DU PONT
PHOTO PRODUCTS DEPARTMENT
TOWANDA, PENNSYLVANIA
LIBRARY

ORGANOPHOSPHORUS REAGENTS IN ORGANIC SYNTHESIS

ORGANOPHOSPHORUS REAGENTS IN ORGANIC SYNTHESIS

Edited by

J. I. G. CADOGAN

Department of Chemistry, University of Edinburgh, Great Britain

1979

ACADEMIC PRESS

LONDON · NEW YORK · TORONTO · SYDNEY · SAN FRANCISCO

A Subsidiary of Harcourt Brace Jovanovich, Publishers

ACADEMIC PRESS INC. (LONDON) LTD.
24/28 Oval Road,
London NW1 7DX

United States Edition published by
ACADEMIC PRESS INC.
111 Fifth Avenue
New York, New York 10003

Copyright © 1979 by
ACADEMIC PRESS INC. (LONDON) LTD.

All Rights Reserved
No part of this book may be reproduced in any form by photostat, microfilm, or any other means, without written permission from the publishers

British Library Cataloguing in Publication Data
Organophosphorus reagents in organic synthesis.
   1. Chemistry, Organic—Synthesis
   2. Organophosphorus compounds
  I. Cadogan, John Ivan George
  547'.2   QD262   79–50307
  ISBN 0-12-154350-1

Printed in Great Britain
by Whitstable Litho Ltd., Whitstable, Kent.

Contributors

R. APPEL
Institute of Inorganic Chemistry, University of Bonn, Gerhard Domagk Strasse 1, D-5300 Bonn 1, West Germany

K. BURGER
Organic Chemical Laboratory, Technical University, Munich, West Germany

J. I. G. CADOGAN
Department of Chemistry, University of Edinburgh, Edinburgh, UK

I. GOSNEY
Department of Chemistry, University of Edinburgh, Edinburgh, UK

M. HALSTENBERG
Institute of Inorganic Chemistry, University of Bonn, Gerhard Domagk Strasse 1, D-5300 Bonn 1, West Germany

R. K. MACKIE
Department of Chemistry, University of St. Andrews, UK

R. RAMAGE
Department of Chemistry, University of Manchester Institute of Science and Technology, UK

A. G. ROWLEY
Department of Chemistry, University of Edinburgh, Edinburgh, UK

D. J. H. SMITH
Department of Chemistry, University of Leicester, UK

B. J. WALKER
Department of Chemistry, The Queen's University, Belfast, UK

E. ZBIRAL
Institute of Organic Chemistry, University of Vienna, Austria

Editor's Preface

The literature of organophosphorus chemistry is very large. This is not surprising. Ever since Clermont in 1854 demonstrated his toughness as well as chemical ability by reporting that tetraethyl pyrophosphate (subsequently shown to be an insecticide) had a burning taste, it has been apparent that organophosphorus compounds are of interest for their own, practical, sake. Add to their industrial and biological importance their increasing value and use as general reagents in organic synthesis and we have all that is necessary for yet another information retrieval problem.

We collaborated in writing this monograph because none devoted to a summary of the use of organophosphorus reagents in general organic synthesis existed. We knew that the field was big but, had we realized just how big was big, this monograph may never have materialized. At the end of the day cost compelled us to be selective rather than comprehensive.

The extraordinarily wide variety of synthetic application of organophosphorus reagents meant that a uniform organizational treatment would be unhelpful. For example, the Wittig reaction, although immense in its scope, is relatively easy to classify: it is simply a beautiful method for the production of carbon–carbon double bonds. On the other hand the reagents Ph_3P-CCl_4 or Ph_3PBr_2, for example, are useful in scores of widely different functional group conversions. We believe that in the first of these cases it is more convenient to the user to classify in terms of reaction type and in the second in terms of type of reagent. With hindsight we can see that the monograph has a subsidiary theme: almost all the reactions described stem from the use of simple tervalent phosphorus reagents at some stage, which surely must distinguish them as the most versatile reagents in organic chemistry.

After an introduction (Chapter 1) to the key types of reactions and reagents used in organophosphorus chemistry, which is brief because this is not intended to be a text book on organophosphorus chemistry *per se*, Chapter 2 is concerned with the Wittig reaction. To cover this comprehensively would require a monograph to itself; to keep this one within manageable size between one set of covers meant selectivity. In the event we decided to concentrate on stereospecific aspects of synthesis, a subject which apart from

Schlosser's review ten years ago, has not received much attention in the review literature.

Chapter 3 is also concerned with synthesis of alkenes via phosphorus-stabilized anions but restricts itself to phosphonyl-activated olefinations of the Horner and Wadsworth–Emmons types. Chapter 4 describes syntheses derived from reactions of phosphorus-stabilized anions with electrophiles other than aldehydes and ketones, while Chapter 5 completes this group with its summary of heterocyclic syntheses via alkylidenephosphoranes, iminophosphoranes, and vinylphosphonium salts. Chapter 6 is also concerned with synthesis of heterocycles, in this case via the deoxygenation of aromatic nitro compounds with phosphorus(III) reagents.

Chapter 7, continuing the theme of deoxygenation, describes a wide variety of functional group conversions brought about by phosphorus(III) reagents both by direct and indirect removal of oxygen. Direct removal, in this case, means direct formation of the product by deoxygenation, whereas indirect deoxygenation can be more complicated, mechanistically, sometimes involving the formation of a second phosphorus reagent from the tervalent starting material, subsequent deoxygenation being only one step in the overall reaction. The most important examples of the latter are the combined use of triphenylphosphine with dialkyl azodicarboxylates and conversions based on the ozone–triphenyl phosphite reagent pair.

Functional group conversions brought about by phosphorus(III)-induced desulphurizations (Chapter 8), combination of phosphorus(III) reagents with polyhalogenomethanes (Chapter 9) and with halides and halogens, including the use of Ph_3PBr_2 (Chapter 10), then follow.

The literature of pentacoordinate phosphoranes is growing very rapidly and reflected in this is the increasing number of applications in general organic synthesis; these are therefore summarized in Chapter 11.

The final chapter deals with the special problems of functional group conversions in peptide synthesis.

It is a pleasure to acknowledge the hard work and cooperation of my co-contributors. Professor Stuart Trippett, Professor Fausto Ramirez, and Dr. Stuart Warren are thanked, too, for helpful comments at an early stage of the project. I am very grateful, also, to Mrs. Christian Ranken whose competence has been a great help in the making of this book, and to Dr. Christopher Pounder for his help with the subject index.

J. I. G. Cadogan
October 1979

Contents

1
Introduction: Types of Organophosphorus Reagents and their Key Reactions
J. I. G. CADOGAN

1. Preamble, 1
2. Tervalent phosphorus compounds, 2
3. Quaternization of tertiary phosphines, 4
4. Quaternization of phosphorus(III) esters, 4
    4.1. The Arbusov reaction, 4
    4.2. The Perkow reaction, 5
5. Phosphorus ylides, 6
    5.1. Structure and reactivity of ylides, 6
    5.2. Formation of ylides from phosphonium salts, 7
    5.3. Reactions of phosphorus ylides: the Wittig reaction, 7
    5.4. Formation of phosphorus ylides from vinylphosphonium salts, 8
6. Dialkyl alkylphosphonates, trialkylphosphine oxides, and vinylphosphine oxides, 8
7. Iminophosphoranes, 9
8. Phosphates, phosphoramidates, and phosphorothioates, 10
9. Pentacoordinate phosphoranes, 11
10. Guide to further reading, 14
11. References, 14

2
Transformations via Phosphorus-stabilized Anions
1: Stereoselective Syntheses of Alkenes via the Wittig Reaction
IAN GOSNEY and (in part) ALAN G. ROWLEY

1. Introduction, 17
2. General aspects, 19
    2.1. Introduction, 19

2.2. Reactivity of the phosphorus ylide, 20
2.3. Choice of base, 23
3. Stereoselectivity in reactions of ylides, 26
  3.1. Introduction, 26
  3.2. Stable ylides, 27
  3.3. Reactive ylides, 42
    3.3.1. trans-Stereoselective alkene synthesis via β-oxido-ylides (Wittig–Schlosser reaction), 55
    3.3.2. α-Substitution plus carbonyl olefination via β-oxido phosphorus ylides ("scoopy" reactions), 56
  3.4. Ylides of intermediate ("moderate") reactivity, 63
    3.4.1. Introduction, 63
    3.4.2. Allylic ylides, 66
    3.4.3. Benzylic ylides, 77
    3.4.4. Propargylic (prop-2-ynyl) ylides, 84
    3.4.5. Summary, 86
4. Applications of stereoselective Wittig reactions in the synthesis of natural products, 89
  4.1. Introduction, 89
  4.2. Naturally occurring long-chain unsaturated fatty acids, 90
    4.2.1. Introduction, 90
    4.2.2. (Z)-Alkenoic acids, 90
    4.2.3. Polyenoic acids, 93
    4.2.4. ω-Hydroxy unsaturated acids, 96
    4.2.5. Polyacetylenic compounds, 97
    4.2.6. Prostaglandins, 99
  4.3. Pheromones, 109
  4.4. Carotenoids and polyenes, 120
  4.5. Rethrolones, 127
  4.6. Sesquiterpenes, 133
  4.7. Steroids, 136
5. References, 142

3
Transformations via Phosphorus-stabilized Anions 2: PO-Activated Olefination

BRIAN J. WALKER

1. Introduction, 155
2. Synthesis of reagents containing the PO group, 159
  2.1. Phosphonates, phosphinates, and phosphine oxides, 159
    2.1.1. The Michaelis–Arbusov reaction, 159
    2.1.2. Via Grignard reagents, 160
    2.1.3. Via alkylation and acylation, 161
  2.2. Phosphoramidates, 162
3. The olefination reaction, 162
  3.1. The carbanion, 162

Contents

- 3.2. The carbonyl group, 164
- 3.3. Reaction conditions, 165
- 3.4. Competing reactions, 166
- 4. Mechanism and stereochemistry, 167
  - 4.1. Mechanism, 167
  - 4.2. Stereochemistry, 168
  - 4.3. Control of stereochemistry, 169
- 5. Applications in synthesis of alkenes, 172
  - 5.1. PO-stabilized carbanions, 172
    - 5.1.1. α,β-Unsaturated ketones, 173
    - 5.1.2. α,β-Unsaturated acid derivatives, 173
    - 5.1.3. Stilbenes and related alkenes, 174
    - 5.1.4. Vinyl halides, 174
    - 5.1.5. Sulphur compounds, 175
    - 5.1.6. Enamines, 176
    - 5.1.7. Allenes and acetylenes, 176
    - 5.1.8. Dienes and polyenes, 177
    - 5.1.9. Prostaglandins, 179
    - 5.1.10. Steroids and related compounds, 180
    - 5.1.11. Carbohydrates, 184
    - 5.1.12. Nitrogen heterocycles, 184
    - 5.1.13. Intramolecular reactions, 186
  - 5.2. Phosphoramidates, 188
  - 5.3. Related reactions, 189
- 6. Examples of experimental procedures, 190
  - 6.1. Diethyl 1-naphthylmethanephosphonate, 191
  - 6.2. 3,6,10-Trimethylundeca-3,9-dien-2-ones, 191
  - 6.3. *trans*-Stilbene, 191
  - 6.4. Ethyl 4-methylpenta-2,4-dienoate, 191
  - 6.5. 2-Phenylbut-2-ene from ethylphosphonic acid bis(dimethylamide), 192
- 7. Summary of PO-stabilized anions, 192
  - 7.1. Phosphine oxides, 192
  - 7.2. Phosphinates, 193
  - 7.3. Phosphonates, 193
  - 7.4. Phosphonamidates, 196
  - 7.5. Phosphoramidates, 196
- 8. References, 197

4
Transformations via Phosphorus-stabilized Anions
3: Reactions with Electrophiles other than Aldehydes and Ketones

D. J. H. SMITH

1. Alkylation reactions, 207
2. Acylation reactions, 211
3. Reactions with epoxides, 214

4. Reactions with carbon–carbon double bonds, 217
5. Reactions with carbon–carbon triple bonds, 218
6. Reactions with carbon–nitrogen multiple bonds, 219
   6.1. Schiff bases, 219
   6.2. Isocyanates and isothiocyanates, 219
   6.3. Nitriles, 220
7. Summary of functional group conversions, 220
8. References, 221

5
Transformations via Phosphorus-stabilized Anions 4: Heterocyclic Synthesis via Alkylidenephosphoranes, Iminophosphoranes, and Vinylphosphonium Salts

ERICH ZBIRAL

1. Syntheses of heterocyclic compounds using alkylidenephosphoranes, 223
   1.1. Syntheses of heterocyclic compounds via carbonyl olefination with alkylidenephosphoranes, 225
      1.1.1. Tetrahydropyrans, thiacyclohexanes, and dibenzo[b,f]oxepins, 225
      1.1.2. Thiophens and thiacyclononatetraenes, 227
      1.1.3. α-Pyrones, coumarins, and 2,3-dihydro-1-benzofurans, 227
      1.1.4. Acridinium betaines, 230
      1.1.5. Pyridines, 231
   1.2. Syntheses via 1,3-dipolar cycloaddition reactions, 231
      1.2.1. 1,2,3-Triazoles, 231
      1.2.2. Pyrazoles, 234
      1.2.3. 1,2-Oxazoles, 234
      1.2.4. 3-Pyrrolines and pyrroles, 234
      1.2.5. Azirines, 236
   1.3. Various syntheses using 2-oxoalkylidenephosphoranes, 236
      1.3.1. γ-Pyrones, 4-(2-oxoalkylidene)pyrans, and 4-phenyliminopyrans, 236
      1.3.2. Fused pyran derivatives, 238
      1.3.3. 1,3-Thiazoles, 239
2. Syntheses of heterocyclic compounds using iminophosphoranes, 241
   2.1. Tetrazoles, 241
   2.2. Pyrazines, 241
   2.3. 1,2,4,5-Tetrazines, 242
   2.4. 1,2,4-Benzotriazines, 242
   2.5. 1,3-Oxazoles, 243
   2.6. Oxadiazoles, 244
   2.7. Pyrazoles, 245
   2.8. 1,3-Thiazoles and 1,3-selenazoles, 246
   2.9. Phthalazines, 246
   2.10. Pyrazoles, pyridazines, and quinolines, 247
   2.11. Reductive cyclization of azido compounds with tervalent phosphorus compounds via iminophosphoranes, 248

Contents

 2.11.1. Quinolines and 1,3-benzoxazoles, 248
 2.11.2. Nigrifactin and 1,2-dehydroproline methylamide, 248
 2.11.3. Aziridines, 249
3. Syntheses of heterocyclic compounds using alk-1-enylphosphonium salts, 250
 3.1. Syntheses using triphenylvinylphosphonium bromide, 250
  3.1.1. Pyrrolizine derivatives, 250
  3.1.2. Furan derivatives, 250
  3.1.3. Chromens and 2,3-dihydro-1-benzoxepins, 252
  3.1.4. Quinoline derivatives, 253
  3.1.5. Pyrroles and 2,3-dihydropyridazines, 254
  3.1.6. Pyrazole derivatives, 254
 3.2. Syntheses using 3-oxoalk-1-enyl phosphonium salts (2-acylvinylphosphonium salts), 256
  3.2.1. 1,2,3-Triazoles, 256
  3.2.2. Pyrazoles, 257
  3.2.3. 1,3-Thiazoles, 258
  3.2.4. Pyrroles and imidazoles, 260
  3.2.5. Heterocyclic-fused imidazoles, 261
4. References, 265

6
Deoxygenation via Phosphorus(III) Reagents
1: Heterocyclic Synthesis using Aromatic Nitro Compounds

J. I. G. CADOGAN

1. Introduction, 269
2. Insertion into adjacent aromatic C–H bonds: formation of carbazoles and heterocyclic analogues, 272
3. Insertion into adjacent olefinic and imino C–H bonds: synthesis of indoles and related compounds, 273
4. Addition to adjacent conjugated C=C bonds: synthesis of indoles and related compounds, 275
5. Insertion into adjacent saturated aliphatic C–H bonds: synthesis of dihydroindoles and related compounds, 276
6. Cyclization at adjacent imino nitrogen (C=N) and azo nitrogen (N=N): synthesis of indazoles, triazoles, and related compounds, 278
7. Interaction with adjacent carbonyl groups, 281
 7.1. Reaction with o-carbonyl groups, 281
 7.2. Reaction with o-α,β-unsaturated carbonyl groups, 282
 7.3. Reaction with adjacent, but not conjugated, carbonyl groups, 285
8. Interaction with adjacent, but not conjugated, aromatic rings: synthesis of phenothiazines and related compounds, 286
9. Ring enlargement reactions: formation of azepines and related compounds, 287

10. Incorporation of phosphorus into the heterocyclic ring: formation of amino(oxy)- and amino(thiyl)-phosphoranes, 288
11. Some conundrums, 290
12. References, 292

7
Deoxygenations using Phosphorus(III) Reagents
1: General Functional Group Conversions

A. G. ROWLEY

1. Direct deoxygenation by tervalent phosphorus reagents, 295
   1.1. Removal of oxygen attached to nitrogen, 295
      1.1.1. Deoxygenation of heterocyclic N-oxides, 295
      1.1.2. Selective monodeoxygenation of heterocyclic di-N-oxides, 297
      1.1.3. Deoxygenation of furoxans and furazans, 298
      1.1.4. Deoxygenation of azoxy compounds, 300
      1.1.5. Deoxygenation of nitrones, 300
      1.1.6. Deoxygenation of nitrile oxides, 300
   1.2. Removal of oxygen attached to sulphur, 301
      1.2.1. Deoxygenation of sulphones and sulphoxides, 301
      1.2.2. Deoxygenation of other oxidized sulphur compounds, 303
   1.3. Deoxygenation of carbonyl compounds, 304
      1.3.1. Principles, 304
      1.3.2. Synthesis of epoxides from aromatic aldehydes, 306
      1.3.3. Synthesis of epoxides from other aroyl derivatives, 308
      1.3.4. Intramolecular condensations of aryl aldehydes: synthesis of aromatic epoxides, 309
      1.3.5. Synthesis of alkenes by reductive dimerization of carbonyl compounds, 310
      1.3.6. Reductive dimerization of dicarboxylic acid anhydrides, 311
      1.3.7. Miscellaneous reductive condensations of carbonyl compounds, 314
      1.3.8. Conversion of aryl trifluoromethyl ketones into alkenes, 314
      1.3.9. Deoxygenation of isocyanates, 315
      1.3.10. Deoxygenation of 1,2-dicarbonyl compounds: synthesis of alkynes, 315
      1.3.11. Other syntheses of alkynes, 316
   1.4. Deoxygenation of epoxides, 316
   1.5. Deoxygenation of other oxygen heterocycles, 317
   1.6. Deoxygenation of peroxy compounds, 318
      1.6.1. Deoxygenation of hydroperoxides, 318
      1.6.2. Deoxygenation of dialkyl peroxides, 320
      1.6.3. Deoxygenation of endocyclic peroxides, 322
      1.6.4. Deoxygenation of diaroyl peroxides, 323
      1.6.5. Deoxygenation of peroxy-esters and peroxy-lactones, 324
      1.6.6. Deoxygenation of peroxydicarbonates, 326
2. Indirect deoxygenation by tervalent organophosphorus reagents, 327

Contents

　2.1. Use of the dialkyl azodicarboxylate–triphenylphosphine reagent (DEADC–TPP), 327
　　2.1.1. Preparation of esters, 328
　　2.1.2. Conversion of alcohols into amines via N-alkylphthalimides, 330
　　2.1.3. Other condensations at nitrogen, 331
　　2.1.4. Alkylation of N-hydroxy groups, 331
　　2.1.5. Synthesis of ethers, 332
　　2.1.6. Synthesis of anhydrides, 333
　　2.1.7. Alkylation of active hydrogen compounds, 333
　　2.1.8. Other nucleophilic substitutions, 333
　　2.1.9. Transesterification of carboxylic esters, 334
　　2.1.10. Intramolecular dehydrations, 335
　2.2. Use of the ozone–triphenyl phosphite adduct, 335
　　2.2.1. 1,4-Addition to dienes, 337
　　2.2.2. The "ene" reaction: formation of allylic hydroperoxides, 338
　　2.2.3. 1,2-Cycloadditions: formation of dioxetans, 340
　　2.2.4. The ozone–triphenyl phosphite adduct as a simple oxidant, 341
　　2.2.5. Other ozone–phosphite adducts, 342
　2.3. Other reactions involving ozone and phosphorus(III) reagents, 343
3. References, 345

8
Desulphurization via Phosphorus(III) Reagents: General Functional Group Conversions

R. K. MACKIE

1. Reactions with thiirans: synthesis of olefins, 351
2. Vinylogous amides and β-diketones from thioimidates and related compounds, 354
3. Alkenes from thionocarbonates and related reactions, 354
4. Alkenes from 1,3-oxathiolan-5-ones and azosulphides, 359
5. Desulphurization of sulphides and thiols, 360
6. Reactions with sulphoxides, sulphenic acids, and sulphenates, 363
7. Desulphurization of di- and poly-sulphides, 365
8. Formation of pyridyl thioesters and related synthons, 373
9. Condensation reactions, including formation of amides, phosphorylation, and application in nucleoside synthesis, 374
10. Reactions involving other sulphenyl compounds, 376
　10.1. Sulphenyl chlorides, 376
　10.2. Compounds containing sulphur–nitrogen bonds, 377
　10.3. Thiolsulphinates and sulphenylthiolsulphinates, 378
　10.4. Miscellaneous sulphenyl compounds, 379
11. Reactions with carbon disulphide, thiobenzophenone, and dicyanosulphonium ylides, 379
12. Miscellaneous reactions, 380
13. References, 382

9
Functional Group Conversions using Phosphorus(III) Reagents and Polyhalogenoalkanes
ROLF APPEL and MECHTHILD HALSTENBERG

1. Introduction, 387
2. Initial steps of the reaction between a tertiary phosphine and a tetrahalogenomethane, 388
    2.1. Triphenylphosphine–CCl_4, 388
    2.2. Aliphatic phosphines–CCl_4, 389
3. Basis of the preparative application of the reagent R_3P–CCl_4 in multicomponent reactions, 391
    3.1. Mechanism of the three-component reaction, 391
4. The three-component reaction: triphenylphosphine–CCl_4–water, 394
    4.1. Chloromethylenephosphoranes and hexaorganylcarbodiphosphoranes from intermediates of the reactions Ph_3P–CCl_4 and Ph_3P–CCl_4–H_2O, 397
5. Chlorinations, 400
    5.1. Chlorinations of O-functional compounds, 400
        5.1.1. Alcohols and thiols, 400
        5.1.2. Aldehydes, ketones, and epoxides, 402
        5.1.3. Carboxylic acids and oxophosphoric acids, 404
    5.2. Chlorinations of –C(X)NH– functional compounds (X = O or S), 405
        5.2.1. Chloroformamidines from trisubstituted ureas, 405
        5.2.2. Imidoyl halides from monosubstituted carboxamides and by Beckmann rearrangement of ketoximes, 406
        5.2.3. Reaction with carbamoyl halides, 407
        5.2.4. Reaction with thiocarbamic O-esters, 408
        5.2.5. N-Phenylchlorothioformimidic esters from thiocarbamic S-esters, 408
6. Dehydrations, 408
    6.1. Nitriles from carboxamides and aldoximes, 408
    6.2. Isocyanides from monosubstituted formamides, 409
    6.3. Carbodiimides from N,N'-disubstituted ureas, 409
    6.4. Ketene imines from α-disubstituted primary acid amides, 410
    6.5. C-Ethoxycarbonyl-C-imidoylketene imines from (aminoalkylene)malonamic esters, 410
    6.6. Nitrogen heterocycles from N-substituted amino-alcohols, 410
    6.7. Removal of water from phosphoramidates and phosphinamides, 411
    6.8. Intermolecular removal of water, 411
7. Reactions providing P–N linkage, 411
    7.1. Aminophosphonium chlorides and iminophosphoranes, 412
    7.2. N-(Triphenylphosphoranediyl)amides of sulphuric acid, sulphonic acids, alkyl sulphates, and fluorosulphamides, and N-(triphenylphosphoranediyl)-phosphoramidic diesters and -phosphinamides, 413
8. Four-component reactions, 414
    8.1. Esterification, 414
    8.2. Acid amides and peptides, 415

8.3. Esters, amidic esters, and amides of phosphinic and phosphoric acids, 415
9. The reagent R_3P–C_2Cl_6, 416
10. Phosphorus(III) esters–CCl_4, 417
    10.1. Two-component reactions, 418
        10.1.1. Reaction products and conditions, 418
        10.1.2. Mechanism, 420
    10.2. Three-component reactions, 420
        10.2.1. OH compounds, 420
        10.2.2. SH compounds, 421
        10.2.3. NH compounds, 421
11. Other reagent combinations, 422
12. Summary and outlook, 423
13. References, 424

10
Functional Group Conversions using Phosphorus(III) Reagents with Halides and Halogens

R. K MACKIE

1. Reactions of phosphorus(III) reagents with halogens, hydrogen halides, and alkyl halides, 433
    1.1. Reactions of phosphites with halogens and hydrogen halides, 433
    1.2. Reactions of phosphites with alkyl halides, 435
    1.3. Syntheses involving phosphine dihalides, 437
    1.4. Reactions involving triphenylphosphine and hydrobromic acid, 446
2. Reactions of phosphorus(III) reagents at positive halogens, 446
    2.1. With compounds containing activated carbon–halogen bonds, 446
    2.2. With compounds containing nitrogen–halogen bonds, 455
    2.3. Elimination from 1,2-dihalogeno compounds, 457
3. Reactions with acyl halides, 459
4. Reactions with aryl halides, 460
5. Summary of functional group conversions, 461
6. References, 462

11
Pentacoordinate Phosphoranes in Synthesis

K. BURGER

1. Introduction, 467
2. 2,2-Dihydro-1,3,2-dioxaphospholens, 469
    2.1. Reaction with oxygen and ozone, 469
    2.2. Hydrogenation, 469

2.3. Hydrolysis, 470
2.4. Alcoholysis, 471
2.5. Reactions with hydrogen halides, 472
2.6. Reactions with acid chlorides and anhydrides, 472
2.7. Bromination, 474
2.8. Reactions with activated carbon–carbon double bonds, 475
2.9. Reactions with carbonyl compounds, 475
2.10. Reactions with heterocumulenes, 476
2.11. Fragmentation reactions, 479
3. 2,2-Dihydro-1,3,2-dioxaphospholans, 481
    3.1. Hydrolysis, 482
    3.2. Alcoholysis, 482
    3.3. Bromination, 484
    3.4. Insertion into P–H bonds, 484
    3.5. Formation of epoxides, 486
    3.6. Formation of carbenes, 487
    3.7. The 2,2-dihydro-1,3,2-dioxaphospholan–2,2-dihydro-1,2-oxaphosphetan transformation, 487
4. 2,2-Dihydro-1,4,2-dioxaphospholans, 488
    4.1. The 2,2-dihydro-1,4,2-dioxaphospholan–2,2-dihydro-1,3,2-dioxaphospholan transformation, 488
    4.2. The 2,2-dihydro-1,4,2-dioxaphospholan–2,2-dihydro-1,2-oxaphosphetan transformation, 488
5. 2,2-Dihydro-1,2-oxaphospholens-4, 488
    5.1. Hydrolysis, 488
    5.2. Thermolysis, 489
    5.3. The 2,2-dihydro-1,2-oxaphospholen–2,2-dihydro-1,2-thiaphospholen transformation, 489
6. 2,2-Dihydro-1,2-oxaphospholans, 490
    6.1. Reactions with electrophiles, 490
    6.2. Thermolysis, 490
7. 3,3-Dihydro-1,3-oxaphospholan-5-ones, 491
8. Trioxaphosphetans [Phosphorus(III)–ozone adducts], 491
9. 2,2-Dihydro-1,3,2-oxathiaphospholans, 492
    9.1. Formation of thiirans, 492
10. 2,2-Dihydro-1,4,2-oxazaphospholens-4 (Δ^4-1,4,2λ^5-oxazaphospholines), 492
    10.1. Thermolysis and photolysis, 492
11. 5.5-Dihydro-1,2,5-oxazaphospholens-2 (Δ^2-1,2,5λ^5-oxazaphospholines), 494
    11.1. Thermolysis, 494
12. 2,2-Dihydro-1,3,2-oxazaphospholans (1,3,2λ^5-oxazaphospholidines), 495
    12.1. Thermolysis, 495
    12.2. Photolysis, 496
13. 5,5-Dihydro-1,2,5-oxazaphospholans (1,2,5λ^5-oxazaphospholidines), 496
14. 2,2-Dihydro-1,3,4,2-oxadiazaphospholens-4 (Δ^4-1,3,4,2λ^5-oxadiazaphospholines), 496
15. 1,1-Dihydrophospholens-3, 497
16. 1,1-Dihydrophospholans, 498
17. 1,1-Dihydrophosphirans, 498
18. Bis-(2,2'-biphenylene)phosphoranes, 499
19. Acyclic phosphoranes, 499

20. Miscellaneous reactions, 501
    20.1. Phosphoranes as intermediates in the Michaelis–Arbusov reaction, 501
    20.2. Formation of copolymers, 501
    20.3. Hexacoordination, 502
21. References, 503

12
Organophosphorus Reagents in the Synthesis of Peptides

R. RAMAGE

1. Amino protection, 511
    1.1. Phosphoric acid derivatives, 513
    1.2. Phosphinic acid derivatives and thio analogues, 515
2. Carbonyl activation and amide formation, 518
    2.1. Anhydrides with phosphoric acid derivatives, 519
    2.2. Anhydrides with phosphinic acid derivatives, 522
    2.3. Anhydrides with phosphorous acid derivatives, 523
    2.4. Combination of phosphorus(III) reagents with amines, 526
    2.5. Activation methods based on R_3PO formation, 528
3. References, 533

Author Index   537

Subject Index   578

1
Introduction: Types of Organophosphorus Reagents and their Key Reactions

J. I. G. CADOGAN

1. Preamble, 1
2. Tervalent phosphorus compounds, 2
3. Quaternization of tertiary phosphines, 4
4. Quaternization of phosphorus(III) esters, 4
5. Phosphorus ylides, 6
6. Dialkyl alkylphosphonates, trialkylphosphine oxides, and vinylphosphine oxides, 8
7. Iminophosphoranes, 9
8. Phosphates, phosphoramidates, and phosphorothioates, 10
9. Pentacoordinate phosphoranes, 11
10. Guide to further reading, 14
11. References, 14

1. Preamble

The art and science of organic chemical synthesis owes much to Nature's decision to make the electronic structure of phosphorus $1s^2 2s^2 2p^6 3s^2 3p^3$. This distribution and the relevant orbital energies lead to well defined families of tri-, tetra-, penta-, and hexa-coordinate derivatives in which the ligands can be organic or inorganic.*

The extraordinary usefulness of organophosphorus reagents in general organic synthesis stems from the ease with which it is possible to progress from the lowest to the highest coordination number $[P(III) \rightarrow P(IV) \rightarrow P(V) \rightarrow P(VI)]$, and occasionally back again [e.g. $P(V) \rightarrow P(IV)$]. Key, overlapping factors in this are: (a) the high nucleophilicity of tervalent phosphorus reagents towards a wide range of electrophiles; (b) the strong bonds that

---

* Throughout this monograph we have followed the advice given in *Handbook for Chemical Society Authors* published by the Chemical Society (1961) whose editors scholarly inform us that we should not mix Greek and Latin roots: hence ter- and quinque-valent, tri- and penta-coordinate.

phosphorus forms with oxygen (particularly P=O), sulphur, nitrogen, the halogens, and carbon; (c) the capability of phosphorus to stabilize adjacent anions, best illustrated by the paramount importance in synthesis of the Wittig ylide.

The need to classify and summarize the large number of reactions of phosphorus reagents of synthetic importance is the rationale of this monograph; in this introductory chapter we seek only to identify certain key reactions and reagents in the hope that this will help link the specialized chapters which follow. So, we look first at tricoordinate phosphorus compounds and their more important transformations, then at tetracoordinate species including phosphonium salts and the large subset $X_3P=Y$, which includes ylides, iminophosphoranes, phosphonates, phosphates, phosphoramidates, and phosphorothioates. This summary then concludes with an introduction to pentacoordinate phosphoranes.

For a detailed discussion of the theoretical principles behind these generalizations, the reader is referred to the many good texts on organophosphorus chemistry now available (Section 10).

2. Tervalent Phosphorus Compounds

These, the workhorses of organophosphorus chemistry, include the phosphines and phosphites, X_3P, where X can be alkyl, aryl, alkoxy, aryloxy, halogeno, alkylthio, amino, alkylamino, or arylamino. These reagents carry a lone pair of electrons which endows basicity and nucleophilicity on the molecule. In general, phosphines are much weaker bases than amines but they are stronger nucleophiles; almost all the reactions of tervalent reagents of value in synthesis depend on this nucleophilicity. As the reader will find, the range of reactions and reactivity is enormous because phosphorus(III) reagents will react by nucleophilic attack at saturated and unsaturated carbon, at oxygen, sulphur, halogen, and nitrogen to give intermediates that either break down *in situ* to desired products or can be used in a second reaction as synthons in their own right. Examples of the first class include Corey and Winter's

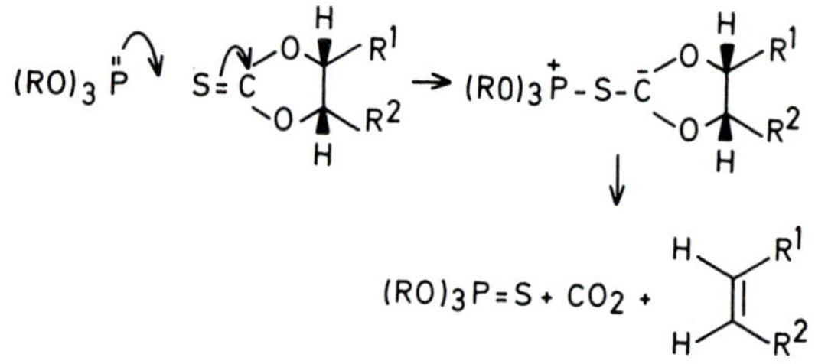

Scheme 1

1 Introduction

stereospecific synthesis of alkenes by desulphurization of cyclic thioncarbonates[1] (Scheme 1) and the very convenient stereoinverted conversion of alcohols into alkyl chlorides by the reaction of triphenylphosphine in carbon tetrachloride (Scheme 2).[2]

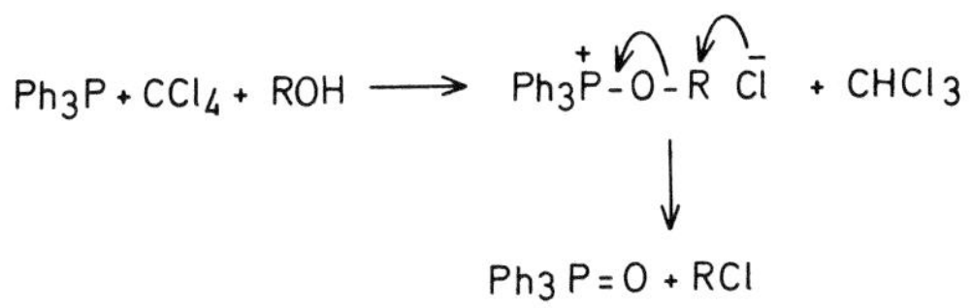

$$Ph_3P + CCl_4 + ROH \longrightarrow Ph_3\overset{+}{P}\text{-}O\text{-}R \ \ Cl^- + CHCl_3$$

$$\downarrow$$

$$Ph_3P=O + RCl$$

Scheme 2

The best example of the second type is the Wittig ylide formed in a two-step process involving nucleophilic attack of the phosphorus(III) reagent on an alkyl halide followed by removal of the acidic proton of the resulting phosphonium salt (Scheme 3).[3] Wittig's demonstration that the ylide so

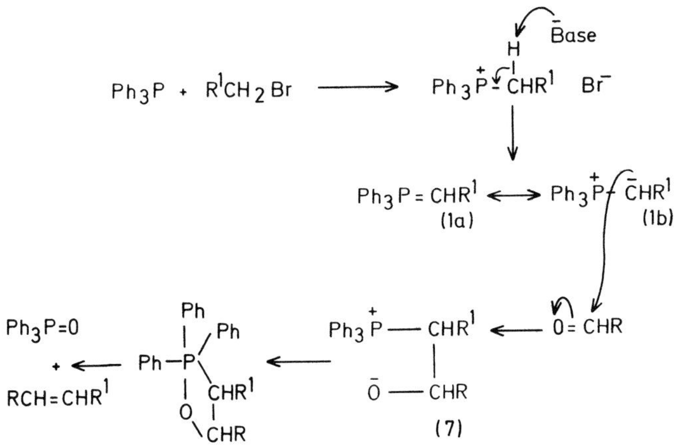

$$Ph_3P + R^1CH_2Br \longrightarrow Ph_3\overset{+}{P}\text{-}CHR^1 \ \ Br^-$$

$$\downarrow$$

$$Ph_3P=CHR^1 \longleftrightarrow Ph_3\overset{+}{P}\text{-}\overset{-}{C}HR^1$$
(1a) \hspace{2em} (1b)

$$Ph_3P=O \ + \ RCH=CHR^1 \longleftarrow Ph\text{-}P\underset{O\text{-}CHR}{\overset{Ph}{\diagdown}}CHR^1 \longleftarrow Ph_3\overset{+}{P}\text{-}CHR^1 \longleftarrow O=CHR$$
$$\hspace{15em} \overset{|}{O}\text{-}CHR$$
$$\hspace{15em} (7)$$

Scheme 3

formed reacts with carbonyl compounds to give alkenes is perhaps the most important discovery in general organic synthesis since that of the Grignard reagent.

The starting point for the Wittig reaction, however, is the simple quaternization, the conversion of tricoordinate phosphorus, via nucleophilic attack on an electrophile, to give tetracoordinate phosphorus:

$$X_3P + YZ \rightarrow [X_3PY]^+Z^-$$

This step or its analogues are important in almost all synthetically valuable reactions of organophosphorus reagents.

3. Quaternization of Tertiary Phosphines

Reagents such as triphenylphosphine undergo a Menshutkin-type reaction with alkyl halides or alkyl-X, where X is a suitable leaving group, to give stable quaternary phosphonium salts:

$$Ph_3P + RCH_2Br \rightarrow Ph_3P^+CH_2R \; Br^-$$

Carried out in a variety of solvents, the reaction displays S_N2 characteristics. As we have seen (Scheme 3) the formation of the phosphonium salt is the first step towards the Wittig reaction.

4. Quaternization of Phosphorus(III) Esters

4.1 The Arbusov Reaction

This is analogous to the corresponding reaction of phosphines but the reaction between alkyl halides and phosphorus(III) reagents that contain alkoxy groups proceeds further. In this modification, known as the Arbusov, or Michaelis–Arbusov, reaction, the first formed quaternary intermediate is a quasiphosphonium salt (2). If the counter anion has a high nucleophilic reactivity towards saturated carbon, such as is shown by chloride, bromide, or iodide, but not by fluoroborate, rapid dealkylation occurs (Scheme 4).

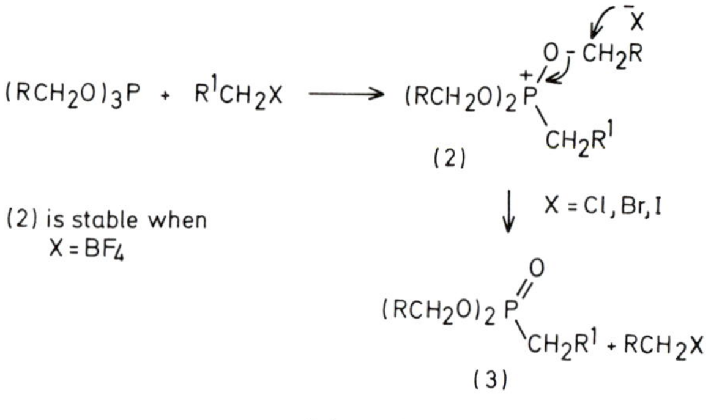

$$(RCH_2O)_3P + R^1CH_2X \longrightarrow (RCH_2O)_2\overset{+}{P}\begin{smallmatrix}\nearrow X \\ O^-CH_2R \\ CH_2R^1\end{smallmatrix}$$

(2) is stable when $X = BF_4$

$$\downarrow X = Cl, Br, I$$

$$(RCH_2O)_2P\overset{O}{\underset{CH_2R^1}{\parallel}} + RCH_2X$$

(3)

Scheme 4

This process is helped by the formation of the strong P=O bond (bond strength *ca.* 630 kJ mol^{-1}; 150 kcal mol^{-1}).

Clearly, from Scheme 4, when $R^1 = R$ in R^1CH_2Br, say, the reaction becomes an isomerization of the trialkyl phosphite to the corresponding dialkyl alkylphosphonate (3), so catalytic quantities, only, of R^1CH_2Br are necessary to bring about the reaction. If R is not equal to R^1 the reaction can

1 Introduction

be used as a method of preparation of "mixed" dialkyl alkylphosphonates (2) useful in PO olefination reactions as discussed below (Section 6) and in detail in Chapter 3.

Triaryl phosphites also undergo the quaternization reaction with alkyl halides, but the intermediate quasiphosphonium salts, e.g. $[(PhO)_3PCH_2R]^+I^-$, having no electrophilic α-alkyl carbon atom available for attack, are stable to about 200°C. Quaternizations of trialkyl or triaryl phosphites by aryl halides occur only on photolysis or under copper or nickel catalysis.

Quasiphosphonium salts have been postulated as intermediates in many deoxygenations (Chapters 6, 7) and desulphurizations (Chapter 8) brought about by phosphorus(III) reagents, e.g.

$$(RO)_3P + ArNO_2 \rightarrow (RO)_3\overset{+}{P}ON(O^-)Ar \rightarrow products$$

4.2 The Perkow Reaction

This is the reaction of trialkyl phosphites with α-halogeno-ketones to give vinyl phosphates (5). It is likely that reaction at the carbonyl carbon atom gives a species that rearranges to the intermediate salt (4) which then dealkylates to (5) (Scheme 5). In this case the reagent eschews nucleophilic attack on positive

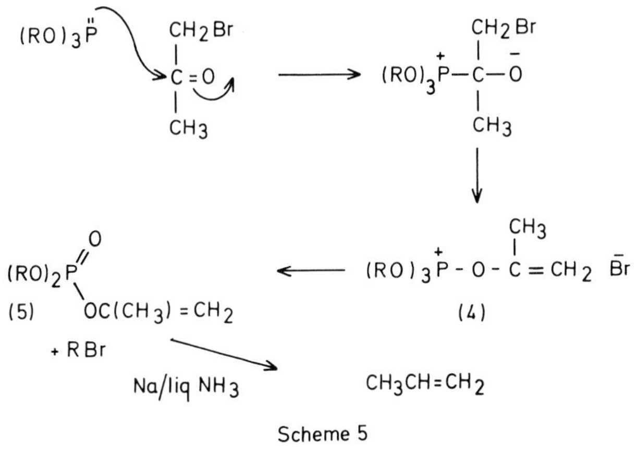

Scheme 5

halogen or at the carbon of the C–Br bond. This is in contrast to reactions of phosphines with α-halogeno-ketones, where attack on positive halogen occurs to give an ion pair (6) (Scheme 6) which may rearrange to the enol phosphonium salt and both ion pairs may then decompose in a variety of ways depending on molecular circumstances.

The main preparative use of the Perkow reaction lies in the conversion of ketones, via the α-bromo derivative, into alkenes by reduction of the resulting

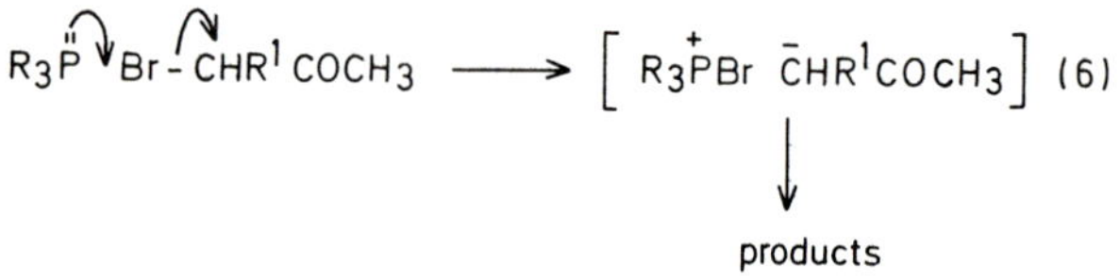

$$R_3\overset{..}{P} \curvearrowright Br \curvearrowleft \bar{C}HR^1 COCH_3 \longrightarrow \left[R_3\overset{+}{P}Br \; \bar{C}HR^1COCH_3 \right] \quad (6)$$

$$\downarrow$$

products

Scheme 6

dialkyl enol phosphate with sodium or lithium in liquid ammonia[4a,b] (Scheme 6) or titanium metal.[4e] This reaction has been extended by the use of the N,N,N',N'-tetramethylphosphorodiamidic protecting group,[4c,d] to provide an alternative to the Wolf–Kishner reduction.

Modifications of the Perkow reaction based on triphenylphosphine are also useful (Chapter 10).

5. Phosphorus Ylides

5.1 Structure and Reactivity of Ylides

Phosphorus ylides have the structure (1a) ↔ (1b) (Scheme 3). Their high stability, compared with nitrogen ylides ($R_3\overset{+}{N}-\bar{C}R_2$), derives from donor $p\pi$–$d\pi$ bonding involving back-donation of the negative charge from the

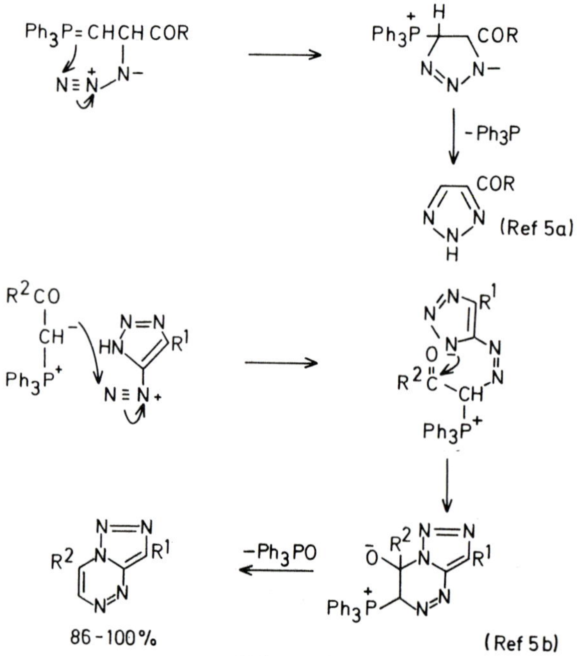

Scheme 7

1 Introduction

ylide carbon into the vacant d-orbital of the phosphorus atom. In accord with this the C–P bond in ylides is shorter (e.g. 1·66 Å) than the single C–P bond (e.g. 1·85 Å), indicating considerable double-bond character. Similar delocalization is unfavoured in the case of nitrogen ylides.

The reactivity of phosphorus ylides varies according to the substituents attached to the carbon end of the dipole. In $Ph_3\overset{+}{P}-\overset{-}{C}HR$, electron-withdrawing groups such as $R = CO_2Et$ lead to delocalization of the carbanionic charge with subsequent loss of carbanionic (and hence ylidic) reactivity. Electron donation by alkyl groups, say, leads to an increase in reactivity. Such reactivity, as we will see, is associated with carbanionic character ($Ph_3\overset{+}{P}-\overset{-}{C}HR$), and most synthetical applications of ylides utilize this, coupled with subsequent loss of triphenylphosphine or its oxide, both of which are good leaving groups, e.g. Scheme 3 and Scheme 7.[5]

5.2. Formation of Ylides from Phosphonium Salts

Phosphorus ylides are usually prepared by removal of the acidic α-hydrogen from a phosphonium salt by a base chosen from amines, ammonia, tertiary amines, pyridine, sodium carbonate, sodium hydroxide, sodium alkoxides, alkyllithiums, and so on. The ylide is usually used *in situ* in solution under nitrogen. The choice of the base can be of great importance in determining the course of subsequent Wittig-type reactions of the ylide (Chapter 2).

5.3. Reactions of Phosphorus Ylides: the Wittig Reaction

As is to be expected from their carbanionic character, phosphorus ylides react with a wide series of electrophiles (Chapter 4), but by far the most important of these is the carbonyl group. The resultant formation of an alkene (the Wittig reaction, Scheme 3) is undoubtedly one of the most valuable chemical reactions ever reported.

The various steps of the reaction beautifully illustrate all the valuable synthetic attributes of organophosphorus reagents:

1. Formation of the desired phosphonium salt utilizing the nucleophilicity of phosphorus(III).
2. Formation of the ylide illustrating the ability of phosphorus to stabilize neighbouring anions.
3. Reaction of the ylide with an electrophilic centre (the carbonyl group) to give the betaine (7) (Scheme 3).
4. Formation of the intermediate pentacoordinate phosphorane by apical attack of the electronegative ligand, oxygen—a favoured reaction—leading to an intermediate in which the four-membered ring easily meets the geometrical demands of phosphorus(V).

5. Decomposition of the phosphorane with formation of the strong P=O bond and elimination of the desired product.

These five processes, which will appear repeatedly in various combinations in the pages of this monograph, are the dominant reactions of organophosphorus reagents in general organic synthesis.

5.4. Formation of Phosphorus Ylides from Vinylphosphonium Salts

These salts are easily prepared, for example by gentle thermolysis of 2-phenoxyethyltriphenylphosphonium bromide (8) (Scheme 8),[6] and are valuable as ylide precursors.

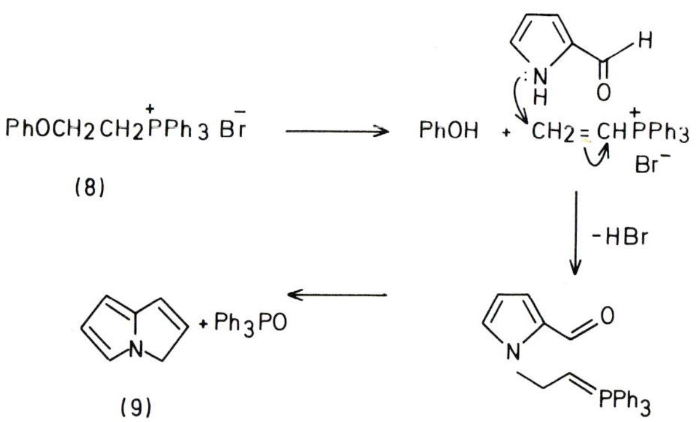

Scheme 8

The terminal vinyl carbon atom is susceptible to attack by a nucleophile, thus creating an ylide *in situ*. Use of nucleophiles containing suitable adjacent carbonyl groups hence provides an elegant method of synthesis of cyclic compounds by what then becomes an intramolecular Wittig reaction.[7]

This is illustrated in Scheme 8 by Schweizer's conversion of pyrrole-2-aldehyde into 3*H*-pyrrolizine (9) (87%) by reaction with triphenylvinylphosphonium bromide (Chapter 5).[7a]

6. Dialkyl Alkylphosphonates, Trialkylphosphine Oxides, and Vinylphosphine Oxides

These reagents are readily prepared (Chapter 3) by the Arbusov reaction (Scheme 4), reactions of Grignard reagents with phosphorus trichloride, or the Michaelis–Becker reaction. The last-named process involves nucleophilic displacement of the dialkyl hydrogen phosphonate anion with alkyl halides (Scheme 9).

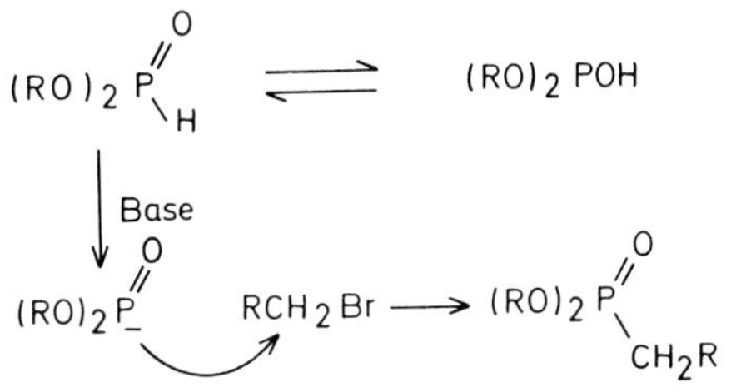

1 Introduction

$$(RO)_2\overset{O}{\underset{H}{P}} \rightleftharpoons (RO)_2POH$$

$$\downarrow \text{Base}$$

$$(RO)_2\overset{O}{\underset{-}{P}} \quad RCH_2Br \longrightarrow (RO)_2\overset{O}{\underset{CH_2R}{P}}$$

Scheme 9

The value of dialkyl alkylphosphonates in synthesis stems from the considerable stabilization that can be imparted by the $>\!\!P\!\!=\!\!O$ group to an adjacent carbanion (Scheme 10) thus providing a synthon with sufficient ylidic character to be very useful in supplementing the Wittig reaction and, by extension to trialkylphosphine oxides and vinylphosphine oxides, in providing an alternative to vinylphosphonium salts (Chapter 5).

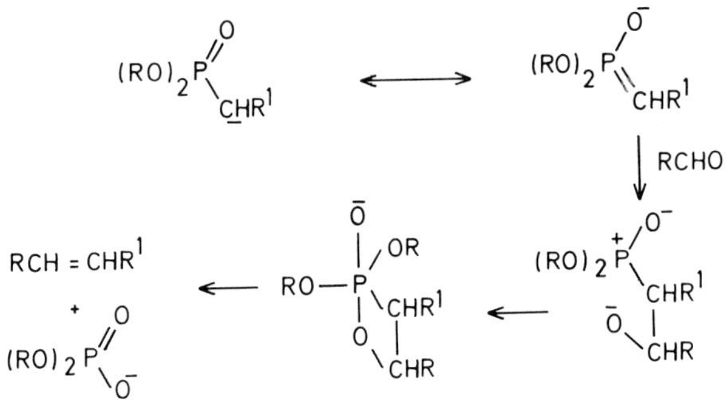

Scheme 10

7. Iminophosphoranes

Readily prepared by reaction of azides with phosphorus(III) reagents (Scheme 11), these compounds are the nitrogen analogues of phosphorus ylides, e.g.

$$Ph_3P\!\!=\!\!NPh \leftrightarrow Ph_3\overset{+}{P}\!\!-\!\!\overset{-}{N}Ph$$

Discovered by Staudinger, more than 50 years ago, they undergo reactions with electrophiles which almost exactly correspond to those of phosphorus ylides giving C=N bonds rather than C=C, as exemplified in Scheme 11. They have been found to be particularly useful in heterocyclic synthesis (Chapter 5).

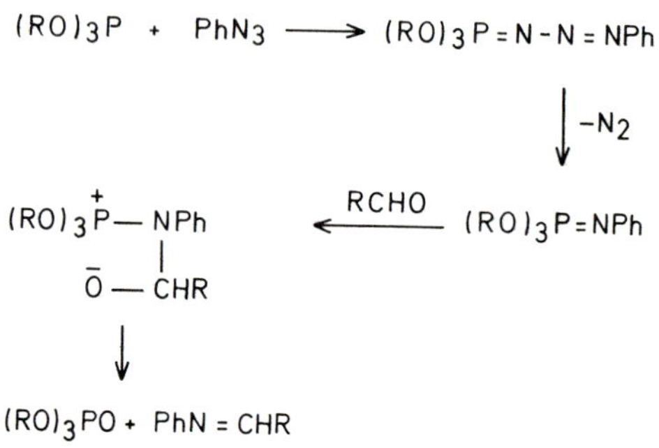

$$(RO)_3P + PhN_3 \longrightarrow (RO)_3P=N-N=NPh$$

$$\downarrow -N_2$$

$$(RO)_3\overset{+}{P}-NPh \quad \overset{RCHO}{\longleftarrow} \quad (RO)_3P=NPh$$
$$\overset{|}{\underset{|}{\bar{O}-CHR}}$$

$$\downarrow$$

$$(RO)_3PO + PhN=CHR$$

Scheme 11

8. Phosphates, Phosphoramidates, and Phosphorothioates

These tetracoordinate species $(RX)_3P=Y$ are exemplified by trialkyl phosphates $[(RO)_3P=O]$, hexamethylphosphorus triamide $[(Me_2N)_3P=O]$, phenyl phosphorodiamidate $[PhO(H_2N)_2P=O]$ and derivatives, and the sulphide (10) (Scheme 14). They do not display a very large range of reactions useful in general organic synthesis, compared with their analogues, the ylides $(R_3P=CHR)$, phosphonates $[R(RO)_2P=O]$, and iminophosphoranes $(R_3P=NR)$. Nevertheless they are important reagents in transferring the moieties R, X, or Y in $(RX)_3P=Y$ to non-phosphorus-containing organic materials. An early example of this is the use of trialkyl phosphates as relatively mild alkylating agents for amines,[8] a reaction readily extended for example to the *N*-alkylation of heterocycles such as phenothiazine,[9] or to the aqueous *S*-methylation of cysteine by trimethyl phosphate.[10]

Phosphoramidates are useful because they can behave as aminating reagents. Phenyl phosphorodiamidate, $PhOP(O)(NH_2)_2$, conveniently converts tautomeric oxo–hydroxy groups directly into amino functions without using the classical but yield-expensive step of formation of a halide first. Examples include conversion of either 6-methyluracil or 6-methylisocytosine into 2,4-diamino-6-methylpyrimidine (Scheme 12),[11a] and the conversion of quinazolin-4-one into 4-aminoquinazoline.[11c] Related aminations, all of which involve attack by an oxygen nucleophile at the phosphoryl centre followed by amino transfer, are brought about by phenyl tetraethylphosphorodiamidate $PhOP(O)(NEt_2)_2$[12a] and hexamethylphospho-

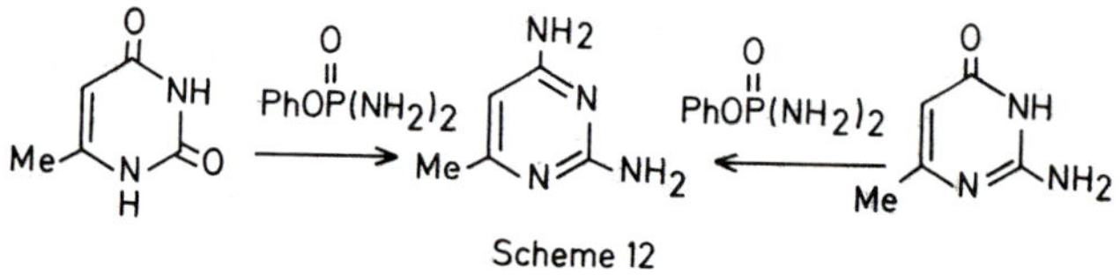

Scheme 12

rus triamide (HMPA or HMPT) $(Me_2N)_3PO$,[12b] and these have been utilized in many heterocyclic syntheses. Thus, reaction of the former diamidate with acetanilide and N,N-diethylformamide leads to 2-diethylaminoquinoline,[12a] while reaction of boiling HMPT with ethyl 2-acetamidothiophen-3-carboxylates gives 4,6-bis(dimethylamino)thieno[2,3-b]pyridines (Scheme 13).[12b]

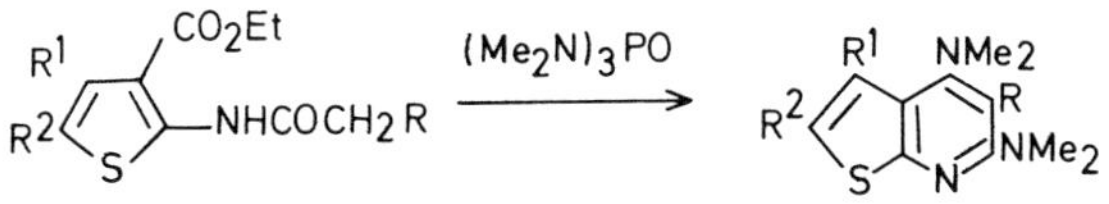

Scheme 13

Regrettably HMPT is carcinogenic, so Pedersen's discovery[12c] that in certain circumstances a preformed mixture of P_2O_5 and dialkylamines behaves chemically in the same way as HMPT is to be welcomed.

Reaction at phosphorus accompanied by thiyl transfer is probably also the important step in thiation reactions brought about by the extraordinarily useful dimer of p-methoxyphenylthionophosphine sulphide (10). Readily prepared from anisole and phosphorus pentasulphide,[13] this reagent has been shown by Lawesson and his coworkers to convert ketones into thioketones,[14a] amides into thioamides,[14b] aliphatic and aromatic esters into the corresponding thiono esters [RC(S)OR],[14c] and tetrahydrothiophen sulphoxide into 1,2-dithian.[14d] No doubt there will soon be many further examples of this thiation reaction.

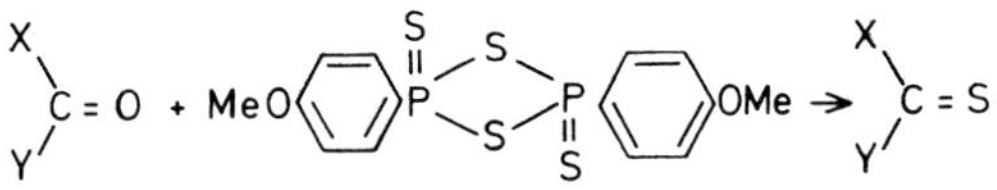

Scheme 14

9. Pentacoordinate Phosphoranes

These are compounds in which phosphorus is covalently linked to five ligands. Most of these possess or are close to trigonal bipyramidal geometry, and most undergo ligand reorganization in a permutational fashion (e.g. pseudorotation). Such isomerization can often be followed by temperature variation n.m.r. studies. The classical example is PF_5. Having trigonal bipyramidal structure this might be expected to exhibit two types of fluorine atoms: equatorial (3 identical P–F bonds) and apical (2 identical P–F bonds) (Scheme 15). In practice, at room temperature, all fluorines appear by n.m.r. to be identical. This suggests that equilibration by ligand reorganization is occurring. As the temperature is lowered the single phosphorus decoupled fluorine

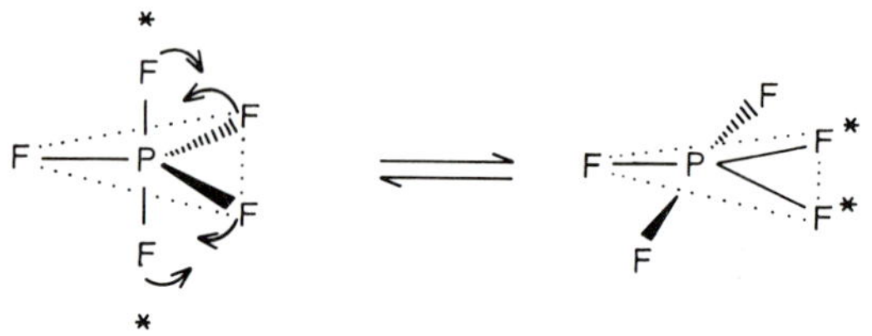

Scheme 15

signal broadens and eventually separates into two signals in the ratio 3:2. This led Berry to suggest the term pseudorotation to explain the phenomenon; it is supposed that the ligands rearrange as in Scheme 15, which is equivalent (because we cannot distinguish the fluorine atoms) to a process in which rotation about a fixed equatorial P–F has occurred (pseudorotation).

The exact locus of the five fluorine atoms during this permutational isomerism is the subject of considerable discussion. It is alternatively argued that the ligands move in groups of three and two (turnstile rotation). What is beyond dispute is that ligands, whether inorganic or organic, do reorganize, a fact that is of importance in understanding the mechanisms of reactions around pentacoordinate phosphorus.

Pentacoordinate phosphoranes have been implicated as short-lived intermediates in a large number of reactions of nucleophiles with tetracoordinate species, e.g. in the hydrolysis of trialkyl phosphates (Scheme 16), where the general rule that electronegative reagents enter and leave apically is accommodated by an intermediate pseudorotatory step (i).

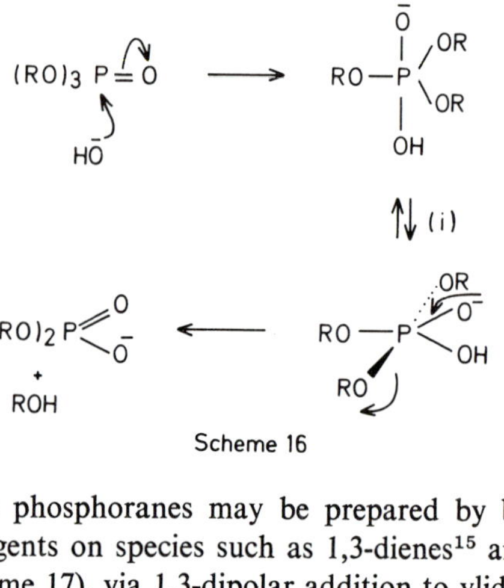

Scheme 16

Thermally stable phosphoranes may be prepared by biphilic attack of phosphorus(III) reagents on species such as 1,3-dienes[15] and 1,2-dicarbonyl compounds[16] (Scheme 17), via 1,3-dipolar addition to ylides[17] (Scheme 18), by reaction of chloramines with diols[18a] or aminophenols[18b] with phos-

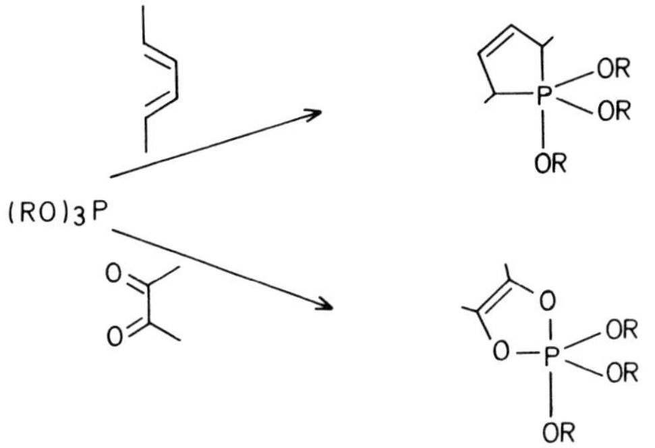

1 Introduction

Scheme 17

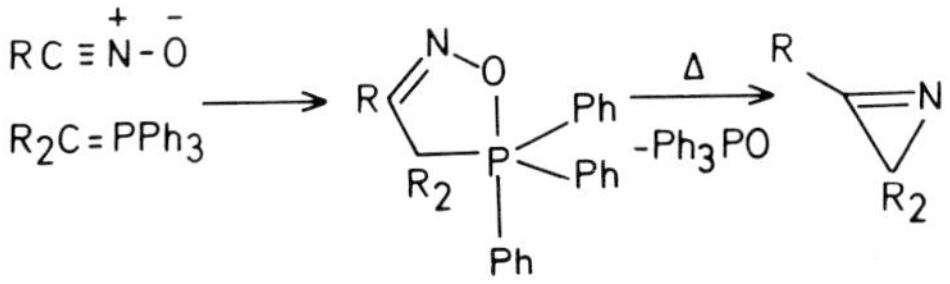

Scheme 18

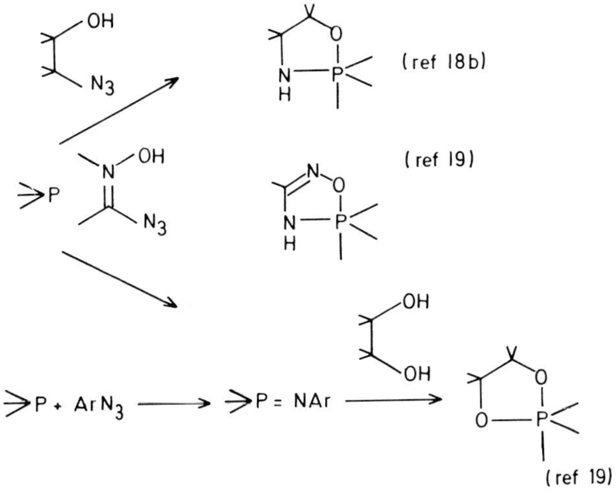

Scheme 19

phorus(III) esters, and more recently by reactions of certain azides[18b,19] (Scheme 19).

Many phosphoranes can be profitably decomposed either thermally or photolytically (Chapter 11), the high bond strength of the resulting P=O or P=S bonds being a major contribution to the driving force of the reaction (Scheme 18).

Our knowledge of this class of compound has expanded greatly in the past few years and we can expect this to increase still further with the increasing availability of ^{31}P Fourier transform n.m.r. instruments. It is now evident (1978) that pentacoordinate phosphoranes are real, rather than imaginary, participants in so many reactions of organophosphorus reagents that large parts of mechanistic organophosphorus chemistry will soon need to be rewritten. Add to this the rapidly growing awareness of the existence and chemistry of hexacoordinate species formed by nucleophilic attack on phosphoranes and we have all the signs of a major growth point in organic chemistry.

10. Guide to Further Reading

Several excellent specialist texts on organophosphorus chemistry exist, and it is to these that we direct the reader seeking a more rigorous and comprehensive treatment of organophosphorus chemistry than space has allowed in this monograph.

The compendia of Kosolapoff and Maier[20] and Houben-Weyl[21] are invaluable sources of detailed and esoteric organophosphorus preparative chemistry, but they are not bedtime reading. It is a pleasure to record that one of our contributors, Brian Walker, has written an excellent, readable, and cheap general text[22] which for its price must be considered a candidate for best buy. Also very good are Emsley and Hall's contribution,[23] and the somewhat older books by Hudson[24] and Kirby and Warren.[25] Very useful too are Trippett's annual volumes on organophosphorus chemistry in the specialist periodical reports published by the Chemical Society. For those with a thirst for the stereochemical complexities of phosphoranes, Luckenbach's monograph[26] is highly recommended.

11. References

1. E. J. Corey and R. A. E. Winter, *J. Amer. Chem. Soc.*, 1963, **85**, 2677.
2. I. M. Downie, J. B. Holmes, and J. B. Lee, *Chem. and Ind.*, 1966, 900.
3. G. Wittig and G. Geissler, *Annalen*, 1953, **580**, 44.
4. (a) R. E. Ireland and G. Pfister, *Tetrahedron Lett.*, 1969, 2145. (b) M. Tefizon, M. Jurian, and N. T. Anh, *J. Chem. Soc., Chem. Comm.*, 1969, 112. (c) R. E. Ireland, D. C. Muchmore, and U. Hengartner, *J. Amer. Chem. Soc.*, 1972, **94**, 5098. (d) R. E. Ireland, R. Giger, and S. Kamata, *J. Org. Chem.*, 1977, **42**, 1276. (e) S. C. Welch and M. E. Walters, *ibid.*, 1978, **43**, 2715.
5. (a) M. Rasberger and E. Zbiral, *Monatsh. Chem.*, 1969, **100**, 64. (b) G. K. Barnes and G. Tennant, to be published.
6. E. E. Schweizer and R. D. Bach, *J. Org. Chem.*, 1964, **29**, 1746.

7. (a) E. E. Schweizer and K. K. Light, *J. Amer. Chem. Soc.*, 1964, **86**, 2963. (b) E. E. Schweizer, S. D. Goff, and W. P. Murray, *J. Org. Chem.*, 1977, **42**, 200.
8. J. H. Billman, A. Radike, and B. W. Mundy, *J. Amer. Chem. Soc.*, 1942, **64**, 2977; J. H. Billman, C. E. Davis, and D. G. Thomas, *ibid.*, 1946, **68**, 895; W. H. C. Rueggburg and J. Chernack, *ibid.*, 1948, **70**, 1802.
9. J. I. G. Cadogan, S. Kulik, C. Thomson, and M. J. Todd, *J. Chem. Soc. (C)*, 1970, 2437.
10. K. Yamaguchi, T. Sugimae, and M. Kinoshita, *Tetrahedron Lett.*, 1977, 1199.
11. (a) E. A. Arutyunyan, V. I. Gunar, E. P. Gracheva, and S. I. Zavyalov, *Izv. Akad. Nauk SSSR, Ser. Khim.*, 1969, 655. (b) E. A. Arutyunyan, V. I. Gunar, and S. I. Zavyalov, *ibid.*, 1970, 1198. (c) A. Rosowsky and N. Papanthanasopoulos, *J. Heterocyclic Chem.*, 1972, **9**, 1235.
12. (a) E. B. Pedersen and D. Carlsen, *Synthesis*, 1977, 890. (b) *Idem, Tetrahedron*, 1977, **33**, 2089 (includes leading references). (c) E. B. Pedersen and D. Carlsen, *Synthesis*, 1978, 844.
13. H. Z. Lecher, R. A. Greenwood, K. C. Whitehouse, and T. H. Chao, *J. Amer. Chem. Soc.*, 1956, **78**, 5018.
14. (a) B. S. Pedersen, S. Schiebye, N. H. Nilsson, and S.-O. Lawesson, *Bull. Soc. Chim. Belg.*, 1978, **87**, 223. (b) S. Schiebye, B. S. Pedersen, and S.-O. Lawesson, *ibid.*, p. 229, p. 299. (c) B. S. Pedersen, S. Schiebye, K. Clausen, and S.-O. Lawesson, *ibid.*, p. 293. (d) J. B. Rasmussen, K. A. Jørgensen, and S.-O. Lawesson, *ibid.*, p. 307.
15. L. D. Quin, "Tervalent Phosphorus Compounds as Dienophiles," in *1,4-Cycloaddition Reactions* (ed. J. Hamer), Academic Press, New York, 1967.
16. F. Ramirez, *Pure Appl. Chem.*, 1964, **9**, 337.
17. R. Huisgen and J. Wulff, *Tetrahedron Lett.*, 1967, 917.
18. (a) S. Antczak, S. A. Bone, J. Brierley, and S. Trippett, *J. Chem. Soc., Perkin I*, 1977, 278. (b) J. I. G. Cadogan, N. J. Stewart, and N. J. Tweddle, *J. Chem. Soc., Chem. Comm.*, 1978, 182.
19. J. I. G. Cadogan, I. Gosney, E. Henry, T. Naisby, B. Nay, N. J. Stewart, and N. J. Tweddle, *J. Chem. Soc., Chem. Comm.*, 1979, 189; J. I. G. Cadogan, N. J. Stewart, and N. J. Tweddle, *ibid.*, p. 191.
20. G. M. Kosolapoff and L. Maier, *Organic Phosphorus Compounds*, Vols. 1–7, Wiley-Interscience, New York, 1973.
21. *Methoden der Organischen Chemie* (Houben-Weyl), Vols. 12/1, 12/2, G. Thieme, Stuttgart, 1963.
22. B. J. Walker, *Organophosphorus Chemistry*, Penguin, London, 1972.
23. J. Emsley and D. Hall, *The Chemistry of Phosphorus*, Harper and Row, London, 1976.
24. R. F. Hudson, *Structure and Mechanism in Organophosphorus Chemistry*, Academic Press, New York, 1965.
25. A. J. Kirby and S. G. Warren, *The Organic Chemistry of Phosphorus*, Elsevier, Amsterdam, 1967.
26. R. Luckenbach, *Dynamic Stereochemistry of Pentacoordinated Phosphorus and Related Elements*, G. Thieme, Stuttgart, 1973.

2
Transformations via Phosphorus-stabilized Anions 1: Stereoselective Syntheses of Alkenes via the Wittig Reaction

IAN GOSNEY AND (in part) ALAN G. ROWLEY

1. Introduction, 17
2. General aspects, 19
3. Stereoselectivity in reactions of ylides, 26
4. Applications of stereoselective Wittig reactions in the synthesis of natural products, 89
5. References, 142

1. Introduction

In 1953 Wittig and Geissler[1] observed that, when methyltriphenylphosphonium bromide (1) was treated with phenyllithium, an ylide was formed which could react with benzophenone to give triphenylphosphine oxide and 1,1-diphenylethylene (2) in 84% yield. This discovery led to the development of a new method[2-6] for the preparation of alkenes which has since found widespread application in synthetic organic chemistry and is now universally known as the Wittig reaction.

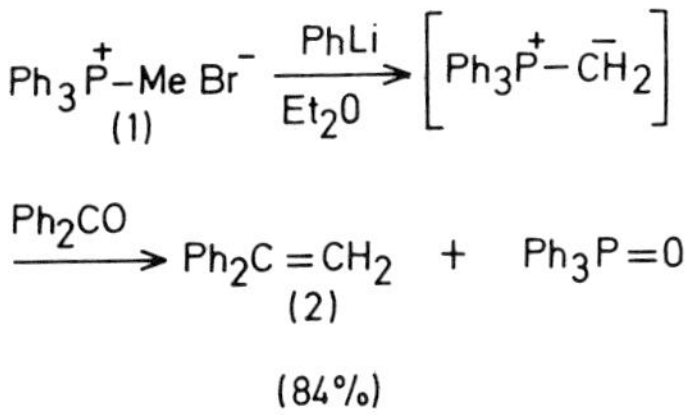

$$Ph_3\overset{+}{P}-Me\ Br^- \xrightarrow[Et_2O]{PhLi} [Ph_3\overset{+}{P}-\overset{-}{C}H_2]$$

$$\xrightarrow{Ph_2CO} Ph_2C=CH_2\ +\ Ph_3P=O$$

(2)

(84%)

Scheme 1 (Ref 1)

One of the main virtues of the Wittig reaction, in marked contrast with other alkene syntheses, is the fact that no ambiguity exists concerning the position of the newly formed carbon–carbon double bond. Even in cases where it occupies an energetically unfavourable position, the double bond always appears at the site of the former carbonyl group. One of the first striking examples of this regioselectivity was the preparation[2] of methylenecyclohexane (3) free of the endocyclic isomer (4).

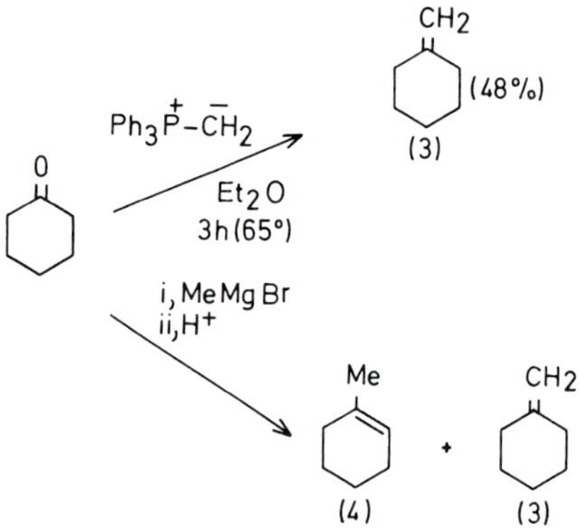

Scheme 2 (Ref 2)

By comparison, the stereochemical course of the reaction is sometimes less predictable, and mixtures of (Z)- and (E)-alkenes are often produced if the ylide and carbonyl compound are both unsymmetrically substituted. This lack of stereoselectivity was first noted by Wittig and Schöllkopf[2] in the reaction of the allylic ylide (5) with benzaldehyde to give a 1:1 mixture of (Z)- (6) and (E)-phenylbutadiene (7). This finding evoked little interest at the time since

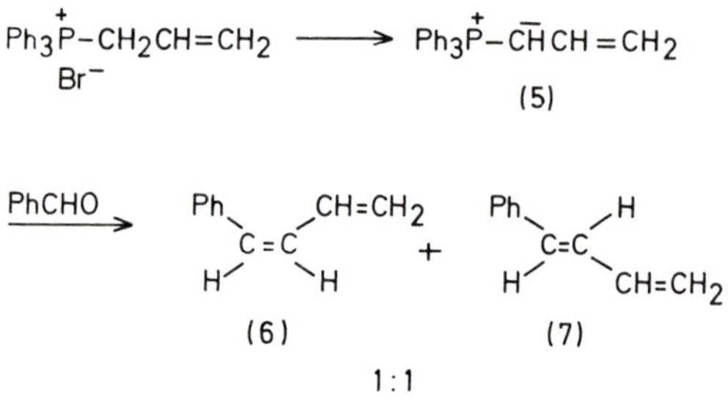

Scheme 3 (Ref 2)

there was no reason to suppose that the Wittig reaction would be intrinsically stereoselective. Subsequent investigations, however, revealed wide variation in the selectivity of the reaction with the direction of coupling,[7] the base used,[2,3] and the reaction temperature.[8] In many cases only the (E)-alkene was formed, a fact that was frequently put to synthetic use. For example, Trippett[9]

2 Wittig Reaction

was able to prepare the first pure sample of all-*trans*-squalene from the condensation of two molecules of geranylacetone (9) with the bis-ylide derived from 1,4-bis(triphenylphosphonio)butane dibromide (8). Whereas steric reasons could be advanced for the preferred formation of the (*E*)-isomer in such cases, no satisfactory explanation could be offered for the occasional predominance of the (*Z*)-isomer.[10-12] This enigma formed the basis

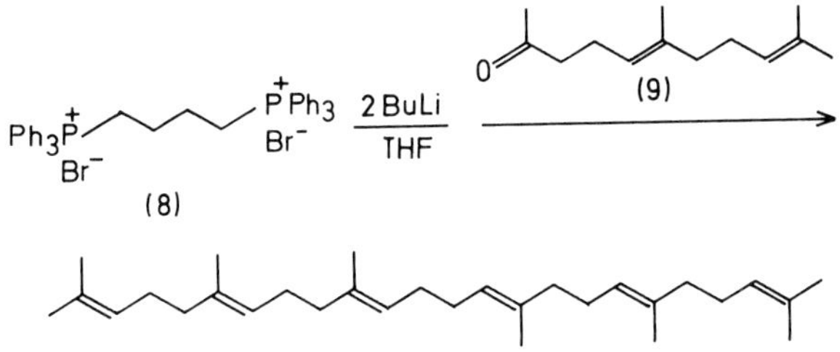

Scheme 4 (Ref 9)

for active study of the stereochemistry of the Wittig reaction, and in particular stimulated a search to see whether conditions could be evolved under which the reaction would become stereospecific.

Bergelson and Shemyakin were the first to explore the *cis*-selectivity in detail and to demonstrate its synthetic utility.[13] Their pioneering work on the effect of solvents and additives on the (*Z*) to (*E*) ratio of alkenes laid the foundations for subsequent mechanistic investigations, notably by Schlosser,[14] which revealed subtle ways to bring about a high degree of stereoselectivity in the Wittig reaction. It is on these developments and their use in synthesis, particularly of natural products, that this chapter will concentrate, but it will commence with a brief discussion of some general aspects of the Wittig reaction. For a more detailed treatment the reader is referred to the several excellent reviews on the subject already available.[15] The applications of the Wittig reaction in specific fields such as those of vitamin A and its derivatives,[16] carbohydrates[17] and industrial practice[18] have also been reviewed elsewhere, as has the use of bis-Wittig reactions in the synthesis of non-benzenoid aromatic ring systems.[19] In addition we recommend reference to a brief survey of the synthesis of cycloalkenes by the intramolecular Wittig reaction.[20]

2. General Aspects

2.1. Introduction

The scope of the Wittig reaction is extremely wide because it allows considerable variation of the carbonyl component (12) of the reactants (Scheme 5);

moreover, the mildness of the reaction conditions permits the inclusion of relatively delicate functionality such as ester or even acetal. The nature of the ylide component (11) is, however, far more critical with regard to both the ease of its preparation and its reactivity. These aspects are discussed in some detail in a later section.

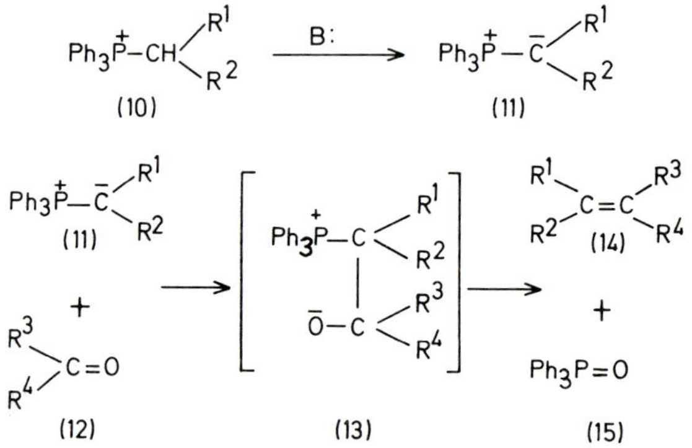

$$Ph_3\overset{+}{P}-CH\begin{smallmatrix}R^1\\R^2\end{smallmatrix} \quad \xrightarrow{B:} \quad Ph_3\overset{+}{P}-\overset{-}{C}\begin{smallmatrix}R^1\\R^2\end{smallmatrix}$$
$$(10) \hspace{5em} (11)$$

$$Ph_3\overset{+}{P}-\overset{-}{C}\begin{smallmatrix}R^1\\R^2\end{smallmatrix} \quad \longrightarrow \quad \left[Ph_3\overset{+}{P}-C\begin{smallmatrix}R^1\\R^2\\|\\\overset{-}{O}-C\begin{smallmatrix}R^3\\R^4\end{smallmatrix}\end{smallmatrix} \right] \quad \longrightarrow \quad \begin{smallmatrix}R^1\\R^2\end{smallmatrix}C=C\begin{smallmatrix}R^3\\R^4\end{smallmatrix}$$
$$(11) \hspace{8em} (13) \hspace{6em} (14)$$
$$+ \hspace{12em} +$$
$$\begin{smallmatrix}R^3\\R^4\end{smallmatrix}C=O \hspace{14em} Ph_3P=O$$
$$(12) \hspace{18em} (15)$$

Scheme 5

The preparation of an alkene by the Wittig reaction involves three stages. First, the phosphonium salt (10) must be prepared, usually from triphenylphosphine. The preparation of phosphonium salts has been comprehensively reviewed.[15d,21] In the second stage of the reaction the salt (10) is treated with a base to convert it into the ylide (11) which is then, third, allowed to react with the carbonyl compound (12) to give the olefinic product (14) and triphenylphosphine oxide (15) via the intermediacy of the betaine (13).

The choice of base for the conversion of (10) into (11) is discussed in Section 2.3, while in Section 2.2 we discuss the factors that influence the reactivity of the ylide (11).

2.2. Reactivity of the Phosphorus Ylide

Phosphorus ylides can be considered as a unique form of carbanion, the charge of which is modified by possible $d\pi-p\pi$ bonding (11a ↔ 11b). The contributing dipolar ylide form (11b) gives the ylide a nucleophilic character which is further modified by the nature of the groups R^1 and R^2. The latter may confer either exceptional stability or abnormal reactivity. Thus groups

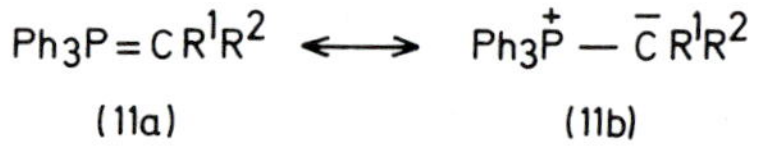

$$Ph_3P=CR^1R^2 \quad \longleftrightarrow \quad Ph_3\overset{+}{P}-\overset{-}{C}R^1R^2$$
$$(11a) \hspace{6em} (11b)$$

Scheme 6

2 Wittig Reaction

that are strongly electron-withdrawing will stabilize the carbanion and consequently reduce the reactivity (nucleophilicity) of the ylide and hence reduce the ease of formation of the betaine (13). In general, ylides with a highly delocalized negative charge are of little or no value in the Wittig reaction. Thus triphenylphosphonium cyclopentadienylide (16) (Scheme 7) does not

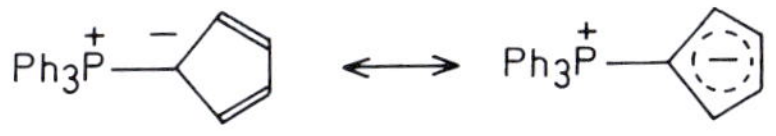

$$Ph_3\overset{+}{P}-\underset{(16a)}{\overset{-}{\bigcirc}} \longleftrightarrow Ph_3\overset{+}{P}-\underset{(16b)}{\bigcirc^{(-)}}$$

Scheme 7

react with carbonyl compounds on heating in ethanol for 120 hours.[22] Furthermore, it is recovered unchanged after prolonged heating with ethanolic potassium hydroxide, a procedure that would convert most ylides into triphenylphosphine oxide and a hydrocarbon. This remarkable stability can be attributed to the importance of the resonance structure (16b), a quasi-aromatic system. By comparison, triphenylphosphonium methylide (11; $R^1 = R^2 = H$), where there is no such interaction, is highly reactive towards all carbonyl compounds and is instantaneously hydrolysed by atmospheric moisture.

The nature of the substituents on phosphorus also affects the reactivity of an ylide. Replacement of the "stationary" phenyl groups on phosphorus by electron-releasing groups, e.g. alkyl, will increase the reactivity of the ylide by stabilizing the contributing dipolar form in the resonance hybrid. Thus tri-n-butylphosphonium fluorenylide (17; $R = Bu^n$) gives much better yields of alkenes with aldehydes and ketones than does its triphenyl analogue (17; $R = Ph$), and unlike the latter is sensitive to water.[23]

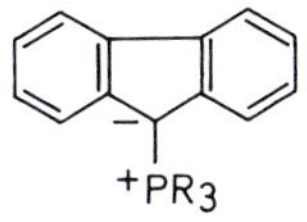

(17)

Scheme 8

Subsequent discussion will show, however, that in spite of reactivity considerations, aliphatic substituents on phosphorus will result in decreased yields in reactions of ylides that are not stabilized by electron-withdrawing groups. Suffice it to mention at this point that in these cases loss of phosphine oxide from the betaine intermediate (Scheme 5) is the critical step in the reaction and increased electron density on phosphorus will have an unfavourable effect upon this step.

In view of the large variation in their reactivity, ylides have been loosely classified as either "reactive" or "stable", with borderline cases sometimes being referred to as "moderate" ylides. Reactive ylides are those that must be prepared in the absence of oxygen and moisture and used immediately, whereas stable ylides can be isolated as relatively hydrolytically stable, characterizable solids.

From the practical point of view of the Wittig reaction the relative reactivities of the different classes of ylides towards carbonyl compounds are of more

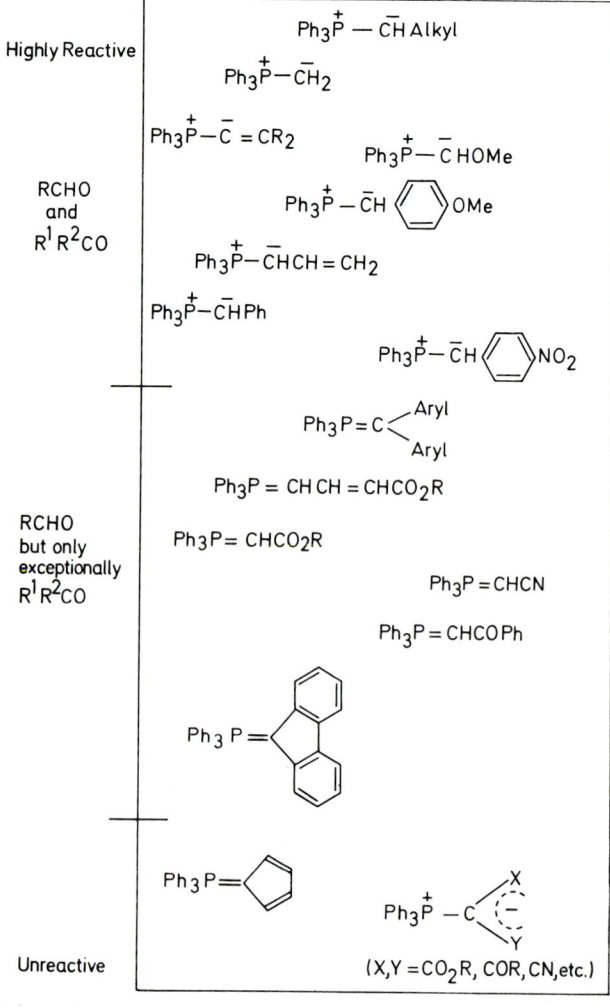

Scheme 9   Approximate reactivities of ylides towards aldehydes and ketones

2 Wittig Reaction

importance than their hydrolytic stability, and Scheme 9 summarizes the relative reactivities of triphenylphosphonium ylides towards aldehydes and ketones. It should be noted that the preparatively important ylides stabilized by an adjacent carbonyl group will react, as a rule, with aldehydes, but not with ketones, except under forcing conditions. A modification of the Wittig reaction, utilizing a phosphonate carbanion, e.g. (18), with the carbonyl compound, has proved of value in such cases.[24,25] Thus (18) (Scheme 10)

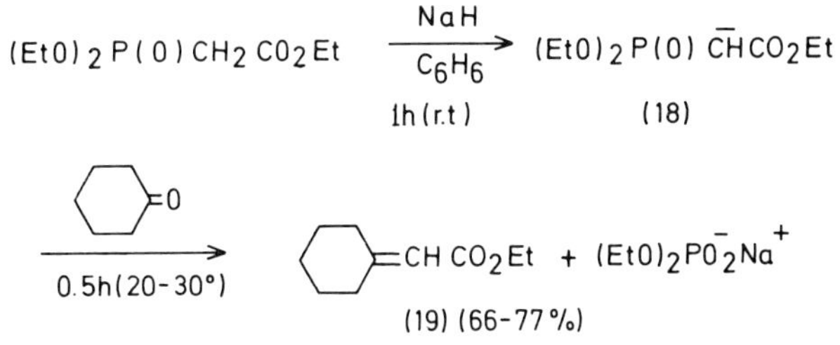

$$(EtO)_2P(O)CH_2CO_2Et \xrightarrow[\text{1h (r.t)}]{\text{NaH}\atop C_6H_6} (EtO)_2P(O)\bar{C}HCO_2Et$$
$$(18)$$

$$\underset{0.5h(20-30°)}{\overset{\text{cyclohexanone}}{\longrightarrow}} \text{cyclohexylidene}=CHCO_2Et + (EtO)_2P\bar{O}_2Na^+$$
$$(19)\ (66-77\%)$$

Scheme 10 (Ref 26)

reacts with cyclohexanone *at room temperature* within a few minutes to give ethyl cyclohexylideneacetate (19) in 70% yield,[26] whereas the corresponding Wittig reaction gives a satisfactory yield only after heating the reactants without solvent at 170° for 10 hours.[27] The greater reactivity of phosphonate anions, as compared to phosphorus ylides, is due to the fact that the phosphonate group is not so effective as the phosphonium group in stabilizing a negative charge, because of "back donation" from oxygen. Generally, phosphonate anions are the reagents of choice for the synthesis of alkenes having electron-withdrawing groups on the α-carbon. In all other cases, phosphorus ylides are to be preferred.

2.3. Choice of Base

The strength of base required to convert a phosphonium salt into an ylide will clearly depend on the acidity of the hydrogen on the α-carbon atom.[15d,28] Table 1 provides a summary of the bases appropriate for the generation of some common types of ylide. Thus phosphonium salts with electron-withdrawing groups on the α-carbon, the precursors to stable ylides, are deprotonated by dilute aqueous alkalis or by neat amines. If there are electron-donating substituents on the α-carbon, for example alkyl groups, then metal alkyls or hydrides are normally required to remove the α-proton.

Intermediate between these two extremes is the case where the α-proton is allylic or benzylic, when alcoholic alkoxide is the base of choice. Any substituents in the α-phenyl groups will of course modify the acidity of the

Table 1. Bases used in the Wittig reaction

Base	Solvent frequently used	Phosphonium salt	Ref.
NH_3 (anh.)	MeCN	$Ph_3P^+CH_2C\equiv CH$	171
NH_4OH	EtOH	$Ph_3\overset{+}{P}$-fluorenyl	365
Et_3N or pyr.	CH_2Cl_2	$Ph_3P^+CH(Br)CO_2R$	366
NaOH	10% aq.	$Ph_3P^+CH(Ph)COMe$	85
		$Ph_3P^+CH_2CO_2Et$	367
	aq. CH_2Cl_2	$Ph_3P^+CH_2Ph$	164, 165
KOH	Pr^iOH	$Ph_3\overset{+}{P}$-$CH_2\diagdown\diagup\diagdown$R	368
epoxide	CH_2Cl_2	$Ph_3P^+CH_2CO_2Et$	45–47
	neat	$Ph_3P^+CH_2alkyl$	47
$Et_4N^+F^-$	MeCN	$Ph_3P^+CH_2COMe$	369
Na_2CO_3	aq. MeOH	$Ph_3P^+CH_2COCH_2Cl$	370
LiOEt	EtOH	$Ph_3\overset{+}{P}$-$CH_2\diagdown\diagup$Ph	371
		$C_6H_4(CH_2$-$\overset{+}{P}Ph_3)_2$	372
NaOEt	EtOH	$Ph_3P^+CH_2CO_2Et$	50, 51
		$Ph_3P^+CH_2aryl$	3, 163, 167, 373
		$Ph_3P^+CH_2OMe$	169
	DMF	$Ph_3P^+CH_2alkyl$	124, 125, 140
$NaOC(Et)Me_2$	C_6H_6	$Ph_3P^+CH_2alkyl$	364
Bu^tOK	Bu^tOH	$Ph_3P^+CH_2Ph$	163
	THF	$Ph_3P^+CH_2alkyl$	130
		$Ph_3P^+CH_2CH=CH_2$	187
		$Ph_3\overset{+}{P}$-CH_2-(morpholinyl)	374
	DMF	$Ph_3P^+CH_2aryl$	375
		$Ph_3\overset{+}{P}$-$CH_2\diagdown\diagup\diagdown$	103
	DMSO	Ph_3P^+Me	354
	DME	$Ph_3P^+CH(Me)OMe$	21
	C_6H_6	$Ph_3P^+CH_2alkyl$	174
NaH^a	DMSO	$Ph_3P^+CH_2alkyl$	105, 113–115, 135, 139, 141, 143, 240, 246
		$Ph_3P^+CH_2aryl$	199
		$Ph_3P^+CH_2CH=CH_2$	168, 376

2 Wittig Reaction

Table 1. *Continued*

Base	Solvent frequently used	Phosphonium salt	Ref.
	DMF	$Ph_3P^+CH_2CH=CHR$	175
	THF	$Ph_3P^+CH_2$alkyl	107, 149
KH	DMSO	$Ph_3P^+CH_2$alkyl	267
$LiNPr^i_2$	THF	$Ph_3P^+CH_2CH=CH_2$	192
$NaNH_2$	liq. NH_3	$Ph_3P^+CH_2$alkyl	33, 377
		$Ph_3P^+CH_2$aryl	166, 201
MeLi	THF	$Ph_3P^+CH_2$alkyl	144
		$Ph_3P^+CH_2C(Pr^i)=CH_2$	360
Bu^nLi	Et_2O	$Ph_3P^+CH_2$alkyl	100, 101, 103, 104, 378
		$Ph_3P^+CH_2CH=CH_2$	173, 379
	toluene	$Ph_3P^+CH_2$alkyl	328
	C_6H_6	$Ph_3P^+CH_2$alkyl	10, 110
		$Ph_3P^+CH_2Ph$	85, 110, 193, 213
	THF	$Ph_3P^+CH_2$alkyl	9, 108, 30, 146, 380
		$Ph_3P^+CH_2CN$	38
		$Ph_3P^+CH_2C{\equiv}CSiMe_3$	204
	dioxan	$Ph_3P^+CH_2Ph$	154
	DMSO	$Ph_3P^+CH_2$alkyl	289
	DMF	$Ph_3P^+CH_2$alkyl	110, 111
		$Ph_3P^+CH_2Ph$	110
	THF–HMPA	$Ph_3P^+CH_2$alkyl	119, 299
PhLi	Et_2O	$Ph_3P^+CH_2$alkyl	1, 102
		$Ph_3P^+CH_2CH=CHR$	2, 190
	THF	$Ph_3P^+CH_2$alkyl	33, 155
		$Ph_3P^+CH_2CH=CHR$	351
K^b	HMPA	$Ph_3P^+CH_2$alkyl	117, 118, 233, 296
$NaN(SiMe_3)_2$	THF	$Ph_3P^+CH_2$alkyl	131, 287, 288, 290, 291
$LiN(SiMe_3)_2$	HMPA	$Ph_3P^+CH_2$alkyl	271

[a] The effective base is dimsyl anion ($MeSOCH_2^-$).
[b] The effective base is Me_2N^- plus $[(Me_2N)_2PO]^-$.

α-proton by their electronic effects. The choice of base may be affected by factors other than the acidity of the α-hydrogen, such as the presence of sensitive functionality in the phosphonium salt. Metal hydrides and alkyls will, for example, reduce carbonyl groups.

The nature of the base, in so far as it affects the overall polarity of the reaction medium, also affects the stereochemical course of the Wittig reaction. We return to this point in some detail in the following section.

3. Stereoselectivity in Reactions of Ylides

3.1. Introduction

Any attempt to rationalize the stereochemistry of the alkene formed in the Wittig reaction hinges on a full understanding of the mechanism of the reaction. Unfortunately, some of the mechanistic aspects are still something of a puzzle, but it is generally accepted that the reaction is a two-stage process initiated by nucleophilic attack of the ylide on the carbonyl carbon atom to give two possible diastereoisomeric betaines (20a) and (20b). Subsequent decomposition is then believed to occur by way of a four-centre cyclic intermediate [29-31] (21) which eliminates phosphine oxide in a *syn* fashion [32,33] to form the (Z)- or (E)-alkene. At least part of the driving force for the second stage of the reaction appears to be the formation of the very strong phosphorus–oxygen bond.

For reasons that are not yet well understood, there appears to be a decided kinetic preference for the formation of the *erythro*-betaine (20a), regardless of

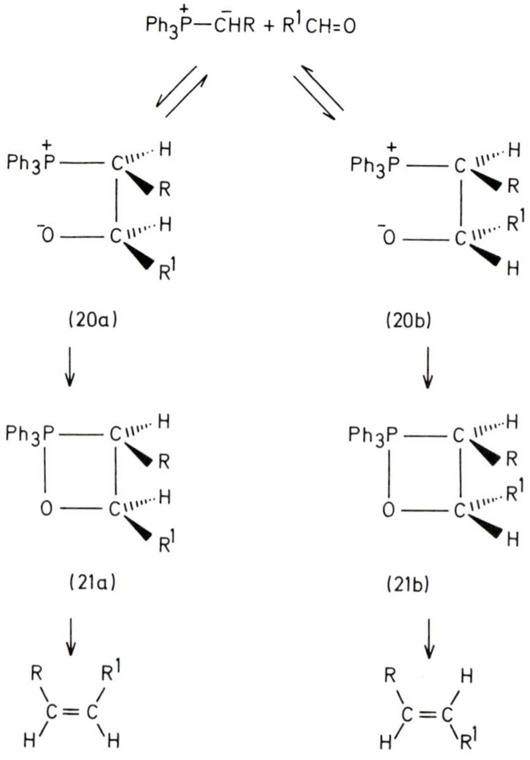

$$Ph_3\overset{+}{P}-\overset{-}{C}HR + R^1CH=O$$

(20a)   (20b)

(21a)   (21b)

Scheme 11

2 Wittig Reaction

the nature of the reactants involved. However, betaine formation is reversible in most, if not all, cases [32,34,35] thus allowing interconversion to the thermodynamically more stable *threo*-isomer (20b). It follows that the stereochemistry of the alkene produced will depend both on structural factors and on reaction conditions that affect the balance between the betaine dissociation and betaine decomposition steps. In attempting to discriminate between these factors and appraise their relative importance it is convenient to deal separately with the three distinct categories of ylides: "stable" ylides, i.e. ylides having a powerful electron-withdrawing substituent in the α-position; "reactive" ylides which do not contain stabilizing α-substituents; and ylides of intermediate reactivity, "moderate" ylides.

3.2. Stable Ylides

In general, reactions employing stabilized ylides favour the formation of the stereoisomer with the activating group at the α-carbon of the alkene *trans* to the larger group at the β-carbon. This stereoselectivity appears to be attributable to the preferential decomposition of the betaine intermediate in the *threo* (20b) rather than the *erythro* (20a) configuration since in the transition state, e.g. (22) leading to (*E*)-alkene, there is less steric interference to conjugative stabilization of the incipient double bond. As a result, the equilibrium between the two betaines is effectively displaced toward the *threo*-isomer, leading to preferential formation of (*E*)-alkene.

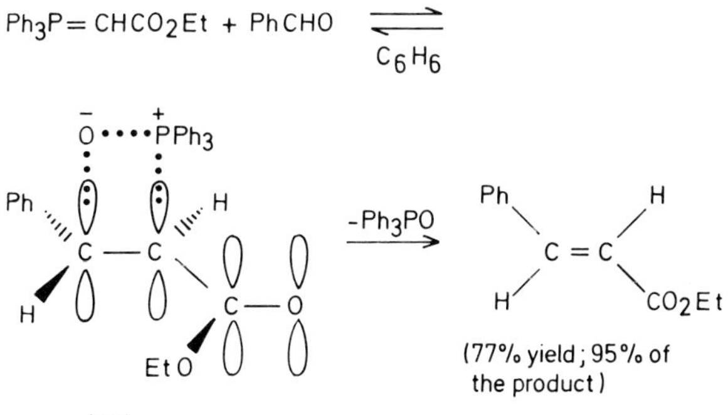

$$Ph_3P=CHCO_2Et + PhCHO \underset{C_6H_6}{\rightleftarrows}$$

(22) → -Ph$_3$PO → Ph, H, C=C, H, CO$_2$Et (77% yield; 95% of the product)

Scheme 12 (Ref 36)

Scheme 13 provides several examples indicating the high degree of stereoselectivity that may be obtained with stabilized ylides. These examples have been chosen to illustrate general experimental procedures. The use of epoxides as the base in a "one pot" synthesis of alkenes by Buddrus is worthy of special note.[45-47]

As many of the examples in Scheme 13 indicate, the (E)-isomer is often produced with almost complete exclusion of the (Z). It should be recognized that this may not be a true reflection of the stereochemistry of the reaction. In many cases, (Z)-alkenes having electron-withdrawing groups at the α-carbon are isomerized to the thermodynamically more stable (E)-isomer by the elevated temperatures required to effect reaction.[48] In addition, the reactants or products may catalyse the isomerization.[49,50]

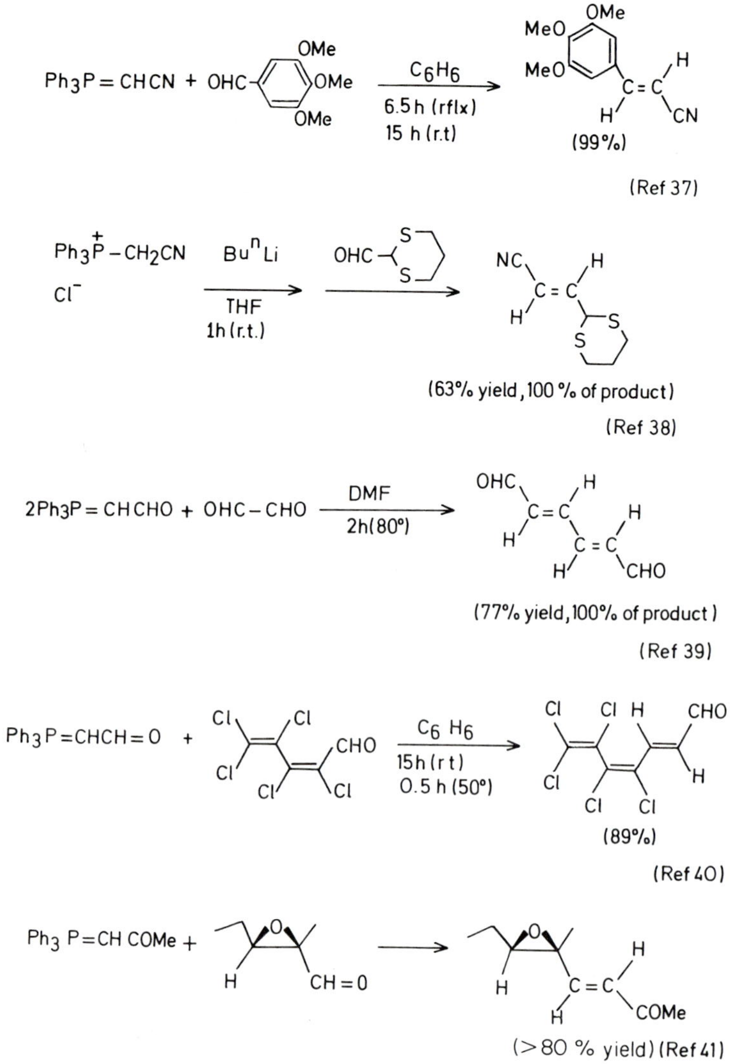

Scheme 13

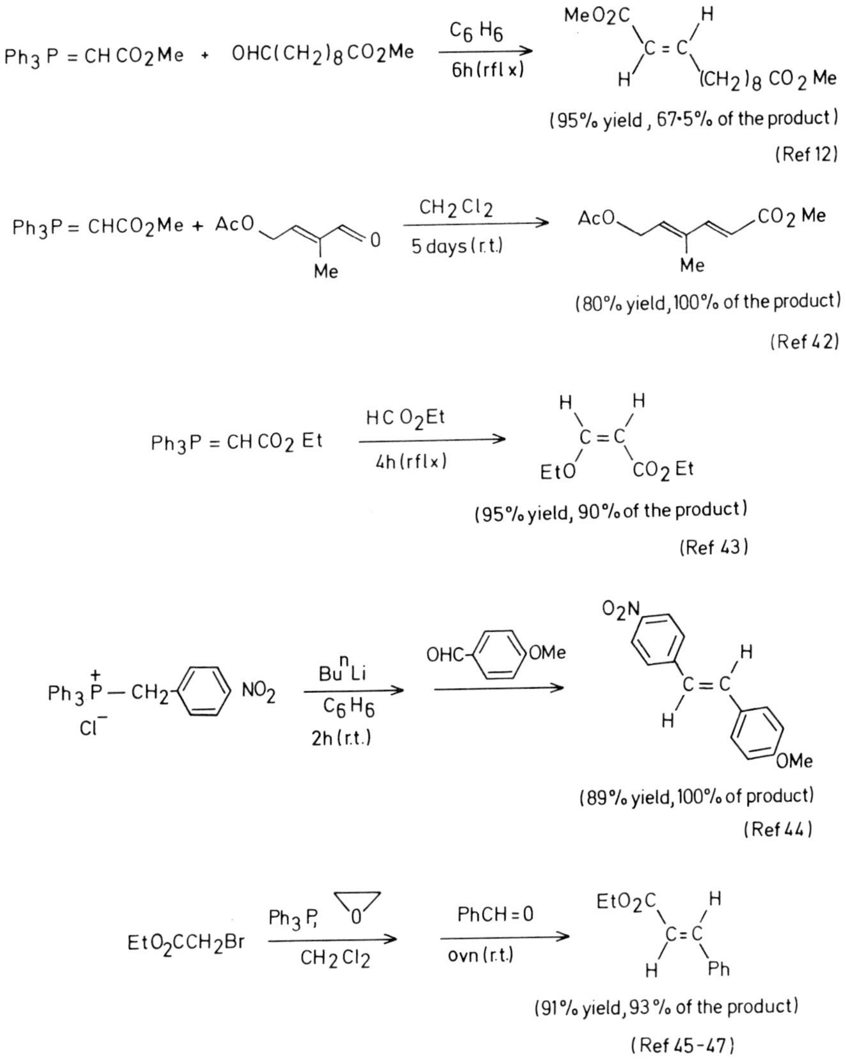

2 Wittig Reaction

$Ph_3P=CHCO_2Me$ + $OHC(CH_2)_8CO_2Me$ $\xrightarrow[6h(rflx)]{C_6H_6}$ MeO_2C, H / $C=C$ / H, $(CH_2)_8CO_2Me$

(95% yield, 67·5% of the product)

(Ref 12)

$Ph_3P=CHCO_2Me$ + $AcO\diagdown\diagup\diagdown\diagup{O}$ (Me) $\xrightarrow[5 days (r.t.)]{CH_2Cl_2}$ $AcO\diagdown\diagup\diagdown\diagup CO_2Me$ (Me)

(80% yield, 100% of the product)

(Ref 42)

$Ph_3P=CHCO_2Et$ $\xrightarrow[4h(rflx)]{HCO_2Et}$ H, H / $C=C$ / EtO, CO_2Et

(95% yield, 90% of the product)

(Ref 43)

$Ph_3\overset{+}{P}-CH_2-\langle\rangle NO_2$ Cl^- $\xrightarrow[\substack{C_6H_6\\2h(r.t.)}]{n\text{-}BuLi}$ $OHC-\langle\rangle-OMe$ $\longrightarrow$ $O_2N-\langle\rangle$, H / $C=C$ / H, $\langle\rangle-OMe$

(89% yield, 100% of product)

(Ref 44)

EtO_2CCH_2Br $\xrightarrow[CH_2Cl_2]{Ph_3P, \triangleleft_O}$ $\xrightarrow[ovn (r.t.)]{PhCH=O}$ EtO_2C, H / $C=C$ / H, Ph

(91% yield, 93% of the product)

(Ref 45-47)

Scheme 13 concluded

The degree of stereoselectivity observed may be influenced by the nature of the substituents on the ylide. A characteristic feature of stabilized ylides is that replacement of phenyl groups on phosphorus by alkyl groups such as butyl[51] or cyclohexyl[51,52] shifts the isomer ratio further in favour of the (E)-product (Scheme 14). To account for this effect it has been suggested that alkyl groups (which are more electron-donating than phenyl) reduce the effective charge at

$$R_3\overset{+}{P}-CH_2CO_2Et \quad X^- \xrightarrow[\text{ii, PhCHO}]{\text{i, NaOEt/EtOH, 25°}} PhCH=CHCO_2Et$$

(X = Br or BPh$_4$)

R	Product (Z) : (E)		Ref
Ph	15	85	50, 51
Bu	5	95	50, 51
n-C$_6$H$_{13}$	2	98	51
n-C$_{10}$H$_{21}$	4	96	51
cyclo-C$_6$H$_{13}$	0	100	51, 52

Scheme 14

phosphorus, thus allowing equilibration of the intermediate betaines. However, it may equally well be due to increased steric interaction preventing the formation of the transition state leading to (Z)-alkene.

As mentioned earlier, substitution of phenyl groups on phosphorus by alkyl groups also results in an increase in the yields of alkene.[23] This is because electron-releasing groups on phosphorus facilitate nucleophilic attack[50] on the carbonyl group in the first (rate-controlling) step of the Wittig reaction by

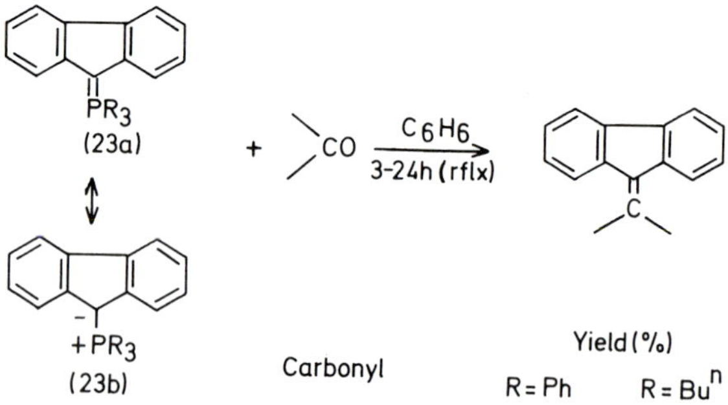

Carbonyl	Yield (%) R = Ph	Yield (%) R = Bu^n
p-O$_2$N C$_6$H$_4$CHO	96	99
p-Cl C$_6$H$_4$CHO	93	96
PhCHO	84	96
p-MeOC$_6$H$_4$CHO	37	94
p-Me$_2$N C$_6$H$_4$CHO	0	94
(p-O$_2$NC$_6$H$_4$)$_2$CO	0	93
p-O$_2$NC$_6$H$_4$CO Me	0	67
p-Cl C$_6$H$_4$CO Me	0	9

Scheme 15 (Ref 23)

stabilizing the contributing dipolar canonical form, e.g. (23b). Thus, fluorenylidenetriphenylphosphorane (23; R = Ph) reacts only with the more reactive (electrophilic) aldehydes, whereas the tri-n-butyl ylide (23; R = Bu^n) gives quantitative yields of alkene with most aldehydes, and good yields with the more reactive ketones, e.g. *p*-nitroacetophenone.

It should be noted that the situation is completely different with the more reactive (non-stabilized) ylides, for which the second step of the Wittig reaction, decomposition of the betaine, is rate-controlling. Here, electron-releasing groups on phosphorus have an unfavourable effect because they slow down betaine decomposition to the point where phosphine oxide elimination is difficult, or does not take place at all.[53-56] For example, the reaction between methylenetrimethylphosphorane (24) and benzophenone stops at the betaine stage (25) (Scheme 16), whereas under similar conditions

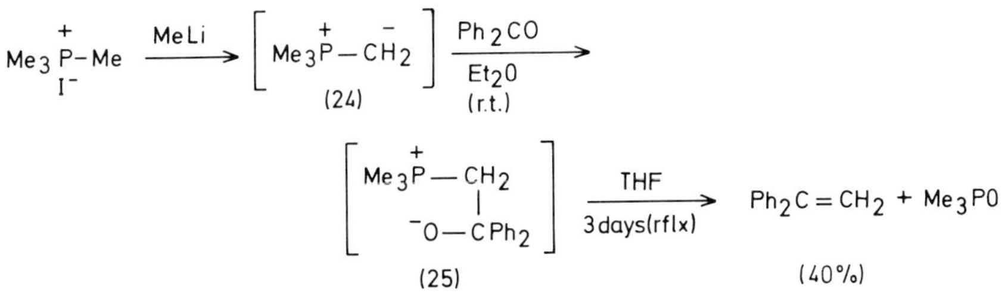

$$Me_3\overset{+}{P}-Me \quad I^- \xrightarrow{MeLi} \left[Me_3\overset{+}{P}-\overset{-}{CH_2} \right] \xrightarrow[Et_2O]{Ph_2CO} \atop (r.t.)$$

$$\left[\begin{array}{c} Me_3\overset{+}{P}-CH_2 \\ | \\ {}^-O-CPh_2 \end{array} \right] \xrightarrow[3\,days(rflx)]{THF} Ph_2C=CH_2 + Me_3PO$$

(24)   (25)   (40%)

Scheme 16 (Ref 53-55)

its triphenyl analogue affords an almost quantitative yield of 1,1-diphenylethylene.[53] It is possible to force the betaine (25) to decompose to 1,1-diphenylethylene by prolonged heating in tetrahydrofuran,[54,55] but even then the yield is only 40%. On the other hand, the betaine from the ylide (26) and benzaldehyde could not be decomposed to phosphine oxide and alkene even at 180°.[56]

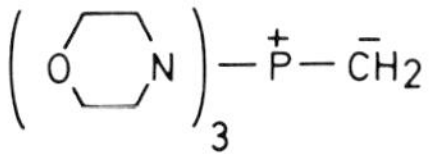

$$\left(O\underset{\smile}{\frown}N \right)_3 -\overset{+}{P}-\overset{-}{CH_2}$$

(26)

Scheme 17 (Ref 56)

The *trans*-stereoselectivity of Wittig reactions of stabilized ylides may be adversely affected by the introduction of substituents at the position α to the phosphorus atom. If the group is sufficiently large, the energy difference between the two diastereoisomers of the intermediate betaine will be lowered. In this case appreciable amounts of (*Z*)-alkenes are formed in addition to the (*E*)-isomers. For example, reaction of phenacylidenetriphenylphosphorane

(27; R = H) with *p*-nitrobenzaldehyde gives entirely (*E*)-alkene, whereas introduction of an α-bromine atom (27; R = Br) leads to the formation of 35% of the (*Z*)-isomer.[35]

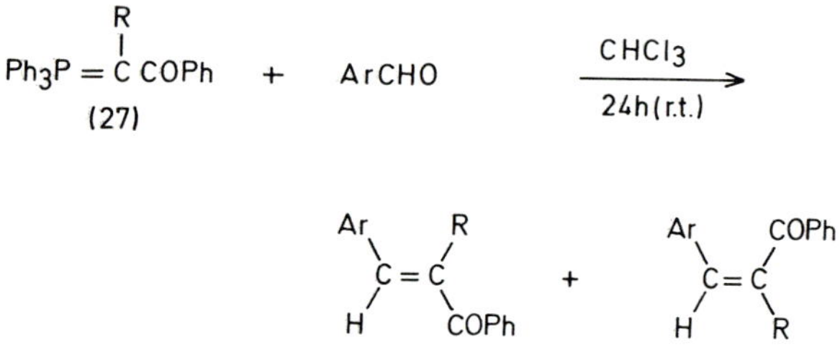

$$Ph_3P = \overset{R}{\underset{(27)}{C}} COPh \quad + \quad ArCHO \quad \xrightarrow{CHCl_3}_{24h(r.t.)}$$

$$\underset{H}{\overset{Ar}{\diagdown}} C = C \underset{COPh}{\overset{R}{\diagup}} \quad + \quad \underset{H}{\overset{Ar}{\diagdown}} C = C \underset{R}{\overset{COPh}{\diagup}}$$

R	Yield (%)	Product composition	
H	72	100	0
Br	64	35	65

(Ar = *p*-O₂NC₆H₄)

Scheme 18 (Ref 35)

A similar situation prevails in the preparation of esters related to the insecticide bioresmethrin (Scheme 19).[57] Condensation of the (*E*)-aldehyde (28) with the ethoxycarbonylmethylide (29; R = H) gives a product containing > 90% of the (*E*)-alkene; on the other hand, when R = Br or Cl, the condensation produces an isomer mixture in which the (*Z*)-product predominates (70–80%).

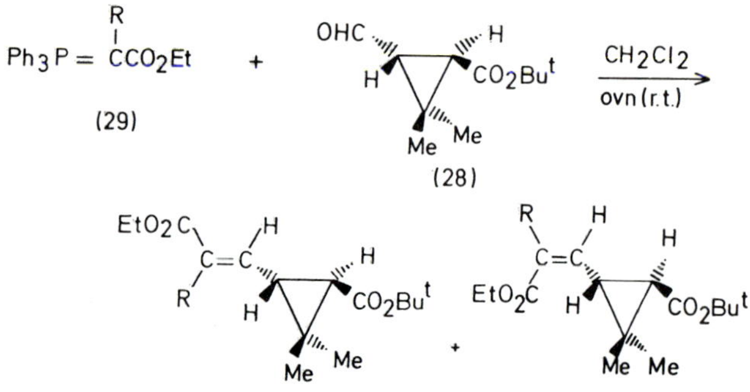

$$Ph_3P = \overset{R}{\underset{(29)}{C}} CO_2Et \quad + \quad (28) \quad \xrightarrow{CH_2Cl_2}_{ovn(r.t.)}$$

R	Product composition		
H	≤10	:	≥90
Cl	70	:	30
Br	80	:	20

Scheme 19 (Ref 57)

2 Wittig Reaction

The effect of α-methyl substituents is of interest. As a rule, the (E)-alkene is formed almost exclusively as illustrated by the reaction[58] of methoxycarbonylethylidenetriphenylphosphorane (30) with acetaldehyde to give practically pure methyl tiglate (31), i.e. the (E)-isomer, and with the acetylenic aldehyde[59] (32) to give a product (33) which was all-*trans* as shown by n.m.r. Similarly, the same ylide with 2-hydroxytetrahydropyran (34)[60,61] afforded methyl 7-hydroxy-2-methyl-(E)hept-2-enoate (35) stereospecifically. How-

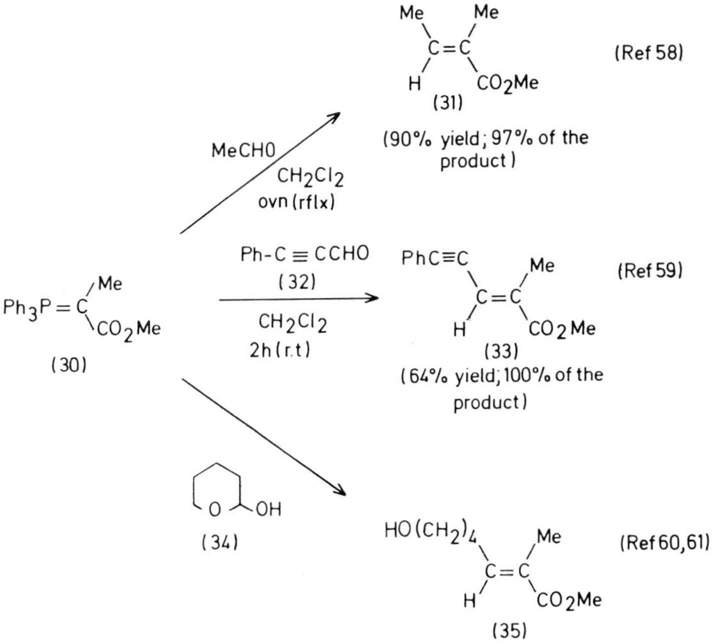

Scheme 20

ever, loss of stereospecificity has occasionally been observed,[62-65] a case in point being the condensation of 2-ethoxycarbonylethylidenetriphenylphosphorane (Scheme 21) with the aldehyde (36) to form a 5:1 mixture of the unsaturated esters (37) and (38) respectively.[63] When reduced the latter gave a mixture of β-santalol (39) and its all-*trans* isomer (40). It is also worth noting[65] that the reaction of cyanoethylidenetriphenylphosphorane (41) with the indole aldehyde (42) is considerably less stereoselective, giving a product mixture, the composition of which displays an interesting dependence on the solvent (Scheme 22). In tetrahydrofuran the thermodynamically unfavourable (Z)-isomer (44) is preferentially formed, whereas in benzene the (E)-isomer (43) becomes the favoured product.

Whereas in the reactions of stabilized ylides with aldehydes the (E)-alkene

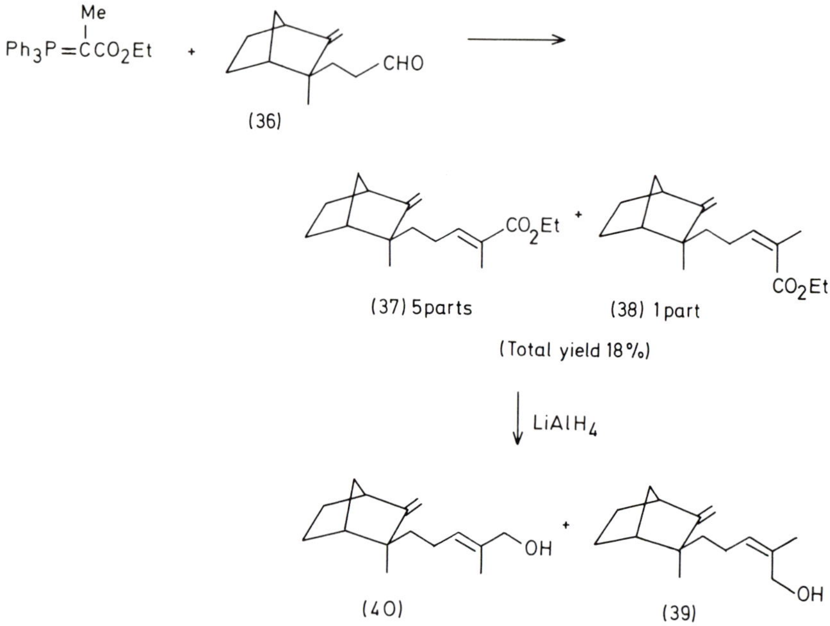

Scheme 21 (Ref 63)

is predominantly formed, interaction with unsymmetrical ketones almost always leads to a considerable amount of the (Z)-isomer. However, condensation often fails unless in the presence of an acid catalyst, e.g. benzoic acid.[66-69] The success of this procedure is attributable to protonation of the carbonyl group making it more electrophilic, and therefore susceptible to nucleophilic attack by the ylide. Typical examples of acid-catalysed reactions with unsymmetrical ketones are illustrated in Scheme 23. It should be noted that catalysis is considerably reduced in hydrogen-bonding solvents, e.g. chloroform.[70]

An alternative, but often better procedure, to acid catalysis involves heating a mixture of the reactants without a solvent to relatively high temperatures (150–200°).[27,71] For example, in the synthesis of abscisic acid (Scheme 24), the condensation between ethoxycarbonylmethylenetriphenylphosphorane (46) and 1-hydroxy-4-oxo-α-ionone (45) was practically complete after 10 min at 140–170° and gave approximately equal amounts of the (E,Z)- and (E,E)-esters (47) and (48) respectively.[72] A comparison of the stereochemical outcome of the procedure with that obtained in toluene (or methanol) after prolonged heating revealed little difference in the isomer ratio[73,74] although increasing the temperature above 170° appeared to increase the amount of (E,E)-isomer.[72]

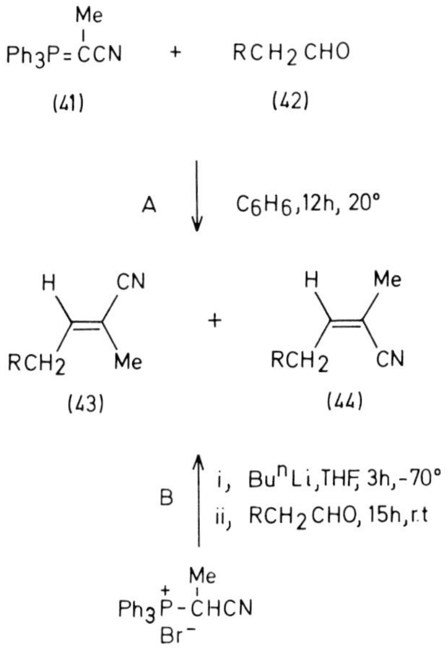

2 Wittig Reaction

$$\text{Ph}_3\text{P=}\overset{\text{Me}}{\underset{|}{\text{C}}}\text{CN} \quad + \quad \text{RCH}_2\text{CHO}$$

(41)   (42)

A ↓ C$_6$H$_6$, 12h, 20°

(43) *(H, CN on one side; RCH$_2$, Me on other)* + (44) *(H, Me on one side; RCH$_2$, CN on other)*

B ↑ i, Bu^{n}Li, THF, 3h, −70°
ii, RCH$_2$CHO, 15h, r.t

$$\text{Ph}_3\overset{+}{\text{P}}-\overset{\text{Me}}{\underset{|}{\text{C}}}\text{HCN} \quad \text{Br}^-$$

Route	Ratio of (43) : (44)	Total yield %
A	63 : 37	54
B	38 : 62	54

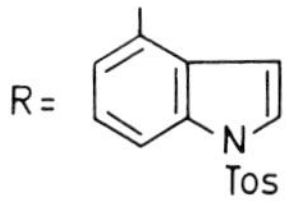

R = (indole substituted at N with Tos)

Scheme 22 (Ref 65)

A third approach which appears to have the greatest synthetic utility makes use of phosphonate anions (49), where R^1 is a group capable of stabilizing the adjacent anion. This procedure, discovered by Horner et al.[75] but also known as the Wadsworth–Emmons reaction,[26] may be regarded as a useful supplement to the Wittig reaction (Scheme 25). The scope, mechanism, and stereochemistry of this reaction are reviewed in Chapter 3 and will not be considered here except by way of comparison. Its greatest merit lies in the fact that phosphonate anions are much more reactive, i.e. nucleophilic, than the corresponding Wittig reagents. This is a result of the fact that the phosphonate group has lower net positive charge and accordingly provides less stabilization for the adjacent carbanion by valence shell expansion, (49a) ↔ (49b). Thus ketones that react sluggishly or not at all with Wittig reagents (stabilized by

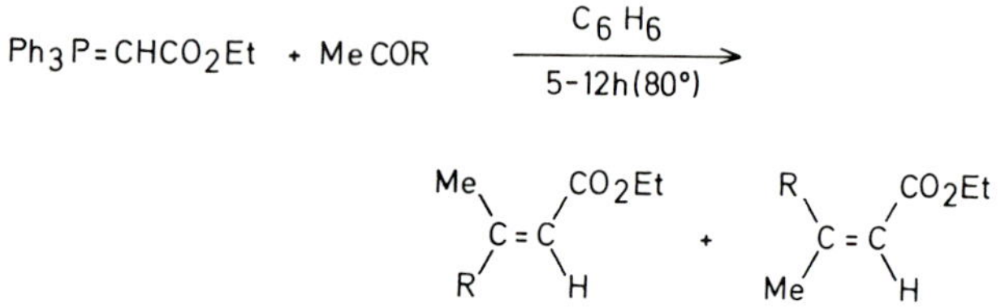

Ph₃P=CHCO₂Et + MeCOR $\xrightarrow[\text{5-12h(80°)}]{C_6H_6}$

$\underset{R}{\overset{Me}{>}}C=C\underset{H}{\overset{CO_2Et}{<}}$ + $\underset{Me}{\overset{R}{>}}C=C\underset{H}{\overset{CO_2Et}{<}}$

R	Yield (%) without PhCOOH	with Ph COOH	Product Composition	
Et	5	89	66	34
Pr	12	83	60.5	39.5

(Ref 67)

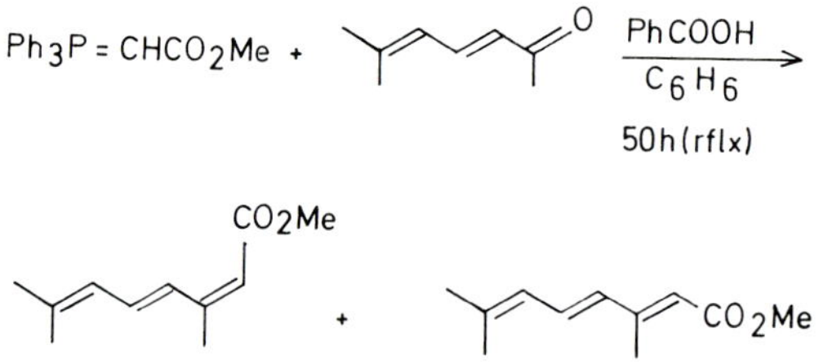

Ph₃P = CHCO₂Me + [dienone] $\xrightarrow[\text{50h(rflx)}]{\text{PhCOOH} \atop C_6H_6}$

[diene-CO₂Me isomer] + [diene-CO₂Me isomer]

2:1

(40%) (Ref 69)

Scheme 23

(45) + Ph₃P = CHCO₂Et (46) →

(47) + (48)

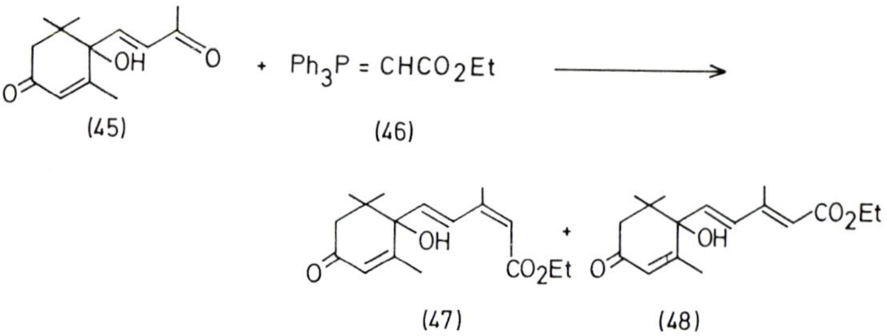

Reaction conditions	Yield (%)	Product Composition		Ref
Melt (10min,140-170°)	83	50	50	72
Toluene (4h,rflx)	77	50	50	72,73
EtOH		38	62	74

Scheme 24

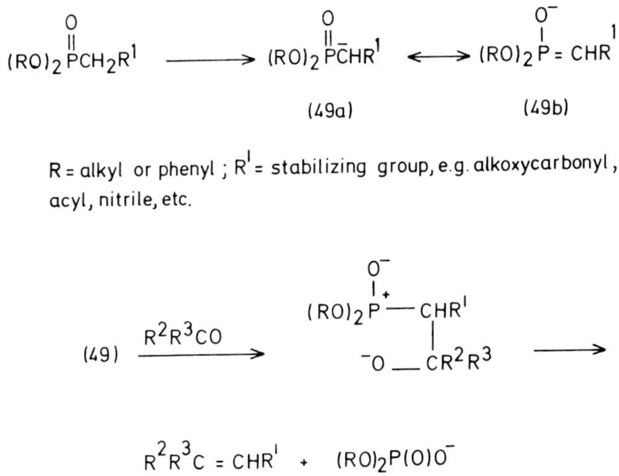

2 Wittig Reaction

$$(RO)_2 \overset{O}{\underset{\|}{P}} CH_2 R^1 \longrightarrow (RO)_2 \overset{O}{\underset{\|}{P}} \bar{C}HR^1 \longleftrightarrow (RO)_2 \overset{O^-}{\underset{|}{P}} = CHR^1$$

$$(49a) \qquad\qquad (49b)$$

R = alkyl or phenyl ; R¹ = stabilizing group, e.g. alkoxycarbonyl, acyl, nitrile, etc.

$$(49) \xrightarrow{R^2R^3CO} \begin{array}{c} O^- \\ | \\ (RO)_2 \overset{+}{P} - CHR^1 \\ | \\ {}^-O - CR^2R^3 \end{array} \longrightarrow$$

$$R^2R^3C = CHR^1 \ + \ (RO)_2P(O)O^-$$

Scheme 25

an alkoxycarbonyl or acyl group)[76] are smoothly converted into the corresponding alkenes by their phosphonate counterparts under very mild conditions. Table 2 summarizes this striking difference in reactivity for the reactions of diethyl ethoxycarbonylmethylphosphonate (50) and its phosphonium analogue with a variety of ketones.

Another important difference in olefin-forming reactions which use phosphonate anions rather than the corresponding Wittig reagents is that the former show a greater selectivity for (E)-alkene formation.[84,85] Even in those cases where mixtures of isomers are almost invariably obtained, as with the more reactive ylides, the phosphonate anions give mainly the (E)-alkene (see Section 3.3), but loss of stereoselectivity will occur when steric interactions in the product become serious.[86-88] For example,[88] in the reactions of diethyl ethoxycarbonylalkylphosphonates (51) with aliphatic aldehydes (52) (Scheme 26), only (E)-ester (53) is formed when R = H. However, with increasing size of R together with branching of the alkyl group (R¹) of the aldehyde, the (Z)-ester (54) becomes the major product of the reaction.

Loss of stereoselectivity is also observed in condensations with unsymmetrical ketones.[89-95] Two examples from syntheses of natural products are illustrative (Scheme 27). Thus, in the first step of the synthesis of racemic β,γ-carotene,[94] condensation of β-ionone (55) with diethyl ethoxycarbonylmethylphosphonate in the presence of sodium methoxide gave an isomer mixture containing 35% of methyl (Z)-β-ionylidene acetate (56b). Similarly, during the synthesis[91,92] of the juvenile hormone isolated from the giant silkworm moth *Hyalophora cecropia*, treatment of the methyl ketone (57)

Table 2. Relative yields of alkenes from the reaction of Wittig and phosphonate reagents with ketones

Ketone	From Ph$_3$P=CHCO$_2$Et Reaction conditions (Yield %)	From (EtO$_2$)$_2$P(O)CH$_2$CO$_2$Et (50)	Alkene	Ref.
Et$_2$CO	Xylene (0) (rflx)	NaOEt–DMF (ca. 65) (20°)	Et$_2$C=CHCO$_2$Et	77
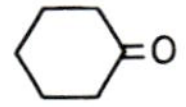 cyclohexanone	C$_6$H$_6$ (rflx) (25)	NaOEt–DMF (ca. 65) (20°)	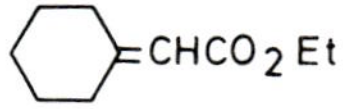 cyclohexylidene-CHCO$_2$Et	77
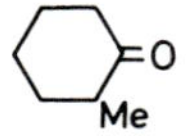 2-methylcyclohexanone	xylene (rflx) (0)	NaOEt–DMF (ca. 65) (20°)	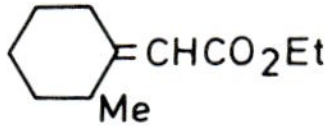 2-methylcyclohexylidene-CHCO$_2$Et	77
EtCOMe	xylene (rflx) (40)	NaNH$_2$–ether (58)	Et(Me)C=CHCO$_2$Et	78, 79
PhCOMe	170°, 10 h (68)	NaNH$_2$–ether (77)	Ph(Me)C=CHCO$_2$Et	27, 79
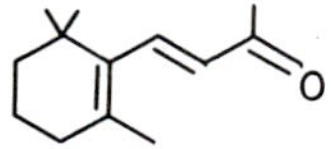 (cyclohexenyl ketone)	toluene (<50) 2 days (rflx)	NaOMe–MeOH (90) 10 h (40°)	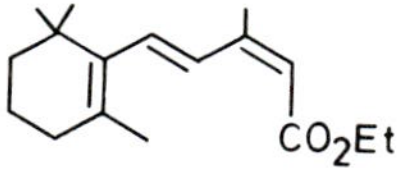 diene-CO$_2$Et ($\underline{Z}$ and $\underline{E}$)	72, 80
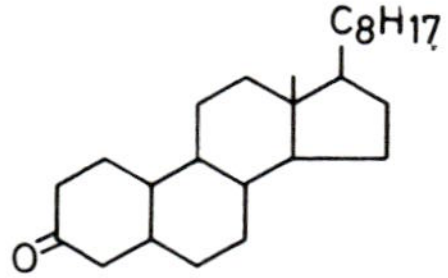 steroid ketone (C$_8$H$_{17}$)	no reaction	NaOEt–DMF (93) 20 h (20°)	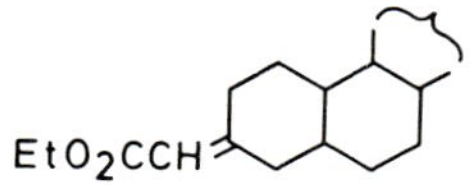 steroid alkene	81
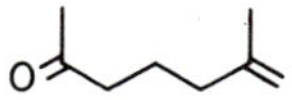 citronellal-type ketone	xylene 6–14 h, rflx (15)	NaOEt–DMF (43) ovn. (30°)	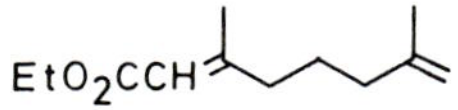 EtO$_2$CCH=...	78
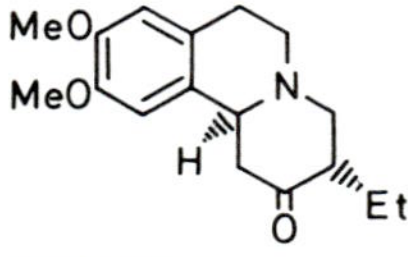 dimethoxy-tetrahydroisoquinoline ketone	152°, 3 h (54)	KOBu^t–DMF 30 h (20°) (99)	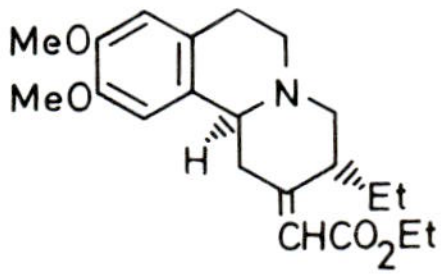 dimethoxy-tetrahydroisoquinoline CHCO$_2$Et	82, 83

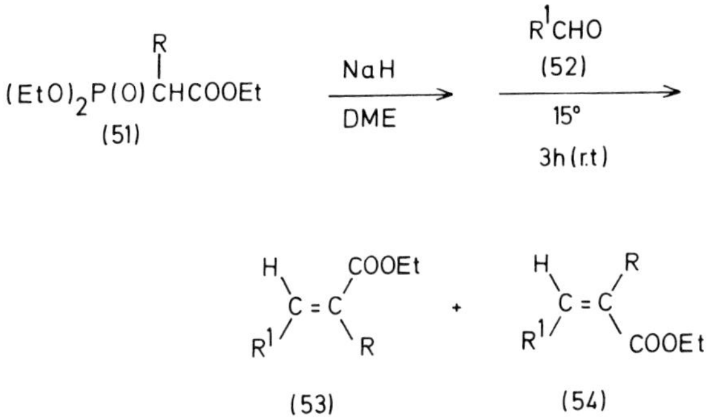

2 Wittig Reaction

$$(EtO)_2P(O)\overset{R}{\underset{|}{C}}HCOOEt \xrightarrow[DME]{NaH} \xrightarrow[\substack{15° \\ 3h\,(r.t)}]{R^1CHO\ (52)}$$
(51)

$$\underset{R^1}{\overset{H}{\diagdown}}C=C\underset{R}{\overset{COOEt}{\diagup}} \quad + \quad \underset{R^1}{\overset{H}{\diagdown}}C=C\underset{COOEt}{\overset{R}{\diagup}}$$

(53)   (54)

R	R^1	% (54) in the product
H	Pr^i	0
H	Bu^t	0
Me	Me	0
Me	Et	16
Me	Pr^i	62-73
Me	Bu^t	67
Et	Me	18
Et	Et	41
Et	Pr^i	84

Scheme 26 (Ref 88)

with the sodium salt of trimethyl phosphonacetate (58) in DME led to a mixture of the (E,E)- (59) and (Z,E)-ester (60) in 90% yield in the ratio 60:40.

Finally, mention should be made of a recent modification of the phosphonate route to alkenes which appears to offer considerable advantage over Wittig reactions employing stabilized ylides in that it provides a method for obtaining almost pure (Z)-alkenes.[96,97] It makes use of the fact that betaines formed from stabilized phosphonate anions with magnesium as the cation are much more stable than comparable sodium or lithium salts (Scheme 28). When hydrolysed, these reaction mixtures afford diastereoisomeric β-hydroxyphosphonates (protonated betaines) from which the *erythro* component may be separated by fractional crystallization and stereospecifically converted into the corresponding (Z)-alkene by heating. For example,[96] in the direct reaction between diphenyl cyanomethylphosphonate (61) and benzaldehyde under normal conditions, the ratio of (Z)- to (E)-cinnamonitriles is 15:85, whereas thermal decomposition of the isolated *erythro* adduct (62a) gives a $(Z):(E)$ ratio of 80:20. This technique is in its infancy and deserves to be fully explored.

Although changes in the structure of the reactant clearly influence the

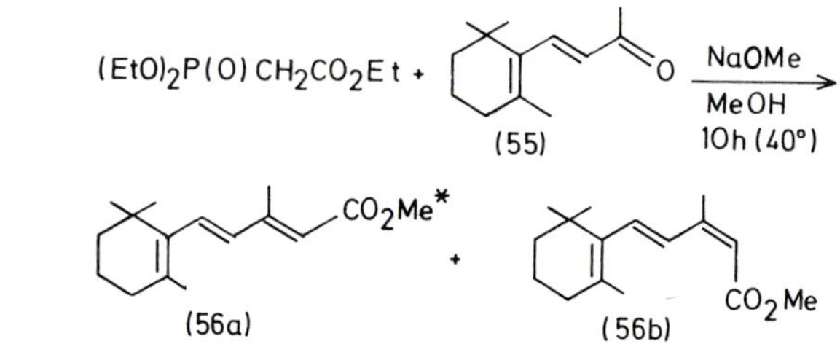

$(EtO)_2P(O)CH_2CO_2Et$ + [compound (55)] $\xrightarrow[MeOH]{NaOMe}$
                                        10h (40°)

(55)

[compound (56a)] CO$_2$Me*

(56a)

+

[compound (56b)] CO$_2$Me

(56b)

65 : 35
(90%)

*ester exchange.   (Ref 94)

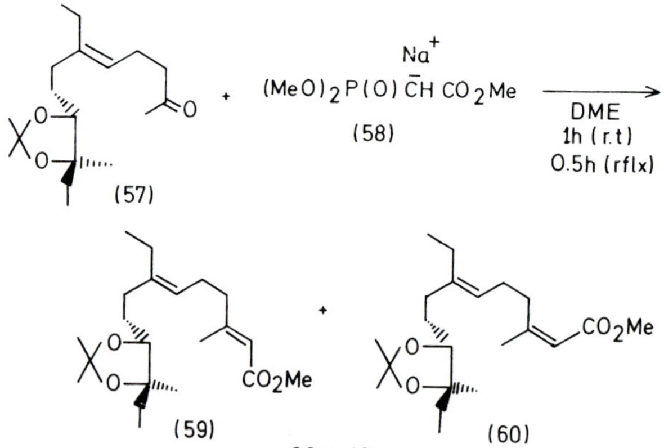

[compound (57)] + $(MeO)_2P(O)\bar{C}HCO_2Me \; Na^+$ $\xrightarrow[\substack{1h\,(r.t) \\ 0.5h\,(rflx)}]{DME}$

(57)   (58)

[compound (59)] CO$_2$Me   +   [compound (60)] CO$_2$Me

(59)   (60)

60 : 40
(90%)

Scheme 27   (Ref 91,92)

proportions of stereoisomers formed in Wittig reactions effected with stabilized ylides, another factor of comparable importance is the nature of the reaction medium. Usually, more (Z)-alkene is obtained when an aprotic solvent is replaced by a protic one, such as methanol, or when soluble lithium salts are added (see Table 3). The reason for this increase is not entirely clear but it has been argued[15b] that in the presence of powerful solvation or complexation the relative stabilities of the intermediate betaines leading to (Z)- and (E)-alkenes are changed, that (63) leading to the (Z)-isomer becoming more stable.

Data concerning the influence of temperature on the stereoselectivity in reactions of stabilized ylides are less plentiful. In the two cases studied[51] one showed a negligible effect, while the other showed a measurable, though minor, decrease in stereospecificity with increasing temperature.

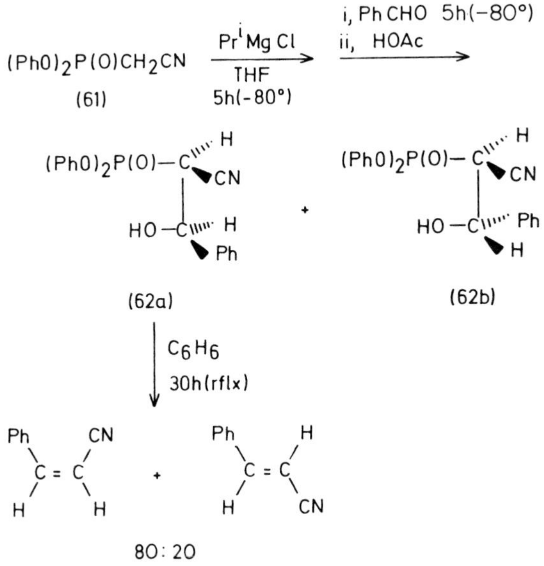

2 Wittig Reaction

$$(PhO)_2P(O)CH_2CN \xrightarrow[\text{THF} \; 5h(-80°)]{Pr^iMgCl} \xrightarrow[\text{ii, HOAc}]{\text{i, PhCHO } 5h(-80°)}$$

(61)

(PhO)$_2$P(O)—C(H)(CN)—C(OH)(H)(Ph)   +   (PhO)$_2$P(O)—C(H)(CN)—C(OH)(Ph)(H)

(62a) (62b)

↓ C$_6$H$_6$, 30h(rflx)

Ph\C=C/CN + Ph\C=C/H
H/   \H   H/   \CN

80:20

Scheme 28 (Ref 96)

Table 3. Effect of reaction medium on yield of (Z)-alkene (25°): Ph$_3$P=CHCOOMe + RCHO → MeCH=CHCOOMe

Reaction medium	(Z)-Isomer in product (%)	Ref.
R = Me		
CH$_2$Cl$_2$, DME or DMF	3–6	
DMF, LiX (X = Br, Cl, ClO$_4$ or NO$_3$)	18–22	85
MeOH	38	
R = Ph		
C$_6$H$_6$	5	98, 99
DMF	6	98, 99
EtOH	15	51
DMF (LiBr)	22	98, 99

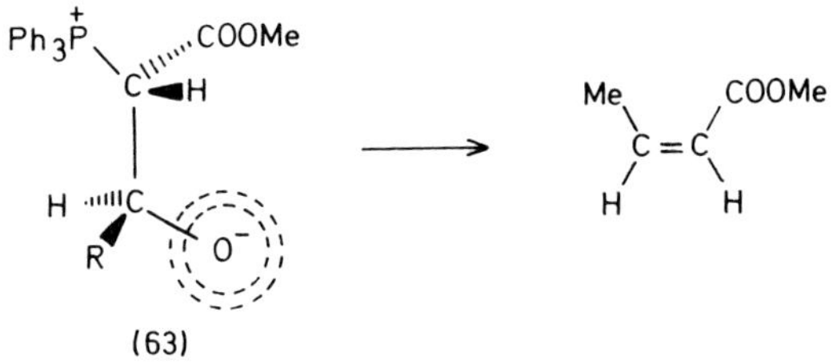

Ph$_3$P$^+$—C(H)(COOMe)—C(H)(R)(O$^-$) → Me\C=C/COOMe with H\ /H

(63)

Scheme 29

3.3. Reactive Ylides

From the foregoing it follows that, in the absence of additives or solvent interactions, the (*E*)-isomer usually forms the major product in Wittig reactions effected with stabilized ylides. This situation changes profoundly in the case of the so-called "reactive" ylides, i.e. ylides that do not contain stabilizing α-substituents. For these the thermodynamically less stable (*Z*)-isomers tend to dominate the mixture of alkenes obtained, as illustrated in Scheme 30.

The examples show that the degree of stereoselectivity varies considerably with the reaction conditions, especially with the nature of the solvent and the

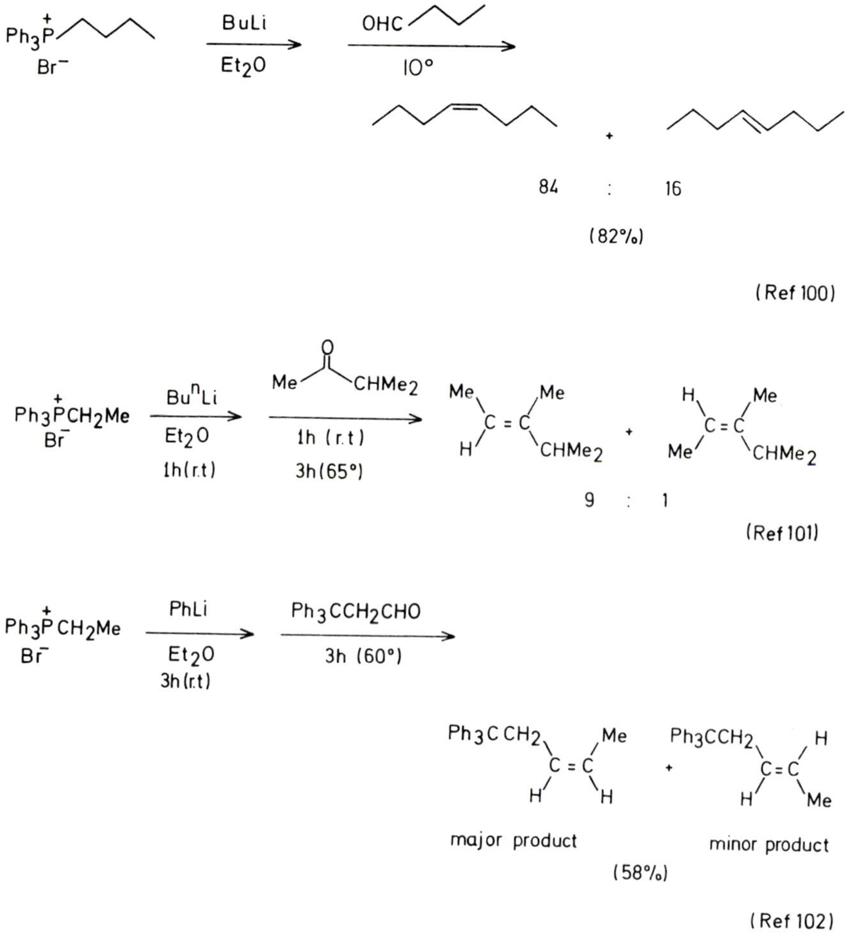

$Ph_3\overset{+}{P}$—CH₂CH₂CH₃ Br⁻ $\xrightarrow{\text{BuLi, Et}_2\text{O}}$ $\xrightarrow{\text{OHC-CH}_2\text{CH}_3, 10°}$

(Z-alkene) + (E-alkene)

84 : 16
(82%)

(Ref 100)

$Ph_3\overset{+}{P}CH_2Me$ Br⁻ $\xrightarrow{\text{Bu}^n\text{Li, Et}_2\text{O, 1h(r.t)}}$ $\xrightarrow{\text{Me-CO-CHMe}_2, \text{1h (r.t), 3h(65°)}}$

$\underset{H}{\overset{Me}{>}}C=C\underset{CHMe_2}{\overset{Me}{<}}$ + $\underset{Me}{\overset{H}{>}}C=C\underset{CHMe_2}{\overset{Me}{<}}$

9 : 1

(Ref 101)

$Ph_3\overset{+}{P}CH_2Me$ Br⁻ $\xrightarrow{\text{PhLi, Et}_2\text{O, 3h(r.t)}}$ $\xrightarrow{\text{Ph}_3\text{CCH}_2\text{CHO, 3h (60°)}}$

$\underset{H}{\overset{Ph_3CCH_2}{>}}C=C\underset{H}{\overset{Me}{<}}$ + $\underset{H}{\overset{Ph_3CCH_2}{>}}C=C\underset{Me}{\overset{H}{<}}$

major product   minor product
(58%)

(Ref 102)

Scheme 30

2 Wittig Reaction

Scheme 30 (Cont.)

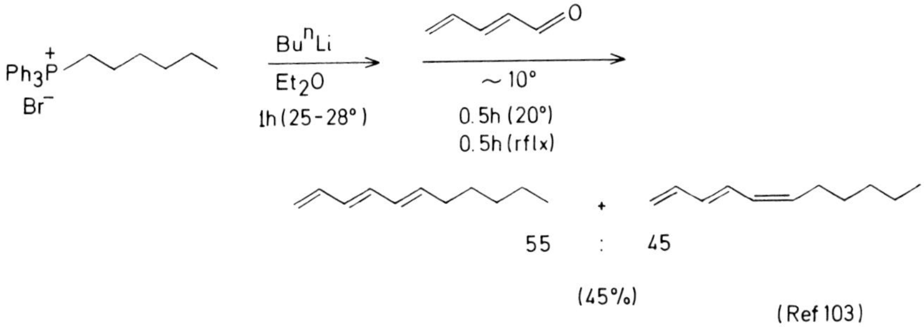

$Ph_3\overset{+}{P}$∼∼∼∼ Br^- → Bu^nLi / Et_2O / 1h (25–28°) → ∼∼CHO / ∼10° / 0.5h (20°) / 0.5h (rflx) →

∼∼∼∼∼∼∼ 55 : 45 ∼∼∼∼∼∼∼

(45%)

(Ref 103)

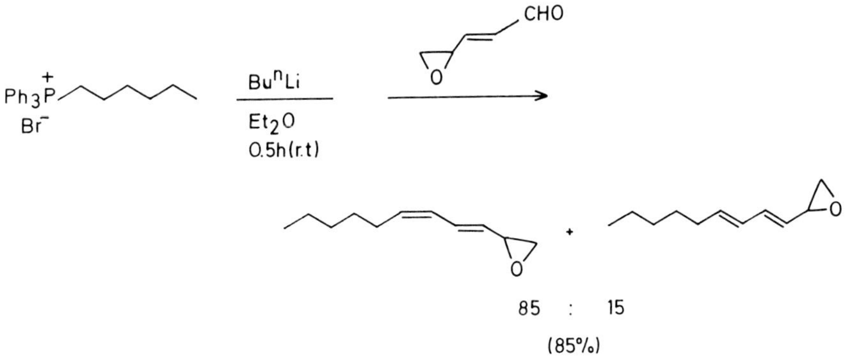

$Ph_3\overset{+}{P}$∼∼∼∼ Br^- → Bu^nLi / Et_2O / 0.5h (r.t) → epoxide-CH=CH-CHO →

85 : 15

(85%)

(Ref 104)

$Ph_3\overset{+}{P}$∼∼∼OTHP Br^- → $Na^+\overset{-}{C}H_2SOMe$ / DMSO → OHC-...-OCOmesityl / r.t. →

THPO∼∼∼ / mesitylOCO∼∼∼

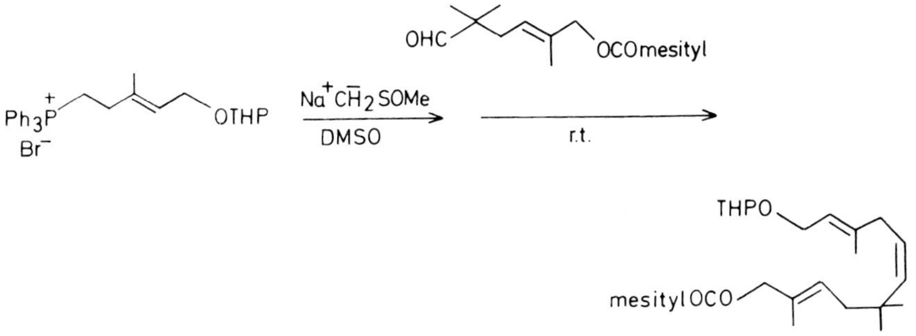

(100% of the product)

(Ref 105)

Scheme 30 (Cont.)

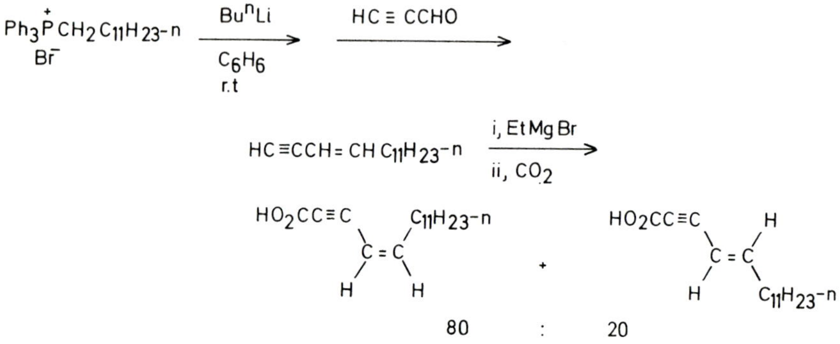

$$Ph_3\overset{+}{P}CH_2C_{11}H_{23-n}\ Br^- \xrightarrow[\text{r.t}]{Bu^nLi,\ C_6H_6} \xrightarrow{HC\equiv CCHO}$$

$$HC\equiv CCH=CHC_{11}H_{23-n} \xrightarrow[\text{ii, } CO_2]{\text{i, EtMgBr}}$$

$HO_2CC\equiv C\diagdown_{H}C=C\diagup^{C_{11}H_{23-n}}_{H}$ + $HO_2CC\equiv C\diagdown_{H}C=C\diagup^{H}_{C_{11}H_{23-n}}$

80 : 20

(Ref 10 cf. 106)

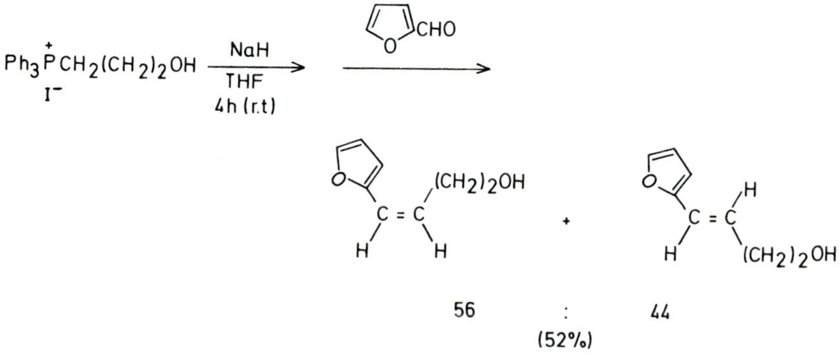

$$Ph_3\overset{+}{P}CH_2(CH_2)_2OH\ I^- \xrightarrow[\text{4h (r.t)}]{NaH,\ THF} \xrightarrow{\text{furfural-CHO}}$$

furyl−CH=CH−(CH$_2$)$_2$OH (Z) + furyl−CH=CH−(CH$_2$)$_2$OH (E)

56 : 44
(52%)

(Ref. 107)

$$Ph_3\overset{+}{P}\text{-}(CH_2)_5\text{-}OH\ Br^- \xrightarrow[\text{20 min (r.t)}]{2\,Bu^nLi,\ THF} \xrightarrow[\text{0.5h, then AcCl}]{\text{Et}_2C=CH\text{-}CH_2CH_2CHO}$$

[structure: Et$_2$C=CH−CH$_2$CH$_2$−CH=CH−(CH$_2$)$_4$−OAc]

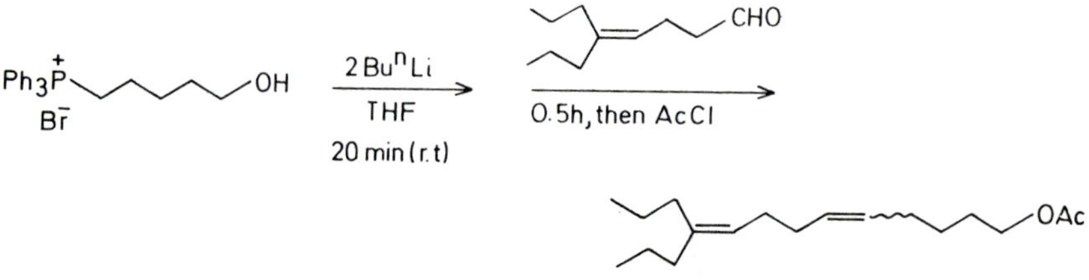

72% (mainly Z-isomer)

(Ref 108)

2 Wittig Reaction

Scheme 30 (Cont.)

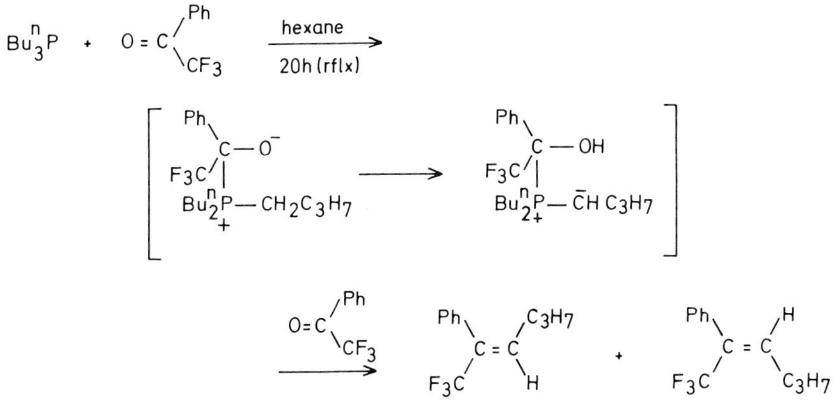

$$Bu_3^nP + O=C\begin{matrix}Ph\\CF_3\end{matrix} \xrightarrow[20h\,(rflx)]{hexane}$$

$$\left[\begin{matrix}Ph\\ C-O^-\\F_3C|\\Bu_2^nP-CH_2C_3H_7\\ +\end{matrix} \longrightarrow \begin{matrix}Ph\\ C-OH\\F_3C|\\Bu_2^nP-\bar{C}HC_3H_7\\ +\end{matrix}\right]$$

$$\xrightarrow{O=C\begin{matrix}Ph\\CF_3\end{matrix}} \begin{matrix}Ph\\ C=C\\F_3C\\ H\end{matrix}\begin{matrix}C_3H_7\\ \\ H\end{matrix} + \begin{matrix}Ph\\ C=C\\F_3C\\ C_3H_7\end{matrix}\begin{matrix}H\\ \\ \end{matrix}$$

$$1 \quad : \quad 3$$
$$(43\%) \qquad (\text{Ref }109)$$

$$Ph_3\overset{+}{P}CH_2C_3H_7\text{-}n \;+\; \text{(furyl)CHO} \;\xrightarrow{12h\,(150°)}\; Br^-$$

$$\begin{matrix}(furyl)\\ C=C\\H\\ H\end{matrix}\begin{matrix}C_3H_7\text{-}n\\ \\ H\end{matrix} + \begin{matrix}(furyl)\\ C=C\\H\\ C_3H_7\text{-}n\end{matrix}\begin{matrix}H\\ \\ \end{matrix}$$

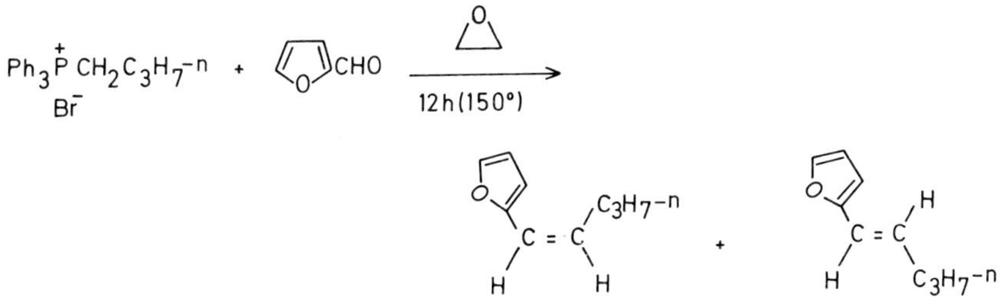

$$60 \quad : \quad 40$$
$$(82\%) \qquad (\text{Ref }47)$$

base used. The decisive influence of dissolved lithium salts is particularly stressed. Until recently it was thought that Wittig reactions of reactive ylides produced largely (E)-alkenes but that the $(Z):(E)$ ratio could be increased by the use of special polar solvents and added lithium salts.

While these conditions have been used to effect striking syntheses of a number of naturally occurring *cis*-unsaturated acids (see Section 4.2) it has become apparent that lithium salts have little or no effect on the stereochemistry of reactions involving reactive ylides in dipolar aprotic solvents. The examples in Scheme 31 illustrate the generalization that reactive ylides normally yield alkenes that are $>90\%$ (Z), irrespective of the presence or absence of lithium salts, if the reaction is carried out in very polar aprotic solvents such as dimethylformamide,[110–112] dimethyl sulphoxide (DMSO),[113–116] or hexamethylphosphoramide (HMPA).[117,118] The addition of DMSO or HMPA as co-solvent to tetrahydrofuran also provides a suitable reaction medium for promoting the formation of (Z)-alkene.[119,120]

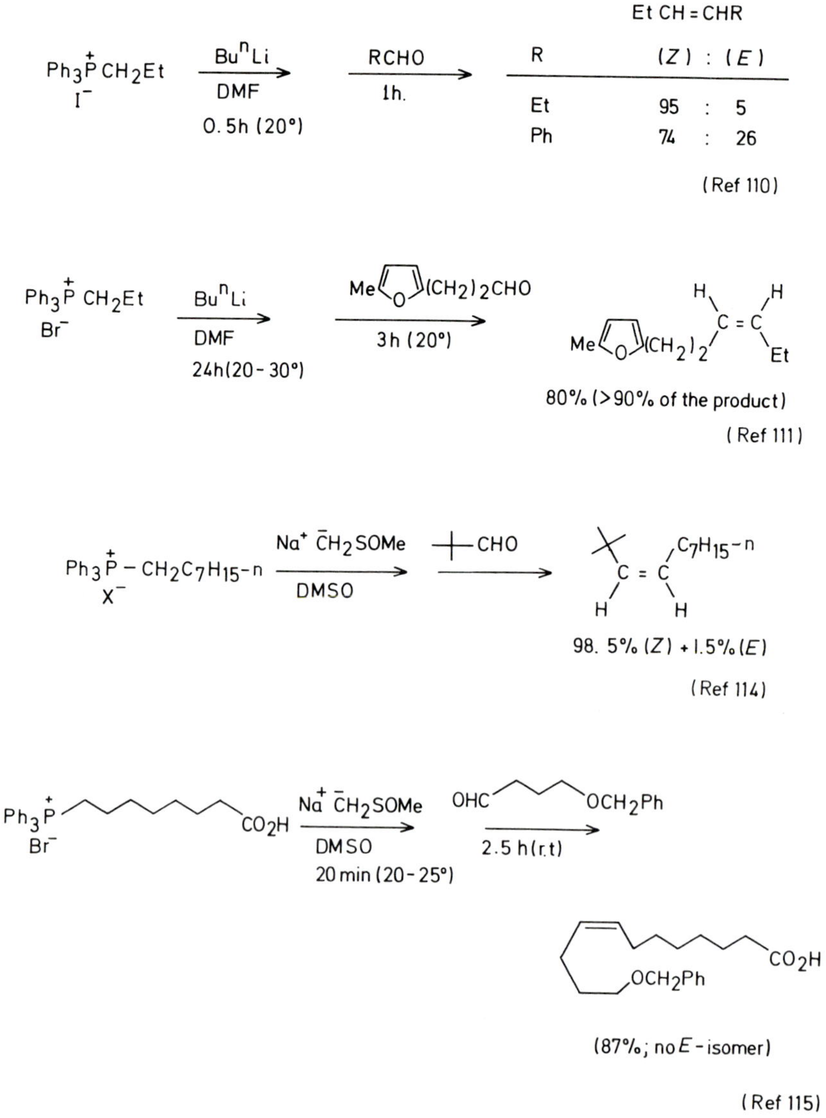

$$Ph_3\overset{+}{P}CH_2Et \cdot I^- \xrightarrow[\text{0.5h (20°)}]{Bu^nLi / DMF} \xrightarrow[\text{1h.}]{RCHO} EtCH=CHR$$

R	(Z) : (E)
Et	95 : 5
Ph	74 : 26

(Ref 110)

$$Ph_3\overset{+}{P}CH_2Et \cdot Br^- \xrightarrow[\text{24h(20-30°)}]{Bu^nLi / DMF} \xrightarrow[\text{3h (20°)}]{Me\text{-furan-}(CH_2)_2CHO} Me\text{-furan-}(CH_2)_2 \overset{H}{\underset{}{C}}=\overset{H}{\underset{Et}{C}}$$

80% (>90% of the product)

(Ref 111)

$$Ph_3\overset{+}{P}-CH_2C_7H_{15}-n \cdot X^- \xrightarrow[\text{DMSO}]{Na^+\ \bar{C}H_2SOMe} \xrightarrow{\text{iPr-CHO}} \underset{H}{\overset{iPr}{C}}=\underset{H}{\overset{C_7H_{15}-n}{C}}$$

98.5% (Z) + 1.5% (E)

(Ref 114)

$$Ph_3\overset{+}{P}\text{-}(CH_2)_n\text{-}CO_2H \cdot Br^- \xrightarrow[\text{20 min (20-25°)}]{Na^+\ \bar{C}H_2SOMe / DMSO} \xrightarrow[\text{2.5 h(r.t)}]{OHC\text{-}(CH_2)_n\text{-}OCH_2Ph}$$

(alkene product with OCH_2Ph and CO_2H groups)

(87%; no E-isomer)

(Ref 115)

Scheme 31

2 Wittig Reaction

Scheme 3[1] (cont)

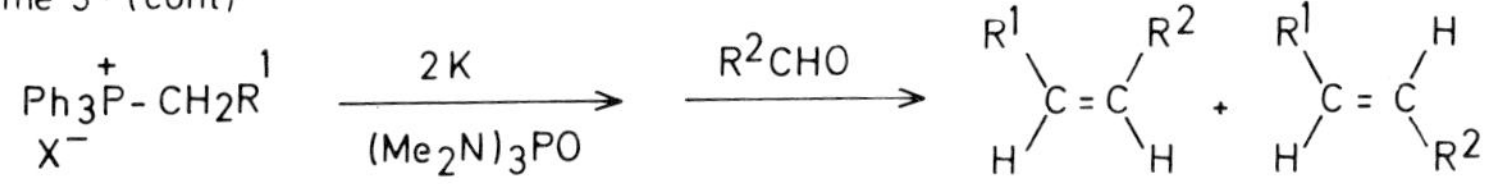

$$Ph_3\overset{+}{P}\text{-}CH_2R^1 \quad X^- \quad \xrightarrow[(Me_2N)_3PO]{2K} \quad \xrightarrow{R^2CHO} \quad \underset{H}{\overset{R^1}{>}}C=C\underset{H}{\overset{R^2}{<}} \quad + \quad \underset{H}{\overset{R^1}{>}}C=C\underset{R^2}{\overset{H}{<}}$$

R^1	R^2	Yield (%)	Product composition
Me	Ph	55	86 : 14
Et	Et	82	96 : 4
n-C$_3$H$_7$	n-C$_3$H$_7$	76	95 : 5

(Ref 117)

It must be stressed, however, that organolithium compounds are unsuitable for use as bases in dimethylformamide since they react with this solvent to give aldehydes[121,122] (Scheme 32) which may co-condense with the ylide.[123]

$$RLi + HCONMe_2 \longrightarrow RCH(\bar{O})NMe_2\overset{+}{Li} \longrightarrow RCHO + Me_2NLi$$

Scheme 32

For this reason sodium hydride is recommended as base if dimethylformamide is used as solvent; metal alkoxides have also enjoyed widespread use. Some typical examples are illustrated in Scheme 33.

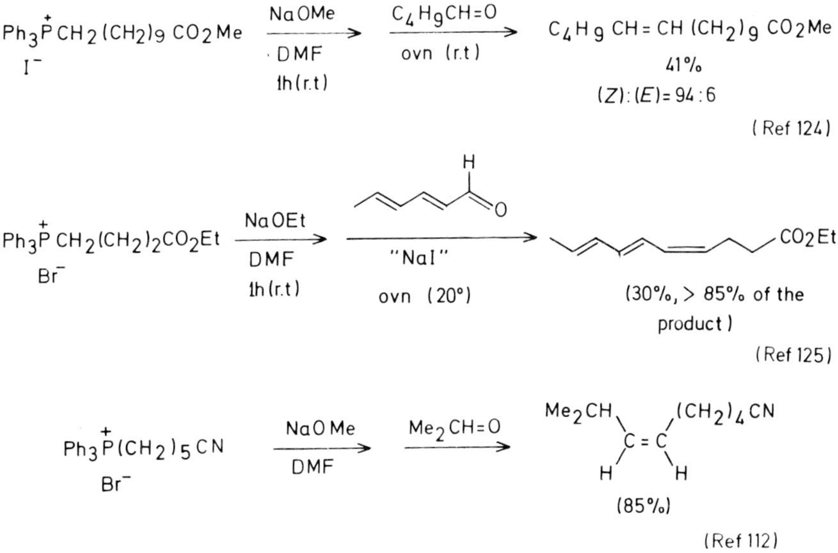

$$Ph_3\overset{+}{P}CH_2(CH_2)_9CO_2Me \quad I^- \quad \xrightarrow[\text{1h (r.t)}]{\text{NaOMe, DMF}} \quad \xrightarrow[\text{ovn (r.t)}]{C_4H_9CH=O} \quad C_4H_9CH=CH(CH_2)_9CO_2Me$$

41%
(Z):(E) = 94:6

(Ref 124)

$$Ph_3\overset{+}{P}CH_2(CH_2)_2CO_2Et \quad Br^- \quad \xrightarrow[\text{1h (r.t)}]{\text{NaOEt, DMF}} \quad \xrightarrow[\text{ovn (20°)}]{\text{"NaI"}} \quad \text{(diene-CO}_2\text{Et product)}$$

(30%, > 85% of the product)

(Ref 125)

$$Ph_3\overset{+}{P}(CH_2)_5CN \quad Br^- \quad \xrightarrow[\text{DMF}]{\text{NaOMe}} \quad \xrightarrow{Me_2CH=O} \quad \underset{H}{\overset{Me_2CH}{>}}C=C\underset{H}{\overset{(CH_2)_4CN}{<}}$$

(85%)

(Ref 112)

Scheme 33

A completely different situation is observed for reactions in non-polar solvents such as benzene and ether. Reactions in solvents of this type show a salt dependence, giving increased amounts of the (E)-alkene on addition of relatively soluble lithium salts such as lithium iodide [33,126] (Scheme 34).

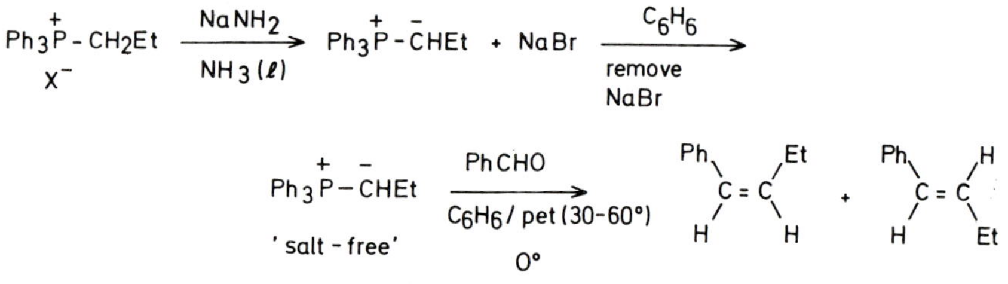

$$Ph_3\overset{+}{P}-CH_2Et \; X^- \xrightarrow[NH_3(\ell)]{NaNH_2} Ph_3\overset{+}{P}-\overset{-}{C}HEt + NaBr \xrightarrow[\text{remove NaBr}]{C_6H_6}$$

$$Ph_3\overset{+}{P}-\overset{-}{C}HEt \text{ 'salt-free'} \xrightarrow[C_6H_6/\text{pet }(30-60°)]{PhCHO} \underset{H}{\overset{Ph}{>}}C=C\underset{H}{\overset{Et}{<}} + \underset{H}{\overset{Ph}{>}}C=C\underset{Et}{\overset{H}{<}}$$

LiX	Yield (%)	(Z : E)
salt-free	88	96 : 4
LiCl	80	90 : 10
LiBr	80	86 : 14
LiI	81	83 : 17
LiBPh$_4$	60	52 : 48

Scheme 34 (Ref 33)

In order to explain these effects in non-polar media, it is argued that when dissolved metal cations, especially Li$^+$, are present in the reaction solution, the decomposition of both diastereoisomeric betaine intermediates to alkene is retarded by complex formation with the cation. As a result, reversal of the betaine formation step becomes important and more of the kinetically favoured *erythro*-betaine is converted into the thermodynamically more stable *threo*-isomer, thus reducing the proportion of (Z)-alkene in the product. The absence of salt effects in more polar media is regarded as being due to destabilization of any such betaine–cation complex due to preferential solvation of the cation.

It is important therefore, when preparing (Z)-alkenes by the Wittig reaction in non-polar media, to ensure that ylide solutions free from dissolved salts are used. Several means of achieving this have been reported. For example, the ylide can be prepared by the use of sodamide in anhydrous ammonia and then extracted into benzene.[127,128] A variation of this method is to prepare the ylide with sodamide in boiling tetrahydrofuran and to remove the insoluble inorganic salts by filtration.[129] Potassium t-butoxide[130] and sodium bis(trimethylsilyl)amide[131] have also been employed as bases in tetrahydrofuran without the tedious necessity for filtration.

2 Wittig Reaction

In another convenient procedure,[132] a mixture of the phosphonium salt and the aldehyde is treated with sodium t-pentoxide dissolved in benzene containing trace amounts of dimethyl sulphoxide. Typical "salt-free" $(Z):(E)$ distributions are also obtained when 18-crown-6-complexes of potassium carbonate or potassium t-butoxide are used in THF.[133] Major advantages of the last two methods are that preforming the ylide is unnecessary. Apart from the obvious advantages of a "one pot" process, *in situ* formation of the ylide gives high yields of alkenes and a minimum of self condensation.

Some examples of the use of these "salt-free" Wittig reactions are shown in Scheme 35.

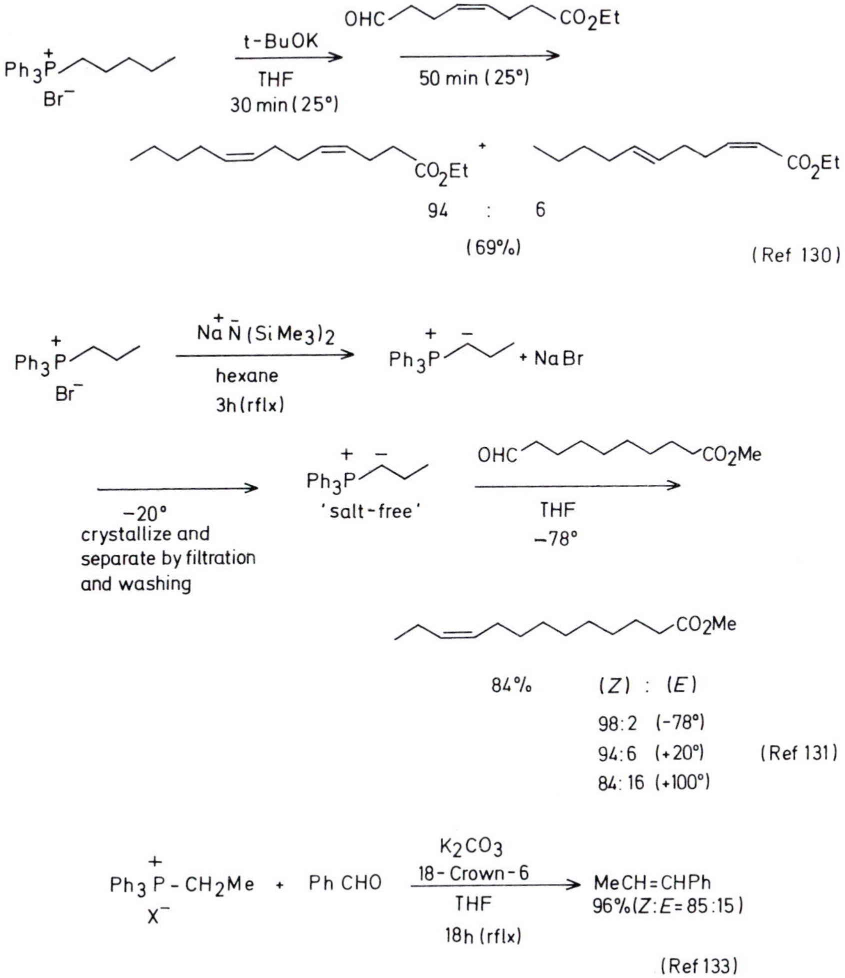

Scheme 35

Although choice of reaction conditions may exert a significant influence on the stereochemical composition of the product of Wittig reactions, other factors of importance are the nature of the substituents both on the carbonyl and on the ylide component. As the data in Table 4 clearly demonstrate, the reaction between homologous unbranched alkylidene ylides and primary aliphatic aldehydes is highly stereoselective in salt-free solution, yielding products in which the proportion of the (Z)-isomer exceeds 90% (in most cases 95%). However, when α,β-unsaturated or aromatic aldehydes serve as the carbonyl component, some loss of stereoselectivity is observed, especially in dipolar aprotic media. This difference in behaviour may best be explained by assuming that betaine formation with aliphatic aldehyde is essentially irreversible (in the absence of lithium salts), whereas with more electrophilic carbonyl compounds such as p-cyanobenzaldehyde, reversibility is more pronounced and the betaine formation and hence the product ratio become thermodynamically controlled.

Table 4. Yields and (Z):(E) ratios of alkenes from "salt-free" Wittig reactions of reactive ylides in benzene, toluene, or tetrahydrofuran at $0°$[14,33,126,134]

$$Ph_3P^+{-}^-CHR + R^1CH{=}O \rightarrow RCH{=}CHR^1 + Ph_3P{=}O$$

	(Z):(E) ratio (yield %)		
R^1	R = Me	R = Et	R = Pr
Et		97:3 (70)	
Pr^n	95:5	95:5 (49)	95:5 (78)
n-C$_5$H$_{11}$	91:9 (66)	96:4	
(E)-PhCH=CH	87:13 (77)		
PhC≡C	73:27		
p-MeOC$_6$H$_4$	90:10 (78)	92:8 (95)	90:10 (80)
p-MeC$_6$H$_4$	89:11 (88)	95:5 (59)	92:8 (71)
Ph	87:13 (98)	96:4 (88)	94:6 (66)
p-ClC$_6$H$_4$	82:18 (77)	93:7 (54)	92:8 (64)
m-ClC$_6$H$_4$	74:26		
p-NCC$_6$H$_4$	74:26		
p-F$_3$C C$_6$H$_4$	69:31		

In the case of ketones the isomer ratio depends on the nature of the substituents at the carbonyl carbon. With alkyl aryl ketones the (Z)-isomer is still favoured[135] as shown in Scheme 36. On the other hand, cis-stereoselectivity is almost totally lost with unsymmetrical aliphatic ketones using a variety of reaction conditions. A few examples will suffice although others could be quoted.[136–138] Thus, in the course of the synthesis[139] of the chemotactic hormone (±)-sirenin, reaction of the ylide derived from the phospho-

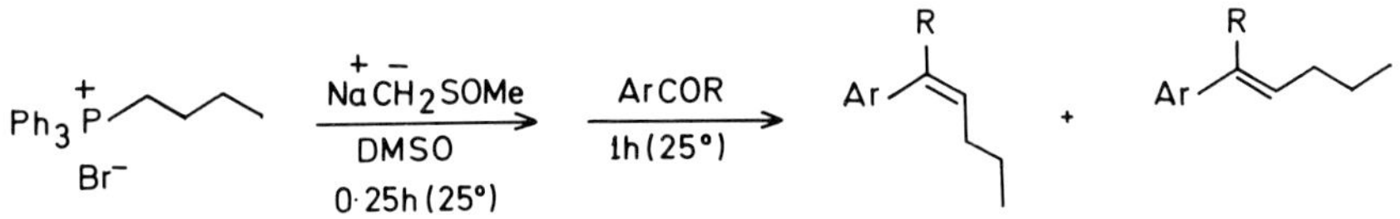

2 Wittig Reaction

Scheme 36 (Ref 135)

R	Yield (%)	Product composition
Me	80	9 : 1
Et	71	9 : 1

(Ar = p-MeOC$_6$H$_4$)

nium salt (64) and excess 6-methylhept-5-en-2-one (65) in dimethyl sulphoxide–tetrahydrofuran (1:1) gave the C_{13} diene acid (66) as a mixture of (Z)- and (E)-isomers in the ratio 3:2.

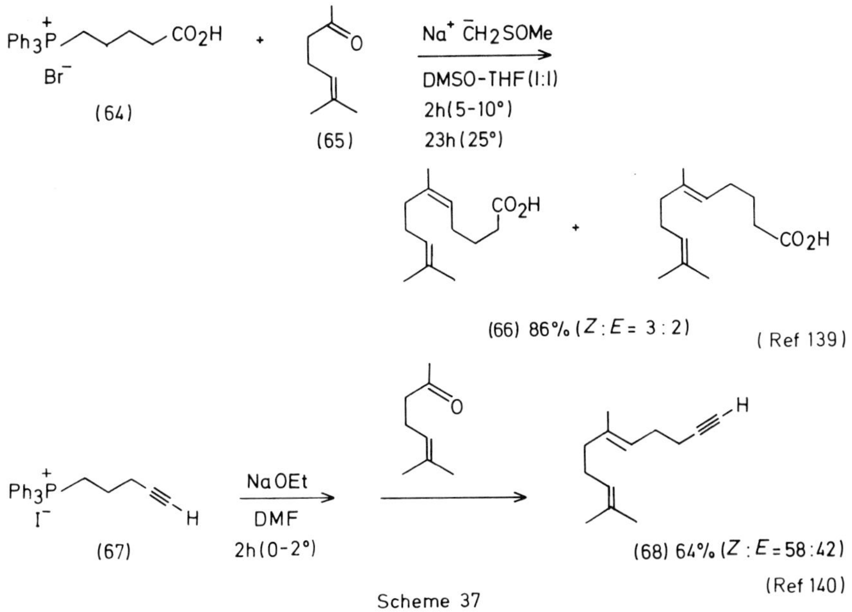

(66) 86% (Z:E = 3:2)   (Ref 139)

(68) 64% (Z:E = 58:42)
(Ref 140)

Scheme 37

A similar (Z):(E) ratio of 58:42 was obtained by condensation of the same ketone with the ylide prepared from (67) with sodium ethoxide in dimethylformamide.[140] Subsequent hydration of the terminal acetylene in the product (68) completed a general method for the elaboration of isoprenoid chains, one unit at a time.

In another instance,[141] a mixture of (Z,Z)- and (E,Z)-isomers (71) was formed in the reaction of the ethyl ketone (70) with the ylide from (69) during a synthesis of the juvenile hormone of *Hyalophora cecropia* (Scheme 38).

Scheme 38 (Ref 141)

Finally, in a single unspecified[142] attempt to synthesize a (Z)-trisubstituted alkene from a methyl ketone and a "salt-free" ylide, the stereoselectivity was again reported to be very poor.

In striking contrast[143] with these observations, treatment of the methyl ester (73) in a polar solvent mixture with the Wittig reagent derived from octadecyltriphenylphosphonium bromide (72) gave, in 48% yield, methyl 8-methylhexacos-8-enoate (74), apparently mainly the (Z)-isomer by g.l.c. analysis (Scheme 39).

Scheme 39 (Ref 143)

R	(Z) : (E)
Me$_3$C	99 : 1
Ph	2 : 1
PhCH$_2$CH$_2$	7 : 3

(Ref 30)

Scheme 40

2 Wittig Reaction

An unexpectedly large tendency to form (Z)-alkenes has been observed[30,144] with aliphatic aldehydes when the alkyl group is branched. As Scheme 40 illustrates, reaction of ethylidenetriphenylphosphorane with bulky aldehydes such as pivalaldehyde and 2,2-dimethylbut-3-enal gives predominantly, if not exclusively, (Z)-alkene, even in the presence of lithium salts.

Steric factors may also be responsible for the exclusive formation[145,146] of the (Z)-alkene (76) from the reaction of hexylidenetriphenylphosphorane with the *endo*-aldehyde (75); in the case of the corresponding *exo*-aldehyde a mixture of (Z)- and (E)-alkenes is obtained in the ratio of 80:20.[147] A possible

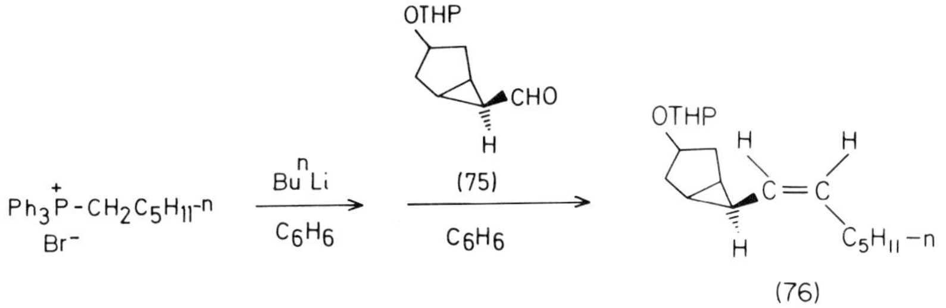

Scheme 41 (Ref 145,146)

explanation for this surprising selectivity can be formulated[145,146] based on initial coordination of carbonyl oxygen to the phosphorus of the ylide *before* C–C bond formation takes place (Scheme 42). Inspection of molecular models indicates that the least sterically hindered position for the bulky aldehyde is as shown in (77) with oxygen in the apical position and with the ylide carbon equatorial (necessary for subsequent bond formation with the aldehyde). Formation of an oxaphosphetan can now result from a rotational movement about the C^+–O bond. If this occurs in an anticlockwise direction, the *erythro* intermediate (78), the immediate precursor of (Z)-alkene, is obtained. While this brings the R^1 and R^2 into an eclipsed conformation, a rotational movement in the opposite direction would lead to a more serious

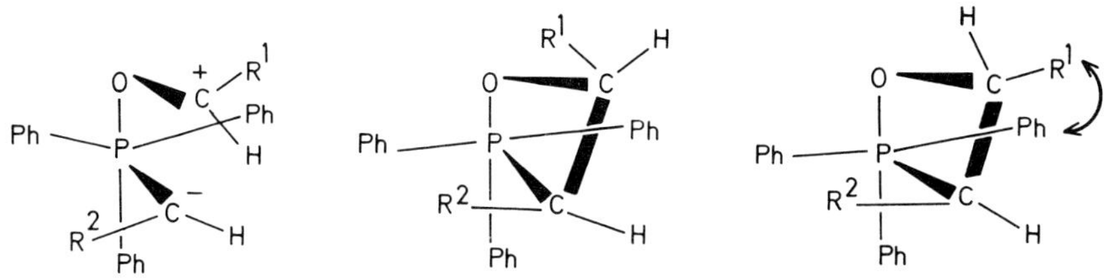

Scheme 42 (Ref 145)

interaction between R¹ and the closest equatorial phenyl group [see arrows in (79)]. As a result the former *erythro* configuration is preferred, resulting in selective (Z)-alkene formation.

This rationale may also explain the loss of *cis*-selectivity in Wittig reactions of reactive ylides having branched alkyl substituents at the α-carbon. The examples in Scheme 43 illustrate this effect.

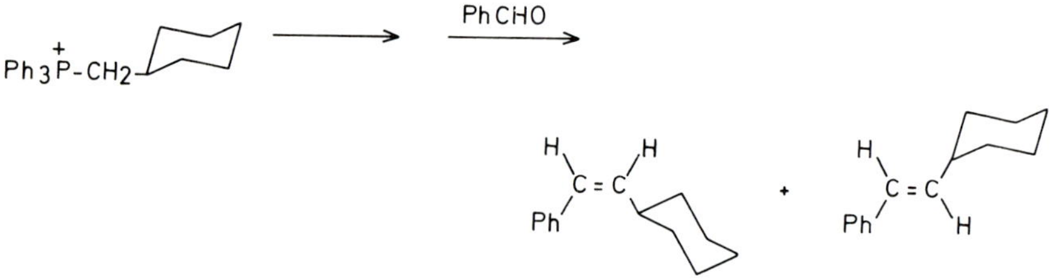

(Ref 148)

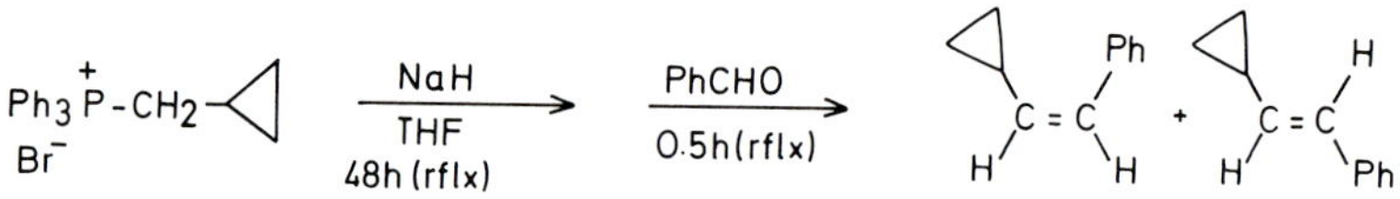

Reaction solvent	Product composition			Yield (%)
THF	66	:	34	75
C₆H₆	69	:	31	30
DMF	61	:	39	84

(Ref 149)

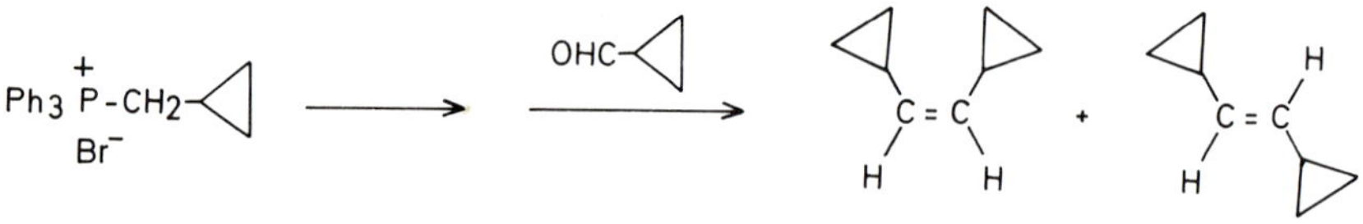

Reaction conditions	Yield (%)	Product composition		
Bu^{n}Li/Et₂O 4h(35°)	70	1	:	1
Na$^+$ $\bar{C}$H₂SOMe/DMSO 4h(25°)	71	2.4	:	1

(Ref 150)

Scheme 43

2 Wittig Reaction

3.3.1. trans-*Stereoselective alkene synthesis via β-oxido-ylides (Wittig–Schlosser reaction)*

As mentioned previously, reactive ylides will afford large amounts of (*E*)-alkenes if equilibration of the *erythro* and *threo* betaines can be accelerated. A method has been described[33,151,152] whereby the initially formed betaine–lithium complex, with predominantly the *erythro* configuration, is α-metallated with an equivalent of an organolithium reagent to form a new β-oxido phosphorus ylide ("betaine ylide") (81) (Scheme 44). Whereas the initially formed *erythro*-complex (80) was relatively stable to inversion, the diastereo-isomeric betaine ylides are rapidly interconverted, even at temperatures as low as −78°. Significantly, the equilibrium lies far to the side of form (81b), so addition of one equivalent of a proton donor occurs with remarkable stereospecificity to form the Wittig intermediate (82) in the *threo* configuration, which decomposes selectively to give pure (*E*)-alkene. It should be noted that protonation of β-oxido ylides prepared from salt-free ylides leads to mixtures of *erythro*- and *threo*-betaines and hence to mixtures of (*Z*)- and (*E*)-alkenes.[153]

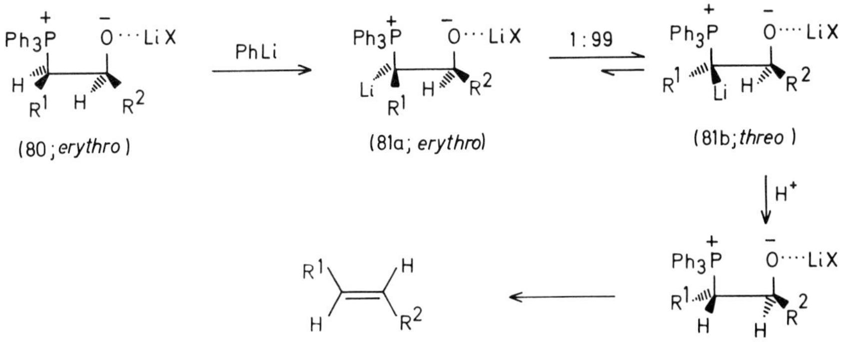

Scheme 44

Table 5 summarizes the application of this method, known as the Wittig–Schlosser reaction, to the preparation of (*E*)-alkenes using reactive ylides in combination with a variety of aliphatic, unsaturated, and aromatic aldehydes. For comparison, the conventional Wittig reactions lead to a (*E*):(*Z*) distribution that varies between 75:25 (for phenylhexene) and 20:80 (for oct-2-ene).

trans-Stereoselective alkene synthesis by means of the Wittig–Schlosser reaction is not limited to aldehydes, as is shown by the reaction of ethylidene-triphenylphosphorane with the unsymmetrical ketone acetophenone to give (*E*)- and (*Z*)-phenylbut-2-enes in the ratio 89:11.[33] When carried out in the

Table 5. *trans*-Stereoselective olefination of aldehydes $R^1CH=O$ with reactive ylides $[Ph_3P^+-^-CHR]$ via the Wittig–Schlosser reaction[33,151,152]

Alkene	R	R^1	Yield (%)	$(E):(Z)$ ratio
Oct-2-ene	Me	n-C_5H_{11}	70	99:1
Oct-2-ene	n-C_5H_{11}	Me	60	96:4
Oct-4-ene	n-C_3H_7	n-C_3H_7	72	98:2
PhCH=CHMe	Me	Ph	69	99:1
"salt-free"			82	76:24
PhCH=CHEt	Et	Ph	72	97:3
PhCH=CH(CH$_2$)$_3$Me	Bu^n	Ph	75	96:4
PhCH=CHCH=CHMe	Me	(E) PhCH=CH	63	97:3

conventional manner, the same reaction led to a ratio of 40:60.[14] In contrast, no detectable increase in stereoselectivity was observed[140] for the reaction of n-butylidenetriphenylphosphorane with the *aliphatic* methyl ketone methylheptenone under the same conditions.

The reaction also proceeds equally well with polymeric ylides.[154] Thus, the polymer (83) with acetophenone gave largely (E)-alkene via the β-oxido-ylide, and almost entirely (Z)-alkene under salt-free conditions.

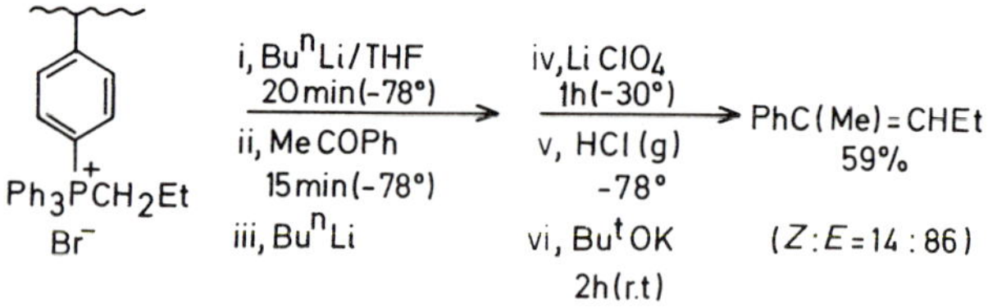

Scheme 45 (Ref 154)

3.3.2. α-*Substitution plus carbonyl olefination via β-oxido phosphorus ylides ("scoopy" reactions).*

β-Oxido-ylides also react with a variety of other electrophiles to produce α-substituted betaines which decompose spontaneously or after "activation" (e.g. treatment with potassium t-butoxide or by heating) to triphenylphosphine oxide and a trisubstituted alkene.[155,156] This procedure, referred to as "α-substitution plus carbonyl olefination via β-oxido phosphorus ylides (scoopy),"[157] is highly stereoselective provided that the preparations are carried out in the presence of soluble lithium salts; only in the case of the unsubstituted methylenetriphenylphosphorane is the stereoselectivity

2 Wittig Reaction

lost.[155,157] Thus, treatment of the β-oxido-ylide (84) with N-chlorosuccinimide followed by elimination of phosphine oxide affords 2-chloro-(E)-non-2-ene (85) as the predominant product, whereas the use of iodobenzene dichloride gives principally the (Z)-isomer (86).[156] The reason for this difference in stereoselectivity is not yet clear (Scheme 46).

Iodo-alkenes, e.g. (87), are also accessible from β-oxido-ylides by an indirect process involving addition of mercuric acetate at −100° followed by

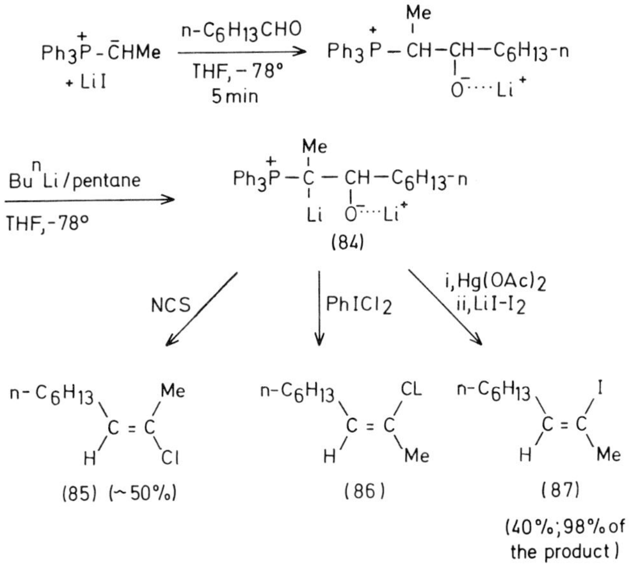

Scheme 46 (Ref 156)

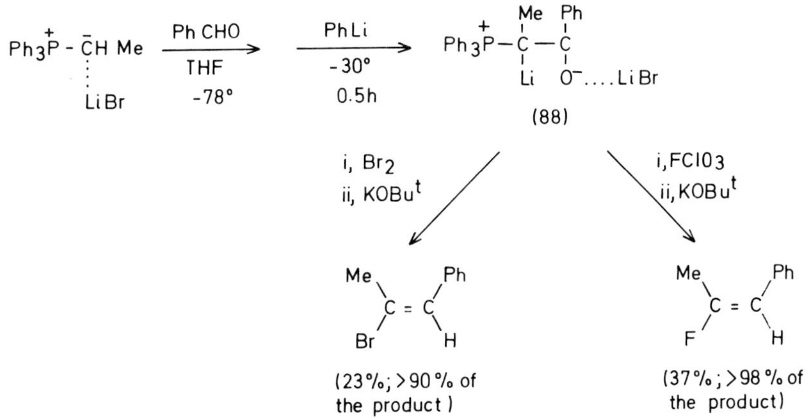

Scheme 47 (Ref 155)

elimination of phosphine oxide at 25° and finally treatment with anhydrous lithium iodide and iodine.[156]

With bromine, the β-oxido-ylide (88) gives β-methyl-(E)-ω-bromostyrene with >90% isomeric purity, albeit in low yield.[155] Direct fluorination of (88) is also possible using perchloryl fluoride; a similar yield and stereoselectivity are obtained.[155] By comparison, the analogous reactions using a methyltriphenylphosphonium salt as starting material give a product mixture containing approximately equal proportions of (Z)- and (E)-isomers.

The stereoselectivity associated with alkylations is significantly lower, but still present.[157,158] For example, 2-methyl-1-phenylpent-1-ene is obtained by a "scoopy" procedure in a 56% yield and a 22:78(Z):(E) ratio (Scheme 48). On the other hand, the unmodified Wittig reaction affords a (Z):(E) mixture of the same alkene in a ratio of 46:54.

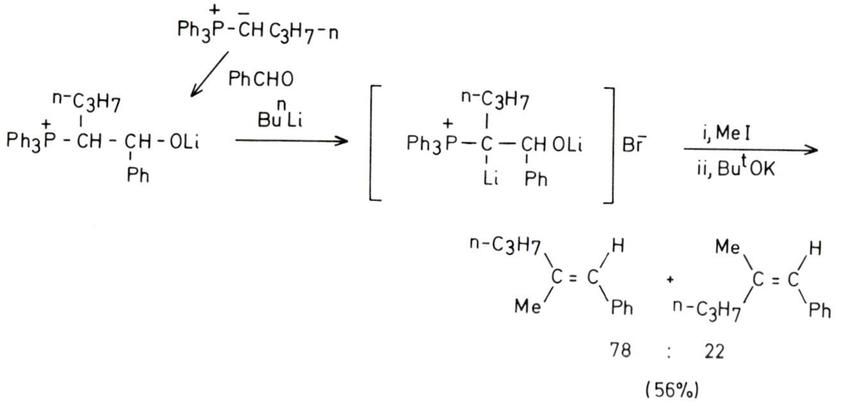

Scheme 48 (Ref 157)

In contrast to the foregoing reactions, electrophilic attack of acyl halides, such as methyl chloroformate or benzoyl chloride, gives only small amounts, if any, of the described products.[14] Seemingly, in these cases, α-substitution favours reversible betaine decomposition and leads to the formation of novel resonance-stabilized ylides.

By far the most important "scoopy" reaction of β-oxido-ylides is the stereospecific conversion of aldehydes into allylic alcohols.[158] Strikingly, the oxygen atom lost is that which originates in the second aldehyde moiety used in the reaction sequence. As shown in Scheme 49, reaction of ethylidene-triphenylphosphorane first with acetaldehyde, and then, after β-oxido-ylide formation, with heptanal yields the alcohol (90) exclusively (67%), whereas the reverse order of aldehyde addition gives only the isomeric alcohol (89) (67%). The above reaction appears to be general. Thus, with acetaldehyde as

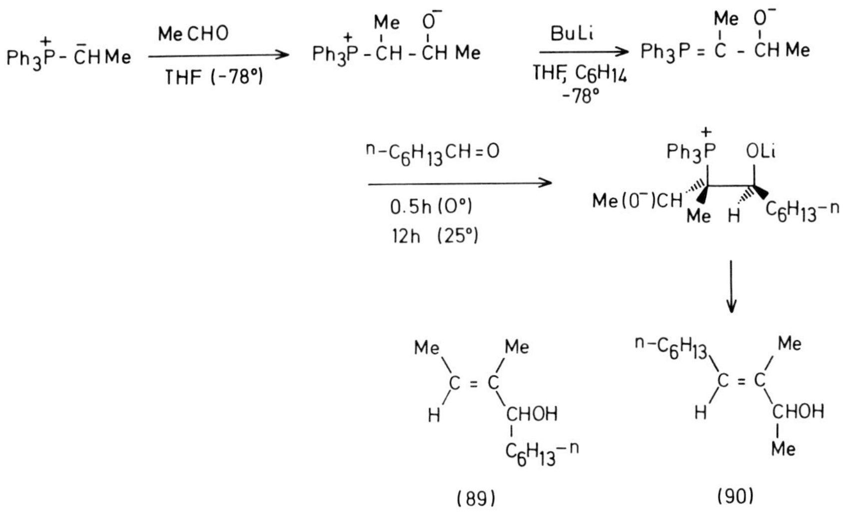

2 Wittig Reaction

$$Ph_3\overset{+}{P}-\overset{-}{C}HMe \xrightarrow[THF\ (-78°)]{MeCHO} Ph_3\overset{+}{P}-\underset{Me}{\overset{O^-}{\underset{|}{CH}}}-\underset{|}{\overset{}{CH}}Me \xrightarrow[THF,\ C_6H_{14}]{BuLi} Ph_3P=\underset{Me}{\overset{O^-}{\underset{|}{C}}}-\overset{}{CH}Me$$

$$\xrightarrow[0.5h\ (0°)]{n-C_6H_{13}CH=O} \quad Me(O^-)CH\underset{Me}{\overset{Ph_3\overset{+}{P}}{\cdots}}\underset{H}{\overset{OLi}{\cdots}}C_6H_{13}-n$$
$$12h\ (25°)$$

↓

(89) Me, Me / C=C / H, CHOH | C₆H₁₃-n

(90) n-C₆H₁₃, Me / C=C / H, CHOH | Me

Scheme 49 (Ref 158)

the second component, the corresponding reaction affords the (*E*)-allylic alcohol (91) and its (*Z*)-isomer in a ratio of 93:7 (65% yield).[159] Both carbonyl compounds may be aromatic aldehydes, as illustrated by the exclusive formation of (*E*)-1,3-diphenyl-2-methylprop-2-en-1-ol (92) using benzaldehyde as the first and second component.[159]

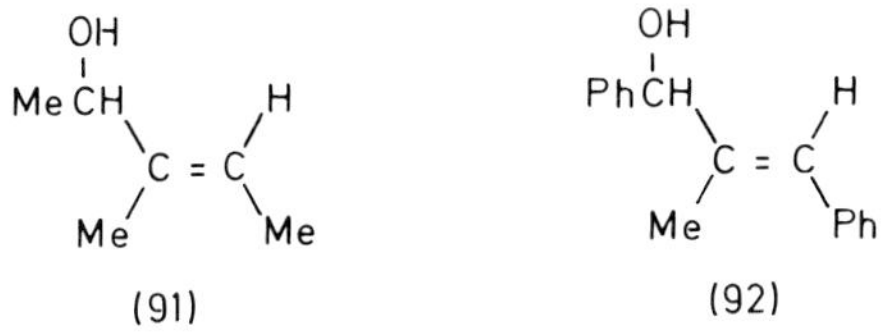

(91) MeCH(OH), H / C=C / Me, Me

(92) PhCH(OH), H / C=C / Me, Ph

Scheme 50 (Ref 159)

A thorough study[159] of these reactions has revealed that preferential formation of the (*E*)-isomer from a racemic adduct of the type (95) determines that the oxygen of the second aldehyde component is removed. What is not clear is why β-oxido-ylides react with aldehydes (or other electrophiles) with such remarkable selectivity. One possible explanation is that β-oxido-ylides possess a cyclic structure with a strong preference for the diastereoisomeric form (93). Reaction of (93) with an aldehyde so as to minimize steric interactions [see (94)] would then lead to the intermediate adduct in a racemic form (95) (Scheme 51).

Some loss of selectivity results if higher temperatures are used in the last step of the reaction.[158] Thus (Scheme 52), reaction of the β-oxido-ylide

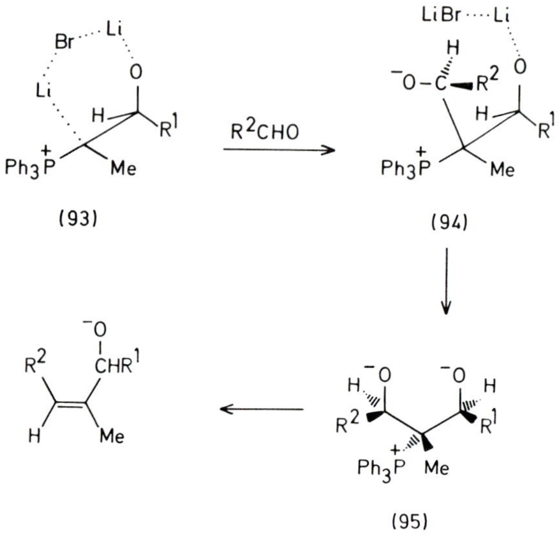

Scheme 51 (Ref 159)

derived from heptanal and ethylidenetriphenylphosphorane in THF at −78° with heptanal followed by treatment with acetic acid at the same temperature produced a mixture of alcohols (96) and (97) in the ratio *ca.* 9:1. On increasing the temperature to 0–20° this ratio decreased to 3:1. The temperature effect presumably reflects the equilibration of the racemic intermediate adduct (see Scheme 51) to the *meso* form, decomposition of which leads mainly to the (Z)-isomer.

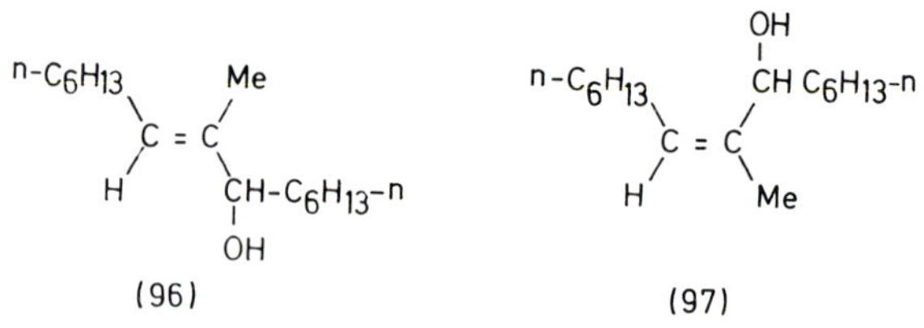

Scheme 52 (Ref 158)

In contrast with the above reactions, the β-oxido-ylide route to allylic alcohols, using paraformaldehyde as the second carbonyl component, gives the product formed by loss of oxygen from the *first* carbonyl component.[158] As Scheme 53 shows, this generates the olefinic linkage of (Z)-stereochemistry with high stereoselectivity. If the first component is a hindered aldehyde or a ketone, oxygen is lost from both carbonyl reagents[160] and mixtures of isomeric allylic alcohols are produced as illustrated in Scheme 54. In extreme cases, e.g. with acetone as the first carbonyl compound, only the product (99),

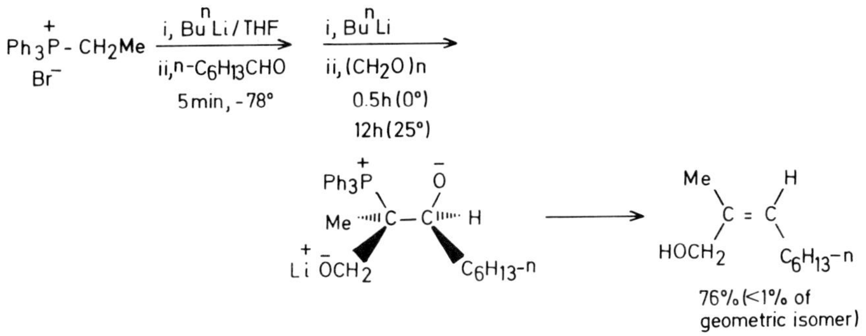

2 Wittig Reaction

$$Ph_3\overset{+}{P}-CH_2Me \cdot Br^- \xrightarrow[\text{ii},n-C_6H_{13}CHO]{i, Bu^nLi/THF} \xrightarrow[\text{ii},(CH_2O)_n]{i, Bu^nLi}$$
5 min, −78°   0.5h (0°)
              12h (25°)

Intermediate:
$$\underset{Li^+ \ \bar{O}CH_2}{\overset{Ph_3\overset{+}{P} \quad \bar{O}}{Me^{...}C-C^{...}H}} \quad \underset{C_6H_{13}-n}{}$$

→ Me, H / HOCH$_2$, C$_6$H$_{13}$-n (alkene)

76% (<1% of geometric isomer)

Scheme 53 (Ref 158)

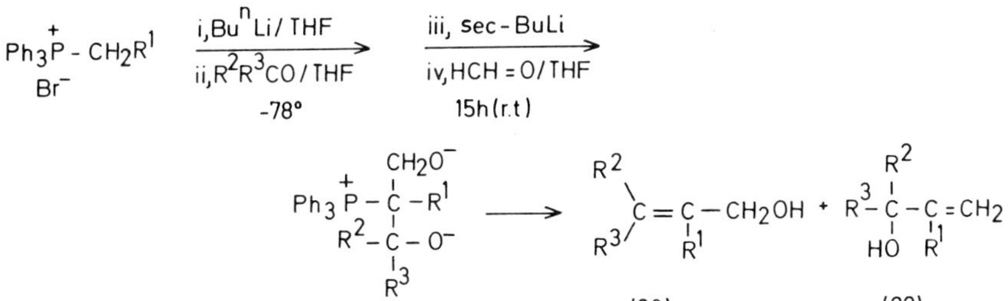

$$Ph_3\overset{+}{P}-CH_2R^1 \cdot Br^- \xrightarrow[\text{ii},R^2R^3CO/THF]{i, Bu^nLi/THF} \xrightarrow[\text{iv},HCH=O/THF]{iii, sec-BuLi}$$
−78°   15h (r.t)

Intermediate:
$$Ph_3\overset{+}{P}-\underset{R^2-C-O^-}{\overset{CH_2O^-}{C-R^1}} \atop R^3$$

→ $\underset{R^3}{\overset{R^2}{C}}=\underset{R^1}{C}-CH_2OH$ + $R^3-\underset{HO}{\overset{R^2}{C}}-\underset{R^1}{C}=CH_2$

  (98)   (99)

R^1	R^2	R^3	% yield (98) (Z):(E)	% yield (99)
Me	n-C$_5$H$_{11}$	H	71 >99:1	0
Me	Ph	H	28 >99:1	12
H	−(CH$_2$)$_4$−		37	10
H	Me	Me	0	41
H	Ph	Me	0	~1

Scheme 54 (Ref 160)

formed by elimination of oxygen from the formaldehyde, is obtained. With acetophenone the reaction fails.

The stereospecificity of the β-oxido-ylide route using paraformaldehyde as one of the aldehyde components is dependent upon the order of use of the aldehydes.[161] Thus, starting from ethylidenetriphenylphosphorane, use of hexanal and paraformaldehyde in that order gave almost pure isomer (100), while their use in the reverse order gave a mixture of the isomers (100) and (101) in the ratio 36:64. Conversely, reactions starting with the unsubstituted

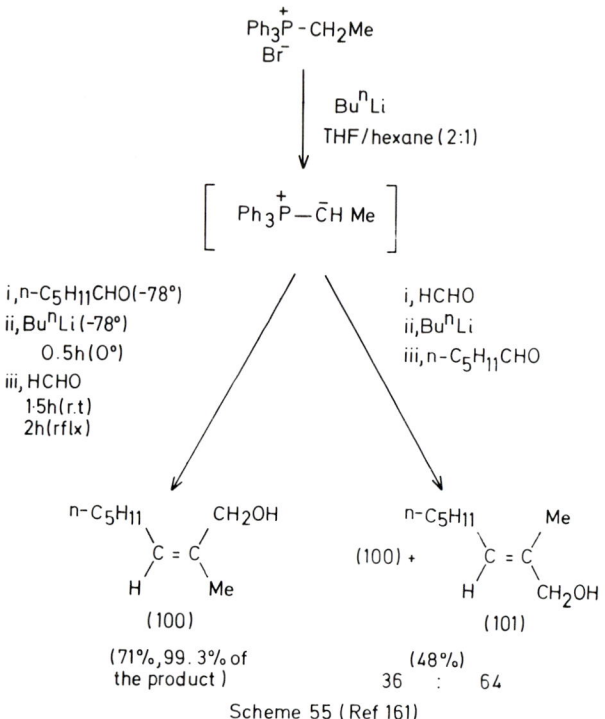

Scheme 55 (Ref 161)

methylenetriphenylphosphorane proceed with greater stereospecificity when paraformaldehyde is the first aldehyde moiety.[161] Thus, the same sequence of reactions as in Scheme 55 produced a mixture of (*E*)- and (*Z*)-isomers (102) and (103) (36% yield) in the ratio 30:70, whereas reverse order of aldehyde addition (paraformaldehyde first) gave (103) specifically, albeit in lower yield (10%) (Scheme 56).

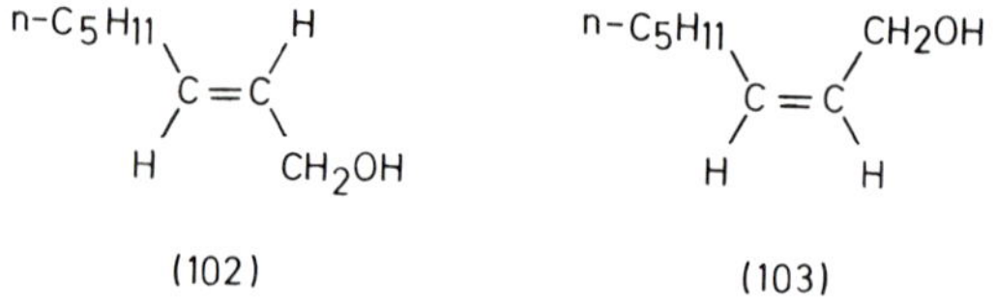

Scheme 56 (Ref 161)

An apparently superior, but as yet little used, alternative to this procedure utilizes 2-hydroxyalkylphosphonium salts as starting materials.[157] Thus, sequential treatment of 2-hydroxyethyltriphenylphosphonium bromide with butyllithium (2 equiv.), benzaldehyde, butyllithium and water affords (*E*)-

2 Wittig Reaction

cinnamic acid in 31% yield and in at least 95% isomeric purity; the only significant by-product is 3-hydroxy-3-phenylpropene (10%) (cf. Scheme 113).

The use in analogous β-oxido-ylide syntheses of a nitrile instead of the second carbonyl reagent leads to α,β-unsaturated ketones,[162] but full experimental details and information on the scope of this reaction have not been reported.

3.4. Ylides of Intermediate ("Moderate") Reactivity

3.4.1. Introduction

As we have seen, ylides fall into two main groups according to their reactivity. The major group are the so-called "reactive" ylides in which the negative charge is localized on the α-carbon, whereas the second group includes the "stable" ylides which are much less reactive owing to extensive delocalization of negative charge through participation of resonance structures. Obviously, a whole spectrum of reactivity lies in between, and some ylides stabilized by α-substituents such as vinyl, aryl, or alkynyl possess an intermediate "moderate" reactivity which is also reflected in the stereochemical outcome of their reactions with carbonyl compounds. The examples in Scheme 57 illustrate the fact that stereoisomeric mixtures of alkenes are usually produced when this preparatively important class of ylides are employed in Wittig reactions under a variety of conditions. It will be noted that in most cases a slight preponderance of the (E)-alkene is favoured.

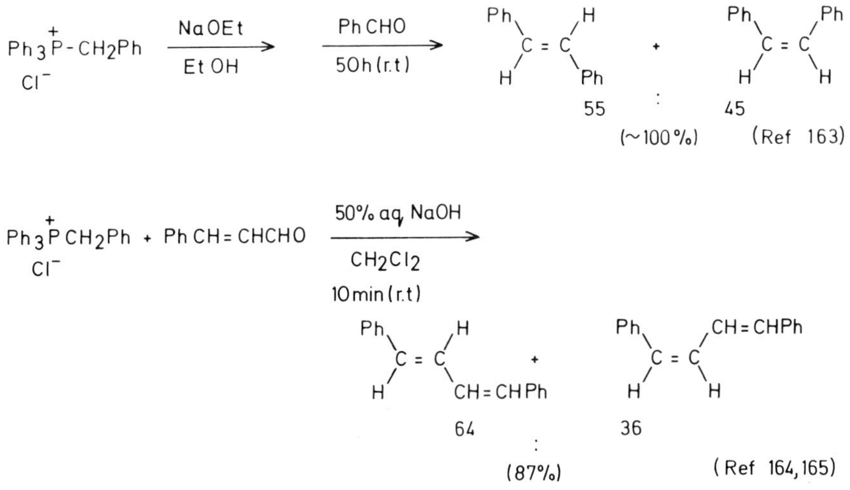

$$Ph_3\overset{+}{P}\text{-}CH_2Ph \;\; Cl^- \xrightarrow[\text{Et OH}]{\text{NaOEt}} \xrightarrow[\text{50h (r.t)}]{\text{Ph CHO}}$$

$$\underset{55}{\overset{Ph}{\underset{H}{>}}C=C\overset{H}{\underset{Ph}{<}}} \;\; + \;\; \underset{45}{\overset{Ph}{\underset{H}{>}}C=C\overset{Ph}{\underset{H}{<}}}$$

(~100%)   (Ref 163)

$$Ph_3\overset{+}{P}CH_2Ph \;\; Cl^- \; + \; Ph CH=CHCHO \xrightarrow[\text{CH}_2\text{Cl}_2]{\text{50\% aq. NaOH}} \text{10 min (r.t)}$$

$$\underset{64}{\overset{Ph}{\underset{H}{>}}C=C\overset{H}{\underset{CH=CHPh}{<}}} \;\; + \;\; \underset{36}{\overset{Ph}{\underset{H}{>}}C=C\overset{CH=CHPh}{\underset{H}{<}}}$$

(87%)   (Ref 164,165)

Scheme 57

Scheme 57 (Cont.)

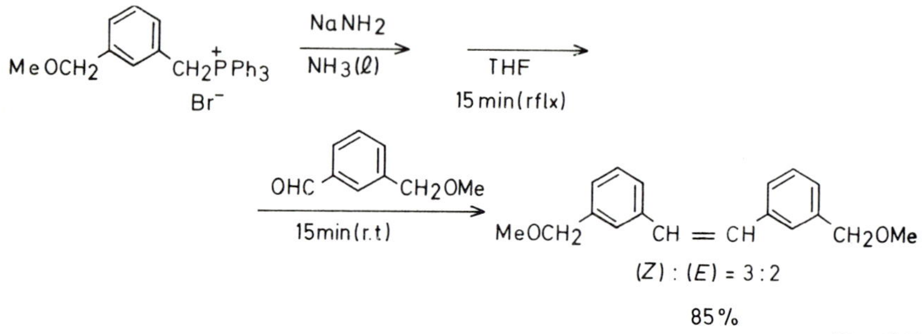
MeOCH2
CH2PPh3
Br⁻
NaNH2
NH3(ℓ)
THF
15 min (rflx)
OHC
CH2OMe
15min (r.t)
MeOCH2
CH = CH
CH2OMe
(Z) : (E) = 3:2
85%

(Ref 166)

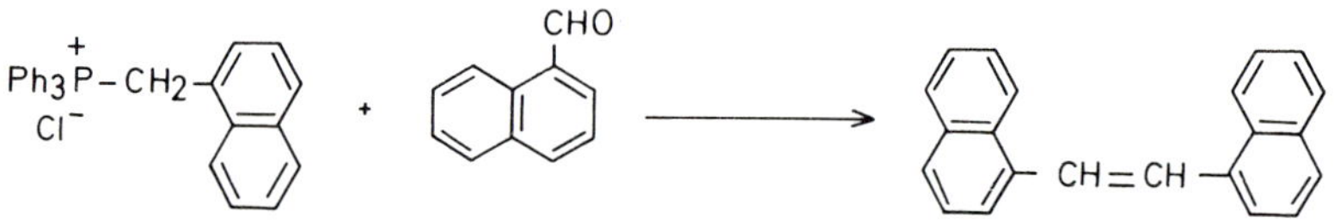
Ph3P−CH2
Cl⁻
CHO
CH=CH

Reaction conditions	Yield (%)	(*Z* : *E*)
NaOMe/MeOH	71	58 : 42
NaOMe/ Et_2O	72	29 : 71
NaOMe/ C_6H_6	88	36 : 64
NaOMe/ DMF	87	37 : 63

(Ref 167)

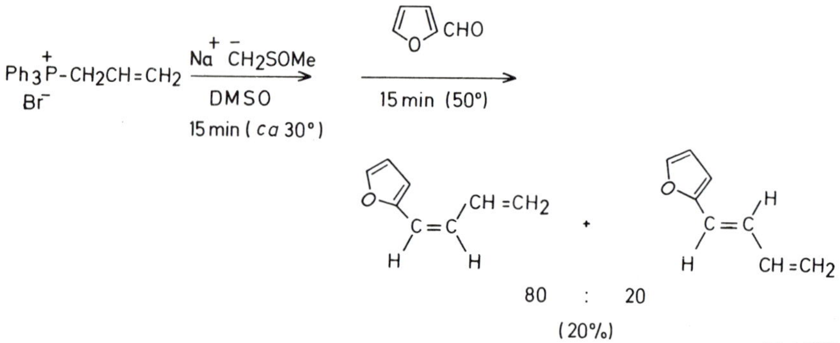
Ph3P-CH2CH=CH2
Br⁻
Na CH2SOMe
DMSO
15 min (ca 30°)
CHO
15 min (50°)
CH=CH2
C=C
H
H
CH=CH2
80 : 20
(20%)

(Ref 168)

2 Wittig Reaction

Scheme 57 (cont.)

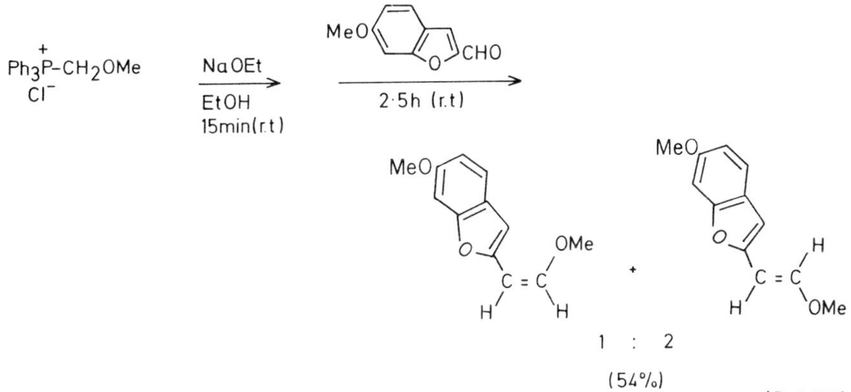

$Ph_3\overset{+}{P}-CH_2OMe \; Cl^-$ $\xrightarrow[\text{15min(r.t)}]{\text{NaOEt}, \text{EtOH}}$ $\xrightarrow[\text{2·5h (r.t)}]{\text{MeO-benzofuran-CHO}}$

MeO-benzofuran-C(H)=C(H)(OMe) + MeO-benzofuran-C(H)=C(OMe)(H)

1 : 2
(54%)
(Ref 169)

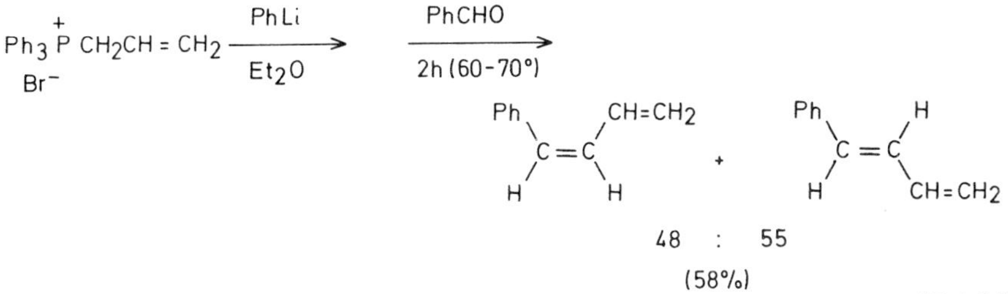

$Ph_3\overset{+}{P}\,CH_2CH=CH_2 \; Br^-$ $\xrightarrow[\text{Et}_2\text{O}]{\text{PhLi}}$ $\xrightarrow[\text{2h (60-70°)}]{\text{PhCHO}}$

Ph(H)C=C(H)(CH=CH_2) + Ph(H)C=C(H)(CH=CH_2)

48 : 55
(58%)
(Ref 2)

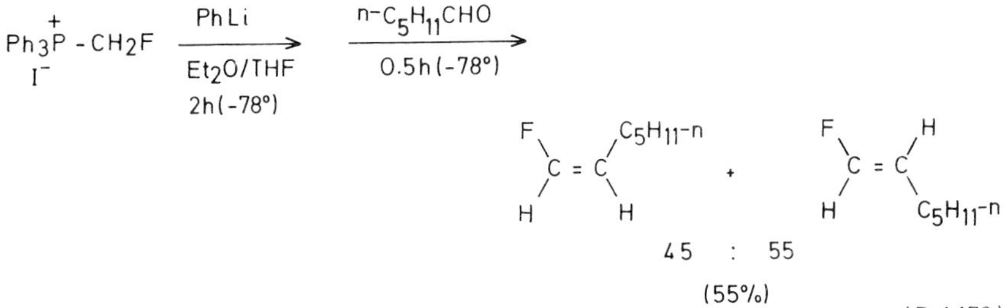

$Ph_3\overset{+}{P}-CH_2F \; I^-$ $\xrightarrow[\substack{\text{Et}_2\text{O/THF}\\ \text{2h(-78°)}}]{\text{PhLi}}$ $\xrightarrow[\text{0.5h(-78°)}]{n\text{-}C_5H_{11}CHO}$

F(H)C=C(C_5H_{11}-n)(H) + F(H)C=C(H)(C_5H_{11}-n)

45 : 55
(55%)
(Ref 170)

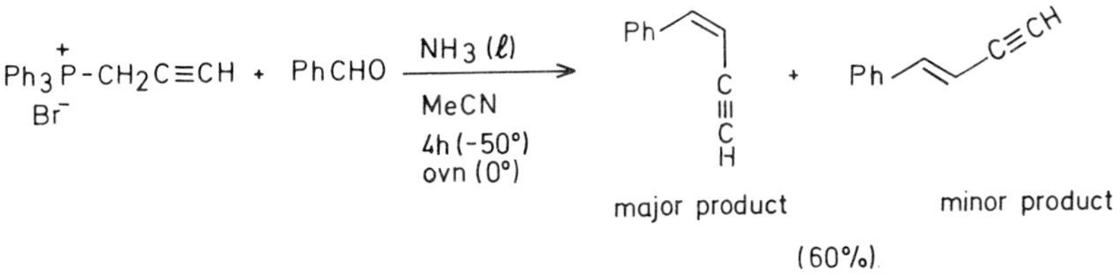

$Ph_3\overset{+}{P}-CH_2C\equiv CH \; Br^-$ + PhCHO $\xrightarrow[\substack{\text{MeCN}\\ \text{4h(-50°)}\\ \text{ovn(0°)}}]{NH_3\,(\ell)}$

Ph-CH=CH-C≡CH (cis, major product) + Ph-CH=CH-C≡CH (trans, minor product)

(60%)
(Ref 171)

Since moderate ylides possess different structural features, certain of which deserve special recognition, it is convenient when discussing stereochemical implications as well as preparative applications to treat separately the three important types of moderate ylides: allylic, benzylic, and prop-2-ynylic (propargylic).

3.4.2. Allylic ylides

In contrast to reactions of reactive ylides, the stereochemistry of Wittig reactions effected with moderately reactive ylides is little affected by the presence of dissolved lithium salts. Scheme 58 shows two examples involving allylic ylides in which the stereochemical outcome is not significantly different

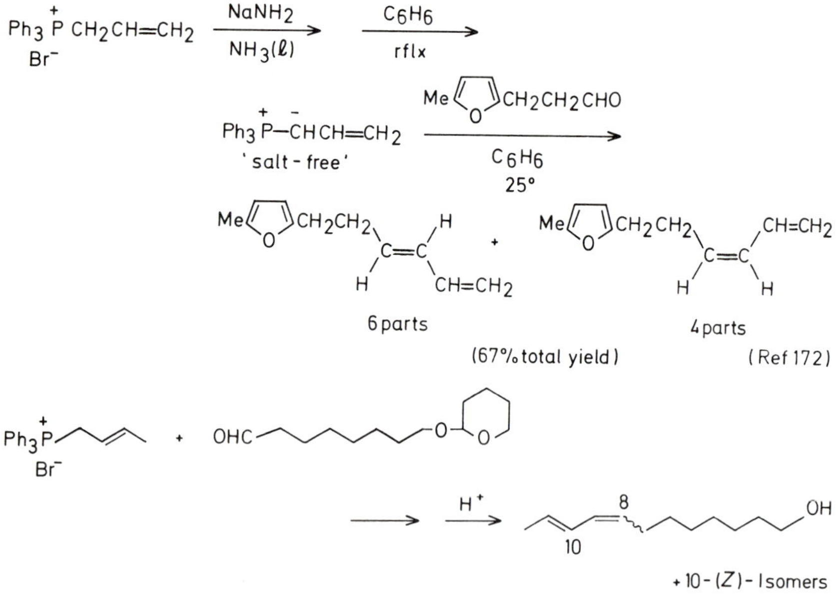

Reaction conditions	Yield (%)	Product composition		
		8E,10E	8Z,10E	10Z
'salt-free'	65	54	28	11 and 7
Bu^nLi/Et_2O	65	50	30	13 and 7
Schlosser's α-metallation procedure using Et_2O as solvent	42	44	36	12 and 8

(Ref 173)

Scheme 58

2 Wittig Reaction

under "salt-free" conditions from that obtained under normal Wittig reaction conditions.[172,173] The second example[173] also illustrates the fact that even the *trans*-olefination procedure of Schlosser, whereby the initially formed Wittig intermediate is α-metallated, loses its stereoselectivity when applied to partly stabilized ylides.

Owing to the intrinsic non-stereoselectivity of allylic ylides, formation of conjugated dienes of definite configuration is best conducted by reaction of a saturated aliphatic reactive ylide with an α,β-unsaturated aldehyde, as exemplified by the synthesis[174] of the (9Z, 11E)-diene (104), a major component of the sex pheromone of the Egyptian cotton leafworm, *Spodoptera littoralis*, by the Wittig reaction of 9-acetoxynonyltriphenylphosphonium bromide with (E)-pent-2-en-1-al in dimethyl sulphoxide using dimsylsodium as the base. This gave a 74% yield of the dienes containing 85–90% of (104) and 10–15% of the (9E, 11E)-isomer (105), whereas the alternative Wittig condensation of (E)-pent-2-enyltriphenylphosphonium bromide with 9-oxonon-1-yl acetate produced a mixture of isomers in various ratios depending on the reaction conditions. The results are recorded in Scheme 59.

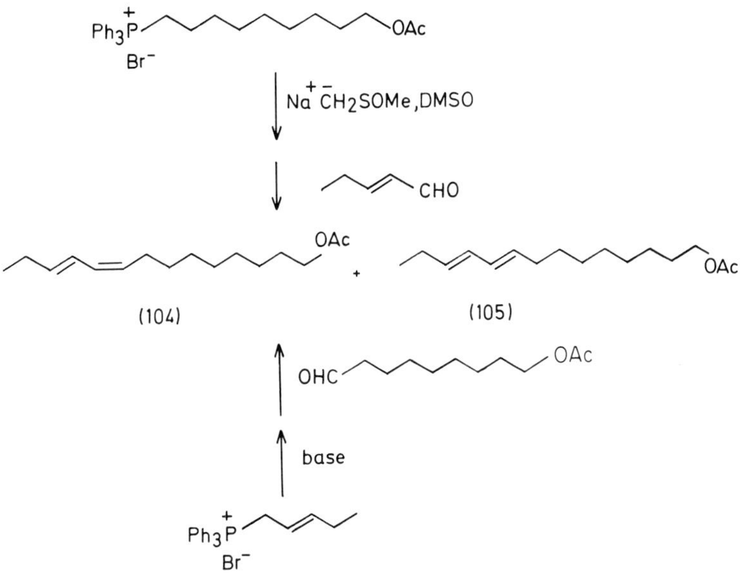

Reaction conditions	(104)	:	(105)
KOBu^t / C$_6$H$_6$	40	:	60
glyme / Bu^{n}Li	50	:	50
Na$^+$ $\bar{C}$H$_2$SOMe / DMSO	60	:	40

Scheme 59 (Ref 174)

Another stereochemical complication sometimes encountered during Wittig reactions with allylic ylides having a terminal substituent is loss of configuration in the allylic double bond. Thus, the reaction[175] between salicylaldehyde and the ylide generated from the crotylphosphonium salt (106) (Scheme 60) gave a mixture of all four geometrical isomers of the diene (107). The same mixture was also obtained starting from pure (E)-crotyl salt.

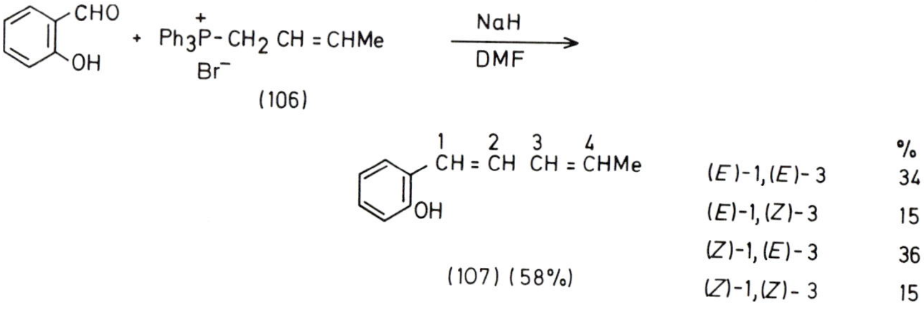

	%
(E)-1,(E)-3	34
(E)-1,(Z)-3	15
(Z)-1,(E)-3	36
(Z)-1,(Z)-3	15

Scheme 60 (Ref 175)

Similarly all four possible dienes were obtained when the ylide from (108) was condensed with benzaldehyde.[176]

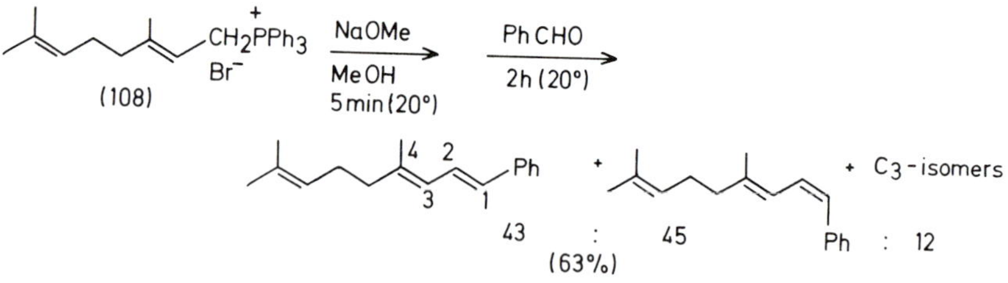

Scheme 61 (Ref 176)

The tendency to isomerize is not, however, a general phenomenon. Preservation of stereochemical integrity has been demonstrated for both (Z)- and (E)-allylic phosphonium salts during ylide formation and subsequent reaction with aldehydes.[12,103] The examples in Scheme 62 are illustrative. Exceptions are to be expected when vigorous reaction conditions or long reaction times are employed. Thus, during the synthesis of triene models relevant to the stereochemistry of the phytoenes (7,8,11,12,7′,8′,11′,12′-octahydro-ψ,ψ-carotenes) (Scheme 63), condensation of the (E)-allylic salt (109) with the (Z)-aldehyde (110) in butylene oxide at 100° gave a mixture (ca 2:3) of the expected (E,E,Z)- and (E,Z,Z)-trienes.[177] However, attempts to prepare the corresponding (Z,Z,Z)- and (Z,E,Z)-isomers by condensing the (Z)-aldehyde with the partly purified (Z)-allylic salt were accompanied by extensive stereomutation of the phosphonium compound, and only traces

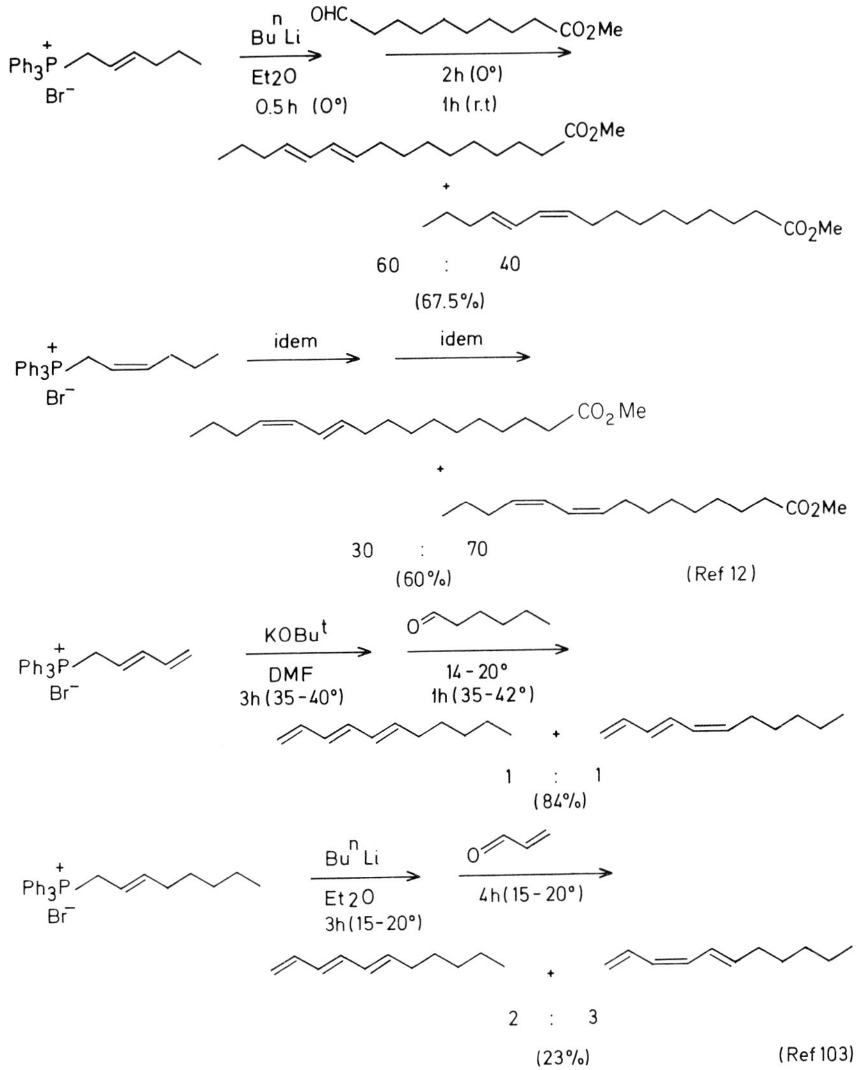

2 Wittig Reaction

Scheme 62

of materials presumed to be the desired products were detected in the reaction mixture.

In some cases, at least, stereomutation may be suppressed by an appropriate choice of reaction conditions. For example, the (Z)-allyltriphenylphosphonium bromide (111) with butyllithium and cyclohexanone in tetrahydrofuran at 40° gave a mixture of dienes containing 20% of the (E)-isomer (113).[178] By

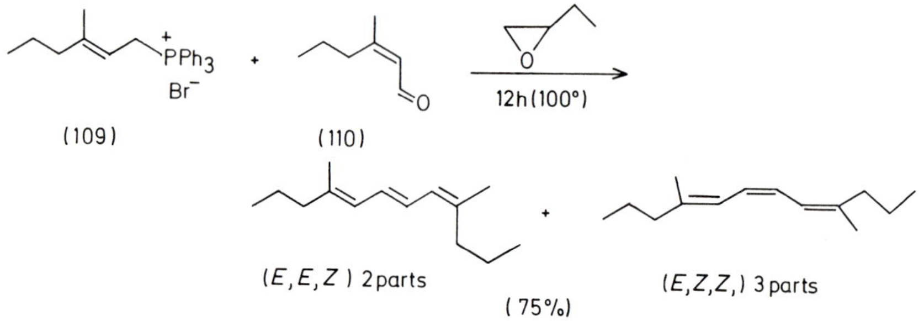

(109) + (110) → 12h (100°)

(E,E,Z) 2 parts (75%)   (E,Z,Z,) 3 parts

Scheme 63 (Ref 177)

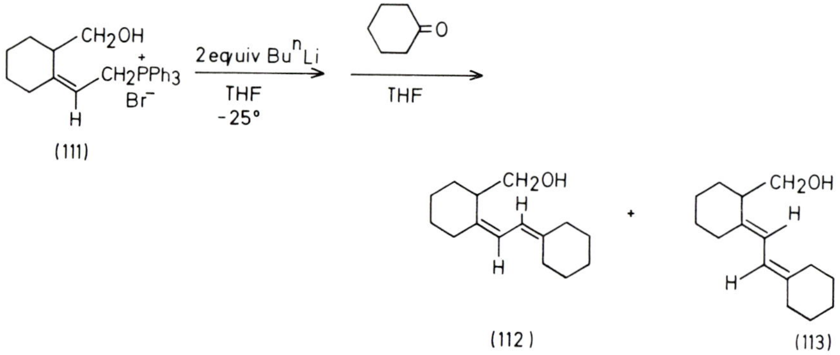

(111) + cyclohexanone, 2 equiv Bu^{n}Li, THF, −25° → THF → (112) + (113)

Scheme 64 (Ref 178)

preparing the ylide at −25° and warming to 5° after addition of the ketone the proportion of (E)-isomer was reduced to < 5%.

Substitution of a methyl group at the β-position usually increases the isomerization about the allylic double bond. For example, treatment of either (E)- or (Z)-ethyl ester of (3-carboxy-2-methylallyl)triphenylphosphonium bromide (114) with base and benzaldehyde gave the same ester mixture (115) in which the sum of (Z)-2-isomers amounted to 57–60%.[179]

In contrast, the ylide from (E)-(3-ethoxycarbonylallyl)triphenylphosphonium bromide (116) reacted with benzaldehyde under the same conditions to produce ethyl 5-phenylpenta-2,4-dienoate (117) that consisted of 94% (E)-2-isomers (Scheme 66).[179] This effect has been attributed to conformational differences in the intermediate allylic ylides. N.m.r. studies revealed[179] that the ylide formed from either isomer of (114) exists as two conformers (118a) and (118b) in rapid equilibrium, with the cisoid form (118a) as the major isomer, whereas the ylide from (116) exists as the two conformers (119a) and (119b), with the transoid form (119a) as the major isomer. These results

2 Wittig Reaction

Scheme 65 (Ref 179)

From structure (114):

PhCH=CH−C(Me)=CHCO₂Et numbered 5 4 3 2
(115) (83%)

from (Z)-ester
22% (E)-2; (E)-4
21% (E)-2; (Z)-4
23% (Z)-2; (E)-4
34% (Z)-2; (Z)-4

Scheme 66 (Ref 179)

From (116):
PhCH=CH−CH=CHCO₂Et numbered 4 3 2
(117) (87%)
(E)-2, (E)-4 39%
(E)-2, (Z)-4 55%
(Z)-2, (E)-4 5%
(Z)-2, (Z)-4 1%

Scheme 67 (Ref 179)

(118a) ⇌ (118b)   70 : 30

(119a) ⇌ (119b)   75 : 25

suggest that (118a) and (118b) must react with similar rate constants with benzaldehyde, but that (119a) may react slightly faster than (119b) with benzaldehyde. In this connection it is of interest that the methyl ester derivative of (116) condenses with 9-anthraldehyde to produce *only* (E)-2-products (120) (Scheme 68).[180] It might also be noted in passing that the

Scheme 68 (Ref 180)

Scheme 69 (Ref 181, 182)

Scheme 70 (Ref 183)

2 Wittig Reaction

(E)-phosphonate ester (121) undergoes little, if any, isomerization upon formation of its anion and subsequent reaction with either benzaldehyde or (Z)-citral.[181,182] On the other hand, condensations of the (Z)-phosphonate ester with the same aldehydes are accompanied by extensive stereomutation, and only 25% of the resulting esters have the (Z)-2 configuration (Scheme 69).

Another difficulty which frequently arises in Wittig reactions of allylic ylides is concurrent condensation at the γ-position. For example, the Wittig reaction of hexanal with the allylic ylide (122) furnished[183] not only all four geometric isomers of methyl 3-methyldeca-2,4-dienoate (123), the normal α-condensation product, but also both geometric isomers of methyl 2-isopropenyloct-2-enoate (124), the γ-condensation product (Scheme 70). The latter could have been formed via the cyclic six-centre phosphorane (125).

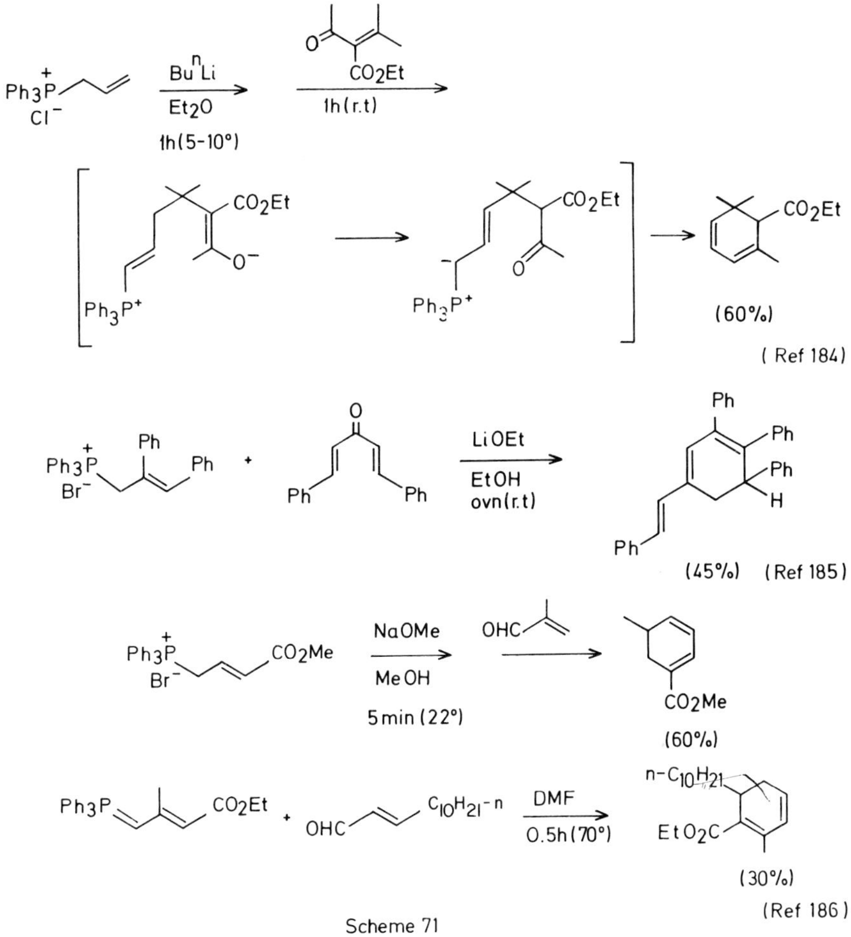

Scheme 71

However, by generating the ylide from its salt, using diisopropylamine in DMF in the presence of cadmium iodide, it is possible to minimize γ-condensation and obtain fair yields of the α-condensation product (123) (36%).

In the case of α,β-unsaturated carbonyl compounds, condensation at the γ-position is highly specific and takes place in a Michael-type fashion to provide a simple and elegant synthesis of cyclohexa-1,3-dienes.[184–186] A possible mechanism for this transformation is illustrated in the first example

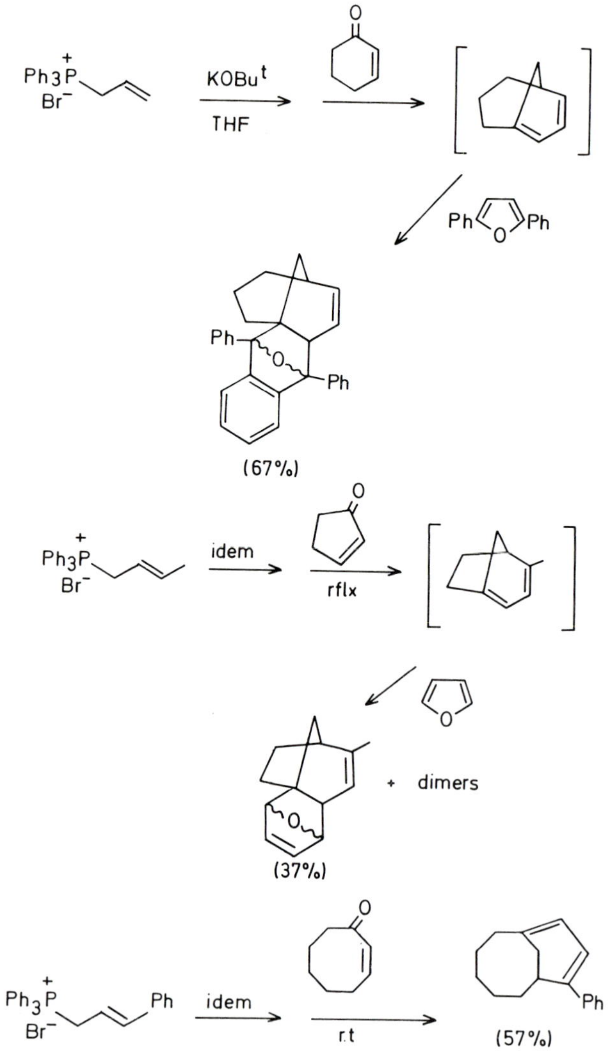

Scheme 72 (Ref 187)

2 Wittig Reaction

in Scheme 71. Cyclic α,β-unsaturated ketones condense in the same way to give bridgehead alkenes which may be highly strained:[187] that from cyclohex-2-enone was trapped as an adduct with 1,3-diphenylisobenzofuran as shown in Scheme 72. Other examples are also indicated. Similar conjugative additions of allylic ylides to dienoic esters lead to bicyclo[4.1.0]heptenes by a mechanism involving loss of triphenylphosphine (Scheme 73).[188]

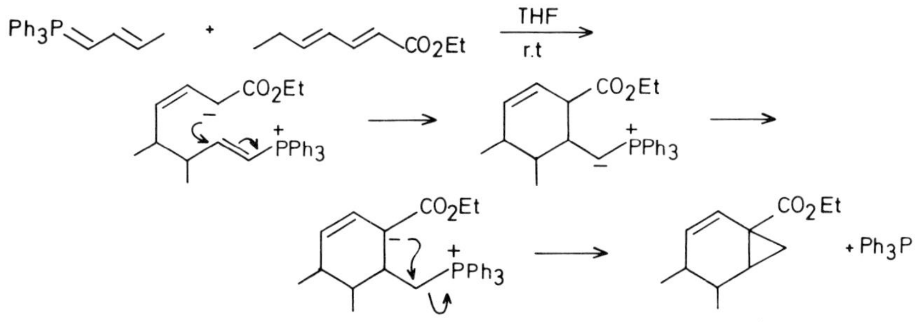

Scheme 73 (Ref 188)

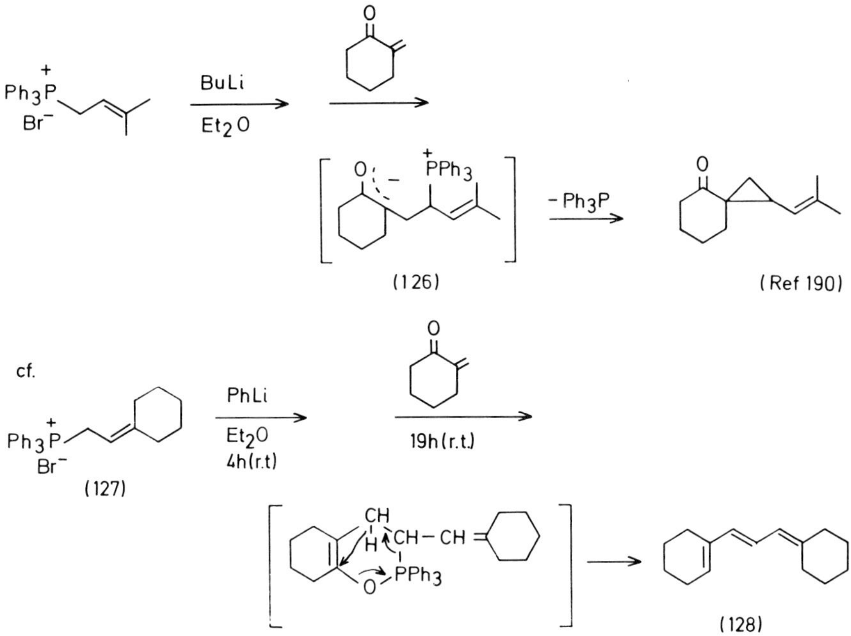

Scheme 74

As expected, increased hindrance at the γ-carbon inhibits γ-condensation.[189-191] Thus, on condensation of 2-methylenecyclohexanone with 3-methyl-2-butylidenetriphenylphosphorane, normal 1,2-addition occurred to give the betaine (126) which subsequently closed to the cyclopropane by ejection of triphenylphosphine.[190] This result is in contrast to a related condensation[191] of 2-methylenecyclohexanone with the ylide derived from the allylic salt (127), which resulted in the formation of 1-(cyclohex-1-enyl)-3-cyclohexylidenepropene (128) via a [1,3]-hydrogen shift as shown in Scheme 74.

Finally, mention should be made of a related "one pot" procedure[192] for the synthesis of conjugated polyene esters or ketones involving the condensation of allylidenetriphenylphosphorane with α,β-unsaturated β-chloro-carbonyl compounds, e.g. (129), in the presence of an aldehyde (Scheme 75). The

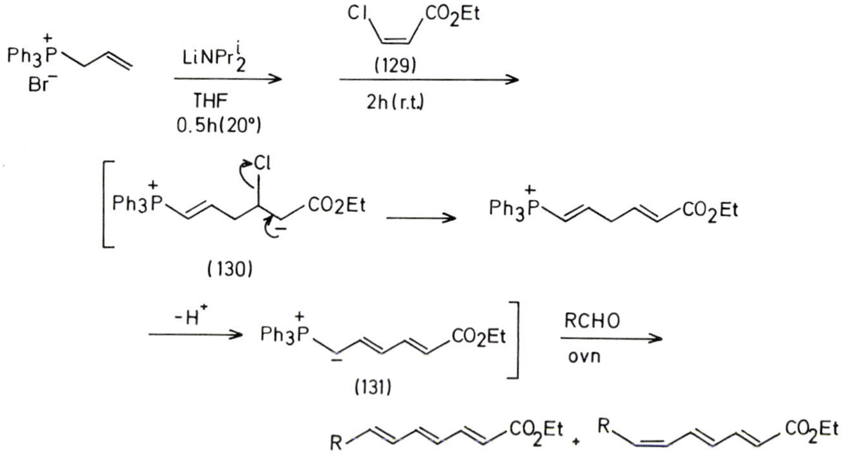

R	Yield(%)	Product composition	
Ph	84	100 : 0	
Me₂CH	67	60 : 40	

Scheme 75 (Ref 192)

reaction is believed to involve analogous conjugate addition to give intermediates such as (130). Subsequent loss of chloride and deprotonation by base convert (130) into the conjugated ylide (131) which is trapped by the aldehyde. Note that trapping the ylide with benzaldehyde gives a high yield of the expected all-(E)-triene whereas the more bulky isobutyraldehyde affords a 40:60 mixture of (Z)- and (E)-isomers. The procedure does not appear to be suitable for the stereoselective synthesis of the isoprenoid polyenes since

2 Wittig Reaction

trapping of the corresponding ylide derived from methyl β-chlorocrotonate with isobutyraldehyde gives comparable amounts of four isomeric alkenes.

3.4.3. Benzylic ylides

For Wittig reactions effected with triphenylphosphonium benzylides, the stereochemistry of the products depends largely on the nature of the substituents in the aryl ring. In keeping with previous stereochemical arguments, electron-releasing groups which enhance the reactivity (nucleophilicity) of the ylide reduce the *trans*-selectivity, whereas the reverse is true for electron-withdrawing groups.[44,193] Important advantage may be gained in synthesis from this influence to produce a higher proportion of (Z)-alkene. For example, in the preparation of 4-nitro-4′-methoxystilbene, the reaction of 4-nitrobenzylidenetriphenylphosphorane (132) with anisaldehyde leads to the predominant formation of the (E)-alkene as shown in Scheme 76. However, by interchanging the ylide and aldehyde substituents, a 1:1 mixture of stereoisomers is obtained, which can be easily separated by fractional crystallization and chromatography on deactivated alumina.

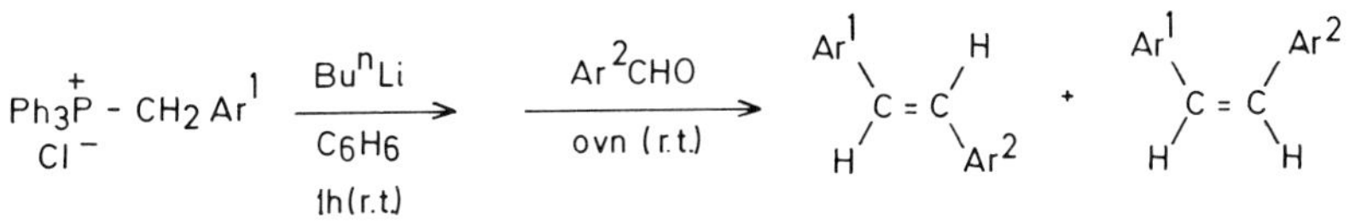

$$Ph_3\overset{+}{P}-CH_2Ar^1 \; Cl^- \xrightarrow[\substack{C_6H_6 \\ \{h(r.t.)\}}]{Bu^nLi} \xrightarrow[\text{ovn (r.t.)}]{Ar^2CHO} \underset{H}{\overset{Ar^1}{\diagup}}C=C\underset{Ar^2}{\overset{H}{\diagup}} + \underset{H}{\overset{Ar^1}{\diagup}}C=C\underset{H}{\overset{Ar^2}{\diagup}}$$

Ylide	Ar^1	Ar^2	Product composition*	
(132)	$p\text{-}O_2NC_6H_4$	$p\text{-}MeOC_6H_4$	74 (100)	: 26 (0)
(133)	$p\text{-}MeOC_6H_4$	$p\text{-}O_2NC_6H_4$	56 (52)	: 44 (48)

*Ref 44 in brackets

Scheme 76 (Ref 193)

Perhaps a more effective means of shifting the isomer ratio in favour of the (Z)-product is through structural alteration of the phosphonium group. As already noted, if the phosphorus atom is rendered more electrophilic by substitution of electron-withdrawing groups, the betaine intermediate has less chance to equilibrate to the more stable *threo* form. As a consequence greater selectivity for (Z)-alkene formation is observed.[193] Table 6 shows that in the synthesis of 4-nitro-4′-methoxystilbene, use of 4-methoxybenzylidenetris-(p-chlorophenyl)phosphorane (134) in place of (133) leads to an increase from 44:56 to 80:20 in the (Z):(E) ratio. In a similar manner, the (Z):(E) ratio increases in the nitrobenzylidene series as the phosphorus substituent is successively changed from n-butyl (135) to phenyl (136) to p-chlorophenyl

Table 6. Relative yields of (Z)- and (E)-4-nitro-4'-methoxystilbenes from the reaction of $R_3P^+CH_2C_6H_4R^1\ Br^-$ with $R^2C_6H_4CHO$ in benzene at room temperature using n-butyllithium as base [193]

Ylide	R	R^1	R^2	(Z):(E) ratio
(133)	Ph	OMe	NO_2	44:56
(134)	p-ClC$_6$H$_4$	OMe	NO_2	80:20
(135)	n-C$_4$H$_9$	NO_2	OMe	17:83
(136)	Ph	NO_2	OMe	26:74
(137)	p-ClC$_6$H$_4$	NO_2	OMe	52:48

(137). As indicated, the course of the reaction may be shifted overwhelmingly in favour of either the (Z)- or the (E)-isomer by the appropriate choice of reactants.

Apart from the effect of the substituents, the degree of stereoselectivity observed may be influenced by changes in the reaction medium. As shown in Tables 7 and 8, formation of (Z)-alkene is favoured by the use of either protic solvents or dipolar aprotic solvents. However, under no circumstances is the effect appreciable, and (E)-alkene usually dominates the product composition even when salt-free solutions of the ylide are employed. This

Table 7. Effect of the reaction medium on the stereoselectivity of the Wittig reaction between benzyltriphenylphosphonium salts and propionaldehyde

Reaction medium (salt present)	PhCH=CHEt (Z):(E) ratio	Yield (%)	Refs.
Bu^{n}Li–C$_6$H$_6$ (LiBr)	27:73	57	181
(LiCl)	23:77	32–83	85
(LiCl)	21:79		110
(LiBr)	27:73		110
(LiI)	41:59		110
Schlosser's α-metallation product	43:57		110
Bu^{n}Li–hexane	22:78		110
"salt-free" C$_6$H$_6$	25:75		33
	18:82	61	110
Et$_2$O–NaOEt	31:69		61
THF–NaOEt	33:67		61
DMF–Bu^{n}Li	44:56	32–44	85
	40–45:60–55	60–74	181
DMF–NaOEt	46:54	66	85
EtOH–NaOEt	46:54	95	85
K$_2$CO$_3$–18-crown-6-ether–THF	25:75	94	133

2 Wittig Reaction

Table 8. Effect of the reaction medium on the stereoselectivity of the Wittig reaction between benzyltriphenylphosphonium salts and benzaldehyde

Reaction medium	PhCH=CHPh $(Z):(E)$ ratio	Yield (%)	Refs.
PhLi–Et$_2$O	30:70	82	2
"salt free" C$_6$H$_6$	34:66		110
	44:56	100	33
KOBu^t–Bu^{t}OH	25:75		163
NaN(SiMe$_3$)$_2$–C$_6$H$_6$	48:52	82	131
NaOEt–EtOH	45:55		163
	53:47	76	3
NaOEt–DMF	53:47		163
THF	42:58	72	194
50% aq.NaOH–CH$_2$Cl$_2$	50:50	88	164

would imply that general solvation of the betaine intermediate is not significantly affecting its formation or reversible decomposition. It has been suggested[14] that *threo*-betaine formation becomes more important under conditions of kinetic control if the ylide side-chain bears bulky ligands such as phenyl (or cyclohexyl). Even so, it should be noted that use of Schlosser's modification, whereby the initially formed betaine–lithium complex is α-metallated, fails to improve the *trans*-selectivity in the reaction with propionaldehyde. A similar situation prevails[154] in the reaction of the corresponding polymeric ylide with acetaldehyde as illustrated in Scheme 77. No change is observed in the product composition when lithium salts are rigorously excluded, or by using the Schlosser *trans*-olefination procedure. However, (Z)-alkene is preferentially formed. Other reports indicate that stereoselectivity differences between the reactions of polymeric and monomeric ylides are small.[194,195]

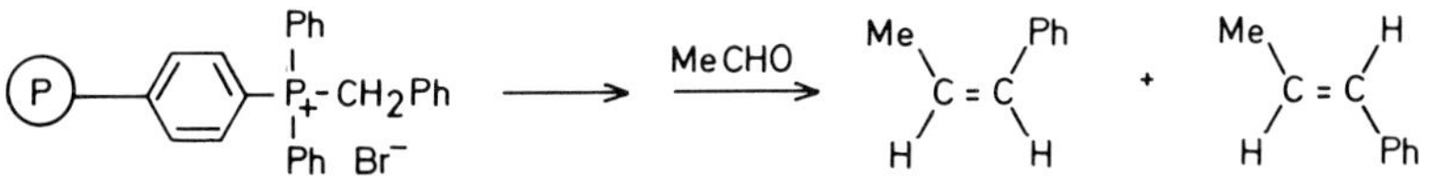

(P)—⟨Ph⟩—P$_+$(Ph)–CH$_2$Ph Br$^-$ $\xrightarrow{\text{MeCHO}}$ (Z) Me,Ph/H,H C=C + (E) Me,H/H,Ph C=C

Reaction conditions	(Z)	:	(E)
Bu^{n}Li/dioxan (60°, 8h)	60	:	40
'salt-free' (Bu^{n}Li/dioxan, washed with THF)	60	:	40
Schlosser's α-metallation of primary Wittig intermediate (THF as solvent)	60	:	40

Scheme 77 (Ref 154)

In contrast to the above cases, an interesting change in selectivity is observed[196] when the reaction solvent in the Wittig condensation between 2-methoxybenzyltriphenylphosphonium bromide and the aldehyde (138) is changed from methanol to dimethylformamide. In methanolic solution (2 h, reflux) with lithium methoxide as base the (E)-stilbene predominated, but in dimethylformamide (1 h, 90°) the (Z)-stilbene predominated. The reason why opposite stereoselectivities are preferred in these solvents is not apparent.

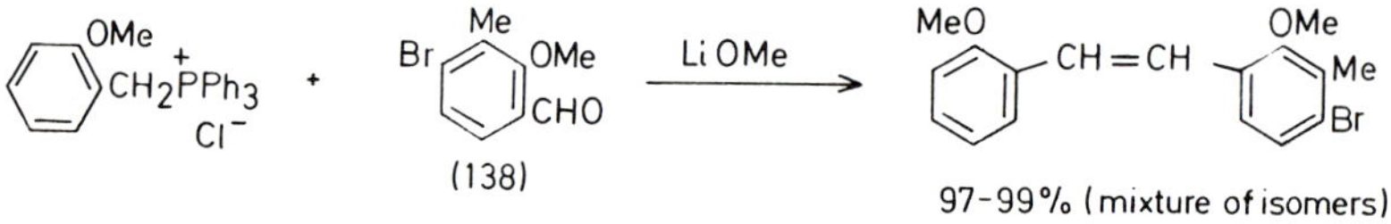

Scheme 78 (Ref 196)

Although extraneous cations appear to have little or no influence on the stereochemistry of Wittig reactions to form stilbenes, some effect is displayed by extraneous inorganic anions.[197] As shown in Table 9, large concentrations of halide ions, particularly of iodide, promote the formation of (E)-stilbene from benzaldehyde and benzyltriphenylphosphonium salts in methanol with methoxide as base. Large concentrations of methoxide ions, on the other hand, slightly favour the formation of the (Z)-isomer. This difference in directive effects its explicable in terms of preferential solvation of P$^+$ by halide ions, leading to greater reversibility of betaine formation, as compared with methoxide ions, which are preferentially solvated by methanol.

Table 9. Effect of addition of inorganic salts on the relative yields of (Z)- and (E)-stilbene from the Wittig reaction of benzyltriphenylphosphonium salts with benzaldehyde in methanol after 1 hour[197]

Base	Salt added (mole l^{-1})	PhCH=CHPh (Z):(E) ratio
NaOMe	—	42:58
	NaI (1)	41:59
	NaI (2)	36:64
	NaI (3)	28:72
	NaOMe (4·5)	54:46
LiOMe	—	42:58
	LiCl (3)	40:60
	LiI (3)	30:70
	LiOMe (2·8)	50:50

2 Wittig Reaction

The isomer ratio may be shifted further in favour of the (*E*)-isomer by replacing the phenyl groups on phosphorus by alkyl groups as illustrated in Table 10 where a comparison is made of the distribution of stilbene isomers obtained in ethanol. It should be noted that substitution of alkyl groups for phenyl groups on phosphorus leads not only to almost pure (*E*)-stilbene, but also to better yields. A similar effect has been noted previously for stabilized ylides (see Section 3.2).

Replacement of phenyl groups on phosphorus by n-butyl groups can lead to unexpected complications.[135] Thus condensation between the salt (139) and 4-methoxyacetophenone in the presence of dimsylsodium gave a mixture of (*Z*)-stilbene and (*E*)-styrene products in stereospecific reactions resulting from dichotomous ylide formation (Scheme 79). Under the same conditions, the reaction with the corresponding triphenylphosphonium salt produced a 15:85 mixture of (*Z*)- and (*E*)-isomers of the stilbene, albeit in low yield (*ca.* 2%).[199] Forcing conditions were required to obtain the stilbenes (140)

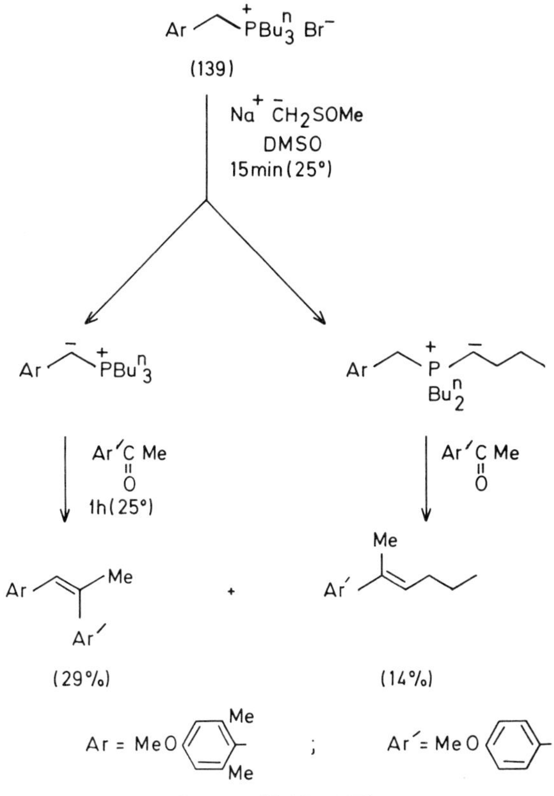

Scheme 79 (Ref 135)

Table 10. Yields and *(Z)* : *(E)* ratios for stilbene resulting from the interaction of different benzylphosphonium salts $R^1R^2R^3P^+CH_2Ph\ X^-$ and benzaldehyde in ethanol–sodium ethoxide

$R^1R^2R^3$	PhCH=CHPh		Ref.
	(Z):(E) ratio	Yield (%)	
Ph	53:47	76	3
Ph$_2$Me	28:72	97	198
PhMe$_2$	13:87	91	198
Bu	9:91		14
cyclohexyl	5:95	92	52

and (141) (yields < 5%) (Scheme 80), which probably reflects steric limitations to the Wittig synthesis of these types of molecules. The striking stereospecific formation of (Z)-stilbene from (139) probably illustrates a similar importance of steric factors[30,145] in controlling the selectivity of the Wittig reaction as discussed earlier in Section 3.3.

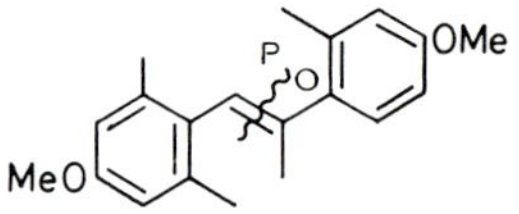

(140) 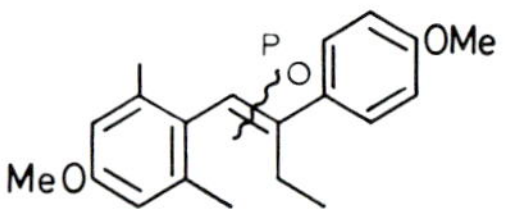　　　　(141)

Scheme 80 (Ref 199)

In general, structural changes in the carbonyl co-reactant exert only a minor effect on the stereochemistry of Wittig reactions of benzylidenetriphenylphosphorane. As illustrated in Table 11 the *trans*-selectivity is preserved when using a variety of aliphatic, unsaturated, and aromatic aldehydes under aprotic conditions. There are some exceptions to this observation, but one in particular deserves special attention.[182] The tabulated examples show that condensations involving α,β-unsaturated aldehydes with a carboxylate ion on the β-carbon atom, whether (Z) or (E), give mixtures of pentadienoic acids, e.g. (142) and (143) (Scheme 81), in which a high proportion (>60%) of the newly formed double bonds have the (Z) configuration. Molecular models suggest that this may well result from preferential stabilization of the *erythro* form of the intermediate betaine by interaction of the positively charged phosphorus atom and the carboxylate ion in the staggered conformation (144) [cf. (145)].[182,200] The latter reaction is also of interest from a preparative viewpoint in that there is complete retention of stereochemistry at the double bond derived from the lactol. We return to this point in Section 4.4.

2 Wittig Reaction

Table 11. *(Z) : (E)* Ratios of alkenes from benzyltriphenylphosphonium salts and aldehydes (RCHO)

R	Reaction medium	PhCH=CHR $(Z):(E)$ ratio	Yield (%)	Ref.
EtO	NaH–DMSO	22:78	90	43
Et	Bu^{n}Li–C$_6$H$_6$	23:77		85
Et	LiI added	25:75		
Et	Bu^{n}Li–C$_6$H$_6$	27:73		181
Et	LiBr added	23:77		
Et	C$_6$H$_6$ (salt-free)	18:82		110
		25:75	61	33
Ph	Bu^{n}Li–C$_6$H$_6$	45:55		110
Ph	C$_6$H$_6$ (salt-free)	34:66		110
		44:56	100	33
p-MeOC$_6$H$_4$	C$_6$H$_6$ (salt-free)	19:81	84	33
p-MeC$_6$H$_4$	C$_6$H$_6$ (salt-free)	36:64	88	33
p-ClC$_6$H$_4$	C$_6$H$_6$ (salt-free)	10:90	72	33
p-NO$_2$C$_6$H$_4$	aq.NaOH–CH$_2$Cl$_2$	56:44	72	164
PhCH=CH–	aq.NaOH–CH$_2$Cl$_2$	36:64	87	
(E)-MeCH=C(Me)	NaOMe–Et$_2$O	34:66	87	182
(E)-MeO$_2$CCH=C(Me)	NaOMe–Et$_2$O	40:60	87	
(Z)-MeO$_2$CCH=C(Me)	NaOMe–Et$_2$O	34:66	70	
(E)-$^-$O$_2$CCH=C(Me)	NaOMe–Et$_2$O	60:40	78	
(Z)-$^-$O$_2$CCH=C(Me)	NaOMe–Et$_2$O	66:34	93	
$^-$O$_2$CC(Me)=CH	NaOMe–Et$_2$O	66:34	90	
MeO$_2$CC(Me)=CH	NaOMe–Et$_2$O	40:60	40	

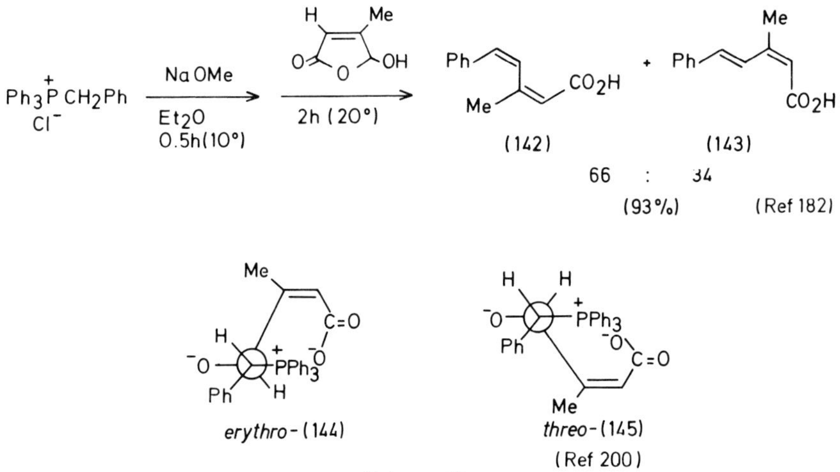

Scheme 81

As might be expected, in some instances only the (E)-isomer is formed, especially when it is strongly favoured on steric grounds. An example is the conversion of indole-3-aldehyde (146) into (E)-1-phenyl-2-(3-indolyl)ethylene (147) in 30–60% yield[201] (Scheme 82). The same ylide also condenses with

Scheme 82 (Ref 201)

9-anthraldehyde to give the (E)-alkene (148) as the sole product.[164] By comparison the corresponding reaction with benzaldehyde affords a 59:41 mixture of (Z)- and (E)-isomers. The simple expedient of using a two-phase system for effecting condensation in the latter reactions is worthy of special note.

Scheme 83 (Ref 164)

3.4.4. Propargylic (prop-2-ynyl) ylides

Only a few reports have appeared on the use of propargylic ylides in Wittig reactions for the introduction of the enyne grouping, presumably because the latter is more easily secured starting from a reactive ylide (for which steric control is possible) and a propargylic aldehyde. One of the obstacles to the use of propargylic ylides lies in their tendency to isomerize prior to Wittig condensation.[202] This possibility is illustrated by the reaction of prop-2-ynyltriphenylphosphonium bromide (149) with cyclohexanealdehyde using

Scheme 84 (Ref 202)

2 Wittig Reaction

butyllithium as base in THF to give the cumulene (151) in low yield, presumably via the cumulene ylide (150). Interestingly, the same salt affords the expected enynes with *conjugated* aldehydes when treated with sodamide in liquid ammonia (cf. Scheme 57).[171]

An indirect, but as yet little used, procedure for the introduction of a *terminal* enyne grouping which avoids cumulene formation involves the initial reaction of the carbonyl compound with the trimethylsilylprop-2-ynyl salt (152)[202,203] (Scheme 85). After condensation the silyl protecting group is removed quantitatively by treatment with either $Bu^n_4N^+F^-$ in THF or $KF \cdot 2H_2O$ in DMF. Scheme 85 provides some representative examples of the preparation of protected enynes. In each given case the process is stereoselective for the (E)- rather than the (Z)-isomer.

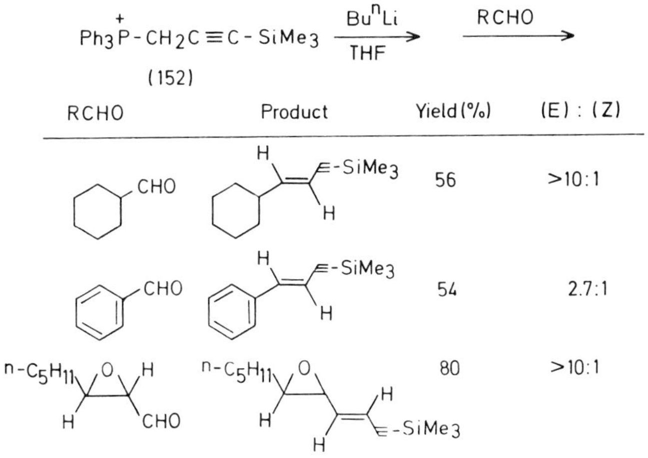

$$Ph_3\overset{+}{P}-CH_2C{\equiv}C-SiMe_3 \quad \xrightarrow[\text{THF}]{Bu^nLi} \quad \xrightarrow{RCHO}$$

(152)

RCHO	Product	Yield (%)	(E) : (Z)
cyclohexyl-CHO	cyclohexyl-CH=C(H)(E-SiMe₃)	56	>10:1
Ph-CHO	Ph-CH=C(H)(E-SiMe₃)	54	2.7:1
n-C₅H₁₁ glyceraldehyde acetonide-CHO	corresponding enyne with E-SiMe₃	80	>10:1

Scheme 85 (Ref 202)

That the ylide derived from the trimethylsilylprop-2-ynyl salt (152) undergoes little if any isomerization during olefination has led[204] to a key intermediate in the synthesis of the polyacetylenic diol (155) obtained from cultures of the fungus *Fistulina pallida* (Berk and Rev.)[204] (Scheme 86). Thus, reaction of (152) with 2,3-O-isopropylidene-D-glyceraldehyde (153) in the presence of butyllithium affords the trimethylsilyl enyne acetonide (154) in an (E):(Z) ratio 1:5 at −78° and 1:1·25 at −40° [cf. the more favoured (E)-double bond formation with other aldehydes in Scheme 85 and the selective (Z)-double bond formation in the skipped enyne system from the corresponding butynyl salt, $Me_3SiC{\equiv}CCH_2CH_2P^+Ph_3\ I^-$ [205]]. In a related study,[206] the corresponding condensation of (152) with tetrahydropyran-2-aldehyde gave the enyne (156) in an (E):(Z) ratio of 1·4:1 at −78° and 3:1 at 0°. The (Z)-enyne [(Z)-(158)] could also be formed in 14% yield from the phosphonium salt

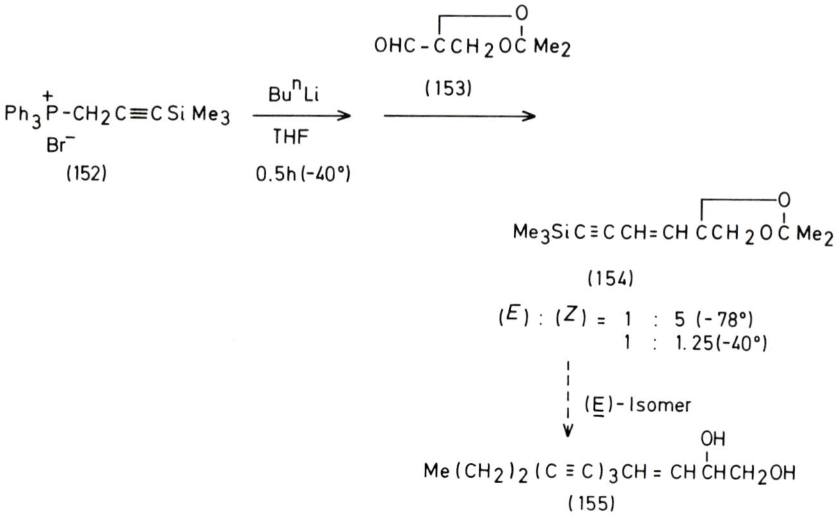

$$\underset{\underset{(152)}{Br^-}}{Ph_3\overset{+}{P}-CH_2C\equiv CSiMe_3} \xrightarrow[\underset{0.5h(-40°)}{THF}]{Bu^nLi} \xrightarrow{\underset{(153)}{OHC-\overset{|}{C}CH_2O\overset{|}{C}Me_2}\overset{\overline{\hspace{1cm}O\hspace{1cm}}}{}}$$

$$Me_3SiC\equiv CCH=CH\overset{\overline{\hspace{1cm}O\hspace{1cm}}}{\overset{|}{C}CH_2O\overset{|}{C}Me_2}$$
(154)

(E) : (Z) = 1 : 5 (-78°)
            1 : 1.25(-40°)

↓ (E)-Isomer

$$Me(CH_2)_2(C\equiv C)_3CH=CH\overset{OH}{\overset{|}{C}H}CH_2OH$$
(155)

Scheme 86 (Ref 204)

$$\underset{\underset{(152)}{Br^-}}{Ph_3\overset{+}{P}-CH_2C\equiv CSiMe_3} \xrightarrow[THF]{Bu^nLi} \xrightarrow{\langle O \rangle-CHO} Me_3SiC\equiv CCH=CH-\langle O \rangle$$

(156) (E) : (Z) = 1.4 : 1 (-78°)
                  3 : 1 (0°)

$$\underset{\underset{(157)}{Br^-}}{Ph_3\overset{+}{P}CH_2\overset{|}{C}H(CH_2)_3\overset{\overline{\hspace{1cm}O\hspace{1cm}}}{C}H_2}} \xrightarrow[\underset{5min(0°)}{THF}]{Bu^nLi} \xrightarrow[0.5h(20°)]{HC\equiv CHCHO} HC\equiv CCH=CH-\langle O \rangle$$

(158) (14%; Z-isomer only)

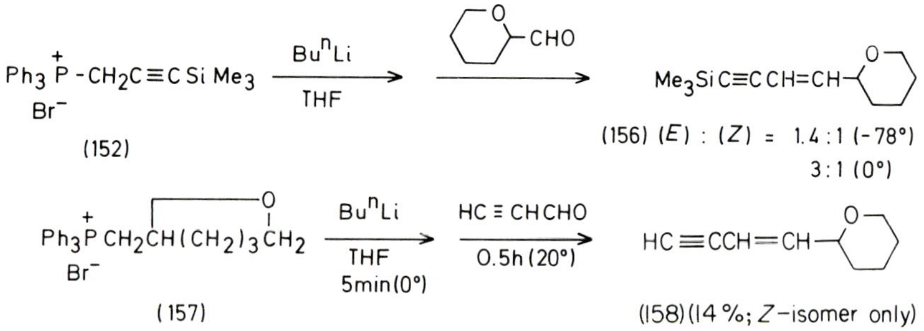

Scheme 87 (Ref 206)

(157), butyllithium, and propiolaldehyde (Scheme 87) (cf. the impossibility of obtaining ylides from similar Wittig salts[207]).

3.4.5. Summary

From the preceding discussion it is clear that Wittig reactions involving ylides of moderate reactivity are in general devoid of marked stereoselectivity. It is also apparent that stereochemical control is limited largely to the selective formation of (E)-alkenes by the correct choice of the phosphorus substituents. For preparative purposes this procedure possesses only marginal utility. One reason is that stereoisomeric mixtures of alkenes having electron-withdrawing groups on the α-carbon can usually be isomerized to the more stable (E)-isomer without much difficulty. An example is the synthesis of vitamin A acetate (160) from β-cyclogeranyltriphenylphosphonium bromide

2 Wittig Reaction

and 8-acetoxy-2,6-dimethylocta-2,4,6-trienal according to Scheme 88.[208,209] This gave a mixture of all-*trans*-(160), together with significant quantities of (7Z)vitamin A acetate (159), which was easily isomerized to (160) by heating with iodine. Scheme 88 also illustrates a different Wittig combination giving a mixture of all-*trans*-(160) and (11Z)-vitamin A acetate (161) from which all-*trans*-(160) could be obtained by subsequent isomerization and crystallization.[210,211] A similar approach has also been used[212] to prepare the all-*trans*-bis(thienylvinyl) derivative (162) as shown in Scheme 89.

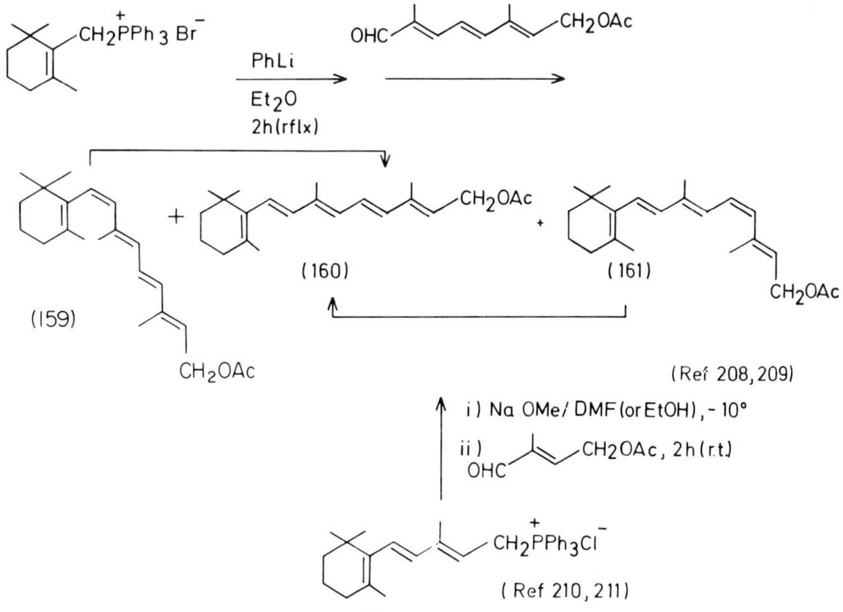

Scheme 88

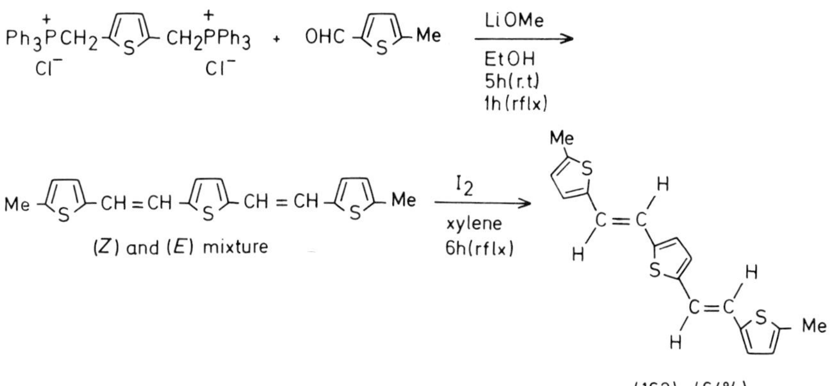

Scheme 89 (Ref 212)

A second reason is that the phosphonate modification of the Wittig reaction offers a more convenient and versatile route for *trans*-stereoselective synthesis of alkenes. The examples[199,213] in Scheme 90 illustrate the fact that (*E*)-alkenes are formed almost exclusively by using phosphonate anions rather than conventional Wittig reagents.

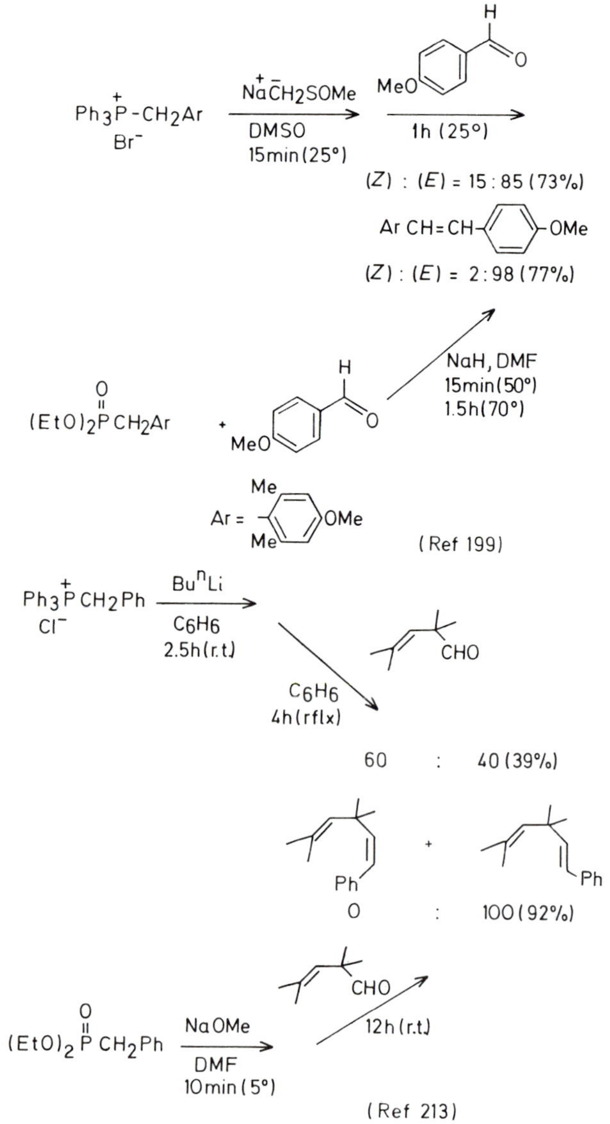

Scheme 90

2 Wittig Reaction

From a preparative point of view, the utility of moderate ylides as Wittig reagents would be greatly enhanced were there available a means for forcing (Z)-alkenes to dominate in the reaction. As yet, this has proved impracticable, but an alternative procedure,[214] utilizing a phosphonamide reagent, e.g. (163), with the carbonyl compound, has been developed for a similar purpose (Scheme 91). The synthesis of pure (Z)-1-phenylpropene is illustrative. Thus condensation of (163) with benzaldehyde in THF–toluene (1:4) at −78° gave the β-hydroxyphosphonamide (164) as a mixture of two diastereoisomers (164a) and (164b) in the ratio 7:2. The *erythro* component (164a) was then

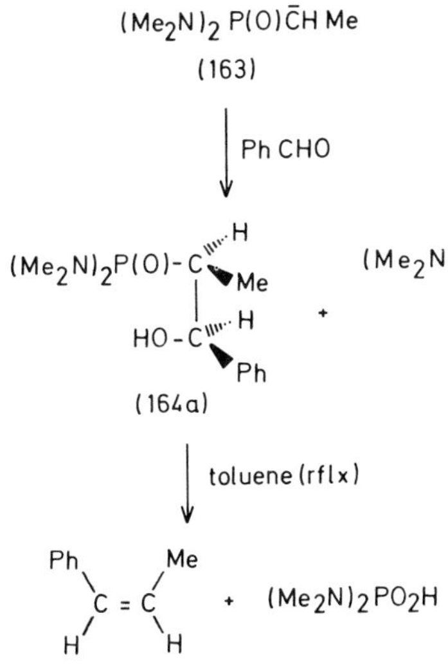

$(Me_2N)_2 P(O)\bar{C}H\ Me$

(163)

↓ Ph CHO

$(Me_2N)_2P(O)-C\overset{H}{\underset{Me}{\cdots}}$ $(Me_2N)_2P(O)-C\overset{H}{\underset{Me}{\cdots}}$
     | |
$HO-C\overset{H}{\underset{Ph}{\cdots}}$ + $HO-C\overset{Ph}{\underset{H}{\cdots}}$

(164a)   (164b)

↓ toluene (rflx)

$\underset{H}{\overset{Ph}{\diagdown}}C=C\underset{H}{\overset{Me}{\diagup}}$ + $(Me_2N)_2PO_2H$

Scheme 91 (Ref 214)

separated by fractional crystallization and stereospecifically converted in excellent yield into (Z)-1-phenylpropene by heating in toluene. This remarkably specific procedure seems to hold much synthetic potential, but so far it has not been applied to moderately reactive ylides.

4. Applications of Stereoselective Wittig Reactions in the Synthesis of Natural Products

4.1. Introduction

The impetus for the development of stereoselective Wittig reactions has come largely from the synthesis of natural products where, of course, the necessity for the introduction of double bonds in a controlled, stereospecific or highly stereoselective manner is often of paramount importance. It is the purpose of this section to survey the application of these reactions in the field of natural product synthesis. No attempt will be made to be encyclopaedic. Much useful information is to be found in the reviews by Reucroft and Sammes[215] and by Faulkner[216] on alternative methods for the stereoselective synthesis of alkenes.

4.2. Naturally Occurring Long-chain Unsaturated Fatty Acids (with A. G. Rowley)

4.2.1. *Introduction*

cis-Stereoselective Wittig reactions first acquired practical importance in the preparation of naturally occurring long-chain unsaturated fatty acids. These syntheses have been extensively reviewed[217,218] by Bergelson and Shemyakin, the most active workers in this field.

The naturally occurring unsaturated fatty acids are found in the lipids of animals, higher plants, and micro-organisms. Before the discovery of ways for controlling the steric course of the Wittig reaction, their syntheses were characterized by low yields, product mixtures (with the exception of acetylenic routes), and/or lengthy procedures.[219]

Since the reviews by Bergelson and Shemyakin are in our opinion definitive and contain detailed experimental instructions, it will suffice here to provide selected examples of synthesis of the principal types of naturally occurring unsaturated fatty acids which seem best to illustrate either stereochemical or structural features of prime significance. Some aspects of the synthesis of related naturally occurring acetylenic compounds, not covered in the above reviews, are also briefly discussed.

4.2.2. *(Z)-Alkenoic acids*

Many of the higher (Z)-alkenoic acids have been synthesized by the Wittig reaction between an ω-alkoxycarbonylalkyltriphenylphosphonium salt and an aliphatic aldehyde in a strongly dipolar aprotic solvent such as dimethylformamide[217] or dimethyl sulphoxide,[116] using sodium methoxide and sodium hydride respectively as base. This procedure generally gives good yields of the alkenoic acid [as the (Z)-isomer] in >90% stereochemical purity. Heptanal and the ylide (165), for example, afford the ethyl ester (166) of (Z)-vaccenic acid stereospecifically in 71% yield (Scheme 92).[181,220] The preparation of ^{14}C-labelled oleic[221,222] and erucic[223] acids in a similar fashion, starting from specifically ^{14}C-labelled pelargonaldehyde, is worthy of special note. The method is especially attractive because the starting

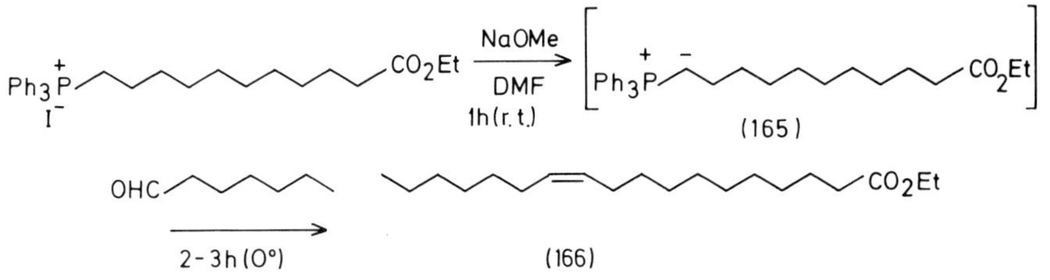

2 Wittig Reaction

Scheme 92 (Ref 181, 220)

materials for the preparation of the phosphonium salts are the readily available ω-chloro-alkanoic acids.

In practice, satisfactory yields of unsaturated esters such as (166) are obtained only when the phosphonium salt is employed in large excess owing to a side-reaction involving base-catalysed elimination of triphenylphosphine from the phosphonium salt.[220] In this connection, it appears to be important to use ω-alkoxycarbonylalkyltriphenylphosphonium iodides rather than the corresponding chlorides since the latter show a greater tendency to decompose.

A limitation to the method is encountered in the case of the phosphonium salt (167) which behaves anomalously and undergoes a favourable acylation to form (168) upon generation of its ylide.[224,225] It is of interest that (168) reacts readily with aldehydes, thereby providing a useful preparative route to α-substituted cyclopentanones, starting from ω-chlorovaleric acid. The formation of (E)-2-benzylidenecyclopentanone (169) from reaction with benzaldehyde is illustrative (Scheme 93).[224]

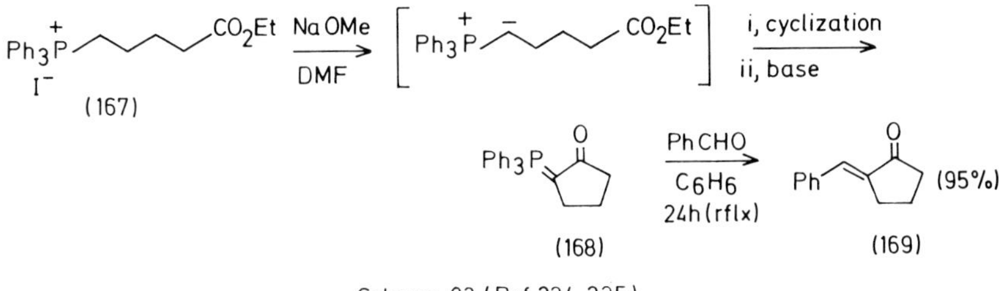

Scheme 93 (Ref 224, 225)

A variant of the method which has been developed for the preparation of (Z)-Δ^5-unsaturated acids consists of condensing an aliphatic non-stabilized ylide with ω-methoxycarbonylbutyraldehyde,[226] as in the synthesis of (Z)-docos-5-enic acid (170) (Scheme 94). This approach also appears to have the greatest synthetic utility for higher unsaturated fatty acids which are found as minor components in plant seeds and in the lipids of marine sponges.

Unlike that with aldehydes, the reaction of ω-alkoxycarbonylalkylidene ylides with ketones is not stereospecific, but *saturated* branched-chain fatty

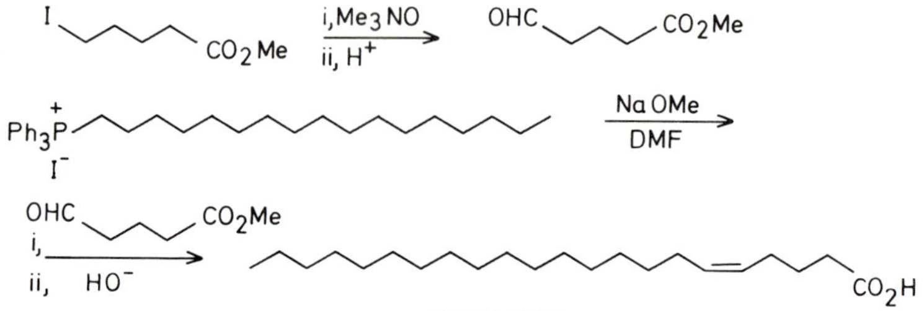

[Scheme 94 (Ref 226)]

acids can be obtained in good yield by catalytic hydrogenation followed by saponification.[227] The synthesis of (*R*,*S*)-tuberculostearic acid (171), isolated from the lipids of tubercle bacilli and other mycobacteria, is an illustration (Scheme 95). The method is of value since alternative routes involve a large number of stages and give only low yields of the desired product.

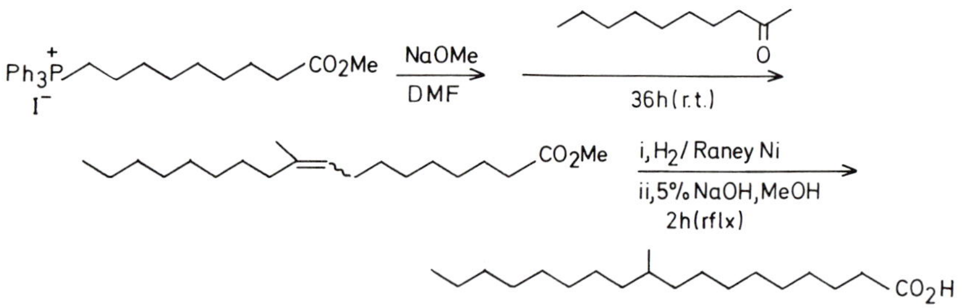

[Scheme 95 (Ref 227)]

Of incidental interest is a reaction related to those previously discussed which provides a versatile route to β,γ-unsaturated acids, starting from 2-carboxyethyltriphenylphosphonium chloride (172).[228] The reaction is experimentally unusual in that rather than the preformed ylide being treated with the carbonyl compound, e.g. *m*-methoxyacetophenone, a mixture of the

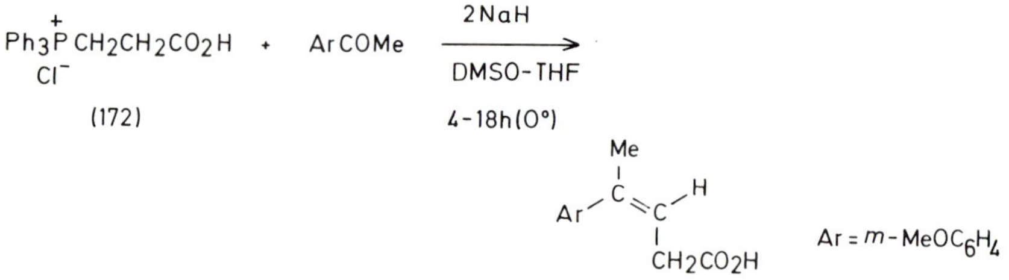

$Ph_3\overset{+}{P}CH_2CH_2CO_2H$ + ArCOMe $\xrightarrow[\text{DMSO-THF}]{2\,NaH}$
Cl^-
(172)   4-18h(0°)

Ar−C(Me)=C(H)−CH$_2$CO$_2$H   Ar = *m*-MeOC$_6$H$_4$

(173) (53-69%; *ca.* 80% Z)

Scheme 96 (Ref 228)

2 Wittig Reaction

reactants dissolved in 1:1 DMSO–tetrahydrofuran is added to the base, in this case sodium hydride under dry nitrogen at 0°. This procedure afforded crude (173), mainly as the (Z)-isomer (ca. 80%), in the range of 53–69% yield with reaction times of 4 and 18 h. By comparison, the use of the reactants in normal order resulted in negligible yields. This is probably due to competitive base-catalysed elimination of triphenylphosphine from (172) to form acrylic acid (see above).

4.2.3. Polyenoic acids

The synthesis of the biologically important fatty acids of the (Z,Z)-divinylmethane type can be effected under similar conditions as illustrated by the preparation of linoleic acid from (Z)-non-3-enyltriphenylphosphonium bromide (174) and ω-methoxycarbonyloctanal[217] (Scheme 97). After chromatography (on neutral alumina), methyl linoleate (175) is obtained in 71% yield containing only 3% of the (E)-isomer, the ester group being subsequently removed by saponification.

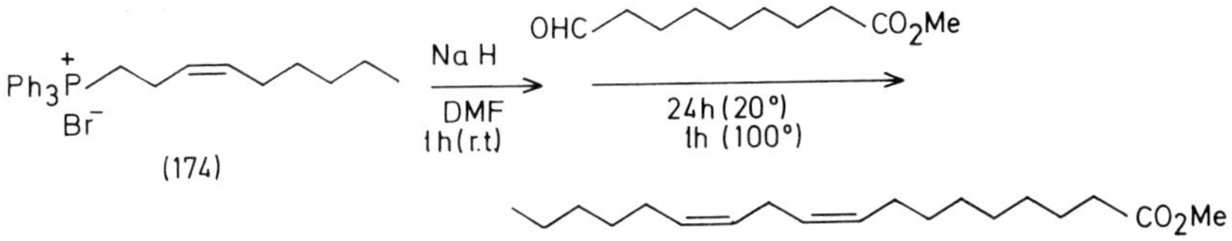

Scheme 97 (Ref 217)

An undesirable side-reaction always encountered in preparations of this type is isomerization of the γ,δ-alkylidenephosphorane to its conjugated isomer, e.g. (176) → (177). This complication may be mitigated, at least in part, by the slow addition of the base to a *cold* solution of the phosphonium salt in the dark.

Scheme 98

The alternative approach starting from (Z)-β,γ-unsaturated aldehydes and ω-alkoxycarbonylalkylidenephosphoranes usually leads to less pure products.[229]

As a consequence of the foregoing difficulties, the Wittig approach is less well suited for the preparation of this type of "skipped" polyenoic acid than procedures based on partial hydrogenation of acetylenic derivatives.[230,231] However, the method can be used for the synthesis of fatty acids containing the corresponding (E,Z) system[232] [cf. the highly stereoselective synthesis of a "skipped" (Z,E)-pheromone by a Wittig condensation using either potassium in HMPA[233] or dimsylsodium in dimethyl sulphoxide;[234] see Scheme 125].

Fatty acids of the (Z,Z)-divinylethane type are also accessible by a general reaction sequence that involves two stereoselective Wittig reactions.[235] The application of this procedure to the preparation of (Z,Z)-hexadeca-9,13-dienoic acid ethyl ester (178) is shown in Scheme 99. The compounds so obtained contain only small amounts of the (E,Z)-isomers (cf. ref. 232).

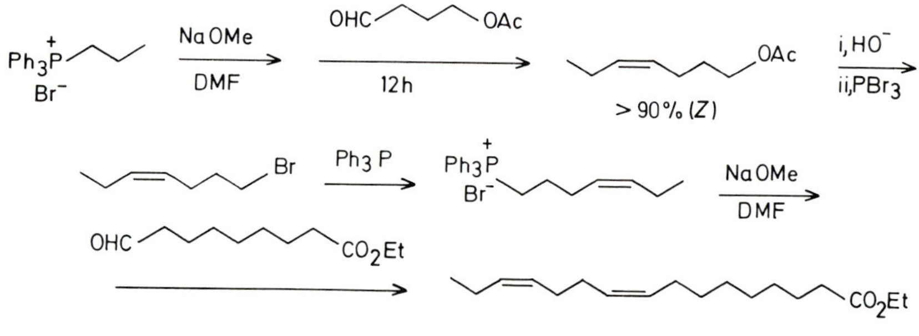

Scheme 99 (Ref 235)

Substances with the (E,Z)-divinylethane grouping can be constructed analogously utilizing a combination of (Z)- and (E)-stereoselective Wittig reactions, as illustrated by the preparation of (E,Z)-nona-2,6-dien-1-ol (179), a constituent of the odoriferous principle of violets (Scheme 100).[217] Another way to the (E,Z)-divinylethane system, which has proved especially valuable in pheromone syntheses, is the reaction of saturated aliphatic aldehydes with (E)-Δ^4-alkenylidenephosphoranes under conditions leading to the selective formation of the (Z)-double bond (see Scheme 119). In the light of this work it is worth considering whether the use of different reaction conditions,

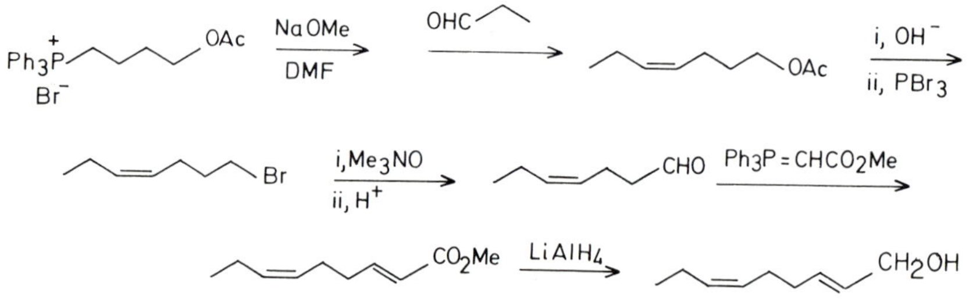

Scheme 100 (Ref 217)

2 Wittig Reaction

namely sodium bis(trimethylsilyl)amide in THF,[131] would improve the selectivity of the foregoing reactions (using sodium methoxide in DMF). Potassium in hexamethylphosphoric triamide may also offer some advantage.[117,118]

Finally, mention should be made of the elegant syntheses of polyenoic acids containing three conjugated double bonds. Thus, α-eleostearic acid (182), the principal fatty acid of tung oil, has been prepared by condensation of (E,E)-nona-2,4-dienal (181) with the ylide derived from 8-ethoxycarbonyl-octyltriphenylphosphonium iodide (180) in dimethylformamide, followed by

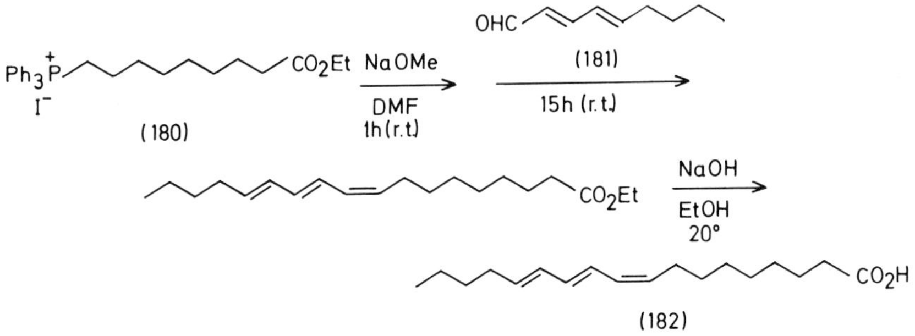

Scheme 101 (Ref 236)

saponification.[124,236] Starting from sorbic aldehyde (183), the same procedure also furnished the triene ester (184), a key intermediate in the synthesis of α-sanshool, containing no more than 15% of the all-*trans*-isomer.[125] It might be noted that the triene ester (184) (Scheme 102) could be obtained with the same purity by carrying out the condensation in THF or DME using sodium hydride as base.

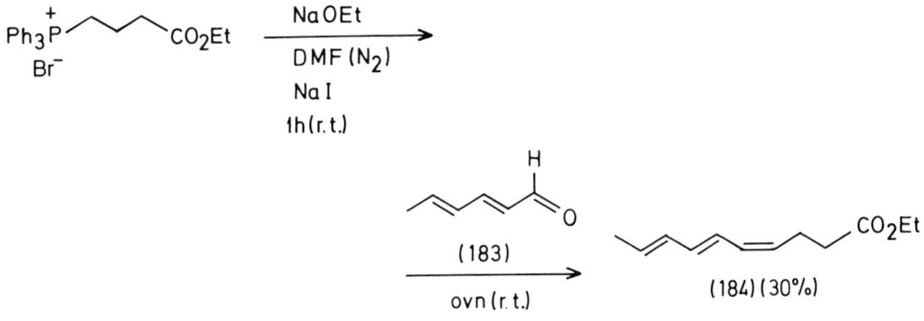

Scheme 102 (Ref 125)

An alternative approach[237] to the synthesis of conjugated polyenoic acids that involves selective formation of an (E)-double bond in the Wittig step is exemplified by the preparation of catalpa acid (185) according to Scheme 103.

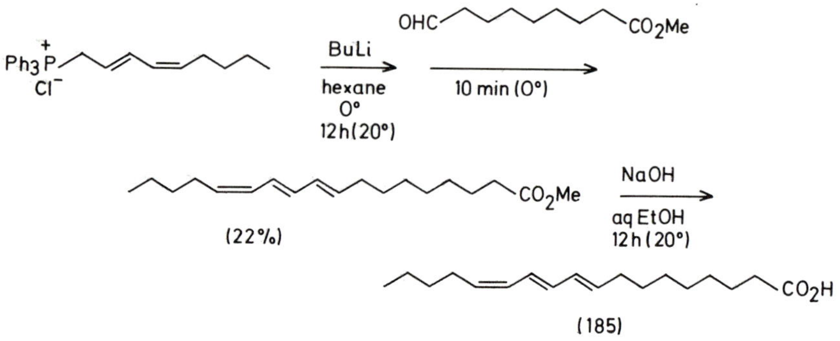

Scheme 103 (Ret 237)

4.2.4. ω-Hydroxy unsaturated acids

In certain cases the Wittig reaction has proved useful for the controlled syntheses of ω-hydroxy unsaturated acids starting from 2-hydroxytetrahydropyran (186a) which may be regarded as the cyclic form of 5-hydroxypentanol (186b) (Scheme 104). For example, condensation of (186) with ethoxycarbonylmethylenetriphenylphosphorane (187) followed by saponification leads to exclusive formation of 7-hydroxy-(E)-hept-2-enoic acid (188), a lower homologue of the royal jelly acid of the honey bee.[238]

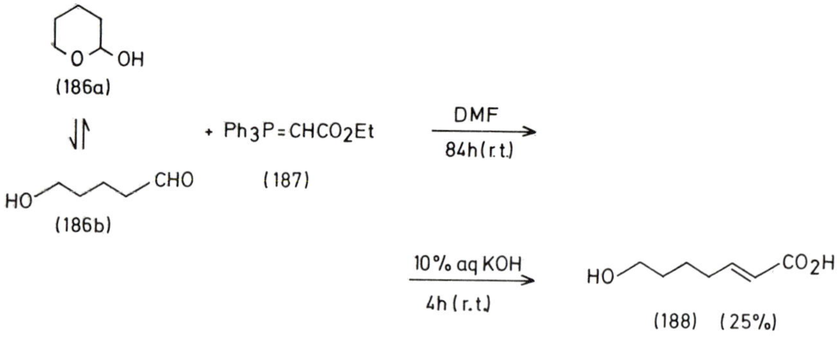

Scheme 104 (Ret 238)

Employment of this method using the vinylogous ylide (189) led to the total synthesis of α-camlolenic acid (190), found in the seed oil of spurge *Euphorbiaceae* (Scheme 105).[239] Note that the final step in the reaction sequence involves the selective introduction of a (Z)-double bond by means of a Wittig reaction.

Surprisingly, attempts to condense (186) with more reactive ylides using dimethylformamide as the reaction solvent failed; decomposition occurred instead. In interesting contrast, however, is the observation that comparable

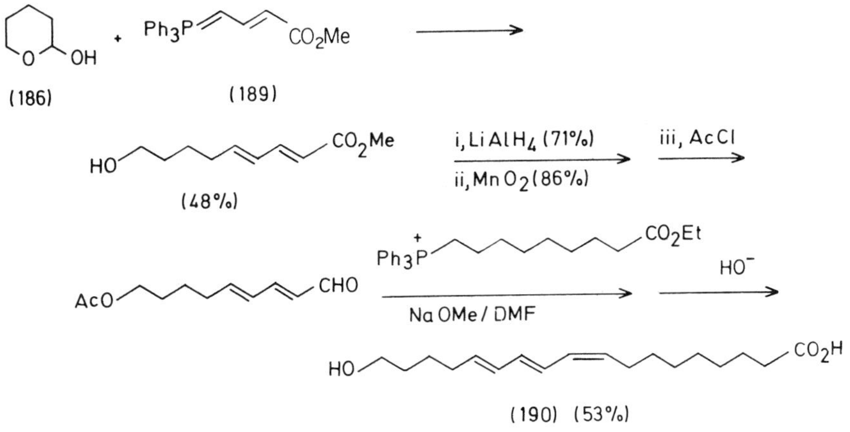

2 Wittig Reaction

Scheme 105 (Ref 239)

reactions proceed efficiently when the ylide is generated with dimsylsodium in DMSO.[240] Under these conditions, for example, octyltriphenylphosphonium bromide (191) and 2-hydroxytetrahydropyran (186) gave (Z)-tridec-5-en-1-ol (192) in 51% yield with no detectable (E)-isomer.

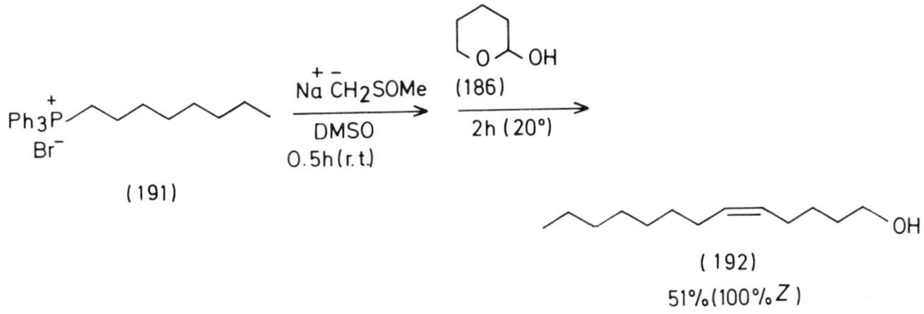

Scheme 106 (Ref 240)

4.2.5. Polyacetylenic compounds

Stereoselective Wittig reactions have also been instrumental in the synthesis of a number of naturally occurring polyacetylenic compounds. An example is the synthesis of crepenynic acid (194), first isolated from seed oils of *Crepis* species, in which the introduction of the (Z)-double bond by a Wittig reaction represented the last and crucial stage (Scheme 107).[241] Of particular interest is the observation that only butyllithium in ether, among a variety of base-solvent combinations (sodium hydride in DMF, sodium methoxide in both methanol and dimethylformamide) tried in the Wittig condensation, gave a satisfactory yield (51%) of methyl crepenynate (193); intriguingly,

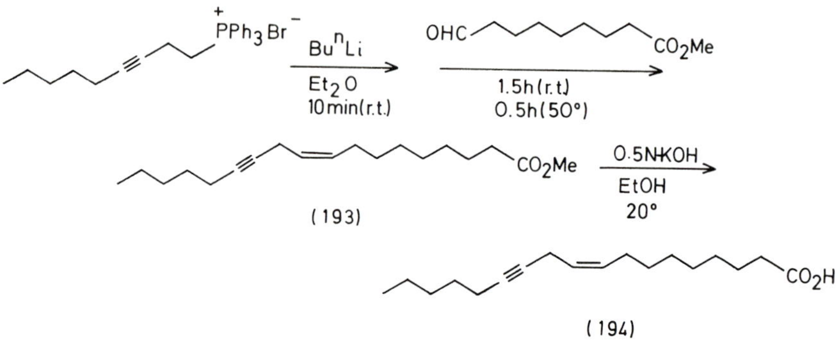

[Scheme 107 diagram]

Scheme 107 (Ref 241)

none of the (E)-isomer could be detected in the reaction mixture. (It will be recalled that these conditions usually lead to a mixture of stereoisomers.)

The same route has been used to prepare both ^{14}C- and ^{3}H-labelled methyl crepenynate and methyl linoleate (by partial hydrogenation of the crepenynate).[221]

A related reaction which utilizes the trimethylsilylphosphonium salt (195) forms the basis of a general route to compounds containing the 1-ene-4,6-diyne and the more highly unsaturated "skipped" ene-yne systems.[205] Scheme 108 illustrates the usefulness of the method for the preparation of a wide variety of C_{18} and C_{16} esters (198) with the (Z)-9-ene-12,14-diyne unsaturation. As indicated, the Wittig step is effected by stirring the salt (195),

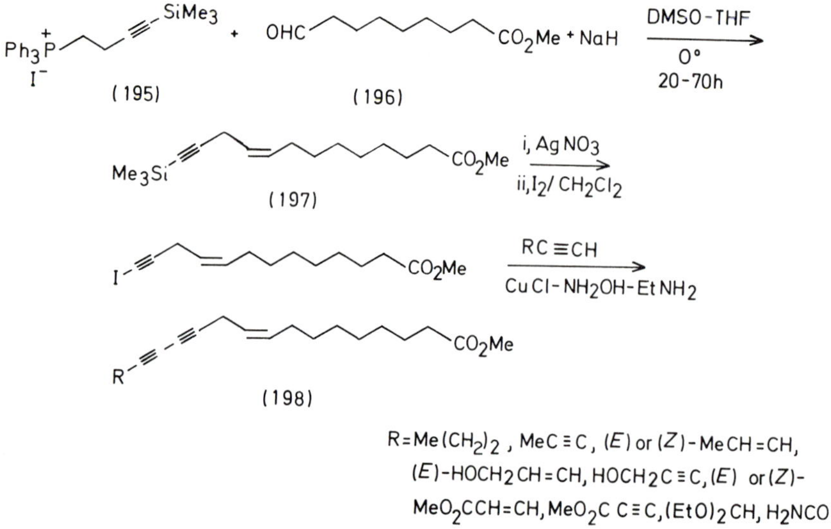

[Scheme 108 diagram]

R = Me(CH$_2$)$_2$, MeC≡C, (E) or (Z)-MeCH=CH, (E)-HOCH$_2$CH=CH, HOCH$_2$C≡C, (E) or (Z)-MeO$_2$CCH=CH, MeO$_2$C C≡C, (EtO)$_2$CH, H$_2$NCO

Scheme 108 (Ref 205)

2 Wittig Reaction

the aldehyde (196), and sodium hydride together in THF–DMSO at 0° for 20–70 h. Under these conditions, the (Z)-isomer (197) is formed almost exclusively in ca. 80% yield; incomplete exclusion of water, however, leads to a loss in the selectivity and also results in the formation of by-products.

Finally, it is of interest that, in the preparation of the fungal metabolite ester (201) from the ylide (199) and the bis-acetylenic aldehyde (200), the usual stereoselectivity associated with stabilized ylides is lost and the (E)-product predominates in the ratio of only 2·5:1 (Scheme 109).[242] The same

$$Ph_3P=CHCO_2Me + OHCC\equiv CC\equiv CCH_2OH \longrightarrow HOCH_2C\equiv CC\equiv CCH=CHCO_2Me$$
$$(199) \qquad (200) \qquad\qquad (201) \quad (E):(Z)=2.5:1$$

Scheme 109 (Ref 242)

phenomenon has been observed in the reaction of (199) with other aldehydes having an α-acetylenic function, e.g. (202),[205] (203),[243] and (204).[244]

$$OHC(C\equiv C)_2CH_2CH=CH(CH_2)_7CO_2Me$$
$$(202) \qquad\qquad (Ref\ 205)$$

$$OHC(C\equiv C)_2CH_2CH_2CO_2Me$$
$$(203) \qquad\qquad (Ref\ 243)$$

$$OHC-C\equiv C-CH=C(SMe)-C\equiv CMe$$
$$(204) \qquad\qquad (Ref\ 244)$$

Scheme 110

4.2.6. Prostaglandins

The selective formation of practically pure (Z)-alkenes from highly reactive ylides by the use of a strongly polar solvent has rendered the Wittig reaction particularly attractive for the introduction of the (5Z)-double bond in naturally occurring C_{20} carboxylic acids of the prostaglandin family. The method first acquired practical importance in the much acclaimed synthesis developed by Corey, by which the entire prostaglandin family together with a number of non-natural analogues can be obtained utilizing a single resolved precursor.[245] The final steps in the synthesis of prostaglandin $F_{2\alpha}$, starting from the bicyclic lactone (205), are typical (Scheme 111).[246] Note that in this sequence the introduction of the lower side-chain and establishment of the (13E)-double bond is accomplished first by using the phosphonate modification of the Wittig reaction. The upper side-chain is then attached by condensation of the lactol (206) with the ylide (207) derived from 4-carboxy-butyltriphenylphosphonium bromide and dimsylsodium in dimethyl sulphoxide. This procedure furnished the bis-tetrahydropyranyl ether (208) of PGF$_{2\alpha}$ stereospecifically in 80% yield [based on the lactol (206)]. Minor modifications of the reaction sequence lead to the synthesis of other primary prostaglandins. For these aspects, the reader is referred to two recent books [247] as well as an earlier review.[248]

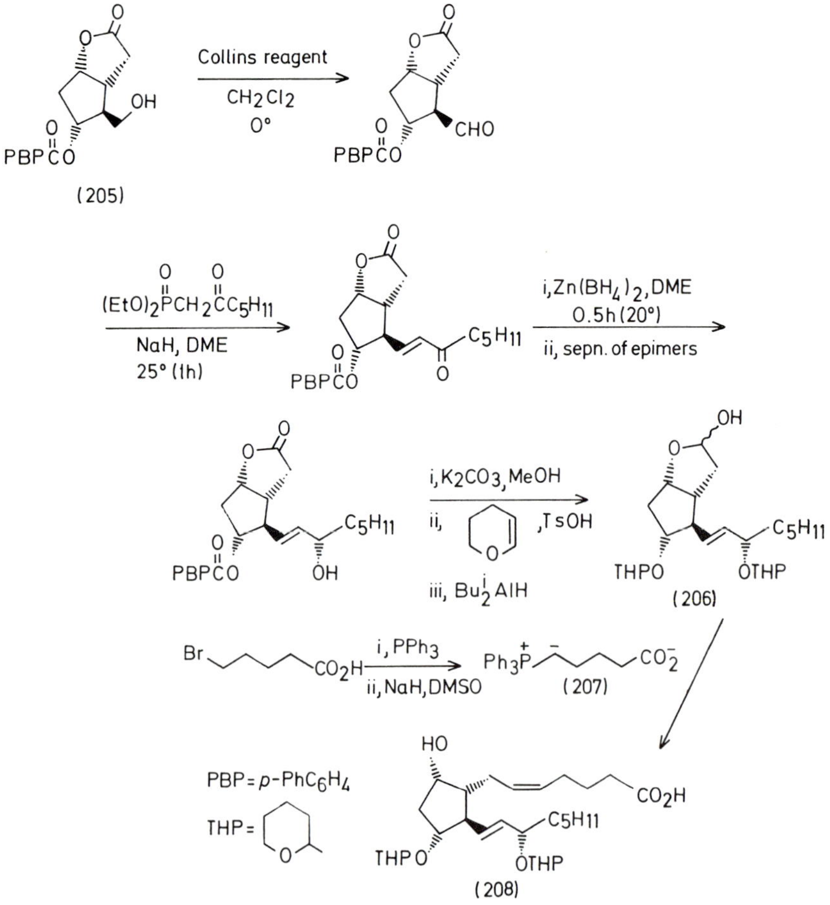

Scheme 111 (Ref 246)

By using this approach it is possible to construct the upper *cisoid* side-chain of a variety of prostaglandin analogues including those containing heteroatoms in the ring. Table 12 lists some of the lactols and aldehydes employed

2 Wittig Reaction

Table 12. Prostaglandin synthesis. Introduction of the (Z)-Δ^5 function via Wittig reaction of lactols and aldehydes with the ylide (207) from 5-triphenylphosphoniopentanoic acid

PG	Reactant	Conditions	Product and yield(s) (%)	Ref.
PGE$_2$	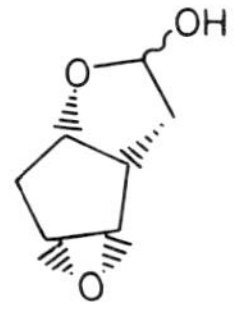		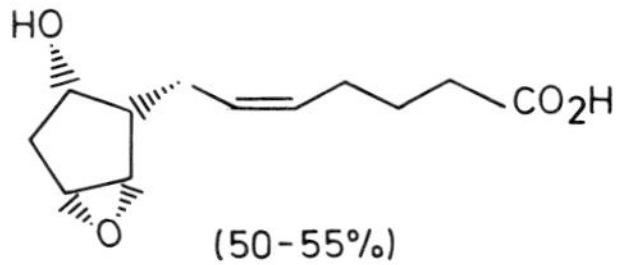 (50–55%)	250
PGA$_2$	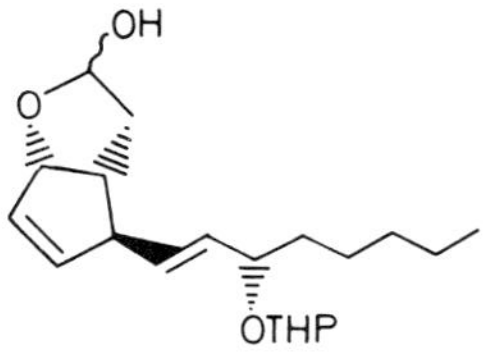	NaH/DMSO	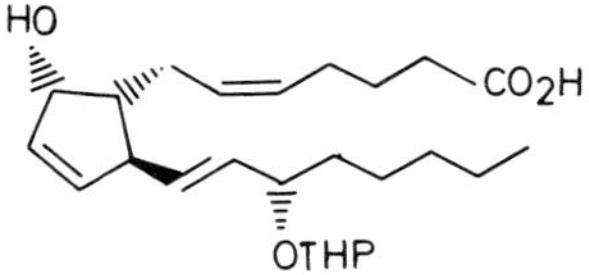	251
PGC$_2$	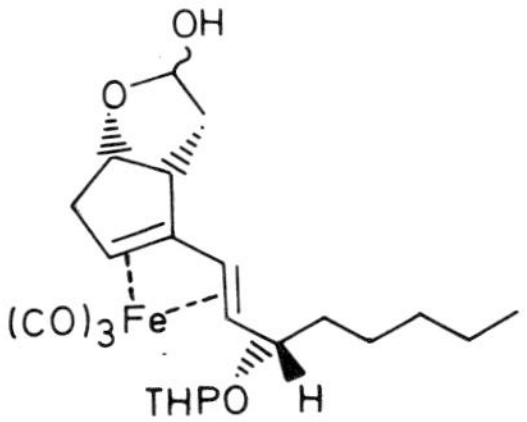	NaH/DMSO	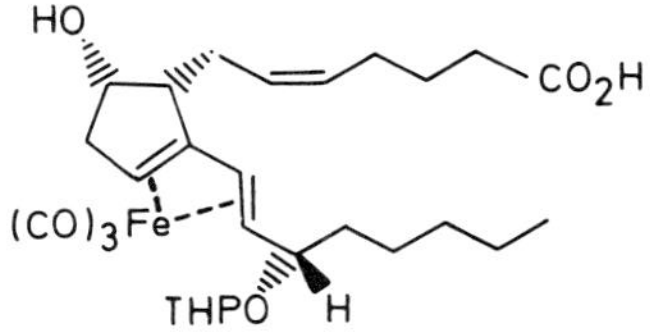 (70%)	252

Table 12. *Continued*

PG	Reactant	Conditions	Product and yield(s) (%)	Ref.
11-deoxy PGE$_2$	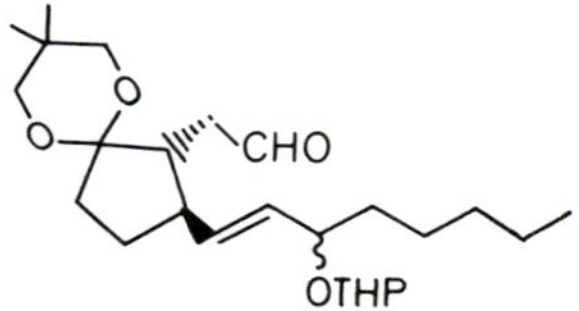	NaH/DMSO	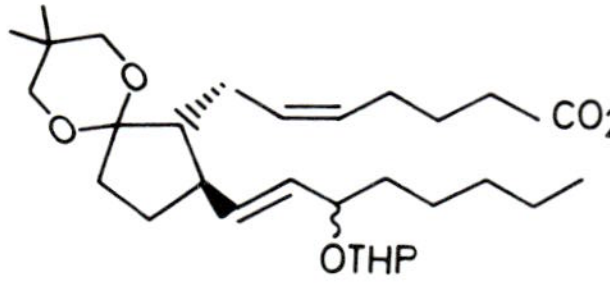	253
12α-fluoro-PGF$_{2\alpha}$	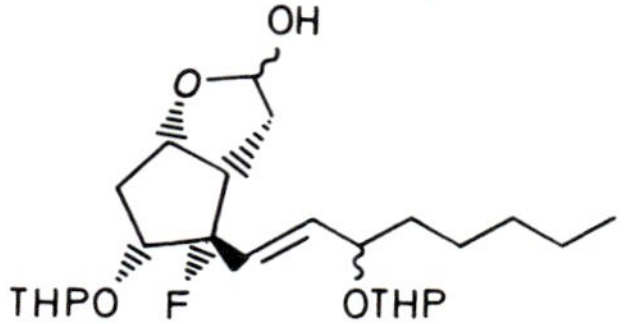	NaH/DMSO	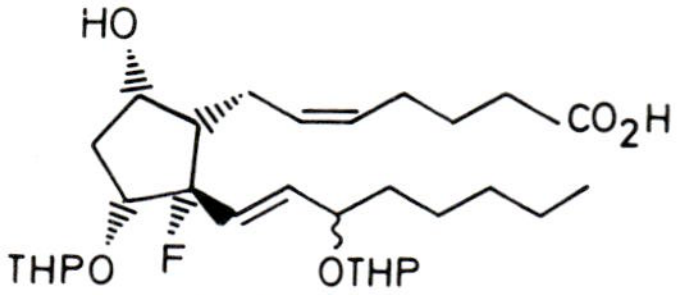 (70%)	254
13-dehydro PGF$_{2\alpha}$	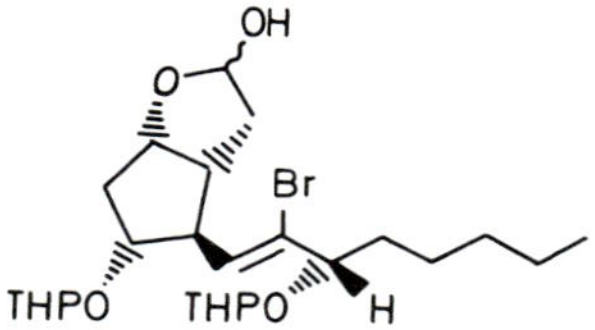		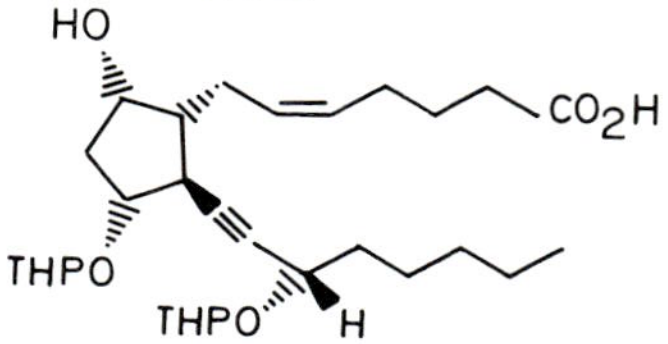	255
	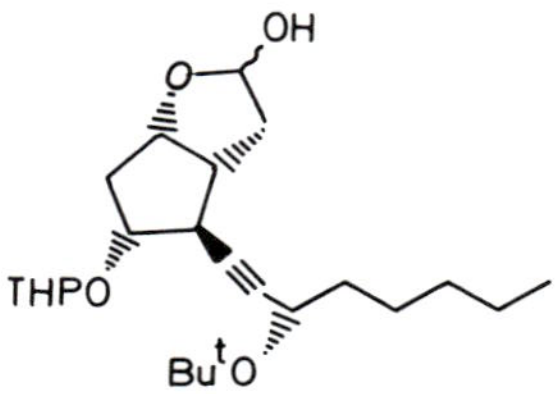		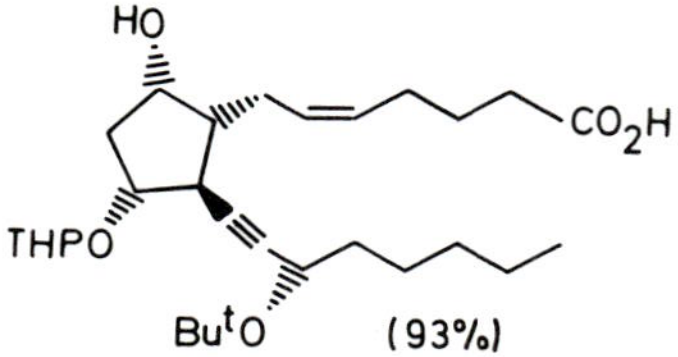 (93%)	256

2 Wittig Reaction

Table 12. *Continued*

PG	Reactant	Conditions	Product and yield(s) (%)	Ref.
16,16-ethano-PGF$_{2\alpha}$	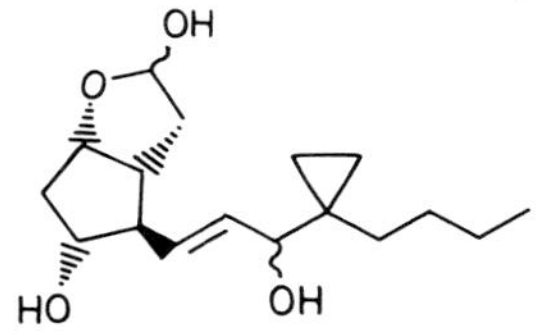	NaH/DMSO 2h (r.t.)	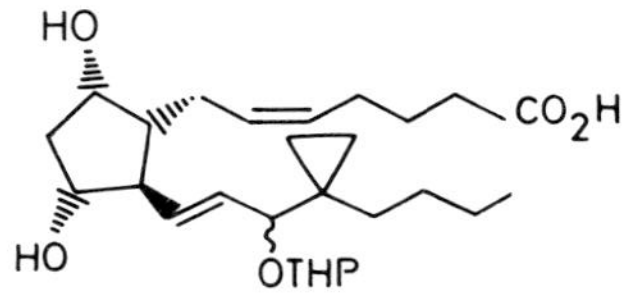 (59%)	257
11-oxa-PGF$_{2\alpha}$	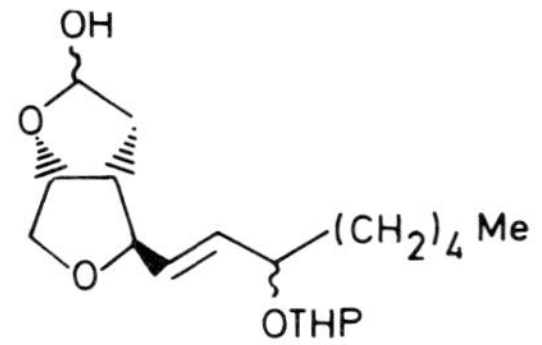	NaH/DMSO	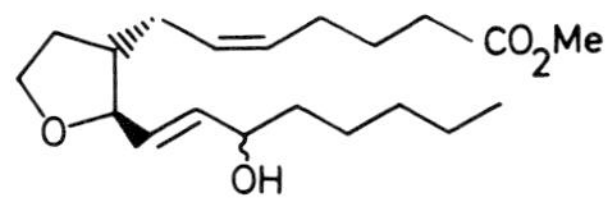	258
	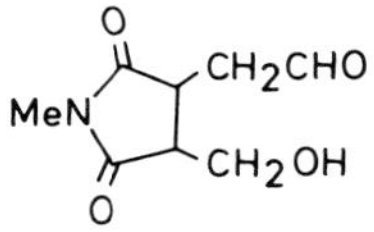		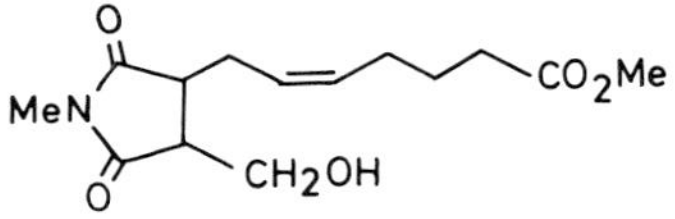	259
	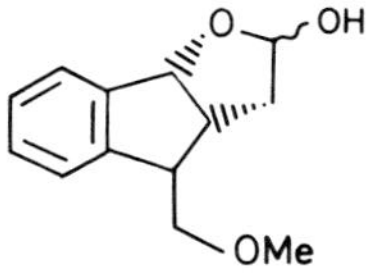	NaH/DMSO (r.t.)	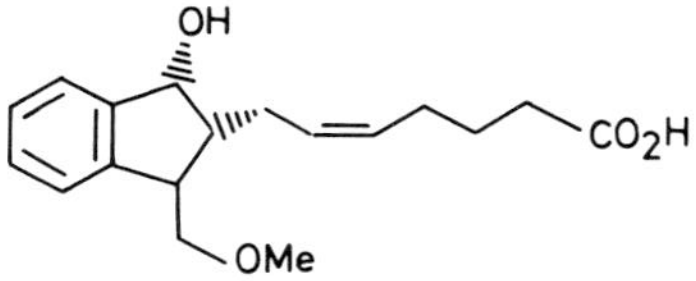	260

Table 12. *Continued*

PG	Reactant	Conditions	Product and yield(s) (%)	Ref.
	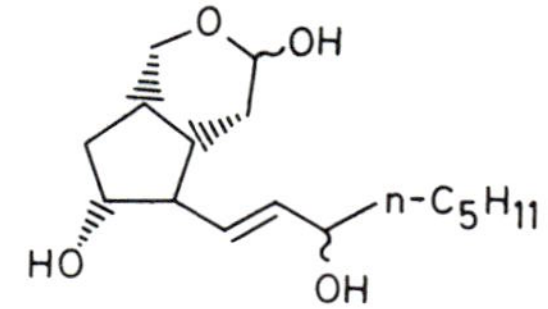		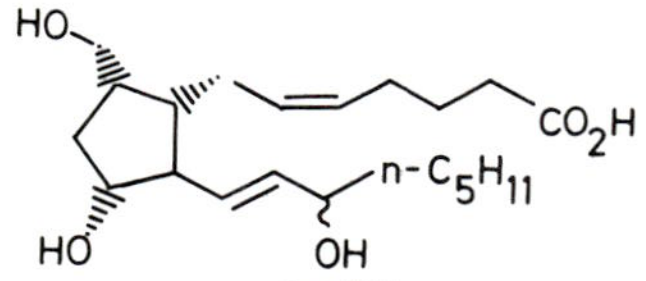	261
6,9-azo-PGH₂ analogue	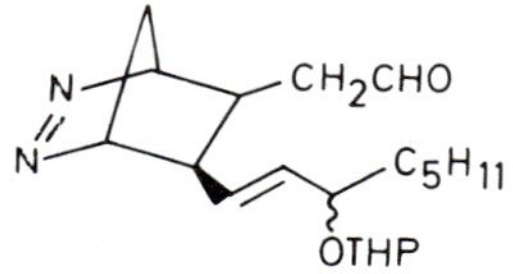	NaH/DMSO	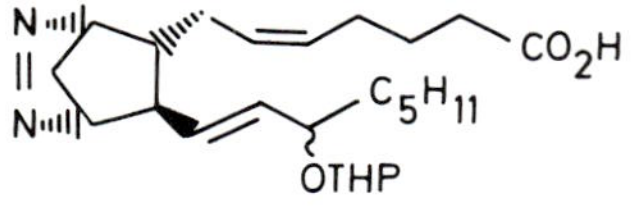 (65%)	262
	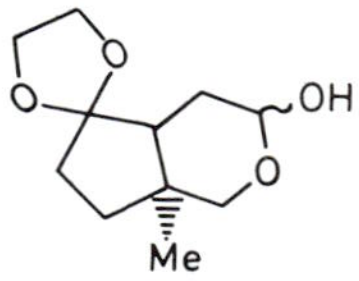	DMSO 3 day (r.t.)	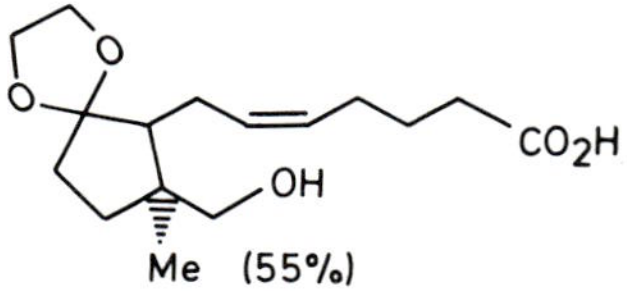 (55%)	263
	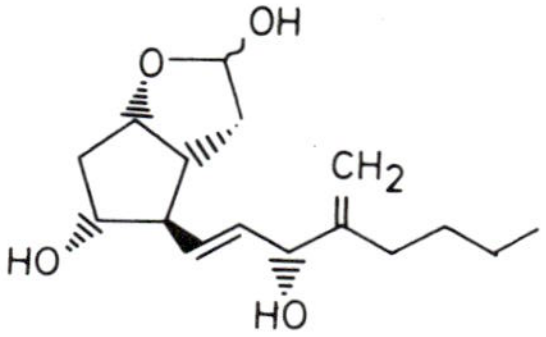 610 mg	NaH/DMSO 2h (r.t)	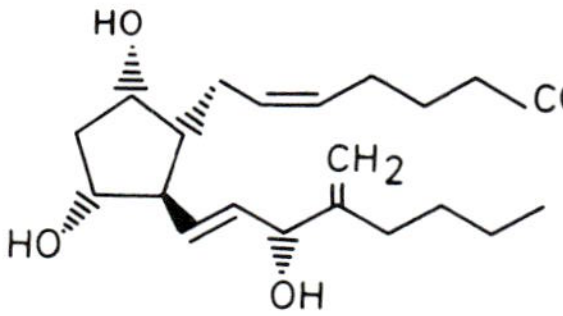 252 mg	264

2 Wittig Reaction

Table 12. *Continued*

PG	Reactant	Conditions	Product and yield(s) (%)	Ref.
	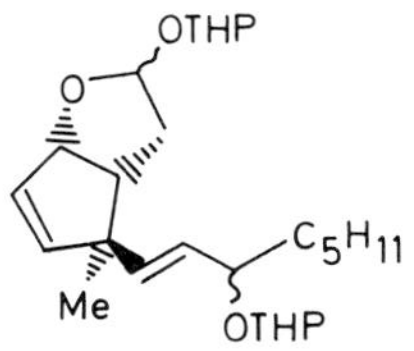	NaH/DMSO	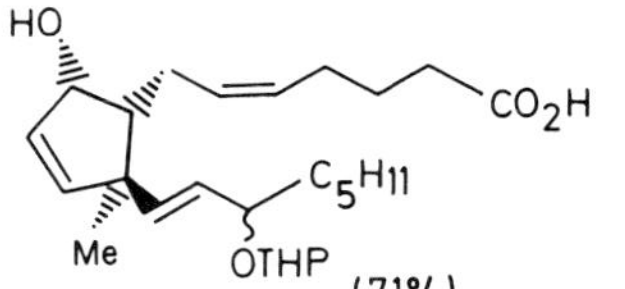 (71%)	265
	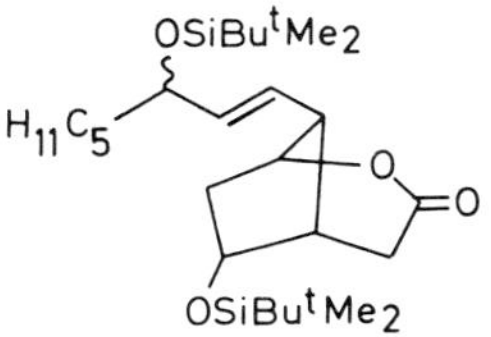	NaH/DMSO	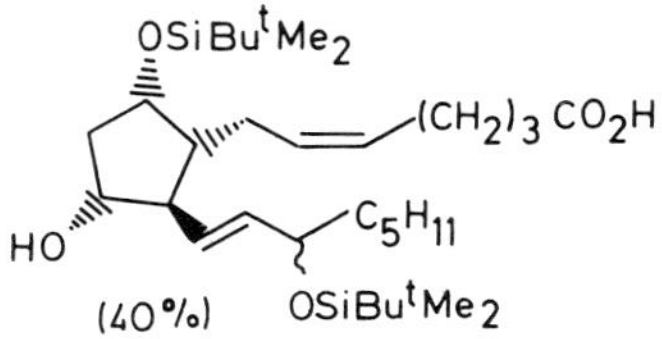 (40%)	266
	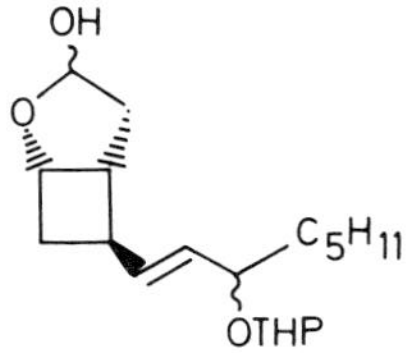	KH/DMSO	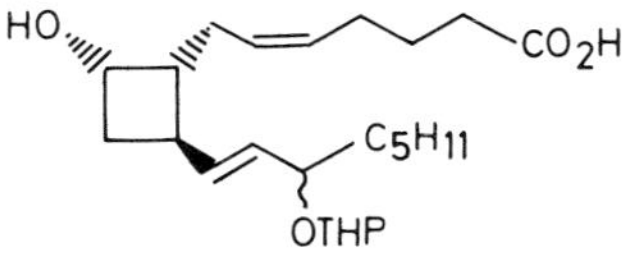	267

Table 12. *Continued*

PG	Reactant	Conditions	Product and yield(s) (%)	Ref.
9,11-etheno PGH$_2$-analogue	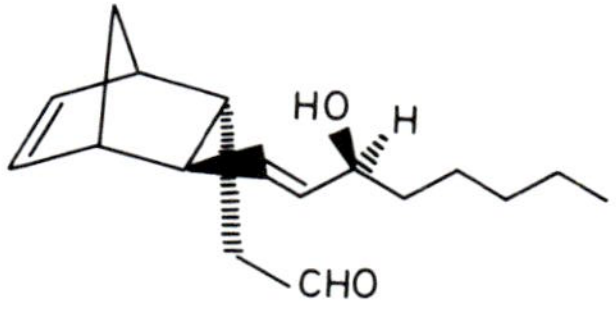 *(structure with norbornene, HO, CHO)*	NaH/DMSO	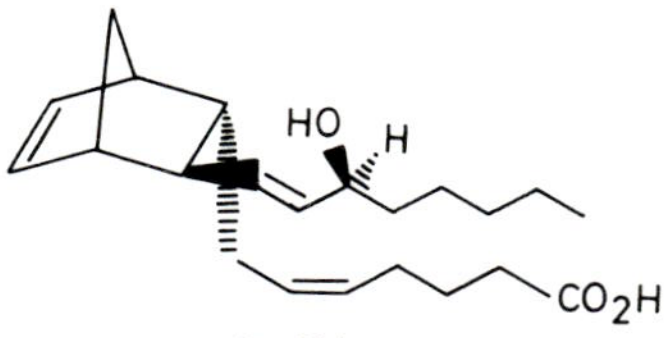 *(structure with norbornene, HO, CO$_2$H)* (88%)	268
TXB$_2$	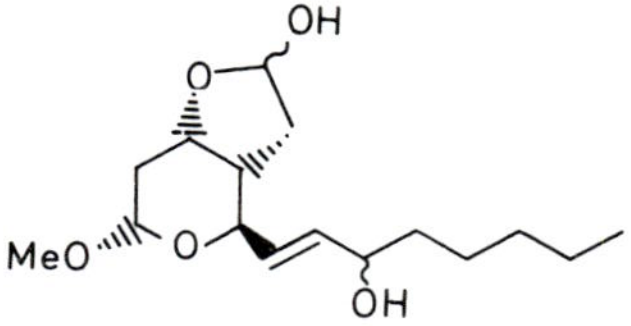 *(pyranose structure with OH, OMe, alkenyl chain)*	NaH/DMSO	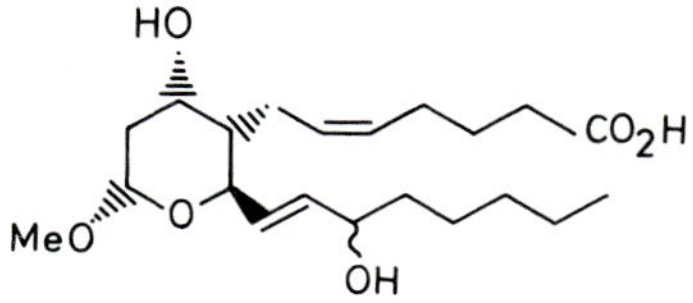 *(pyranose structure with HO, OMe, CO$_2$H)* (34%)	269, 270
TXB$_2$	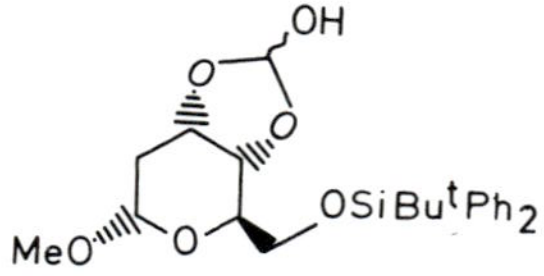 *(dioxolane-pyranose structure with OSiBu^{t}Ph$_2$, OMe)*	LiN(SiMe$_3$)$_2$ HMPA	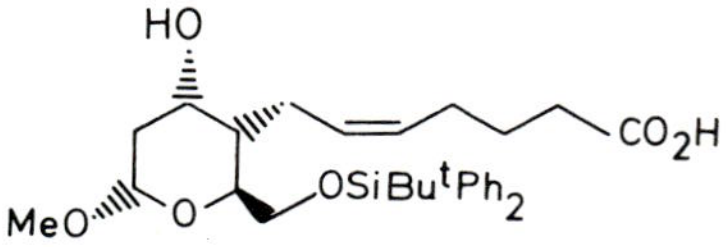 *(pyranose structure with HO, OMe, OSiBu^{t}Ph$_2$, CO$_2$H)* (67–71%)	271

2 Wittig Reaction

Table 13. Introduction of the (Z)-Δ⁵ olefinic bond in prostaglandin analogues via Wittig condensation of lactols with $Ph_3P^+R\ X^-$

Reagent R	Lactol	Conditions	Product and yield(s) (%)	Ref.
$-(CH_2)_2CH(CO_2H)_2$	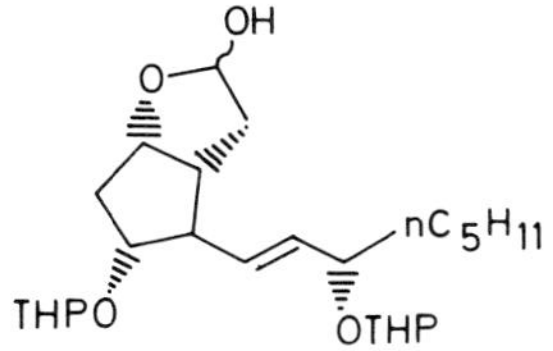	NaH/DMSO 2h (r.t.)	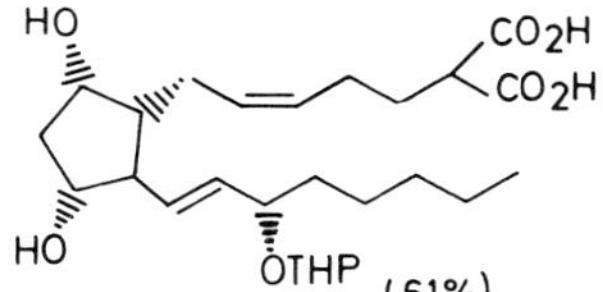 (61%)	272
 $-CH_2-\triangle-CO_2H$	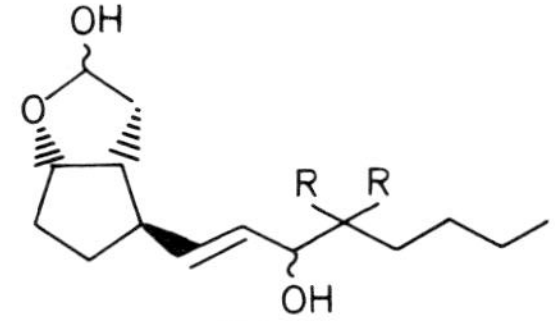		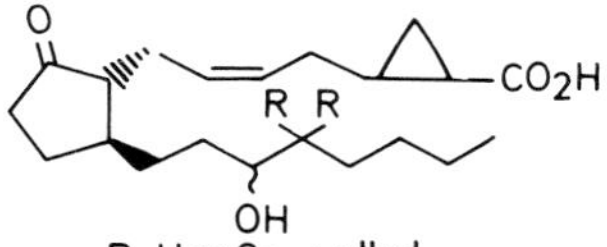 R=H or C_{1-4} alkyl	273
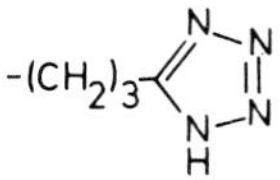 $-(CH_2)_3-\underset{N\underset{H}{\overset{N=N}{\mid}}}{\overset{N}{\mid}}$	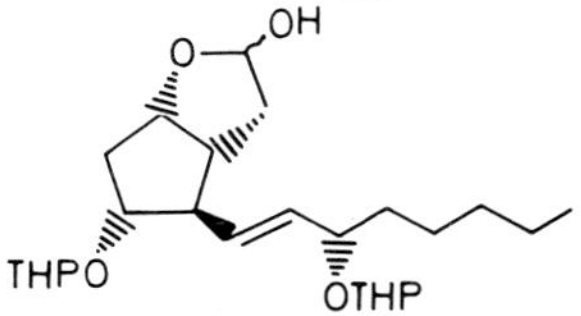	NaH	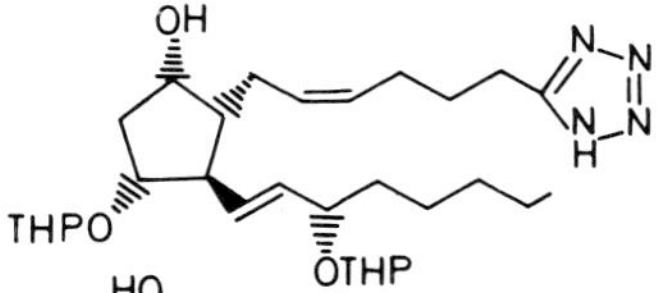	274
$-(CH_2)_3SO_3H$	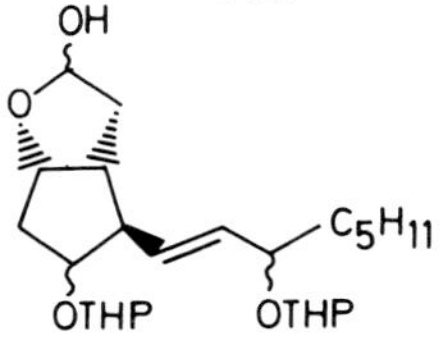	NaH/DMSO 3h (30°)	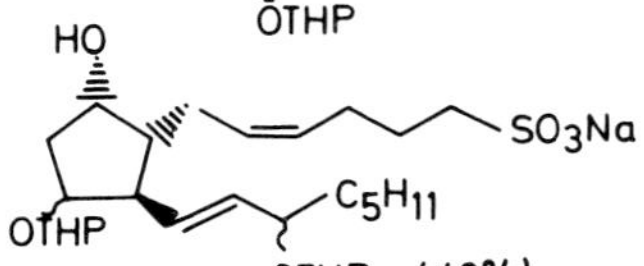 (48%)	275

as the carbonyl co-reactant; the last two examples are of interest as intermediates for the synthesis of thromboxane B_2, one of a new class of compounds closely related to prostaglandins.[249]

At the present time, dimsylsodium, prepared from sodium hydride and dimethyl sulphoxide (70°, *ca.* 1 h) with excess of the latter as solvent, is the preferred base–solvent combination for these condensations. However, it is worth noting that in one instance[267] the use of dimsylpotassium (KH–DMSO, 25°, <5 min) gave more reproducible results. The combination of lithium bis(trimethylsilyl)amide and hexamethylphosphoramide may also offer some advantage.[271]

Table 13 lists some other ylides used with lactols to complete the synthesis of the upper side-chain of prostaglandin analogues.

While the formation of predominantly *trans* unsaturation makes the use of phosphonate carbanions the method of choice for introduction of the lower side-chain of prostaglandins, stabilized ylides have been used in a few cases as illustrated by the selected examples in Scheme 112. As noted earlier, and also shown in Scheme 112, ylides derived from tributylphosphine rather than triphenylphosphine offer the advantage of milder reaction conditions.

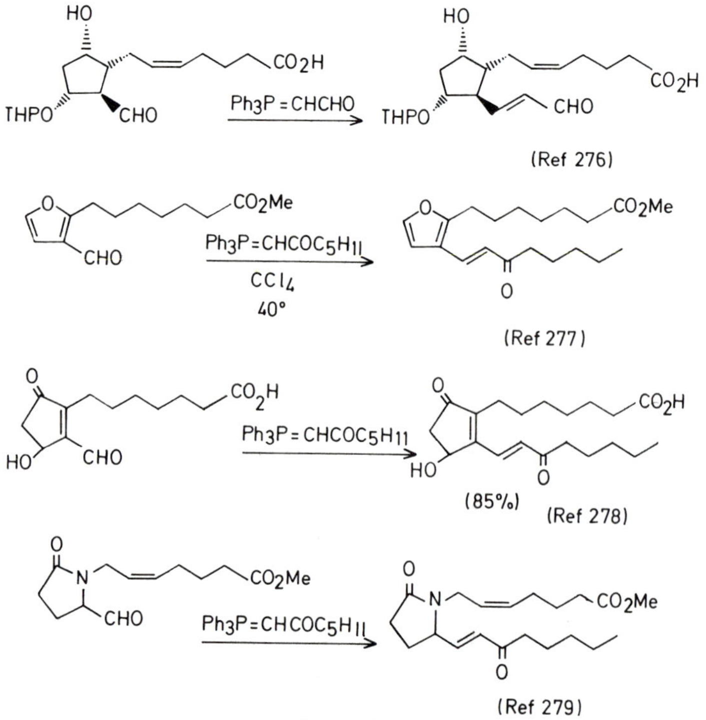

(Ref 276)

(Ref 277)

(85%)   (Ref 278)

(Ref 279)

Scheme 112

2 Wittig Reaction

Scheme 112 (Cont.)

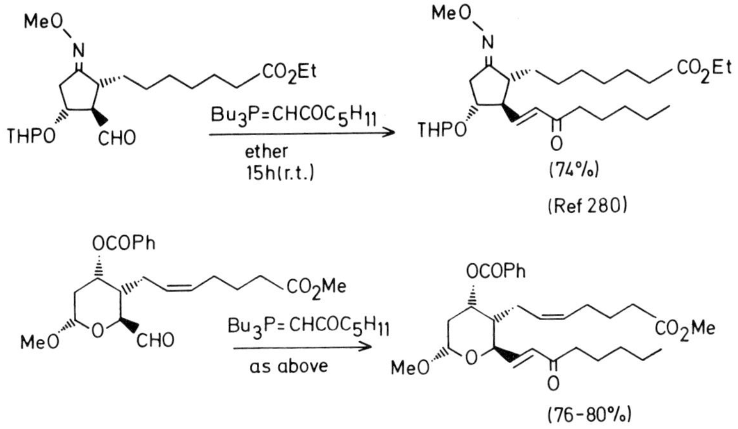

Scheme 112

Finally mention should be made of an innovative approach recently applied to the stereospecific total syntheses of prostaglandins E_3 and $F_{3\alpha}$.[281] In the course of synthesis of the latter, the (E)-double bond in the lower side-chain together with the 15α-hydroxy function were stereospecifically introduced *in one step* by condensing the optically active aldehyde (210) with the β-oxido-ylide reagent derived from the hydroxy-phosphonium salt (S)-$(+)$-(209). The desired alcohol (211) was isolated in 35% yield, and none of the epimeric allylic alcohol could be detected in the reaction mixture.

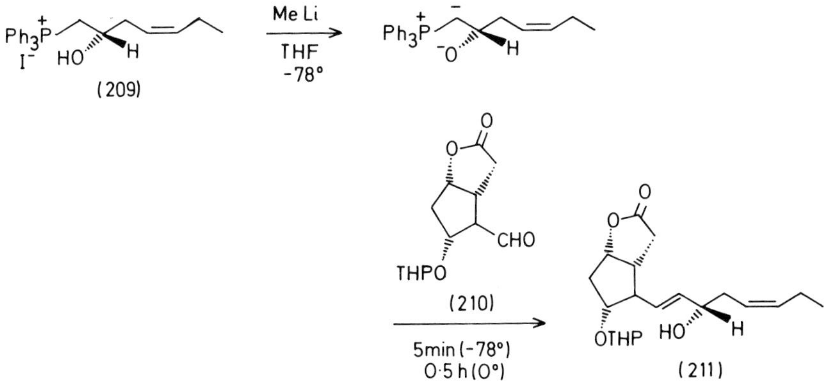

Scheme 113 (Ref 281)

4.3. Pheromones

In no field have stereoselective Wittig procedures achieved importance as rapidly as in the area of insect hormone synthesis. Recent progress in the use of synthetic sex pheromones has emphasized the need to have stereochemically pure alkenes for maximum effect.[282] One notable example of stereospecific chemoreception is bombykol, the sex attractant of the female silk moth, *Bombyx mori*, identified as (E,Z)-hexadeca-10,12-dien-1-ol (212), which is 10^9 times more active than other stereoisomers of the same structure (Scheme 114).[283]

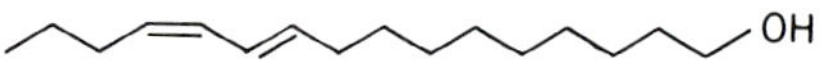

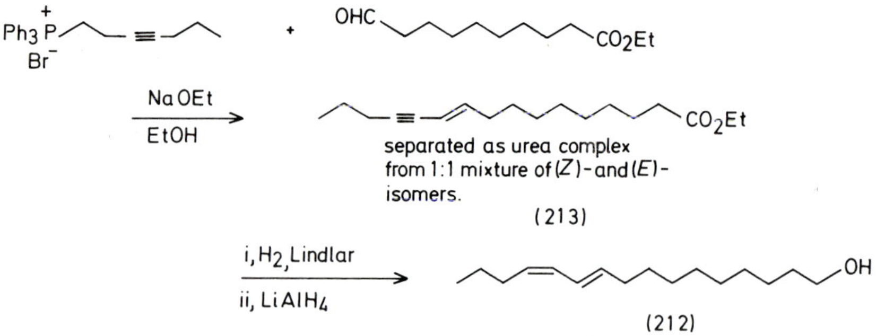

~~~~~~~~~OH
(212)

Scheme 114

Bombykol itself has been synthesized by a number of alternative routes utilizing non-stereoselective Wittig reactions.[12,284,285] These syntheses are of little use for preparing large quantities of bombykol, but are of interest in showing the development of more efficient and practical routes. For example, in the approach outlined in Scheme 115,[284] the key Wittig step afforded a mixture of (E)- and (Z)-isomers in about equal amounts. The required (E)-isomer (213) was separated by repeated recrystallization of its urea clathrate complex and converted into *impure* bombykol by partial hydrogenation followed by lithium aluminium hydride reduction. Apart from the necessity to separate the desired Wittig product from the mixture, this route suffers

Scheme 115 (Ref 284)

from the added drawback that partial hydrogenation of *conjugated* enynes such as (213) is not very selective and usually gives mixtures containing low yields of the required diene.[286] However, by using a judicious combination of stereoselective Wittig reactions these complications may be avoided. Such a route is outlined in Scheme 116.[287] Thus, condensation of methyl 10-oxodecanoate (215) with the stabilized ylide (214) furnished almost exclusively the (E)-aldehyde (216) which was converted into the conjugated (E,Z)-diene ester (218) with greater than 96% selectivity, by reaction with the ylide generated from the phosphonium salt (217) using sodium bis(trimethylsilyl)-amide in tetrahydrofuran. Reduction with lithium aluminium hydride then gave a product containing, as shown by g.l.c. analysis, 92% of the required (*E,Z*)-isomer (212). This highly stereoselective sequence of reactions seems to hold much potential as a general route to (*E,Z*)-conjugated dienes.

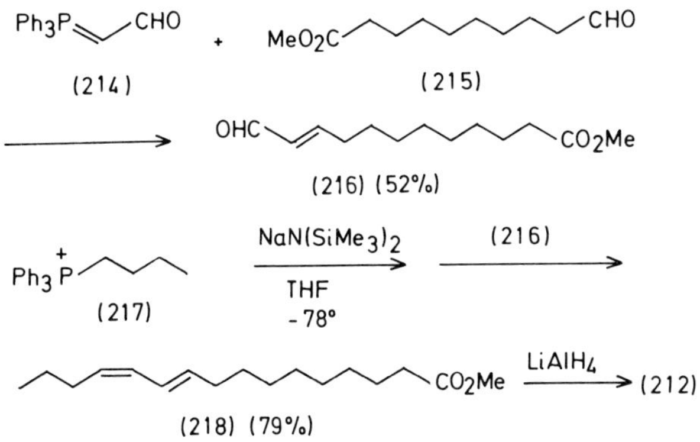

Scheme 116 (Ref 287)

cis-Stereoselective Wittig reactions have been applied with great success to the synthesis of many other lepidopterous pheromones. The higher order of *cis*-stereoselectivity and experimental simplicity of using sodium bis(trimethylsilyl)amide in THF has made it the most satisfactory procedure for the preparation of reactive ylides for use in these reactions.[131] The stereoselectivity is temperature dependent, but by operating at $-78°$ the (*Z*)-alkene obtained contains <2·0% of the (*E*)-isomer. Several of the aliphatic (*Z*)-monoenic acetate pheromones have been synthesized under these conditions.[288] As an example of the general method used to prepare attractants of this type, the synthesis of (*Z*)-tetradec-11-en-1-yl acetate (219), the main component of the

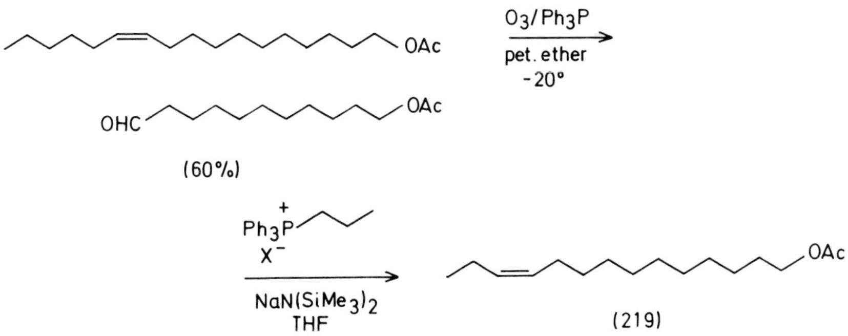

Scheme 117 (Ref 288)

sex pheromone of the European corn borer, and of the redbanded leafroller *Argyrotaenia vebitinana*, is outlined in Scheme 117.

The pheromone emitted by the female gypsy moth *Porthetria dispar*, cis-7,8-epoxy-2-methyloctadecane (221), has also been synthesized by a Wittig reaction route in which (Z)-alkene formation was used prior to epoxidation (Scheme 118).[118,289,290] As will be noted, the intermediate alkene, (Z)-2-methyloctadec-7-ene (220), is obtained with greatest purity (>98%) when the

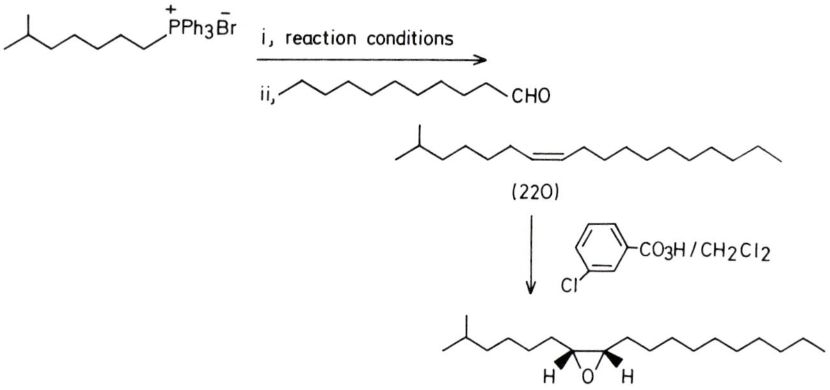

Reaction conditions	Yield (%)	(Z) : (E) of (220)	Ref.
Bu^{n}Li / DMSO, 15-20° then 10-20°	91	88 : 12	289
K / HMPA		94 : 6	118
Na$^+$ N̄(SiMe$_3$)$_2$ / THF 2h reflux, then aldehyde at -78°(1h), 12h (r.t.)	75	98 : 2	290

Scheme 118

ylide is generated with sodium bis(trimethylsilyl)amide in THF. The latter conditions also proved very effective for the three Wittig condensations used in a route (Scheme 119)[291] to gossyplure, the sex pheromone of the female pink vollworm moth *Pectinophora gossypiella*, which is a *ca* 1:1 mixture of (Z,Z)- and (Z,E)-hexadeca-7,11-dien-1-yl acetate, (222) and (224) respectively. By using this approach, it is possible to prepare the (Z,Z)-isomer (222) in 94·5% purity, but the (Z,E)-isomer (224) can be obtained in only 83·7% purity owing to loss of selectivity in the preparation of the phosphonium salt (223). The necessity for isomerically pure alkenes is greatly emphasized in this case by field tests which show that the presence of as little as 10% of the two other isomers greatly diminishes the potency of the mixture.[292]

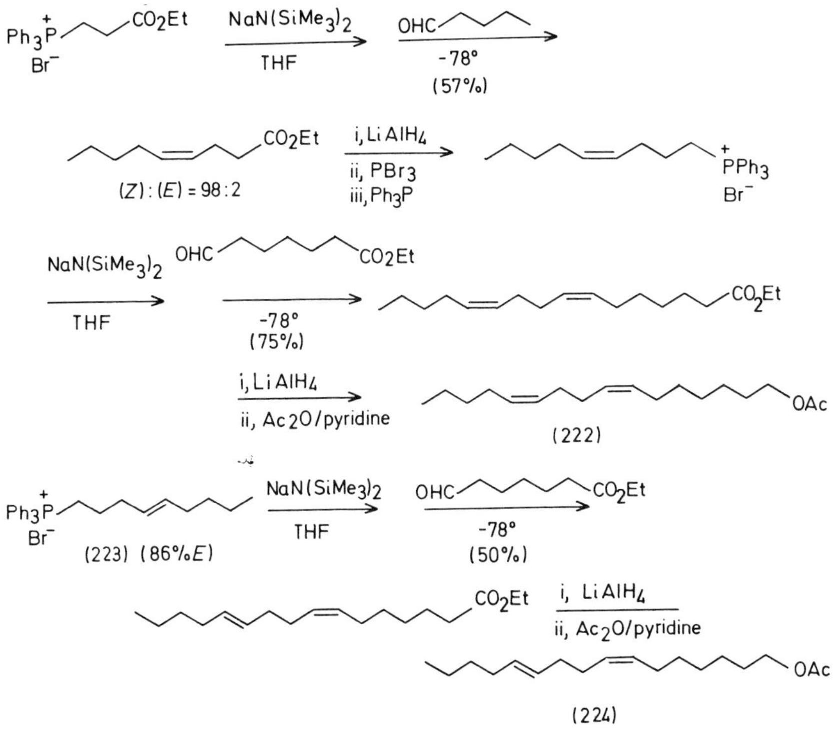

2 Wittig Reaction

Scheme 119 (Ref 291)

Two other approaches to gossyplure are worthy of attention for their use of stereoselective Wittig reactions to introduce the critical olefinic linkages. The first synthesis,[120] illustrated in Scheme 120, emphasizes the use of hexamethylphosphoramide (HMPA) as a co-solvent with tetrahydrofuran for obtaining (Z)-alkenes with >95% selectivity from the reactions of reactive ylides with aldehydes.[119] An interesting feature of this approach is that attempts to prepare the (E)-isomer of (226) from the phosphonium salt (225) by Schlosser's modification to the Wittig reaction failed. Even a change to a less polar solvent (toluene), whilst improving the *trans*-selectivity, still gave the (Z)-isomer as the major product.

The second approach[293] to gossyplure is notable for the ingenious procedure specifically developed to prepare the required 1:1 mixture of the isomers (229) and (230) by a single synthetic scheme involving stereochemical control of the Wittig reaction of the primary aldehyde (228) and the ylide from the phosphonium salt (227) (Scheme 121). Thus, while a normal Wittig reaction in ether in the presence of lithium bromide gave a 78:22 ratio of (229) and (230) respectively, the judicious addition of ethanol to the reaction

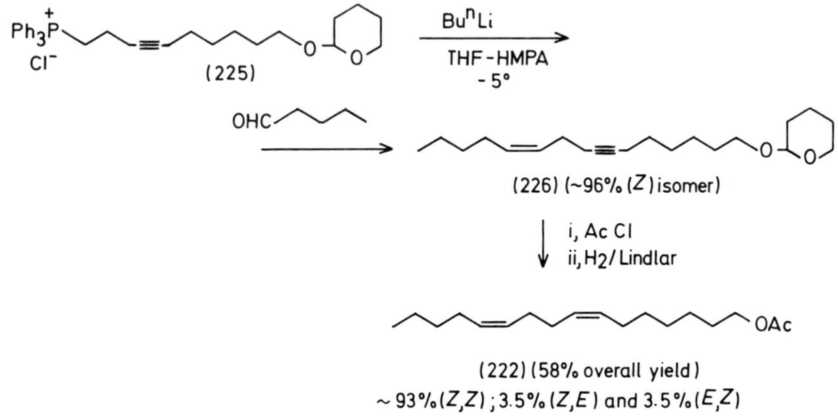

Scheme 120 (Ref 120)

mixture at −40° gave directly a ratio of 49:51. The effect of the alcohol is presumably to promote reversible betaine dissociation to reactants in a manner analogous to the role of lithium salts, but to a greater extent, and thus increase the proportion of the (*E*)-isomer (230) in the product. This approach to the stereochemical control of the Wittig reaction to give specific mixtures of (*Z*)- and (*E*)-alkenes promises to be particularly useful in pheromone synthesis, where such specific ratios of isomers are critically important.

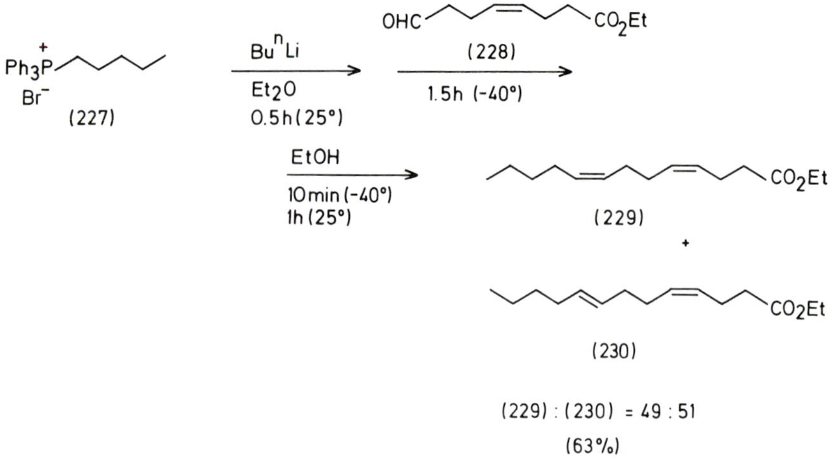

Scheme 121 (Ref 293)

cis-Stereoselective Wittig reactions have also been instrumental in the synthesis of a number of other sex pheromones, or their precursors for which alternative procedures such as Lindlar hydrogenation proved to be less

2 Wittig Reaction

successful. For example, a salt-free ylide has been used to obtain more than 98% of the (Z)-intermediate (231) in the course of the synthesis of multifiden (232), the biologically active main constituent of the androgamete attractant of *Cutleria multifida*.[294] In another instance, (Z)-tetradec-9-en-1-yl acetate

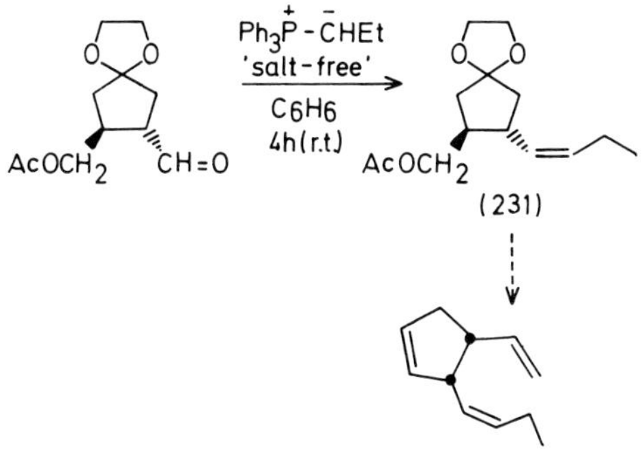

Scheme 122 (Ref 294)

(235), a synthetic pheromone which shows strong activity in the electrophysiological test for various species of *Noctuidae* (*Lepidoptera*), has been obtained in 50% yield and greater than 95% isomeric purity by the reaction of 9-acetoxynonanal (234) with n-pentylidenetriphenylphosphorane (233) in salt-free solution (Scheme 123).[295] This same reaction can also be performed

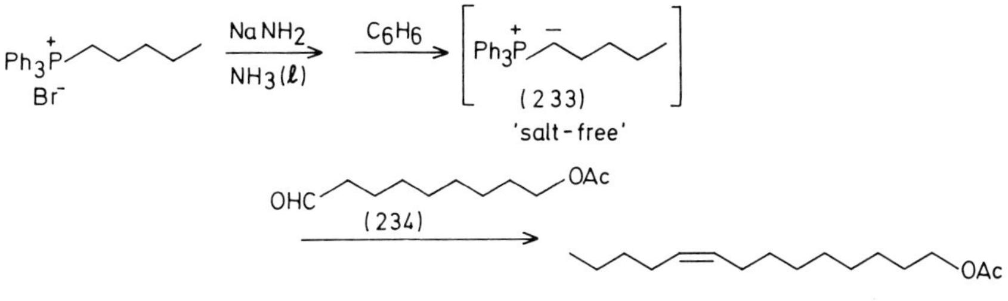

Scheme 123 (Ref 295)

with equal stereoselectivity, but in a less tedious way, by using potassium in HMPA to generate the ylide.[296] Moreover, with the exception of sodium bis(trimethylsilyl)amide in THF, this latter method has proved more effective for the selective formation of the (Z)-isomer, as illustrated by its use in the synthesis of *cis*-7,8-epoxy-2-methyloctadecane (221) (see Scheme 118),[118] the insect attractant of the female gypsy moth *Porthetria dispar*, and (Z)-tricos-9-ene (236), the most active hydrocarbon isolated from the cuticular lipids of

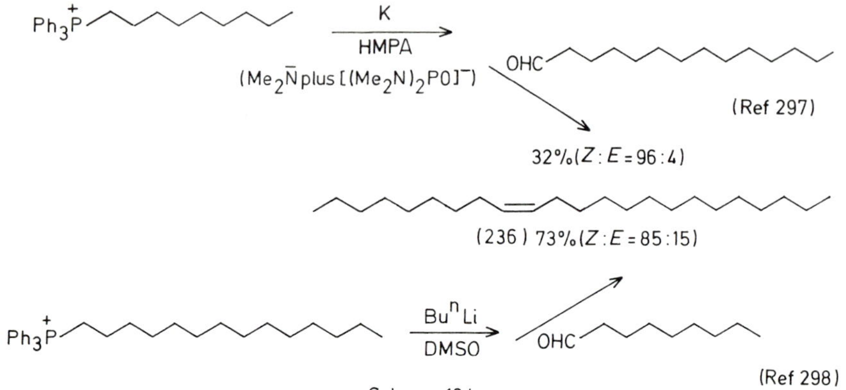

Scheme 124

female house flies *Musca domestica* (Scheme 124).[297,298] Another example is encountered in the preparation of (Z,E)-tetradeca-9,12-dien-1-yl acetate (237), the major component of the sex pheromone of the Indian meal moth *Plodia interpunctuella*, as shown in Scheme 125.[233,234]

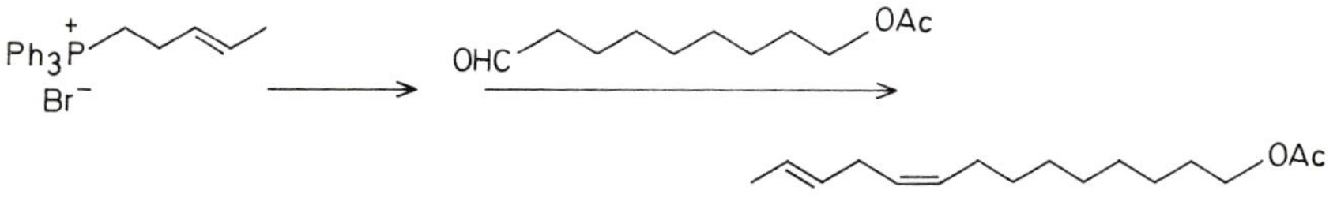

Reaction conditions	Yield (%)	(9Z,12E) : (9E,12E)	Ref
K, HMPA	40	95 : 5	233
NaH, DMSO	53	91 : 9	234

Scheme 125

As pointed out earlier, (Z)-alkenes of the same purity are also obtained when hexamethylphosphoramide (or dimethyl sulphoxide) is used as co-solvent with tetrahydrofuran. Since smaller quantities of the more expensive polar solvents are required and n-butyllithium (which is more convenient to use than the potassium compound) serves as the base, this procedure should be particularly attractive for the large-scale preparation of insect sex pheromones.[119,299] The degree to which the *cis*-selectivity is affected by a small amount of HMPA or DMSO is summarized in Table 14 for the preparation of (Z)-nonacos-11-ene, found in the cuticular hydrocarbons of both male and female face flies *Musca autumnalis*.[119]

Of the various base–solvent combinations available for effecting selective formation of (Z)-alkenes from reactive ylides, the use of dimsylsodium in

2 Wittig Reaction

Table 14. *(Z)*-Content of nonacos-11-ene from various modifications of the Wittig reaction between $Ph_3P^+-{}^-CH(CH_2)_{16}Me$ and $Me(CH_2)_9CHO$[119]

Solvent[a]	$Me(CH_2)_9CH=CH(CH_2)_{17}Me$	
	Yield (%)	(Z) (%)
THF	84	84
THF–DMSO (2:1)	87	94
THF–HMPA (2:1)	81	96
HMPA	81	95
DMF (NaOEt)	46	92
DMF (NaH)	39	91

[a] Bu^{n}Li employed as the base unless otherwise stated.

dimethyl sulphoxide has proved to be, contrary to expectations,[114] the least satisfactory for the preparation of unsaturated aliphatic compounds related to pheromones. For example, in the course of the synthesis of (Z,E)-tetradeca-9,11-dien-1-yl acetate (239), the most active component of the sex pheromone for the Egyptian cotton leafworm *Spodoptera littoralis*, the critical (9Z) configuration in the Wittig product (238) was obtained[300] with only 80% stereoselectivity by using these conditions (Scheme 126) (cf. the high stereoselectivity obtained in the related reaction used in Scheme 116). Nevertheless,

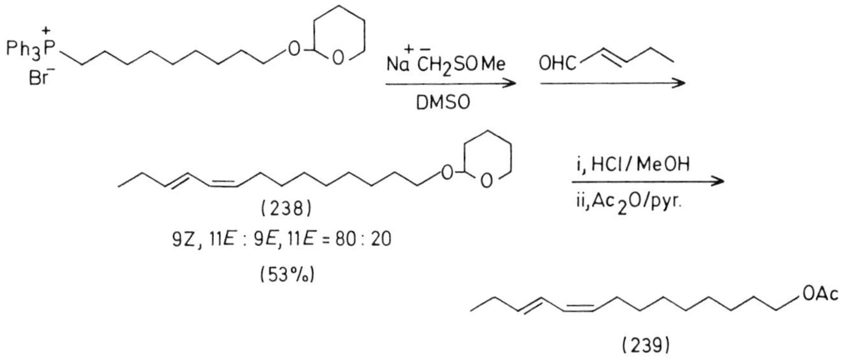

Scheme 126 (Ref 300)

it is noteworthy that, under the same conditions, the alternative Wittig condensation between (E)-pent-2-enyltriphenylphosphonium bromide (240) and 9-oxonon-1-yl acetate (241) is even less stereoselective, giving a mixture of (239) and its (9E,11E)-isomer in the ratio 60:40.[174,301] (See Section 3.4.2; cf. Scheme 59.)

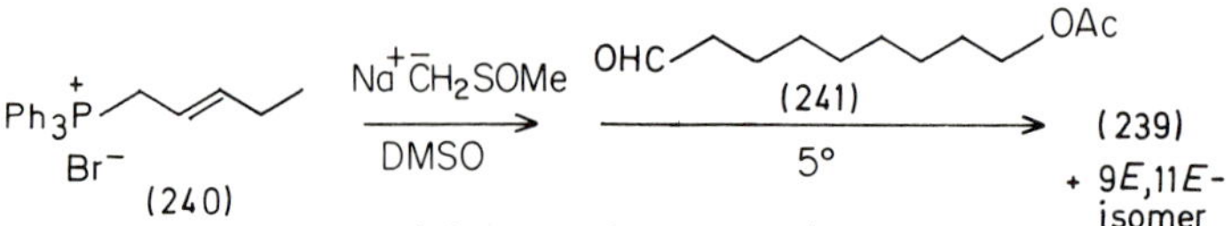

Scheme 127 (Ref 174, 301)

Stereochemical integrity can also be lost under these conditions, as illustrated by the synthesis of (Z,Z)-pentadeca-5,10-dien-1-ol (245) which utilized a two-fold Wittig reaction with 2-hydroxytetrahydropyran (242) functioning as the bifunctional C_5 building block common to both steps (Scheme 128).[240] In contrast to the first reaction, which afforded (Z)-dec-5-en-1-ol (243) in 93% isomeric purity, the second reaction of (242) and (244) gave a 10:5:1 mixture of (Z,Z)-isomer (245), (E,Z)-isomer (246), and (E,E)-isomer (247) in 36% yield, indicating that some isomerization of the (Z)-linkage in (244) had occurred during ylide formation.

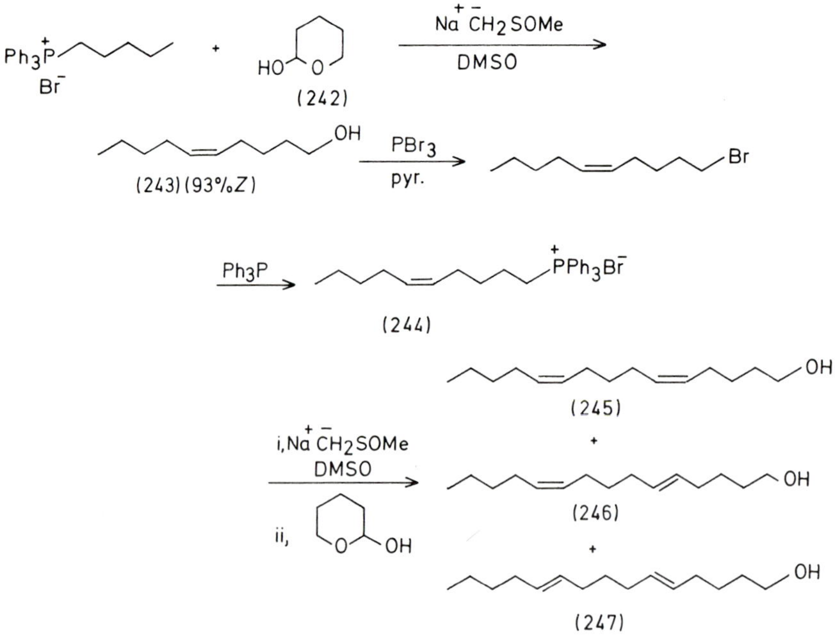

Scheme 128 (Ref 240)

Since insect sex pheromones are often long-chain (Z)-alkenols and alkenol acetates[282,302] it is to be noted that the Wittig reaction of 2-hydroxytetrahydropyran (242) with phosphorus ylides having a saturated aliphatic side-chain is a general method for the preparation of alk-5-en-1-ols with high cis-selectivity. As might be expected, this preference is lost when an additional double bond is in conjugation with the P=C of the ylide (see Table 15).

2 Wittig Reaction

Table 15. Wittig reaction of 2-hydroxytetrahydropyran (242) with various phosphonium salts in DMSO using dimsylsodium as base[240]

Phosphonium salt $Ph_3P^+CH_2R$	Product	Yield (%)	(Z):(E) ratio
R = MeCH₂	 ⁓⁓⁓OH	56	93:7
Me(CH₂)₄	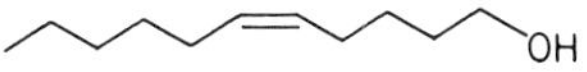 ⁓⁓⁓⁓OH	53	98:2
Me(CH₂)₇	 ⁓⁓⁓⁓⁓OH	47	100:0
(E) MeCH=CH	 ⁓⁓⁓⁓OH	34	55:45
Me₂C=CH (branched)	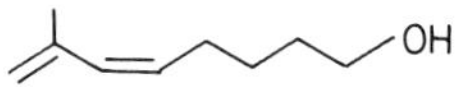 ⁓⁓⁓OH	11	50:50
Me₂C=CH	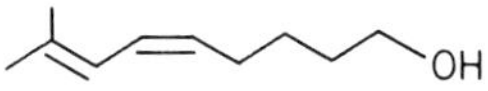 ⁓⁓⁓OH	25	50:50

Likewise, 2-hydroxytetrahydrofuran lends itself to a one-step stereoselective preparation of (Z)-alk-4-en-1-ols, e.g. (248), as shown in Scheme 129.[240] As noted earlier, an analogous transformation via γ-lactols has been extensively employed to construct the C_8 side-chain in prostaglandins (see Section 4.2.6).

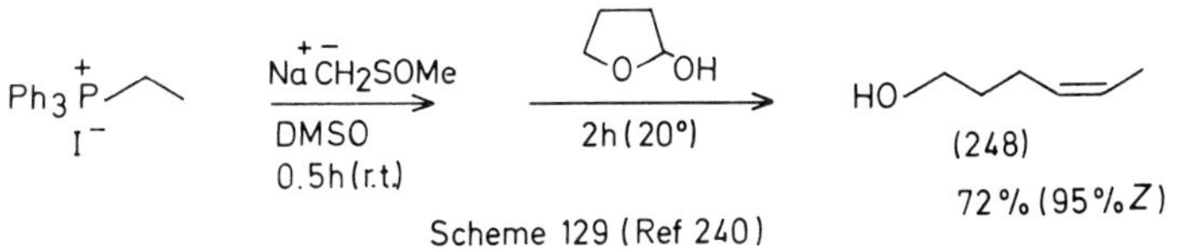

$Ph_3\overset{+}{P}\diagdown \quad I^- \quad \xrightarrow[\text{0.5h (r.t.)}]{\text{Na}^+\text{CH}_2\text{SOMe} \text{ DMSO}} \quad \xrightarrow[\text{2h (20°)}]{\text{tetrahydrofuran-OH}} \quad HO\diagup\diagdown= \quad (248) \quad 72\% (95\% Z)$

Scheme 129 (Ref 240)

Finally, mention should be made of the use of polymer-bound Wittig reagents to produce insect pheromones. This method of synthesis gives yields and stereoselectivity comparable to those of previous methods and has the potential, at least on a small scale, to be adapted to an automated procedure.[303] For example, treatment of the trityl-bound phosphonium salt (249) with n-butyllithium followed by addition of butyraldehyde gave the polymer-bound alken-1-ol (250) (Scheme 130). Subsequent hydrolysis and acetylation then produced tetradec-10-en-1-yl acetate (251) containing more than 91% of the (Z)-isomer, which is the sex attractant for the oak leaf roller *Archips semiferanus*. That this process should be so highly stereoselective is particularly noteworthy since no precautions, i.e. "salt-free" conditions, were taken to promote (Z)-alkene formation. In addition, the synthesis in the reverse sense, whereby a polymer-bound aldehyde is condensed with a Wittig reagent, gives a mixture of (Z)- and (E)-isomers in the ratio 67:36.

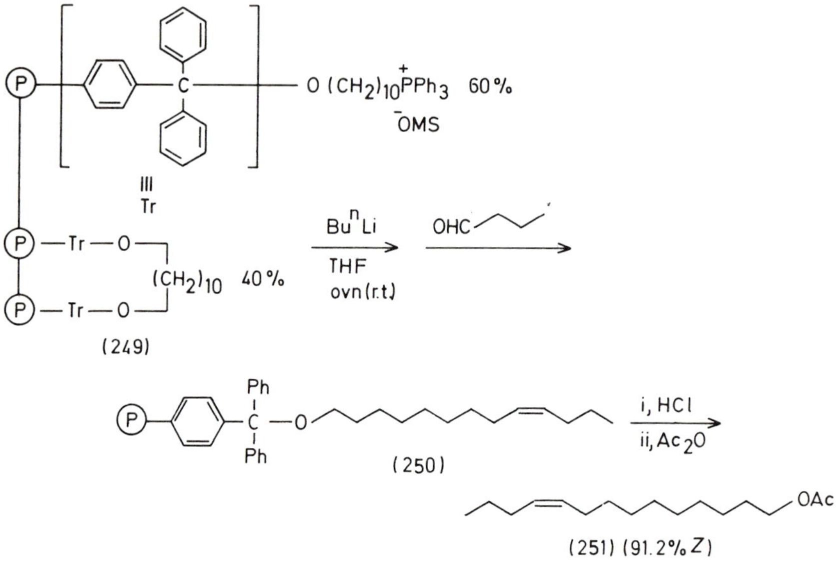

Scheme 130 (Ref 303)

One drawback to this approach is that about 40% of the starting diol is actually bound to the polymer at both ends [see (249)], and although this doubly-bound material does not interfere in the chemical synthesis its presence means that it is liberated in the final cleavage step from the polymer and must be separated from the desired product. Attempts to overcome this problem by using difunctionalized polymers containing proximally located trityl alcohol and benzoic acid groups have been reported.[304]

4.4. Carotenoids and Polyenes

Of the various procedures used for building up the carbon skeleton of carotenoids and related polyenes, none has proved more successful than the Wittig reaction. In the 25 years since its discovery, more than 150 carotenoids have been synthesized in this way, often using more than one approach. It is outside the scope of this review to describe this tremendous amount of work in detail. For a comprehensive survey, the reader is referred to the excellent review by Mayer and Isler[305] covering the literature up to 1971, and to a monograph by Weedon[306] for more recent developments. To cite one exemplary case will suffice here. Table 16 summarizes the six different building schemes devised to prepare all-*trans*-β-carotene. These consist in combining two components with identical carbon skeletons, or three components, two of which have identical carbon skeletons, in a single Wittig step.

2 Wittig Reaction

Table 16. Synthesis of all-*trans*-β-carotene

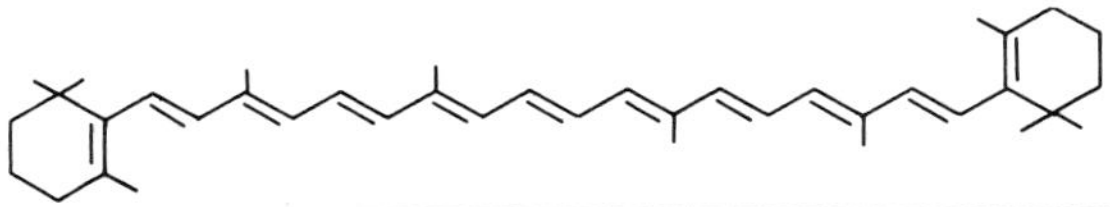

Building scheme	Reagent	Reactant	Base/Solvent	Ref.
$C_{20} + C_{20} = C_{40}$	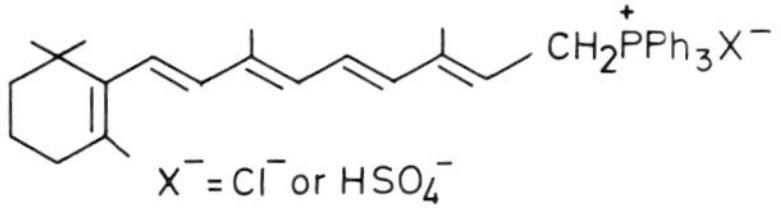 $CH_2\overset{+}{P}Ph_3 X^-$ $X^- = Cl^-$ or HSO_4^-	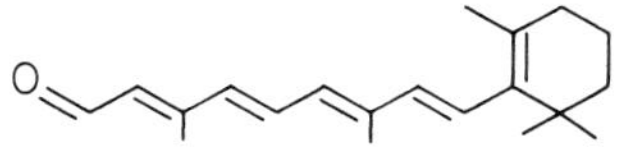	KOH or NaOMe in MeOH	307
				307d
$C_{18} + C_4 + C_{18} = C_{40}$	$\overset{+}{P}Ph_3 Br^-$ 2		NaOMe or NaC≡CH in DMF or MeCN	210, 308
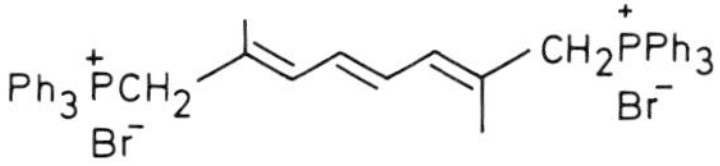 $C_{15} + C_{10} + C_{15} = C_{40}$	$CH_2\overset{+}{P}Ph_3 Cl^-$ 2 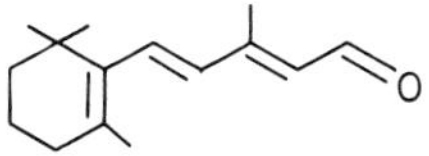		PhLi	309
	$Ph_3\overset{+}{P}CH_2\ Br^-$ ... $CH_2\overset{+}{P}Ph_3\ Br^-$			
$C_{13} + C_{14} + C_{13} = C_{40}$	$\overset{+}{P}Ph_3 Cl^-$ 2 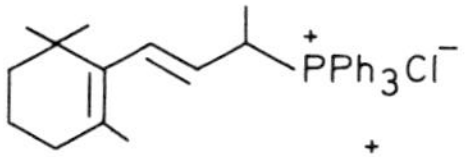	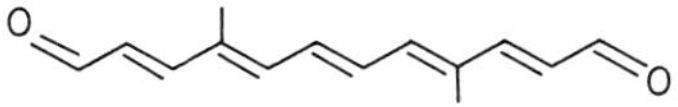	NaOMe in DMF or MeLi in THF	310
$C_{10} + C_{20} + C_{10} = C_{40}$	$CH_2\overset{+}{P}Ph_3 Br^-$ 2 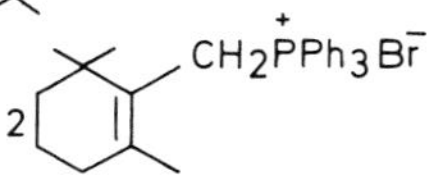	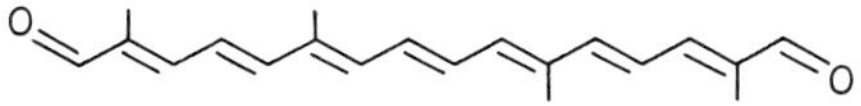	NaOMe in DMF	311

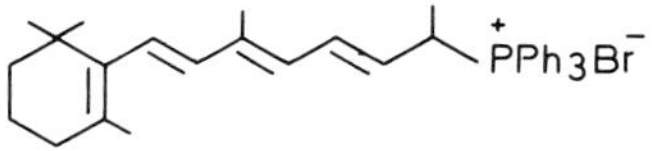

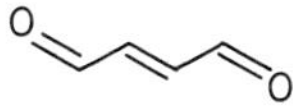

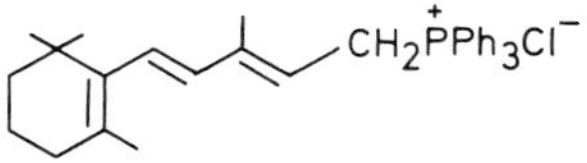

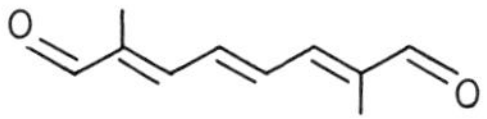

Much of the work devoted to the total synthesis of polyisoprenoids has been directed to the preparation of the all-*trans* isomers. However, in common with other methods, routes based on the Wittig reaction often lead to mixtures of geometrical isomers as might be expected from previous considerations. Fortunately, most (Z)-isomers of carotenoids can be converted into the more stable (E)-product by heating in heptane or by irradiation in the presence of catalytic amounts of iodine. Often the conditions of the Wittig condensation are sufficient for an effective isomerization. Interestingly, the ease of isomerization is dependent upon the position of the double bond in the polyene chain. Most difficulties are encountered with carotenoids having a (Z) configuration about the 9,10-double bond. For example, in the preparation of vitamin A derivatives by the Wittig reaction of β-ionyltriphenylphosphonium bromide (252) with C_7 diene aldehydes (253), a mixture of products is obtained in which the (9Z)-isomer predominates in the ratio *ca.* 60:40 (Scheme 131).[209,312] Although the yields are good, the method is of little practical importance because the (9Z)-isomers cannot be forced into the (E) configuration by the methods described above (see Scheme 88).

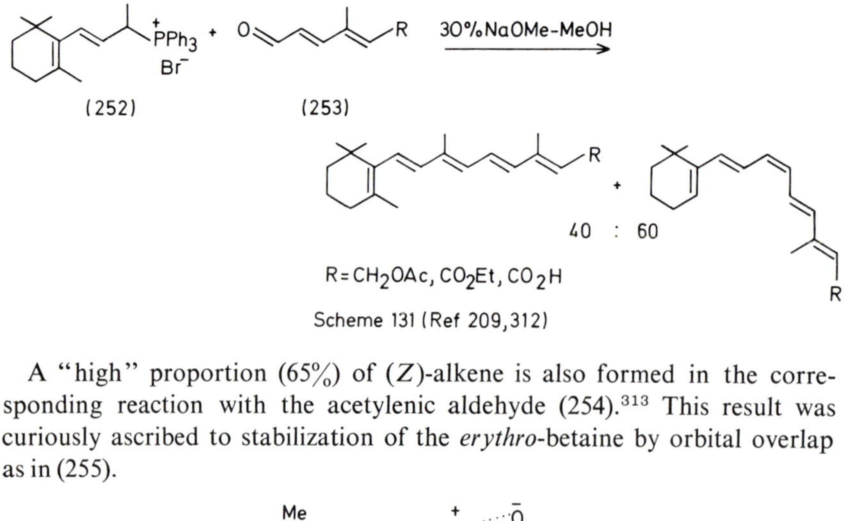

R = CH_2OAc, CO_2Et, CO_2H

Scheme 131 (Ref 209, 312)

A "high" proportion (65%) of (Z)-alkene is also formed in the corresponding reaction with the acetylenic aldehyde (254).[313] This result was curiously ascribed to stabilization of the *erythro*-betaine by orbital overlap as in (255).

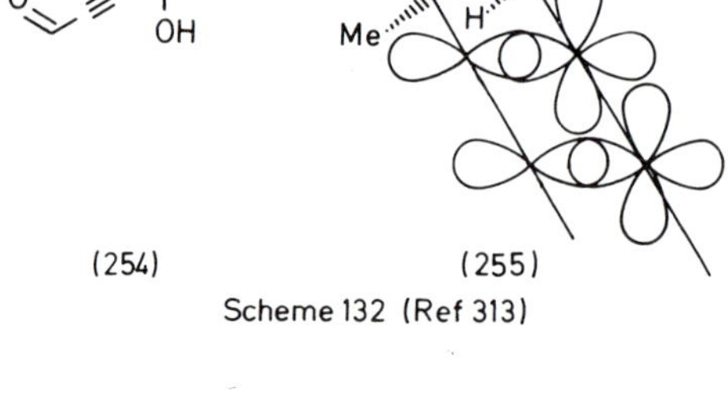

(254)   (255)

Scheme 132 (Ref 313)

2 Wittig Reaction

A similar situation is observed[310] in the synthesis of all-*trans*-β-carotene according to the $C_{13} + C_{14} + C_{13}$ building scheme depicted in Table 16. The reaction product consisted of a mixture of several geometrical isomers, mainly the (9Z,9'Z)-isomer, which was relatively stable and difficult to isomerize to the all-*trans*-β-carotene. A variant of this procedure using the corresponding central acetylenic C_{14} dialdehyde also gave a similar mixture of 15,15'-didehydro-β-carotene isomers, from which the all-*trans* form was subsequently obtained by crystallization.[310b,314]

By comparison, polyene carotenoids with a *cis* double bond in the 7,8-, 11,12-, 13,14-, or 15,15'-positions undergo isomerization relatively easily.[315] Scheme 133 illustrates the general isomerization pattern for β-carotene.[316] A likely explanation for the difference in stabilities has been given based on the severity of steric interactions of the types illustrated in Scheme 134.[315]

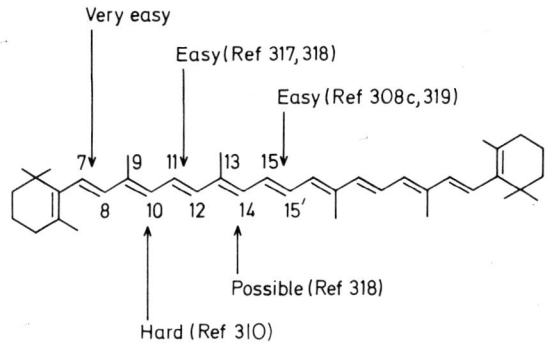

Scheme 133. Isomerization of (Z)-β-carotenes into their (E)-isomers (from Kienzle [316])

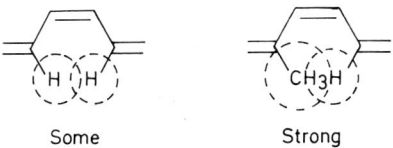

Scheme 134 Steric interference of hydrogen atoms in CH-CH=CH-CH, and of methyl and hydrogen in C(CH₃)-CH=CH-CH, with (Z) configurations (from Weedon [315])

The Wittig reaction also appears to be acquiring practical importance in the carotenoid and vitamin A fields for the preparation of specific (Z)-isomers, in particular those with methylated (Z)-double bonds since these are not accessible by acetylenic routes. Because stereochemical control is not feasible in the Wittig step, the routes developed have usually incorporated the desired (Z) configuration in the starting ylide or its carbonyl co-reactant. This approach was first applied[318] to the synthesis of (13Z)-β-carotene (258) by condensing β-retinyltriphenylphosphonium sulphate (256) with (13Z)-retinol (257) in the presence of potassium hydroxide in methanol. No indication of stereomutation was reported (Scheme 135).

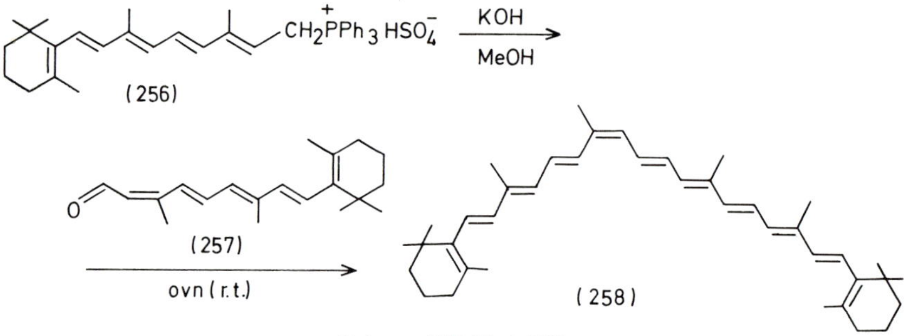

Scheme 135 (Ref 318)

A similar approach was adopted in the final step of the synthesis of (Z)-methylbixin (261) by the route outlined in Scheme 136.[42] Under carefully

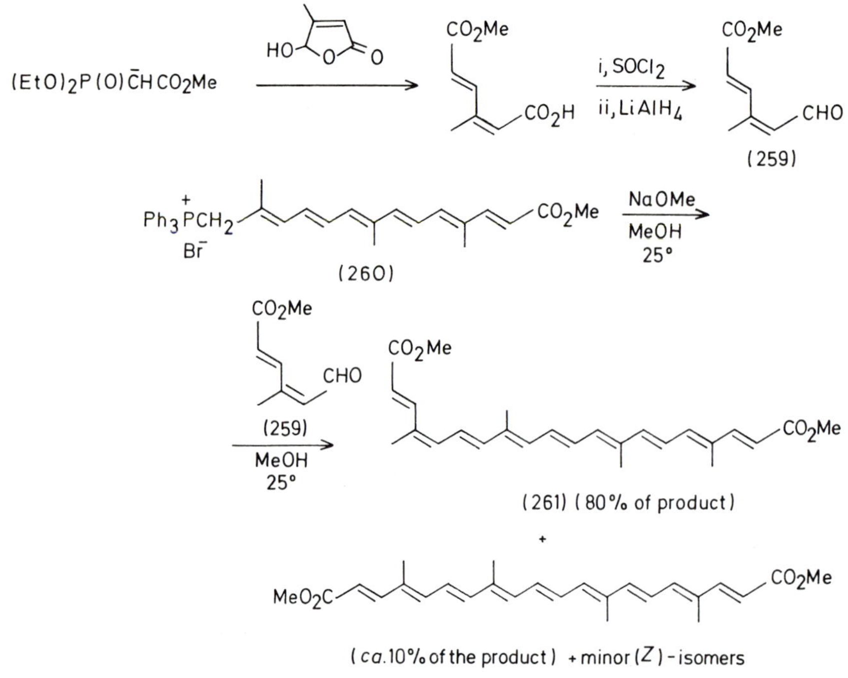

Scheme 136 (Ref 42)

2 Wittig Reaction

controlled conditions, the condensation between the labile (Z,E)-aldehyde (259) and the C_{17} phosphonium salt (260) proceeded with almost complete retention of stereochemistry in (259) and led to methyl natural bixin (261) as the major product (*ca.* 80%).

Retention of stereochemistry about the (Z)-double bond is also observed in comparable syntheses involving the (Z)-C_{15} phosphonium salt (262).[320] Thus condensation of the ylide from (262) with the trienedial (263) gave the (9Z)-C_{25} apo-12-alloxanthinal (264) which could be further condensed to give the (9Z,9′Z)-carotenoid (265), identical with the principal geometrical isomer from the iodine-catalysed stereomutation of natural alloxanthin (Scheme 137). The corresponding condensation with α-apo-12′-carotenal

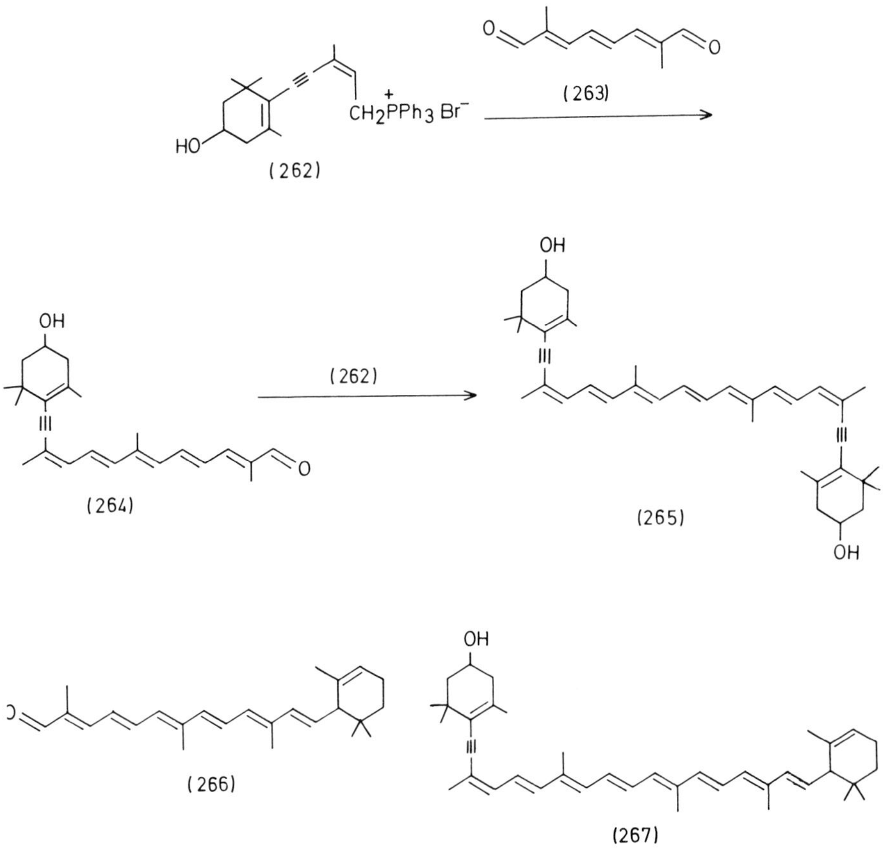

Scheme 137 (Ref 320)

(266) similarly led to the (9Z)-carotenoid (267) which was identified with the main stereomutation product of natural crocoxanthin. An interesting feature of this reaction is that, under the conditions required to condense the (E)-analogue of (264) with the ylide from (262), the product underwent extensive stereomutation to (9Z,9'Z)-alloxanthin (265). It has been suggested[320] that, in contrast to their polyene analogues, these acetylenic (Z)-carotenoids are less sterically hindered, and therefore more stable than their all-*trans* forms (cf. Scheme 133).

Perhaps the most striking example of the stereochemical control that can be exercised in this way is the synthesis of the (Z,Z,Z)-pentaene (271), a "poly-*cis*" C_{20} analogue of phytoene, by the condensation of the (Z)-α-unsaturated phosphonium salt (268) with (Z)-citral (269) in THF with Bu^{n}Li as base.[321] The product consisted of a 68:32 mixture of (Z,E,Z)- and (Z,Z,Z)-isomers (270) and (271), respectively, showing retention of configuration about

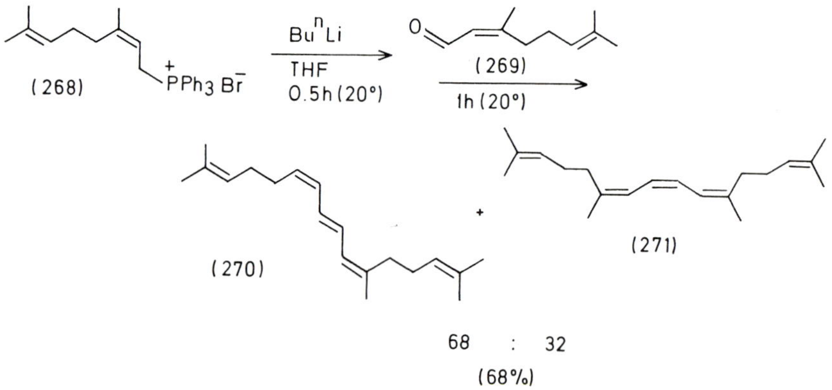

Scheme 138 (Ref 321)

the double bonds of the starting materials. A similar condensation of the (Z)-salt (268) with (E)-citral produced a mixture (*ca.* 60:40) of the corresponding (Z,Z,E)- and (Z,E,E)-pentaenes. In each case preparative thin-layer chromatography on $AgNO_3$–silica gel gave homogeneous samples of each isomer. It should be noted that similar condensations in 1,2-epoxybutane were accompanied by extensive stereomutation in the starting materials.[177]

Finally, mention should be made of the reactions of 4-hydroxy-3-methyl-but-2-en-4-olide (273)* with allylic Wittig reagents to give polyene acids with complete retention of (Z) stereochemistry about the α,β-double bond and a high proportion of (Z) configuration about the newly formed carbon–carbon double bond[181,182,200] (see p. 82). The procedure has particular merit,

---

* The lactol can be regarded as the cyclic form of (Z)-β-formylcrotonic acid which cannot be isolated as such owing to spontaneous cyclization.

2 Wittig Reaction

especially in the vitamin A field. For example, condensation of (273) with the Wittig reagent from (2-ionylideneethyl)triphenylphosphonium bromide (272) yielded a mixture of (13Z)-(275) and the hitherto unknown (11Z,13Z)-vitamin A acid (274), in which the latter predominated (>60%).[181,200] When isomerized with iodine under mild conditions, the (Z,Z)-isomer gave the (13Z)-acid (275) in excellent yield (Scheme 139).[200]

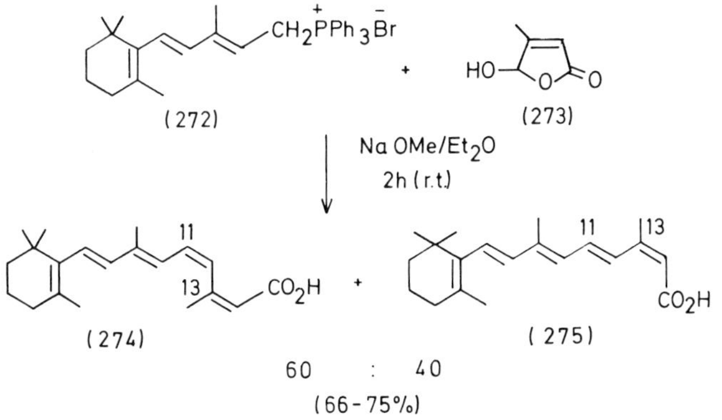

Scheme 139 (Ref 181, 200)

(9Z)-Vitamin A acid has been prepared in basically the same way, starting from the (9Z)-analogue of (272) and ethyl β-formylcrotonate.[210,311a]

4.5. Rethrolones

cis-Stereoselective Wittig reactions have also proved to be very suitable for the synthesis of cinerolone (276a), jasmolone (276b), and pyrethrolone (276c), known collectively as rethrolones, which are the alcohol components of the insecticidal "pyrethrin" esters found in *Chrysanthemum cinerariaefolium* (Scheme 140). In the case of cinerolone and jasmolone the (Z)-side-chain

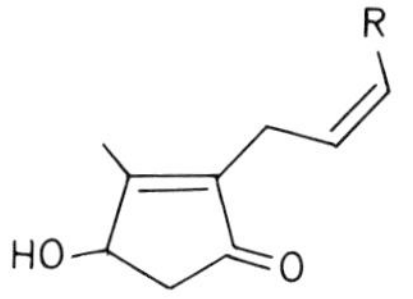

(276)

a; R = Me
b; R = Et
c; R = CH=CH

Scheme 140

may be introduced by partial hydrogenation of acetylenic intermediates,[322,323] but with pyrethrolone, which contains a sensitive (Z)-vinyl diene system, the hydrogenation of conjugated enyne precursors is not selective.[323] Indeed the overall yield by this route is low (0·2%). In contrast, the Wittig reaction between the reactive ylide (277) and acraldehyde, under "salt-free" conditions, affords the (Z)-diene (278) almost exclusively, which can be transformed by a series of subsequent steps into highly pure ($\pm$)-(Z)-pyrethrolone (21%) (Scheme 141).[324] In view of the ease with which acraldehyde polymerizes in the presence of strong base, it is important to note that the yield of diene (278) produced is increased by passing the vapour of acraldehyde in nitrogen over a stirred solution of the ylide for long periods. In this way ca. 50% yields of diene are consistently obtained (cf. references 7 and 100).

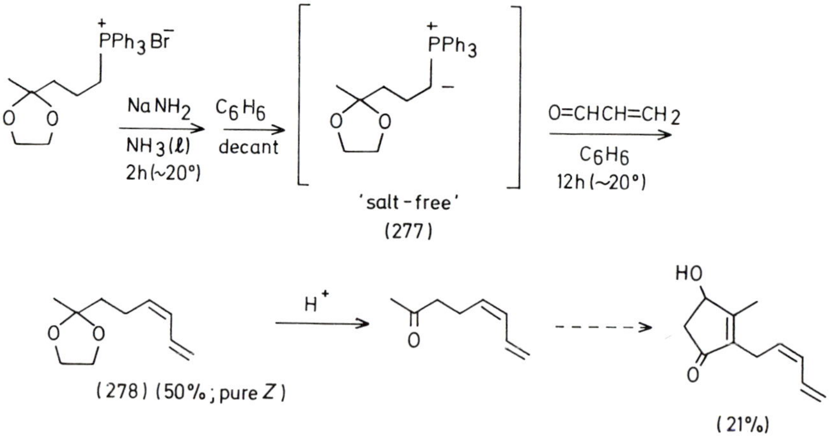

Scheme 141 (Ref 324)

The corresponding Wittig reactions with either acetaldehyde or propionaldehyde under "salt-free" conditions also provide the (Z)-precursors (279a) and (279b) for cinerolone (276a) and jasmolone (276b) syntheses, respectively; here, ca. 6–8% of the corresponding (E)-alkenes are produced concurrently.

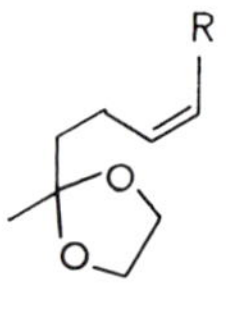

(279) a; R = Me
b; R = Et

Scheme 142 (Ref 324)

The (Z)-double bond present in jasmolone (276b) has also been introduced stereoselectively and in 23% yield by treatment of the unstable ketoaldehyde (280) with the ylide derived from propyltriphenylphosphonium bromide in dimethyl sulphoxide, using dimsylsodium as base (Scheme 143).[325] A similar

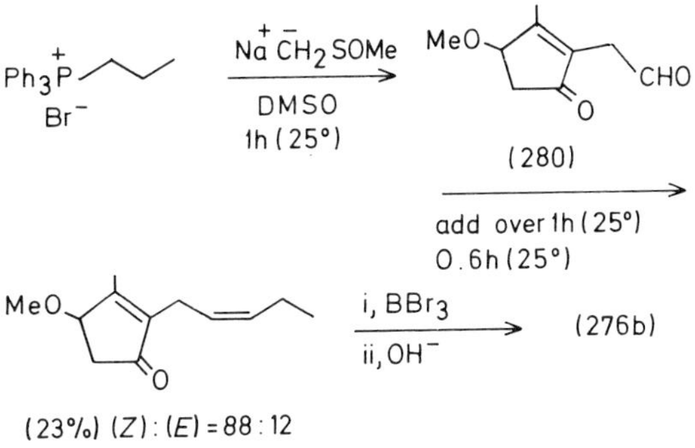

Scheme 143 (Ref 325)

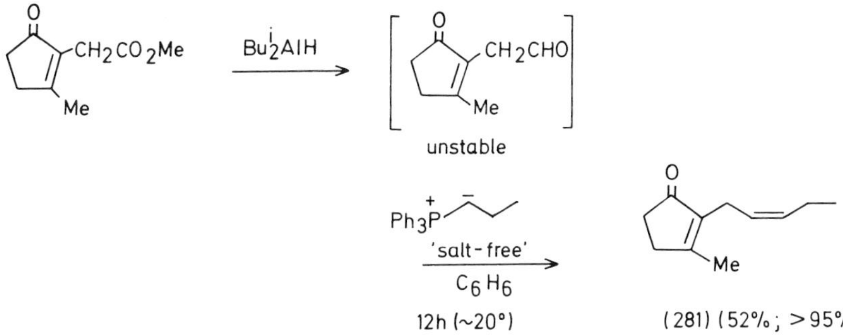

(281) (52%; >95% Z)
(Ref 326)

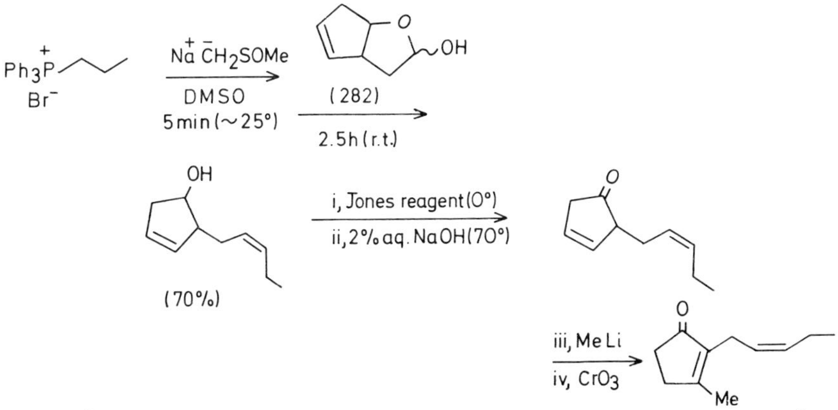

(281)

Scheme 144 (Ref 327)

approach[326] was used for the synthesis of the corresponding rethrone, (Z)-jasmone (281), an important perfumery constituent, as shown in Scheme 144. Also illustrated is another route to (Z)-jasmone using the same mode of assembly of the side-chain, but employing a hemiacetal (282) as the aldehyde component.[327]

Other intermediates for the synthesis of (Z)-jasmone that have been prepared using stereoselective Wittig reactions include (283)[111,172] and (285).[328] Note that in the case of the former the (E)-isomer is obtained almost exclusively[329] when no special precautions, i.e. removal of lithium salt, are taken to ensure (Z)-selectivity. However, in the case of the latter, the chemoselective Wittig reaction with the tricarbonyl compound (284) proceeds in 54% yield and with >90% cis-stereoselectivity under comparable conditions.

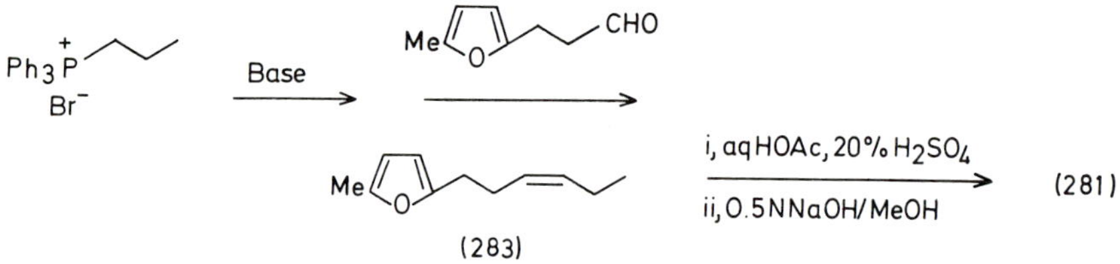

Reaction conditions	% Yield	% (Z)	Ref
Bu^nLi/DMF, 3h, 20°	80	>90	111
'salt-free' C_6H_6	32	ca. 88	172
Bu^nLi/Et_2O, 24h, rflx	66	<10	329

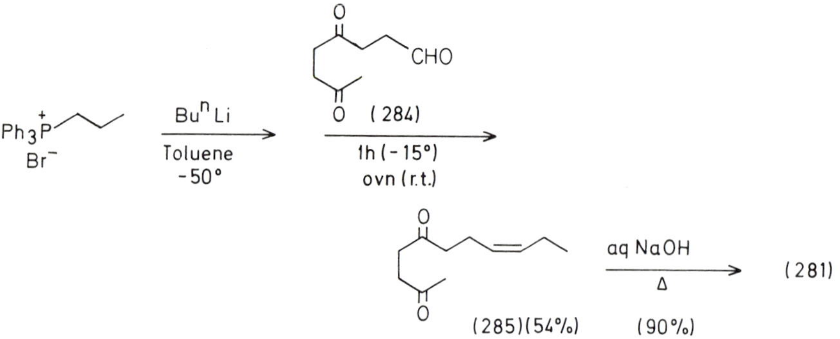

Scheme 145

trans-Stereoselective alkene formation by means of the Wittig reaction has found application in the synthesis of pyrethric acid (289), another constituent of pyrethrum, with a specific ^{14}C label.[330] Thus, condensation of the stabilized ylide (286) with the (+)-(E)-aldehyde (287) and hydrolysis of the resulting dimethyl ester (288) gave the ^{14}C-labelled diacid (289), identical with that

2 Wittig Reaction

obtained from hydrolysis of pyrethrum extract (Scheme 146). Model experiments with unlabelled ylide (286) and the ($\pm$)-(E)-aldehyde indicated that none of the corresponding (Z)-olefination product was formed in the Wittig condensation. In contrast, the analogous phosphonate anion, $(EtO)_2P(O)C^-(Me)CO_2Me$, condenses with the (E)-aldehyde to give the (E,E)-product (288) with only 83% stereoselectivity.[331] Similar losses of *trans*-selectivity have been reported[332] for other condensations of phosphonate anions that have an α-methyl group (see also reference 25).

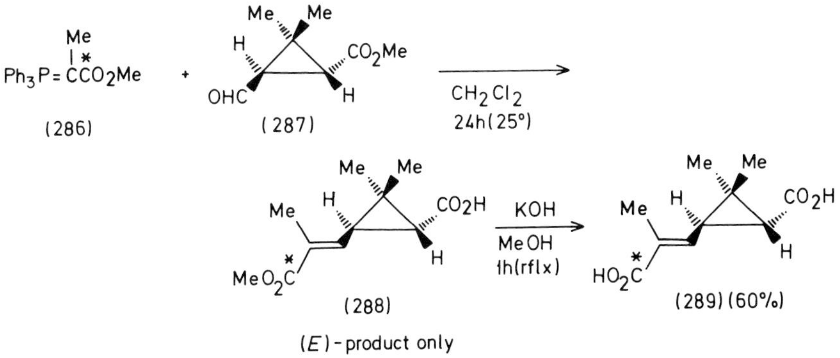

Scheme 146 (Ref 330)

The (E)-aldehyde (287) has also been condensed with a number of reactive ylides under "salt-free" conditions.[330] For example, (287) and ethylidenetriphenylphosphorane gave methylnorchrysanthemate (290) containing not less than 88% of the (Z)-isomer. The acid of (290) is reported to be the principal acidic product from the pyrolytic decomposition of pyrethric acid (289).

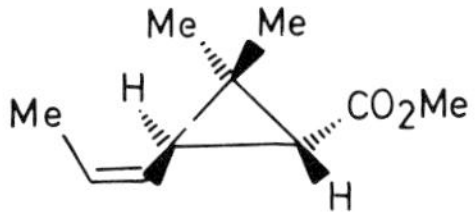

Scheme 147

Surprisingly, condensation of (287) with the ylide derived from ^{14}C-labelled (291) under "salt-free" conditions, followed by hydrolysis, gave a negligible yield (ca. 2%) of ^{14}C-labelled (+)-(Z)-chrysanthemic acid (292) (Scheme 148).[330] The use of n-butyllithium as base also proved unsuccessful. However, by carrying out the reaction in dimethyl sulphoxide using dimsylsodium as base, (292) was obtained in 64% yield.

It should be noted that methyl (E)-chrysanthemate (294) can be obtained stereospecifically in more than 60% yield *by simply mixing* THF solutions of

Scheme 148 (Ref 330)

methyl (E)-4-oxobutenoate (293) (1 equiv.) with isopropylidenetriphenylphosphorane (2·4 equiv.).[333,334] Other ylides react in a similar manner to produce the corresponding chrysanthemate analogues. That these reactions occur by an initial Wittig condensation followed by addition to the activated

Scheme 149 (Ref 333,334)

double bond is shown by the selective formation of the corresponding dienoic esters when only 1 equivalent of the ylide is used. Hexylidenetriphenylphosphorane (295), for instance, reacts with ethyl (E)-4-oxobutenoate (296) to give ethyl (E,Z)-deca-2,4-dienoate (297), identified as part of the odoriferous principle of Bartlett pears, in 63% purified yield [70% of the (2E,4Z)-isomer].[333] This same reaction under "salt-free" conditions is reported to occur with 85% stereoselectivity (Scheme 150).[335]

Scheme 150 (Ref 335)

In a related approach[104] to the "pear ester" (297), condensation of hexylidenetriphenylphosphorane with the vinylogous epoxy-aldehyde (298) under normal Wittig reaction conditions gave the desired (Z)-linkage to the extent of 85%, but subsequent oxidative degradation of the epoxy-diene (299) to "pear ester" reduced the (Z):(E) ratio to 75:25.

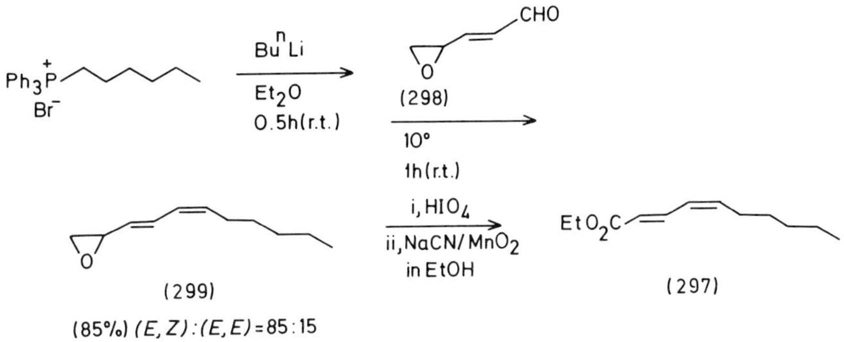

2 Wittig Reaction

Scheme 151 (Ref 104)

4.6. Sesquiterpenes

The considerable interest in the possible utilization of insect juvenile hormones as a non-insecticidal means of insect control is reflected in the growing number of different synthetic approaches to the juvenile hormones of the giant silkworm moth *Hyalophora cecropia*. Not surprisingly, some of these syntheses have involved the production of a plethora of isomers from which the desired substance has been separated with varying degrees of difficulty. A pertinent example, whereby three consecutive Wittig reactions are utilized for building the carbon skeleton of $(\pm)$-C_{18} juvenile hormone (300), is summarized in Scheme 152.[336] Stereochemically, the synthesis provides little prospect for control, since, as pointed out in Section 3.3, about equal amounts of (Z)- and (E)-isomers are normally obtained when methyl ketones are allowed to react with alkylidenetriphenylphosphoranes under "selective" conditions.* In the event, the synthesis required separation and chemical assignment of the desired isomers after introduction of each double bond. This and related approaches[92,141,337] to juvenile hormone (JH) are to be contrasted with the following method[338,339] for the assembly of the basic JH chain *in a single step* from three components specifically in the correct stereochemical form. The essential *trans* stereochemistry of the 2,3- and 6,7-double bonds was established by taking advantage of the previously described method for stereospecific synthesis of trisubstituted alkenes from β-oxido phosphonium ylides (see Section 3.3.2). Scheme 153 outlines the reactions utilized to prepare the unsaturated alcohol derivative (301) uncontaminated by stereoisomeric or other impurities in 50% yield. Conversion of this alcohol into $(\pm)$-C_{17} JH (302; R = Me) and $(\pm)$-C_{18} JH (302; R = Et) was then accomplished by a sequence of straightforward steps.

* Also note that an increase in selectivity for (E)-trisubstituted alkene formation cannot be accomplished using Schlosser's modification to the Wittig reaction (see Section 3.3.1).

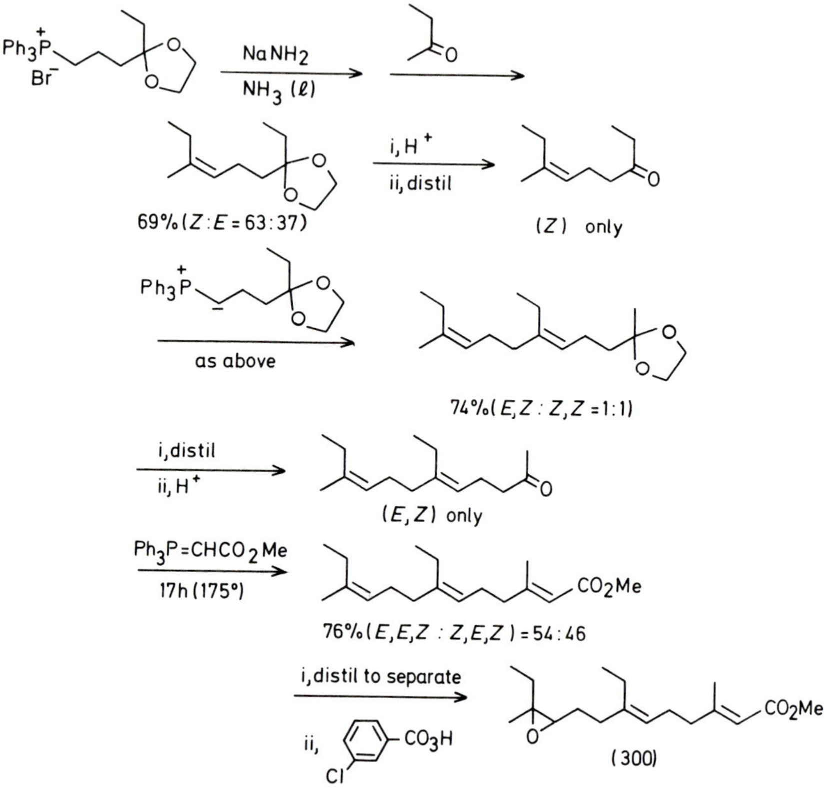

Scheme 152 (Ref 336)

The stereochemical control and the flexibility inherent in the β-oxido-ylide approach makes it a potentially powerful method for the synthesis of numerous structures of the acyclic triisoprenoid type. Farnesol itself (306) has been synthesized in *two steps*, stereospecifically, from the phosphonium salt (303), the aldehyde (304), and paraformaldehyde, as shown in Scheme 154.[340] The hydroxylated farnesol derivative (305) could also be converted in a highly selective manner into (307), a biologically active positional isomer of C_{17} *Cecropia* juvenile hormone.

The procedure has also been used with great effect to prepare the important essential oil α-santalol (309) in a single step stereospecifically and efficiently, starting from tricycloekasantalal (308) (Scheme 155).[158] The simplicity of this route is to be contrasted with a previous synthesis of α-santalol from the same aldehyde by a process involving a conventional Wittig reaction which was stereoselective for the undesired isomer of (309) as shown in Scheme 156.[62] (See also Scheme 21 for a related non-stereoselective synthesis of β-santalol to which the β-oxido-ylide modification could probably be applied.)

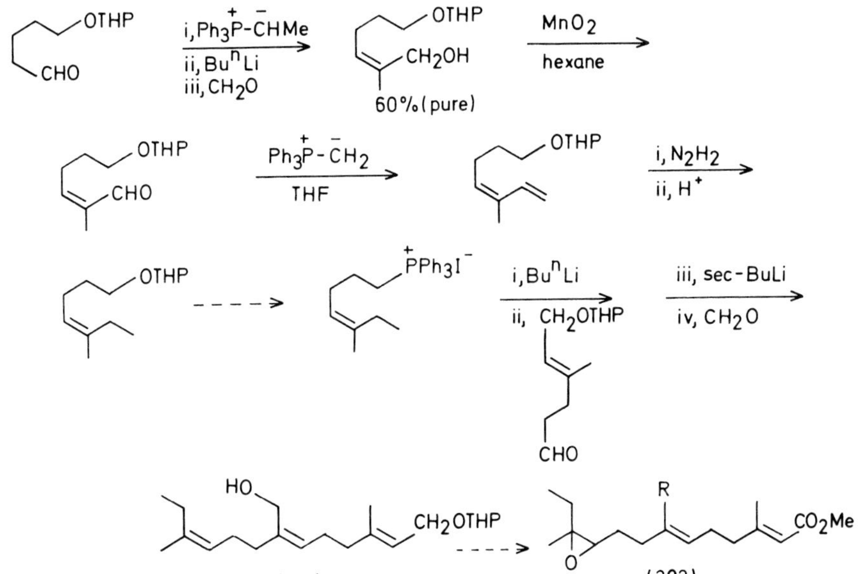

2 Wittig Reaction

Scheme 153 (Ref 338, 339)

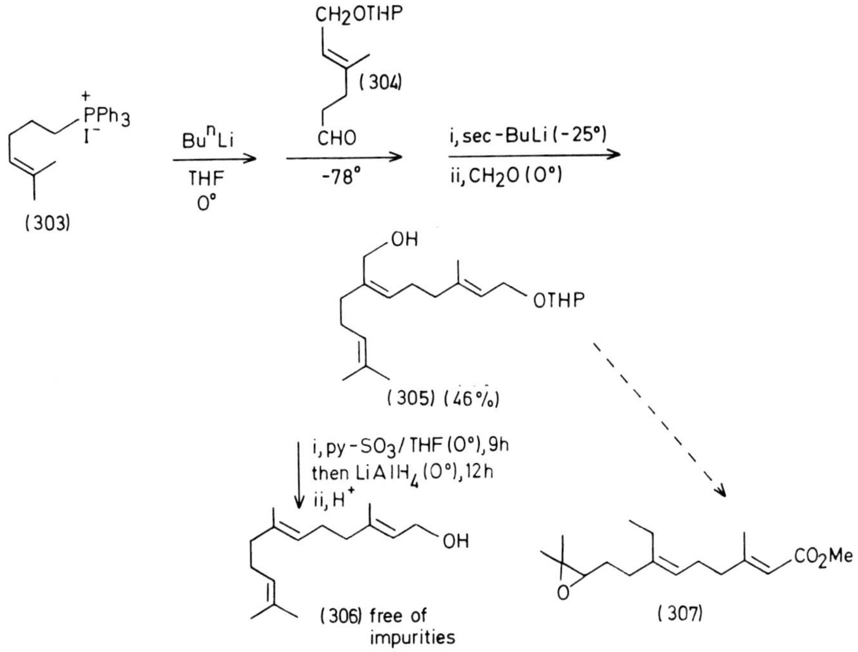

Scheme 154 (Ref 340)

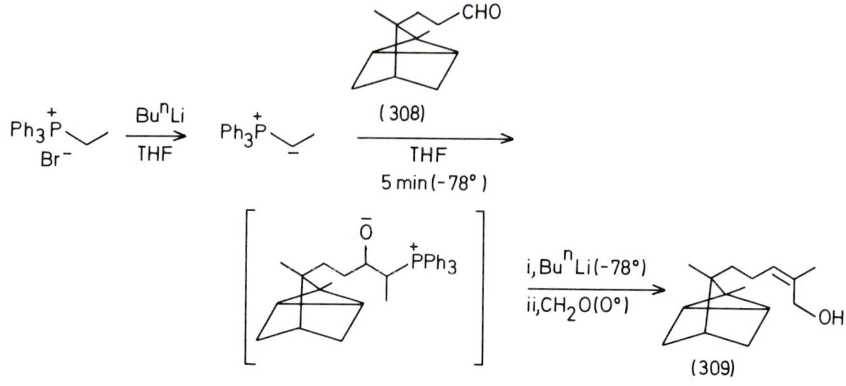

Scheme 155 (Ref 158)

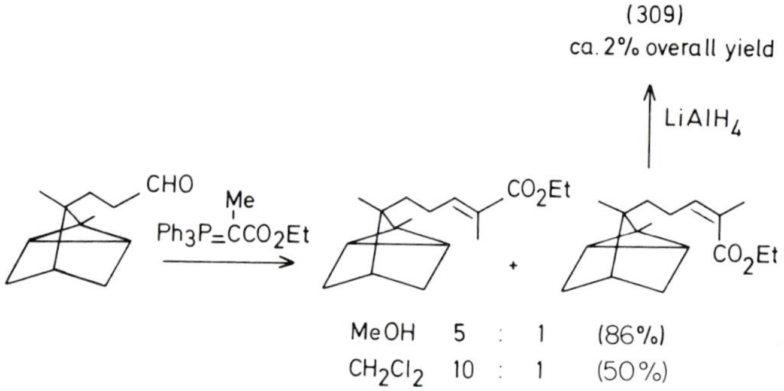

Scheme 156 (Ref 62)

MeOH	5 : 1	(86%)
CH_2Cl_2	10 : 1	(50%)

Finally, it is worth noting that, in the synthesis of acyclic sesquiterpenes using the Wittig reaction, the direction of coupling can have a profound effect on the stereoselectivity of trisubstituted alkene formation. A key step in the synthesis of β-sinensal (313), an ingredient of the essential oil of the Chinese orange, is illustrative (Scheme 157).[341] Thus condensation of the methyl ketone (311) with the ylide derived from the phosphonium salt (310) gave a predominance of the wrong isomer (312b), whereas the alternative condensation between (314) and (315) under the same conditions furnished a mixture of stereoisomers containing 60% of the desired isomer (316a).

4.7. Steroids

The Schlosser modification of the Wittig reaction, whereby (E)-alkenes are obtained from β-oxido-ylides by stereospecific protonation, has been used with notable success to prepare key intermediates required for the synthesis

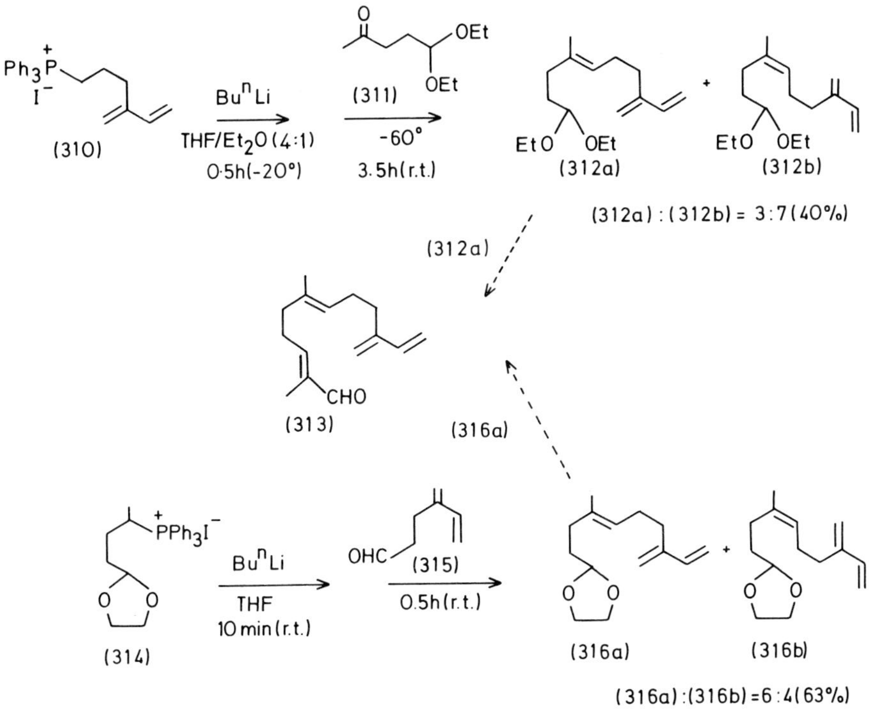

2 Wittig Reaction

Scheme 157 (Ref 341)

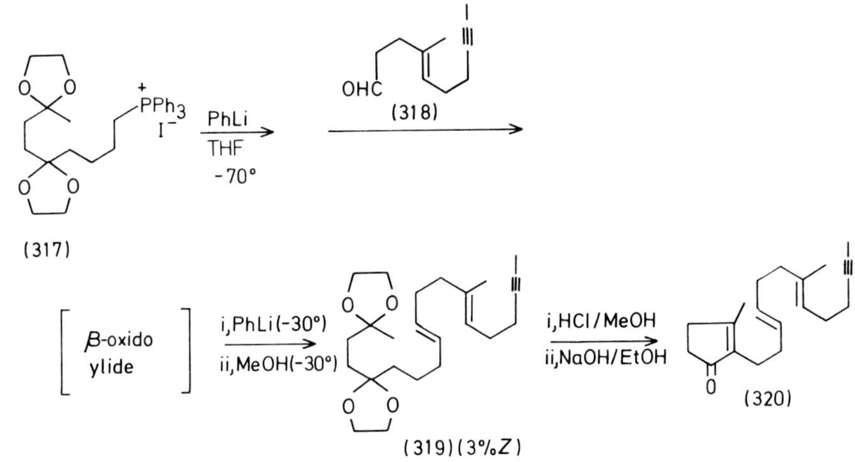

Scheme 158 (Ref 343)

of steroids via biomimetic polyene cyclizations. The purity of such compounds is of critical importance since the corresponding (*Z*)-isomers fail to give tetracyclic material.[342] An example is the preparation of the (±)-progesterone precursor (320) which involved, as the key (convergent) step, condensation of the phosphonium salt (317) with the aldehyde (318) under Schlosser conditions to give the ketal (319) containing only 3% of the (*Z*)-isomer.[343]

Other cyclization substrates that have been prepared in this way include (321), [344] (322),[345] (323),[346] and (324).[347] In order to achieve high *trans*-selectivity in the case of (324) some modification of Schlosser's original procedure was necessary. The reader is recommended to consult this reference for the details.

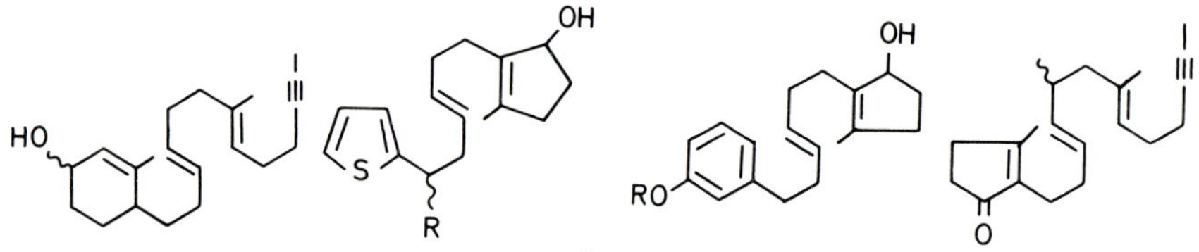

(321) (322); R = Me or Bu^t (323); R=Me or H (324)

Scheme 159 (Ref 344-347)

The use of a reaction sequence involving the coupling of an allylic bromide with an allylidenephosphorane followed by reduction of the resulting phosphonium salt also provides a useful route to a number of other precursors for biogenetic-type syntheses.[348] For example, coupling of the

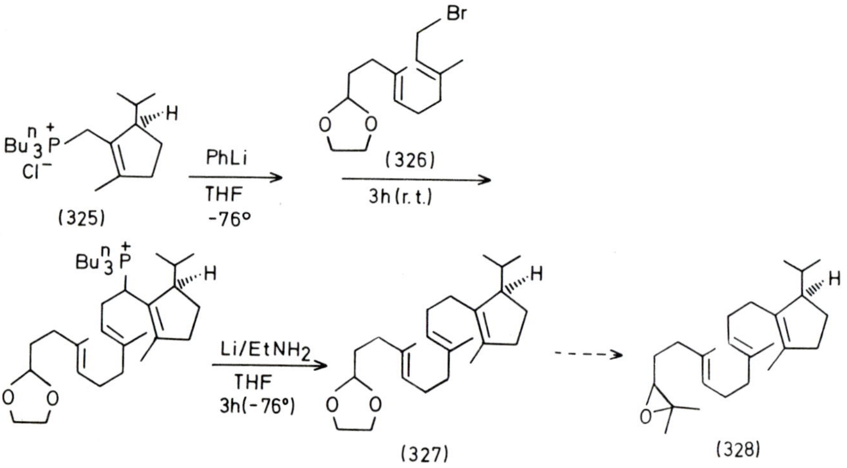

Scheme 160 (Ref 349)

2 Wittig Reaction

ylide from the phosphonium chloride (325) with farnesyl bromide trisnor-acetal (326) gave, on reduction with lithium–ethylamine, the monocyclic acetal (327) which was convertible into the epoxide (328), the cyclization substrate for the isoeuphenol system.[349] This coupling procedure, which serves as a general method for the preparation of 1,5-dienes,[350] has been applied in the synthesis of all-*trans*-squalene as shown in Scheme 161.[351]

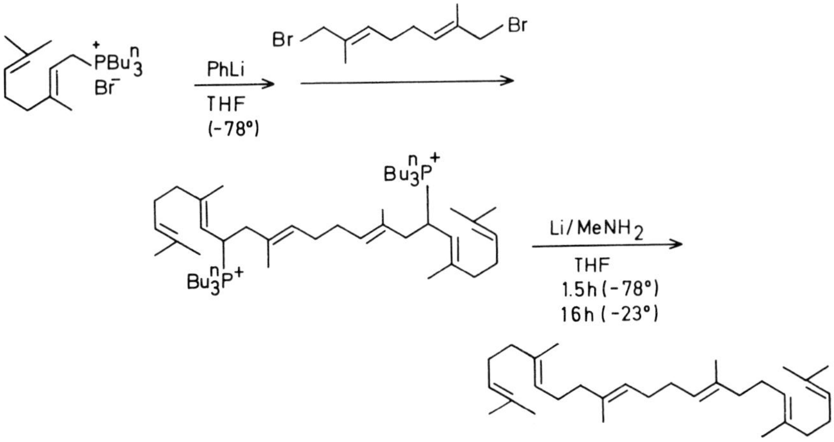

Scheme 161 (Ref 351)

Increasing attention is being given to the use of stereoselective Wittig reactions for the elaboration of steroidal side-chains. The method first acquired practical importance in the stereospecific conversion of 17-oxo-steroids into 20α-hydroxy-steroids by a reaction sequence involving (Z)-alkene formation followed by hydroboration.[352,353] In this way, for example, 5α-pregnane-3β,20α-diol (331) has been prepared from isoandrosterone (329) (or its acetate[354]), where a Wittig reaction with ethylidenetriphenylphosphorane in DMSO gave the (Z)-$\Delta^{17(20)}$-alkene (330) in a highly stereoselective manner (Scheme 162).[352]

Surprisingly, the product from the analogous reaction between the cyclopentanone (332) and the ethylidene Wittig reagent is the cyclohexane derivative (333) and not the expected ethylidenecyclopentane.[355] The mechanism of this novel rearrangement is by no means clear (Scheme 163).

The synthesis of 20-methylcholesterol (337) which started from the *i*-steroid aldehyde (334) also involved a stereoselective Wittig reaction (Scheme 164).[356] Thus, methylation of (334) at C-20 via its carbanion, followed by a Wittig condensation with isopentyltriphenylphosphonium bromide (335) in DMSO using dimsylsodium as base, gave the (Z)-sterol (336) as the sole intermediate from which 20-methylcholesterol was obtained. Under similar

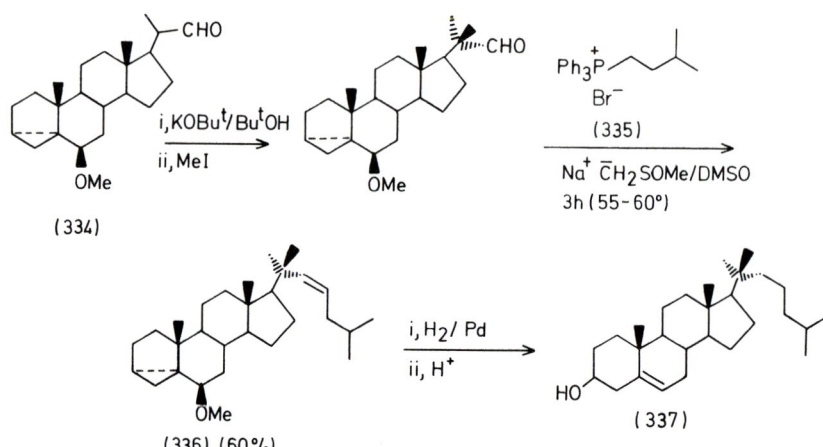

Scheme 162 (Ref 352)

Scheme 163 (Ref 355)

Scheme 164 (Ref 356)

2 Wittig Reaction

conditions, related transformations have produced the (Z)-isomer predominantly, if not exclusively.[357,358] The aldehyde (334) also served as the starting material for Wittig synthesis of a series of (Z)- and (E)-Δ^{22}-sterols with modified side-chains, ranging from the C_{23} to the C_{27} series, and including various patterns of chain-branching.[359]

Whilst, in general, the phosphonate-olefin formation method is preferred,[24,25] the Wittig reaction has on occasion been employed for the introduction of *trans* unsaturation into steroidal side-chains. For example, in a partial synthesis of the yeast sterol, ergosta-7,22,24(28)-trien-3β-ol, a conventional Wittig reaction between the pure (20S)-aldehyde (339) and the 2-isopropyl-allylidenephosphorane (338) gave an 85% yield of the required sterol acetate (340) stereospecifically.[360] That the (E)-isomer is formed exclusively is somewhat surprising, since the Wittig reaction might be expected to give a

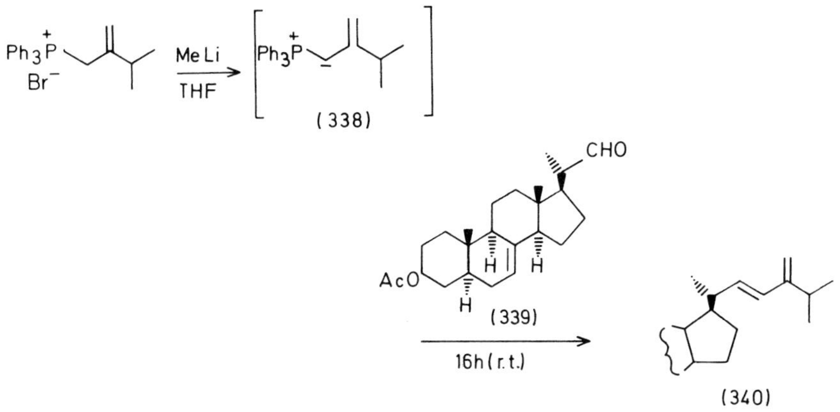

Scheme 165 (Ref 360)

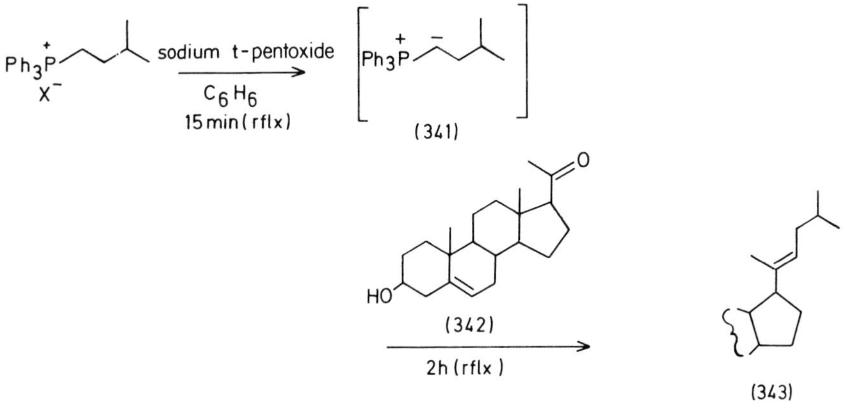

Scheme 166 (Ref 364)

mixture of (Z)- and (E)-isomers under these conditions (see Section 3.4.2), as has been observed in similar syntheses.[361-363]

Another example is the conventional Wittig reaction of pregnenolone (342) with various alkylidenephosphoranes to give a range of $\Delta^{20(22)}$-alkenes containing only the (E)-isomers.[364] Thus the isopentylidenephosphorane (341) gave cholesta-5,20(22)-dien-3β-ol (343), the acetate of which yielded cholesterol acetate on selective hydrogenation.[364] Interestingly, prolongation of the Wittig reaction (48 h) resulted in the predominant formation of 3β-phenoxy derivatives; a control reaction showed that the triphenylphosphonium salts were responsible for the O-phenylation.

5. References

1. G. Wittig and G. Geissler, *Annalen*, 1953, **580**, 44.
2. G. Wittig and U. Schöllkopf, *Chem. Ber.*, 1954, **87**, 1318.
3. G. Wittig and W. Haag, *Chem. Ber.*, 1955, **88**, 1654.
4. G. Wittig, *Angew. Chem.*, 1956, **68**, 505.
5. G. Wittig, *Experientia*, 1956, **12**, 41.
6. G. Wittig, *Festschr. Arthur Stoll*, 1957, 48 (*Chem. Abs.*, 1958, **52**, 15413).
7. F. Bohlmann and H.-J. Mannhardt, *Chem. Ber.*, 1955, **88**, 1330.
8. A. Mondon, *Annalen*, 1957, **603**, 115.
9. S. Trippett, *Chem. and Ind.*, 1956, 80.
10. P. C. Wailes, *Chem. and Ind.*, 1958, 1086.
11. F. Bohlmann, E. Inhoffen, and P. Herbst, *Chem. Ber.*, 1957, **90**, 1661.
12. E. Truscheit and K. Eiter, *Annalen*, 1962, **658**, 65.
13. L. D. Bergelson and M. M. Shemyakin, *Pure Appl. Chem.*, 1964, **9**, 271.
14. M. Schlosser, in *Topics in Stereochemistry* (ed. E. L. Ellel and N. L. Allinger), Vol. 5, p. 1, Wiley-Interscience, New York, 1970.
15. (a) S. Trippett, *Quart. Rev.*, 1963, **17**, 406. (b) S. Trippett, *Pure Appl. Chem.*, 1964, **9**, 255. (c) A. Maercker, *Org. Reactions*, 1965, **14**, 270. (d) A. W. Johnson, *Ylide Chemistry*, Academic Press, New York, 1966. (e) H. O. House, *Modern Synthetic Reactions*, 2nd Ed., Ch. 10, Benjamin, New York, 1972. (f) J. Mathieu and J. Weill-Raynal, in *Introduction of a Carbon Chain or an Aromatic Ring*, Vol. 2, p. 608, G. Thieme, Stuttgart, 1975.
16. H. Pommer, *Angew. Chem.*, 1960, **72**, 811, 911; H. Freyschlag, H. Grassner, A. Nürrenbach, H. Pommer, W. Reif, and W. Sarnecki, *Angew. Chem. Internat. Ed.*, 1965, **4**, 287.
17. Yu. A. Zhdanov, Yu. E. Alexeev, and V. G. Alexeeva, in *Advances in Carbohydrate Chemistry and Biochemistry*, Vol. 27, p. 227, Academic Press, New York, 1972.
18. H. Pommer, *Angew. Chem. Internat. Ed.*, 1977, **16**, 423.
19. K. P. C. Vollhardt, *Synthesis*, 1975, 765.
20. K. B. Becker, *Helv. Chim. Acta*, 1977, **60**, 68.
21. P. Beck, in *Organic Phosphorus Compounds* (ed. G. M. Kosolapoff and L. Maier), Vol. 2, Ch. 4, John Wiley and Sons, 1972; H. J. Bestmann, *Angew. Chem. Internat. Ed.*, 1965, **4**, 583.

22. F. Ramirez and S. Levy, *J. Amer. Chem. Soc.*, 1957, **79**, 6167.
23. A. W. Johnson and R. B. LaCount, *Tetrahedron*, 1960, **9**, 130.
24. J. Boutagy and R. Thomas, *Chem. Rev.*, 1974, **74**, 87.
25. W. S. Wadsworth, Jr., *Org. Reactions*, 1977, **25**, 73.
26. W. S. Wadsworth, Jr., and W. D. Emmons, *J. Amer. Chem. Soc.*, 1961, **83**, 1733; *Org. Synth.*, 1965. **45**, 44.
27. G. Fodor and I. Tomoskozi, *Tetrahedron Lett.*, 1961, 579.
28. H. J. Bestmann and R. Zimmermann, in *Organic Phosphorus Compounds* (ed. G. Kosolapoff and L. Maier), Vol. 3, Ch. 5A, John Wiley and Sons, 1972.
29. G. H. Birum and C. N. Matthews, *J. Chem. Soc., Chem. Comm.*, 1967, 137.
30. E. Vedejs and K. A. J. Snoble, *J. Amer. Chem. Soc.*, 1973, **95**, 5778.
31. M. Schlosser, A. Piskala, C. Tarchini, and H. B. Tuong, *Chimia* (Switz.), 1975, **29**, 341.
32. M. Schlosser and K. F. Christmann, *Angew. Chem. Internat. Ed.*, 1965, **4**, 689.
33. M. Schlosser and K. F. Christmann, *Annalen*, 1967, **708**, 1.
34. S. Fliszar, R. F. Hudson, and G. Salvadori, *Helv. Chim. Acta*, 1963, **46**, 1580.
35. A. J. Speziale and K. W. Ratts, *J. Amer. Chem. Soc.*, 1963, **85**, 2790.
36. L. D. Bergelson, V. A. Vaver, L. I. Barsukov, and M. M. Shemyakin, *Izv. Akad. Nauk SSSR, Otd. Khim. Nauk*, 1963, 1053 (*Chem. Abs.*, 1963, **59**, 8783).
37. G. P. Schiemenz and H. Engelhard, *Chem. Ber.*, 1961, **94**, 578.
38. A. I. Meyers and R. C. Strickland, *J. Org. Chem.*, 1972, **37**, 2579.
39. G. Koszmehl and B. Bohn, *Chem. Ber.*, 1974, **107**, 710.
40. A. Roedig, G. Märkl, and H. Schaller, *Chem. Ber.*, 1970, **103**, 1022.
41. S. Masamune, C. U. Khim, K. E. Wilson, G. O. Spessard, P. E. Georghiou and G. S. Bates, *J. Amer. Chem. Soc.*, 1975, **97**, 3512.
42. G. Pattenden, J. E. Way, and B. C. L. Weedon, *J. Chem. Soc.* (*C*), 1970, 235.
43. V. Subramanyam, E. H. Silver, and A. H. Soloway, *J. Org. Chem.*, 1976, **41**, 1272.
44. R. Ketcham, D. Jambotkar, and L. Martinelli, *J. Org. Chem.*, 1962, **27**, 4666.
45. J. Buddrus, *Angew. Chem. Internat. Ed.*, 1968, **7**, 536.
46. J. Buddrus, *Angew. Chem. Internat. Ed.*, 1972, **11**, 1041.
47. J. Buddrus, *Chem. Ber.*, 1974, **107**, 2050.
48. K. Mackenzie, in *The Chemistry of Alkenes* (ed. S. Patai), Ch. 7, Interscience, 1964; Vol. 2, Ch. 3, 1970.
49. A. T. Blomquist and V. J. Hruby, *J. Amer. Chem. Soc.*, 1964, **86**, 5041.
50. A. J. Speziale and D. E. Bissing, *J. Amer. Chem. Soc.*, 1963, **85**, 3878.
51. D. E. Bissing, *J. Org. Chem.*, 1965, **30**, 1296.
52. H. J. Bestmann and O. Kratzer, *Chem. Ber.*, 1962, **95**, 1894.
53. G. Wittig and M. Rieber, *Annalen*, 1949, **562**, 177.
54. S. Trippett and D. M. Walker, *Chem. and Ind.*, 1960, 933.
55. S. Trippett and D. M. Walker, *J. Chem. Soc.*, 1961, 1266.
56. G. Wittig, H.-D. Weigmann, and M. Schlosser, *Chem. Ber.*, 1961, **94**, 676.
57. M. Elliot, N. F. Janes, and D. A. Pulman, *J. Chem. Soc., Perkin I*, 1974, 2470.
58. H. O. House and G. H. Rasmusson, *J. Org. Chem.*, 1961, **26**, 4278.
59. G. Pattenden and B. C. L. Weedon, *J. Chem. Soc.* (*C*), 1968, 1997.

60. L. D. Bergelson, E. V. Dyatlovitskaya, and M. M. Shemyakin, *Izv. Akad. Nauk SSSR, Ser. Khim.*, 1964, 2003 (*Chems. Abs.*, 1965, **62**, 8998).
61. L. D. Bergelson and M. M. Shemyakin, *Tetrahedron*, 1963, **19**, 149.
62. R. G. Lewis, D. Gustafson, and W. F. Erman, *Tetrahedron Lett.*, 1967, 401.
63. H. Kretschmar and W. F. Erman, *Tetrahedron Lett.*, 1970, 41.
64. Y. Badar, W. J. S. Lockley, T. P. Toube, B. C. L. Weedon, and L. R. Guy Valadon, *J. Chem. Soc., Perkin I*, 1973, 1416.
65. H. Plieninger, E. Meyer, F. Sharif-Nassirian, and E. Weidmann, *Annalen*, 1976, 1475.
66. C. Rüchardt, S. Eichler, and P. Panse, *Angew. Chem. Internat. Ed.*, 1963, **2**, 619.
67. C. Rüchardt, P. Panse, and S. Eichler, *Chem. Ber.*, 1967, **100**, 1144.
68. A. K. Bose, M. S. Manhas, and R. M. Ramer, *J. Chem. Soc. (C)*, 1969, 2728.
69. F. Bohlmann and H.-J. Bax, *Chem. Ber.*, 1974, **107**, 1773.
70. S. Fliszar, R. F. Hudson, and G. Salvadori, *Helv. Chim. Acta*, 1964, **47**, 159.
71. H. T. Openshaw and N. Whittaker, *Proc. Chem. Soc.*, 1961, 454.
72. D. L. Roberts, R. A. Heckman, B. P. Hege, and S. A. Bellin, *J. Org. Chem.*, 1968, **33**, 3566.
73. J.-C. Bonnafous and M. Mousseron-Canet, *Bull. Soc. Chim. France*, 1971, 4551.
74. T. Oritani and K. Yamashita, *Tetrahedron Lett.*, 1972, 2521.
75. L. Horner, H. Hoffmann, and H. G. Wippel, *Chem. Ber.*, 1958, **91**, 61.
76. S. Sugasawa and H. Matsuo, *Chem. Pharm. Bull.* (Japan), 1960, **8**, 819 (*Chem. Abs.*, 1961, **55**, 20901).
77. S. Trippett and D. M. Walker, *Chem. and Ind.*, 1961, 990.
78. L. A. Yanovskaya and V. F. Kucherov, *Izv. Akad. Nauk SSSR, Ser. Khim.*, 1964, 1341 (*Chem. Abs.*, 1964, **61**, 11887).
79. M. Kirilov and G. Petrov, *Monatsh. Chem.*, 1972, 1651.
80. A. G. Andrewes and S. Liaaen-Jensen, *Acta Chem. Scand.*, 1973, **27**, 1401.
81. A. K. Bose and R. T. Dahill, Jr., *J. Org. Chem.*, 1965, **30**, 505.
82. C. Szantay, L. Toke, and P. Kolonits, *J. Org. Chem.*, 1966, **31**, 1447.
83. H. T. Openshaw and N. Whittaker, *J. Chem. Soc.*, 1963, 1461.
84. D. H. Wadsworth, O. E. Schupp, III, E. J. Seus, and J. A. Ford, Jr., *J. Org. Chem.*, 1965, **30**, 680.
85. H. O. House, V. K. Jones, and G. A. Frank, *J. Org. Chem.*, 1964, **29**, 3327.
86. K. Sasaki, *Bull. Chem. Soc. Japan*, 1968, **41**, 1252, and references cited therein.
87. D. E. McGreer and N. W. K. Chiu, *Canad. J. Chem.*, 1968, **46**, 2225.
88. T. H. Kinstle and B. Y. Mandanas, *J. Chem. Soc., Chem. Comm.*, 1968, 1699.
89. L. A. Yanovskaya and V. F. Kucherov, *Izv. Akad. Nauk SSSR, Ser. Khim.*, 1965, 1504 (*Chem. Abs.*, 1965, **63**, 16387).
90. K. H. Dahm, B. M. Trost, and H. Röller, *J. Amer. Chem. Soc.*, 1967, **89**, 5292.
91. G. W. K. Cavill and P. J. Williams, *Austral. J. Chem.*, 1969, **22**, 1737.
92. G. W. K. Cavill, D. G. Laing, and P. J. Williams, *Austral. J. Chem.*, 1969, **22**, 2145.
93. K. Mori, T. Mitsui, J. Fukami, and T. Ohtaki, *Agric. Biol. Chem.*, 1971, **35**, 1116.
94. A. G. Andrewes and S. Liaaen-Jensen, *Acta Chem. Scand.*, 1973, **27**, 1401.
95. G. Jones and R. F. Maisey, *J. Chem. Soc., Chem. Comm.*, 1968, 543.

96. B. Deschamps, G. Lefebvre, and J. Seyden-Penne, *Tetrahedron*, 1972, **28**, 4209.
97. T. Bottin-Strzalko, *Tetrahedron*, 1973, **29**, 4199.
98. L. D. Bergelson, V. A. Vaver, and M. M. Shemyakin, *Izv. Akad. Nauk SSSR, Otd. Khim. Nauk*, 1961, 729 (*Chem. Abs.*, 1961, **55**, 22196).
99. L. D. Bergelson, V. A. Vaver, L. I. Barsukov, and M. M. Shemyakin, *Dokl. Akad. Nauk SSSR*, 1962, **143**, 111 (*Chem. Abs.*, 1962, **57**, 7298).
100. C. F. Hauser, T. W. Brooks, M. L. Miles, M. A. Raymond, and G. B. Butler, *J. Org. Chem.*, 1963, **28**, 372.
101. J. P. Dusza, *J. Org. Chem.*, 1960, **25**, 93.
102. G. Wittig and D. Wittenberg, *Annalen*, 1957, **606**, 1.
103. F. Näf, R. Decorzant, W. Thommen, B. Willhalm, and G. Ohloff, *Helv. Chim. Acta*, 1975, **58**, 1016.
104. G. Ohloff and M. Pawlak, *Helv. Chim. Acta*, 1973, **56**, 1176.
105. E. J. Corey and E. Hamanaka, *J. Amer. Chem. Soc.*, 1967, **89**, 2758.
106. F. Bohlmann, E. Inhoffen, and P. Herbst, *Chem. Ber.*, 1957, **90**, 1661.
107. A. R. Hands and A. J. H. Mercer, *J. Chem. Soc.* (*C*), 1968, 2448.
108. A. I. Meyers and E. W. Collington, *Tetrahedron*, 1971, **27**, 5979.
109. D. J. Burton, F. E. Herkes, and K. J. Klabunde, *J. Amer. Chem. Soc.*, 1966, **88**, 5042.
110. L. D. Bergelson, L. I. Barsukov, and M. M. Shemyakin, *Tetrahedron*, 1967, **23**, 2709, and references cited therein.
111. T. Moroe, M. Indo, T. Matsui, and K. Tsuchiya, Brit. Pat. 1,202,343 (*Chem. Abs.*, 1970, **73**, 120490).
112. R. Rangoonwala and G. Seitz, *Deut. Apoth. Ztg.*, 1970, **110**, 1946 (*Chem. Abs.*, 1971, **74**, 99541).
113. R. Greenwald, M. Chaykovsky, and E. J. Corey, *J. Org. Chem.*, 1963, **28**, 1128.
114. E. J. Corey and G. T. Kwiatkowski, *J. Amer. Chem. Soc.*, 1966, **88**, 5653.
115. M. I. Dawson and M. Vasser, *J. Org. Chem.*, 1977, **42**, 2783; see also C.-L. Yeh and M. Dawson, *Tetrahedron Lett.*, 1977, 4257.
116. A. S. Kovaleva, V. M. Bulina, L. L. Ivanov, Yu. B. Pyatnova, and R. P. Evstigneeva, *Zh. Org. Khim.*, 1974, **10**, 696 (*Chem. Abs.*, 1974, **81**, 37206).
117. H. J. Bestmann and W. Stransky, *Synthesis*, 1974, 798.
118. H. J. Bestmann and O. Vostrowsky, *Tetrahedron Lett.*, 1974, 207.
119. P. E. Sonnet, *Org. Prep. Proced. Int.*, 1974, **6**, 269.
120. P. E. Sonnet, *J. Org. Chem.*, 1974, **39**, 3793.
121. E. A. Evans, *J. Chem. Soc.*, 1956, 4691.
122. E. A. Evans, *Chem. and Ind.*, 1957, 1596.
123. J. O. Currie, Jr., R. A. LaBar, R. D. Breazeale, and A. G. Anderson, Jr., *Annalen*, 1973, 166.
124. N. Petragnani and G. Schill, *Chem. Ber.*, 1964, **97**, 3293.
125. P. E. Sonnet, *J. Org. Chem.*, 1969, **34**, 1147.
126. M. Schlosser, G. Müller, and K. F. Christmann, *Angew. Chem. Internat. Ed.*, 1966, **5**, 667.
127. G. Wittig, H. Eggers, and P. Duffner, *Annalen*, 1958, **619**, 10.
128. H. J. Bestmann, *Angew. Chem. Internat. Ed.*, 1965, **4**, 583.
129. R. Köster, D. Simic, and M. A. Grassberger, *Annalen*, 1970, **739**, 211.
130. R. J. Anderson and C. A. Henrick, *J. Amer. Chem. Soc.*, 1975, **97**, 4327.

131. H. J. Bestmann, W. Stransky, and O. Vostrowsky, *Chem. Ber.*, 1976, **109**, 1694.
132. M. Fétizon and J. Schalbar, *Fr. Ses Parfums*, 1969, **12**, 330 (*Chem. Abs.*, 1970, **72**, 42886).
133. R. M. Boden, *Synthesis*, 1975, 784.
134. M. Schlosser, *Bull. Soc. Chim. France*, 1971, 453.
135. B. G. James and G. Pattenden, *J. Chem. Soc., Perkin I*, 1976, 1476.
136. W. Bondinell, C. D. Snyder, and H. Rapoport, *J. Amer. Chem. Soc.*, 1969, **91**, 6889.
137. M. Baumann, W. Hoffmann, and H. Pommer, *Annalen*, 1976, 1626.
138. S. Isoe, Y. Hayase, and T. Sakan, *Tetrahedron Lett.*, 1971, 3691.
139. J. J. Plattner, U. T. Bhalerao, and H. Rapoport, *J. Amer. Chem. Soc.*, 1969, **91**, 4933; U. T. Bhalerao, J. J. Plattner, and H. Rapoport, *ibid.*, 1970, **92**, 3429.
140. K. Sato, S. Inoue, and S. Ota, *J. Org. Chem.*, 1970, **35**, 565.
141. J. A. Findlay and W. D. Mackay, *J. Chem. Soc., Chem. Comm.*, 1969, 733.
142. S. F. Brady, M. A. Ilton, and W. S. Johnson, *J. Amer. Chem. Soc.*, 1968, **90**, 2882.
143. A. W. Burgstahler, L. O. Weigel, W. J. Bell, and M. K. Rust, *J. Org. Chem.*, 1975, **40**, 3456.
144. R. G. Salomon and N. El Sanadi, *J. Amer. Chem. Soc.*, 1975, **97**, 6214.
145. W. P. Schneider, *J. Chem. Soc., Chem. Comm.*, 1969, 785.
146. U. Axen, F. H. Lincoln, and J. L. Thompson, *J. Chem. Soc., Chem. Comm.*, 1969, 303.
147. G. Just, C. Simonovitch, F. H. Lincoln, W. P. Schneider, U. Axen, G. B. Spero, and J. E. Pike, *J. Amer. Chem. Soc.*, 1969, **91**, 5364.
148. G. Heublein, unpublished work quoted in ref. 14.
149. E. E. Schweizer, J. G. Thompson, and T. A. Ulrich, *J. Org. Chem.*, 1968, **33**, 3082.
150. T. Teraji, I. Moritani, E. Tsuda, and S. Nishida, *J. Chem. Soc. (C)*, 1971, 3252.
151. M. Schlosser and K. F. Christmann, *Angew. Chem. Int. Ed.*, 1966, **5**, 126.
152. M. Schlosser, in *Methodicum Chimicum* (ed. F. Korte, K. Niedenzu, and H. Zimmer), Vol. 7, p. 551, Thieme, Stuttgart/Academic Press, N.Y., 1976.
153. M. Schlosser, K. F. Christmann, and A. Piskala, *Chem. Ber.*, 1970, **103**, 2814.
154. W. Heitz and R. Michels, *Annalen*, 1973, 227.
155. M. Schlosser and K. F. Christmann, *Synthesis*, 1969, 38.
156. E. J. Corey, J. I. Shulman, and H. Yamamoto, *Tetrahedron Lett.*, 1970, 447.
157. M. Schlosser, K. F. Christmann, A. Piskala, and D. Coffinet, *Synthesis*, 1971, 29.
158. E. J. Corey and H. Yamamoto, *J. Amer. Chem. Soc.*, 1970, **92**, 226.
159. E. J. Corey, P. Ulrich, and A. Venkateswarlu, *Tetrahedron Lett.*, 1977, 3231.
160. M. Schlosser and D. Coffinet, *Synthesis*, 1972, 575.
161. M. Schlosser and D. Coffinet, *Synthesis*, 1971, 380.
162. M. Schlosser, D. Coffinet, and H. B. Tuong, unpublished work in ref. 160.
163. O. H. Wheeler and H. N. Batlle de Pabon, *J. Org. Chem.*, 1965, **30**, 1473.
164. G. Märkl and A. Merz, *Synthesis*, 1973, 295.
165. W. Tagaki, I. Inoue, Y. Yano, and T. Okonogi, *Tetrahedron Lett.*, 1974, 2587.
166. H. Blaschke, C. E. Ramey, I. Calder, and V. Boekelheide, *J. Amer. Chem. Soc.*, 1970, **92**, 3675.

167. G. Drefahl, D. Lorenz, and G. Schnitt, *J. Prakt. Chem.*, 1964, **23**, 143.
168. J. W. Van Reijendam, G. J. Heeres, and M. J. Janssen, *Tetrahedron*, 1970, **26**, 1291.
169. J. D. Brewer and J. A. Elix, *Austral. J. Chem.*, 1972, **25**, 545, and references cited therein.
170. M. Schlosser and M. Zimmermann, *Synthesis*, 1969, 75; *Chem. Ber.*, 1971, **104**, 2885.
171. K. Eiter and H. Oediger, *Annalen*, 1965, **682**, 62.
172. L. Crombie, P. Hemesley, and G. Pattenden, *J. Chem. Soc. (C)*, 1969, 1024.
173. C. Descoins and C. A. Henrick, unpublished results quoted in C. A. Henrick, *Tetrahedron*, 1977, **33**, 1845.
174. G. Goto, T. Shima, H. Masuya, Y. Masuoka, and K. Hiraga, *Chem. Lett.*, 1975, 103.
175. R. Hug, H.-J. Hansen and H. Schmid, *Helv. Chim. Acta*, 1972, **55**, 1828; see also E. E. Schweizer, E. T. Shaffer, C. T. Hughes, and C. J. Berninger, *J. Org. Chem.*, 1966, **31**, 2907.
176. J. E. Johansen and S. Liaaen-Jensen, *Acta Chem. Scand.*, 1974, **B28**, 301.
177. N. Khan, D. E. Loeber, T. P. Toube, and B. C. L. Weedon, *J. Chem. Soc., Perkin I*, 1975, 1457.
178. I. T. Harrison and B. Lythgoe, *J. Chem. Soc.*, 1958, 843.
179. R. K. Howe, *J. Amer. Chem. Soc.*, 1971, **93**, 3457.
180. G. Kresze, J. Firl, and H. Braun, *Tetrahedron*, 1969, **25**, 4481.
181. G. Pattenden and B. C. L. Weedon, *J. Chem. Soc. (C)*, 1968, 1984.
182. G. Pattenden and B. C. L. Weedon, *J. Chem. Soc. (C)*, 1968, 1997.
183. E. J. Corey and B. W. Erickson, *J. Org. Chem.*, 1974, **39**, 821.
184. G. Büchi and H. Wüest, *Helv. Chim. Acta*, 1971, **54**, 1767.
185. A. Padwa and L. Brodsky, *J. Org. Chem.*, 1974, **39**, 1318.
186. F. Bohlmann and C. Zdero, *Chem. Ber.*, 1973, **106**, 3779.
187. W. G. Dauben and J. Ipaktschi, *J. Amer. Chem. Soc.*, 1973, **95**, 5088.
188. W. G. Dauben and A. P. Kozikowski, *Tetrahedron Lett.*, 1973, 3711.
189. E. Vedejs, J. P. Bershas, and P. L. Fuchs, *J. Org. Chem.*, 1973, **38**, 3625.
190. W. G. Dauben, D. J. Hart, J. Ipaktschi, and A. P. Kozikowski, *Tetrahedron Lett.*, 1973, 4425.
191. H. H. Inhoffen, K. Brückner, G. F. Domagk, and H.-M. Erdmann, *Chem. Ber.*, 1955, **88**, 1415.
192. E. Vedejs and J. P. Bershas, *Tetrahedron Lett.*, 1975, 1359.
193. A. W. Johnson and V. L. Kyllingstad, *J. Org. Chem.*, 1966, **31**, 334.
194. S. V. McKinley and J. W. Rakshys, Jr., *J. Chem. Soc., Chem. Comm.*, 1972, 134.
195. F. Camps, J. Castells, and F. Vela, *Anales de Quin*, 1974, **70**, 374 (*Chem. Abs.*, 1974, **81**, 63015).
196. T. M. Cresp, R. G. F. Giles, M. V. Sargent, C. Brown, and D. O'N. Smith, *J. Chem. Soc., Perkin I*, 1974, 2435.
197. T. Bottin-Strzalko, J. Seyden-Penne, and B. Tchoubar, *Compt. Rend. (C)*, 1971, **272**, 778.
198. M. E. Jones and S. Trippett, *J. Chem. Soc. (C)*, 1966, 1090.
199. B. G. James, G. Pattenden, and L. Barlow, *J. Chem. Soc., Perkin I*, 1976, 1466.
200. C. F. Garbers, D. F. Schneider, and J. P. van der Merwe, *J. Chem. Soc. (C)*, 1968, 1982.
201. R. S. Tewari and K. C. Gupta, *Indian J. Chem.*, 1976, **14B**, 419.

202. E. J. Corey and R. A. Ruden, *Tetrahedron Lett.*, 1973, 1495.
203. N. N. Belyaev, M. D. Stadnichuk, A. A. Petrov, and A. N. Belyaev, *J. Gen. Chem. (U.S.S.R.)*, 1972, **42**, 710 (*Chem. Abs.*, 1972, **77**, 75252).
204. M. Ahmed, G. C. Barley, M. T. W. Hearn, E. R. H. Jones, V. Thaller, and J. A. Yates, *J. Chem. Soc., Perkin I*, 1974, 1981.
205. A. G. Fallis, M. T. W. Hearn, E. R. H. Jones, V. Thaller, and J. L. Turner, *J. Chem. Soc., Perkin I*, 1973, 743.
206. R. K. Bentley, C. A. Higham, J. K. Jenkins, E. R. H. Jones and V. Thaller, *J. Chem. Soc., Perkin I*, 1974, 1987.
207. F. Bohlmann and P. Herbst, *Chem. Ber.*, 1959, **92**, 1319.
208. H. Pommer and G. Wittig, Ger. Pat. 951,212 (1956) (*Chem. Abs.*, 1959, **53**, 437).
209. H. Pommer, *Angew. Chem.*, 1960, **72**, 811.
210. H. Pommer and W. Sarnecki, Ger. Pat. 1,068,710 (1959) (*Chem. Abs.*, 1961, **55**, 12446).
211. H. Pommer and W. Sarnecki, Ger. Pat. 1,059,900 (1959) (*Chem. Abs.*, 1961, **55**, 14511).
212. G. Manecke and M. Härtel, *Chem. Ber.*, 1973, **106**, 655.
213. H. E. Zimmermann, P. Baeckstrom, T. Johnson, and D. W. Kurtz, *J. Amer. Chem. Soc.*, 1974, **96**, 1459.
214. E. J. Corey and G. T. Kwiatkowski, *J. Amer. Chem. Soc.*, 1968, **90**, 6816.
215. J. Reucroft and P. G. Sammes, *Quart. Rev.*, 1971, 135.
216. D. J. Faulkner, *Synthesis*, 1971, 175.
217. L. D. Bergelson and M. M. Shemyakin, *Angew. Chem. Internat. Ed.*, 1964, **3**, 250; also in *Newer Methods of Preparative Organic Chemistry*, Vol. 5, p. 154, 1968.
218. L. D. Bergelson and M. M. Shemyakin, in *The Chemistry of Carboxylic Acids and Esters*, p. 295, Interscience, London, 1969.
219. W. J. Gensler, *Chem. Rev.*, 1957, **57**, 191.
220. L. D. Bergelson, V. A. Vaver, V. Yu. Kovtun, L. B. Senyavina, and M. M. Shemyakin, *Zh. Obshch. Khim.*, 1962, **32**, 1802 (*Chem. Abs.*, 1963, **58**, 4415).
221. G. C. Barley, E. R. H. Jones, V. Thaller, and R. A. Vere Hodge, *J. Chem. Soc., Perkin I*, 1973, 151.
222. L. Pichat, J. P. Guermont, and J. C. Levron, *Bull. Soc. Chim. France*, 1969, 1198.
223. L. Pichat, J. C. Levron, and J. P. Guermont, *Bull. Soc. Chim. France*, 1969, 1200.
224. H. O. House and H. Babad, *J. Org. Chem.*, 1963, **28**, 90.
225. L. D. Bergelson, V. A. Vaver, L. I. Barsukov, and M. M. Shemyakin, *Izv. Akad. Nauk SSSR, Otd. Khim. Nauk*, 1963, 1134 (*Chem. Abs.*, 1963, **59**, 8607).
226. L. D. Bergelson, V. A. Vaver, L. I. Barsukov, and M. M. Shemyakin, *Izv. Akad. Nauk SSSR, Ser. Khim.*, 1963, 1417 (*Chem. Abs.*, 1963, **59**, 15176).
227. L. D. Bergelson, V. A. Vaver, A. A. Bezzubov, and M. M. Shemyakin, *Zh. Obshch. Khim.*, 1962, **32**, 1807 (*Chem. Abs.*, 1963, **58**, 4416).
228. H. S. Corey, Jr., J. R. D. McCormick, and W. E. Swensen, *J. Amer. Chem. Soc.*, 1964, **86**, 1884.
229. L. D. Bergelson, V. A. Vaver, and M. M. Shemyakin, *Izv. Akad. Nauk SSSR, Otd. Khim. Nauk*, 1962, 1894 (*Chem. Abs.*, 1963, **58**, 7825).
230. W. H. Kunau, *Chem. Phys. Lipids*, 1973, **11**, 254.

231. F. D. Gunstone, *Chem. and Ind.*, 1976, 243, and references cited therein.
232. L. D. Bergelson, A. A. Bezzubov, V. A. Vaver, and M. M. Shemyakin, *Izv. Akad. Nauk SSSR, Ser. Khim.*, 1968, 2558 (*Chem. Abs.*, 1969, **70**, 57954).
233. H. J. Bestmann, O. Vostrowsky, and A. Plenchette, *Tetrahedron Lett.*, 1974, 779.
234. J. S. Reed and P. S. Beevor, *J. Stored Prod. Res.*, 1976, **12**, 55.
235. L. D. Bergelson, V. A. Vaver, A. A. Bezzubov, and M. M. Shemyakin, *Izv. Akad. Nauk SSSR, Ser. Khim.*, 1964, 1453 (*Chem. Abs.*, 1966, **64**, 14086).
236. L. D. Bergelson, V. D. Solodovnik, and M. M. Shemyakin, *Izv. Akad. Nauk SSSR Otd. Khim. Nauk*, 1962, 1315 (*Chem. Abs.*, 1963, **58**, 7824).
237. L. D. Bergelson, V. D. Solodovnik, and M. M. Shemyakin, *Izv. Akad. Nauk SSSR, Ser. Khim.*, 1967, 843 (*Chem. Abs.*, 1967, **67**, 99692).
238. L. D. Bergelson, E. V. Dyatlovitskaya, and M. M. Shemyakin, *Izv. Akad. Nauk SSSR, Otd. Khim. Nauk*, 1963, 506 (*Chem. Abs.*, 1963, **59**, 3766).
239. L. D. Bergelson, E. V. Dyatlovitskaya, and M. M. Shemyakin, *Izv. Akad. Nauk SSSR, Otd. Khim. Nauk*, 1963, 388 (*Chem. Abs.*, 1963, **58**, 13783h).
240. G. Ohloff, C. Vial, F. Näf, and M. Pawlak, *Helv. Chim. Acta*, 1977, **60**, 1161.
241. R. W. Bradshaw, A. C. Day, E. R. H. Jones, C. B. Page, V. Thaller, and R. A. Vere Hodge, *J. Chem. Soc.* (C), 1971, 1156.
242. I. W. Farrell, V. Thaller, and J. L. Turner, *J. Chem. Soc., Perkin I*, 1977, 1886.
243. M. T. W. Hearn, E. R. H. Jones, M. G. Pellatt, V. Thaller, and J. L. Turner, *J. Chem. Soc., Perkin I*, 1973, 2785.
244. F. Bohlmann and W. Skuballa, *Chem. Ber.*, 1970, **103**, 1886.
245. E. J. Corey, *Ann. New York Acad. Sci.*, 1971, **180**, 24.
246. E. J. Corey, T. K. Schaaf, W. Huber, U. Koelliker, and N. M. Weinshenker, *J. Amer. Chem. Soc.*, 1970, **92**, 397.
247. J. S. Bindra and R. Bindra, *Prostaglandin Synthesis*, Academic Press, New York, 1977; A. Mitra, *The Synthesis of Prostaglandins*, Wiley, 1977.
248. P. H. Bentley, *Chem. Soc. Rev.*, 1973, **2**, 29.
249. For a recent review on the synthesis of thromboxanes see: K. C. Nicolaou, P. Crabbé, A. Guzman, and M. Vera, *Tetrahedron Lett.*, 1973, 3021.
250. M. Brawner Floyd, *Synth. Commun.*, 1974, **4**, 317.
251. E. J. Corey and G. Moinet, *J. Amer. Chem. Soc.*, 1973, **95**, 6831.
252. E. J. Corey and G. Moinet, *J. Amer. Chem. Soc.*, 1973, **95**, 7185; see also P. Crabbé, A. Guzman, and M. Vera, *Tetrahedron Lett.*, 1973, 3021.
253. W. Bartmann, G. Beck, and U. Lerch, *Tetrahedron Lett.*, 1974, 2441.
254. C.-L. J. Wang, P. A. Grieco, and F. J. Okuniewicz, *J. Chem. Soc., Chem. Comm.*, 1976, 468, 939.
255. C. Gandolfi, R. Pellegata, E. Dradi, A. Forgione, and E. Pella, *Farmaco. Ed. Sc.*, 1976, **31**, 763 (*Chem. Abs.*, 1977, **86**, 55017).
256. J. Fried, M. S. Lee, B. Gaede, J. C. Sih, Y. Yoshikawa, and J. A. McCracken, in *Advances in Prostaglandin and Thromboxane Research* (ed. B. Samuelsson and R. Paoletti), Vol. 1, p. 173, Raven, New York, 1976.
257. H. Miyake, S. Iguchi, S. Kori, and M. Hayashi, *Chem. Lett.*, 1976, 211.
258. G. J. Lourens and J. M. Koekemoer, Ger. Offen. 2,618,861 (*Chem. Abs.*, 1977, **86**, 89589); see also *Tetrahedron Lett.*, 1975, 3719.
259. Hoechst A. G., Neth. Appl. 7512,794 (*Chem. Abs.*, 1977, **86**, 89596).
260. A. Sugie, H. Shimomura, J. Katsube, and H. Yamamoto, *Tetrahedron Lett.*, 1977, 2759.
261. K. Kojima and K. Sakai, *Tetrahedron Lett.*, 1976, 101.

262. E. J. Corey, K. Narasaka, and M. Shibasaki, *J. Amer. Chem. Soc.*, 1976, **98**, 6417.
263. N. Nakamura and K. Sakai, *Tetrahedron Lett.*, 1976, 2049.
264. M. Hayashi, S. Kori, and H. Miyake, Japan Kokai 75 49,259 (*Chem. Abs.*, 1975, **83**, 205824).
265. P. A. Grieco, C. S. Pogonowski, and M. Miyashita, *J. Chem. Soc., Chem. Comm.*, 1975, 592.
266. M. J. Dimsdale, R. F. Newton, D. K. Rainey, C. F. Webb, T. V. Lee, and S. M. Roberts, *J. Chem. Soc., Chem., Comm.*, 1977, 716.
267. A. E. Greene, J. P. Deprès, M. C. Meana, and P. Crabbé, *Tetrahedron Lett.*, 1976, 3755.
268. E. J. Corey, M. Shibasaki, K. C. Nicolaou, C. Malmsten, and B. Samuelsson, *Tetrahedron Lett.*, 1976, 737.
269. N. A. Nelson and R. W. Jackson, *Tetrahedron Lett.*, 1976, 3275.
270. E. J. Corey, M. Shibasaki, J. Knolle, and T. Sugahara, *Tetrahedron Lett.*, 1977, 785; see also E. J. Corey, M. Shibasaki, and J. Knolle, *ibid.*, 1977, 1625.
271. S. Hanessian and P. Lavallee, *Canad. J. Chem.*, 1977, **55**, 562.
272. M. Hayashi and H. Miyake, Japan Kokai 74 116,068 (*Chem. Abs.*, 1975, **82**, 170200).
273. P. Bollinger, Ger. Offen. 2,431,930 (*Chem. Abs.*, 1975, **83**, 58248).
274. H. J. E. Hess, L. J. Czuba, and T. K. Schaaf, U.S. Pat. 3,928,391 (*Chem. Abs.*, 1976, **84**, 121292).
275. Y. Iguchi, S. Kori, and M. Hayashi, *J. Org. Chem.*, 1975, **40**, 521; M. Hayashi, S. Kori, and Y. Iguchi, Japan Kokai 75 35,133 (*Chem. Abs.*, 1976, **84**, 43430).
276. M. Hayashi, S. Kori, H. Wakatsuka, M. Kawamura, and Y. Konishi, Japan Kokai 75 137,961 (*Chem. Abs.*, 1976, **84**, 164263).
277. G. Ambrus, I. Barta, G. Cseh, P. Tolney, and C. Mehesfahvi, Hung. Teljes 11,745 (*Chem. Abs.*, 1977, **86**, 16352).
278. M. Miyano, C. R. Dorn, F. B. Colton, and W. J. Marsheck, *J. Chem. Soc., Chem. Comm.*, 1971, 425; see also M. Miyano and M. A. Stealy, *J. Org. Chem.*, 1975, **40**, 1748, and references cited therein.
279. J. Himizu, S. Saijo, K. Noguchi, M. Wada, Y. Harigaya, and O. Takaichi, Japan Kokai 76 01461 (*Chem. Abs.*, 1976, **85**, 123751).
280. N. Finch and J. J. Fitt, *Tetrahedron Lett.*, 1969, 4639; N. Finch, L. D. Vecchia, J. J. Fitt, R. Stephani, and I. Vlattas, *J. Org. Chem.*, 1973, **38**, 4412.
281. E. J. Corey, H. Shirahama, H. Yamamoto, S. Terashima, A. Venkateswarlu, and T. K. Schaaf, *J. Amer. Chem. Soc.*, 1971, **93**, 1490.
282. See C. A. Henrick, *Tetrahedron*, 1977, **33**, 1845, for pertinent references.
283. A. Butenandt, R. Beckmann, D. Stamm, and E. Hecker, *Z. Naturforsch.*, 1959, **14b**, 283.
284. A. Butenandt and E. Hecker, *Angew. Chem.*, 1961, **73**, 349.
285. A. Butenandt, E. Hecker, M. Hopp, and W. Koch, *Annalen*, 1962, **658**, 39.
286. F. D. Gunstone and M. L. K. Jie, *Chem. Phys. Lipids*, 1970, **4**, 1; S. G. Morris, S. F. Herb, P. Magidman, and F. E. Luddy, *J. Amer. Oil Chem. Soc.*, 1972, **49**, 92; L. Garanti, A. Marchesini, U. M. Pagnoni, and R. Trave, *Gazz. Chim. Ital.*, 1976, **106**, 187.
287. H. J. Bestmann, O. Vostrowsky, H. Paulus, W. Billmann, and W. Stransky, *Tetrahedron Lett.*, 1977, 121.
288. H. J. Bestmann, I. Kantardjiew, P. Rösel, W. Stransky, and O. Vostrowsky. *Chem. Ber.*, 1978, **111**, 248.

289. B. A. Bierl, M. Beroza, and C. W. Collier, *Science*, 1970, **170**, 87; B. A. Bierl, M. Beroza, and C. W. Collier, *J. Econ. Entomol.*, 1972, **65**, 659.
290. H. J. Bestmann, O. Vostrowsky, and W. Stransky, *Chem. Ber.*, 1976, **109**, 3375.
291. H. J. Bestmann, K. H. Koschatzky, W. Stransky, and O. Vostrowsky, *Tetrahedron Lett.*, 1976, 353.
292. B. A. Bierl, M. Beroza, R. T. Staten, P. E. Sonnet, and V. E. Adler, *J. Econ. Entomol.*, 1974, **67**, 211.
293. R. J. Anderson and C. A. Henrick, *J. Amer. Chem. Soc.*, 1975, **97**, 4327; U.S. Pat. 3,919,329; 3,953,532; 3,989,729.
294. L. Jaenicke and W. Boland, *Annalen*, 1976, 1135.
295. H. J. Bestmann, P. Range, and R. Kunstmann, *Chem. Ber.*, 1971, **104**, 65.
296. H. J. Bestmann, W. Stransky, O. Vostrowsky, and P. Range, *Chem. Ber.*, 1975, **108**, 3582.
297. H. J. Bestmann, O. Vostrowsky, and H. Platz, *Chem.-Ztg.*, 1974, **98**, 161.
298. D. A. Carlson, M. S. Mayer, D. L. Silhacek, J. D. James, M. Beroza, and B. A. Bierl, *Science*, 1971, **174**, 76; D. A. Carlson, R. E. Doolittle, M. Beroza, W. M. Rogoff, and G. H. Gretz, *J. Agric. Food Chem.*, 1974, **22**, 194.
299. E. C. Uebel, P. E. Sonnet, R. W. Miller, and M. Beroza, *J. Chem. Ecol.*, 1975, **1**, 195; P. E. Sonnet, E. C. Uebel, and R. W. Miller, *Environ. Entomol.*, 1975, **4**, 761.
300. D. R. Hall, P. S. Beevor, R. Lester, R. G. Poppi, and B. F. Nesbitt, *Chem. and Ind.*, 1975, 216.
301. B. F. Nesbitt, P. S. Beevor, R. A. Cole, R. Lester, and R. G. Poppi, *Nature New Biol.*, 1973, **244**, 208.
302. M. Jacobson, *Insect Sex Pheromones*, Academic Press, New York, 1972.
303. C. C. Leznoff, T. M. Fyles, and J. Weatherston, *Canad. J. Chem.*, 1977, **55**, 1143.
304. T. M. Fyles, C. C. Leznoff, and J. Weatherston, *Canad. J. Chem.*, 1978, **56**, 1031.
305. H. Mayer and O. Isler, in *Carotenoids* (ed. O. Isler), Ch. 6, Birkhauser, Basel, 1971.
306. B. C. L. Weedon, *Pure Appl. Chem.*, 1976, **47**, 161.
307. (a) H. Pommer and W. Sarnecki, Ger. Pat. 1,068,709 (1959) (*Chem. Abs.*, 1961, **55**, 13472). (b) W. Sarnecki, A. Nürrenbach, and W. Reid, Ger. Pat. 1,158,505 (1963) (*Chem. Abs.*, 1964, **60**, 5570). (c) Badische Anilin und Soda-Fabrik, Neth. Appl. 6,909,081 (1969); Derwent Farmdoc. No. 41,597 (1970). (d) H. Pommer, *Angew. Chem.*, 1960, **72**, 911.
308. (a) G. Wittig and H. Pommer, Ger. Pat. 954,247 (1956) (*Chem. Abs.*, 1959, **53**, 2279). (b) H. Pommer and W. Sarnecki, Ger. Pat. 1,068,705 (1959) (*Chem. Abs.*, 1962, **56**, 1487). (c) H. Pommer and W. Sarnecki, Ger. Pat. 1,068,703 (1959) (*Chem. Abs.*, 1961, **55**, 13473).
309. J. D. Surmatis and A. Ofner, *J. Org. Chem.*, 1961, **26**, 1171.
310. (a) G. Wittig and H. Pommer, Ger. Pat. 971,986 (1959); Brit. Pat. 813,539 (1959) (*Chem. Abs.*, 1960, **54**, 15320). (b) H. Pommer and W. Sarnecki, Ger. Pat. 1,068,704 (1959) (*Chem. Abs.*, 1961, **55**, 13473).
311. (a) H. Pommer and W. Sarnecki, U.S. Pat. 3,006,939 (1961) (*Chem. Abs.*, 1962, **56**, 12960). (b) F. Hoffmann-La Roche and Co. Ltd., Brit. Pat. 787,497 (1957) (*Chem. Abs.*, 1958, **52**, 10189).
312. H. Pommer and W. Sarnecki, Ger. Pat. 1,050,763 (1959) (*Chem. Abs.*, 1961, **55**, 5572).

313. J.-L. Olivé, M. Mousseron-Canet, and J. Dornand, *Bull. Soc. Chim. France*, 1969. 3247.
314. H. Pommer and W. Sarnecki, Ger. Pat. 1,068,706 (1959) (*Chem. Abs.*, 1962, **56**, 512).
315. B. C. L. Weedon, in *Carotenoids* (ed. O. Isler), Ch. 5, Birkhauser, Basel, 1971.
316. F. Kienzle, *Pure Appl. Chem.*, 1976, **47**, 183.
317. O. Isler, L. H. Chopard-dit-Jean, M. Montavon, R. Rüegg, and P. Zeller, *Helv. Chim. Acta*, 1957, **40**, 1256; F. Hoffman-La Roche and Co. Ltd., Brit. Pat. 812,267 (1959) (*Chem. Abs.*, 1959, **53**, 18982).
318. J. D. Surmatis, U.S. Pat. 3,367,985 (1968) (*Chem. Abs.*, 1968, **68**, 114786).
319. O. Isler, H. Lindlar, M. Montavon, R. Rüegg, and P. Zeller, *Helv. Chim. Acta*, 1956, **39**, 249.
320. B. C. L. Weedon, *Rev. Pure Appl. Chem.* (Australia), 1970, **20**, 51.
321. L. Barlow and G. Pattenden, *J. Chem. Soc., Perkin I*, 1976, 1029.
322. L. Crombie and S. H. Harper, *J. Chem. Soc.*, 1950, 1152.
323. L. Crombie, S. H. Harper, and F. C. Newman, *J. Chem. Soc.*, 1956, 3963.
324. L. Crombie, P. Hemesley, and G. Pattenden, *J. Chem. Soc.* (*C*), 1969, 1016.
325. G. Pattenden and R. Storer, *J. Chem. Soc., Perkin I*, 1974, 1603.
326. A. J. Birch, K. S. Keogh, and V. R. Mamdapur, *Austral. J. Chem.*, 1973, **26**, 2671.
327. P. A. Grieco, *J. Org. Chem.*, 1972, **37**, 2363.
328. P. Bakuzis and M. L. F. Bakuzis, *J. Org. Chem.*, 1977, **42**, 2362.
329. N. S. Zefirov, P. V. Kostetskii, and Yu. K. Yur'ev, *Zh. Obshch. Khim.*, 1964, **34**, 1069 (*Chem. Abs.*, 1964, **61**, 581).
330. L. Crombie, C. F. Doherty, and G. Pattenden, *J. Chem. Soc.* (*C*), 1970, 1076.
331. K. Ueda and M. Matsui, *Agric. Biol. Chem.*, 1970, **34**, 1119.
332. E. J. Corey, K. Achiwa, and J. A. Katzenellenbogen, *J. Amer. Chem. Soc.*, 1969, **91**, 4318.
333. M. J. Devos, L. Hevesi, P. Bayet, and A. Krief, *Tetrahedron Lett.*, 1976, 3911.
334. A. Krief and H. Laszlo, Belg. Pat. 827,651 (*Chem. Abs.*, 1976, **85**, 177686).
335. M. Baumann and W. Hoffmann, *Synthesis*, 1977, 681; Ger. Pat. (DOS) 253 4859.
336. H. Schulz and I. Sprung, *Angew. Chem. Internat. Ed.*, 1969, **8**, 271.
337. B. H. Braun, M. Jacobson, M. Schwarz, P. E. Sonnet, N. Wakabayashi, and R. M. Waters, *J. Econom. Entomol.*, 1968, **61**, 866.
338. E. J. Corey, H. Yamamoto, D. K. Herron, and K. Achiwa, *J. Amer. Chem. Soc.*, 1970, **92**, 6635.
339. E. J. Corey and H. Yamamoto, *J. Amer. Chem. Soc.*, 1970, **92**, 6636.
340. E. J. Corey and H. Yamamoto, *J. Amer. Chem. Soc.*, 1970, **92**, 6637.
341. E. Bertele and P. Schudel, *Helv. Chim. Acta*, 1967, **50**, 2445.
342. A. Corvers, J. H. Van Mil, M. M. E. Sap, and H. M. Buck, *Rec. Trav. Chim. Pays-Bas*, 1977, **96**, 18.
343. W. S. Johnson, M. B. Gravestock, and B. E. McCarry, *J. Amer. Chem. Soc.*, 1971, **93**, 4332.
344. R. L. Markezich, W. E. Willy, B. E. McCarry, and W. S. Johnson, *J. Amer. Chem. Soc.*, 1973, **95**, 4414.
345. A. A. Macco, R. J. de Brouwer, and H. M. Buck, *J Org. Chem.*, 1977, **42**, 3196.
346. P. A. Bartlett and W. S. Johnson, *J. Amer. Chem. Soc.*, 1973, **95**, 7501.

347. W. S. Johnson and G. E. DuBois, *J. Amer. Chem. Soc.*, 1976, **98**, 1038.
348. E. E. Van Tamelen, *Accounts Chem. Res.*, 1975, **8**, 152, and references therein.
349. E. E. Van Tamelen, G. M. Milne, M. I. Suffness, M. C. Rudler Chauvin, R. J. Anderson, and R. S. Achini, *J. Amer. Chem. Soc.*, 1970, **92**, 7202.
350. E. Axelrod, G. Milne, and E. E. Van Tamelen, *J. Amer. Chem. Soc.*, 1970, **92**, 2139.
351. U. T. Bhalerao and H. Rapoport, *J. Amer. Chem. Soc.*, 1971, **93**, 5311.
352. A. M. Krubiner and E. P. Oliveto, *J. Org. Chem.*, 1966, **31**, 24.
353. A. M. Krubiner, N. Gottfried, and E. P. Oliveto, *J. Org. Chem.*, 1968, **33**, 1715.
354. G. Drefahl, K. Ponsold, and H. Schick, *Chem. Ber.*, 1965, **98**, 604.
355. E. G. Brain, F. Cassidy, A. W. Lake, P. J. Cox, and G. A. Sim, *J. Chem. Soc., Chem. Comm.*, 1972, 497.
356. Y. Letourneux, G. Büjüktür, M. T. Ryzlak, A. K. Banerjee, and M. Gut, *J. Org. Chem.*, 1976, **41**, 2288.
357. S. G. Wyllie and C. Djerassi, *J. Org. Chem.*, 1968, **33**, 305.
358. R. F. N. Hutchins, M. J. Thompson, and J. A. Svoboda, *Steroids*, 1970, **15**, 113.
359. Y. M. Sheikh and C. Djerassi, *Steroids*, 1975, **26**, 129.
360. D. H. R. Barton, P. J. Davies, U. M. Kempe, J. F. McGarrity, and D. A. Widdowson, *J. Chem. Soc., Perkin I*, 1972, 1231.
361. D. H. R. Barton, T. Shioiri, and D. A. Widdowson, *J. Chem. Soc. (C)*, 1971, 1968.
362. M. Fryberg, A. C. Oehlschlager, and A. M. Unrau, *Tetrahedron*, 1971, **27**, 1261.
363. M. Fryberg, A. C. Oehlschlager, and A. M. Unrau, *J. Chem. Soc., Chem. Comm.*, 1971, 1194.
364. J. P. Schmit, M. Piraux, and J. F. Pilette, *J. Org. Chem.*, 1975, **40**, 1586.
365. L. A. Pinck and H. E. Gilbert, *J. Amer. Chem. Soc.*, 1947, **69**, 723.
366. D. B. Denney and S. T. Ross, *J. Org. Chem.*, 1962, **27**, 998.
367. O. Isler, H. Gutmann, M. Montavon, R. Rüegg, G. Ryser, and P. Zeller, *Helv. Chim. Acta*, 1957, **40**, 1242.
368. M. Akhtar, A. E. Faruk, C. J. Harris, G. P. Moss, S. W. Russell, and B. C. L. Weedon, *J. Chem. Soc., Perkin I*, 1978, 1511; see also ref. 177.
369. G. P. Schiemenz, J. Becke, and J. Stöckigt, *Chem. Ber.*, 1970, **103**, 2077.
370. R. F. Hudson and P. A. Chopard, *J. Org. Chem.*, 1963, **28**, 2446.
371. R. N. McDonald and T. W. Campbell, *Org. Synth.*, 1960, **40**, 36; 1973, Coll. Vol. 5, 499.
372. C. E. Griffin, K. R. Martin, and B. E. Douglas, *J. Org. Chem.*, 1962, **27**, 1627.
373. G. Witschard and C. E. Griffin, *J. Org. Chem.*, 1964, **29**, 2335.
374. G. R. Malone and A. I. Meyers, *J. Org. Chem.*, 1974, **39**, 623.
375. G. Koszmehl and B. Bohn, *Angew. Chem. Internat. Ed.*, 1973, **12**, 237.
376. J. G. Atkinson, M. H. Fisher, D. Horley, A. T. Morse, R. S. Stuart, and E. Synnes, *Canad. J. Chem.*, 1965, **43**, 1614.
377. H. J. Bestmann and E. Kranz, *Chem. Ber.*, 1969, **102**, 1802.
378. G. Wittig and U. Schöllkopf, *Org. Synth.*, 1960, **40**, 66; 1973, Coll. Vol. 5, 751.
379. R. F. Heck, *J. Amer. Chem. Soc.*, 1963, **85**, 3387.
380. R. Raap, C. G. Chin, and R. G. Micetich, *Canad. J. Chem.*, 1971, **49**, 2143.

3
Transformations via Phosphorus-stabilized Anions 2: PO-activated Olefinations

BRIAN J. WALKER

1. Introduction, 155
2. Synthesis of reagents containing the PO group, 159
3. The olefination reaction, 162
4. Mechanism and stereochemistry, 167
5. Applications in synthesis of alkenes, 172
6. Examples of experimental procedures, 190
7. Summary of PO-stabilized anions, 192
8. References, 197

1. Introduction

In the twenty years since the original report by Horner and Wippel[1] of the use of PO-stabilized carbanions in olefin synthesis [equation (1)] the reaction has come to challenge the Wittig reaction as the synthetic method of choice when a specific alkene is required.

$$R_2\overset{O}{\overset{\|}{P}}CH_2C_6H_5 \xrightarrow[2.(C_6H_5)_2CO]{1.NaNH_2/\text{Benzene}} C_6H_5CH=C(C_6H_5)_2 + R_2P\overset{O}{\underset{O^-}{\diagup}} \quad (1)$$

$R = C_6H_5, OEt$

The present chapter covers the literature through December 1977, with a number of references to papers published in 1978. The popularity of PO-activated olefin syntheses has necessitated a strict selection of references (although this has been made easier by the repetition in applications in many cases). In Section 8 brief reference is made to a majority of publications and only representative papers for the synthesis of each olefin type are discussed in any detail.

Surprisingly, no name or combination of names has become generally accepted to describe PO-activated olefin synthesis, and Horner–Wittig,

Horner, Emmons, Wadsworth–Emmons, etc. are all frequently used in the literature. A choice from these is not easy; while Horner[1] clearly published the first example of the reaction, mechanistically it is closely related to the Wittig reaction,[2] and Wadsworth and Emmons[3] soon published an excellent and definitive paper which indicates almost every future application of the reaction. However, Wittig has given his name to the original olefin synthesis involving phosphorus, so some combination of these names seems most appropriate.

The possibility of a PO-activated olefin synthesis can be traced back to 1927 when the elder Arbusov[4] showed that the phosphonate group would stabilize an adjacent carbanion (1), which in turn could be alkylated by a

$$(EtO)_2\overset{O}{\overset{\|}{P}}CH_2CO_2Et \xrightarrow[K]{Na \text{ or}} (EtO)_2\overset{O}{\overset{\|}{P}}\overset{-}{C}H\ CO_2Et \quad M^+$$
(1)

variety of reagents. Horner, Hoffman and Wippel's original observations[1,5,6] of the olefination reaction involved the reaction of phosphonate- (2) or phosphonyl- (3) stabilized carbanions with benzophenone to give the corresponding olefin by elimination of dialkyl phosphate or diarylphosphinate, respectively. It is interesting that the latter reaction is one of the few examples of an *unsubstituted* PO-stabilized carbanion yielding an olefin on reaction with a carbonyl compound.

$$(EtO)_2\overset{O}{\overset{\|}{P}}\overset{-}{C}HC_6H_5 + (C_6H_5)_2CO \longrightarrow (C_6H_5)_2C=CHC_6H_5 + (EtO)_2P\overset{O}{\underset{O}{\diagup}}{}^{-}$$
(2)

$$(C_6H_5)_2\overset{O}{\overset{\|}{P}}\overset{-}{C}H_2 + (C_6H_5)_2CO \longrightarrow (C_6H_5)_2C=CH_2 + (C_6H_5)_2P\overset{O}{\underset{O}{\diagup}}{}^{-}$$
(3)

Patai and Schwartz[7] also observed olefin formation from the ester phosphonate (4) and aldehydes, but failed to appreciate the significance of their results, mainly owing to a competing condensation reaction in some cases to give vinyl phosphonates.

$$(EtO)_2\overset{O}{\overset{\|}{P}}CH_2CO_2Et$$
(4)

Ohta and his coworkers[8] showed that olefination was possible with a variety of phosphonates, and following their detailed investigation[3] of the scope of Horner's original reaction, Wadsworth and Emmons investigated

3 PO-Activated Olefinations

the largely analogous reactions of phosphoramidate anions (5) to prepare imines (from aldehydes) and a variety of heteroallenes.[9] Other modifications include the use of phosphinates (6),[10] phosphonamides (7),[11-14] and thiophosphonates (8),[15] and each is claimed to have specific advantages. Phosphi-

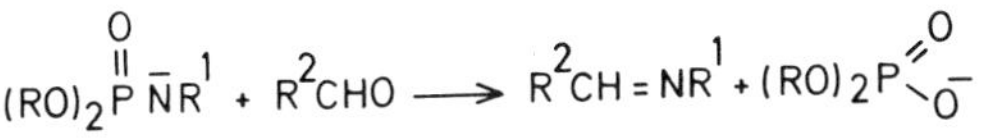

$$(RO)_2\overset{O}{\overset{\|}{P}}\bar{N}R^1 + R^2CHO \longrightarrow R^2CH=NR^1 + (RO)_2P\overset{O}{\underset{O^-}{\diagdown}}$$

(5)

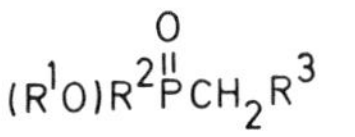

$$(R^1O)R^2\overset{O}{\overset{\|}{P}}CH_2R^3 \qquad (R^1O)_2\overset{S}{\overset{\|}{P}}CH_2R^2 \qquad (R^1_2N)_2\overset{O}{\overset{\|}{P}}CH_2R^2$$

(6) 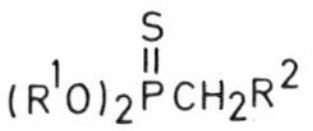   (8) 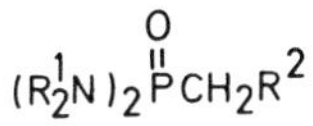   (7)

$$\left[(EtO)_2\overset{O}{\overset{\|}{P}}CH_2\right]_2\overset{O}{\overset{\|}{P}}(OEt) + RCHO \longrightarrow \begin{array}{c} RCH=CH\overset{O}{\overset{\|}{P}}(OEt)_2 \\ O^+ \quad O \\ (EtO)_2\overset{\|}{P}CH_2\overset{\|}{P}(OEt) \\ | \\ O^- \end{array}$$

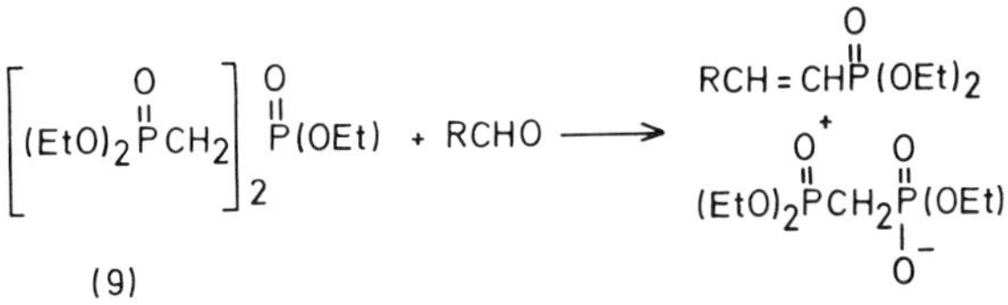

(9)

nates have been little used, although there is some evidence[16] that they can compete successfully with phosphonates (9). The range of PO- and PS-activated reagents used in olefin synthesis is summarized in Scheme 1.

$$R^1R^2\overset{X}{\overset{\|}{P}}-\bar{C}\diagup \qquad\qquad (RO)_2\overset{X}{\overset{\|}{P}}-\bar{N}-$$

$R^1=R^2=$ OAlkyl or OAryl   X = O or S
$R^1=$ OAlkyl, $R^2=$ Alkyl or Aryl
$R^1=R^2=$ Alkyl or Aryl
$R^1=R^2=$ NAlkyl$_2$
X = O or S

Scheme 1

The major advantage of both the Wittig and PO-activated reactions over other olefin syntheses is that the double bond can be introduced regio-specifically in almost every case. However, when compared with the Wittig reaction the latter reaction has a number of other advantages.

(a) PO-stabilized carbanions are much more nucleophilic than the corresponding phosphonium ylides, so they react with a wider range of carbonyl compounds under milder conditions; for example, β-keto- (10) and ester- (11) stabilized ylides require prolonged refluxing to form alkenes, even with reactive carbonyl compounds, whereas the corresponding phosphonate carbanions (12)[3] and (13)[17] react exothermically with both aldehydes and ketones.

$$(C_6H_5)_3 \overset{+}{P} \overset{-}{C}H \, CO \, C_6H_5 \qquad (C_6H_5)_3 \overset{+}{P} \overset{-}{C}H \, CO_2R$$
$$(10) \qquad\qquad\qquad (11)$$

$$(EtO)_2 \overset{O}{\overset{\|}{P}} \overset{-}{C}H \, CO \, C_6H_5 \qquad (EtO)_2 \overset{O}{\overset{\|}{P}} \overset{-}{C}H \, CO_2R$$
$$(12) \qquad\qquad\qquad (13)$$

At first sight the results from competition reactions can be misleading. Horner[18] showed that the reaction of an equimolar mixture of the phosphine oxide (14) and the phosphonium salt (15) with half an equivalent of benzaldehyde and half an equivalent of potassium t-butoxide gave olefin (68%) mainly by the Wittig route; triphenylphosphine oxide (53%) and diphenylphosphinic acid (10%) were formed. However, this is due to the greater acidity of the phosphonium salt and in the presence of excess base the PO-activated reaction predominates.

$$(C_6H_5)_2 \overset{O}{\overset{\|}{P}} CH_2 C_6H_5 \; + \; (C_6H_5)_3 \overset{+}{P} CH_2 C_6H_5 \; Br^-$$
$$(14) \qquad\qquad\qquad\qquad (15)$$
$$\downarrow C_6H_5 CHO, \; KOBu^t$$
$$(C_6H_5)_3 PO \; + \; (C_6H_5)_2 P\!\!\overset{\nearrow O}{\underset{\searrow O^-}{}} \; + \; C_6H_5 CH = CH C_6H_5$$

A further advantage of their greater nucleophilicity is the ease of alkylation and acylation of PO-stabilized carbanions, which offers a convenient route to α-substituted reagents.[19] However, the lesser stabilizing effect of the phosphoryl and phosphonyl groups means that electron-withdrawing α-substituents are generally required at the carbanion centre before preparative yields of olefin can be obtained.

(b) One major problem in the Wittig reaction is separation of the alkene and phosphine oxide products. The phosphinic, phosphonic, and phosphoric acid derivatives obtained from PO-activated syntheses are all soluble in water, so separation from the olefin is easily achieved.

(c) While phosphonium ylides require relatively expensive phosphine starting materials, phosphonates are cheaply and conveniently prepared from trialkyl phosphites by the Michaelis–Arbusov reaction (see p. 159). This does not apply to phosphine oxides, and since these reagents are less reactive towards carbonyl compounds, but more sensitive to air than phosphonate carbanions, they have been less generally used.[20]

(d) As we shall see (p. 168) statements that PO-activated oefin syntheses always lead to *trans* stereochemistry are very much an oversimplification, and although it is true to say that in most cases the *trans* product predominates,

3 PO-Activated Olefinations

recent developments [21,22] suggest that considerable control of stereochemistry is possible.

(e) Although side-reactions do occur (see p. 168) these are generally less frequent than in the corresponding Wittig reaction.[23]

2. Synthesis of Reagents Containing the PO Group

Numerous methods are available for the preparation of compounds containing the P=O bond, and comprehensive descriptions of these are found in the indispensable *Methoden der Organischen Chemie* (Houben-Weyl)[24] and in the more recent series by Kosolapoff and Maier.[25] However, in practice the requirement for cheap, convenient methods from readily available starting materials has restricted these considerably, and in this section only those generally useful methods will be discussed.

2.1. Phosphonates, Phosphinates, and Phosphine Oxides

Representatives of these reagents are available from a number of chemical suppliers.

2.1.1. *The Michaelis–Arbusov reaction*

This reaction has been known for over eighty years[26] and has been extensively reviewed.[27] In its most general form it involves the reaction of a tervalent phosphorus alkyl ester with alkylating or acylating agents to give an intermediate phosphonium salt which rapidly decomposes to the corresponding P=O compound (16). Obviously this reaction could be used to prepare any PO reagent, but in practice its overwhelming use is in the synthesis of phosphonates (17) from trialkyl phosphites since the latter are directly available at low cost from phosphorus trichloride.

$$R^2_2POR^1 + RX \longrightarrow R^2_2\overset{+}{P}R(OR^1)\ X^- \longrightarrow R^2_2\overset{O}{\overset{\|}{P}}R + R^1X \quad (16)$$
$$R^2 = OR, NR_2, Alkyl, Aryl, SR$$

$$PCl_3 + 3R^1OH \xrightarrow{\text{Base}} (R^1O)_3P \xrightarrow{RCH_2X} (R^1O)_2\overset{O}{\overset{\|}{P}}CH_2R \quad (17)$$

The use of the Michaelis–Arbusov reaction to prepare β-keto-phosphonates (18) from α-halogenoketones has the disadvantage that the Perkow reaction,[28,29] to give enol phosphates (19), often competes, and although this can be minimized by varying the reaction conditions, β-keto-phosphonates are best

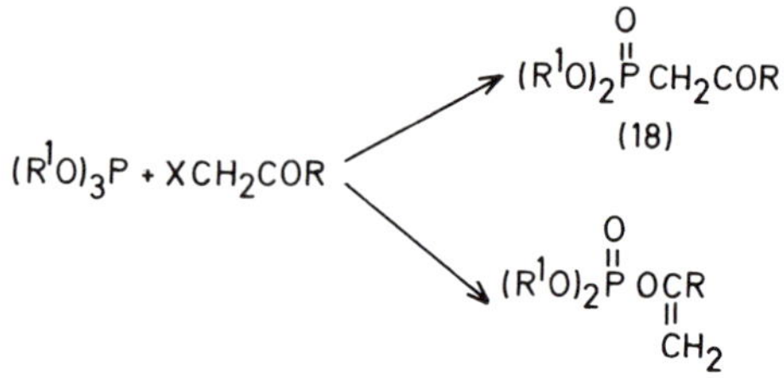

$$(R^1O)_3P + XCH_2COR \begin{array}{c} \nearrow (R^1O)_2\overset{O}{\overset{\|}{P}}CH_2COR \\ (18) \\ \searrow (R^1O)_2\overset{O}{\overset{\|}{P}}O\underset{\|}{\overset{}{C}}R \\ CH_2 \\ (19) \end{array}$$

prepared by acylation of the corresponding alkylphosphonate (see p. 162). Phosphinates (20) are also normally prepared by the Michaelis–Arbusov reaction, but the phosphonite starting material is much less readily available.

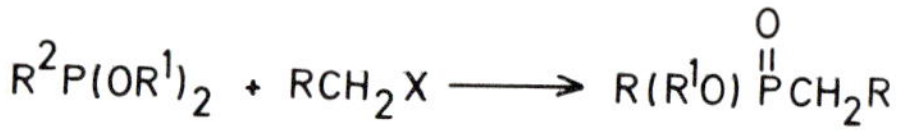

$$R^2P(OR^1)_2 + RCH_2X \longrightarrow R(R^1O)\overset{O}{\overset{\|}{P}}CH_2R$$
$$(20)$$

The Michaelis–Becker reaction,[30] involving alkylation of dialkyl phosphites or their salts [equation (2)], has been less used, although dialkyl phosphites

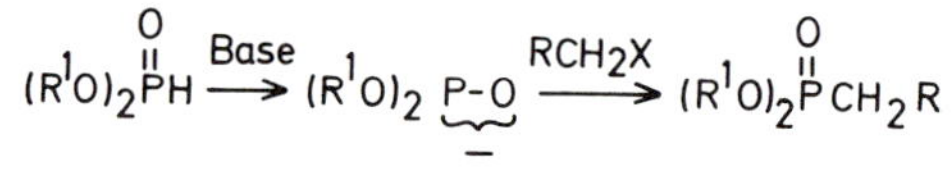

$$(R^1O)_2\overset{O}{\overset{\|}{P}}H \xrightarrow{\text{Base}} (R^1O)_2\underline{P\text{-}O} \xrightarrow{RCH_2X} (R^1O)_2\overset{O}{\overset{\|}{P}}CH_2R \qquad (2)$$

are readily available. Unfortunately this reaction cannot be effectively used for the synthesis of β-keto-phosphonates because, while the Perkow reaction does not occur, the major product is the 1,2-epoxyalkylphosphonate (21).[31]

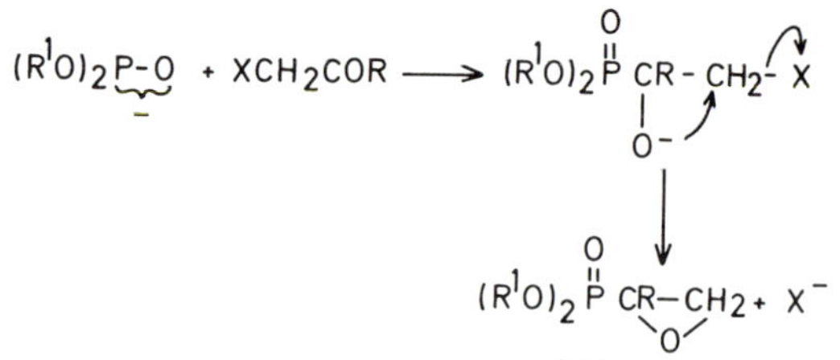

$$(R^1O)_2\underline{P\text{-}O} + XCH_2COR \longrightarrow (R^1O)_2\overset{O}{\overset{\|}{P}}\underset{\underset{O^-}{|}}{CR}\text{-}CH_2\text{-}X$$

$$\downarrow$$

$$(R^1O)_2\overset{O}{\overset{\|}{P}}\underset{\diagdown O \diagup}{CR\text{-}CH_2} + X^-$$
$$(21)$$

2.1.2. Via Grignard reagents

Reaction of the appropriate alkyl Grignard reagent with halogenophosphines or phosphonyl halides provides the most general route to tertiary phosphine oxides (22) suitable for PO olefin synthesis.[24,32,33] The halogenophosphine is usually more readily available (specifically Ph_2PCl) and the phosphine product is conveniently oxidized to the phosphine oxide.

The route could be made somewhat more cost effective by preparing the trialkylphosphine oxide (23) from phosphorus trichloride; however, the

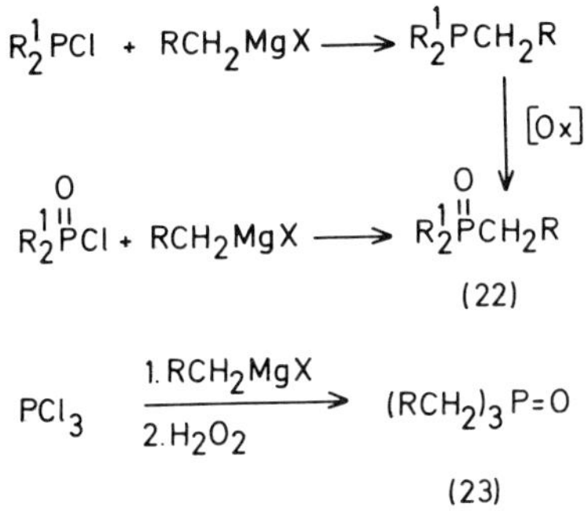

3 PO-Activated Olefinations

$$R^1_2PCl + RCH_2MgX \longrightarrow R^1_2PCH_2R$$
$$\downarrow [Ox]$$
$$R^1_2\overset{O}{\underset{||}{P}}Cl + RCH_2MgX \longrightarrow R^1_2\overset{O}{\underset{||}{P}}CH_2R$$
(22)

$$PCl_3 \xrightarrow[2.\,H_2O_2]{1.\,RCH_2MgX} (RCH_2)_3P=O$$
(23)

effect on olefin synthesis of varying substituents on phosphorus would require further study.

A variety of other, but generally less convenient, routes to phosphine oxides are discussed in references 24, 32, and 33.

2.1.3. Via alkylation and acylation

The formation and nucleophilic properties of phosphonate and phosphine oxide carbanions, and the availability of alkylphosphonates from the Michaelis–Arbusov reaction, offer a useful route to a variety of substituted reagents.[34]

Alkylation can be achieved with reactive and moderately reactive alkylating agents,[35] although in the case of phosphonates use of the diethyl ester (24) is

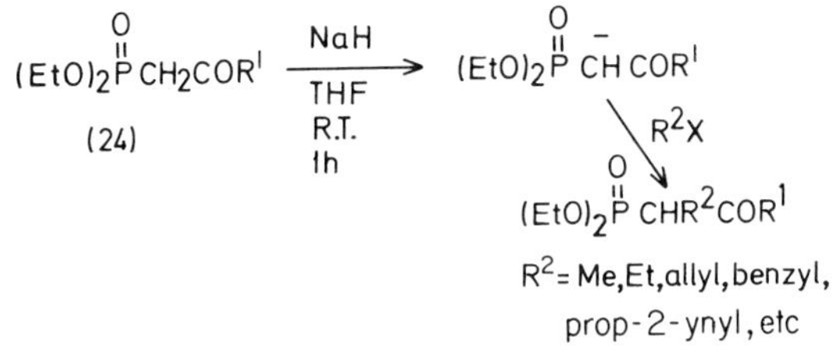

$$(EtO)_2\overset{O}{\underset{||}{P}}CH_2COR^1 \xrightarrow[\substack{THF\\R.T.\\1h}]{NaH} (EtO)_2\overset{O}{\underset{||}{P}}\overset{-}{C}HCOR^1$$
(24)
$$\searrow R^2X$$
$$(EtO)_2\overset{O}{\underset{||}{P}}CHR^2COR^1$$

R^2 = Me, Et, allyl, benzyl, prop-2-ynyl, etc

advisable owing to competing demethylation in the dimethyl case. A general synthesis of 2-oxoalkylphosphonates (useful in prostaglandin synthesis; see p. 179) is available[36] through alkylation of the dianion (25), which predictably takes place entirely at the γ-position. (See also reference 37.)

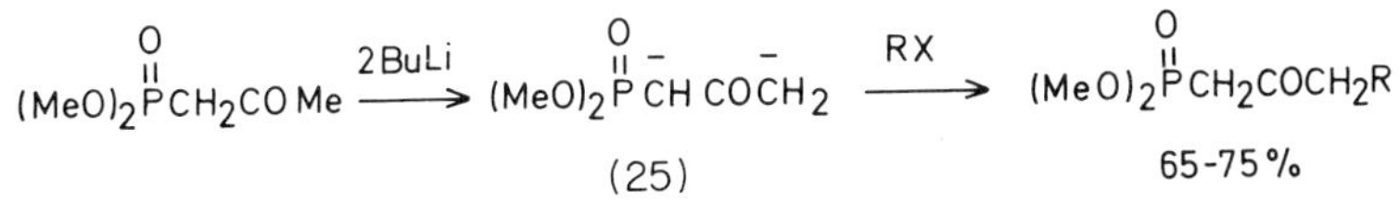

$$(MeO)_2\overset{O}{\underset{||}{P}}CH_2COMe \xrightarrow{2\,BuLi} (MeO)_2\overset{O}{\underset{||}{P}}\overset{-}{C}H\overset{-}{C}OCH_2 \xrightarrow{RX} (MeO)_2\overset{O}{\underset{||}{P}}CH_2COCH_2R$$
(25)
65-75%

Phase-transfer catalysts and two-phase systems have been effectively employed in the alkylation of carbanions of low reactivity, e.g. (26),[38] and to achieve specific monoalkylation[39] in cases, e.g. (27), where mixtures of products are obtained by conventional methods.

$$(EtO)_2\overset{O}{\underset{\|}{P}}\overset{-}{C}HCN \qquad (Me_2N)_2\overset{O}{\underset{\|}{P}}CH_2CN$$

$$(26) \qquad\qquad (27)$$

β-Keto-phosphine oxides (28)[40,41] and β-keto-phosphonates (29)[42] (which are not generally available via the Michaelis–Arbusov reaction; see p. 159) can be prepared in high yield by acylation of the corresponding alkyl carbanion.

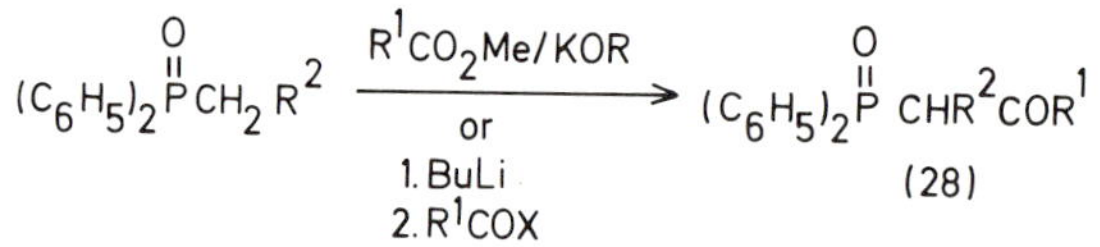

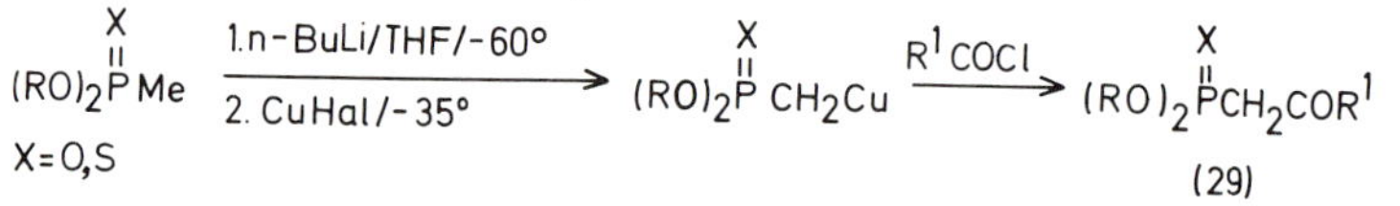

$$(C_6H_5)_2\overset{O}{\underset{\|}{P}}CH_2R^2 \xrightarrow[\substack{\text{or}\\ \text{1. BuLi}\\ \text{2. }R^1COX}]{R^1CO_2Me/KOR} (C_6H_5)_2\overset{O}{\underset{\|}{P}}CHR^2COR^1 \quad (28)$$

$$(RO)_2\overset{X}{\underset{\|}{P}}Me \xrightarrow[\text{2. CuHal}/-35°]{\text{1. n-BuLi/THF}/-60°} (RO)_2\overset{X}{\underset{\|}{P}}CH_2Cu \xrightarrow{R^1COCl} (RO)_2\overset{X}{\underset{\|}{P}}CH_2COR^1$$

X = O, S

$$(29)$$

60 – 100%

Very recently[43] copper(I)-catalysed replacement of α-halogen atoms has been used to prepare α-substituted phosphonates.

2.2. Phosphoramidates

Phosphoramidates (31) can be prepared[9] from dialkyl phosphorochloridates (30) or from the more readily available dialkyl phosphites.

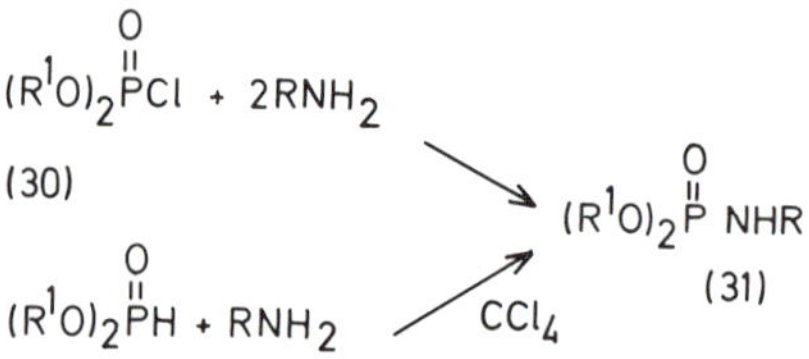

$$(R^1O)_2\overset{O}{\underset{\|}{P}}Cl + 2RNH_2$$
$$(30)$$
$$\searrow$$
$$(R^1O)_2\overset{O}{\underset{\|}{P}}NHR$$
$$(31)$$
$$\nearrow \text{CCl}_4$$
$$(R^1O)_2\overset{O}{\underset{\|}{P}}H + RNH_2$$

3. The Olefination Reaction

3.1. The Carbanion

Although phosphonates, phosphinates, and phosphine oxides have been studied using a wide range of physical methods,[44] similar investigations of

3 PO-Activated Olefinations

the corresponding carbanions have been few.[45,46] ^{31}P n.m.r. chemical shifts of the β-keto-carbanions (32) are around -38 p.p.m. and are largely independent of the cation.[45] However, the situation is much more complex for ester-stabilized carbanions[46] which appear to exist as a cation-dependent mixture of the two forms (33) and (34) in t-butanol–THF.

$$\text{(EtO)}_2\overset{O}{\overset{\|}{P}}\overset{-}{C}H\,COMe\quad M^+$$
(32)

$$(EtO)_2P\underset{C}{\overset{O\cdot\cdot\cdot\overset{M^+}{\cdot\cdot}\cdot\cdot O}{\diagup\quad\diagdown}}\overset{}{\underset{OMe}{\diagdown}}$$
(33)

$$(EtO)_2\,\overset{\delta-}{P}\!-\!\overset{|}{C}HCO_2Me\\ \quad\ \delta\text{-}O\text{----}M^+$$
(34)

Results from various spectroscopic studies suggest[45] that the phosphonate group is about as effective as a carbonyl group for stabilizing an adjacent carbanion and rather less effective than a phosphonium group, and this is supported by their relative reactivities. However, by an appropriate choice of base, phosphonate carbanions with almost any α-substituent can be prepared (a comprehensive list of these is given in Section 7).

This does not mean that all of these carbanions react with carbonyl compounds to give olefins. Phosphonate- or phosphine oxide-stabilized carbanions with hydrogen or alkyl groups as the only α-substituent ("unactivated" carbanions) react rapidly with carbonyl compounds[1,47,48] or with carbon dioxide,[49] but in most cases at best give only poor yields of olefins, presumably because the intermediate alkoxide (35) requires "activating"

$$R^1_2\overset{O}{\overset{\|}{P}}\,\underset{\underset{O^-}{|}}{C}HR^2\\ \qquad\quad \overset{\diagdown}{C}HR^3$$
(35)

substituents before it will decompose under the reaction conditions (see p. 165). The catalytic effect of silver salts in certain cases is interesting in this connection.[47]

Some other α-substituents, e.g. alkoxy,[50,51] are sufficient to allow olefin formation from (35; $R^2 = OR$), but only under vigorous conditions and this has led to contradictory reports. Further confusion is generated by the more rapid decomposition of intermediates (35) derived from phosphine oxides[1,50] than those derived from phosphonates.[47,48,51]

The use of lithium bases leads to rather similar effects since the rate of olefin formation from (35) is cation-dependent. Horner[5] obtained no olefin from benzaldehyde and the phosphine oxide (36) using phenyllithium as the base, but did isolate a high yield of the intermediate (37) after the addition of water. A similar reaction in the presence of potassium t-butoxide, or treatment of (37) with the same base, gave stilbene in high yield.

$$(C_6H_5)_2\overset{O}{\overset{\|}{P}}CH_2C_6H_5 + C_6H_5\,CHO \xrightarrow[\text{2. }H_2O]{\text{1. }C_6H_5Li} (C_6H_5)_2\overset{O}{\overset{\|}{P}}\underset{HOCHC_6H_5}{CHC_6H_5}$$

(36)   (37)

The relative stability of these intermediate alkoxides when associated with lithium cations has been used[11–14,52] to isolate β-hydroxy-phosphine oxides and phosphonates; again the alkoxides derived from phosphine oxides decompose to alkene more rapidly.[53,54] Similar cation effects have been observed in the Wittig reaction.[55]

Steric effects also reduce the reactivity of the carbanions, and often tertiary anions (38) will not form olefins with ketones,[56] hence the synthesis of tetrasubstituted alkenes by this route is restricted.

$$(RO)_2\overset{O}{\overset{\|}{P}}\overset{-}{C}R^1R^2$$

(38)

3.2. The Carbonyl Group

Ketones are generally less reactive towards phosphonate carbanions than aldehydes[57] and frequently require much more vigorous conditions for olefination.[58] This is primarily due to steric effects and electronic effects seem less important,[59] although trifluoromethyl ketones have enhanced reactivity.[60]

Long-chain aldehydes are also very unreactive, and that this is not due to the stability of the alkoxide intermediate is demonstrated by the ease of formation of olefin (in >90% yield) from (40) prepared by reduction of the ketone (39) and base treatment.[61]

$$(EtO)_2\overset{O}{\overset{\|}{P}}CH\underset{COC_{17}H_{35}}{\overset{CO_2Et}{\diagup}} \xrightarrow[\text{2. NaOEt/EtOH}]{\text{1. NaBH}_4} \left[(EtO)_2\overset{O}{\overset{\|}{P}}-\underset{\overset{|}{\overset{-}{O}}\diagdown CHC_{17}H_{35}}{CH\,CO_2Et} \right]$$

(39)   (40)

Readily enolizable ketones often give poor yields of olefin[62,63] (cyclopentanone and acetophenone are common examples), although this appears to be less important with more acidic phosphonates.[64]

Ketenes[3,65] react normally with phosphonate anions to give allenes, although they have been little studied. Similar reactions with ester carbonyls are rare, but oxalate esters do give vinyl esters (41).[66]

3 PO-Activated Olefinations

$$(RO)_2\overset{O}{\overset{\|}{P}}\bar{C}H\,CO_2R^1 + R^2OCO\,CO_2R^2 \longrightarrow R^1OCOCH=C(OR^2)CO_2R^2$$

(41)

Imines react to give alkenes in an analogous way to aldehydes and ketones,[67] although an excess of base is required to avoid formation of the β-aminophosphonate (42). The formation of imines by the reaction of phosphonate carbanions with nitroso groups has also been reported.[68,69]

$$(RO)_2\overset{O}{\overset{\|}{P}}CHR^1CHR^2NHR^3$$

(42)

3.3. Reaction Conditions

Typically the phosphorus compound, dissolved in a suitable solvent and under nitrogen, is treated with a base to generate the carbanion, and finally the ketone or aldehyde is added. The reaction mixture is poured into water or on to ice, and the alkene is easily separated from the phosphorus acid anion by solvent extraction (this is sometimes facilitated by adjustment of the pH[49]). In many early examples of the reaction the carbanion was generated in the presence of the carbonyl compound; however, this is unsatisfactory for aldehydes and base-sensitive ketones.[3] (Typical experimental procedures are given in Section 6.)

An almost unlimited range of solvents and bases have been used depending on, among other factors, solubility, the basicity of the carbanion to be formed,[70] the reactivity of the carbonyl compound and any other functional groups present, the reaction temperature, and most recently the stereochemistry of olefin required (see p. 169). A comprehensive list of solvents and bases used is given in Section 7.

Cation effects on the intermediate alkoxide (43) have already been mentioned (p. 163), and effects on the stereochemistry of the olefin formed will be discussed later (see p. 169).

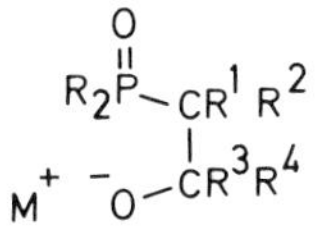

$$\begin{array}{c} \overset{O}{\overset{\|}{R_2P}}\!\!-\!\!CR^1R^2 \\ M^+ \;\; ^-O\!-\!\overset{|}{C}R^3R^4 \end{array}$$

(43)

Recently phase-transfer catalysts[71] have been used in PO-activated olefin synthesis.[72-76] Advantages claimed include improved yields,[72] one-step reactions,[76] and variations in alkene stereochemistry.[73-75] However, in at least one case[77] similar results were obtained in a two-phase system in the *absence* of a phase-transfer catalyst!

3.4. Competing Reactions

With α,β-unsaturated aldehydes and ketones Michael addition competes with olefin formation[78,79] and in the case of ketones often predominates,[80] although this depends to some extent on the reaction conditions.[81]

Allylic phosphonates often react normally to give dienes, but γ-addition can compete[82] depending on the substituents present. It seems likely that γ-addition (44) commonly occurs, but providing it is reversible the expected

$$(RO)_2\overset{O}{\overset{\|}{P}}\text{-CH-CH-CH}_2 \;\rightleftharpoons\; (RO)_2\overset{O}{\overset{\|}{P}}\text{CH=CHCH}_2\overset{O^-}{\underset{|}{C}}R^1R^2 \quad (44)$$

$$+ R^1R^2CO \;\searrow\;$$

$$(RO)_2\overset{O}{\overset{\|}{P}}\text{-CHCH=CH}_2$$
$$\underset{O}{\overset{|}{\underset{\diagdown}{}}}\overset{|}{C}R^1R^2$$

$$\downarrow$$

$$R^1R^2C\text{=CH-CH=CH}_2$$

alkene is the final product. This is supported by the reactions of 3-halogeno-alkylphosphonate anions (45) with ketones,[83] which give various products from α- and γ-addition depending on the conditions, while replacement of halide by a poor leaving group leads to normal olefination (46).

$$(RO)_2\overset{O}{\overset{\|}{P}}\text{CH-CR}^1\text{-CR}^2X \xrightarrow[X=OR,SR,NR_2]{R^3R^4CO} R^3R^4C\text{=CH CR}^1\text{=CR}^2X$$
$$(45) \qquad\qquad\qquad\qquad\qquad (46)$$

$$R^3R^4CO \downarrow X=Cl,Br$$

$$R^3R^4C\text{=CH CR}^1\text{=CR}^2X$$
$$+$$
$$(RO)_2\overset{O}{\overset{\|}{P}}\text{CH=CH CR}^2\text{-CR}^3R^4$$
$$\underset{\diagdown O\diagup}{}$$

Although PO-activated olefination is normally regiospecific excess base should be avoided since thermodynamically controlled isomerization can take place[84,85] (see p. 185).

Anomalous olefin formation through dehydration of intermediate β-hydroxy-phosphonates (47)[7] or directly from phosphonate carbanions[59] can

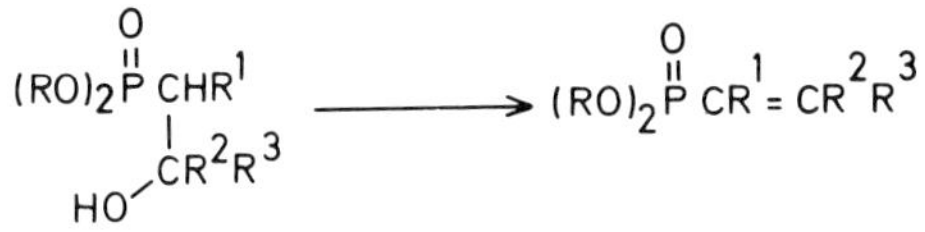

3 PO-Activated Olefinations

$$(RO)_2\overset{O}{\overset{\|}{P}}\underset{\underset{HO}{\overset{|}{CR^2R^3}}}{CHR^1} \longrightarrow (RO)_2\overset{O}{\overset{\|}{P}}CR^1=CR^2R^3$$

(47)

occur, and a variety of condensation products are formed from phosphonates having more than one highly acidic site.[86]

4. Mechanism and Stereochemistry

Since the present chapter is intended primarily as an aid to synthesis, only a summary of the mechanism of PO-activated olefination will be presented (an in-depth discussion can be found in reference 87 and the bibliography at the beginning of Section 8). However, a more detailed account will be given of the aspects of mechanism relating to stereochemistry.

4.1. Mechanism

The accepted mechanism[1,3] for PO-activated olefin synthesis is closely analogous to that of the Wittig reaction.[88] The phosphorus-stabilized carbanion and the carbonyl compound react in a reversible step[61,89-94] to form an alkoxide intermediate (48) which irreversibly decomposes to alkenes by

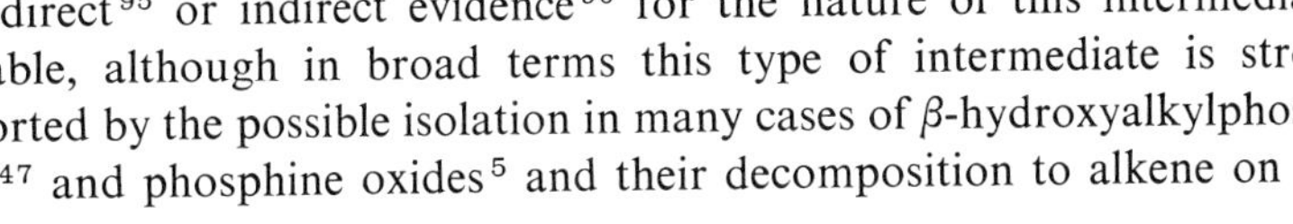

$$R_2\overset{O}{\overset{\|}{P}}-\bar{C}\!\!<\; +\; R^1R^2CO \;\rightleftharpoons\; \left[R_2\overset{O}{\overset{\|}{P}}-\underset{\underset{|}{\overset{|}{O-C-}}}{C}\!\!<\right] \;\longrightarrow\; R_2P\overset{\nearrow O}{\underset{\searrow O^-}{}} \;+\; \overset{\diagdown}{\underset{\diagup}{C}}\overset{\diagup}{\underset{\diagdown}{\|}}\overset{}{C}$$

R = OAlkyl, OAryl, alkyl, aryl, NR$_2$ (48)

the *cis*-elimination of a phosphorus acid anion. Unlike the Wittig reaction, little direct[95] or indirect evidence[96] for the nature of this intermediate is available, although in broad terms this type of intermediate is strongly supported by the possible isolation in many cases of β-hydroxyalkylphosphonates[47] and phosphine oxides[5] and their decomposition to alkene on base-treatment.

Good evidence for a large build-up of negative charge in the T.S. on the carbon atom α to phosphorus is provided by the need for resonance-stabilizing

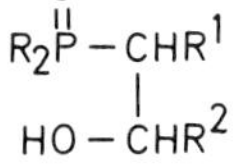

$$R_2\overset{O}{\overset{\|}{P}}-\underset{\underset{HO-CHR^2}{|}}{CHR^1}$$

(49)

α-substituents[97] (see p. 163). In the absence of these substituents β-hydroxyalkyl compounds (49) are isolated, and generally these will not give alkenes even on base treatment, except under the most vigorous conditions.[47] An exception to this are the β-hydroxyalkyl phosphoramidates (49; $R = NR_2$).[11-14]

4.2. Stereochemistry

Early reports claim exclusive *trans* stereochemistry for the olefins formed using PO reagents. This is hardly surprising, since the systems investigated generally contained groups capable of conjugating with the incipient double bond *and* clear-cut steric factors favouring *trans* olefin formation (see p. 169). Reports of the formation of some *cis*-olefin soon appeared,[98-100] although those relating to tri- and tetra-substituted olefins[98] were often based on semantics rather than unexpected stereochemistry. However, some experiments, for example equations (3),[99] (4),[100] and (5),[101] showed unambiguous

$$(MeO)_2\overset{O}{\overset{\|}{P}}CH_2CO_2Me \xrightarrow[\text{2. Me}_2\text{CH CHO R.T.}]{\text{1. NaH/glyme } 15-35°} Me_2CHCH=CHCO_2Me \quad (3)$$
$$14\% \text{ cis}$$
$$32\% \text{ trans}$$

$$(EtO)_2\overset{O}{\overset{\|}{P}}CH_2CO_2Et \xrightarrow[\text{2. HCO CO}_2\text{Et}]{\text{1. NaH/DME}} EtOCOCH=CHCO_2Et \quad (4)$$
$$38\% \text{ cis}$$
$$20\% \text{ trans}$$

$$(Me_2N)_2\overset{O}{\overset{\|}{P}}CH_2OR^1 \xrightarrow[\text{2. R}^2\text{CHO}]{\text{1. NaH}} R^2CH=CH\ OR^1 \quad (5)$$
$$50-70\% \text{ cis}$$

formation of significant amounts of *cis*-alkene, although the explanations of its formation often left much to be desired!

The intermediate (50), formed by the addition of the carbanion to an aldehyde or ketone, can exist in two diastereoisomeric forms (50a) and (50b), and each of these decomposes to give a specific olefin isomer.[102,103] This is analogous to the Wittig reaction and is supported by a variety of evidence[87] including the isolation of the pure *erythro* (50a; $R^2 = H$) and *threo* (50b; $R^2 = H$) intermediates as their conjugate acids[104] and their base-catalysed decomposition to the appropriate olefin.[90,103,105]

The stereochemistry of the alkene obtained in a given reaction depends on the relative rates of formation and of decomposition of the intermediates (50a) and (50b)[91] (sometimes referred to as the "reversibility factor") and to the extent of direct interconversion of these intermediates; the last has been demonstrated in a number of cases.[90,92-94]

3 PO-Activated Olefinations

$$R_2\overset{O}{\overset{\|}{P}}\bar{C}HR^1 + R^2R^3CO$$

$$\begin{array}{c}R_2\overset{O}{\overset{\|}{P}}\diagdown_{C\overset{H}{\underset{R^1}{\blacktriangleleft}}}\\ {}^-O\diagup^{C\overset{\text{\tiny///}R^2}{\underset{R^3}{\blacktriangleleft}}}\end{array} \rightleftharpoons \begin{array}{c}R_2\overset{O}{\overset{\|}{P}}\diagdown_{C\overset{H}{\underset{R^1}{\blacktriangleleft}}}\\ {}^-O\diagup^{C\overset{\text{\tiny///}R^3}{\underset{R^2}{\blacktriangleleft}}}\end{array}$$

(50a)   (50b)

$$\underset{R^1}{\overset{H}{\diagdown}}C=C\underset{R^3}{\overset{R^2}{\diagup}} \qquad \underset{R^1}{\overset{H}{\diagdown}}C=C\underset{R^4}{\overset{R^3}{\diagup}}$$

When groups capable of conjugating with the incipient double bond are absent, the relative steric bulk of R^2 and R^3, and hence their interaction with R^1 in (50), can often be seen to control olefin stereochemistry.[106–108] However, when conjugating groups are present the alkene in which these groups are *trans* usually predominates,[109] although in the presence of particularly large opposing steric effects this may be reversed.[100]

4.3. Control of Stereochemistry

The development of an understanding of the factors affecting stereochemistry in PO-activated olefin synthesis naturally led to attempts to control this stereochemistry.[110] Extensive investigations of the effect of changing the base, solvent, temperature, and the nature of the cation associated with (50) have been carried out.[89,91,98,104] In some cases quite notable effects have been achieved within specific systems; examples are given in Scheme 2.

In at least some cases solvent effects are derived from the extent of betaine reversibility; for example, the reaction of diethyl cyanomethylphosphonate anion (51) with aromatic aldehydes in benzene, DME, or THF is thermo-

$$(EtO)_2\overset{O}{\overset{\|}{P}}\bar{C}H\ CN$$

(51)

dynamically controlled; however, in HMPT the reaction can be thermodynamically or kinetically controlled depending on the proportions and order of addition of the reagents.[112] Similar information, more directly related to the intermediate (50), has been obtained[113] through isolation of the *erythro*-isomer (53a) by addition of acid to the reaction of the phosphonate (52) with

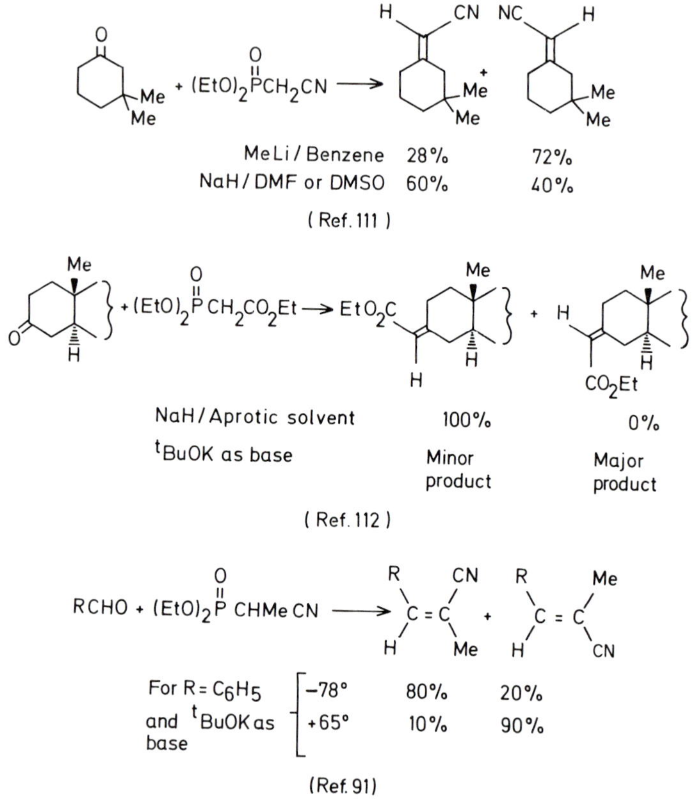

Scheme 2

$$\text{cyclohexanone-Me,Me} + (EtO)_2\overset{O}{\underset{\|}{P}}CH_2CN \longrightarrow$$ (E/Z isomers with CN)

	28%	72%
MeLi / Benzene	28%	72%
NaH / DMF or DMSO	60%	40%

(Ref. 111)

$$\text{decalone} + (EtO)_2\overset{O}{\underset{\|}{P}}CH_2CO_2Et \longrightarrow EtO_2C\text{-isomer} + \text{isomer with }CO_2Et$$

NaH / Aprotic solvent   100%   0%
^{t}BuOK as base   Minor   Major
                           product   product

(Ref. 112)

$$RCHO + (EtO)_2\overset{O}{\underset{\|}{P}}CHMe\,CN \longrightarrow \underset{H}{\overset{R}{>}}C=C\underset{Me}{\overset{CN}{<}} + \underset{H}{\overset{R}{>}}C=C\underset{CN}{\overset{Me}{<}}$$

For R = C$_6$H$_5$ $-78°$ 80%   20%
and ^{t}BuOK as $+65°$ 10%   90%
base

(Ref. 91)

Scheme 2

benzylideneaniline in ether at $-33°$. In ether at 10° or in liquid ammonia a similar addition of acid leads to isolation of the *threo*-isomer (53b).

Unfortunately the effect of changing cations on the stereochemistry of olefins from PO-activated synthesis is far less predictable than the equivalent effect in the Wittig reaction[114] (however, see references 75 and 92).

The isolation and separation of the diastereoisomeric betaines (50a) and (50b) in the form of their conjugate acids provides a further method of stereochemical control. Base-catalysed decomposition of (53a) and (53b) should lead specifically to the appropriate olefin, provided that betaine reversibility is slow relative to betaine decomposition (this is frequently *not* the case). This has been accomplished for the β-hydroxyalkylphosphonates (54),[90] the β-hydroxyalkylphosphine oxides (55),[105] and the β-hydroxyalkyl-phosphoramidates (56).[11–14] In the last case base-treatment is not required since the conjugate acids themselves decompose to alkene on heating. Unfortunately in many cases one betaine is formed to only a small extent,

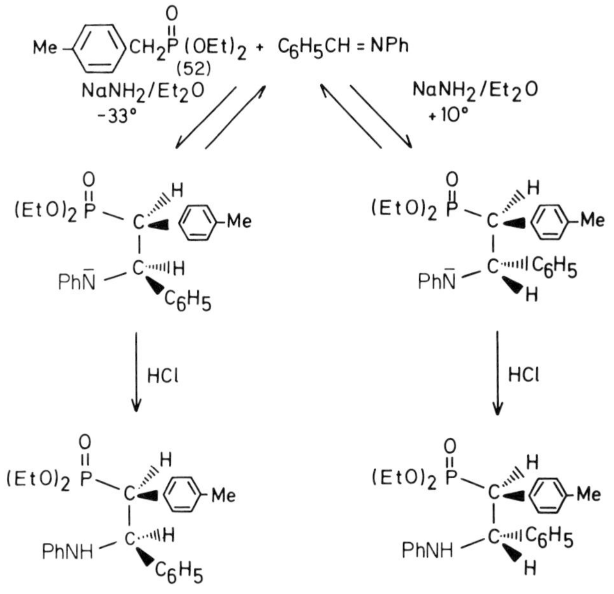

3 PO-Activated Olefinations

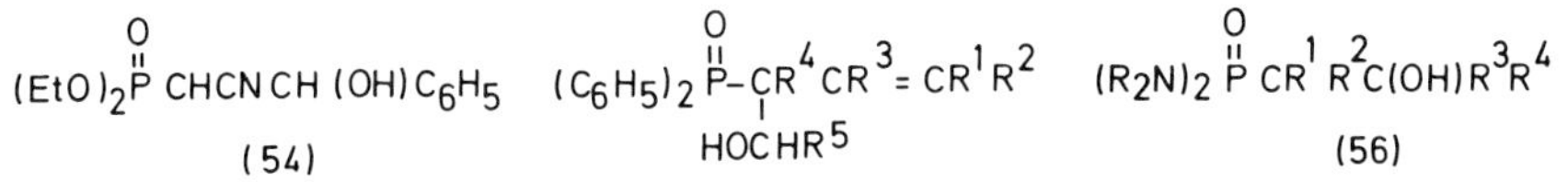

[Scheme showing reaction of Me-C₆H₄-CH₂P(O)(OEt)₂ (52) + C₆H₅CH=NPh with NaNH₂/Et₂O at −33° and +10°, giving diastereomeric intermediates that on treatment with HCl yield (53a) and (53b).]

(EtO)₂P(O)CHCNCH(OH)C₆H₅   (C₆H₅)₂P(O)-CR⁴CR³=CR¹R²   (R₂N)₂P(O)CR¹R²C(OH)R³R⁴
(54)   HOCHR⁵   (56)
                      (55)

and this effectively precludes any practical synthesis of its derived olefin. Whitham[115] has used this type of approach to interconvert olefin isomers (Scheme 3).

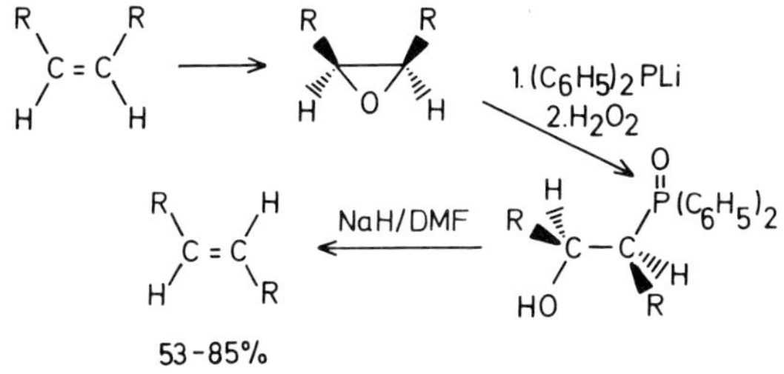

[Scheme 3: cis-olefin → epoxide → 1.(C₆H₅)₂PLi, 2.H₂O₂ → β-hydroxyphosphine oxide → NaH/DMF → trans-olefin, 53-85%]

Scheme 3

A recent and very interesting development is the use of cyclic phosphonates[21] and phosphine oxides.[22] It was reasoned that the use of the phosphonate (57) should (a) lead to more rapid closure to a five-coordinate intermediate (58) (release of strain in the five-membered ring) and (b) reduce the rate of

(57)

(58)

reversion to phosphonate and carbonyl compound for similar reasons. This in turn should lead to a kinetically controlled reaction and so to a higher proportion of *cis*-olefin; this was clearly true in every case studied. Similar but smaller effects were noted for the phosphine oxide (59).[22] The Wittig reaction of the phosphole ylide (60) also gives a higher proportion of *cis*-olefin than the acyclic analogue (61), possibly for the same reasons.[116]

(59)

(60)

$(C_6H_5)_3P{=}CHCO_2Et$

(61)

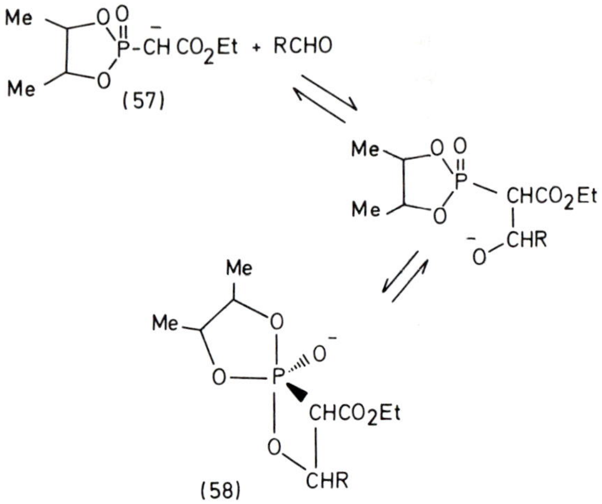

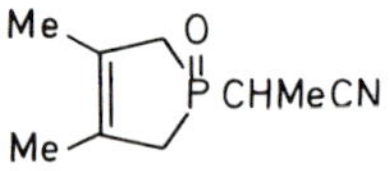

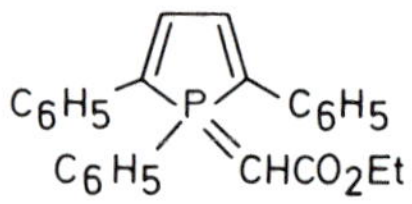

5. Applications in Synthesis of Alkenes

5.1. PO-stabilized Carbanions

For reasons already discussed, the synthesis of alkenes that do not contain at least one activating group is not normally possible by this method. However, Corey has prepared a variety of olefins of this type, e.g. *trans*-2,2-dimethylundec-2-ene,[12] 4-methylene-t-butylcyclohexane,[13] and *cis*- and *trans*-but-2-ene,[14] using alkylphosphonic acid bis(dialkylamides) (62) and the derived betaines (see Section 6).

$$(R_2N)_2\overset{O}{\overset{\|}{P}}CHR^1R^2$$

(62)

3 PO-Activated Olefinations

5.1.1. α,β-Unsaturated ketones

β-Ketoalkylphosphonate anions have been used[3,98] to prepare a variety of unsaturated ketones [equation (6)], and have the advantage over the equiv-

$$(EtO)_2\overset{O}{\underset{\|}{P}}CH_2COMe \xrightarrow[2.\ R^1R^2CO]{1.\ NaH/THF\ or\ DME\ or\ DMF/10°} R^1R^2C=CHCOMe \quad (6)$$
$$16-90\%$$

alent Wittig reagent of much greater reactivity. However, poor yields have been noted in some cases,[8] and phosphonates (63)[117] and (64),[118] with masked keto functions, have been used. A similar reaction with the phosphonate (63; $R^2 = H$) provides a route to α,β-unsaturated aldehydes in excellent yield.[119] (See also prostaglandins, p. 179.)

$$R^1_2\overset{O}{\underset{\|}{P}}CH=CR^2NHR^3 \xrightarrow[2.\ R^4_2CO]{1.\ NaH/THF} R^4_2C=CH\overset{NR^3}{\underset{\|}{C}}R^2 + R^1_2P\overset{O}{\underset{OH}{\diagup}}$$
(63)

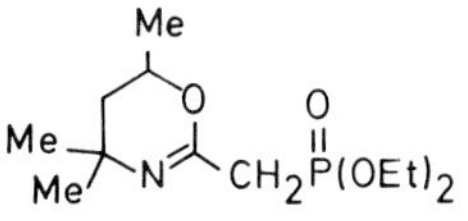

(Structure 64: a tetrahydrooxazine with Me, Me, Me substituents and N–CH₂P(OEt)₂ group)

(64)

5.1.2. α,β-Unsaturated acid derivatives

Unsaturated acids,[49,118,120] esters,[100,107] amides,[56,121] nitriles,[107,111] and isocyanides[122] have been prepared in excellent yields from the corresponding phosphonates (in the case of carboxylic acids two mole equivalents of base are required). Detailed investigations of the factors controlling the stereochemistry of the nitriles[22,91,106] and esters[21,98,107] formed have also been carried out.

Royal jelly (exclusively *trans*) has been synthesized[123] from the protected aldehyde (65), although general syntheses of fatty acids from ester phosphonates have been hindered by the low reactivity of long-chain aldehydes.[61]

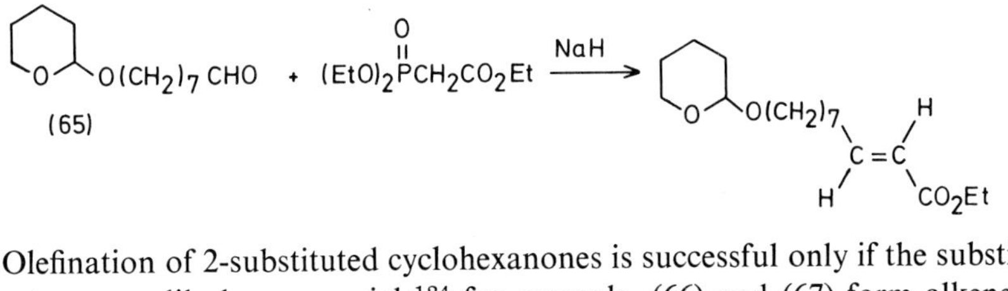

$$\underset{(65)}{\text{THP-O}(CH_2)_7CHO} + (EtO)_2\overset{O}{\underset{\|}{P}}CH_2CO_2Et \xrightarrow{NaH} \text{THP-O}(CH_2)_7\overset{H}{\underset{H}{\diagdown}}C=C\overset{H}{\underset{CO_2Et}{\diagdown}}$$

Olefination of 2-substituted cyclohexanones is successful only if the substituent can readily become axial;[124] for example, (66) and (67) form alkenes

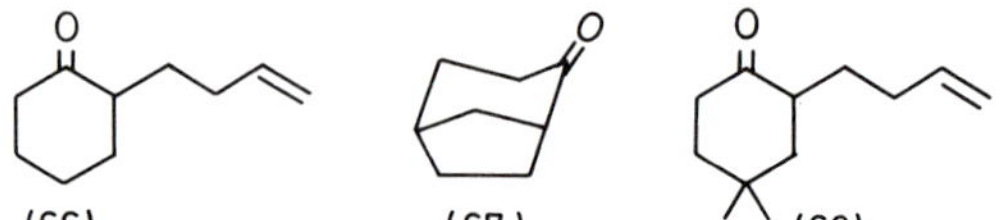

(66) (67) (68)

but (68) does not, presumably owing to 1,3-diaxial interactions. However, 4-substituted cyclohexanones react normally, and equation (7) provides one of the few examples of asymmetric induction in phosphonate olefin syntheses.[125]

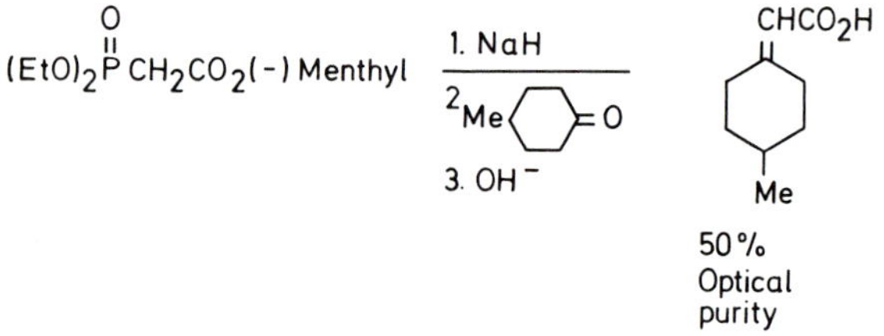

$$(EtO)_2\overset{O}{\overset{\|}{P}}CH_2CO_2(-)Menthyl \xrightarrow[\substack{2. Me\text{-cyclohexanone}\\ 3.\ OH^-}]{1.\ NaH} \text{4-Me-cyclohexylidene-CHCO}_2H \quad (7)$$

50% Optical purity

Olefinations of the cyanoalkylphosphonate (69) have been used as the basis of one-carbon homologation of carbonyl compounds through conversion of vinyl ether (70) into acids, esters, and amides.[126]

$$(EtO)_2\overset{O}{\overset{\|}{P}}CH(CN)OBu^t \xrightarrow[2.\ R^1R^2CO]{1.\ NaH} R^1R^2C=C(CN)OBu^t$$

(69)   (70)

5.1.3. Stilbenes and related alkenes

The reactions of benzylphosphonate[1,102] and phosphine oxide[5,103] carbanions with aromatic aldehydes to give stilbenes were amongst the first reported examples of PO-activated olefination, and a variety of similar olefins prepared in this way are shown in Scheme 4. *trans*-Alkene is formed exclusively in virtually every case, presumably owing to the extensive conjugation with the incipient double bond in the transition state.

5.1.4. Vinyl halides

α-Bromo- and α-chloro-alkylphosphonates are readily available from halogenation[134] of unsubstituted alkylphosphonate carbanions, and the former compounds undergo normal olefination reactions with aldehydes and ketones[135] to give isomeric mixtures which depend on the other substituents.[136]

Acetylenes can be obtained directly from these reactions in the presence of excess base [equation (8)].[137]

$$(C_6H_5O)_2\overset{O}{\overset{\|}{P}}CHCl\,Ar + RCHO \xrightarrow[THF]{2\ NaH} ArC\equiv CR \quad (8)$$

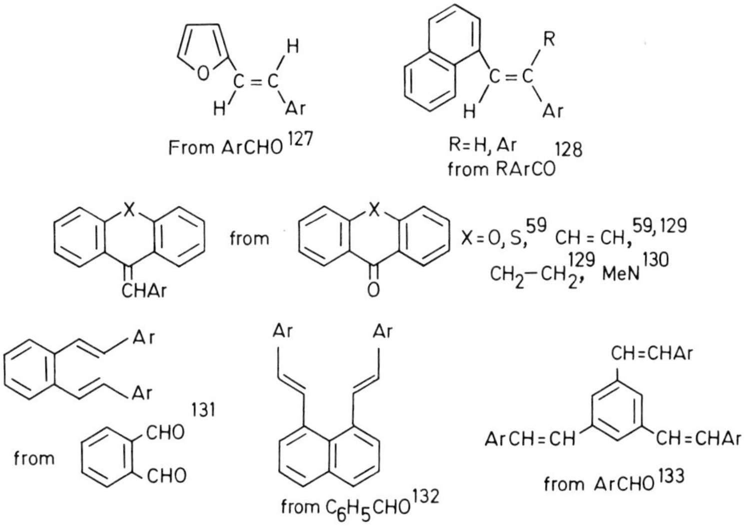

3 PO-Activated Olefinations

[Scheme 4 with structures:
- From ArCHO 127 (furan-CH=CH-Ar)
- R=H, Ar 128 from RArCO (naphthyl-C(R)=CH-Ar)
- Xanthene =CHAr from xanthone (X=O, S,⁵⁹ CH=CH,⁵⁹,¹²⁹ CH₂–CH₂,¹²⁹ MeN¹³⁰)
- 131 from o-phthalaldehyde (naphthalene with two CH=CHAr)
- 132 from C₆H₅CHO (1,8-bis(styryl)naphthalene)
- 133 from ArCHO (1,3,5-tris(styryl)benzene)]

Scheme 4

5.1.5. Sulphur compounds

Vinyl sulphides [63] and ketene thioacetals [76,138] can be prepared in high yield from the phosphonates (71) and (72) respectively. Unsaturated sulphones (largely E)[58,109,139] and sulphoxides (E and Z)[64] have been obtained by similar methods.

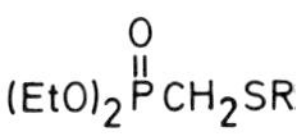

$$(EtO)_2\overset{O}{\overset{\|}{P}}CH_2SR \qquad (EtO)_2\overset{O}{\overset{\|}{P}}CH(SR)_2$$
$$(71) \qquad\qquad (72)$$

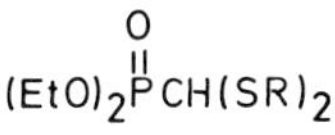

Sulphonium-substituted phosphonates react with aldehydes to give *cis*- and *trans*-vinylsulphonium salts (73) rather than epoxyalkylphosphonates,[140] although similar reactions with acrylic esters lead to cyclopropylphosphonates (74).

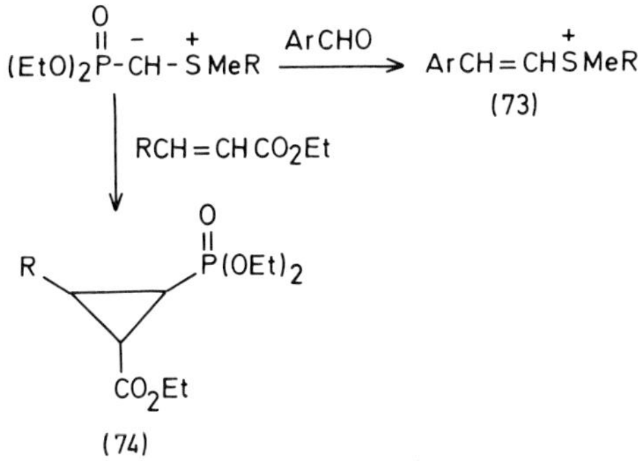

$$(EtO)_2\overset{O}{\overset{\|}{P}}\text{-}\overset{-}{CH}\text{-}\overset{+}{S}MeR \xrightarrow{ArCHO} ArCH=CH\overset{+}{S}MeR \quad (73)$$

$$\downarrow RCH=CHCO_2Et$$

[cyclopropane structure with R, P(O)(OEt)₂, and CO₂Et substituents]
(74)

Thioalkylphosphonates and phosphine oxides have also been used to prepare precursors for hetero Cope reactions [141] and as acyl anion equivalents [equation (9)].[142]

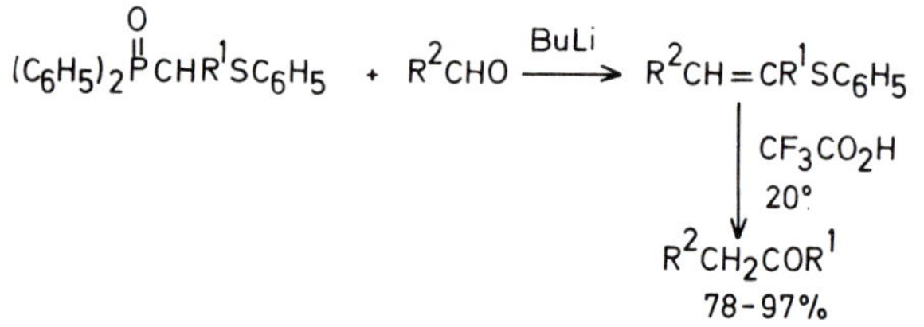

$$(C_6H_5)_2\overset{O}{\overset{\|}{P}}CHR^1SC_6H_5 + R^2CHO \xrightarrow{BuLi} R^2CH=CR^1SC_6H_5 \quad (9)$$

$$\downarrow \begin{array}{c} CF_3CO_2H \\ 20° \end{array}$$

$$R^2CH_2COR^1$$
$$78\text{-}97\%$$

5.1.6. Enamines

α-Aminoalkylphosphonates have been used to prepare amidines [143] and enamines,[144] and consequently substituted ketones,[145] substituted aldehydes,[146] α- and β-diketones,[147] and carboxylic acids,[148] often under very mild conditions and in high yields.

5.1.7. Allenes and acetylenes

Ketenes undergo "normal" olefination reactions with PO-stabilized carbanions to give allenes [79,149] [equation (10)], and this reaction is preferred to

$$(EtO)_2\overset{O}{\overset{\|}{P}}\bar{C}R^1CO_2Et + R^2R^3C=C=O \longrightarrow R^2R^3C=C=CR^1CO_2Et \quad (10)$$

the Wittig route owing to the mild conditions required. However, similar reactions with isocyanates and isothiocyanates *do not* lead to ketenimines.[150]

Allenes of up to 12% optical purity can be obtained from reactions with optically active phosphinates.[151] Conjugate acids (75), related to the presumed

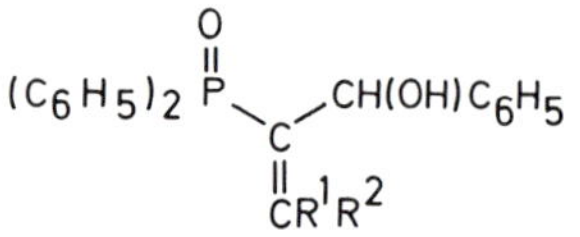

$$(C_6H_5)_2\overset{O}{\overset{\|}{P}}\diagdown\underset{\underset{CR^1R^2}{\|}}{C}\diagup CH(OH)C_6H_5$$

(75)

reaction intermediates, can be prepared from the corresponding acetylene phosphine oxide and give high yields of allenes on base treatment.[152]

Under severe conditions keto-stabilized anions (76)[41] eliminate phosphate

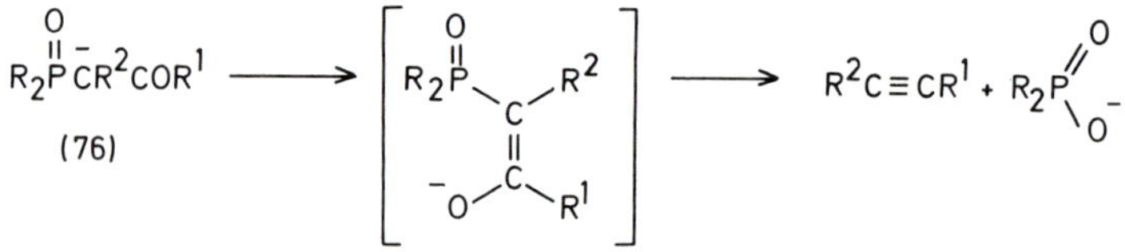

$$R_2\overset{O}{\overset{\|}{P}}\bar{C}R^2COR^1 \longrightarrow \left[R_2\overset{O}{\overset{\|}{P}}\diagdown\underset{\underset{-O}{\overset{\|}{C}}\diagdown R^1}{\overset{R^2}{C}}\right] \longrightarrow R^2C\equiv CR^1 + R_2\overset{O}{\overset{\diagup\!\!\diagup}{P}}\diagdown O^-$$

(76)   (77)

or phosphinate to give acetylenes, presumably via the enolate (77)[153] (see also reference 154).

5.1.8. *Dienes and polyenes*

Dienes can be prepared from saturated[72,155] [equation (11)] or unsaturated[83] [equation (12)] phosphorus compounds, and competing reactions can take

$$(RO)_2\overset{O}{\overset{\|}{P}}CH_2CO_2Et \xrightarrow[2. R^1CH=CHCOR^2]{1. Base} R^1CH=CH\ CR^2=CHCO_2Et \quad (11)$$

$$(RO)_2\overset{O}{\overset{\|}{P}}CH_2CH=CHR^1 \xrightarrow[2. R^1R^2CO]{1. Base} R^1CH=CH\ CH=CR^1R^2 \quad (12)$$

place in both cases (see p. 166). However, yields of alkene are often better than those from the corresponding Wittig reaction.[156]

The stereochemistry of the double bond formed is controlled by the factors already discussed[105] (see p. 169), and since these include conjugation with the incipient double bond, that isomer usually predominates which has the phosphonate activating group and the preformed double bond *trans*.[54,108,157]

The stereochemistry of any double bonds present in the carbonyl substrate is invariably retained,[108] and this is often true of double bonds in the phosphorus compound[53,54,158a] [equation (12)]. In this connection, use of diphenylphosphinoyl [Ph$_2$P(O)] as a migrating functional group shows considerable promise in general stereoselective syntheses of 1,3-dienes.[158a-c] However, when the unsaturated phosphonate carries a strongly stabilizing group in the γ-position, e.g. (78), base-catalysed isomerization of the least stable isomer usually takes place during the reaction.[159,160]

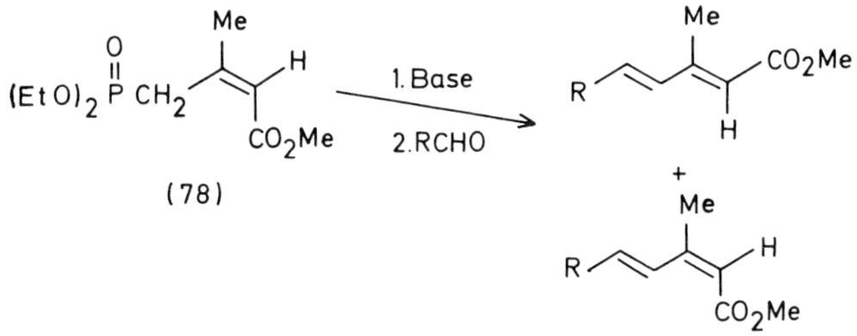

Similar methods have been applied to polyene synthesis[159,161] including 3-deoxy-vitamin D$_2$ (79),[53] precursors (80) of trisporic acid,[162] β-carotene,[163] and 7-*cis*,9-*cis*-retinoates.[164] In the case of readily polymerized compounds the use of alkoxide–alcohol base-system seems advantageous.[165] A variety of

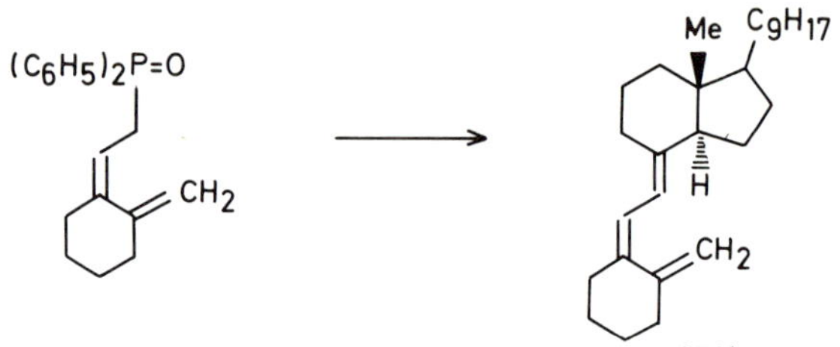

$(C_6H_5)_2P=O$

[structure with CH₂]

→

[structure (79) with Me, C₉H₁₇, H, CH₂]

(79)

[structure (80) with Me, Me, Me, CO₂Me, OMe]

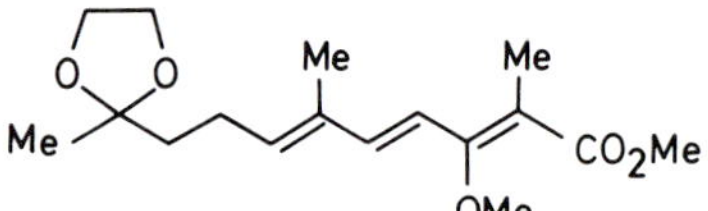

(80)

polyene polymers have been prepared from bisphosphonates and difunctional aldehydes.[166]

Extensive use has been made of phosphonate carbanions in the synthesis of insect metabolites. *Cecropia* (81)[167,168] and *cecropia* dehydro[169] juvenile

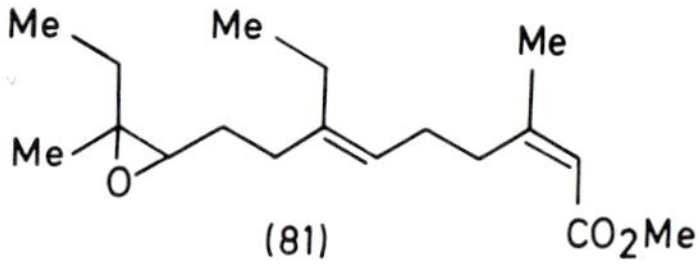

[structure (81) with Me, Me, Me, Me, CO₂Me]

(81)

hormones and their dioxa-[170] and fluorinated[60] analogues have been prepared in this way. Labelled juvenile hormone has been synthesized from methoxycarbonyl[^{14}C]methylphosphonate.[177] Phosphonate carbanions appear to react specifically with carbonyl functions in the presence of epoxy groups,[172] and in view of this it is surprising that this type of approach has not been applied to juvenile hormone synthesis.

The allenic sex pheromone (82) of the dried bean beetle has been prepared[173] as a racemate, but stereospecifically *trans*. However, attempts[79] to prepare the corresponding amides gave only poor yields.

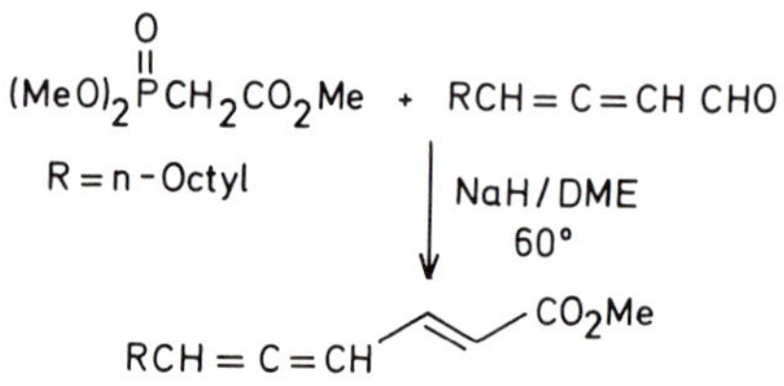

$$(MeO)_2\overset{O}{\overset{\|}{P}}CH_2CO_2Me + RCH=C=CH\ CHO$$

R = n-Octyl

↓ NaH/DME
60°

$$RCH=C=CH\diagdown\hspace{-0.3em}\diagup CO_2Me$$

(82)

The use of phosphonate carbanions in terpene synthesis has been extensively studied by Vig and his coworkers.[174] The reaction has been used to

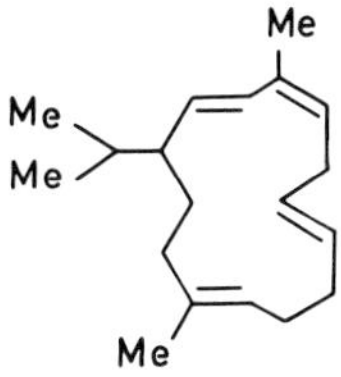

3 PO-Activated Olefinations

(83)

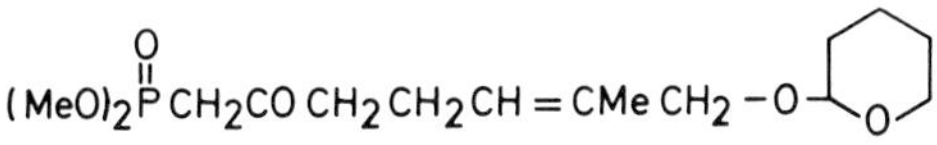

(84) $(MeO)_2\overset{O}{\overset{\|}{P}}CH_2CO\,CH_2CH_2CH=CMe\,CH_2-O-\text{(THP)}$

prepare ^{3}H- and ^{14}C-labelled stereoisomers of farnesol and geraniol,[175] and cembrene (83) has been obtained using the phosphonate (84).[176] Vinyl sulphides (85) and (87), prepared from methylthioalkylphosphonates, have proved useful in the synthesis of (±)-occidentalol (86)[177] and nootkatane sesquiterpenes, e.g. (88),[178] respectively.

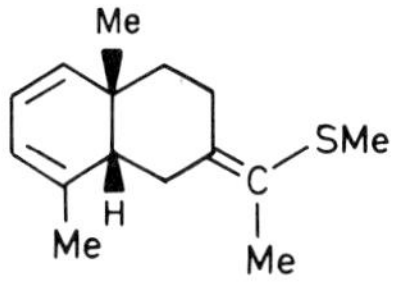

(85) 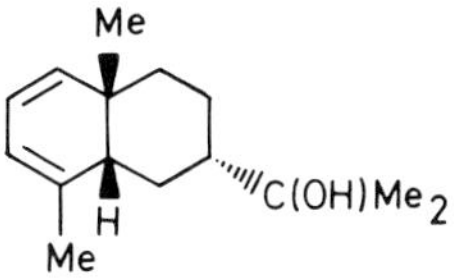 (86) 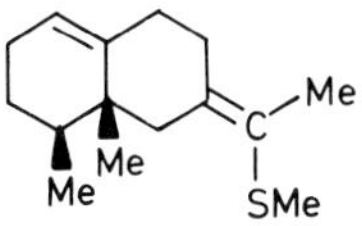 (87) 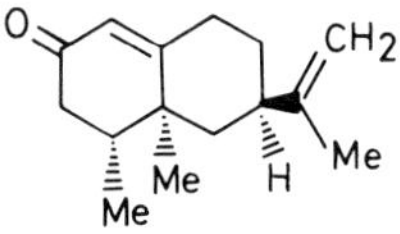 (88)

5.1.9. Prostaglandins

The reaction of 2-oxoheptylphosphonate carbanions with the appropriate aldehyde, followed by reduction of the keto group (e.g. 10,10-dimethylprostaglandin E_1; Scheme 5),[179] has become the method of choice for the

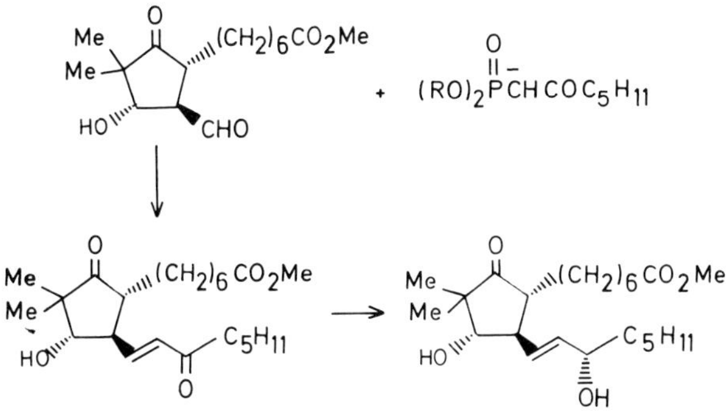

$(RO)_2\overset{O}{\overset{\|}{P}}\overset{-}{CH}COC_5H_{11}$

Scheme 5

introduction of the C_8 side-chain in prostaglandins and has been used to synthesize compounds of the E_1, F_1,[180] E_2, and F_2[181] series, 12-methyl-prostaglandins,[182] and cyclobutane,[183] aryl,[184] and aza analogues.[185] It appears that phosphonate anions are more reactive in this connection than the corresponding phosphonamide anions (89).[186] Modified C_8 side-chains, e.g. (90), have been introduced[187] in a similar way, and recently triethyl phosphonosorbate (91) has been used to attach the C_7 side-chain.[188]

(90)

$$(Me_2N)_2\overset{O}{\overset{\|}{P}}\bar{C}HCOR$$

(89)

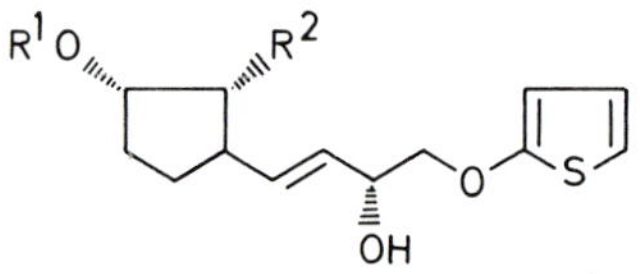

$$(EtO)_2\overset{O}{\overset{\|}{P}}CH_2CH{=}CH\,CH{=}CH\,CO_2Et$$

(91)

5.1.10 *Steroids and related compounds*

The phosphonate olefin synthesis has been used extensively to prepare potentially pharmacologically active alkenyl derivatives, for example digitoxigenin analogues (92),[189] cardiotonics (93),[190] and antigonadotrophic and anabolic compounds (94)[191] from steroidal ketones[192] and aldehydes.

(92)

(93)

(94)

(95)

(96)

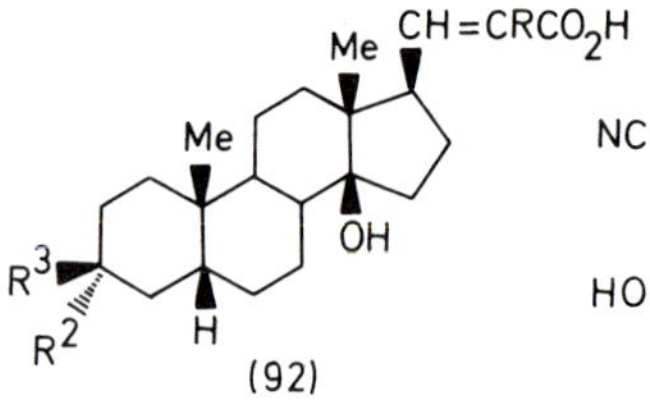

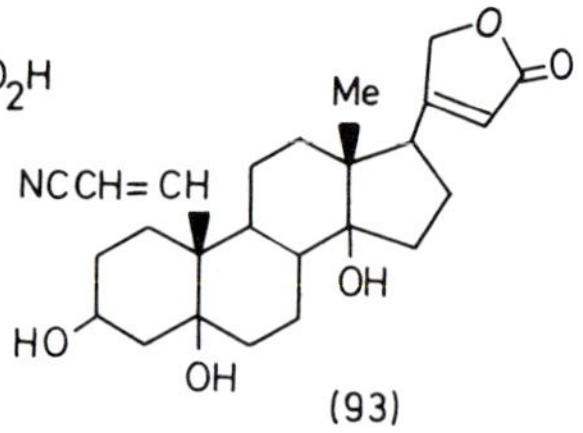

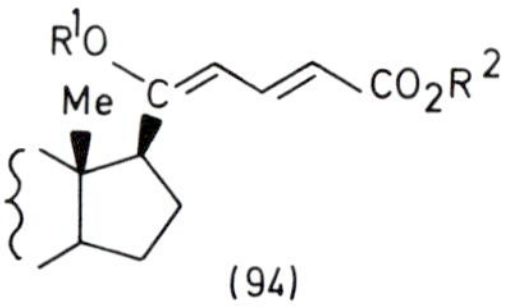

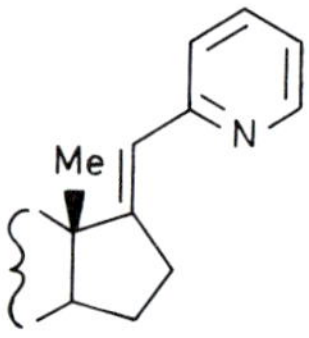

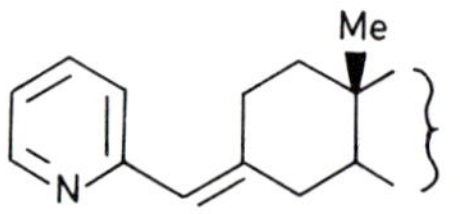

3 PO-Activated Olefinations

The 3-keto group appears to react preferentially, especially with ester-stabilized phosphonates;[193] however, 17-ones and 20-ones[194] will also form alkenes, e.g. (95) and (96).[195] Under the appropriate conditions reaction with the 3-keto group can even compete with side-chain aldehyde groups (Scheme 6).[196]

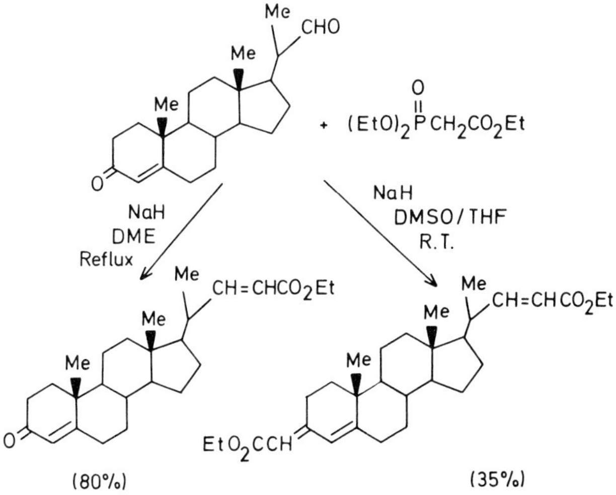

Scheme 6

The stereochemistry of the alkenes formed from 3-oxo-[112b,197] and 17-oxo-steroids[198] depends on the reaction conditions; two examples of this are given in Scheme 7. At higher temperatures the isomeric alkene (97) is formed[112b] from reactions of 3-oxo-steroids, and it is interesting that this *does not* arise by isomerization of the expected product (Scheme 6).

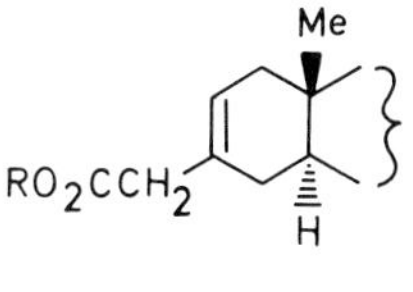

(97)

Various approaches to the synthesis of bufadienolides and isobufadienolides have involved reactions with phosphonate carbanions. One route involves cyclization of the diene (99), which can be obtained stereospecifically by reaction of the aldehyde (98) with phosphonotriester.[199] A similar reaction with diethyl cyanomethylphosphonate gives a mixture of isomers, and the authors suggest that this is due to the lower steric requirement of nitrile compared with ester. Optically active bufadienolides (100) of perhydrostilbenes have been prepared by analogous methods.[200]

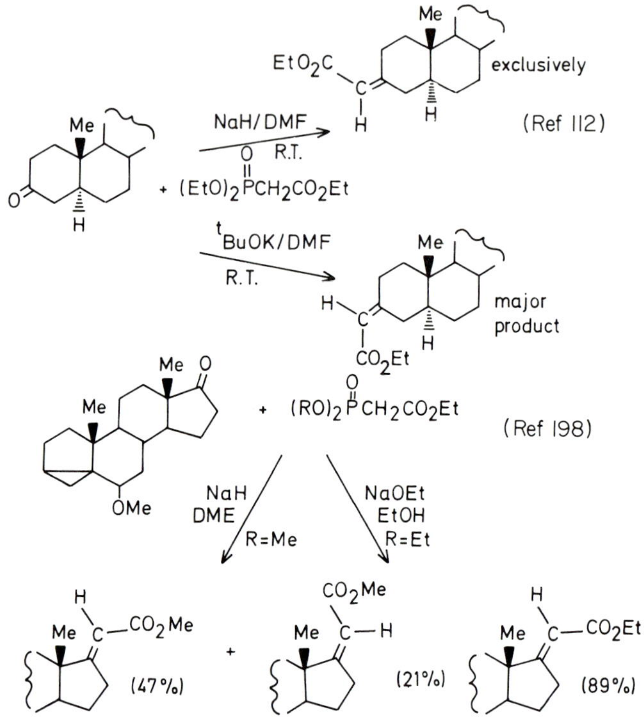
Me
EtO2C
exclusively
NaH/DMF
R.T.
(Ref 112)
+ (EtO)2PCH2CO2Et
tBuOK/DMF
R.T.
major product
CO2Et
Me
+ (RO)2PCH2CO2Et
(Ref 198)
OMe
NaH
DME
R=Me
NaOEt
EtOH
R=Et
CO2Me
(47%)
(21%)
(89%)

Scheme 7

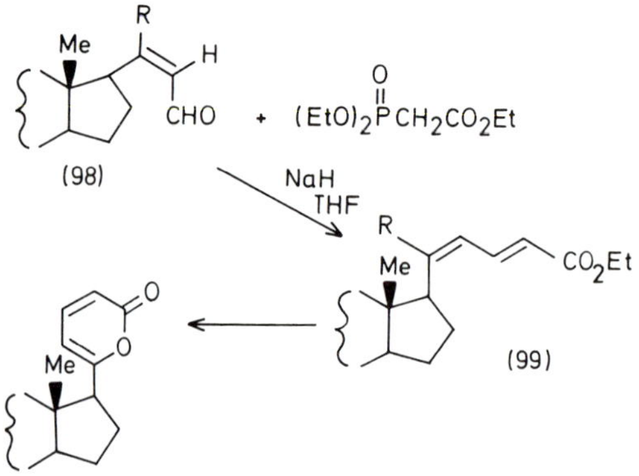
R
Me
H
CHO
+ (EtO)2PCH2CO2Et
(98)
NaH
THF
CO2Et
(99)

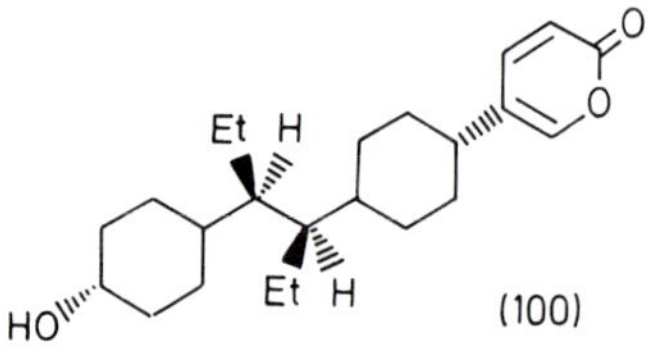
Et H
Et H
HO

(100)

3 PO-Activated Olefinations

Olefination in the presence of an epoxy group offers an alternative route to bufadienolides, and this has been used [201] in the synthesis of scillarenin (Scheme 8) (see also reference 202).

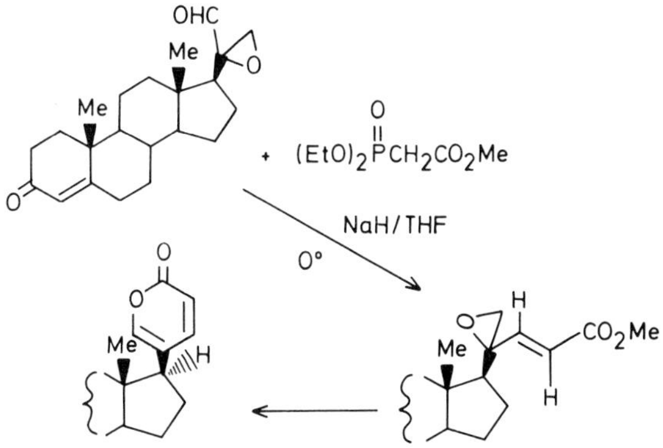

Scheme 8

Phosphonate carbanions have also been used in the synthesis of a variety of cardenolides [203] and cardenolide analogues.[204] A common approach has involved esterification of 14β-hydroxy-steroid aglycones with phosphonate acetic acids followed by intramolecular olefination [205] (Scheme 9), although direct olefination of the carbonyl group followed by intramolecular ester exchange (Scheme 10) has also been used.[194]

Cyclic α,β-unsaturated ketones have been prepared by the reaction of

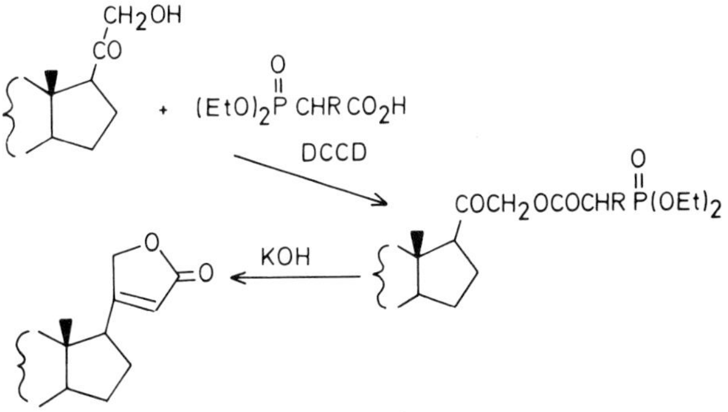

Scheme 9

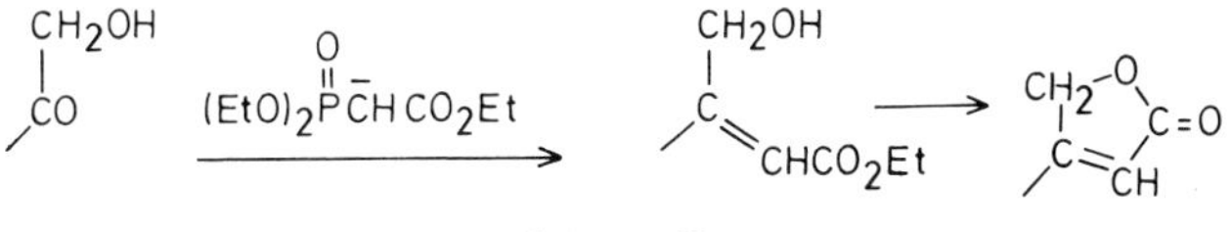

Scheme 10

phosphonate anions with cyclic enol lactones (Scheme 11), and the reaction has been applied to the synthesis of steroid A and D rings.[206]

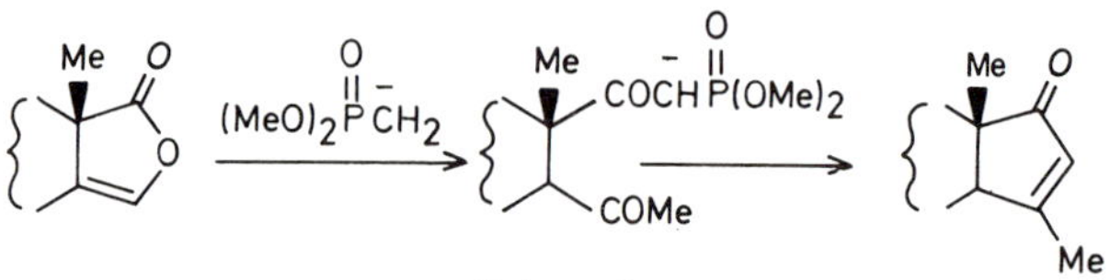

Scheme 11

5.1.11. Carbohydrates

Various branched-chain sugars have been synthesized by phosphonate olefination of the corresponding ketone followed by hydrogenation.[207,208] More interesting is the use of methylenebiphosphonates both as a chain extension method and to prepare phosphonate analogues of purine ribonucleotides[209] and vinylphosphonates[210] from aldehydo sugars. The last case gives entirely *trans* product (101), but novel variations using the silylmethylphosphonate (102) or the phosphonate ylide (103)[211] give predominantly *cis*.

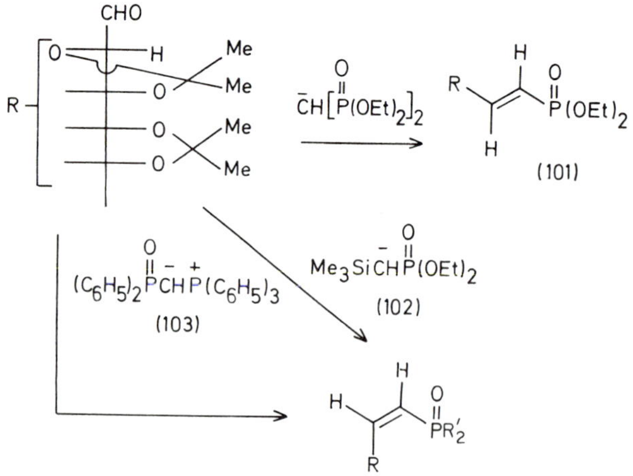

5.1.12. Nitrogen heterocycles

Phosphonate carbanions have been used in a standard way to prepare hetero analogues (104) of methylenecyclobutane[212] and substituted benzimidazoles (105).[213]

Good yields of the expected olefin (107), as a mixture of geometrical isomers, have been obtained from benzo[a]quinolizine (106) and diethyl ethoxycarbonylmethylphosphonate,[214] and used in a synthesis of emetine.

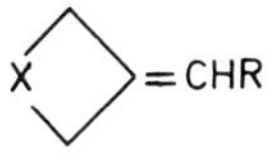

3 PO-Activated Olefinations

[Structures (104), (105), (106), (107), (108) shown with reagents (EtO)$_2$PCH$_2$CO$_2$Et, ^{t}BuOK/DMF or EtOH R.T., and NaOEt/EtOH reflux]

X = C$_6$H$_5$CH$_2$N, S
R = CO$_2$Me, CN, Cl
(104)

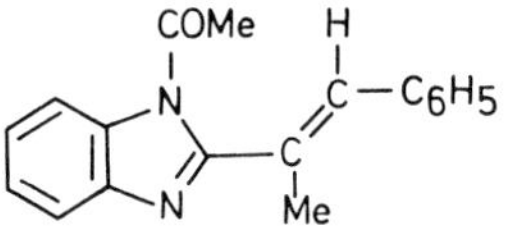

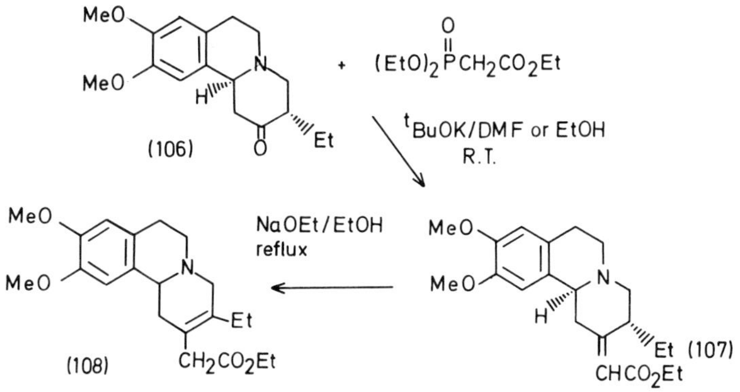

In the presence of excess base some epimerization occurs, and refluxing the product in sodium ethoxide–ethanol gives the endocyclic isomer (108). Similar results have been obtained from the model compounds (109),[84a] (110),[84b] and (111),[84c] although if diethyl cyanomethylphosphonate is used in place of the ester phosphonate only exocyclic olefin is obtained irrespective of the conditions.

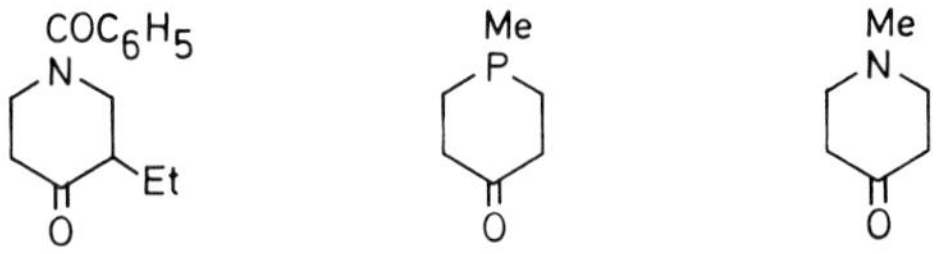

[Structures (109), (110), (111) shown]

In the case of 1-acetyl-3-indolinones the endocyclic olefin (112) is the only product from a reaction with cyanomethylphosphonate carbanion,[85] and ester- and keto-stabilized phosphonate anions did not react at all.

The reaction of the hydroxy-lactam (113) with ester phosphonate carbanion has been used in a biomimetic asymmetric synthesis of pyrrolidine

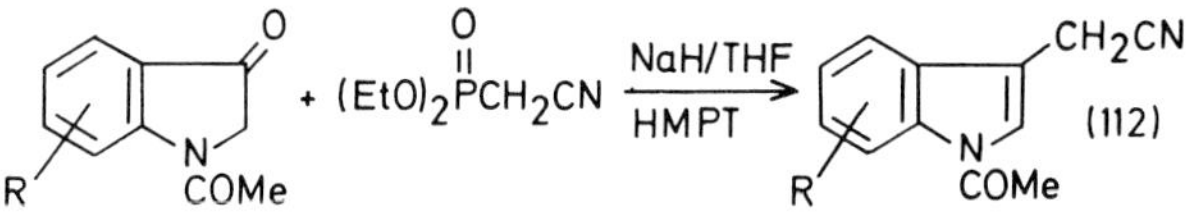

[Reaction scheme with (EtO)$_2$PCH$_2$CN, NaH/THF, HMPT giving structure (112)]

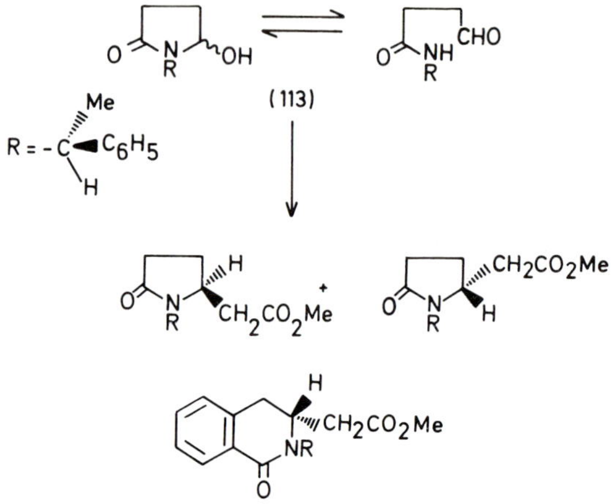

derivatives.[215] 3-Substituted dihydroisocarbostyril derivatives (114) have been prepared by an analogous route.[216]

The reactions of phosphonate carbanions with 4-amino-5-nitrosopyrimidine have been used as a new synthesis of pteridine derivatives (Scheme 12).[69]

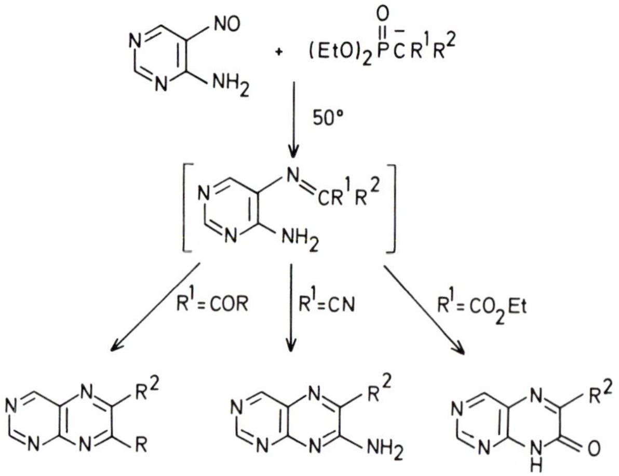

Scheme 12

5.1.13. Intramolecular reactions

The cyclization of phosphonates (115) under very mild conditions to give unsaturated lactones and the application of the reaction to cardenolide synthesis has already been discussed (p. 183). Analogous cyclization of β,ε-diketo-phosphonates to cyclopentenones, e.g. *cis*-jasmone (116), requires

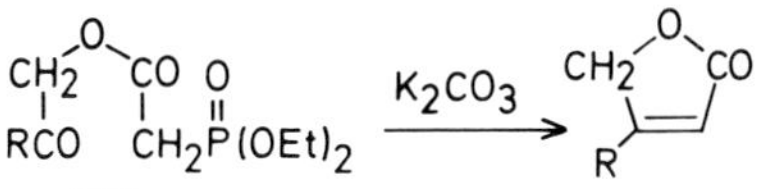

3 PO-Activated Olefinations

[Structure (115): CH$_2$-O-CO with RCO and CH$_2$P(OEt)$_2$ groups] → K$_2$CO$_3$ → [cyclic product with R substituent]

[Structure showing diketone phosphonate] → NaH, THF → [cyclopentenone (116)] (68%)

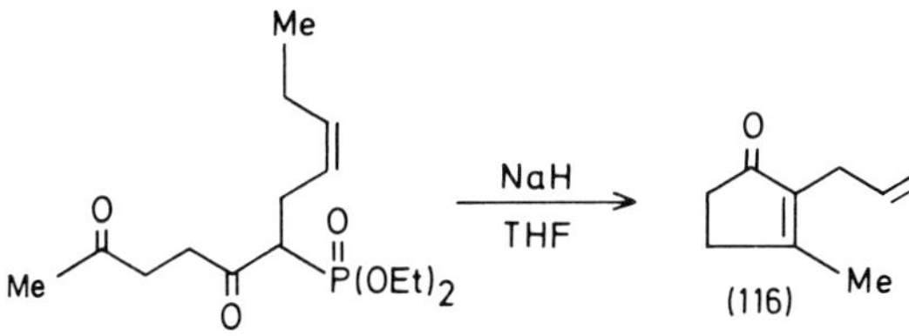

somewhat more vigorous conditions,[217] as does synthesis of alkylpyrrolones and dihydropyridones from the corresponding amino-ketones (Scheme 13).[218] Cyclohexenones (117) can be prepared by an analogous method.[219]

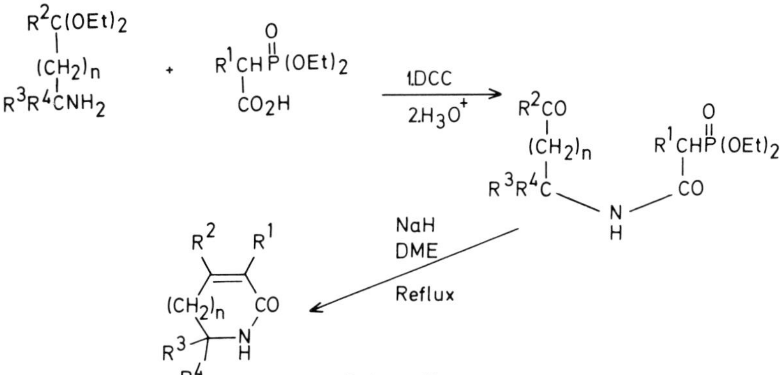

Scheme 13:

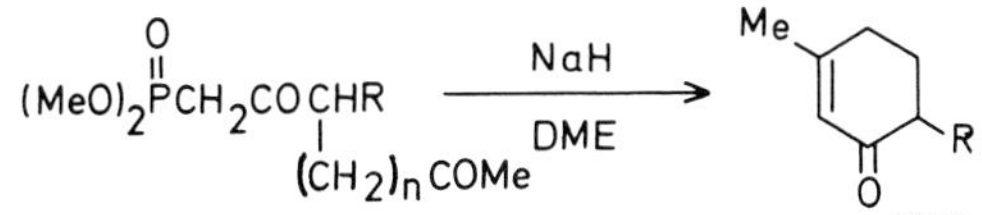

$R^2C(OEt)_2$, $(CH_2)_n$, $R^3R^4CNH_2$ + $R^1CHP(OEt)_2$, CO_2H → (1.DCC, 2.H$_3$O$^+$) → R^2CO, $(CH_2)_n$, R^3R^4C-NH-CO-$CHR^1P(OEt)_2$ → (NaH, DME, Reflux) → cyclic enamine product with R^1, R^2, $(CH_2)_n$, R^3, R^4, N-H, CO

$(MeO)_2PCH_2COCHR(CH_2)_nCOMe$ → NaH, DME → (117) [cyclohexenone with Me and R]

Christensen and his coworkers[220] have used the same principle in the preparation of thiazines [equation (13)], and the extremely mild conditions required allow application of the reaction to β-lactam antibiotic synthesis, e.g. (±)-cefoxitin (118).

Olefination of diketones with bisphosphonates has been used to prepare cyclic dienes, e.g. (119), but the yields are often poor.[221]

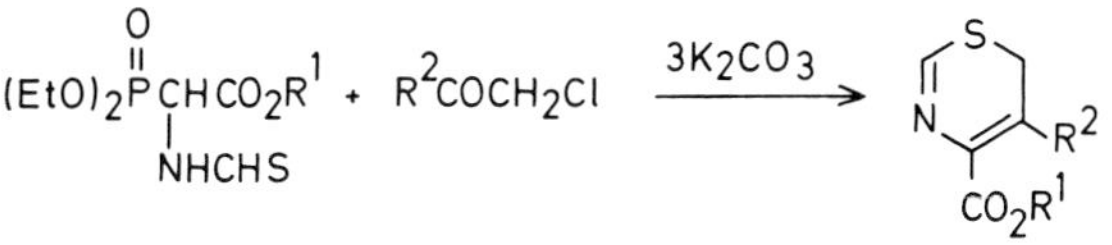

$(EtO)_2PCHCO_2R^1$ (with NHCHS) + R^2COCH_2Cl → 3K$_2$CO$_3$ → [thiazine with S, N, R^2, CO_2R^1]   (13)

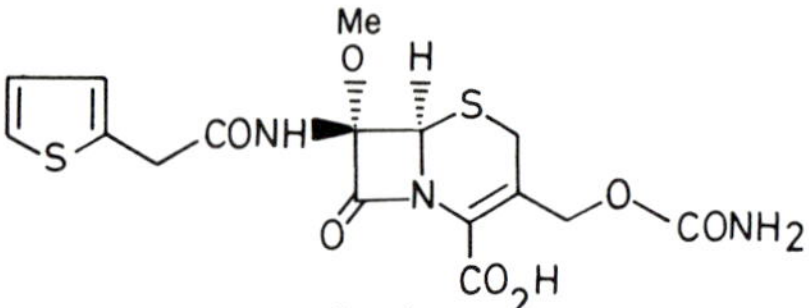

(118)

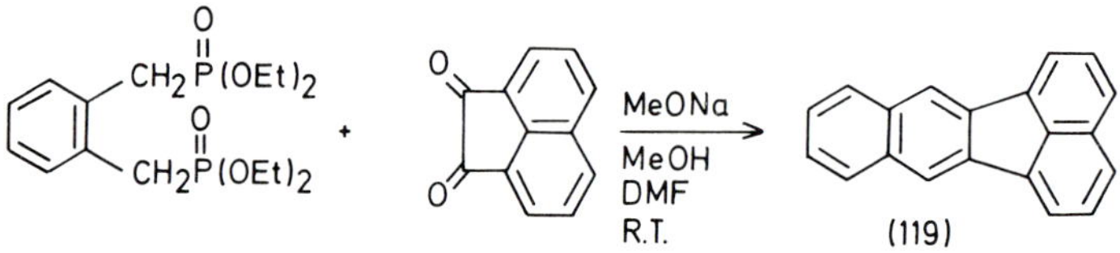

(119)

5.2. Phosphoramidates

Wadsworth and Emmons have shown[9] that phosphoramidate anions (120) react with carbonyl groups in an analogous way to phosphonate carbanions and have used the reaction to prepare a variety of imines and heteroallenes under mild conditions and in good yields (Scheme 14). The advantages of this reaction over those with phosphine imines (121) are similar to those of phosphonate carbanions over phosphonium ylides (see p. 157). In spite of this the reaction has been little used, although the same authors have

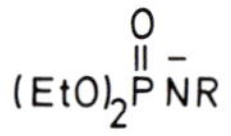

$(EtO)_2\overset{O}{\overset{\|}{P}}\overset{-}{N}R$ $(C_6H_5)_3P=NR$

(120) 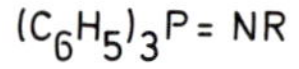   (121)

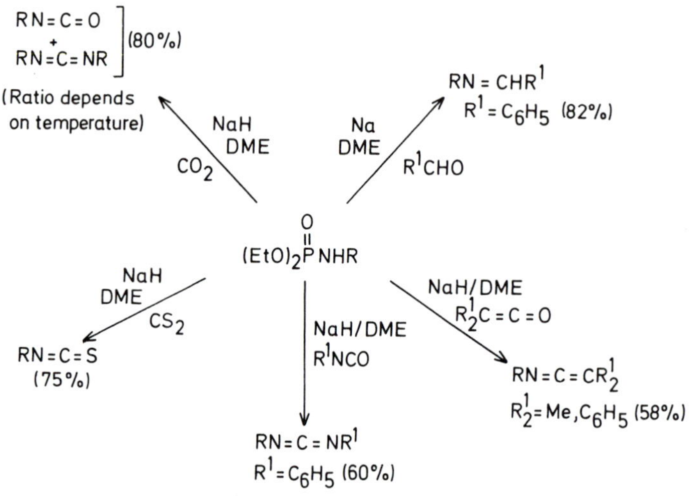

Scheme 14

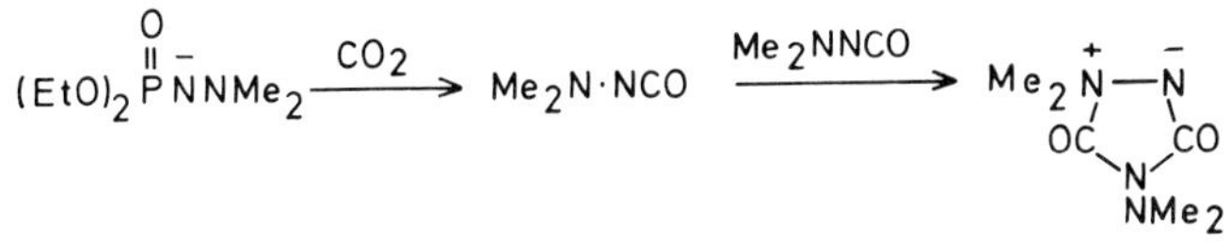

3 PO-Activated Olefinations

$$(EtO)_2\overset{O}{\overset{\|}{P}}\overset{-}{N}NMe_2 \xrightarrow{CO_2} Me_2N \cdot NCO \xrightarrow{Me_2NNCO} Me_2\overset{+}{N}-\overset{-}{N}\underset{\underset{NMe_2}{N}}{\overset{OC\diagdown\diagup CO}{}}$$

(122)

reported[222] the synthesis of dimethylaminoisocyanates, which rapidly dimerize to triazoline-3,5-dione ylides (122).

5.3. Related Reactions

In addition to olefin formation with carbonyl compounds, PO-stabilized carbanions undergo a number of mechanistically related reactions of some synthetic potential.

The direct formation of stilbene[223] from carbanion (123) on treatment with a deficiency of oxygen involves initial formation of benzaldehyde followed by a "normal" olefination; in the presence of an excess of oxygen benzophenone is formed in excellent yield from (124). It is interesting that this report pre-dates that of the analogous reaction of phosphonium ylides.[224]

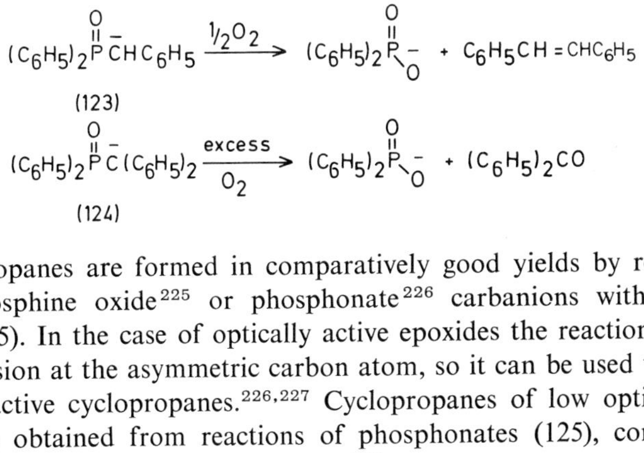

$$(C_6H_5)_2\overset{O}{\overset{\|}{P}}\overset{-}{C}HC_6H_5 \xrightarrow{\frac{1}{2}O_2} (C_6H_5)_2\overset{O}{\overset{\|}{P}}{}^-_{\diagdown O} + C_6H_5CH=CHC_6H_5$$

(123)

$$(C_6H_5)_2\overset{O}{\overset{\|}{P}}\overset{-}{C}(C_6H_5)_2 \xrightarrow[O_2]{excess} (C_6H_5)_2\overset{O}{\overset{\|}{P}}{}^-_{\diagdown O} + (C_6H_5)_2CO$$

(124)

Cyclopropanes are formed in comparatively good yields by reaction of either phosphine oxide[225] or phosphonate[226] carbanions with epoxides (Scheme 15). In the case of optically active epoxides the reaction proceeds with inversion at the asymmetric carbon atom, so it can be used to prepare optically active cyclopropanes.[226,227] Cyclopropanes of low optical purity have been obtained from reactions of phosphonates (125), containing a chiral alkyl group, with racemic epoxides.[228]

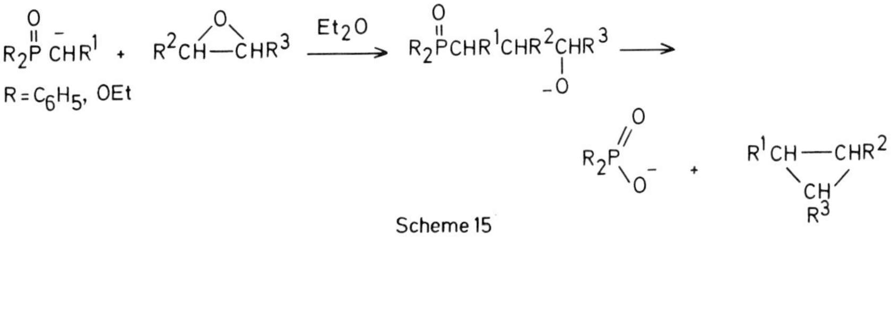

$$R_2\overset{O}{\overset{\|}{P}}\overset{-}{C}HR^1 + R^2CH\overset{O}{\diagup\diagdown}CHR^3 \xrightarrow{Et_2O} R_2\overset{O}{\overset{\|}{P}}CHR^1CHR^2\underset{\underset{-O}{|}}{C}HR^3 \longrightarrow$$

$R = C_6H_5$, OEt

$$R_2P\overset{O}{\diagup}{}_{\diagdown O^-} + \underset{R^3}{R^1CH-CHR^2\diagdown CH\diagup}$$

Scheme 15

$$(EtO)_2\overset{O}{\overset{\|}{P}}CH_2CO_2R^1 \qquad R = (-)\,Menthyl, (-)Octyl$$

(125)

Recently the reaction of nitrones with phosphonate carbanions has been extensively investigated.[229] The proportions of aziridine (126) and alkene (127) in the product depend on the nitrone substrate and the reaction conditions.

$$\underset{\overset{-}{O}}{\overset{R^1}{\underset{R^2}{\Large\diagdown}}\!\!\overset{+}{N}\!\!\diagup} + (EtO)_2\overset{O}{\overset{\|}{P}}CH_2X \xrightarrow[DME]{NaH} \underset{(126)}{\overset{R^1}{\underset{R^2}{\Large\diagdown}}\!\!N\!\!\diagup\!\!\diagdown_X} + \underset{(127)}{\overset{R^1}{\underset{R^2}{\Large\diagdown}}\!\!\underset{H}{N}\!\!\diagup\!\!CHX}$$

Diethyl phosphoroselenoate (128) and diethyl phosphorotelluroate (129) have been used to deoxygenate sulphoxides[230] and epoxides,[231] respectively, and in both cases the driving force is formation of diethyl phosphate.

$$(EtO)_2\overset{O}{\overset{\|}{P}}XH \qquad X = Se\ (128)$$
$$X = Te\ (129)$$

6. Examples of Experimental Procedures

In this section a fairly typical Michaelis–Arbusov reaction is described; however, the conditions required vary considerably depending on the nature of the halide used. The advantages of using trimethyl or triethyl phosphite are their ready availability and the volatility of methyl and ethyl halides. Although the procedure involving alkylphosphonic acid bis(dimethylamides) appears lengthy, an example has been included because it provides the only practicable route to unactivated alkenes using PO-activated reagents.

The conditions required for carbanion formation vary considerably, depending on its stability. Generally the carbanion is best generated before addition of the carbonyl component (especially in the case of aldehydes); however, with highly reactive carbanions, generation in the presence of the ketone may be preferable and a nitrogen atmosphere is invariably required.

Solvent choice does not seem to be especially important (however, see p. 169), but bases that provide a gaeous by-product offer a convenient method of estimating the extent of conversion into carbanion.

Finally, high temperatures are normally unnecessary and in fact should be avoided since they frequently reduce yields.

3 PO-Activated Olefinations

6.1. Diethyl 1-Naphthylmethanephosphonate [240]

A mixture of 43 g of 1-chloromethylnaphthalene and 41 g of triethyl phosphite is heated for 4 h at 150–160°. Distillation of the mixture gives 58 g (87%) of the title phosphonate, b.p. 205–206°/5 mm.

6.2. 3,6,10-Trimethylundeca-3,9-dien-2-ones [108]

To a suspension of 1·56 g (0·04 mol) of sodium amide in 30 ml of dry DMF there was added a solution of 8·33 g (0·04 mol) of diethyl 1-methyl-2-oxopropylphosphonate in 20 ml of dry DMF with stirring and cooling in an ice–salt bath. Stirring and cooling were continued and dry nitrogen gas was passed into the mixture until the evolution of ammonia had ceased. The resultant solution was cooled below $-10°C$ and treated dropwise with 6·17 g (0·04 mol) of β-citronellal in 20 ml of dry DMF. After the solution had stood overnight the solvent was removed under vacuum and the residue was poured into water and extracted with ether. The ethereal extract was washed with water and dried over sodium sulphate. The solvent having been evaporated, the residue was distilled under reduced pressure to yield the product (82%; 94·5% *trans*, 5·5% *cis*, by g.l.c.) as a colourless oil, b.p. 74·5–78·0°/0·15 mm, ν_{max} (liquid film) 1689 (*cis*) and 1670 (*trans*) cm^{-1}.

6.3. *trans*-Stilbene [3]

Diethyl benzylphosphonate (11·4 g, 0·05 mol), 50% sodium hydride (2·4 g, 0·05 mol), and benzaldehyde (5·3 g, 0·05 mol) were added to 100 cc of dry 1,2-dimethoxyethane. The mixture was heated slowly to 85° with stirring. At 70° there was a large evolution of gas and the appearance of a semi-solid precipitate. The solution was refluxed for 0·5 h, cooled, and then taken up in a large excess of water. The aqueous solution was filtered and the precipitate recrystallized twice from alcohol, giving 5·6 g (62·6%) of white crystals, m.p. 124·5°. The infrared spectrum of the product was identical with that of an authentic sample of *trans*-stilbene.

6.4. Ethyl 4-Methylpenta-2,4-dienoate [155]

Sodium (4·6 g, 0·2 g-atom) was added to absolute ethanol (100 ml). After the metal had dissolved, the solution was cooled in an ice bath and triethyl phosphonoacetate (44·8 g, 0·20 mol) was added. After the resulting solution had been stirred at 0° for about 10 min, a solution of 2-methylacrolein (14·0 g, 0·20 mol) in ethanol (20 ml) was added slowly. The reaction was strongly exothermic and the rate of addition was controlled so that the solution temperature did not exceed 20°. When addition of the aldehyde was complete, the ice bath was removed and the yellow solution was stirred at room temperature for 15–30 min. The reaction mixture was then poured into brine and the mixture extracted with hexane. The hexane was dried over magnesium sulphate, concentrated at reduced pressure, and distilled to give the title diene (52%), b.p. 81–83°/17 mm, and ethyl 4-methyl-5-ethoxypent-2-enoate (4%), b.p. 120–122°/17 mm.

6.5. 2-Phenylbut-2-ene from Ethylphosphonic Acid Bis(dimethylamide)[13]

A stirred solution containing 1·0 g (6·09 mmol) of ethylphosphonic acid bis-(dimethylamide) in 15 ml dry THF was treated, at −78° and under nitrogen, with 3·9 ml (6·25 mmol) of 1·6M n-butyllithium in hexane. The mixture required stirring at −50° for 3 h to effect complete metallation. This solution was cooled to −70° and 0·75 g (0·60 mmol) of acetophenone was added with stirring. After stirring at −70° for 2 h and at −70° to +25° for 1 h, 10 ml of water was added. The mixture was extracted with ether; the ether extracts were washed with water, dried, and evaporated under vacuum to afford an oil. The oil, shown by infrared spectroscopy to contain acetophenone, was crystallized from pentane–ether (10:1) to give a 1·30 g (74%) mixture of two diastereoisomers, m.p. 99–111°. A solution of this mixture and 0·2 g of silica gel in 1·5 ml of toluene was heated at reflux for 15 h. G.l.c. analysis of the product after dilution with ether and filtration indicated a 1·09:1·0 ratio of *cis*- and *trans*-2-phenylbut-2-ene in 90% yield. The n.m.r. spectrum of the olefinic mixture after removal of the solvent was in agreement with that reported for *cis*- and *trans*-2-phenylbut-2-ene.

7. Summary of PO-Stabilized Anions

7.1. Phosphine Oxides

Anion	Conditions	Reference
$Ph_2P(O)\bar{C}H_2$	$NaNH_2$/benzene	1
$Ph_2P(O)\bar{C}HCH_2P(O)Ph_2$	Bu^tOK/benzene or toluene	23
$Ph_2P(O)\bar{C}HMe$	$BuLi/(Me_2NCH_2)_2$	142, 225
$Ph_2P(O)\bar{C}HOMe$	$Pr^i_2NLi/THF/20°$	50
$\begin{array}{c}Me\\ \\ Me\end{array}\!\!\!\!\!\!\bigg\rangle\!\!P(O)\bar{C}MeCN$	$BuLi$ or Bu^tOK/THF	22
$Ph_2P(O)\bar{C}HCH=CH_2$	$BuLi/Et_2O/-78°$	54, 105
$Ph_2P(O)\bar{C}HCH=CRMe$	$BuLi/THF$	53, 158
	$NaOMe/MeOH/DMG/20°$	159

3 PO-Activated Olefinations

Anion	Conditions	Reference
Ph$_2$P(O)C̄HPh	BuLi/Et$_2$O	41, 54
	C$_6$H$_5$Li/Et$_2$O	5, 103
	Bu^{t}OK/benzene or toluene	5, 6, 18, 40
Ph$_2$P(O)C̄Me(C$_6$H$_{10}$) 	BuLi/Et$_2$O	105
Ph$_2$P(O)C̄Ph$_2$	Bu^{t}OK/benzene or toluene	223

7.2. Phosphinates

Anion	Conditions	Reference
Ph(MeO)P(O)C̄HCO$_2$Me	Benzene/Et$_2$O	151
Ph(EtO)P(O)C̄HPh	Bu^{t}OK/benzene or toluene	10

7.3. Phosphonates

Anion	Conditions	Reference
(RO)$_2$P(O)C̄H$_2$	Bu^{n}Li/THF/$-78°$	47, 48, 49
(RO)$_2$P(S)C̄H$_2$	Bu^{n}Li/THF/$-70°$ also Cu compound	42, 206
[(EtO)$_2$P(O)]$_2$C̄H	NaH/DME	210
(EtO)$_2$P(O)C̄HP(O)(OEt)CH$_2$P(O)(OEt)$_2$	NaH/benzene	16
(EtO)$_2$P(O)C̄Br$_2$	Pr$^i_2$NLi/Et$_2$O/THF	134
(EtO)$_2$P(O)C̄Cl$_2$	BuLi/THF/$-80°$	135, 136
	BuLi/CCl$_4$/$-75°$ [from (EtO)$_2$P(O)CH$_2$Cl]	236
(MeO)$_2$P(O)C̄N$_2$	BuLi/THF/$-80°$	153
(RO)$_2$P(S)C̄HMe	BuLi/THF/$-78°$	47
(EtO)$_2$P(O)C̄HSMe	NaH/DME/HMPA/$62°$	177, 178
(EtO)$_2$P(O)C̄HSO$_2$R	BuLi/THF	139
(EtO)$_2$P(O)C̄ClR	BuLi/THF/Et$_2$O/$-90°$	52
(EtO)$_2$P(O)C̄HCN	NaNH$_2$/THF	8
	NaH/DME	106, 112
	NaH/DMF or DMSO	111
	NaH/benzene or THF	112
	NaH/THF/HMPT	85
	MeLi/benzene	111
	Na or K/Et$_2$O/dioxan	233
	Mg/NH$_3$(l)	233
	NaOH/H$_2$O/phase-transfer compound	38
(EtO)$_2$P(O)C̄RNC R = H, Ph	BuLi/THF/$-70°$	122
	NaCN/EtOH	122
(EtO)$_2$P(O)C̄HCO$_2^-$	2BuLi/hexane/THF/$-60°$	49
(PhCH$_2$O)$_2$P(O)C̄HCO$_2$	2Pr$^i_2$NLi/THF/$-80°$	120

Anion	Conditions	Reference
$(EtO)_2P(O)\bar{C}HCOMe$	BuLi/THF	36
	NaH/THF/10°	35, 98, 108
	$NaNH_2$/THF	8
	$NaNH_2$/DMF	108
	NaOEt/EtOH	78
	Zn^{2+} salts	45
$(EtO)_2P(O)\bar{C}MeCN$	Bu^tOM/various solvents (M = Li, Na, K)/−78° to +65°	91, 104
$(MeO)_2P(O)\bar{C}HCO_2Me$	NaH/glyme/15–35°	99
$(EtO)_2P(O)\bar{C}MeSMe$	NaH/DME/HMPT/62°	177, 178
$(EtO)_2P(O)\bar{C}HSO_2Et$	NaH/DME or Et_2O or toluene	109
$(EtO)_2P(O)\bar{C}HS^+RMe$	NaH/THF/−20°	140
$(PhO)_2P(O)\bar{C}R^1NHR^2$	KOH/EtOH/−30°	143
$[(EtO)_2P(O)]_2\bar{C}NMe_2$	NaH/dioxan/20°	148
$(MeO)_2P(O)\bar{C}HCH=CClMe$	NaH/DME	83
$(EtO)_2P(O)\bar{C}HCO_2Et$	NaH/DME	3, 84, 98, 100, 107, 167
	NaH/diglyme	81, 123
	NaH/benzene/20°	164
	$NaNH_2/Et_2O$	81
	$NaNH_2$/THF or DMF	108
	BuLi/THF	46
	Bu^tOK/THF	46
	EtONa/EtOH	84, 155, 165
	Na or K/benzene	4
	piperidine/EtOH	7
$(EtO)_2P(S)\bar{C}HCO_2Et$	Na	78
	Bu^tOK/benzene	15
$\begin{array}{c}Me-O\\ \diagdown\\ P(O)-\bar{C}HCO_2Et\\ \diagup\\ Me-O\end{array}$	NaH/DME/20°	21
$(EtO)_2P(O)\bar{C}HCONR^1R^2$	NaH/DME	79
	EtONa/EtOH	165, 170
$(EtO)_2P(O)\bar{C}HSCH_2CH=CH_2$	BuLi/THF/−78°	141
$(EtO)_2P(O)\bar{C}\!\!\begin{array}{c}S\!-\!\!\\ \diagdown\\ \diagup\\ S\!-\!\!\end{array}$	BuLi/THF	138
	KOH/EtOH	138
$(EtO)_2P(O)\bar{C}=CR^1NHR^2$	NaH/THF/0°	117, 119
$(EtO)_2P(O)\bar{C}R^1N=CHR^2$	NaH/THF	145
$(EtO)_2P(O)\bar{C}HSiMe_3$	BuLi/THF	210
$(EtO)_2P(O)\bar{C}R^1CO_2Et$	Bu^tOK/DME	205
	K_2CO_3	205
	KOH/H_2O	205
$R^1 = Me$	EtONa/EtOH	98
	BuLi/THF	46
$(EtO)_2P(O)\bar{C}HCH=CHCO_2Me$	MeONa/MeOH/DMF	157, 160

3 PO-Activated Olefinations

Anion	Conditions	Reference
$(EtO)_2P(O)\bar{C}HCR^1=CR^2X$ $X = OEt, OC_6H_5, SPh, NR_2,$ Br, Cl	NaH/DME/HMPT	83
$(EtO)_2P(O)\bar{C}HCH(OEt)_2$	$NaNH_2$/THF	8
$(EtO)_2P(O)-\bar{C}H-N\bigcirc$	BuLi/THF/$-78°$	146
$(RO)_2P(O)\bar{C}(COR^3)NR^1R^2$	NaH/THF	147
$(EtO)_2P(O)\bar{C}(CO_2R^1)NHCHS$	K_2CO_3	220
$(EtO)_2P(O)\bar{C}(CN)OBu^t$	NaH/THF	126
$(EtO)_2P(O)\bar{C}(CO_2Et)OEt$	NaH/THF	200
$(EtO)_2P(O)\bar{C}HCMe=CHCO_2Me$	Pr^i_2NLi/THF/HMPT/$-65°$	169
$(EtO)_2P(O)\bar{C}HCH=CR(OEt)$		232
$(EtO)_2P(O)\bar{C}HPh$	$NaNH_2$/benzene	1, 102
	NaOMe/DMF	235
$(EtO)_2P(O)\bar{C}FPh$	NaH/DMF/$60°$	237
$(PhO)_2P(O)\bar{C}ClPh$	NaH/DMSO	137
$(EtO)_2P(O)\bar{C}H\langle\text{F-C}_6\text{H}_4\text{-F}\rangle$	NaOMe/DMF	131
$(EtO)_2P(O)\bar{C}Br(p-NO_2C_6H_4)$	NaOR/ROH	238
$[(PhO)_2P(O)\bar{C}Cl]_2p-C_6H_4$	NaH/THF	239
$(MeO)_2P(O)\bar{C}HCO(CH_2)_4Me$	NaH/THF	179, 180
	NaH/DME	182
$(EtO)_2P(O)\bar{C}(CN)CR^1=CMe_2$	NaH	86
$(EtO)_2P(O)\bar{C}BuCO_2Et$	NaH/DME	3
$(EtO)_2P(O)\bar{C}H(CH=CH)_2CO_2Et$	NaH/DME	188
$(EtO)_2P(O)\bar{C}HCONRPh$ $R = alkyl$		56
$(EtO)_2P(O)\bar{C}H(p-MeC_6H_4)$	$NaNH_2/Et_2O$ or $NH_3/-33°$	113
$(EtO)_2P(O)\bar{C}H(p-CNC_6H_4)$	NaH/DMSO/$100°$	129
$(EtO)_2P(O)\underset{(EtO)_2P(O)\bar{C}H}{\bar{C}H}\langle\text{C}_6\text{H}_4\rangle$	NaOMe/MeOH/DMF	221
	Bu^tK/toluene/$110°$	166
$(EtO)_2P(O)\bar{C}H\langle\text{Me-O-N(Me)-Me}\rangle$	NaH/DME	118
$(EtO)_2P(O)\bar{C}RCH=NC_6H_{11}$	NaH/THF/$0°$	234
$(EtO)_2P(O)\bar{C}HR$ $R = $ 2-pyridyl, 2-quinolyl	NaH/toluene	195
$(EtO)_2P(O)\bar{C}H-\alpha$-naphthyl	MeONa/DMF	128
$(PhO)_2P(O)-\bar{C}\langle\text{Ph, N-morpholino}\rangle$	KOH/MeOH	144

Anion	Conditions	Reference
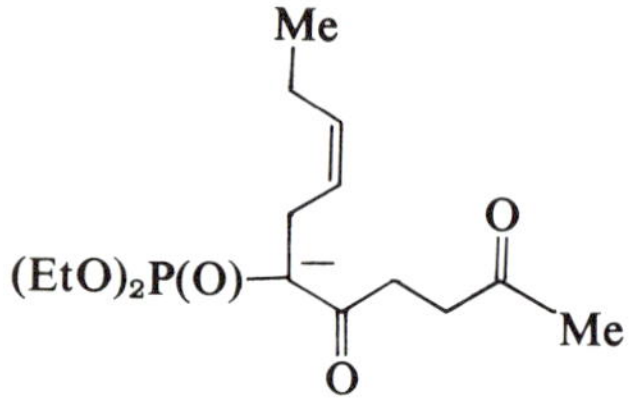 (EtO)$_2$P(O)–CH(CH$_2$CH=CHMe)–CH$_2$C(O)Me (with Me group as shown)	NaH/THF	217
(EtO)$_2$P(O)C̄(CO$_2$Et)(CH$_2$)$_n$-CO$_2$Et	NaH/THF (poor yields of olefin with ketones)	57
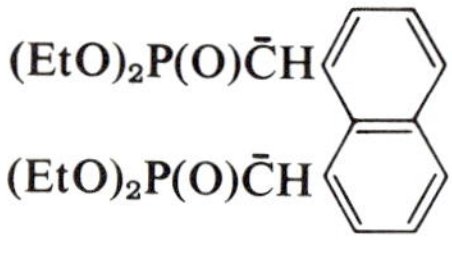 (EtO)$_2$P(O)C̄H(naphthyl)	NaOMe/DMF	132
(EtO)$_2$P(O)C̄H(naphthyl)		
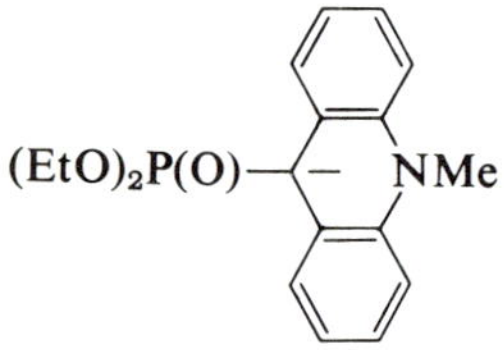 (EtO)$_2$P(O)–CH–(N-methylacridinyl) NMe	NaH/DME/70°	130
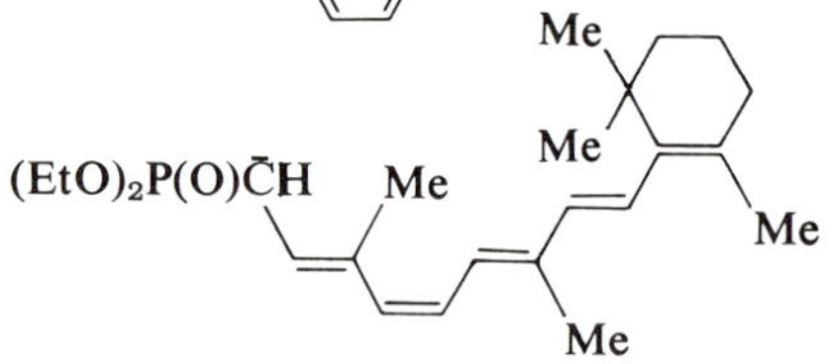 (EtO)$_2$P(O)C̄H–(retinyl-type polyene with Me substituents and trimethylcyclohexenyl group)	MeONa/pyridine/0°	163

7.4. Phosphonamides

Anion	Conditions	Reference
(R$_2$N)$_2$P(O)C̄R^{1}R^2	BuLi/THF/–78°	11–13
(Me$_2$N)$_2$P(O)C̄HOR		101
(Me$_2$N)$_2$P(O)C̄HCH=CH$_2$	BuLi/THF (α-coupling with ketones)	14
[Me-N, Me-N]P(O)C̄HCH=CH$_2$	BuLi/THF (α-coupling with ketones)	14
(Me$_2$N)$_2$P(O)C̄HCOEt		186

7.5. Phosphoramidates

Anion	Conditions	Reference
(EtO)$_2$P(O)N̄R	NaH/DME	9
(EtO)$_2$P(O)N̄NMe$_2$	NaH/benzene/30°	222

8. References

Bibliography for further reading:

W. S. Wadsworth, Jr., "Synthetic Applications of Phosphoryl-Stabilised Anions," *Org. Reactions*, 1978, **25**, 73.
A. W. Johnson, *Ylid Chemistry*, Academic Press, New York, 1966.
J. Boutagy and R. Thomas, "Olefin Synthesis with Organic Phosphate Carbanions," *Chem. Rev.*, 1974, **75**, 87.
A. Hantz, "Utilization of Compounds Activated with PO Groups as Reagents in the Synthesis of Olefins," *Stud. Cercet. Chim.*, 1968, **16**, 665 (Romanian).
A. V. Dombrovskii and V. A. Dombrovskii, "Olefin Synthesis with PO Activated Reagents," *Russ. Chem. Rev.*, 1966, **35**, 733.
H. O. Huisman, "The Wittig Reaction and the Utility of Phosphonate Carbanions in Olefin Synthesis," *Chem. Weekblad*, 1963, **59**, 133.
L. Horner, "Preparative Phosphorus Chemistry," *Fortschr. Chem. Forsch.*, 1966, **7**, 1.
S. Trippett, "Ylides and Related Compounds," Specialist Periodical Reports, *Organophosphorus Chemistry*, Ch. 9; Vol. 1 appeared in 1970.

1. L. Horner, H. Hoffmann, and H. G. Wippel, *Chem. Ber.*, 1958, **91**, 61.
2. G. Wittig and U. Schöllkopf, *Chem. Ber.*, 1954, **87**, 1318.
3. W. S. Wadsworth and W. D. Emmons, *J. Amer. Chem. Soc.*, 1961, **83**, 1733.
4. A. E. Arbusov and A. A. Dunin, *Ber. Deut. Chem. Ges.*, 1927, **60**, 291; A. E. Arbusov and A. I. Razumov, *J. Russ. Phys. Chem. Soc.*, 1929, **61**, 623 (*Chem. Abs.*, 1929, **23**, 4444).
5. L. Horner, H. Hoffmann, H. G. Wippel, and G. Klahre, *Chem. Ber.*, 1959, **92**, 2499.
6. H. Hoffmann, *Angew. Chem.*, 1959, **71**, 379.
7. S. Patai and A. Schwartz, *J. Org. Chem.*, 1960, **25**, 1232.
8. H. Takahashi, K. Fugiwara, and M. Ohta, *Bull. Chem. Soc. Japan*, 1962, **35**, 1498.
9. W. S. Wadsworth and W. D. Emmons, *J. Amer. Chem. Soc.*, 1962, **84**, 1316; *J. Org. Chem.*, 1964, **29**, 2816.
10. L. Horner, H. Hoffmann, W. Klink, A. Ertel, and V. G. Toscano, *Chem. Ber.*, 1962, **95**, 581.
11. E. J. Corey and G. T. Kwiatkowski, *J. Amer. Chem. Soc.*, 1964, **88**, 5652.
12. E. J. Corey and G. T. Kwiatkowski, *J. Amer. Chem. Soc.*, 1966, **88**, 5653.
13. E. J. Corey and G. T. Kwiatkowski, *J. Amer. Chem. Soc.*, 1968, **90**, 6816.
14. E. J. Corey and D. E. Cane, *J. Org. Chem.*, 1969, **34**, 3053.
15. J. Michalski and S. Musierowicz, *Tetrahedron Lett.*, 1964, 1187.
16. W. F. Gilmore and J. W. Huber, tert., *J. Org. Chem.*, 1973, **38**, 1423.
17. S. Trippett and D. M. Walker, *Chem. and Ind.*, 1961, 990.
18. L. Horner, W. Klink, and H. Hoffmann, *Chem. Ber.*, 1963, **96**, 3133.
19. A. W. Johnson, *Ylid Chemistry*, pp. 203–4, Academic Press, New York, 1966.
20. D. Gloyna, K. G. Bernot, H. Koeppel, and H. G. Henning, *J. Prakt. Chem.*, 1976, **318**, 327.
21. E. Breuer and D. M. Bannet, *Tetrahedron Lett.*, 1977, 1141.
22. B. Deschamps, J. P. Lampin, and F. Mathey, *Tetrahedron Lett.*, 1977, 1137.

23. L. Horner, H. Hoffmann, W. Klink, H. Ertel, and V. G. Toscano, *Chem. Ber.*, 1962, **95**, 581.
24. K. Sasse, in *Methoden der Organischen Chemie* (Houben-Weyl) (ed. E. Muller), Vol. 12, Part 1, pp. 135–65, 247–61, 423–522, G. Thieme, Stuttgart, 1964.
25. *Organic Phosphorus Compounds* (ed. G. M. Kosolapoff and L. Maier), Wiley, New York, 1972.
26. A. Michaelis and R. Kaehne, *Ber. Deut. Chem. Ges.*, 1898, **31**, 1048.
27. E.g. G. R. Harvey and E. R. De Sombre, in *Topics in Phosphorus Chemistry*, Vol. 1, Interscience, New York, 1964; reference 25, Vol. 7, p. 23; B. A. Arbusov, *Pure Appl. Chem.*, 1964, **9**, 307.
28. W. Perkow, K. Ullerich, and F. Meyer, *Naturwiss.*, 1952, **39**, 353.
29. B. J. Walker, in *Organophosphorus Chemistry* (ed. S. Trippett), Specialist Periodical Reports, The Chemical Society, London, Volumes 1–10, Ch. 5.
30. A. Michaelis and T. Becker, *Ber. Deut. Chem. Ges.*, 1892, **30**, 1003.
31. G. W. Kenner, *Proc. Chem. Soc.*, 1957, 136; G. Sturtz, *Bull. Soc. Chim. France*, 1964, 2333.
32. K. D. Berlin and G. B. Butler, *Chem. Rev.*, 1960, **60**, 243.
33. Reference 25, Vol. 1, Ch. 1; Vol. 3, Ch. 6.
34. A. W. Johnson, *Ylid Chemistry*, p. 204, Academic Press, New York, 1966.
35. E.g. R. D. Clark, L. G. Kozar, and C. H. Heathcock, *Synthesis*, 1975, 635.
36. P. A. Grieco and C. S. Pogonowski, *J. Amer. Chem. Soc.*, 1973, **95**, 3071.
37. M. Baboulène, A. Belbéoch, and G. Sturtz, *Synthesis*, 1977, 240.
38. E.g. E. D'Incan and J. Seyden-Penne, *Synthesis*, 1975, 516.
39. J. Blanchard, N. Collignon, P. Savignac, and H. Normant, *Synthesis*, 1975, 655.
40. H. Hoffmann, *Angew. Chem.*, 1959, **71**, 379; H. G. Henning, *Z. Chem.*, 1974, **14**, 209.
41. S. T. D. Gough and S. Trippett, *J. Chem. Soc.*, 1962, 2333.
42. P. Savignac and F. Mathey, *Tetrahedron Lett.*, 1976, 2829.
43. J. Villieras, A. Reliquet, and J. F. Normant, *Synthesis*, 1978, 27.
44. J. C. Tebby, in *Organophosphorus Chemistry* (ed. S. Trippett), Specialist Periodical Reports, The Chemical Society, London, Volumes 1–10, Ch. 11.
45. F. A. Cotton and R. A. Schunn, *J. Amer. Chem. Soc.*, 1963, **85**, 2394.
46. T. Bottin-Strzalko, J. Seyden-Penne, and M-P Simonnin, *Chem. Comm.*, 1976, 905; see also T. A. Albright and E. E. Schweizer, *J. Org. Chem.*, 1976, **41**, 1168.
47. E. J. Corey and G. T. Kwiatkowski, *J. Amer. Chem. Soc.*, 1966, **88**, 5654.
48. H. Paulsen and W. Bartsch, *Chem. Ber.*, 1975, **108**, 1229.
49. P. Coutrot, M. Snoussi, and P. Savignac, *Synthesis*, 1978, 133.
50. C. Earnshaw, C. J. Wallis, and S. Warren, *Chem. Comm.*, 1977, 314.
51. E. Schaumann and F. Grabley, *Annalen*, 1977, 88.
52. P. Perriot, J. Villieras, and J. F. Normant, *Synthesis*, 1978, 33.
53. B. Lythgoe, T. A. Moran, M. E. N. Nambudiry, and S. Ruston, *J. Chem. Soc., Perkin I*, 1976, 2386; B. Lythgoe, T. A. Moran, M. E. N. Nambudiry, S. Ruston, J. Tideswell, and P. W. Wright, *Tetrahedron Lett.*, 1975, 3863.
54. M. Schlosser and T. B. Huynh, *Chimia*, 1976, **30**, 197; M. Schlosser and H. B. Tuong, *ibid.*, p. 197.
55. M. Schlosser and K. F. Christmann, *Angew. Chem. Int. Ed.*, 1964, **3**, 636.
56. I. Shahak, J. Almog, and E. D. Bergmann, *Israel J. Chem.*, 1969, **7**, 585.

3 PO-Activated Olefinations

57. E.g. K. Sato, M. Hirayama, T. Inoue, and S. Kikuchi, *Bull. Chem. Soc. Japan*, 1969, **42**, 250.
58. I. Shahak, *Synthesis*, 1969, 170; I. Shahak and J. Almog, *ibid.*, 1970, 145.
59. E. D. Bergmann and A. Solomonovici, *Synthesis*, 1970, **2**, 183.
60. F. Camps, J. Coll, A. Messegner, and A. Roca, *Tetrahedron Lett.*, 1976, 791.
61. G. Durrant and J. K. Sutherland, *J. Chem. Soc., Perkin I*, 1972, 2582.
62. B. G. Kovalev, E. M. Al'tmark, and E. S. Lavrinenko, *Zh. Org. Khim.*, 1970, 2187 (*Chem. Abs.*, 1971, **74**, 41827).
63. M. Green, *J. Chem. Soc.*, 1963, 1324; E. J. Corey and J. I. Shulman, *J. Org. Chem.*, 1970, **35**, 777.
64. M. Mikolajczyk, S. Grzejszczak, and A. Zatorski, *J. Org. Chem.*, 1975, **40**, 1979.
65. S. D. Andrews, A. C. Day, and R. N. Inwood, *J. Chem. Soc. (C)*, 1969, 2443.
66. W. Grell and H. Machleidt, *Annalen*, 1966, **693**, 134.
67. M. Kirilov, J. Petrova, S. Momchilova, and B. Galunski, *Chem. Ber.*, 1976, **109**, 1684; M. Kirilov and J. Petrova, *ibid.*, 1968, **101**, 3467.
68. J. A. Maassen, T. A. J. Wajer, and T. J. de Boer, *Rec. Trav. Chim. Pays-Bas*, 1969, **88**, 5.
69. R. D. Youssefyeh and A. Kalmus, *Chem. Comm.*, 1970, 1371; E. C. Taylor and B. E. Evans, *ibid.*, 1971, 189.
70. Reference 19, p. 195.
71. K. Nara, *Kagaku To Kogyo* (Osaka), 1976, **50**, 413 (*Chem. Abs.*, 1977, **87**, 151184k).
72. C. Piechucki, *Synthesis*, 1976, 187.
73. C. Piechucki, *Synthesis*, 1974, 869.
74. E. D'Incan, *Tetrahedron*, 1977, **33**, 951.
75. M. Mikolajczyk, S. Grzejszczak, W. Midura, and A. Zatorski, *Synthesis*, 1975, 278.
76. M. Mikolajczyk, S. Grzejszczak, A. Zatorski, and B. Mlotkowska, *Tetrahedron Lett.*, 1976, 2731.
77. M. Mikolajczyk, S. Grzejsczak, W. Midura, and A. Zatorski, *Synthesis*, 1976, 396.
78. A. N. Pudovik and N. M. Lebedeva, *Zh. Obshch. Khim.*, 1955, **25**, 1920; A. N. Pudovik and N. M. Lebedeva, *Dokl. Akad. Nauk SSSR*, 1953, **90**, 799; J. Michalski and S. Musierowicz, *Tetrahedron Lett.*, 1964, 1187.
79. P. D. Landor, S. R. Landor, and O. Odyek, *J. Chem. Soc., Perkin I*, 1977, 93.
80. M. Cossentini, B. Deschamps, N. T. Anh, and J. Seyden-Penne, *Tetrahedron*, 1977, **33**, 409; B. Deschamps and J. Seyden-Penne, *ibid.*, 1977, **33**, 413.
81. E. D. Bergmann and A. Solomonovici, *Tetrahedron*, 1971, **27**, 2675.
82. A. H. Davidson, C. Earnshaw, J. I. Grayson, and S. Warren, *J. Chem. Soc., Perkin I*, 1977, 1452.
83. G. Lavielle, *Compt. Rend. (C)*, 1970, **270**, 86; G. Lavielle and G. Sturtz, *Bull. Soc. Chim. France*, 1970, 1369.
84. (a) R. J. Sundberg and F. O. Holcombe, *J. Org. Chem.*, 1969, **34**, 3273. (b) L. D. Quin, J. W. Russell, Jr., R. D. Price, and H. E. Shook, Jr., *J. Org. Chem.*, 1971, **36**, 1495. (c) R. Borne and H. Aboul-Enein, *J. Heterocyclic Chem.*, 1972, **9**, 869.
85. A. Buzas and C. Herisson, *Synthesis*, 1977, 129.
86. D. Danion and R. Carrie, *Tetrahedron Lett.*, 1971, 3219.
87. W. S. Wadsworth, Jr., *Org. Reactions*, 1978, **25**.

88. Reference 19, pp. 152–71.
89. T. Bottin-Strzalko, *Tetrahedron*, 1973, **29**, 4199; D. Danion and R. Carrie, *Compt. Rend. (C)*, 1968, **267**, 735.
90. G. Lefebvre and J. Seyden-Penne, *Chem. Comm.*, 1970, 1308.
91. A. Redjel and J. Seyden-Penne, *Tetrahedron Lett.*, 1974, 1733.
92. G. Lavielle, M. Carpentier, and P. Savignac, *Tetrahedron Lett.*, 1973, 173.
93. B. Deschamps, G. Lefebvre, and J. Seyden-Penne, *Tetrahedron*, 1972, **28**, 4209.
94. D. Danion and R. Carrie, *Tetrahedron*, 1972, **28**, 4223.
95. E. Vedejs and K. A. J. Snoble, *J. Amer. Chem. Soc.*, 1973, **95**, 5778.
96. A. J. Speziale and D. E. Bissing, *J. Amer. Chem. Soc.*, 1963, **85**, 3878; H. Goetz, F. Nerdel, and H. Michaelis, *Naturwiss.*, 1963, **50**, 496.
97. G. Sturtz, *Bull. Soc. Chim. France*, 1964, 2349.
98. H. Normant and G. Sturtz, *Compt. Rend. (C)*, 1963, **256**, 1800; G. Gallagher, Jr., and R. L. Webb, *Synthesis*, 1974, 122.
99. D. E. McGreer and N. W. K. Chiu, *Canad. J. Chem.*, 1968, **46**, 2225.
100. R. K. Huff, C. E. Moppett, and J. K. Sutherland, *J. Chem. Soc. (C)*, 1968, 2725.
101. G. Lavielle and D. Reisdorf, *Compt. Rend. (C)*, 1971, **272**, 100.
102. D. H. Wadsworth, O. E. Schupp, E. J. Seus, and J. A. Ford, Jr., *J. Org. Chem.*, 1965, **30**, 680.
103. L. Horner and W. Klink, *Tetrahedron Lett.*, 1964, 2467; L. Horner and H. Winkler, *ibid.*, p. 3265.
104. B. Deschamps, G. Lefebvre, A. Redjel, and J. Seyden-Penne, *Tetrahedron*, 1973, **29**, 2437; T. Bottin-Strzalko and J. Seyden-Penne, *Tetrahedron Lett.*, 1972, 1945.
105. A. H. Davidson and S. Warren, *Chem. Comm.*, 1975, 148; A. H. Davidson and S. Warren, *J. Chem. Soc., Perkin I*, 1976, 639.
106. G. Jones and R. F. Massey, *Chem. Comm.*, 1968, 543.
107. T. H. Kinstle and B. Y. Mandanas, *Chem. Comm.*, 1968, 1699; M. Gernayova, J. Kovac, M. Dandarova, B. Hasova, and R. Palovcik, *Coll. Czech. Chem. Comm.*, 1976, **41**, 764; D. Danion and R. Carrie, *Bull. Soc. Chim. France*, 1974, 2065.
108. K. Sasaki, *Bull. Chem. Soc. Japan*, 1968, **41**, 1252; 1967, **40**, 2967, 2968; 1966, **39**, 2703.
109. I. C. Popoff, J. L. Dever, and G. R. Leader, *J. Org. Chem.*, 1969, **34**, 1128.
110. M. J. Gallagher and I. D. Jenkins, *Topics Stereochem.*, 1968, **3**, 38.
111. J. H. Babler and T. R. Mortell, *Tetrahedron Lett.*, 1972, 669.
112. (a) J. Seyden-Penne and M. G. Lefebvre, *Compt. Rend. (C)*, 1969, **269**, 48.
    (b) H. Kaneko and M. Okazaki, *Tetrahedron Lett.*, 1966, 219.
113. M. Kirilov and J. Petrova, *Tetrahedron Lett.*, 1970, 2129.
114. M. Schlosser and K. F. Christmann, *Annalen*, 1967, **708**, 1.
115. A. J. Bridges and G. H. Whitham, *Chem. Comm.*, 1974, 142.
116. P. A. Chopard, R. F. Hudson, and R. J. G. Searle, *Tetrahedron Lett.*, 1965, 2357; M. B. Hocking, *Canad. J. Chem.*, 1966, **44**, 1581.
117. N. A. Portnoy, C. J. Morrow, M. S. Chattha, J. C. Williams, and A. M. Aguiar, *Tetrahedron Lett.*, 1971, 1401; M. S. Chattha and A. M. Aguiar, *ibid.*, p. 1419.
118. G. R. Malone and A. I. Meyers, *J. Org. Chem.*, 1974, **39**, 623.

3 PO-Activated Olefinations

119. W. Nagata and Y. Hayase, *J. Chem. Soc. (C)*, 1969, 460.
120. G. A. Koppel and M. D. Kinnick, *Tetrahedron Lett.*, 1974, 711.
121. T. H. Sellstedt, *J. Org. Chem.*, 1975, **40**, 1508.
122. U. Schöllkopf, R. Schröder, and D. Stafforst, *Annalen*, 1974, 44.
123. O. P. Vig, A. K. Vig, J. S. Mann, and K. C. Gupta, *J. Indian Chem. Soc.*, 1975, **52**, 538; O. P. Vig, A. K. Vig, M. S. Grewal, and K. C. Gupta, *ibid.*, p. 543; O. P. Vig, A. K. Vig, A. L. Ganba, and K. C. Gupta, *ibid.*, p. 541.
124. K. E. Harding and C.-Y. Tseng, *J. Org. Chem.*, 1975, **40**, 929.
125. I. Tömösközi and G. Janzso, *Chem. and Ind.*, 1962, 2085.
126. S. E. Dinizo, R. W. Freerksen, W. E. Pabst, and D. S. Watt, *J. Amer. Chem. Soc.*, 1977, **99**, 182.
127. C. Rivalle, J. A. Louisfert, and E. Bisagni, *Tetrahedron*, 1976, **32**, 829.
128. W. E. Witold and J. Zimnicki, *Rocz. Chem.*, 1969, **43**, 95 (*Chem. Abs.*, 1969, **71**, 3169p).
129. N. Rabinovitz, A. Solomnovici, and H. Weiler-Feilchenfeld, *J. Chem. Soc., Perkin II*, 1972, 1836.
130. D. Redmore, *J. Org. Chem.*, 1969, **34**, 1420.
131. E. Muller, M. Sauerbier, D. Streichfuss, R. Thomas, W. Winter, G. Zountsas, and J. Heiss, *Annalen*, 1970, **735**, 99.
132. J. Meinwald and J. W. Young, *J. Amer. Chem. Soc.*, 1971, **93**, 725.
133. L. Ya Malkes, L. Ya Kheifets, and N. P. Kovalenko, *Stsintill. Org. Lyuminofory*, 1972, 32 (*Chem. Abs.*, 1975, **83**, 96820).
134. P. Savignac and P. Coutrot, *Synthesis*, 1976, 197.
135. D. Seyferth and R. S. Marmor, *J. Organometal. Chem.*, 1973, **59**, 237.
136. J. Villieras, P. Perriot, and J. F. Normant, *Synthesis*, 1978, 29.
137. H. Zimmer, K. R. Hickey, and R. J. Schumacher, *Chimia*, 1974, **28**, 656.
138. C. G. Kruse, N. L. J. M. Broekhof, A. Wijsman, and A. Van der Gen, *Tetrahedron Lett.*, 1977, 885.
139. G. H. Posner and D. J. Brunelle, *J. Org. Chem.*, 1972, **37**, 3547.
140. K. Kando, Y. Liu, and D. Tunemoto, *J. Chem. Soc., Perkin I*, 1974, 1279.
141. E. J. Corey and J. I. Shulman, *J. Amer. Chem. Soc.*, 1970, **92**, 5522.
142. P. Blatcher, J. I. Grayson, and S. Warren, *Chem. Comm.*, 1976, 547.
143. H. Zimmer, P. J. Bercz, and G. E. Heuer, *Tetrahedron Lett.*, 1968, 171.
144. M. Hattori and S. Sato, *Nippon Kagaku Kaishi*, 1975, 1780 (*Chem. Abs.*, 1976, **84**, 4079q).
145. A. Dehnel, J. P. Finet, and G. Lavielle, *Synthesis*, 1977, 474.
146. S. F. Martin and R. Gompper, *J. Org. Chem.*, 1974, **39**, 2814.
147. H. Gross and W. Buerger, *J. Prakt. Chem.*, 1969, **311**, 395.
148. H. Gross and B. Costisella, *Angew. Chem. Internat. Ed.*, 1968, **7**, 391.
149. W. Runge, G. Kresze, and E. Ruch, *Annalen*, 1975, 1361.
150. Z. Hamlet and W. Mychajlowskij, *Chem. and Ind.*, 1974, 829.
151. S. Musierowicz, A. Wróblewski, and H. Krawczyk, *Tetrahedron Lett.*, 1975, 437.
152. M. B. Marszak, M. Simalty, and A. Seuleiman, *Tetrahedron Lett.*, 1974, 1905.
153. E. W. Colvin and B. J. Hamill, *Chem. Comm.*, 1973, 151.
154. H. Eckes and M. Regitz, *Tetrahedron Lett.*, 1975, 447.
155. R. J. Sundberg, P. A. Bukowick, and F. O. Holcombe, *J. Org. Chem.*, 1967, **32**, 2938.
156. H. Machleidt, V. Hartmann, and H. Bünger, *Annalen*, 1963, **667**, 35.

157. R. S. Burden and L. Crombie, *J. Chem. Soc.* (*C*), 1969, 2477.
158. (a) A. H. Davidson and S. Warren, *Chem. Comm.*, 1976, 181. (b) P. Blatcher, J. I. Grayson, and S. Warren, *Chem. Comm.*, 1978, 657. (c) I. Fleming, A. Pearce, and R. L. Snowden, *Chem. Comm.*, 1976, 182.
159. G. Pattenden and B. C. L. Weedon, *J. Chem. Soc.* (*C*), 1968, 1984; 1968, 1997.
160. G. Pattenden, *J. Chem. Soc.* (*C*), 1970, 1404; R. N. Gedge, K. C. Westaway, P. Arora, R. Bisson, and A. H. Khalil, *Canad. J. Chem.*, 1977, **55**, 1218.
161. G. M. Peters, F. A. Stuber, and H. Ulrich, *J. Org. Chem.*, 1975, **40**, 2243; H. de Koning, G. N. Mallo, A. Springer-Fidder, K. E. C. Subramanian-Erhart, and H. O. Huisman, *Rec. Trav. Chim. Pays-Bas*, 1973, **92**, 683, 237.
162. J. A. Edwards, V. Schwarz, J. Fajkos, M. L. Maddoz, and J. H. Fried, *Chem. Comm.*, 1971, 292.
163. J. D. Surmatis and R. Thommen, *J. Org. Chem.*, 1969, **34**, 559; M. Ito, R. Masahara, and K. Tsukida, *Tetrahedron Lett.*, 1977, 2767.
164. A. E. Asato and R. S. H. Liu, *J. Amer. Chem. Soc.*, 1975, **97**, 4128.
165. K. Hejno and V. Jarolim, *Coll. Czech. Chem. Comm.*, 1973, **38**, 3511.
166. G. Drefahl, H. H. Hoerhold, and H. Wildner, E. Ger. Pat. 51,436 (*Chem. Abs.*, 1966, **66**, 105346h); G. A. Lapitskii, S. M. Makin, E. K. Lyapina, and A. S. Chebotarev, U.S.S.R. Pat. 275,408 (*Chem. Abs.*, 1971, **74**, 13593y; 1972, **76**, 127488q; 1969, **71**, 30745e).
167. G. W. K. Cavill and P. J. Williams, *Austral. J. Chem.*, 1969, **22**, 1737; G. W. K. Cavill, D. G. Laing, and P. J. Williams, *ibid.* p. 2145; K. Mori, J. Mitsui, J. Fukami, and T. Ohtaki, *Agric. Biol. Chem.*, 1971, **35**, 1116.
168. K. H. Dahm, B. M. Trost, and H. Roller, *J. Amer. Chem. Soc.*, 1967, **89**, 5292.
169. E. J. Corey, J. A. Katzenellenbogen, S. A. Roman, and N. W. Gilman, *Tetrahedron Lett.*, 1971, 1821.
170. V. Jarolim and F. Sorm, *Coll. Czech. Chem. Comm.*, 1977, **42**, 1894.
171. W. Hafferl, R. Zurflueh, and L. Dunham, *J. Labelled Compounds*, 1971, **7**, 331.
172. N. Bensel, H. Marschall, and P. Weyerstahl, *Tetrahedron Lett.*, 1976, 2293; L. S. Stanishevskii, I. G. Tishchenko, V. I. Tyvorskii, and T. I. Prikota, *Zh. Org. Khim.*, 1973, **9**, 1369.
173. P. D. Landor, S. R. Landor, and S. Mukasa, *Chem. Comm.*, 1971, 1638.
174. O. P. Vig, O. P. Chugh, V. K. Handa, and A. K. Vig, *J. Indian Chem. Soc.*, 1975, **52**, 161; O. P. Vig, B. Ram, and J. Kaur, *ibid.*, 1972, **49**, 1181; O. P. Vig, A. Lal, G. Singh, and K. L. Matta, *ibid.*, 1967, **5**, 475; 1968, **6**, 431; 1970, **47**, 851, 999.
175. M. G. Peter, W.-D. Woggon, and H. Schmid, *Helv. Chim. Acta*, 1977, **60**, 2288.
176. W. G. Dauben, G. H. Beasley, M. D. Broadhurst, B. Muller, D. J. Peppard, P. Pesnelle, and C. Suter, *J. Amer. Chem. Soc.*, 1974, **96**, 4724.
177. D. S. Watt and E. J. Corey, *Tetrahedron Lett.*, 1972, 4651.
178. H. M. McGuire, H. C. Odom, Jr., and A. R. Pinder, *J. Chem. Soc., Perkin I*, 1974, 1879.
179. O. G. Plantema, H. de Koning, and H. O. Huisman, *J. Chem. Soc., Perkin I*, 1978, 304.
180. E. J. Corey, I. Vlattas, N. H. Anderson, and K. Harding, *J. Amer. Chem. Soc.*, 1968, **90**, 3247; D. Taub, R. D. Hoffsommer, C. H. Kuo, H. L. Slates, Z. S. Zelawski, and N. L. Wendler, *Chem. Comm.*, 1970, 1258.

3 PO-Activated Olefinations

181. E. J. Corey, Z. Arnold, and J. Hutton, *Tetrahedron Lett.*, 1970, 307.
182. P. A. Grieco, C. S. Pogonowski, S. D. Burke, M. Nishizawa, M. Miyashita, Y. Masaki, C.-L. J. Wang, and G. Majetich, *J. Amer. Chem. Soc.*, 1977, **99**, 4111.
183. D. Reuschling, K. Kühlein, and A. Linkies, *Tetrahedron Lett.*, 1977, 17.
184. M. R. Johnson, H. J. E. Hess, T. K. Schaaf, and J. S. Bindra, Ger. Offen. 2,353,159 (*Chem. Abs.*, 1975, **83**, 58094p).
185. P. A. Zoretic, B. Branchaud, and N. D. Sinha, *J. Org. Chem.*, 1977, **42**, 3201; J. W. Bruin, H. de Koning, and H. O. Huisman, *Tetrahedron Lett.*, 1975, 4599; R. Aries, Fr. Demande 2,258,376 (*Chem. Abs.*, 1976, **84**, 121288t); P. A. Zoretic and F. Bracelos, *Tetrahedron Lett.*, 1977, 529.
186. W. G. Dauben, G. H. Beasley, M. D. Broadhurst, B. Muller, D. J. Peppard, P. Pesnelle, and C. Suter, *J. Amer. Chem. Soc.*, 1975, **97**, 4973.
187. J. Buendia and J. Schalbar, *Tetrahedron Lett.*, 1977, 4499.
188. T. A. Eggelte, H. de Koning, and H. O. Huisman, *Chem. Lett.*, 1977, 433; T. A. Eggelte, H. de Koning, and H. O. Huisman, *J. Roy. Netherlands Chem. Soc.*, 1977, **96**, 271.
189. J. Boutagy and R. Thomas, *Austral. J. Pharm. Sci.*, 1972, 67; 1973, 9; J. Boutagy and R. Thomas, *Austral. J. Chem.*, 1971, **24**, 2723.
190. Farbwerke Hoechst A.G., Fr. Demande 2,013,358 (*Chem. Abs.*, 1971, **74**, 3797).
191. J. B. Dusza, J. P. Joseph, and S. Berustein, U.S. Pat. 3,351,638 (*Chem. Abs.*, 1968, **68**, 114864).
192. E.g. G. Drefahl, K. Ponsold, and H. Schick, *Chem. Ber.*, 1964, **97**, 2011; Farbwerke Hoechst A. G., Neth. Appl. 6,607,315 (*Chem. Abs.*, 1967, **66**, 115884); Fr. Demande 1,498,548 (*Chem. Abs.*, 1968, **69**, 97001); Fr. Demande 1,519,931 (*Chem. Abs.*, 1969, **71**, 13279); Ger. Pat. 1,302,787 (*Chem. Abs.*, 1971, **74**, 31903); H. Kaneko and M. Okazaki, Jap. Pat. 6,821,059 (*Chem. Abs.*, 1969, **70**, 58136).
193. A. K. Bose and R. T. Dahill, *J. Org. Chem.*, 1965, **30**, 505; idem, *Tetrahedron Lett.*, 1963, 959.
194. W. Fritsch, U. Stache, and H. Ruschig, *Annalen*, 1966, **699**, 195.
195. W. Gustowski, M. Kocor, and J. Michalski, *Bull. Acad. Pol. Sci., Ser. Sci. Chim.*, 1973, **21**, 887 (*Chem. Abs.*, 1974, **80**, 60068q).
196. E. D. Bergmann and A. Solomonovici, *Steroids*, 1976, **27**, 431.
197. A. K. Bose and R. M. Ramer, *Steroids*, 1968, **11**, 27; idem, *J. Chem. Soc. (C)*, 1969, 2728.
198. J. Wicha, K. Bal, and S. Pickut, *Synth. Comm.*, 1977, **7**, 215 (*Chem. Abs.*, 1977, **87**, 152457p).
199. J. C. Knight, G. R. Pettit, and C. L. Herald, *Chem. Comm.*, 1967, 445; G. R. Pettit, J. C. Knight, and C. L. Herald, *J. Org. Chem.*, 1970, **35**, 1393.
200. W. Kreiser and G. Neef, *Annalen*, 1974, 1279.
201. K. Radscheit, U. Stache, W. Haede, W. Fritsch, and H. Ruschig, *Tetrahedron Lett.*, 1969, 3029; U. Stache, K. Radscheit, W. Fritsch, W. Haede, H. Kohl, and H. Ruschig, *Annalen*, 1971, **750**, 149.
202. C. R. Popplestone and A. M. Unrau, *Canad. J. Chem.*, 1973, **51**, 1223.
203. E.g. G. R. Pettit, C. L. Herald, and J. P. Yardley, *J. Org. Chem.*, 1970, **35**, 1389; Farbwerke Hoechst A.G., Fr. Demande 1,491,081 (*Chem. Abs.*, 1968, **69**, 77606).
204. W. Kreiser, H.-U. Warnecke, and G. Neef, *Annalen*, 1973, 2071.

205. H. G. Lechmann and R. Wiechert, *Angew. Chem. Internat. Ed.*, 1968, **7**, 300; G. Kruger, *Canad. J. Chem.*, 1974, **52**, 4139; W. Eberlein, J. Nickl, G. Dahms, and H. Machleidt, *Chem. Ber.*, 1972, **105**, 3686.
206. C. A. Henrick, E. Bohme, J. A. Edwards, and J. H. Fried, *J. Amer. Chem. Soc.*, 1968, **90**, 5926.
207. A. Rosenthal and L. Nguyen, *Tetrahedron Lett.*, 1967, 2393; A. Rosenthal and P. Catsoulacos, *Canad. J. Chem.*, 1968, **46**, 2868; A. Rosenthal and D. A. Baker, *Tetrahedron Lett.*, 1969, 397.
208. Yu. A. Zhdanov, Yu. E. Alekseev, and V. G. Alekseeva, *Adv. Carbohydrate Chem.*, 1972, **27**, 227.
209. J. A. Montgomery and K. Hewson, *Chem. Comm.*, 1969, 15.
210. H. Paulsen and W. Bartsch, *Chem. Ber.*, 1975, **108**, 1732; F. A. Carey and A. S. Court, *J. Org. Chem.*, 1972, **37**, 939; D. J. Peterson, *J. Org. Chem.*, 1968, **33**, 780.
211. H. Paulsen, W. Bartsch, and J. Thiem, *Chem. Ber.*, 1971, **104**, 2545.
212. G. Seitz and H. Hoffmann, *Chem.-Ztg.*, 1976, **100**, 440.
213. T. Sasaki and K. Ando, *Yuki Gosei Kagaku Kyokaishi*, 1968, **26**, 70 (*Chem. Abs.*, 1968, **69**, 43859d).
214. C. Szantay, L. Toke, and P. Kolonits, *J. Org. Chem.*, 1966, **31**, 1447.
215. T. Wakabayashi and M. Saito, *Tetrahedron Lett.*, 1977, 93; see also J. J. J. de Boer and W. N. Speckamp, *Tetrahedron Lett.*, 1975, 4039.
216. T. Wakabayashi and K. Watanabe, *Tetrahedron Lett.*, 1977, 4595.
217. R. D. Clark, L. G. Kozar, and C. H. Heathcock, *Synth. Comm.*, 1975, **5**, 1.
218. G. Stork and R. Matthews, *Chem. Comm.*, 1970, 445.
219. P. A. Grieco and C. S. Pogonowski, *Synthesis*, 1973, 425.
220. R. W. Ratcliffe and B. G. Christensen, *Tetrahedron Lett.*, 1973, 4645; 1973, 4649; 1973, 4653; R. N. Guthikonda, L. D. Cama, and B. G. Christensen, *J. Amer. Chem. Soc.*, 1974, **96**, 7584.
221. H. W. Whitlock, Jr., *J. Org. Chem.*, 1964, **29**, 3129.
222. W. S. Wadsworth, Jr., and W. D. Emmons, *J. Org. Chem.*, 1967, **32**, 1279.
223. L. Horner, H. Hoffmann, G. Klahre, V. G. Toscano, and H. Ertel, *Chem. Ber.*, 1961, **94**, 1987.
224. H. J. Bestmann and O. Kratzer, *Chem. Ber.*, 1963, **96**, 1899.
225. L. Horner, H. Hoffmann, and V. G. Toscano, *Chem. Ber.*, 1962, **95**, 536.
226. I. Tömösközi, *Tetrahedron*, 1963, **19**, 1969; 1966, **22**, 179; M. Baboulene and G. Sturtz, *Phosphorus*, 1973, **2**, 195.
227. I. Tömösközi, *Chem. and Ind.*, 1965, 689; R. A. Izydore and R. G. Ghirardelli, *J. Org. Chem.*, 1973, **38**, 1790.
228. I. Tömösközi, *Angew. Chem. Internat. Ed.*, 1963, **2**, 261.
229. E. Breuer and S. Zbaida, *J. Org. Chem.*, 1977, **42**, 1904; E. Breuer, S. Zbaida, J. Pesso, and I. Ronen-Braunstein, *Tetrahedron*, 1977, **33**, 1145; D. S. C. Black and V. C. Davis, *Austral. J. Chem.*, 1976, **29**, 1735.
230. D. L. J. Clive, W. A. Kiel, S. M. Menchen, and C. K. Wong, *Chem. Comm.*, 1977, 657.
231. D. L. J. Clive and S. M. Menchen, *Chem. Comm.*, 1977, 658.
232. G. Sturtz and G. Lavielle, *Compt. Rend. (C)*, 1965, **261**, 2679.
233. M. Kirilov and G. Petrov, *Chem. Ber.*, 1971, **104**, 3073.
234. W. Nagata and Y. Hayase, *Tetrahedron Lett.*, 1968, 4359.
235. F. W. Batchelor, A. A. Loman, and L. R. Snowdon, *Canad. J. Chem.*, 1970, **48**, 1554.

236. P. Savignac, J. Petrova, M. Dreux, and P. Coutrot, *Synthesis*, 1975, 535; P. Coutrot, C. Laurenco, J. Petrova, and P. Savignac, *ibid.*, 1976, 107.
237. E. D. Bergmann, I. Shahak, and J. Appelbaum, *Israel J. Chem.*, 1968, **6**, 73.
238. A. Yamaguchi and M. Okazaki, *Nippon Kagaku Kaishi*, 1972, 2103 (*Chem. Abs.*, 1973, **78**, 29372a).
239. H. Zimmer and H. R. Hickey, *Angew. Chem. Internat. Ed.*, 1971, **10**, 867.
240. G. M. Kosolapoff, *Org. Reactions*, 1951, **6**, 273.

4
Transformations via Phosphorus-stabilized Anions 3: Reactions with Electrophiles other than Aldehydes and Ketones

D. J. H. SMITH

1. Alkylation reactions, 207
2. Acylation reactions, 211
3. Reactions with epoxides, 214
4. Reactions with carbon–carbon double bonds, 217
5. Reactions with carbon–carbon triple bonds, 218
6. Reactions with carbon–nitrogen multiple bonds, 219
7. Summary of functional group conversions, 220
8. References, 221

The study of reactions of phosphorus ylides and similar species with electrophilic reagents has quite rightly been concentrated on the Wittig and related reactions, that is, reactions with aldehydes and ketones. However, reactions with other electrophiles are well known and many have been studied in some detail. Phosphorus-stabilized carbanions are powerful nucleophilic reagents and react with a range of functional groups. Although many of these reactions lead to rather esoteric phosphorus compounds, some are useful in general organic synthesis and it is these that will be described in this chapter.

Groups such as phosphonium, phosphonate ester, or simple phosphine oxide moieties can be used to activate molecules to enable new bonds to be formed via carbanionic intermediates. These activating groups can then be removed by a variety of techniques including hydrolysis, thermolysis, and electrolysis.

1. Alkylation Reactions

Alkylidenephosphoranes form *C*-alkylated phosphonium halides when treated with alkyl halides. These reactions were first described by Wittig[1] who showed that methylenetriphenylphosphorane could be alkylated by methyl iodide. Phosphonium salts such as t-butyltriphenylphosphonium iodide,[2] not easily prepared by conventional routes, may be obtained in this way. Similarly, reaction with allyl bromide can lead to phosphonium bromides of the type $RCH=CHCH(PPh_3)^+CH_2CH=CH_2$, reductive cleavage of which provides a general route to 1,5-dienes.[2b]

Iminophosphoranes, phosphonate anions, and similar species react very much like phosphorus ylides as indicated in Table 1.

Table 1. Alkylation of phosphorus-stabilized carbanions by alkyl halides

$Ph_3P=CH_2$ + MeI	$\longrightarrow$	$Ph_3\overset{+}{P}-CH_2CH_3$ I^-
$Ph_3P=C\begin{smallmatrix}Me\\Me\end{smallmatrix}$ + MeI	$\longrightarrow$	$Ph_3\overset{+}{P}-Bu^t$ I^-
$Ph_3P=NR^1$ + R^2X	$\longrightarrow$	$Ph_3\overset{+}{P}NR^1R^2 X^-$
$(EtO)_2\overset{O}{\underset{\|}{P}}\underset{}{\overset{O}{\underset{\|}{C}}}R^1 + R^2X$	$\longrightarrow$	$(EtO)_2\overset{O}{\underset{\|}{P}}\underset{R^2}{\overset{O}{\underset{\|}{C}}}R^1$ a
$(MeO)_2\overset{S}{\underset{\|}{P}}-\bar{N}R^1 + R^2X$	$\longrightarrow$	$(MeO)_2\overset{S}{\underset{\|}{P}}-NR^1R^2$ b
$(EtO)_2\overset{O}{\underset{\|}{P}}-\bar{N}(CH_2)_2NHR^1 + R^2X$	$\longrightarrow$	$(EtO)_2\overset{O}{\underset{\|}{P}}NR^2(CH_2)_2NHR^1$ c

^a R. D. Clark, L. G. Kozar, and C. H. Heathcock, *J. Org. Chem.*, 1975, **40**, 1979.
^b B. Miller and T. P. O'Leary, *J. Org. Chem.*, 1962, **27**, 3382.
^c P. Savignac, G. Lavielle, and M. Dreux, *J. Organometal. Chem.*, 1974, **72**, 361.

Complications do arise with stabilized ylides, since treatment with an alkyl halide may result in *C*- or *O*-alkylation. *O*-Alkylated products are generally formed from acylphosphoranes (1). It should be noted, however, that the initial product from the reaction with benzyl iodide rearranges on heating to the *C*-alkylated salt.[3]

There is a delicate balance between the products formed in these reactions. Alkylation of ester-stabilized ylides usually leads to *C*-alkylated salts (2) if the alkyl halide contains an electron-donating group. If the alkyl halide bears an electron-withdrawing group this salt will be deprotonated in a transylidation reaction to produce a new ylide (3). These reactions form a useful ester synthesis since the phosphorane (3) can be hydrolysed with boiling water.[4]

$$Ph_3P=CHCOR + PhCH_2I \longrightarrow Ph_3\overset{+}{P}-CH=C\underset{R}{\overset{OCH_2Ph}{\diagup}} \longrightarrow Ph_3\overset{+}{P}-\underset{\underset{O}{\overset{CH_2Ph}{|}}}{CHCR} \quad I^-$$

(1)

$$Ph_3P=CHCO_2Et + RX \longrightarrow Ph_3\overset{+}{P}-CHRCO_2Et \quad X^- \quad (2)$$
$$\downarrow$$
$$Ph_3PO + RCH_2CO_2Et \longleftarrow Ph_3P=CRCO_2Et$$

(3)

Since the phosphonate group can also be easily removed, by reduction with lithium aluminium hydride, alkylation of allylphosphonates can be used to synthesize alkenes. It is noteworthy that pure *trans*-alkenes are isolated even when the phosphonates (4) are not isomerically pure.[5]

$$MeR^1C=CHCH_2\underset{\overset{\|}{O}}{P}(OEt)_2 \xrightarrow[2.\ R^2X]{1.\ BuLi} MeR^1C=CH\underset{\underset{O}{\overset{\|}{R^2}}}{CHP(OEt)_2} \longrightarrow \underset{H}{\overset{MeR^1CH}{\diagdown}}C=C\underset{R^2}{\overset{H}{\diagup}}$$

(4)

The phosphonium salts obtained from nucleophilic substitution of alkyl halides with phosphacumulene ylides, e.g. (5), undergo cycloaddition reactions with another mole of ylide to yield a four-membered ylide phosphonium salt (6). In all cases studied the second step is faster than the first. Compounds of the type (6) have been successfully used in the synthesis of β-oxo-esters and acetylenecarboxylates.[6]

$$Ph_3P=C=C=O \quad \underset{RX}{\overset{+}{\longrightarrow}} \quad \underset{R}{\overset{Ph_3\overset{+}{P}}{\diagdown}}C=C=O \quad X^- \longrightarrow \left[\begin{array}{c} Ph_3\overset{+}{P} \quad R \quad O^- \\ \square \\ O \quad \overset{+}{P}Ph_3 \end{array} \right] X^-$$

(5)   (6)

The direct introduction of alkyl substituents into heterocyclic nuclei can be achieved by treatment of nuclear-chlorinated heterocycles with reactive ylides. The ylides obtained, e.g. (7), can be hydrolysed or used in Wittig reactions. Similar reactions have been used in the synthesis of quinine and related alkaloids.[7]

[quinoline with Cl at 4, Me at 2] + 2Ph₃P=CH₂ → [quinoline with CH=PPh₃ at 4, Me at 2] (7) → [quinoline with Me at 4, Me at 2]

Intramolecular *C*-alkylation of alkylidenephosphoranes is well suited to the formation of cyclic compounds and lends itself to a number of synthetic possibilities. For example, phosphonium salts (8) are easily obtained, and when treated with base form the corresponding ylides which will yield cyclic phosphonium salts. These salts can be converted into cyclic alkanes or alkenes by hydrolysis or thermolysis respectively (Scheme 1), besides being useful for Wittig reactions.

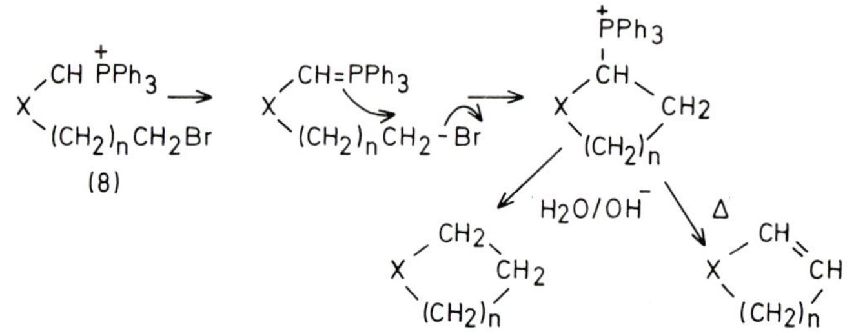

Scheme 1

Recent examples of these reactions include the synthesis of the hydrocarbon (9) of known absolute configuration from optically active dibromide,[8] and the preparation of the phosphonium salt (10).[9]

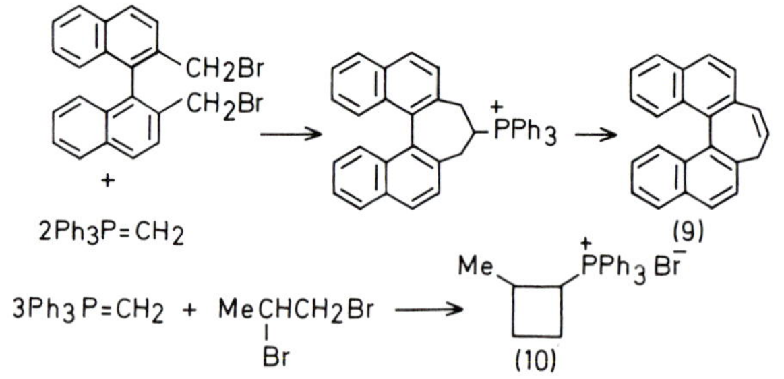

These cyclizations need not be intramolecular, as shown by the alkylation of the bis-ylide (11) with dibromomethane. The salt obtained may be converted into a cyclohexenone by sequential hydrolysis.[10]

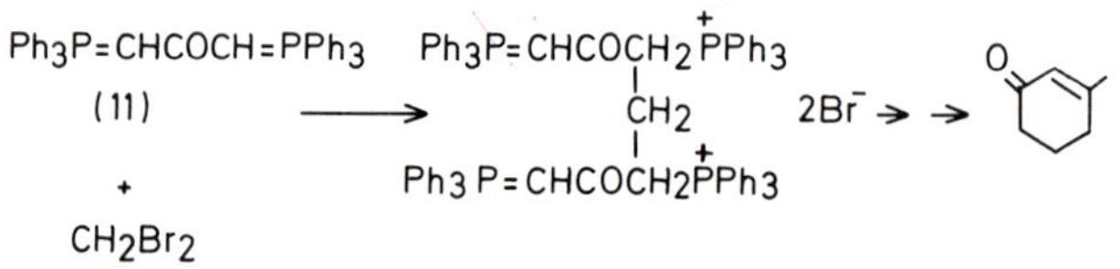

Although reactive ylides may form alkenes when treated with α-halogenocarbonyl compounds, stable alkylidenephosphoranes yield γ-keto-phosphonium salts which subsequently lead to new ylides by a transylidation reaction (Scheme 2). The hydrogen atom in the β-position to the phosphorus atom is strongly activated, and such ylides readily decompose by a Hofmann elimination to unsaturated compounds and a phosphine. Labelling experiments indicate that the final step involves an intermolecular hydrogen atom transfer.[11] This process makes it possible to prepare β-acylacrylic acid esters from ylides (12) and α-halogeno-ketones,[12] as shown in Scheme 2.

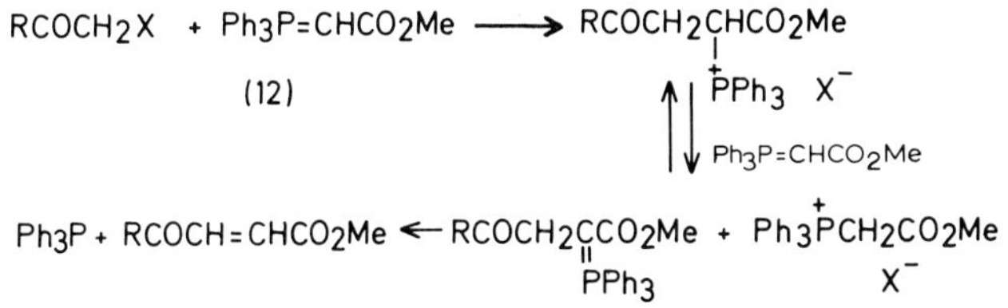

$$RCOCH_2X + Ph_3P=CHCO_2Me \longrightarrow RCOCH_2CHCO_2Me$$
$$(12) \qquad\qquad\qquad \overset{+}{P}Ph_3 \ X^-$$
$$\Updownarrow Ph_3P=CHCO_2Me$$

$$Ph_3P + RCOCH=CHCO_2Me \longleftarrow RCOCH_2\underset{\underset{PPh_3}{||}}{C}CO_2Me + Ph_3\overset{+}{P}CH_2CO_2Me \quad X^-$$

Scheme 2

In analogous fashion alkylidenephosphoranes and α-bromo- or α-iodo-acetates form salts (13) which can decompose to α,β-unsaturated esters via the ylide (as previously) or directly from the phosphonium salt.[11]

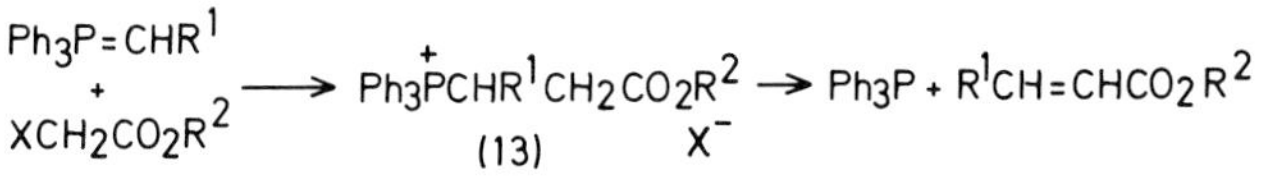

$$\begin{array}{c} Ph_3P=CHR^1 \\ + \\ XCH_2CO_2R^2 \end{array} \longrightarrow Ph_3\overset{+}{P}CHR^1CH_2CO_2R^2 \longrightarrow Ph_3P + R^1CH=CHCO_2R^2$$
$$(13) \quad X^-$$

Reaction of halogenoacetates with acylmethylenetriphenylphosphoranes leads to isolable ylides (14). These decompose on heating, but because of the high temperatures necessary it is preferable to convert them into benzoate salts before undertaking the elimination. The alkaline hydrolysis of (14) gives good yields of γ-keto-acids.[13]

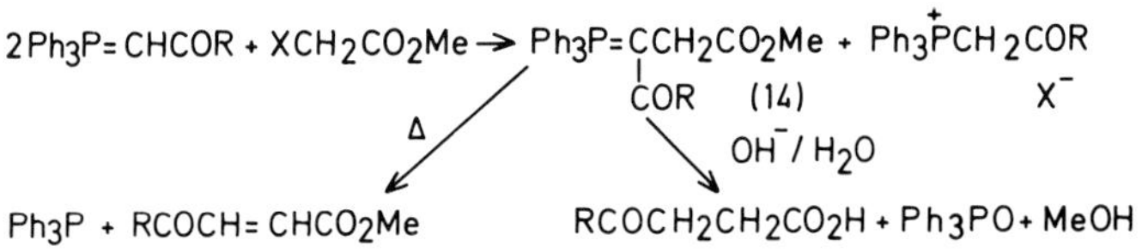

$$2Ph_3P=CHCOR + XCH_2CO_2Me \longrightarrow Ph_3P=\underset{\underset{COR}{|}}{C}CH_2CO_2Me + Ph_3\overset{+}{P}CH_2COR \quad X^-$$
$$(14)$$
$$\swarrow \Delta \qquad\qquad \searrow OH^-/H_2O$$
$$Ph_3P + RCOCH=CHCO_2Me \qquad RCOCH_2CH_2CO_2H + Ph_3PO + MeOH$$

Strangely, the use of methyl *chloro*acetate produces the *trans*-cyclopropane (15) directly, probably via the intermediate formation of dimethyl fumarate.[14]

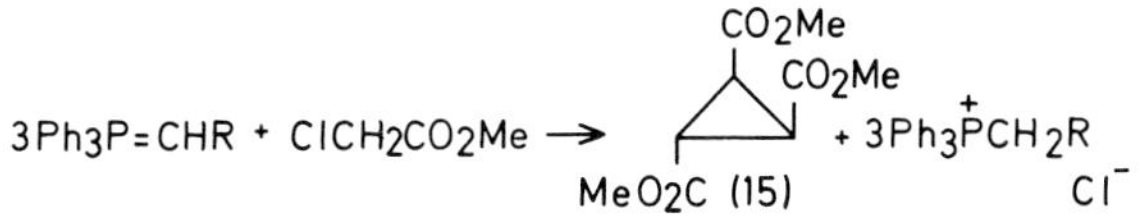

$$3Ph_3P=CHR + ClCH_2CO_2Me \longrightarrow \underset{MeO_2C \ (15)}{\triangle\!\!\!\!\!\!\!\!\overset{CO_2Me}{}\!\!\!\!\!\!\overset{CO_2Me}{}} + 3Ph_3\overset{+}{P}CH_2R \quad Cl^-$$

2. Acylation Reactions

Acyl halides, thioesters, and *N*-acylimidazoles react with alkylidenephosphoranes to form β-keto-phosphonium salts which undergo transylidation with a second mole of ylide (Scheme 3).

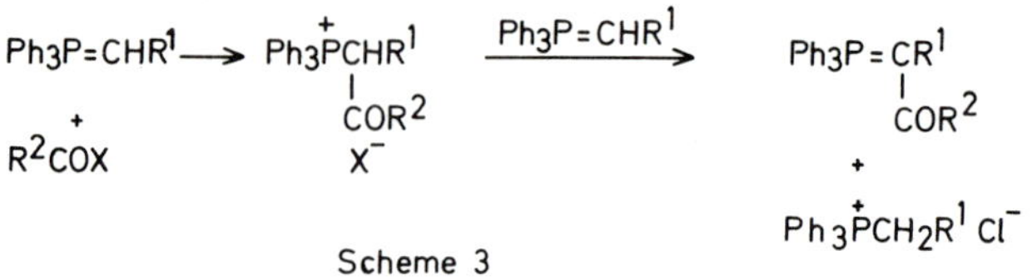

$$Ph_3P=CHR^1 \longrightarrow Ph_3\overset{+}{P}CHR^1 \xrightarrow{Ph_3P=CHR^1} Ph_3P=CR^1$$
$$+ \qquad\qquad |\qquad\qquad\qquad\qquad\qquad |$$
$$R^2COX \qquad\quad COR^2 \qquad\qquad\qquad\qquad\quad COR^2$$
$$\qquad\qquad\qquad X^- \qquad\qquad\qquad\qquad\qquad +$$
$$\qquad\qquad\qquad\qquad\qquad\qquad\qquad\qquad Ph_3\overset{+}{P}CH_2R^1\, Cl^-$$

Scheme 3

The acylation of alkylidenephosphoranes is of synthetic importance since the preparation of acyl ylides is straightforward in only a few cases.[15] However, two moles of starting ylide are required, a disadvantage that can be removed by the use of thioesters because the salt (16) formed usually eliminates mercaptan on heating to give the acyl ylide.

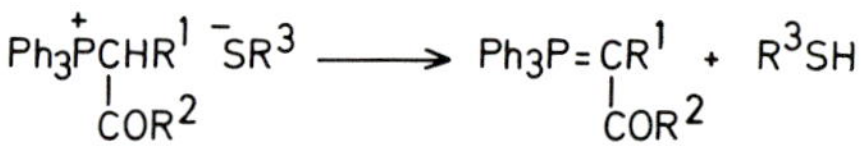

$$Ph_3\overset{+}{P}CHR^1\;\bar{S}R^3 \longrightarrow Ph_3P=CR^1 + R^3SH$$
$$\quad\; |\qquad\qquad\qquad\qquad\qquad\quad |$$
$$\;\; COR^2 \qquad\qquad\qquad\qquad\quad COR^2$$

(16)

In general, salt-free solutions of ylides do not react with ethyl or methyl esters. In the presence of lithium salts reactions do occur but the yields are poor. Better results can be obtained with more reactive esters.

Reactions of other phosphorus-stabilized carbanions such as iminophosphoranes[16] and phosphonate anions[17] occur in an analogous manner. In contrast, the more reactive phosphine oxide carbanions can be acylated with carboxylic acid esters as well as with acyl chlorides.[17]

β-Ketoalkylidenephosphoranes are *O*-acylated by aromatic acid chlorides and dehydrohalogenate aliphatic acid chlorides.[18] Ester-stabilized ylides are *C*-acylated by all acid chlorides. Acylation also occurs with anhydrides to form phosphonium salts which eliminate carboxylic acid to form a new acylated ylide (17). It has been suggested that *O*-acylation is the kinetically controlled reaction and that *C*-acylation is the product of thermodynamic control.[19]

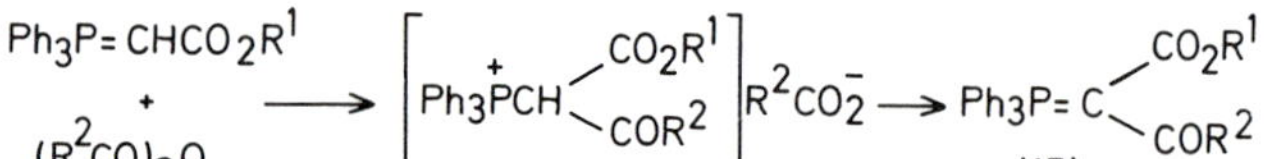

$$Ph_3P=CHCO_2R^1 \qquad\qquad\qquad\qquad CO_2R^1 \qquad\qquad\qquad\qquad CO_2R^1$$
$$+ \qquad\qquad \longrightarrow \left[Ph_3\overset{+}{P}CH\diagup_{COR^2}\right] R^2CO_2^- \longrightarrow Ph_3P=C\diagup_{COR^2}$$
$$(R^2CO)_2O \qquad\qquad\qquad\qquad\qquad\qquad\qquad\qquad\qquad (17)$$

The transylidation process can be prevented by using only one equivalent of ylide and working at low temperatures. The acylated salts obtained, e.g. (18), may be electrolysed to produce α-branched β-keto-esters.[20]

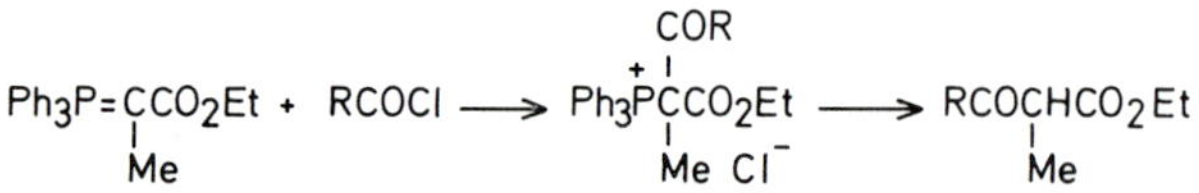

$$\qquad\qquad\qquad\qquad\qquad\quad COR$$
$$\qquad\qquad\qquad\qquad\qquad\quad +\;|$$
$$Ph_3P=CCO_2Et + RCOCl \longrightarrow Ph_3\overset{+}{P}CCO_2Et \longrightarrow RCOCHCO_2Et$$
$$\quad\;\; | \qquad\qquad\qquad\qquad\qquad\qquad | \qquad\qquad\qquad\qquad |$$
$$\quad\;\, Me \qquad\qquad\qquad\qquad\qquad\quad Me\; Cl^- \qquad\qquad\qquad\;\, Me$$

(18)

4 P-Stabilized Anions

Acylation of the phosphonate anions (19) has proved to be difficult as a result of competing lithium–halogen exchange. This problem has been overcome by the addition of one mole of n-butyllithium before the addition of the acid chloride. Any exchange then occurs with the n-butyllithium.[21]

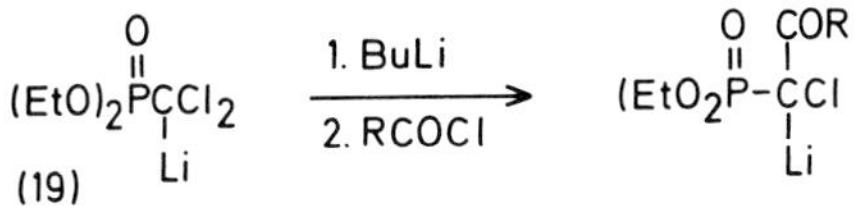

$$(EtO)_2\overset{O}{\overset{\|}{P}}\underset{Li}{\overset{|}{C}}Cl_2 \xrightarrow[\text{2. RCOCl}]{\text{1. BuLi}} (EtO_2\overset{O}{\overset{\|}{P}}-\underset{Li}{\overset{\overset{COR}{|}}{C}}Cl$$

(19)

Ketones of the general structure (20) are easily synthesized from acylating agents by a procedure involving acylation of an alkylidenephosphorane followed by electrolysis or reductive cleavage with zinc and acetic acid.[22]

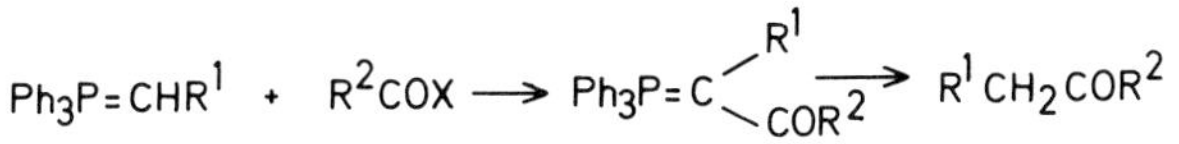

$$Ph_3P=CHR^1 + R^2COX \longrightarrow Ph_3P=C\overset{R^1}{\underset{COR^2}{\diagdown}} \longrightarrow R^1CH_2COR^2$$
(20)

Ylides, e.g. (21), which are obtained by the C-alkylation of methoxycarbonylmethylenetriphenylphosphorane with acid chlorides, generate acetylenic carboxylates on thermal decomposition.[23] In a similar way acyliminophosphoranes (22) are known to decompose at about 200°C to phosphine oxides and nitriles.[24]

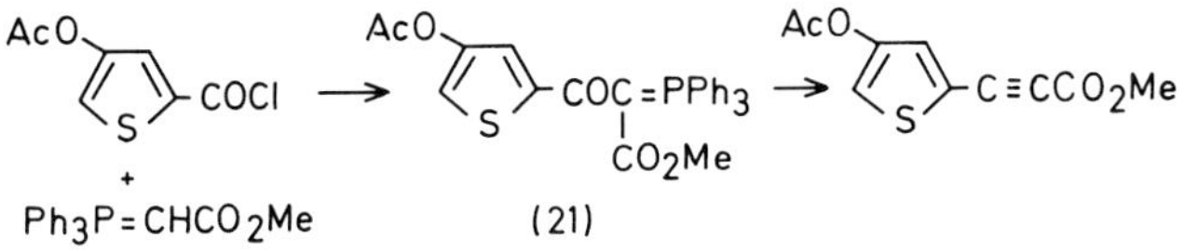

AcO—⟨S⟩—COCl → AcO—⟨S⟩—COC=PPh₃ → AcO—⟨S⟩—C≡CCO₂Me
              |
              CO₂Me
+
Ph₃P=CHCO₂Me   (21)

Ph₃P=NCOR ⟶ Ph₃PO + RCN
(22)

The phosphorane (23) is also C-alkylated by acid chlorides, but in the absence of an α-hydrogen subsequent thermal reactions lead to allenic esters.[25] Optically active allenic esters may be produced in this way since partial asymmetric synthesis results if an optically active acid chloride is treated with a racemic alkylidenephosphorane.[26]

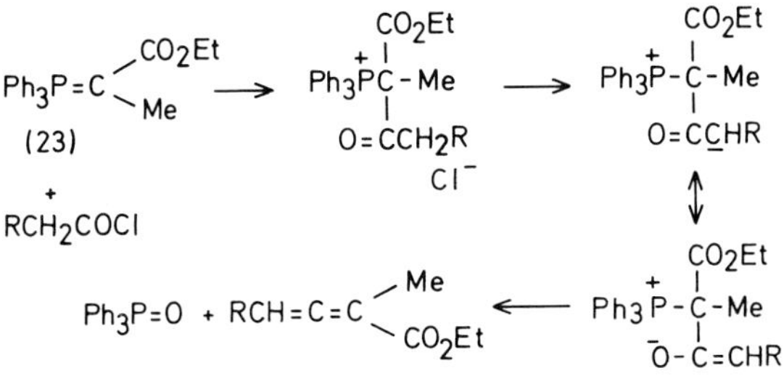

$$Ph_3P=C\overset{CO_2Et}{\underset{Me}{\diagdown}} \longrightarrow Ph_3\overset{+}{P}\overset{\overset{CO_2Et}{|}}{C}-Me \longrightarrow Ph_3\overset{+}{P}-\overset{\overset{CO_2Et}{|}}{C}-Me$$
(23)   O=CCH₂R   O=CCHR
                   Cl⁻   ↕
+
RCH₂COCl
                                               CO_2Et
                                                + |
Ph₃P=O + RCH=C=C⟨Me / CO₂Et⟩ ⟵ Ph₃P-C-Me
                                                 ⁻|
                                                 O-C=CHR

Bestmann has recently shown[27] that treatment of hexaphenylcarbodiphosphorane with aromatic acid chlorides leads to a salt (24) which eliminates triphenylphosphine oxide on thermolysis.

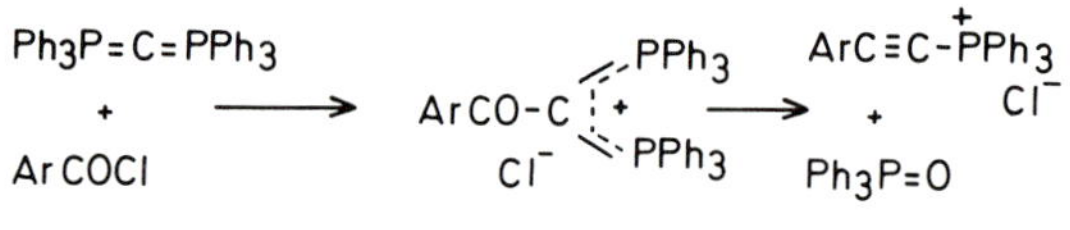

$$Ph_3P=C=PPh_3 + ArCOCl \longrightarrow \underset{(24)}{ArCO-C\overset{PPh_3}{\underset{PPh_3}{\cdot}}\ Cl^-} \longrightarrow ArC\equiv\overset{+}{C}-PPh_3\ Cl^- + Ph_3P=O$$

The lack of reactivity of the phosphonate carbanion (25) with long-chain aldehydes can be overcome by a procedure that involves initial acylation followed by reduction and treatment with base.[28]

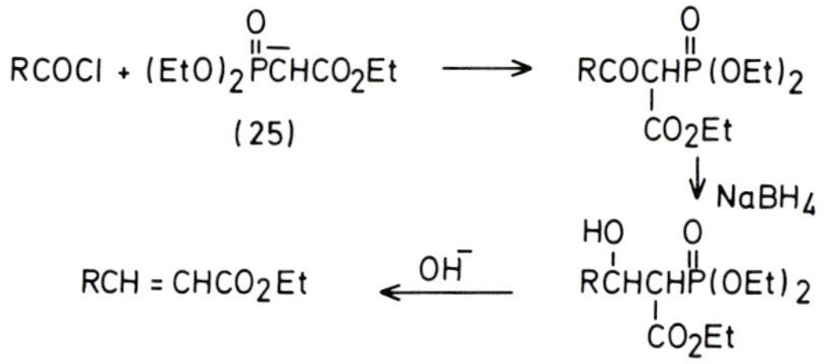

$$RCOCl + \underset{(25)}{(EtO)_2\overset{O}{\overset{\|}{P}}\bar{C}HCO_2Et} \longrightarrow \underset{CO_2Et}{RCOCH\overset{O}{\overset{\|}{P}}(OEt)_2}$$

$$\downarrow NaBH_4$$

$$RCH=CHCO_2Et \xleftarrow{OH^-} \underset{CO_2Et}{\underset{|}{R\overset{HO}{\overset{|}{C}}HCH\overset{O}{\overset{\|}{P}}(OEt)_2}}$$

Ylides react quite readily with chloroformate esters. This is a useful synthetic procedure since the alkoxycarbonylalkylidenephosphoranes formed are easily transformed into carboxylic acids by hydrolysis. Cyclopropylacetic acid (26) can be formed in this way.[29]

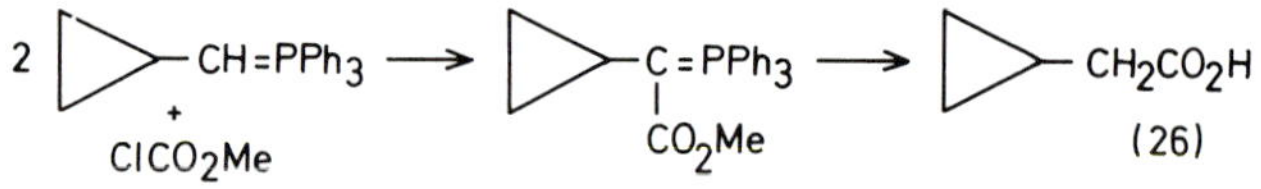

$$2\ \triangleright\!\!-CH=PPh_3 + ClCO_2Me \longrightarrow \triangleright\!\!-\underset{CO_2Me}{\overset{|}{C}=PPh_3} \longrightarrow \underset{(26)}{\triangleright\!\!-CH_2CO_2H}$$

3. Reactions with Epoxides

A variety of products can be isolated from the reactions of phosphorus-stabilized carbanions with epoxides. Generally these depend upon the substituents present and can be accounted for by a comprehensive mechanism suggested by Trippett[30] involving the initial formation of a zwitterionic intermediate as shown in Scheme 4.

When electron-donating alkyl groups are attached to phosphorus, i.e. reducing the electrophilic character of that atom, a reaction path involving elimination of phosphine and a 1,3 hydrogen atom transfer is favoured. Thus McEwen has shown that substantial amounts of a ketone are produced from the reaction of the phosphorane (27).[31] However, when styrene oxide is treated with benzylidenetriphenylphosphorane only a small amount of the

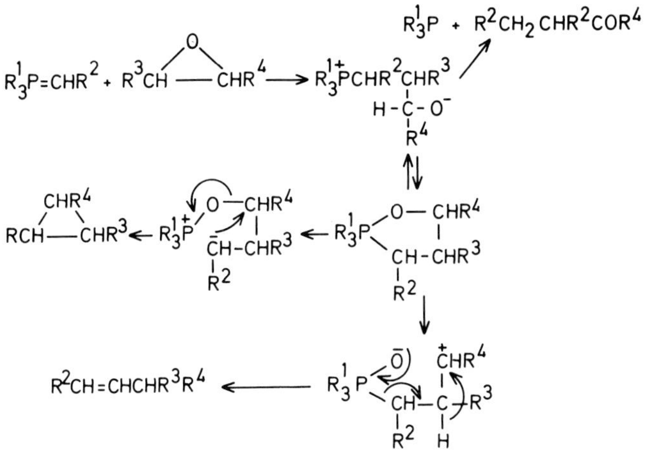

4 P-Stabilized Anions

$$R^1_3P=CHR^2 + R^3CH\overset{O}{-\!\!\!-\!\!\!-}CHR^4 \longrightarrow R^1_3\overset{+}{P}CHR^2CHR^3 \longrightarrow R^1_3P + R^2CH_2CHR^2COR^4$$

Scheme 4

ketone is formed, the major product being a mixture of cis- and trans-1,3-diphenylpropene. In a similar way ester-stabilized ylides form α,β-unsaturated esters in reactions with epoxides.

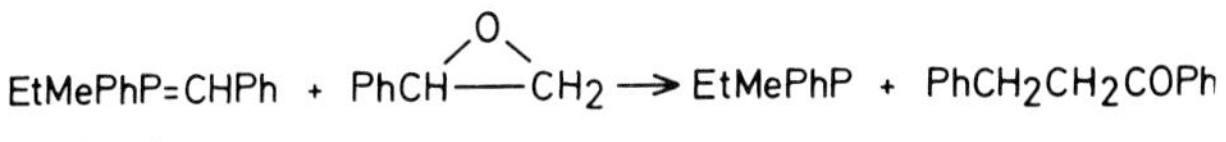

$$\text{EtMePhP=CHPh} + \text{PhCH}\overset{O}{-\!\!\!-\!\!\!-}\text{CH}_2 \longrightarrow \text{EtMePhP} + \text{PhCH}_2\text{CH}_2\text{COPh}$$
(27)

With stabilized ylides only the end products can be isolated, but when strongly basic ylides are used cyclic intermediates may be isolated. Huisgen[32] was able to isolate the compound (28) from methylenetriphenylphosphorane and styrene oxide, which on pyrolysis gave propiophenone and triphenylphosphine.

With cyclic epoxides, such as cyclohexane oxide, ring-contraction occurs from an intermediate phosphorane (29).[33]

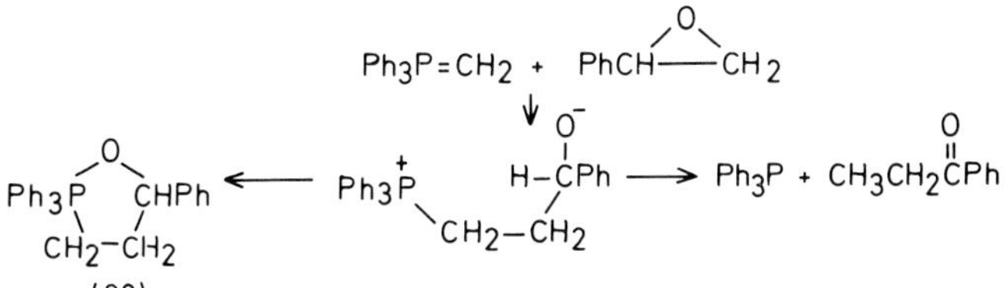

$$Ph_3P=CH_2 + PhCH\overset{O}{-\!\!\!-\!\!\!-}CH_2$$

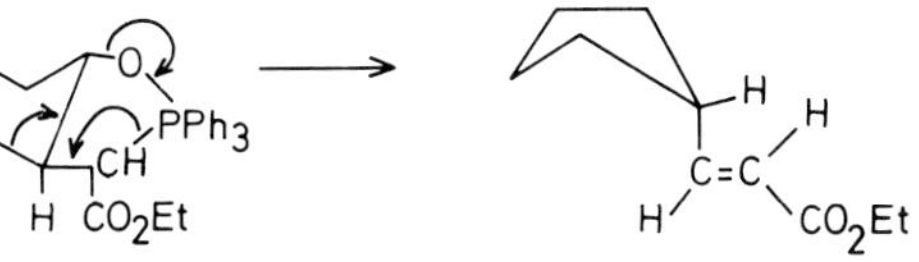

Synthetically the most useful reaction is the formation of cyclopropanes. This will occur if R^2 (Scheme 4) is capable of stabilizing a carbanion. For instance the reaction of styrene oxide with ethoxycarbonyltriphenylphosphorane yields a cyclopropane (30).

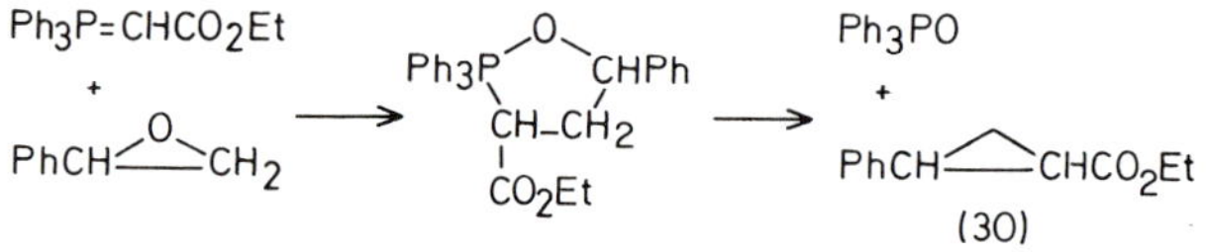

$$Ph_3P=CHCO_2Et + PhCH\overset{O}{-\!\!-\!\!-}CH_2 \longrightarrow \underset{\underset{CO_2Et}{|}}{\overset{Ph_3P\diagup^O\diagdown CHPh}{CH-CH_2}} \longrightarrow Ph_3PO + PhCH\triangle CHCO_2Et$$
(30)

Reaction of phosphonate carbanions with epoxides invariably leads to cyclopropanes. Since the *cis*-epoxide (31) is converted into the corresponding *cis*-cyclopropane using a phosphonate carbanion, it is assumed that inversion of configuration occurs at both carbon atoms.[34] This is in contrast to the report that phosphine oxide carbanions react with optically active styrene oxide with retention of configuration at the asymmetric carbon atom.[35] However, it should be noted that lithium metallated phosphine oxides and epoxides give only lithium alkoxides whereas the corresponding potassium salts lead to cyclopropanes.[36]

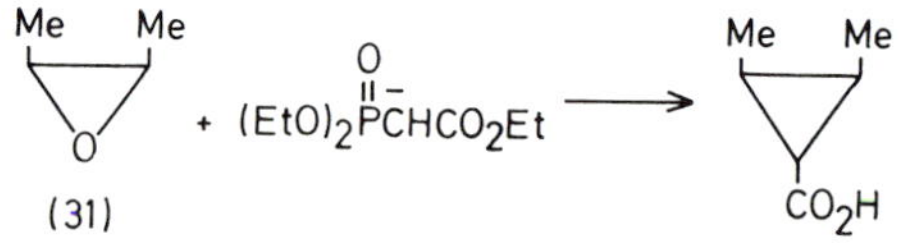

$$\underset{(31)}{\overset{Me\ \ Me}{\triangle_O}} + (EtO)_2\overset{O}{\overset{\|}{P}}\bar{C}HCO_2Et \longrightarrow \underset{CO_2H}{\overset{Me\ \ Me}{\triangle}}$$

Aziridines can be conveniently formed from iminophosphoranes and epoxides. In some instances the phosphorane intermediate, e.g. (32), has been isolated and shown to decompose thermally to a phosphine oxide and an aziridine.[37]

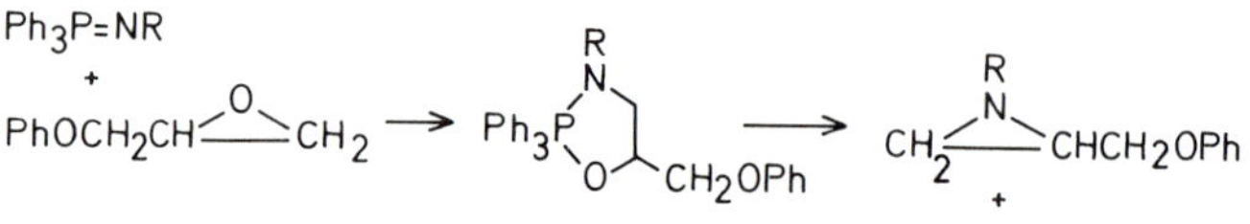

$$Ph_3P=NR + PhOCH_2CH\overset{O}{-\!\!-\!\!-}CH_2 \longrightarrow \underset{(32)}{Ph_3\overset{\overset{R}{|}}{\underset{O}{P}}\diagdown_{CH_2OPh}} \longrightarrow CH_2\overset{\overset{R}{N}}{\triangle}CHCH_2OPh + Ph_3PO$$

Aziridines themselves are cleaved by ylides. One can obtain a new ylide or a pyrroline derivative depending on whether the ylide used has an α-hydrogen atom present[38] as indicated in Scheme 5.

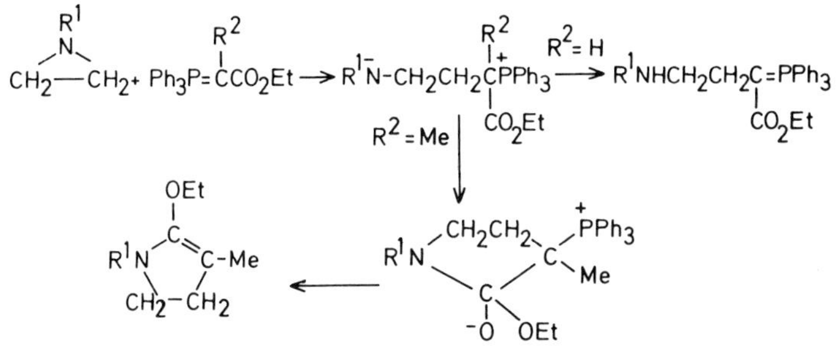

4 P-Stabilized Anions

Scheme 5

4. Reactions with Carbon–Carbon Double Bonds

Alkylidenephosphoranes add to activated carbon–carbon double bonds to give an intermediate zwitterion which may then decompose in one of three ways, depending upon the substituents present, as shown in Scheme 6.

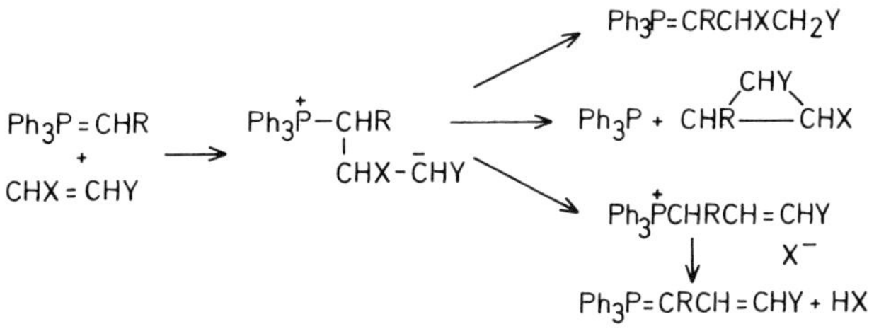

Scheme 6

When the substituent R is electron-withdrawing a 1,3-proton transfer occurs to form a new stable ylide, e.g. (33).[39] If the group X is capable of forming a stable anion, the intermediate zwitterion will eliminate HX to form a more stable phosphorane. For instance, the addition of cyanomethylene-triphenylphosphorane to the malonate (34) generates a new ylide.[40]

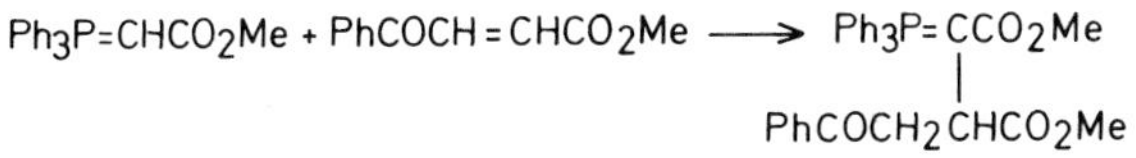

$$Ph_3P=CHCO_2Me + PhCOCH=CHCO_2Me \longrightarrow Ph_3P=CCO_2Me$$
$$\qquad\qquad\qquad\qquad\qquad\qquad\qquad\qquad\qquad\qquad\quad |$$
$$\qquad\qquad\qquad\qquad\qquad\qquad\qquad\qquad\qquad PhCOCH_2CHCO_2Me$$
$$(33)$$

$$Ph_3P=CHCN \atop + \atop EtOCH=C(CO_2Et)_2 \longrightarrow Ph_3\overset{+}{P}-CHCN \atop | \atop EtO\overset{-}{CH}C(CO_2Et)_2 \longrightarrow Ph_3P=CCN \atop | \atop CH=C(CO_2Et)_2$$

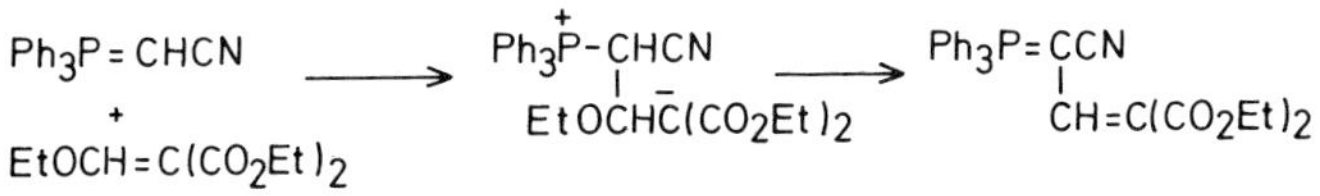

(34)

If neither of the above features is present the intermediate may eliminate a phosphine to form a cyclopropane. This process, first described by Sondheimer[41] for the preparation of the cyclopropane (35), is a very useful synthetic procedure. For instance, chrysanthemate esters can be obtained stereospecifically by simply mixing solutions of the unsaturated aldehyde (36) with isopropylidenetriphenylphosphorane.[42] Similar processes can occur on additions to α,β-unsaturated ketones, provided that the carbonyl group is highly hindered, as in compound (37).[43]

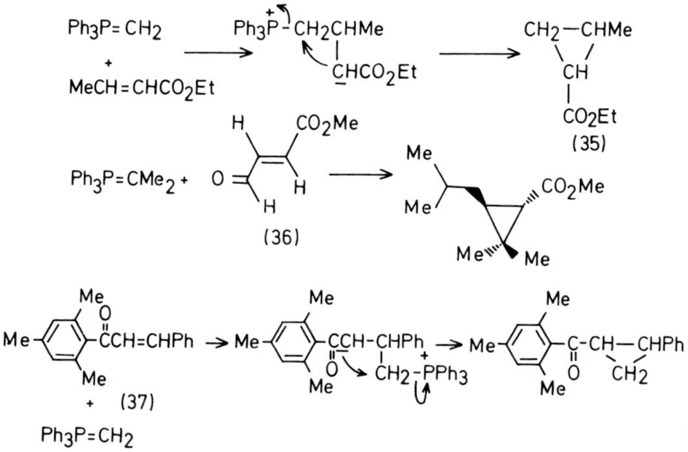

5. Reactions with Carbon–Carbon Triple Bonds

The mechanism of the addition of methylenephosphoranes to acetylenes is determined by steric and electronic factors.[44] Phosphoranes that contain an α-proton can undergo a Michael addition in protic solvents or a cycloaddition in aprotic solvents.

The reaction of ylides such as (38) with dimethyl acetylenedicarboxylate leads to stable ylides via a phosphacyclobutene.[45] Such processes are also

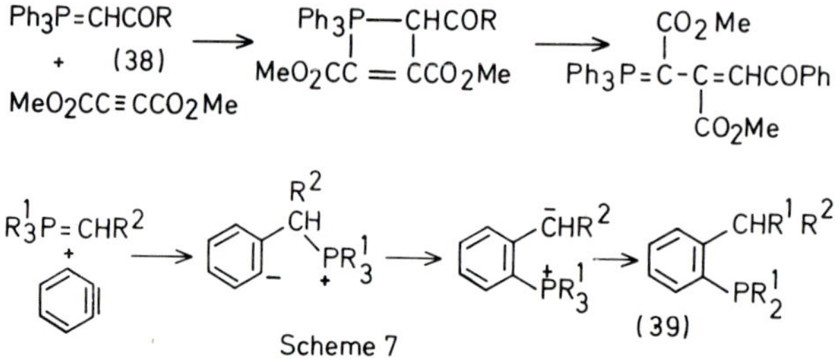

Scheme 7

known for iminophosphoranes although electron-withdrawing groups on nitrogen prevent this reaction.[44]

The interaction of ylides with benzyne leads to phosphines (39) by a multi-step rearrangement as shown in Scheme 7.[46]

6. Reactions with Carbon–Nitrogen Multiple Bonds

6.1. Schiff Bases

Alkenes (40) are generally formed by a Wittig-type reaction when Schiff bases are treated with ylides. However, these reactions are synthetically much more useful when the ylide used has a β-CH$_2$ group present. In these cases the first-formed betaine (41) decomposes on warming to generate an allene.[47]

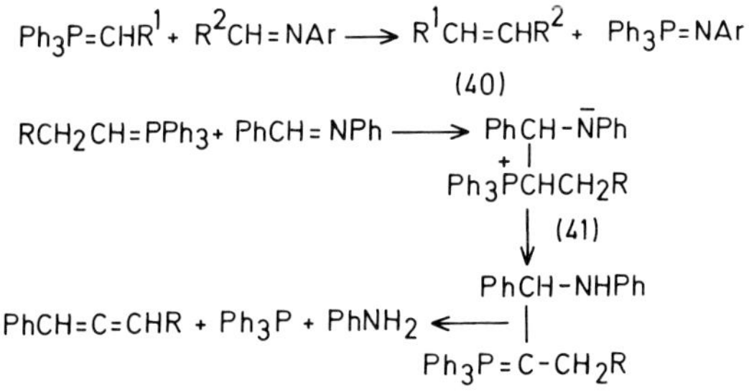

$$Ph_3P=CHR^1 + R^2CH=NAr \longrightarrow R^1CH=CHR^2 + Ph_3P=NAr$$
$$(40)$$

$$RCH_2CH=PPh_3 + PhCH=NPh \longrightarrow \begin{array}{c} PhCH-\bar{N}Ph \\ + | \\ Ph_3\overset{+}{P}CHCH_2R \end{array}$$
$$\downarrow (41)$$
$$PhCH=C=CHR + Ph_3P + PhNH_2 \longleftarrow \begin{array}{c} PhCH-NHPh \\ | \\ Ph_3P=C-CH_2R \end{array}$$

6.2. Isocyanates and Isothiocyanates

Phosphorus-stabilized carbanions react with isocyanates and isothiocyanates in a normal Wittig reaction to yield cumulenes. Thus ylides and phosphonate carbanions will produce ketenimines,[48] and iminophosphoranes and phosphoramidate anions will generate carbodiimides.[15] Complications sometimes arise if an α-proton is available on the intermediate betaine (42), particularly if R exhibits an electron-withdrawing effect. In these cases the preferred reaction is migration of a proton to the nitrogen atom to form a new ylide.

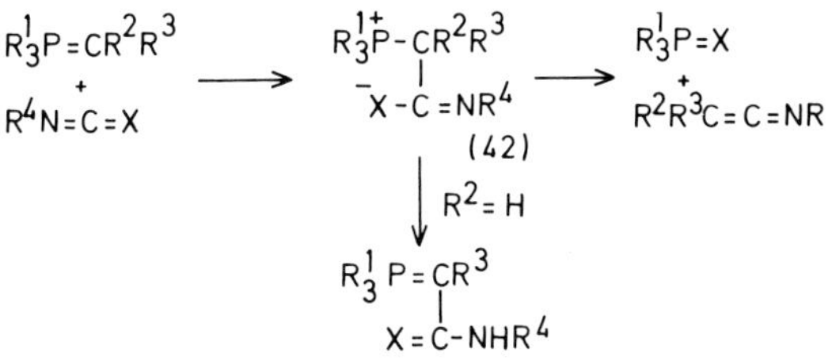

$$\begin{array}{c} R_3^1P=CR^2R^3 \\ + \\ R^4N=C=X \end{array} \longrightarrow \begin{array}{c} R_3^{1+}P-CR^2R^3 \\ | \\ {}^-X-C=NR^4 \\ (42) \end{array} \longrightarrow \begin{array}{c} R_3^1P=X \\ + \\ R^2R^3C=C=NR \end{array}$$
$$\downarrow R^2=H$$
$$\begin{array}{c} R_3^1P=CR^3 \\ | \\ X=C-NHR^4 \end{array}$$

6.3. Nitriles

It has been demonstrated that benzylidenetriphenylphosphorane and benzonitrile interact to give an iminophosphorane (43) in a process similar to reactions with acetylenes. Stabilized ylides, which do not react with benzonitrile, form iminophosphoranes with activated nitriles.[49]

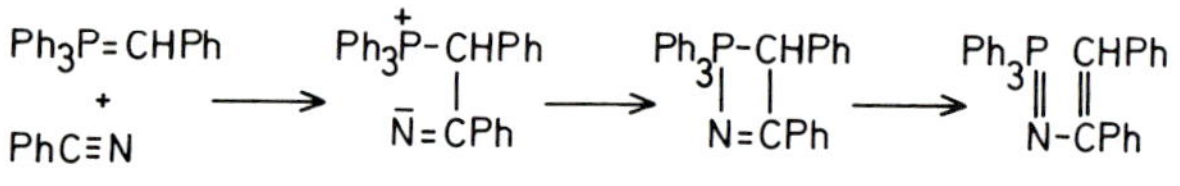

$$Ph_3P=CHPh + PhC\equiv N \longrightarrow \underset{\underset{N=CPh}{|}}{Ph_3\overset{+}{P}-CHPh} \longrightarrow \underset{\underset{N=CPh}{|}}{\overset{|}{Ph_3P-CHPh}} \longrightarrow \underset{\underset{N-CPh}{||}}{\overset{||}{Ph_3P\ CHPh}}$$

(43)

A useful way of converting nitriles into ketones is to hydrolyse the initially formed betaine (44). It is thought that two competing reactions occur on hydrolysis, one path being direct attack of water on the phosphorus atom and the other being hydration of the imino group and intramolecular collapse (Scheme 8). These reactions apparently occur simultaneously as shown by studies with optically active ylides.[50]

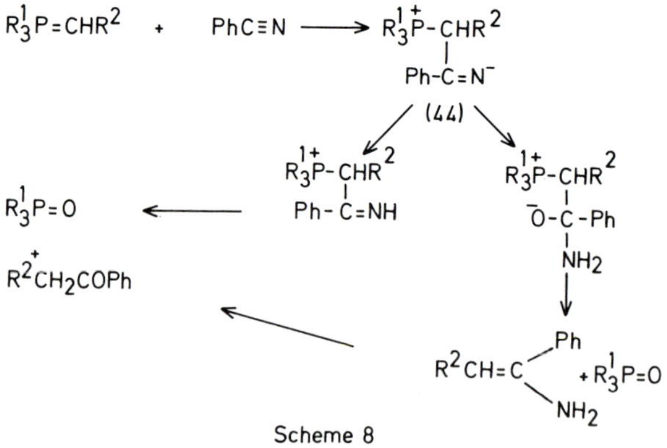

Scheme 8

7. Summary of Functional Group Conversions

Products	From	Page
Acetylenecarboxylates	alkyl halides	209
	acid chlorides	213
β-Acylacrylic acids	halogeno-ketones	211
Alkenes	epoxides	215
trans-Alkenes	allyl halides	209
Allenes	imines	219
Allenic esters	acid chlorides	213
Aziridines	amines	216
α-Branched β-keto-esters	acid chlorides	212

Products	From	Page
Carbodiimides	alkyl halides	219
Carboxylic acids	chloroformate esters	214
Cyclic alkanes	alkyl dihalides	210
Cyclic alkenes	alkyl dihalides	210
Cyclopropanes	α,β-unsaturated esters	218
	epoxides	216
1,5-Dienes	allyl halides	208
Esters	alkyl halides	209
N-Heterocycles		216, 219
Ketenimines	amines	219
γ-Keto-acids	α-halogenoacetates	211
Ketones	acyl chlorides, thioesters	212
	epoxides	215
	nitriles	219
Nitriles	acid halides	212
β-Oxo-esters	alkyl halides	209
α,β-Unsaturated esters	halogenoacetates	211
	acid chlorides	214

8. References

1. G. Wittig and M. Rieber, *Annalen*, 1949, **562**, 177.
2. (a) C. T. Eyles and S. Trippett, *J. Chem. Soc. (C)*, 1966, 67. (b) E. Axelrod, G. Milne, and E. E. van Tamelen, *J. Amer. Chem. Soc.*, 1970, **92**, 2139.
3. N. A. Nesmeyanov, S. T. Berman, and D. A. Reutov, *Bull. Acad. Sci., U.S.S.R.*, 1975, **24**, 2737.
4. H. J. Bestmann and H. Schulz, *Chem. Ber.*, 1962, **95**, 2921.
5. K. Kondo, A. Negishi, and D. Tunemoto, *Angew. Chem. Internat. Ed.*, 1974, **13**, 407.
6. H. J. Bestmann, *Angew. Chem. Internat. Ed.*, 1977, **16**, 349.
7. E. C. Taylor and S. F. Martin, *J. Amer. Chem. Soc.*, 1972, **94**, 2874, 6218.
8. H. J. Bestmann and W. Both, *Chem. Ber.*, 1974, **107**, 2926.
9. J. E. Baldwin and R. H. Fleming, *J. Amer. Chem. Soc.*, 1972, **94**, 2140.
10. A. Hercouet and M. Le Corre, *Tetrahedron Lett.*, 1974, 2491.
11. H. J. Bestmann, H. Haberlein, and I. Pils, *Tetrahedron*, 1964, **20**, 2079.
12. H. J. Bestmann and H. Schulz, *Angew. Chem.*, 1961, **73**, 620.
13. H. J. Bestmann, G. Graf, and H. Hartung, *Annalen*, 1967, **706**, 68.
14. H. J. Bestmann, H. Dornauer, and K. Rostock, *Annalen*, 1970, **735**, 52.
15. H. J. Bestmann and R. Zimmermann, in *Organic Phosphorus Compounds* (ed. G. M. Kosolapoff and L. Maier), Vol. 3, Ch. 5, Wiley-Interscience, New York, 1972.
16. R. Appel and A. Hauss, *Chem. Ber.*, 1960, **93**, 405.
17. N. Kreutzkamp, *Chem. Ber.*, 1955, **88**, 195.
18. S. T. D. Gough and S. Trippett, *J. Chem. Soc.*, 1962, 2333.
19. P. A. Chopard, R. J. G. Searle, and F. H. Devitt, *J. Org. Chem.*, 1965, **30**, 1015.
20. H. J. Bestmann, G. Graf, S. Kolewa, and E. Vilsmeier, *Chem. Ber.*, 1970, **103**, 2794.

21. J. Villieras, P. Perriot, and J. F. Normant, *Synthesis*, 1978, 29.
22. H. J. Bestmann, *Angew. Chem. Internat. Ed.*, 1965, **4**, 645.
23. F. Bohlmann and W. Skuballa, *Chem. Ber.*, 1973, **106**, 497.
24. H. Staudinger and E. Hauser, *Helv. Chim. Acta*, 1921, **4**, 861.
25. H. J. Bestmann and H. Hartung, *Chem. Ber.*, 1966, **99**, 1198.
26. H. J. Bestmann and I. Tömösközi, *Tetrahedron*, 1968, **24**, 3299.
27. H. J. Bestmann and W. Kloesters, *Angew. Chem. Internat. Ed.*, 1977, **16**, 45.
28. G. Durrant and J. K. Sutherland, *J. Chem. Soc., Perkin I*, 1972, 2582.
29. A. Maercker and W. Theysohn, *Annalen*, 1972, **759**, 132.
30. S. Trippett, *Quart. Rev.*, 1963, **17**, 406.
31. W. E. McEwen, A. Bladé-Font, and C. A. Vanderwerf, *J. Amer. Chem. Soc.*, 1962, **84**, 677.
32. R. Huisgen and J. Wulff, *Chem. Ber.*, 1969, **102**, 1841.
33. R. M. Gerkin and B. Rickborn, *J. Amer. Chem. Soc.*, 1967, **89**, 5850.
34. R. A. Izydore and R. G. Ghirardelli, *J. Org. Chem.*, 1973, **38**, 1790.
35. I. Tömösközi, *Chem. and Ind.*, 1965, 689.
36. L. Horner, H. Hoffmann, and V. G. Toscano, *Chem. Ber.*, 1962, **95**, 536.
37. R. Appel and M. Halstenberg, *Chem. Ber.*, 1976, **109**, 814.
38. H. W. Heine, G. B. Lowrie, and K. C. Irving, *J. Org. Chem.*, 1970, **35**, 444.
39. H. J. Bestmann and F. Seng, *Angew. Chem. Internat. Ed.*, 1962, **1**, 116.
40. S. Trippett, *J. Chem. Soc.*, 1962, 4733.
41. R. Mechoulan and F. Sondheimer, *J. Amer. Chem. Soc.*, 1958, **80**, 4386.
42. M. J. Devos, L. Hevesi, P. Bayet, and A. Krief, *Tetrahedron Lett.*, 1976, 3911. See also P. A. Grieco and R. S. Finkelhor, *ibid.*, 1972, 3781.
43. J. P. Freeman, *Chem. and Ind.*, 1959, 1254.
44. G. W. Brown, *J. Chem. Soc.*, 1967, 2018.
45. J. B. Hendrikson, R. Rees, and J. F. Templeton, *J. Amer. Chem. Soc.*, 1964, **86**, 107.
46. E. Zbiral, *Tetrahedron Lett.*, 1964, 3963.
47. H. Daniel and J. Paetsch, *Chem. Ber.*, 1968, **101**, 1451.
48. H. Staudinger and E. Hauser, *Helv. Chim. Acta*, 1921, **4**, 887.
49. E. Ciganek, *J. Org. Chem.*, 1970, **35**, 3631.
50. R. G. Barhardt and W. E. McEwen, *J. Amer. Chem. Soc.*, 1967, **89**, 7009.

5
Transformations via Phosphorus-stabilized Anions 4: Heterocyclic Synthesis via Alkylidenephosphoranes, Iminophosphoranes, and Vinylphosphonium Salts

ERICH ZBIRAL

1. Syntheses of heterocyclic compounds using alkylidenephosphoranes, 223
2. Syntheses of heterocyclic compounds using iminophosphoranes, 241
3. Syntheses of heterocyclic compounds using alk-1-enylphosphonium salts, 250
4. References, 265

This chapter summarizes methods for the synthesis of heterocyclic systems offered by the use of three types of organophosphorus compounds: alkylidenephosphoranes, iminophosphoranes, and vinylphosphonium salts. Previously, this topic has been reviewed only in part and from a different point of view.[1] Related surveys include: (i) heterocyclic synthesis via pentacoordinate phosphoranes,[2] aspects of which are covered in Chapter 11; (ii) formation of N-heterocycles by reductive cyclization of aromatic nitro compounds[3] (Chapter 6); (iii) formation of heterocycles via phosphacumulenes and phosphaallenylidenes. The last named area has been the subject of an excellent review by Bestmann;[4a] he and his group have been responsible for many of the most notable advances. A selection of applications of the method is given in Table 1. Many of the products are themselves ylidic and can be used, in turn, as synthons.

1. Syntheses of Heterocyclic Compounds using Alkylidenephosphoranes

Heterocyclic synthesis using alkylidenephosphoranes can be subdivided into three classes according to the reaction pattern.
1. A carbonyl olefination takes place, which leads to the heterocyclic compound either directly or via an appropriate reactive intermediate.

2. A 2-oxoalkylidenetriphenylphosphorane undergoes a primary reaction with a 1,3-dipolarophile, followed by formation of the heterocyclic compound with elimination of triphenylphosphine oxide.
3. A 2-oxoalkylidenetriphenylphosphorane reacts with a suitable partner to form an intermediate from which the heterocyclic secondary product is formed either directly or by a further synthetic operation.

Table 1. Some cycloaddition reactions of phosphacumulene ylides ($Ph_3P=C=C=X$)

Reactant	Product	
O=Z-A-Y-H	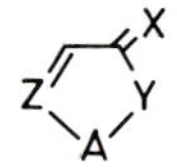 [ring with X, Z, A, Y]	
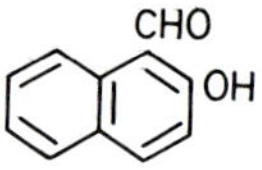 naphthalene-CHO, OH	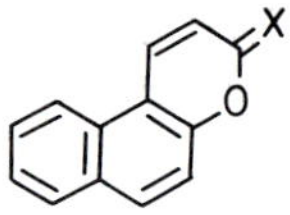 naphtho-pyran with X	X=O, NPh
Ph CO CH(OH) Ph	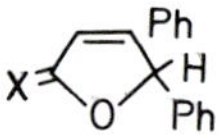 furan ring with Ph, H, Ph, X	X=O, NPh
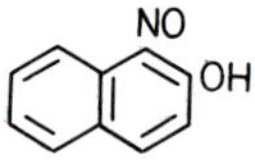 naphthalene-NO, OH	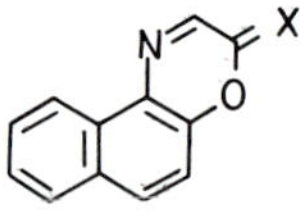 naphtho ring with N, X, O	X=NPh
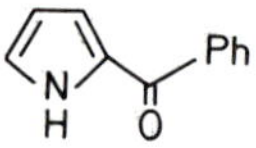 pyrrole-C(O)Ph (N-H)	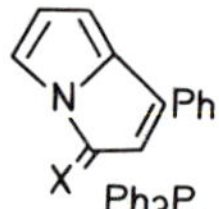 pyrrolizine with Ph, X	X=O, NPh
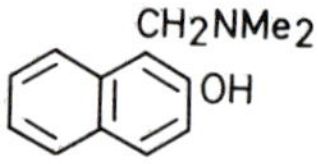 naphthalene-CH$_2$NMe$_2$, OH	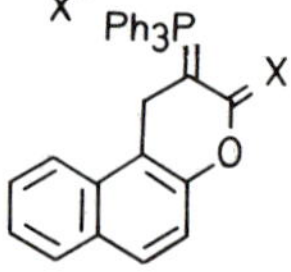 naphtho ring with Ph_3P, X, O	X=O, S
RN=C=O	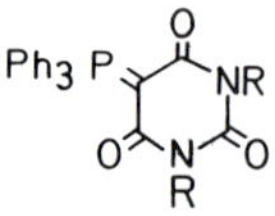 Ph_3P ring with O, NR, N-R, O	X=O
RN=C=S	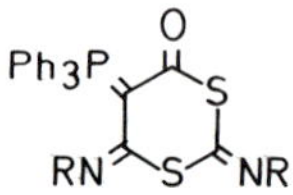 Ph_3P ring with O, S, NR, S, NR	X=O
CO_2	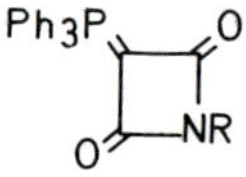 Ph_3P ring with O, NR, O	X=NR

5 Ylides in Heterocyclic Synthesis

Table 1. Continued

Reactant	Product	
Y=C=S (Y=O,S,NR)	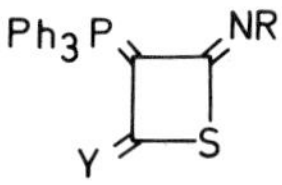 Ph₃P-C(NR)-Y-C(S) (4-membered ring)	X=NR
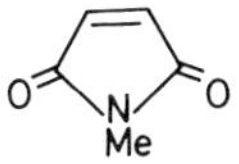 maleimide (O=C-N(Me)-C=O, cyclic)	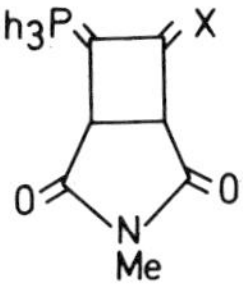 Ph₃P-C(X)-C-C(O)-N(Me)-C(O) (bicyclic)	X=NR
Tos N₃	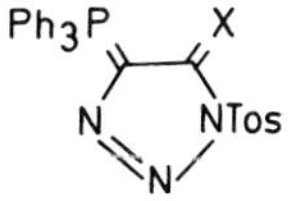 Ph₃P-C(X)-N=N-NTos ring	X=O
R₂CN₂	Ph₃P-C(X)-N=N-CR₂ ring	X=NPh
R¹COCH=CHR²	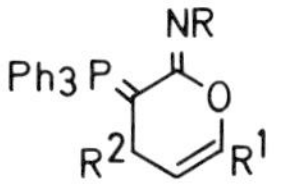 Ph₃P-C(NR)-O-C(R¹)=C-R² (6-membered)	X=NR

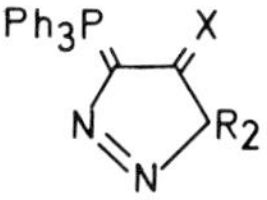

1.1. Syntheses of Heterocyclic Compounds via Carbonyl Olefination with Alkylidenephosphoranes

1.1.1. *Tetrahydropyrans, thiacyclohexanes, and dibenzo[b,f]oxepins*

Compounds of these types can be obtained by applying the standard reactions of phosphorane chemistry to appropriate difunctional ethers or thioethers respectively (Schemes 1 and 2). The use of bis-Wittig reactions in this way has been reviewed.[4b] Thus di-(2-bromoethyl) ether (1a) and di-(2-bromoethyl) sulphide (1b) can be converted into their respective monophosphonium salts by treatment with methylenetriphenylphosphorane; elimination of hydrogen bromide, followed by intramolecular alkylation and a second elimination of hydrogen bromide affords the phosphorane (2) which contains the desired heterocyclic ring; oxidation of (2) yields tetrahydropyran-4-one (3a) or thiacyclohexan-4-one (3b) respectively, while carbonyl olefination of benzaldehyde with (2) yields the corresponding 4-benzylidene derivatives (4a,b).[5] By an analogous reaction sequence 3-benzylidene-9-oxabicyclo[3.3.1]nonane (6) can be obtained from 2,6-di-(tosyloxymethyl)tetrahydropyran (5).[6]

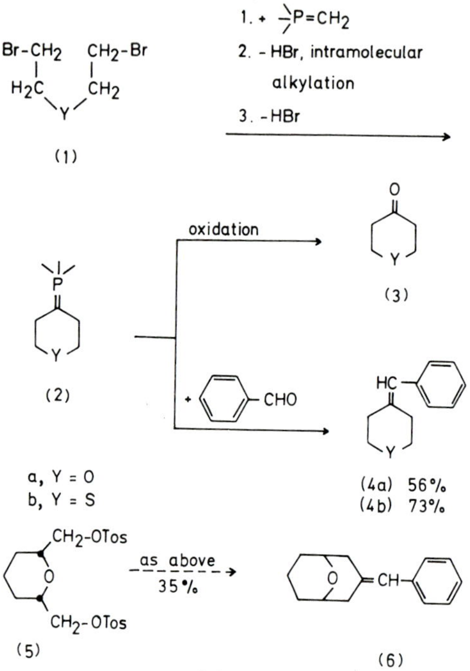

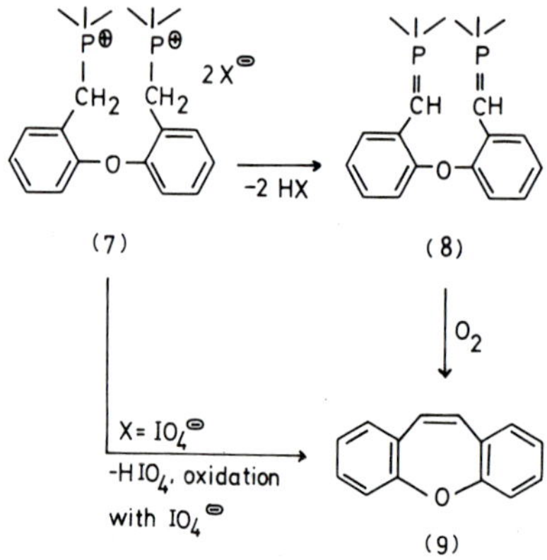

The formation of dibenzo[b,f]oxepin (9) from the bis-phosphonium salt (7) (Scheme 2) represents another type of reaction. Air oxidation of the diphosphorane (8) obtained from (7) leads to intramolecular carbonyl olefination of the initially formed monocarbonyl compound and ring closure to (9).[7] Alternatively, if the diphosphorane is formed from the periodate of (7), the periodate ion functions as oxidizing agent, followed by ring closure to (9).[8] This reaction is a special case of the general oxidative coupling of ylides first discovered by Bestmann[7b,c] which has found use in many other areas of polyene chemistry, e.g. in the synthesis of β-carotene.[8a]

1.1.2. Thiophens and thiacyclononatetraenes

Carbonyl olefination of suitable *vic*-diketones with bis(triphenylphosphoranylidenemethyl) sulphide (10) offers a route to thiophen derivatives,[8] the scope of which has probably not been fully exploited (Scheme 3). From cyclobutenedione, 3-thiabicyclo[3.2.0]heptatriene (11) can be prepared, while 2,3-dihydrobenzocyclobutenedione affords the tricyclic benzo[3,4]cyclobuta[1,2-*c*]thiophen (12).

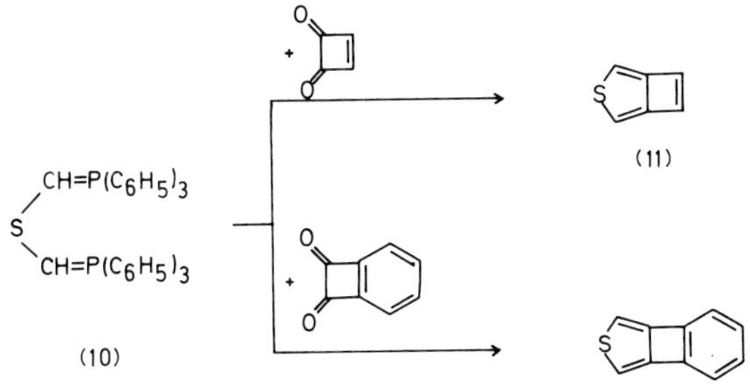

Scheme 3

In a reaction which represents an inversion of the type illustrated in Scheme 3, di-(2-formylcyclohexenyl) sulphide (13) and 2,3-bis(triphenylphosphoranylidene)-2,3-dihydrobenzocyclobutene (14) yield the *cis–trans* isomeric thiacyclononatetraene derivatives (15a) and (15b) (the latter in only small amounts).[9]

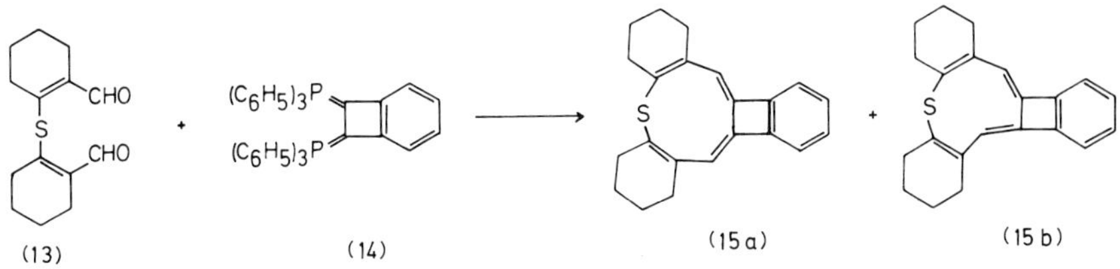

Scheme 4

1.1.3. α-Pyrones, coumarins, and 2,3-dihydro-1-benzofurans

Alkoxycarbonylmethylenetriphenylphosphoranes (17) do not undergo a double Wittig reaction with *o*-benzoquinones (16). Instead, monocarbonyl olefination takes place, followed by Michael addition of a second phosphorane molecule to the intermediate (18). The resulting phosphonium betaine (19) eliminates triphenylphosphine to yield the *o*-hydroxyphenylfumaric acid ester (20), which cyclizes with loss of one molecule of alcohol to the coumarin-4-carboxylic acid ester (21) (Scheme 5).[10] In a similar reaction, 2-methoxy-1,4-naphthoquinone (22) affords the coumarin derivative (21f), 6-methoxy-1-methoxycarbonyl-3-oxo-3H-naphtho[2,1-b]pyran.[10]

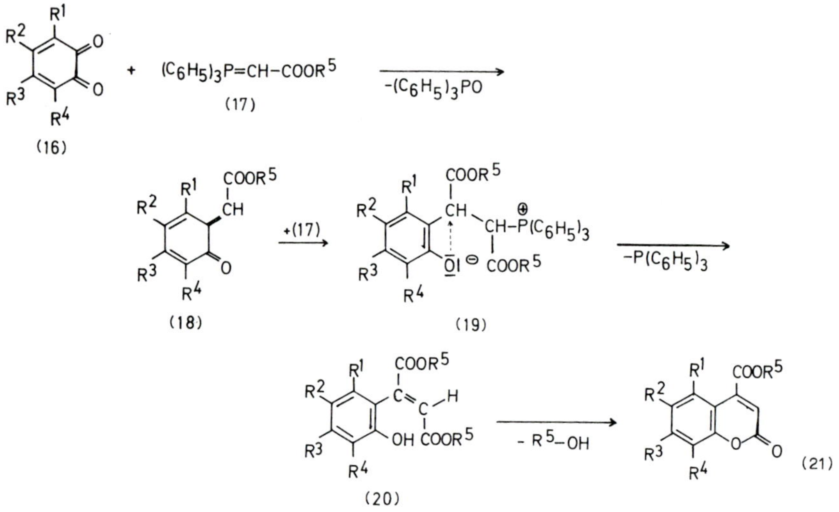

	R^1	R^2	R^3	R^4	R^5	Yield (%)
(21 a)	H	H	—(CH=CH)$_2$—		CH$_3$	85
(21 b)	H	H	H	H	C$_2$H$_5$	73
(21 c)	—(CH=CH)$_2$—		—NH—C$_6$H$_5$	H	CH$_3$	46
(21 d)	—(CH=CH)$_2$—		—(CH=CH)$_2$—		CH$_3$	76
(21 e)	—(CH=CH)$_2$—		—(CH=CH)$_2$—		C$_2$H$_5$	56

Scheme 5

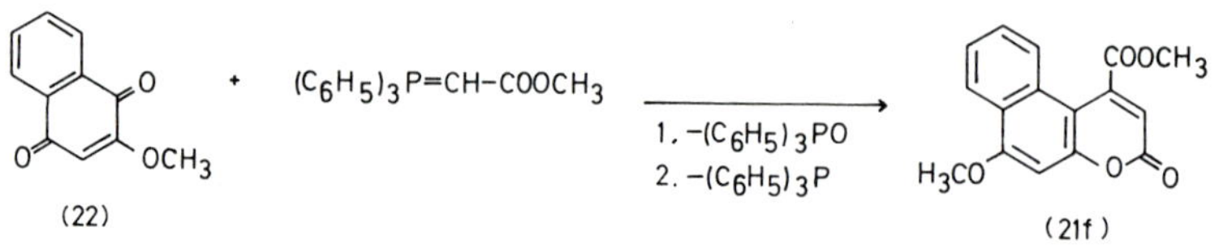

Scheme 6

In the reactions of o-benzoquinones (16) and 2-methoxy-1,4-naphthoquinones (22) with benzylidenetriphenylphosphorane (23) the first steps are analogous to those of the corresponding reactions with alkoxycarbonyltriphenylphosphoranes (Schemes 5, 6); the 1,2-diphenyl-1-(2-hydroxyphenyl)-ethylenes (24) formed as intermediates then undergo the expected cyclization to 2,3-diphenylcoumaran derivatives (25) and (26) (Schemes 7 and 8).[10] The

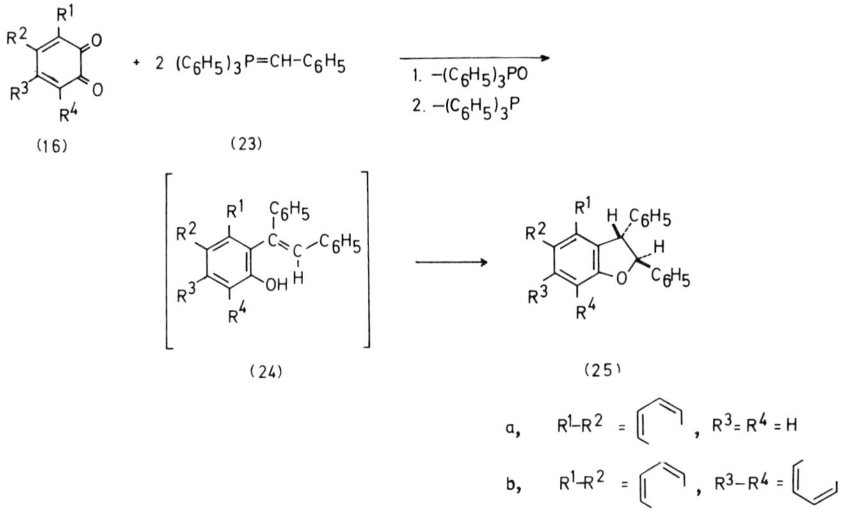

Scheme 7

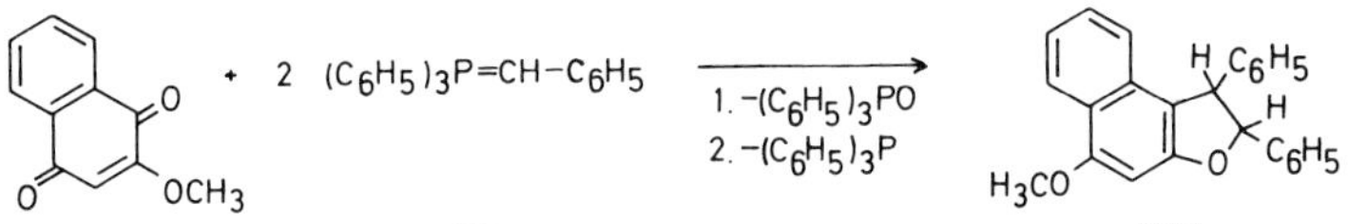

(22) (23) (26)

Scheme 8

coumaran derivatives thus obtained, for example the 2,3-diphenyl-2,3-dihydrophenanthro[9,10-*b*]furans (27),[11] can be oxidized with *N*-bromo-succinimide to the corresponding condensed furans, e.g. (28).

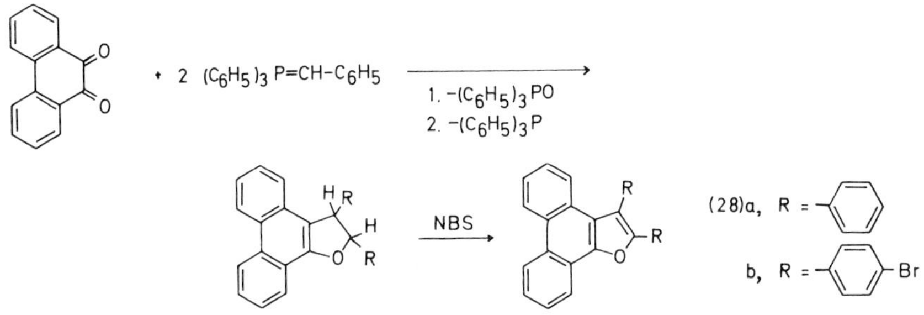

(27)

Scheme 9

An alternative route to coumarin derivatives, e.g. (29), (30), is the carbonyl olefination of aromatic *o*-hydroxy-aldehydes (in Scheme 10, 2-hydroxy-naphthalene-1-aldehyde) with alkoxycarbonylmethylenetriphenylphosphoranes.[12]

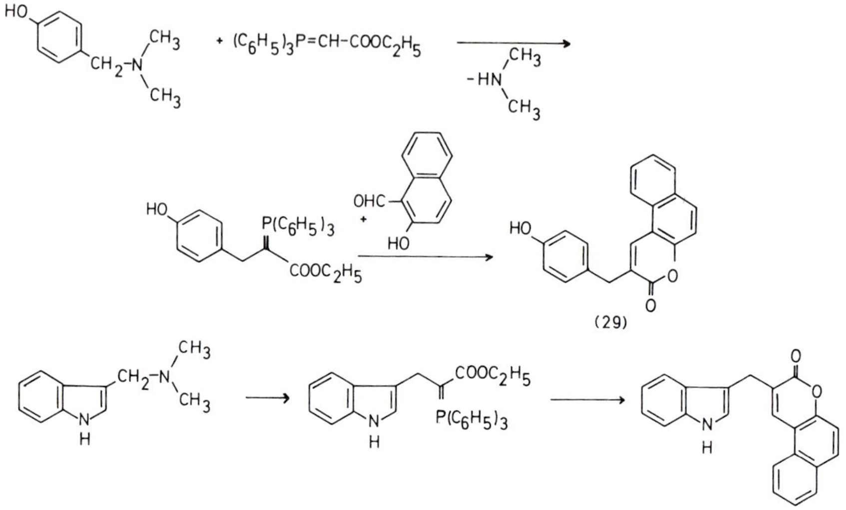

Scheme 10

Non-condensed α-pyrones (31) can be prepared by reaction of 1,3-diketones bearing aromatic substituents with alkoxycarbonylmethylenetriphenylphosphoranes,[13] although in poor yield (Scheme 11). An analogous triphenylphosphorane and 2-benzoylcyclohexanone form 3-oxo-1-phenyl-5,6,7,8-tetrahydro-3H-[2]benzopyran (32).

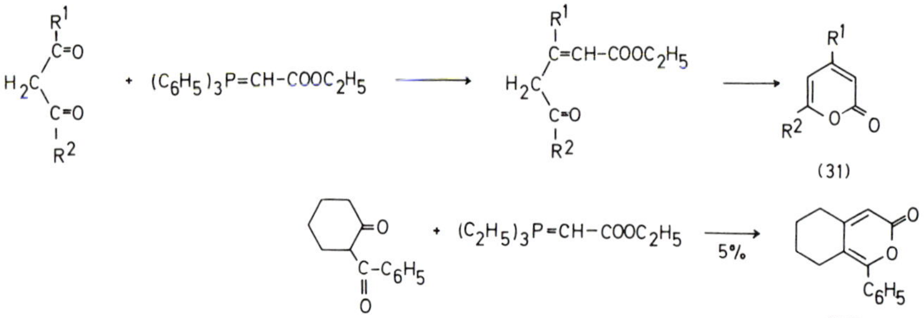

Scheme 11

1.1.4. Acridinium betaines

Application of the reaction outlined in Scheme 6 to N-substituted 2-amino-1,4-naphthoquinones (33) makes possible the direct syntheses of 7-substituted 5-oxybenzo[a]acridinium betaines (34) (Scheme 12).[14]

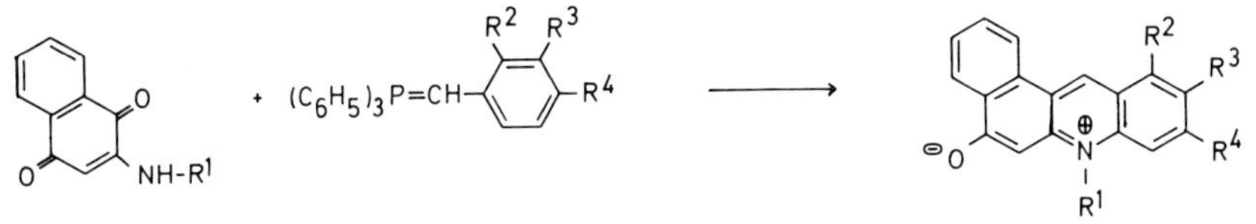

5 Ylides in Heterocyclic Synthesis

[Scheme 12 structures]

	R^1	R^2	R^3	R^4	
a,	$R^1 = C_6H_5$,	$R^2 = R^3 = R^4 = H$			(20 %)
b,	$R^1 = C_6H_5$,	$R^2 = R^3 = H$,	$R^4 = OCH_3$		(11 %)
c,	$R^1 = C_6H_5$,	$R^2 = R^3 = \langle \rangle$,	$R^4 = H$		(42 %)
d,	$R^1 = CH_3$,	$R^2 = R^3 = \langle \rangle$,	$R^4 = H$		(46 %)
e,	$R^1 = C_2H_5$,	$R^2 = R^3 = \langle \rangle$,	$R^4 = H$		(11 %)

Scheme 12

1.1.5. *Pyridines*

An interesting method for the synthesis of 2-substituted and 2,5-disubstituted pyridine heterocycles (34a) with a chiral alkyl substituent in position 5 has recently been published (Scheme 13).[15]

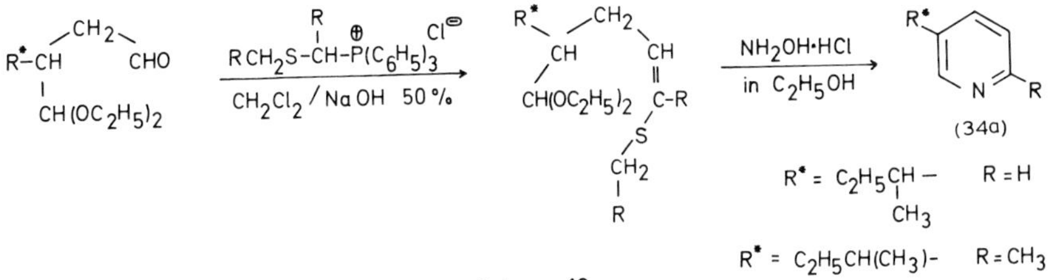

[Scheme 13 structures]

$R^* = C_2H_5\underset{CH_3}{\overset{|}{C}H}-$ $R = H$

$R^* = C_2H_5CH(CH_3)-$ $R = CH_3$

Scheme 13

1.2. Syntheses via 1,3-Dipolar Cycloaddition Reactions

1.2.1. *1,2,3-Triazoles*

The reaction of 2-oxoalkylidenephosphoranes (35) with azido compounds or acid azides as shown in Scheme 14 affords a versatile general synthesis of 1,2,3-triazoles (36). The reaction can be visualized as a 1,3-dipolar cycloaddition of the azido group to the enol betaine form of the phosphorane (35). Originally described for phosphoranes (35) in which $R^1 = H$,[16] the method was later broadened in scope[17-21] to include phosphoranes in which $R^1 =$ alkyl (Scheme 15). The reaction applied to carboxylic acid azides has been shown to yield primary products (36) in which the acyl group is attached to N-1 of the triazole, but migration to N-2 can occasionally occur.[19,21]

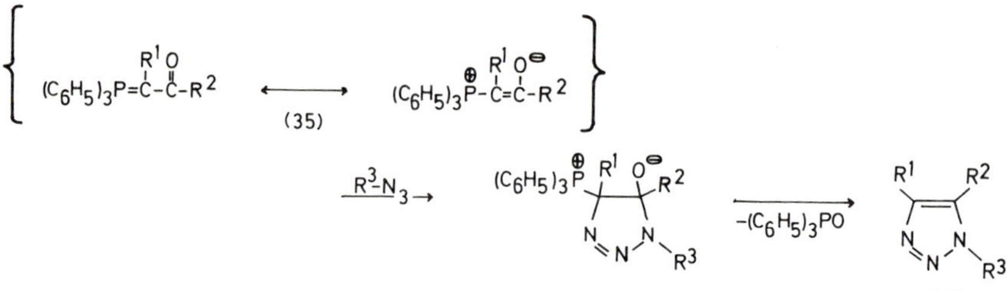

Scheme 14

$R^1 = H$, $R^2 = CH_3$ resp. Aryl, $R^3 = $ Aryl, Ar-SO_2- and ArCO

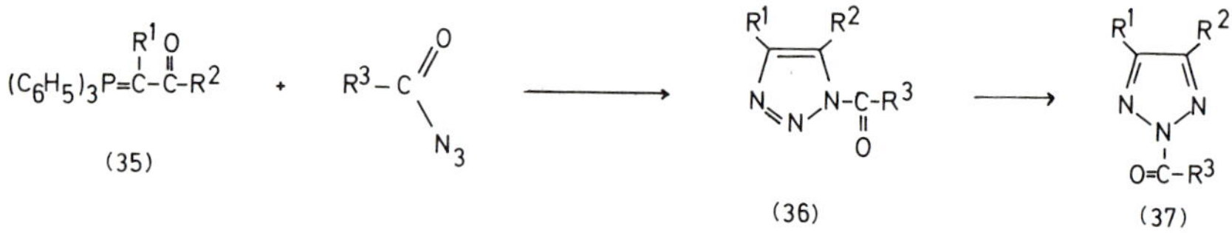

Scheme 15

The azido group is occasionally found to undergo a 1,3-dipolar cycloaddition to the P=C "double bond" of the phosphorane (35) as shown in Scheme 16.[16,20,24] The preparation of 2-oxoalkylidenetriphenylphosphoranes

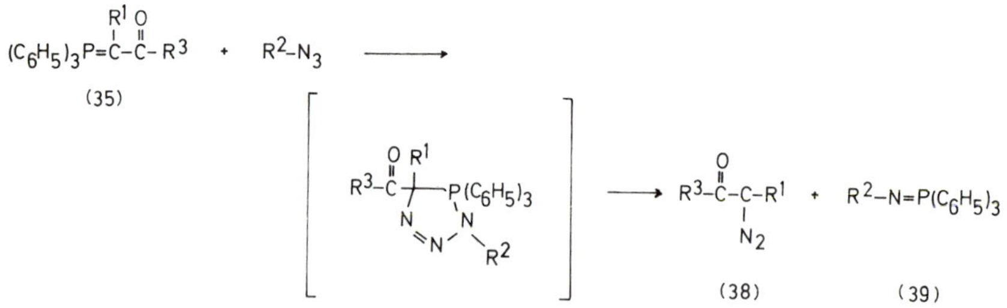

$R^1 = H$ resp. CH_3, $R^2 = $ Ar-SO_2, ArCO, Aryl and -$P(OC_2H_5)_2$,
$R^3 = OC_2H_5$, -$N(C_2H_5)_2$ $\|$
                                                            O

Scheme 16

from 1H-alkylidenetriphenylphosphoranes and acyl halides, and the stepwise synthesis of 1,2,3-triazoles which this reaction offers, have also been realized.[22] The N-acyl-1,2,3-triazoles described in this paper were hydrolysed without further characterization to the corresponding 1,2,3-triazoles (40) (Scheme 17).

The highly regiospecific reaction between aryl azides and 2-oxoalkylidenephosphoranes (35) affords 5-substituted 1-aryl-1,2,3-triazoles (41) in yields ranging from 54 to 98% (Scheme 18).[23,24]

The reaction formulated in Scheme 14 is also applicable to vinyl azides (Scheme 19),[25] although extremely long reaction times are often necessary,

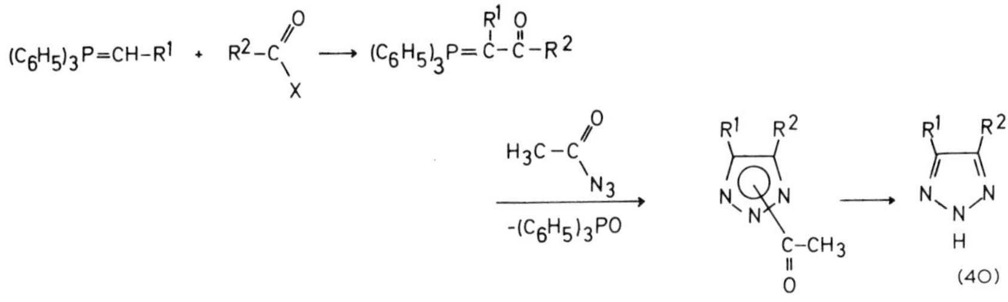

5 Ylides in Heterocyclic Synthesis

$(C_6H_5)_3P=CH-R^1$ + $R^2-C(=O)X$ → $(C_6H_5)_3P=C(R^1)-C(=O)-R^2$

$\xrightarrow[-(C_6H_5)_3PO]{H_3C-C(=O)N_3}$

[pyrazole with R^1, R^2, N, N, $\overset{|}{N}$-C(=O)-CH$_3$] → [triazole R^1, R^2, N, N, N, H] (40)

R^1 = Alkyl, Alkenyl. R^2 = Alkyl, Cycloalkyl, and C_2H_5O (60–90%)

Scheme 17

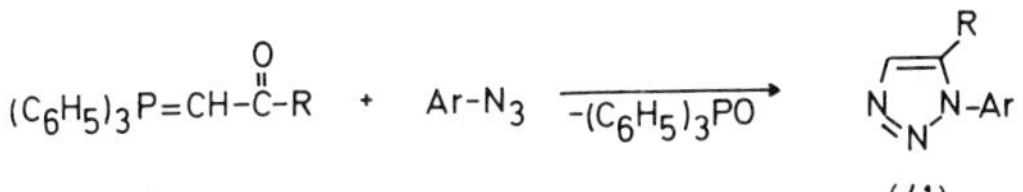

$(C_6H_5)_3P=CH-C(=O)-R$ + $Ar-N_3$ $\xrightarrow{-(C_6H_5)_3PO}$ [pyrazole R, N, N-Ar]

(35)   (41)

Scheme 18

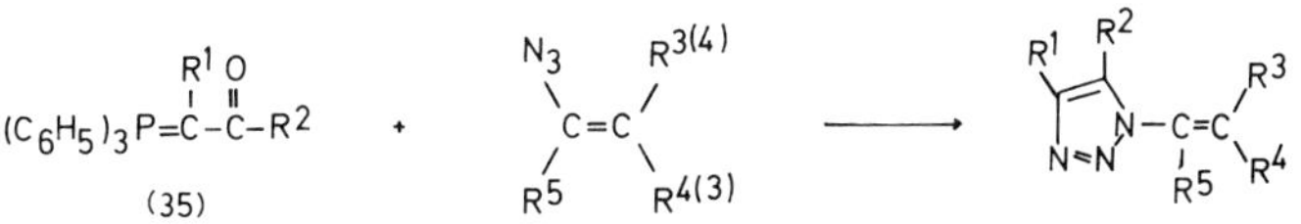

$(C_6H_5)_3P=C(R^1)-C(=O)-R^2$ + $N_3-C(R^{3(4)})=C(R^{4(3)})R^5$ → [triazole R^1, R^2, R^3, R^4, R^5]

(35)   (42)

Scheme 19

[ribose derivative with ØCOO, O, N$_3$, ØCOO, OCOØ] $\xrightarrow{\overset{\cdot}{P}=CH-COR}$ [N-triazolyl ribose with R]

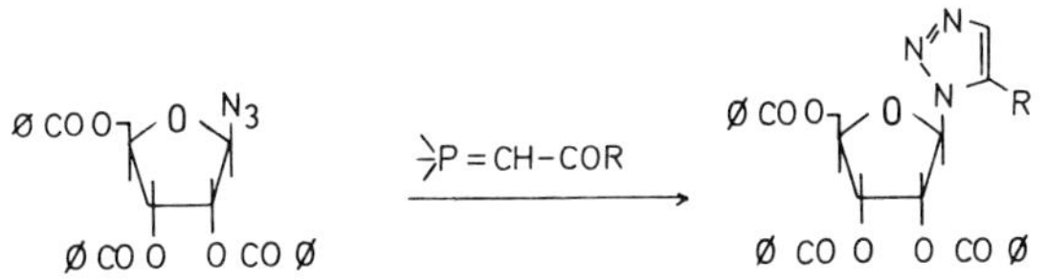

R = H, Alkyl, COOC$_2$H$_5$ and -CONH$_2$ (30–90%)

Scheme 20

satisfactory yields are obtained, and considerable latitude in the choice of substituents is permissible. Recently, this method has been applied to the synthesis of *N*-triazolylriboside analogues of nucleosides (Scheme 20).[26]

1.2.2. Pyrazoles

Nitrileimines as 1,3-dipolar components can undergo cycloaddition to 2-oxo-alkylidenephosphoranes (35), yielding pyrazoles (43) (Scheme 21),[27-29] a reaction closely analogous to that discussed in Section 1.2.1.

An alternative method for the preparation of pyrazoles in yields of 60–80% is offered by the cycloaddition of nitrileimines to allylenephosphoranes with electron-accepting substituents in the γ-position. The final step in this interesting reaction is the relatively uncommon ylide fragmentation shown in Scheme 22.[30]

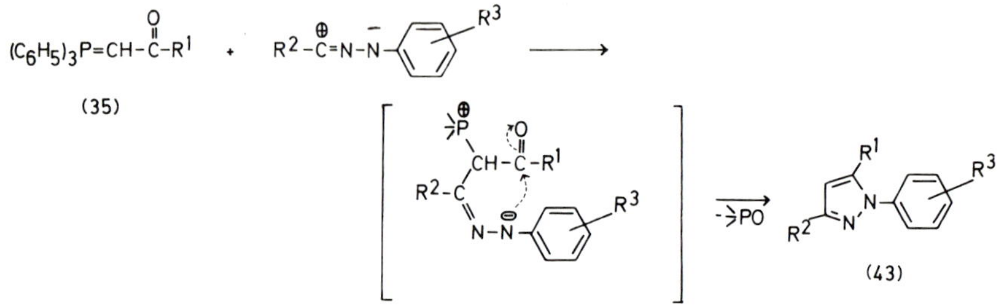

R^2 = Alkoxycarbonyl, CH_3CO and C_6H_5
R^3 = H resp. 4- and 2-NO_2-C_6H_4- (40-80%)

Scheme 21

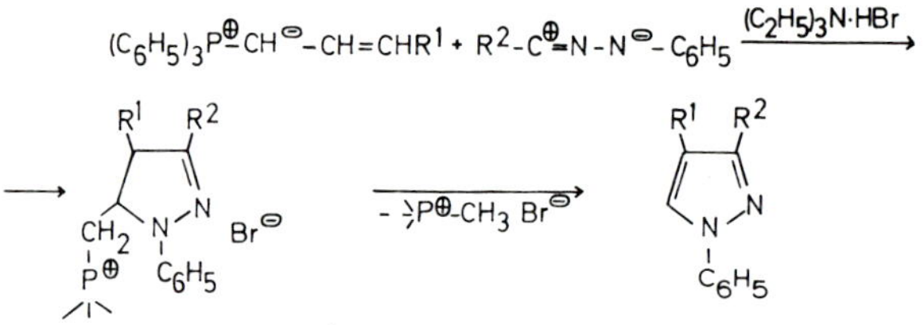

R^1 = CHO, COOCH$_3$, CN
R^2 = C_6H_5, COOCH$_3$

Scheme 22

1.2.3. *1,2-Oxazoles*

If, instead of a nitrileimine, a nitrile oxide is taken as the 1,3-cycloaddend, reaction with an allylenephosphorane takes place in a manner analogous to that in Scheme 22 to yield an isoxazole (Scheme 23).[30]

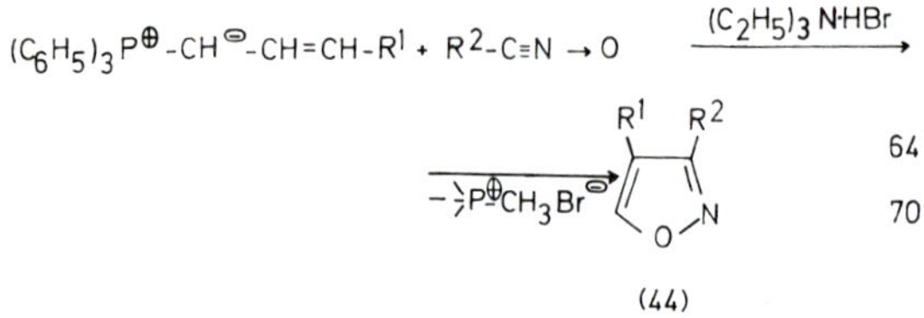

64% (44a) R^1 = CN, R^2 = 4-NO_2-C_6H_4-
70% (44b) R^1 = COOC$_2$H$_5$, R^2 = 4-NO_2-C_6H_4-

Scheme 23

1.2.4. *3-Pyrrolines and pyrroles*

An interesting route to 3-pyrrolines or pyrroles is the reaction of certain phosphoranes with aziridines.[31] At elevated temperatures 2-alkoxycarbonylaziridines are in equilibrium with the corresponding azomethine ylides. Cyanomethylenephosphorane and methoxycarbonylmethylenephosphorane

5 Ylides in Heterocyclic Synthesis

both behave as nucleophiles towards these azomethine ylides. A following intramolecular Wittig reaction leads to 3-pyrrolines (45) and (46) respectively. The 3-pyrrolines formed in variable yields have not always been isolated as such, but as their oxidation products, the pyrroles (47) and (48) (Scheme 24). This reaction is much faster if the aziridine contains two geminal ester groups. Aziridines with $R^2 \neq H$ yield mixtures of diastereoisomeric 3-pyrrolines. The classical reaction pattern of a 1,3-dipolarophilic cycloaddition, which *a priori* one would expect in this reaction, has been observed only with benzoylmethylenetriphenylphosphorane ($R^1 = COC_6H_5$), which yields 3-pyrrolines with a substitution sequence excluding the mechanism in Scheme 24.[31b]

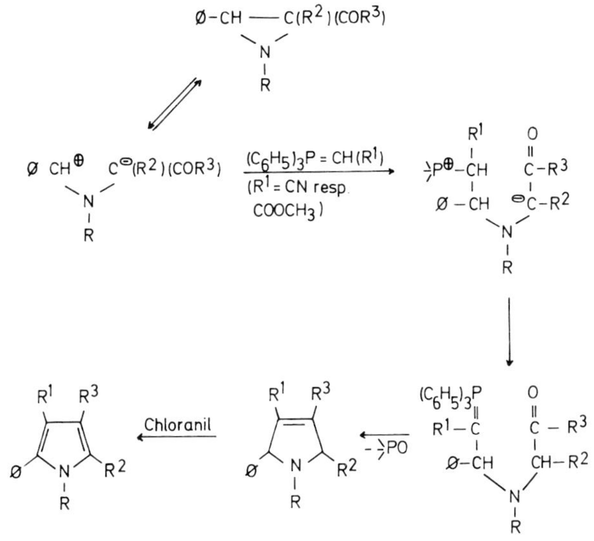

(47) R^1 = CN
(48) R^1 = COOCH$_3$

(45) R^1 = CN
(46) R^1 = COOCH$_3$

Scheme 24

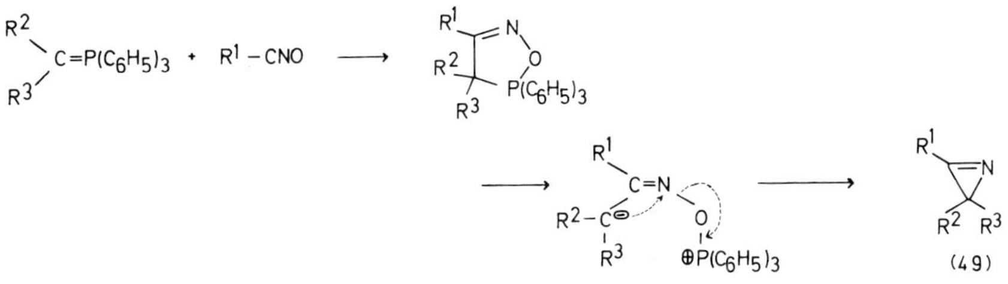

R^1 = Aryl, -COOC$_2$H$_5$ R^2 = H, CH$_3$, C$_6$H$_5$. R^3 = H, Alkyl, COOC$_2$H$_5$, C$_6$H$_5$ (30-90 %)

Scheme 25

1.2.5. Azirines

Nitrile oxides undergo a 1,3-dipolar addition to alkylidenetriphenylphosphoranes to yield 4,5-dihydro-1,2,5-oxazaphosph(v)oles. The latter can, if suitably substituted, eliminate triphenylphosphine oxide to form azirines (49) (Scheme 25).[27,33]

1.3. Various Syntheses using 2-Oxoalkylidenephosphoranes

1.3.1. γ-Pyrones, 4-(2-oxoalkylidene)pyrans, and 4-phenyliminopyrans

2-Oxoalkylidenephosphoranes can react with active methylene compounds such as β-keto-esters, malonic acid esters, or diacylacetic acid esters to form

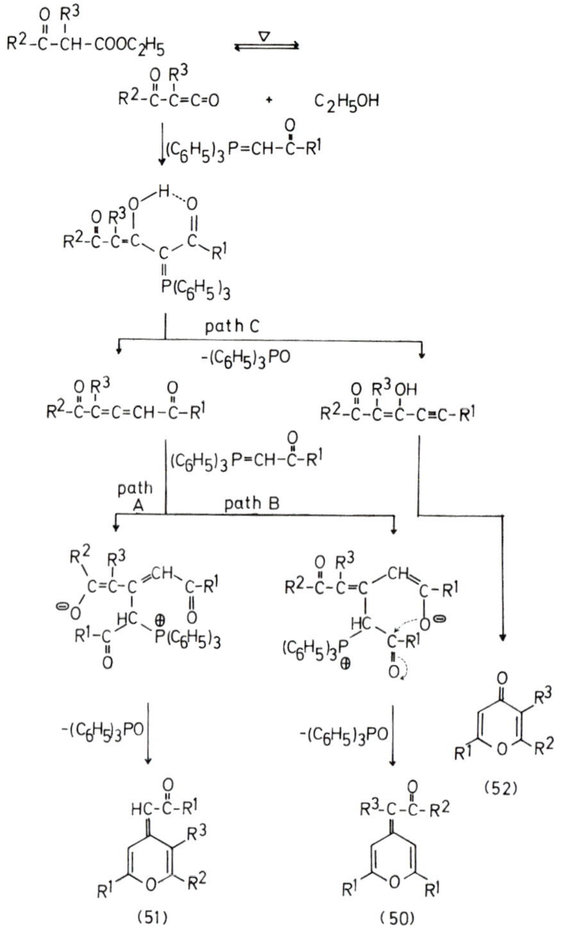

Scheme 26

γ-pyrones (52) and/or 4-(2-oxoalkylidene)pyrans (50, 51). The formation of γ-pyrones requires one mole of phosphorane for each mole of the active methylene compound, while two moles are necessary for 4-(2-oxoalkylidene)-pyrans to be formed (Scheme 26).[34,37] Some reactions described in the literature take paths B and C concurrently (Scheme 27),[35] some take path C exclusively (Scheme 28),[35,36] while others take path B to the exclusion of C (Scheme 29).[38] Path A seems to be preferred when the reaction is applied to

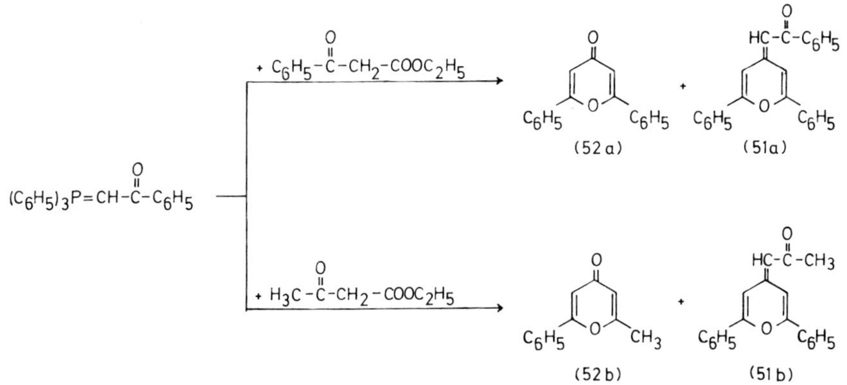

Scheme 27

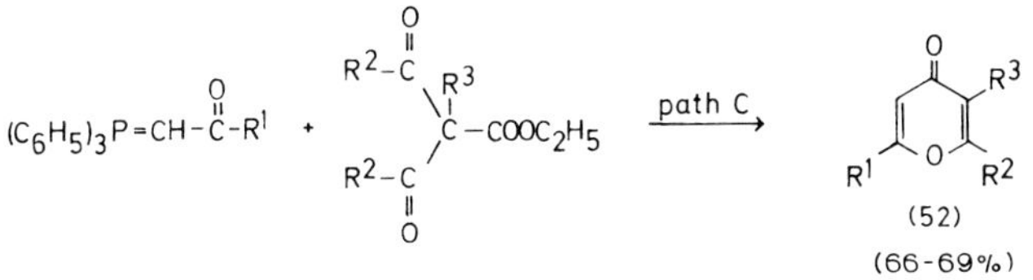

Scheme 28

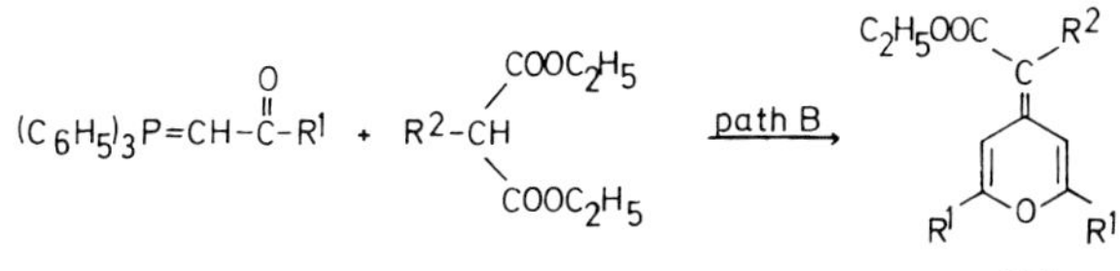

Scheme 29

$R^1 = C_6H_5$, $R^2 = H$ 76% $R^1 = C_6H_5$, $R^2 = CH_3$ 41%
$R^1 = CH_3$, $R^2 = H$ 58% $R^1 = CH_3$, $R^2 = CH_3$ 30%

alkyl acetoacetic esters (Scheme 30).[37] It is appropriate to point out in this connection that the 4-(2-oxoalkylidene)pyrans (50) and (51) can be readily transformed into pyrylium salts,[35] and these latter into pyridine derivatives.

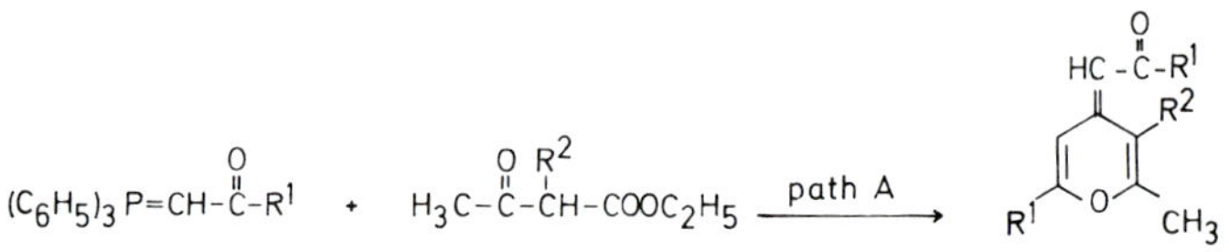

Scheme 30

The discovery of the reactions outlined in Scheme 26 resulted from the observation that aryl 1-diazoalkyl ketones, when heated with 2-oxoalkylidenephosphoranes, yield compounds of type (50).[39–41] The first step of this latter transformation is elimination of nitrogen from the diazo-ketones followed by Schmidt rearrangement to ketenes.

Isocyanates, which are isoelectronic with ketenes, undergo an analogous reaction with 2-oxoalkylidenephosphoranes to yield 4-iminopyrans[42] (Scheme 31). The same products [type (53)] are obtained from the reaction of N-substituted acylketene imines with 2-oxoalkylidenephosphoranes.[41]

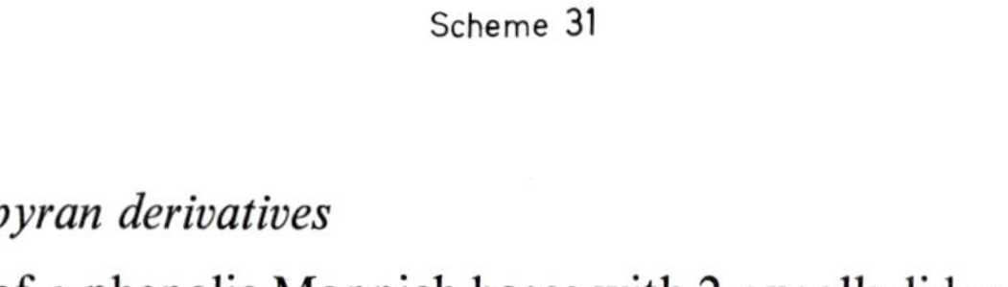

Scheme 31

1.3.2. Fused pyran derivatives

The reaction of o-phenolic Mannich bases with 2-oxoalkylidenephosphoranes represents an interesting approach to the construction of annellated pyrans (54, 55) and dihydro-α-pyrones [lactones such as (57)] (Scheme 32).[12] When subjected to these conditions, o-dimethylaminomethylphenols are first converted by loss of the dimethylamino group into an o-quinone methide, which then reacts with a second molecule of the phosphorane to yield the pyran derivative.

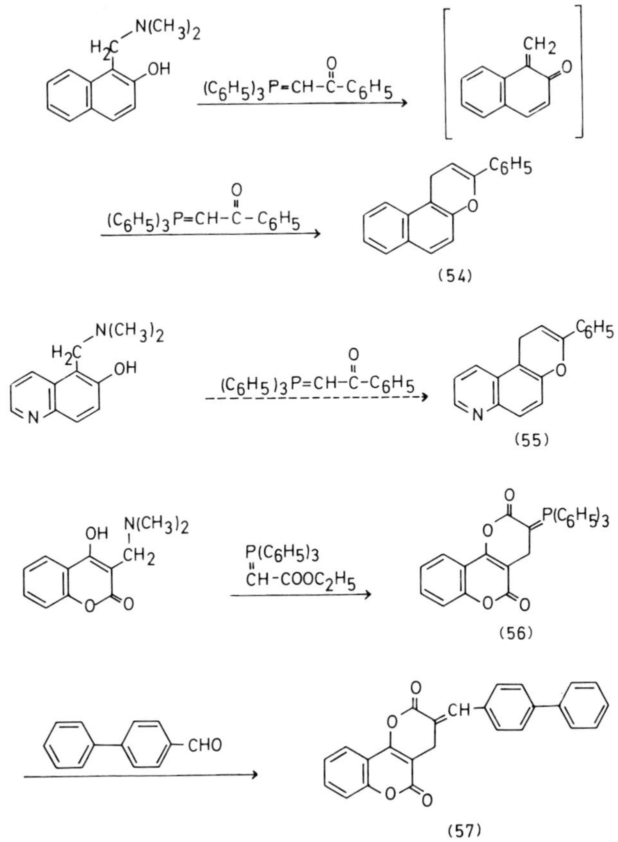

5 Ylides in Heterocyclic Synthesis

Scheme 32

1.3.3. *1,3-Thiazoles*

The readily obtainable 2-oxoalkylidenephosphoranes of type (58) can be caused to react with dicyano disulphide to yield 1-isothiocyanato-2-thiocyanatoalkenes (59), a class of compound for which no alternative synthesis is available to date.

In a subsequent step, one of two possible cyclization reactions occurs, depending on the nucleophile made available to the 1-isothiocyanato-2-thiocyanatoalkene (Scheme 33). Reaction with hydroxide ion is followed by elimination of cyanate ion and ring closure to 2-mercapto-1,3-thiazoles (61), while reaction with amines leads to elimination of thiocyanate ion and formation of 2-amino-1,3-thiazoles (60).[43]

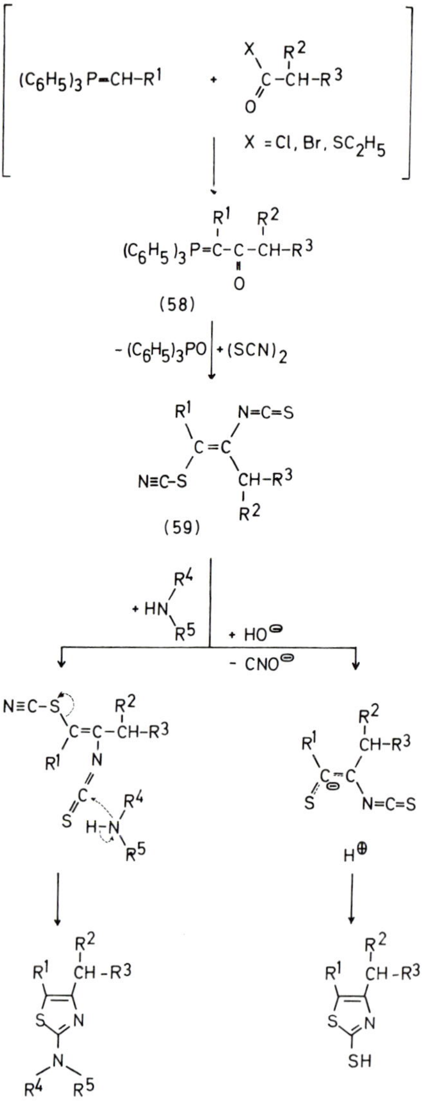

(60)a, $R^1 = n\text{-}C_3H_7$, $R^2 = R^3 = R^4 = R^5 = CH_3$ (61)a, $R^1 = n\text{-}C_3H_7$
b, $R^1 = n\text{-}C_3H_7$, $R^2 = R^3 = CH_3$, $R^2 = R^3 = CH_3$
$R^4 = C_6H_5$, $R^5 = H$

Scheme 33

2. Syntheses of Heterocyclic Compounds using Iminophosphoranes

2.1. Tetrazoles

The possible application of iminophosphoranes (P–N ylides) to the synthesis of heterocyclic systems has attracted some attention in recent years. One example is the synthesis of tetrazoles (62) from iminophosphoranes, acyl halides, and sodium azide.[44] This reaction can follow two different pathways as shown in Scheme 34. Although it is often difficult to distinguish between the two, this is not a matter of great importance with respect to the synthetic goal. Because the reaction proceeds under very mild conditions, it offers a useful alternative to other tetrazole syntheses.

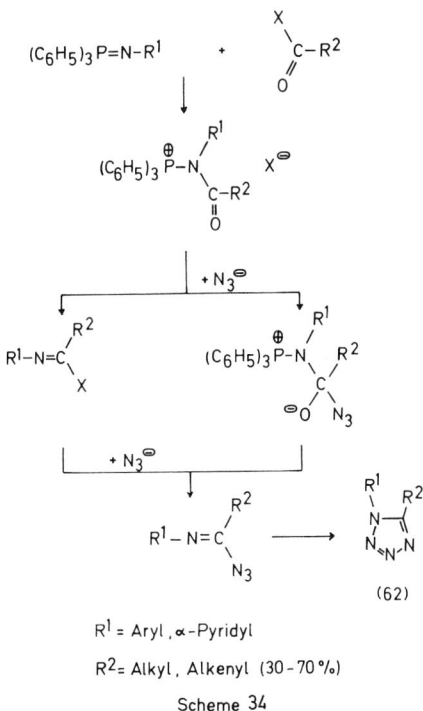

R^1 = Aryl, α-Pyridyl
R^2 = Alkyl, Alkenyl (30–70%)

Scheme 34

2.2. Pyrazines

The reaction of α-azido-αH-ketones (63) with triphenylphosphine (Scheme 35) offers a convenient route to the pyrazine system. A Staudinger reaction in the first step, probably giving rise to an N-(2-oxoalkyl)iminophosphorane as reactive intermediate, is followed by intermolecular condensation with elimination of triphenylphosphine oxide to yield a 2,5-dihydropyrazine (64). This latter is converted into the pyrazine (65) either by symmetry-permitted 1,4-elimination of hydrogen or by oxidative dehydrogenation in the presence of oxygen.[45]

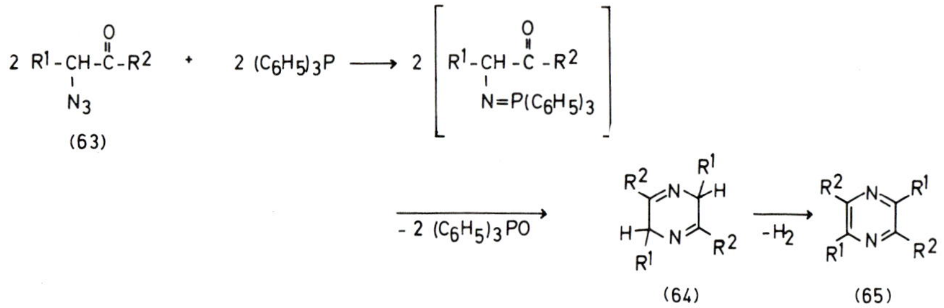

$$2\ R^1\text{-CH-}\underset{\underset{N_3}{|}}{\overset{\overset{O}{\|}}{C}}\text{-}R^2\ +\ 2\ (C_6H_5)_3P\ \longrightarrow\ 2\left[R^1\text{-CH-}\underset{\underset{N=P(C_6H_5)_3}{|}}{\overset{\overset{O}{\|}}{C}}\text{-}R^2\right]$$

(63)

$$\xrightarrow{-2\ (C_6H_5)_3PO}\ \underset{(64)}{\text{dihydropyrazine}}\ \xrightarrow{-H_2}\ \underset{(65)}{\text{pyrazine}}$$

$R^1 = H$, R^2 = Aryl, Alkyl, and $R^1\text{-}R^2 = -(CH_2)_4-$ (25-75%)

Scheme 35

2.3. 1,2,4,5-Tetrazines

Closely related to the foregoing pyrazine ring closure is the synthesis of 1,2,4,5-tetrazines (66) from *N*-acylaminoiminophosphoranes via an intermolecular Wittig reaction.[46] Here too the final product is presumably formed from the corresponding intermediate dihydro compound.

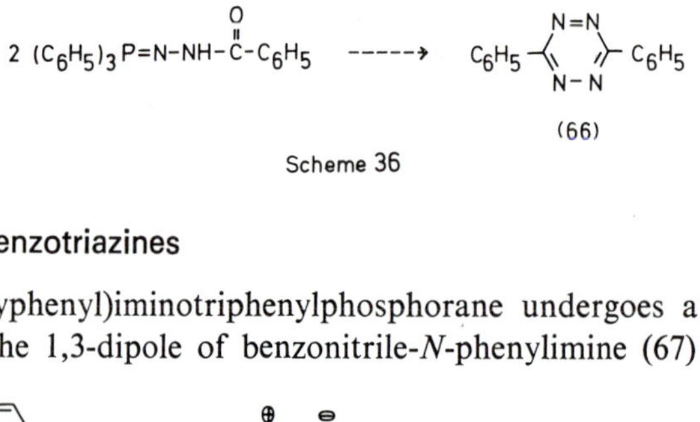

$$2\ (C_6H_5)_3P=N-NH-\overset{\overset{O}{\|}}{C}-C_6H_5\ \dashrightarrow\ \underset{(66)}{C_6H_5\text{-tetrazine-}C_6H_5}$$

Scheme 36

2.4. 1,2,4-Benzotriazines

N-(4-Methoxyphenyl)iminotriphenylphosphorane undergoes a stepwise reaction with the 1,3-dipole of benzonitrile-*N*-phenylimine (67) as shown in

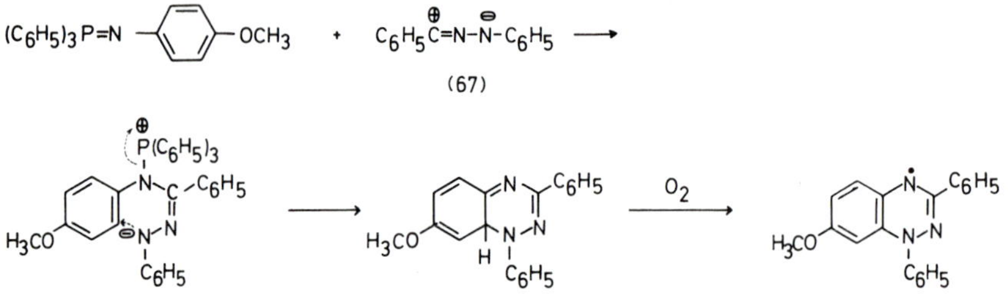

$$(C_6H_5)_3P=N-\langle\rangle-OCH_3\ +\ C_6H_5\overset{\oplus}{C}=N-\overset{\ominus}{N}-C_6H_5\ \longrightarrow$$

(67)

Scheme 37 (68)

5 Ylides in Heterocyclic Synthesis

Scheme 37 to yield as final product the radical 1,3-diphenyl-7-methoxy-1,2,4-benzotriazin-4-yl (68).[27]

2.5. 1,3-Oxazoles

A three-component reaction between α-azido-αH-ketones (63) with acyl halides (69) and triphenylphosphine yields 1,3-oxazoles (72) as shown in Scheme 38. The acyl halide (69) reacts with the intermediate iminophosphorane (cf. Scheme 35) either to give the phosphonium salt (70)[47] or

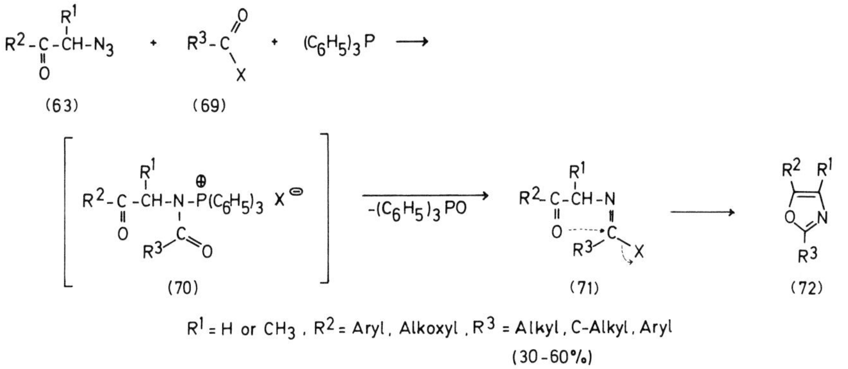

R^1 = H or CH_3, R^2 = Aryl, Alkoxyl, R^3 = Alkyl, C-Alkyl, Aryl
(30–60%)

Scheme 38

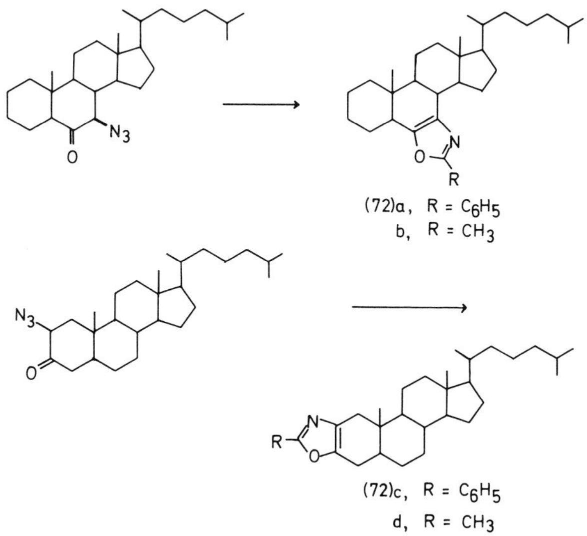

(72)a, R = C_6H_5
b, R = CH_3

(72)c, R = C_6H_5
d, R = CH_3

Scheme 39

directly[48] to give the imide halide (71) which can undergo immediate cyclization to the 1,3-oxazole (72). The method is widely applicable, since the substituents R^1, R^2, and R^3 can contain functional groups that are not compatible with other 1,3-oxazole syntheses but are unchanged under the conditions of the Staudinger reaction.[47]

One application of the reaction outlined in Scheme 38 has been to the synthesis of steroid derivatives containing a 1,3-oxazole ring (Scheme 39).[49]

2.6. Oxadiazoles

Dehydrohalogenation of N'-acylhydrazinotriphenylphosphonium halides with triethylamine, proton abstraction from the resulting N-acylaminoiminotriphenylphosphorane (73) with phenyllithium, and subsequent reaction with benzoyl chloride lead to the formation, with loss of triphenylphosphine oxide, of 5-substituted 2-phenyl-1,3,4-oxadiazoles (74) (Scheme 40).[45]

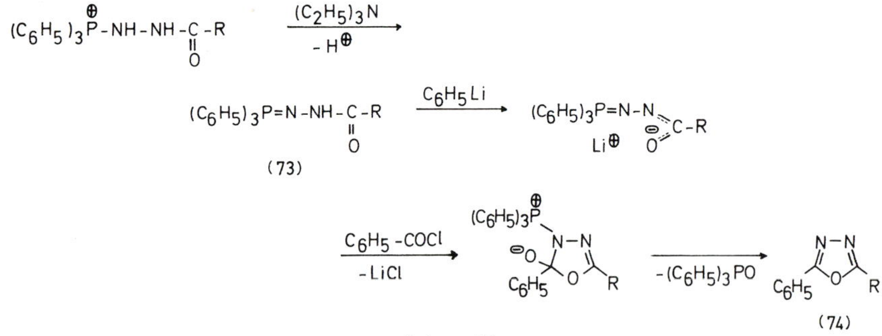

Scheme 40

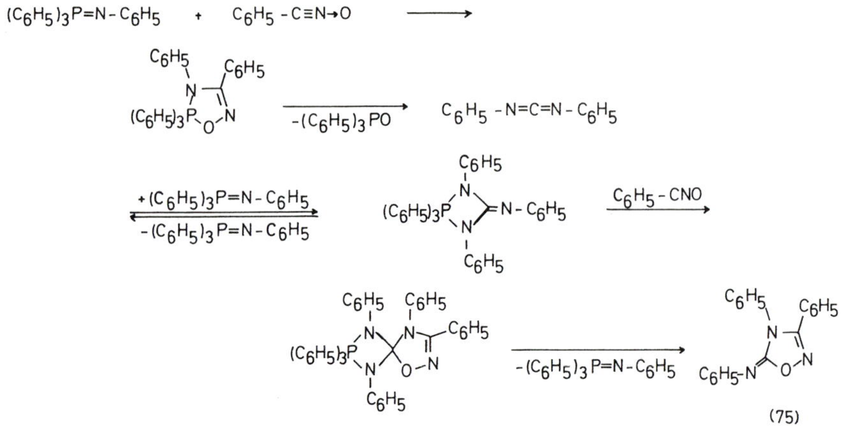

Scheme 41

5 Ylides in Heterocyclic Synthesis

A specific example of the formation of a 1,2,4-oxadiazole system is afforded by the synthesis of 3,4-diphenyl-5-phenylimino-4,5-dihydro-1,2,4-oxadiazole (75) by the reaction of N-phenyliminotriphenylphosphorane with benzonitrile oxide (Scheme 41).[27]

2.7. Pyrazoles

Two specific examples of pyrazole syntheses with the aid of phosphoranes are, firstly, the synthesis of 3,4,5-trialkoxycarbonylpyrazoles (77) from acetylenedicarboxylic esters and N-alkoxycarbonylmethyleneazinotriphenylphosphoranes (76) (from triphenylphosphine and diazoacetic esters),[50] and secondly, the synthesis of 3-ethoxy-5-ethoxycarbonyl-4-ethoxycarbonylaminopyrazole (80) from ethyl N-ethoxycarbonylmethylene carbamate (78) and ethyl diazoacetate, in which the diazo compound (79) cyclizes in the presence of triphenylphosphine (Scheme 42).[51]

A series of pyrazoles (R, R^1, and R^2 = alkyl and/or aryl groups) have been prepared by the reaction of bis(iminophosphoranes) with 1,3-dicarbonyl compounds in yields in the range 50–90% (Scheme 43).[52]

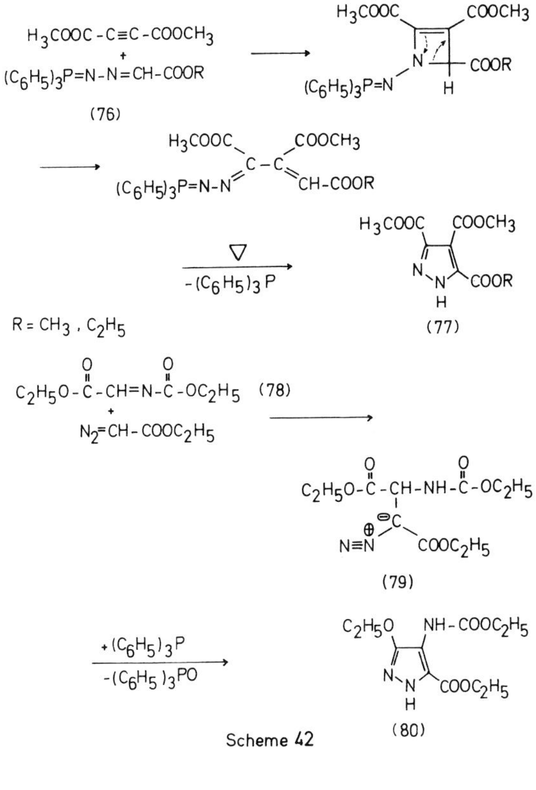

Scheme 42

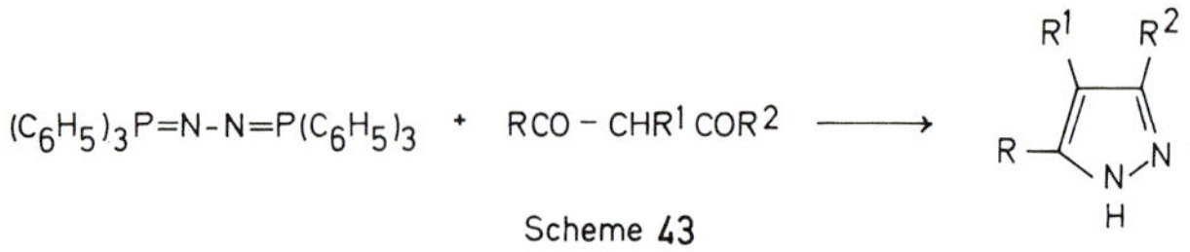

$(C_6H_5)_3P=N-N=P(C_6H_5)_3$ + $RCO-CHR^1COR^2$ ⟶ [pyrazole with R^1, R^2, R, N-N-H]

Scheme 43

2.8. 1,3-Thiazoles and 1,3-Selenazoles

A reaction in which N-aryliminophosphoranes are undoubtedly formed as intermediates is the oxidation of 2-naphthylaminotriphenylphosphonium thiocyanate or selenocyanate with lead tetraacetate. The first step is an acid–base reaction in which one of the acetate groups of the lead tetraacetate and the pseudohalide ion undergo exchange, the latter being then transferred to the nucleophilic *ortho* position (Scheme 44). The N-aryliminophosphorane intermediates undergo the subsequent reaction steps shown, to yield 2-aminonaphtho[2,1-*d*]-1,3-thiazole (81) or -1,3-selenazole (82) respectively.[53]

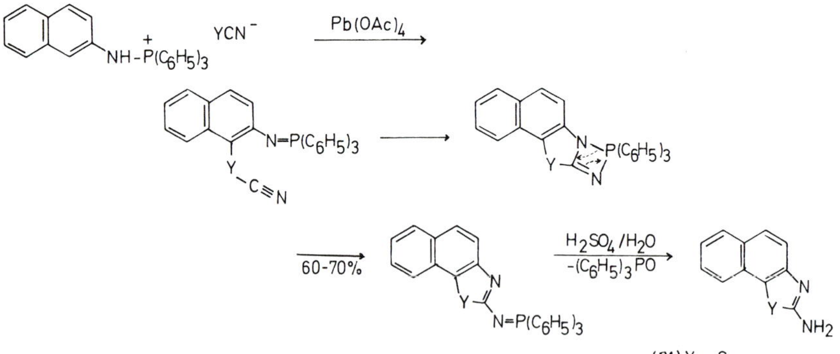

Scheme 44

(81) Y = S
(82) Y = Se   70-80%

2.9. Phthalazines

The reaction of *o*-phthaldialdehyde or 1,2-dibenzoylbenzene with azinobis(triphenylphosphorane) yields phthalazine (83a) or 1,4-diphenylphthalazine (83b) respectively (Scheme 45).[54]

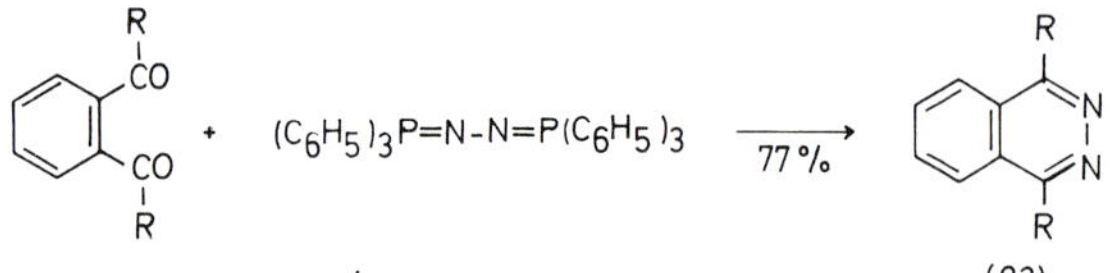

(83a) R = H
(83b) R = C_6H_5

(83)

Scheme 45

2.10. Pyrazoles, Pyridazines, and Quinolines

Treatment of *N*-susbstituted iminotriphenylphosphoranes with prop-2-ynyl-triphenylphosphonium bromide affords a phosphorane of type (85)[55] [Scheme 46: in the author's opinion, (85) is formed via the 4-membered ring intermediate (84), which undergoes a valence isomerization reaction to yield (85)].

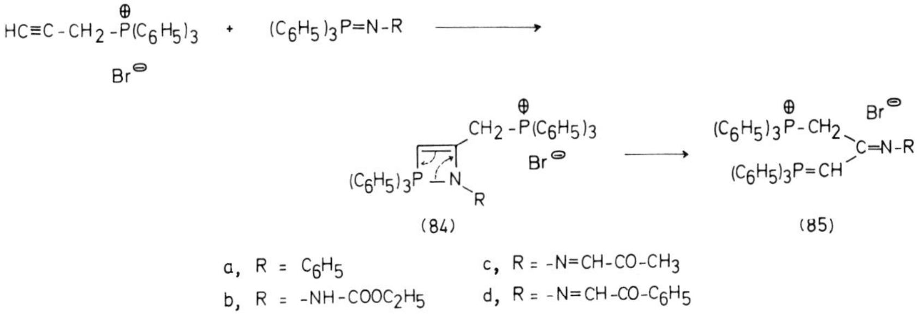

$HC≡C-CH_2-\overset{\oplus}{P}(C_6H_5)_3 \;\; Br^{\ominus}$ + $(C_6H_5)_3P=N-R$ ⟶

(84)   (85)

a, R = C$_6$H$_5$
b, R = -NH-COOC$_2$H$_5$
c, R = -N=CH-CO-CH$_3$
d, R = -N=CH-CO-C$_6$H$_5$

Scheme 46

Compounds of type (85) are useful starting materials for syntheses of pyrazoles, e.g. (86), (86a), and pyridazines, e.g. (87) (Scheme 47).

The reaction of prop-2-ynyltriphenylphosphonium bromide with 2-aminobenzophenone or with esters of 2-aminobenzoic acid yields phosphonium salts of type (88). Removal of a proton from the latter is presumably followed

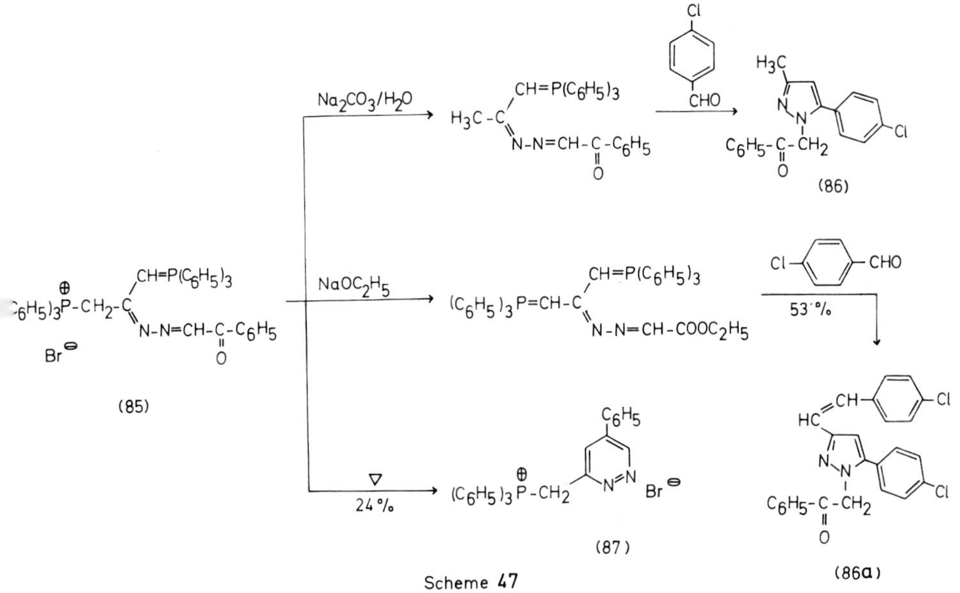

Scheme 47

by the formation of an intermediate 2-iminoalkylidenephosphorane (vinylogous P–N ylide), which cyclizes in an intramolecular Wittig reaction to the quinoline (89) (Scheme 48).

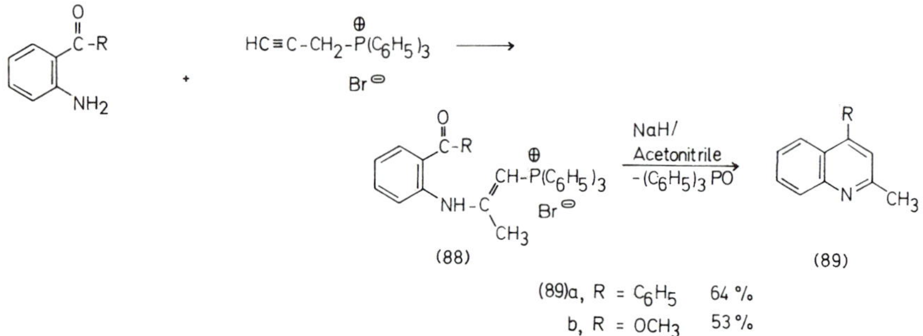

Scheme 48

(89)a, R = C$_6$H$_5$ 64 %
b, R = OCH$_3$ 53 %

2.11. Reductive Cyclization of Azido Compounds with Tervalent Phosphorus Compounds via Iminophosphoranes

A number of closely related syntheses of nitrogen heterocycles are based on the conversion of an ω'-functionally substituted ω-azido compound into an ω'-functionally substituted iminophosphorane followed by intramolecular reaction of the P=N group with the ω'-functional group.

2.11.1. Quinolines and 1,3-benzoxazoles

o-Azidocinnamic acid derivatives (90) react with triethyl phosphite to yield the corresponding iminophosphoranes, which can be subjected to photochemical cyclization to the quinoline (91) (Scheme 49),[56a] involving as a second step trans–cis isomerization about the double bond. Closely related is the synthesis of 1,3-benzoxazoles (70%) from o-azidophenyl acetate or benzoate.[56b]

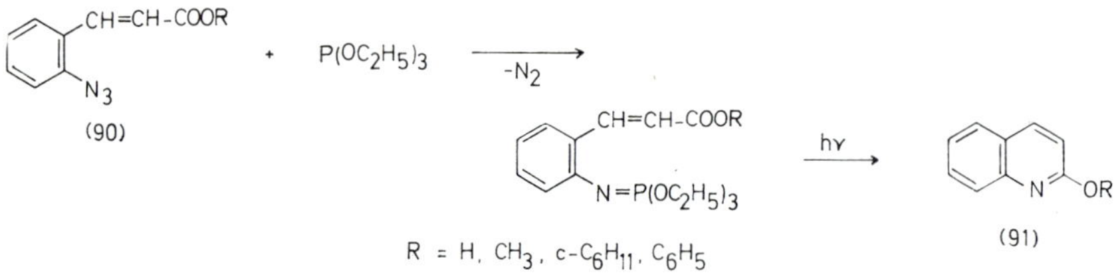

R = H, CH$_3$, c-C$_6$H$_{11}$, C$_6$H$_5$

Scheme 49

2.11.2. Nigrifactin and 1,2-dehydroproline methylamide

The final step in a synthesis of nigrifactin (92) is the cyclization shown in Scheme 50.[57] 1,2-Dehydroproline methylamide is similarly obtained from 5-azido-2-oxovaleric methylamide.[99]

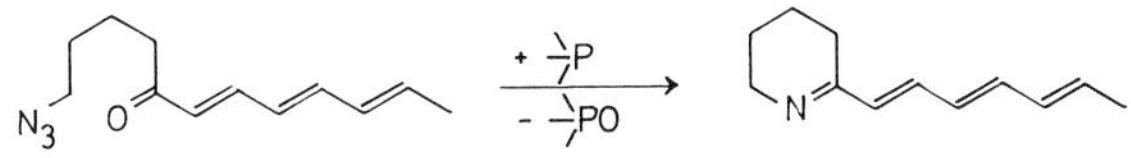

5 Ylides in Heterocyclic Synthesis

[Scheme 50 showing azide reacting via +$\frac{1}{2}$P / -$\frac{1}{2}$PO to give (92)]

Scheme 50

2.11.3. Aziridines

vic-Azidoiodoalkanes (93, *threo* and *erythro*) undergo Staudinger reactions with phosphines or phosphites followed by cyclization to aziridinophosphonium salts (94) or dialkylaziridine-N-phosphonates (dialkylphosphoroaziridates) (95) (Scheme 51).[58] The formation of (96) can be explained in a similar way.

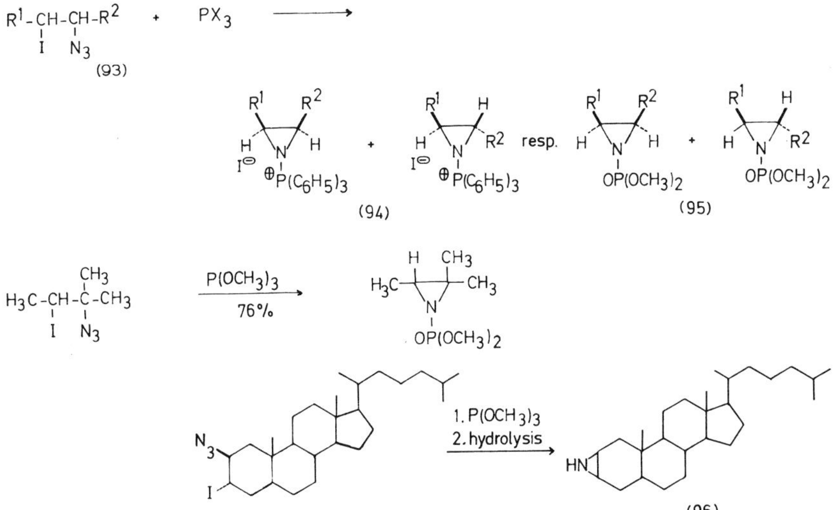

[Scheme 51 showing reaction of (93) R^1-CH(I)-CH(N_3)-R^2 with PX_3 giving aziridinium salts (94) with $\overset{\oplus}{P}(C_6H_5)_3$ and $I^{\ominus}$ counterion, and phosphonates (95) with $OP(OCH_3)_2$; also shown: H_3C-CH(I)-C(N_3)(CH_3)-CH_3 with $P(OCH_3)_3$ giving 76% of trimethylaziridine N-phosphonate; and cholesteryl azidoiodide with 1. $P(OCH_3)_3$ 2. hydrolysis giving aziridine (96)]

Scheme 51

N-Substituted aziridines (98) are obtained from the reaction of 2-aminoalcohols (97) with dibromotriphenylphosphorane (Scheme 52).[59] A phosphine imide is undoubtedly formed as intermediate. Hexahydropyrazines, formed by intermolecular substitution, are sometimes obtained as by-products in these reactions.

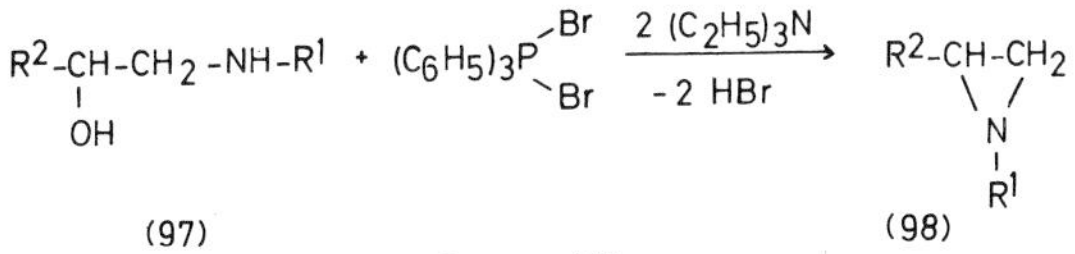

[Scheme 52: R^2-CH(OH)-CH_2-NH-R^1 (97) + $(C_6H_5)_3P(Br)_2$ $\xrightarrow[-2\,HBr]{2\,(C_2H_5)_3N}$ aziridine (98)]

Scheme 52

3. Syntheses of Heterocyclic Compounds using Alk-1-enyl-phosphonium Salts

3.1. Syntheses using Triphenylvinylphosphonium Bromide

A striking difference between vinylphosphonium salts and analogous ammonium salts is the ease with which the C=C bond of the former reacts with nucleophiles or dienes. Both tributylvinyl-[60] and triphenylvinyl-phosphonium salts[61,62] display this characteristic, which was also described earlier for diphenylvinylphosphine oxide.[63]

In the synthesis of heterocyclic compounds, a carbonyl compound that bears a nucleophilic centre Y in a β- or γ-position relative to the carbonyl group is treated with a vinylphosphonium salt. Following an initial Michael addition, an intramolecular Wittig reaction leads to the formation of the heterocycle (Scheme 53).

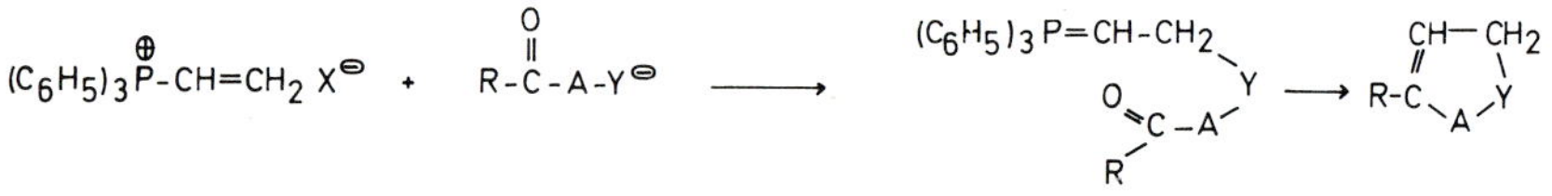

Y = nucleophilic reaction centre
A = a unit with 2 or 3 C-atoms

Scheme 53

3.1.1. *Pyrrolizine derivatives*

The reaction type outlined in Scheme 53 can be applied to 2-acylpyrroles or 2-formylindole respectively, which react with triphenylvinylphosphonium bromide to yield 3*H*-pyrrolizines (99) and pyrrolo[1,2-*a*]indole (101) respectively.[61,62] Catalytic hydrogenation of (99) yields the hexahydro derivative (100), and of (101) the tetrahydro derivative (102) (Scheme 54).

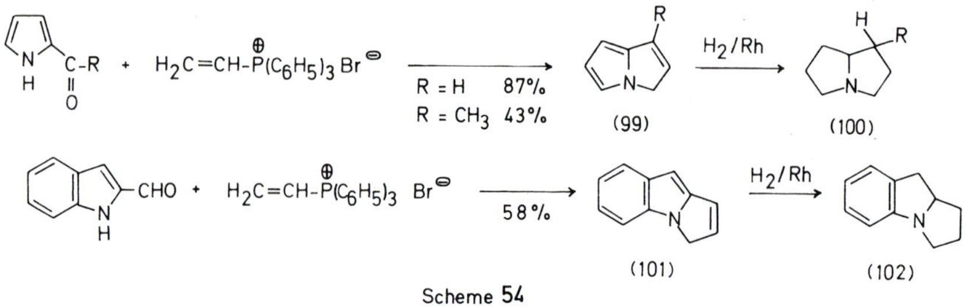

Scheme 54

3.1.2. *Furan derivatives*

In a reaction analogous to that of Scheme 54, 2,5-dihydrofurans can be obtained from metallated acyloins and triphenylvinylphosphonium bromide (Scheme 55).[62,64,65]

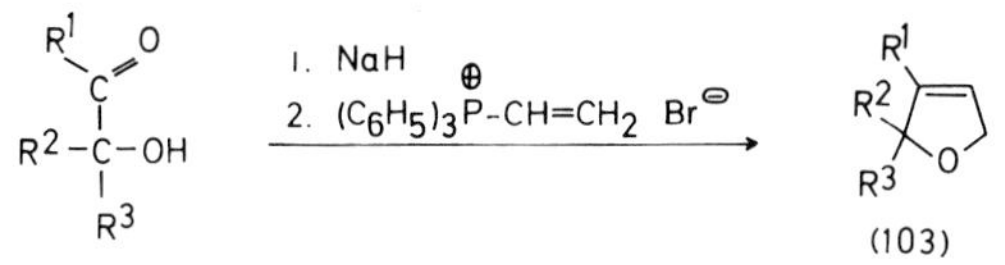

5 Ylides in Heterocyclic Synthesis

Scheme 55

Treatment of the readily available 2-ethoxyvinyltriphenylphosphonium bromide with α-hydroxy-ketones represents an elegant modification of this approach to furan derivatives. The initial product, a dihydrofuran analogous to (103) with an α-ethoxy group, undergoes acid-catalysed elimination to yield the furan derivative (104) (Scheme 56).[66]

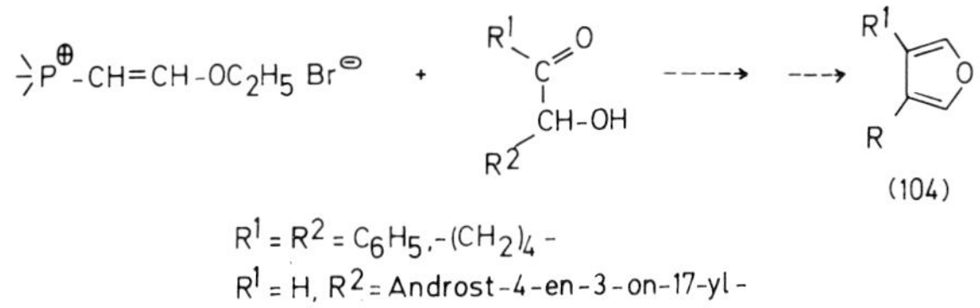

$R^1 = R^2 = C_6H_5, -(CH_2)_4-$
$R^1 = H, R^2 =$ Androst-4-en-3-on-17-yl-

Scheme 56

The preparation of 2,3-dihydrofurans (105) by treatment of 1-ethoxy-carbonylcyclopropyltriphenylphosphonium bromide with carboxylates is a further variation of this approach to furans (Scheme 57).[67] This reaction has been exploited in a synthesis of the furoquinoline alkaloid dictamnin.[67]

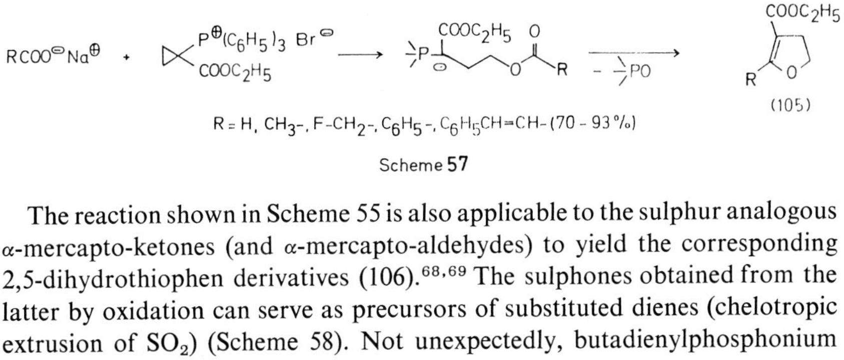

$R = H, CH_3-, F-CH_2-, C_6H_5-, C_6H_5CH=CH-$ (70 – 93 %)

Scheme 57

The reaction shown in Scheme 55 is also applicable to the sulphur analogous α-mercapto-ketones (and α-mercapto-aldehydes) to yield the corresponding 2,5-dihydrothiophen derivatives (106).[68,69] The sulphones obtained from the latter by oxidation can serve as precursors of substituted dienes (chelotropic extrusion of SO_2) (Scheme 58). Not unexpectedly, butadienylphosphonium

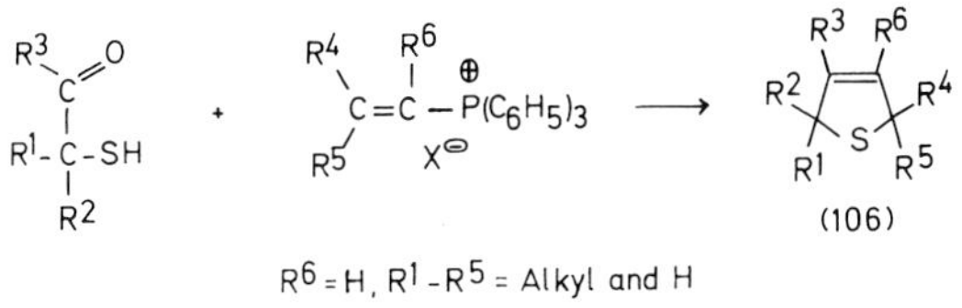

$R^6 = H, R^1-R^5 =$ Alkyl and H

Scheme 58

salts and α-mercaptocarbonyl compounds undergo a Hofmann-type reaction with formation of vinylthiophens, rather than a Wittig-type cyclization (Scheme 59).[70] A discussion of the stereochemical course of the reaction has been presented.[71,72] If R^6 represents a substituent other than H, the reaction usually follows a complex pattern.

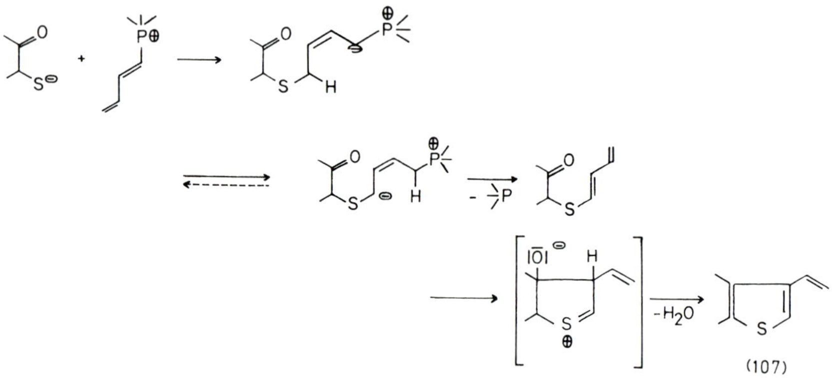

Scheme 59

3.1.3. Chromens and 2,3-dihydro-1-benzoxepins

2H-Chromens (109) are formed by the reaction of triphenylvinylphosphonium bromide with the sodium salt of a 2-hydroxybenzaldehyde or a 2-hydroxyphenyl ketone (108) (Scheme 60).[61,73]

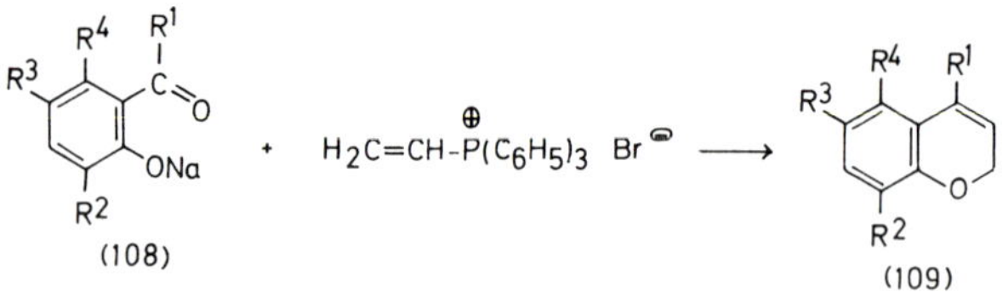

R^1 = H, Alkyl; R^2 = H, OCH$_3$; R^3 = H, Cl, NO$_2$; R^4 = H, or R^3-R^4 =

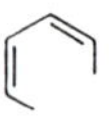

Scheme 60

If the sodium salt of 2-hydroxybenzaldehyde is allowed to react with cyclopropyltriphenylphosphonium bromide instead of triphenylvinylphosphonium bromide, a mixture of 2,3-dihydro-1-benzoxepin (110) and 2-methyl-2H-chromen (111) is formed (Scheme 61).[74] This dual reaction pathway has been studied by subjecting the phosphorane derived from the phosphonium salt (112) to cyclization conditions. Correct choice of reaction conditions enabled (110) or (111) to be obtained exclusively as the product of cyclization.[74,75]

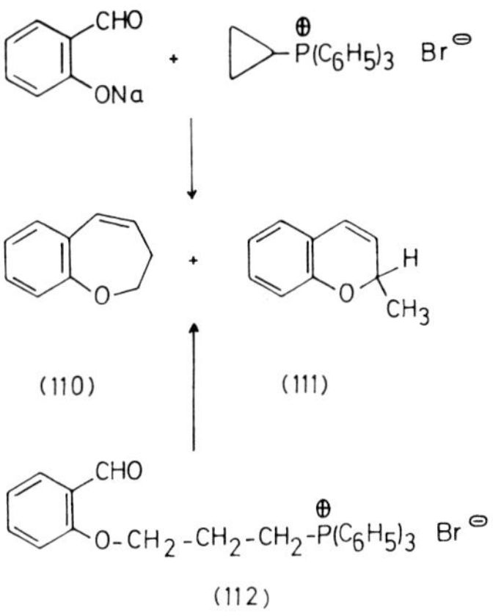

5 Ylides in Heterocyclic Synthesis

[Scheme 61 diagram showing reaction of salicylaldehyde sodium salt with cyclopropyltriphenylphosphonium bromide yielding (110) and (111), and intermediate (112)]

Scheme 61

3.1.4. Quinoline derivatives

The reaction type described above can be adapted to the synthesis of 1,2-dihydroquinolines (113) from o-aminophenyl carbonyl compounds and triphenylvinylphosphonium bromide (Scheme 62).[76]

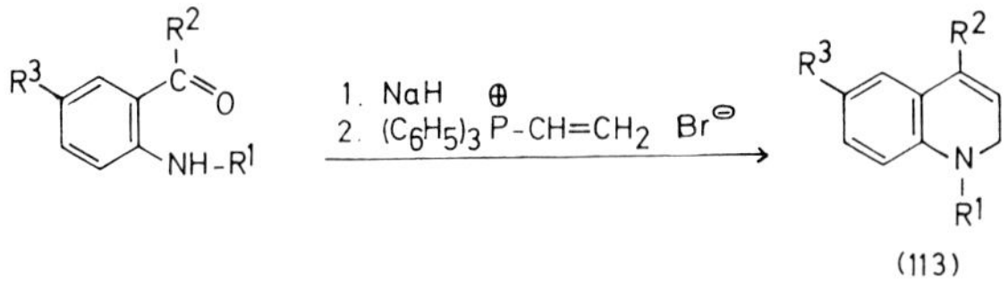

[Scheme 62 diagram]

R^1 = Tos, Ac, R^2 = H, CH_3, C_6H_5, R^3 = H, Cl, NO_2

Scheme 62

A versatile method for the preparation of quinoline derivatives in yields of 30–60% involves the reaction of prop-2-ynyltriphenylphosphonium bromide, which is in tautomeric equilibrium with allenyltriphenylphosphonium bromide, with a wide range of amino compounds (Scheme 63).[77,78] The fate

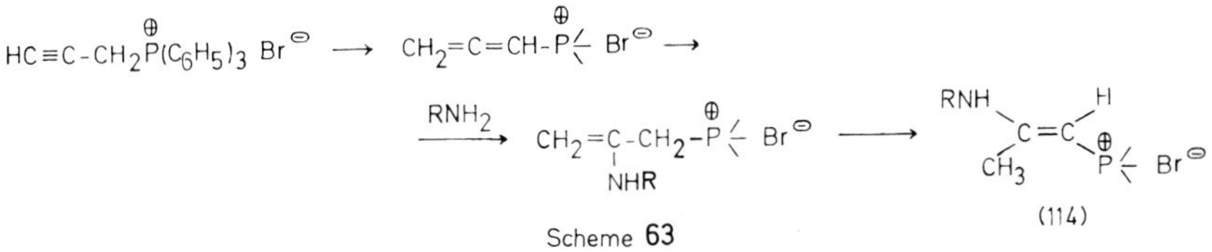

[Scheme 63 diagram showing intermediate (114)]

Scheme 63

of the intermediate (114) depends on the nature of the substituent R, and this determines which of several types of cyclization reaction takes place in the next step (Scheme 64).[77] Cyclization of type (c) is also observed when the

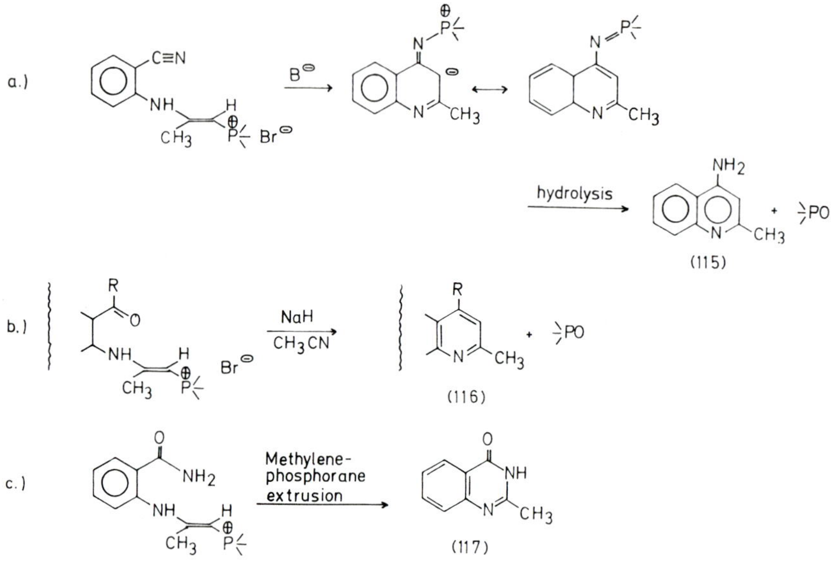

Scheme 64

amide function is replaced by a hydroxy or thio group, and gives rise to a benzoxazole or benzothiazole respectively.[78]

3.1.5. *Pyrroles and 2,3-dihydropyridazines*

If monoximes (118) or monohydrazones (120) of 1,2-dicarbonyl compounds are treated with triphenylvinylphosphonium bromide according to the generalized reaction sequence in Scheme 53, the net reaction represents a novel pathway to 1-hydroxypyrroles (119), 1-aminopyrroles (122), and 2,3-dihydropyridazines (121) respectively (Scheme 65). It should be noted that the product obtained from the hydrazones (120), a dihydropyridazine (121) or a pyrrole (122), depends on the configuration of the hydrazone. In the case of the monoximes (118), only the (*E*)-isomer yields the 1-hydroxypyrrole (119). A photochemical cycloaddition reaction in which pyrrole derivatives (123) are formed is observed when a mixture of a 3-phenyl-2*H*-azirine and triphenylvinylphosphonium bromide is irradiated[80] (Scheme 66).

3.1.6. *Pyrazole derivatives*

The well known pyrazoline synthesis from diazo compounds and activated alkenes is also applicable to alkenes in which the C=C bond is activated by positive phosphorus (Scheme 67).[81,82] The resulting pyrazolinyltriphenylphosphonium bromides (124) can be cleaved with alkali to the corresponding

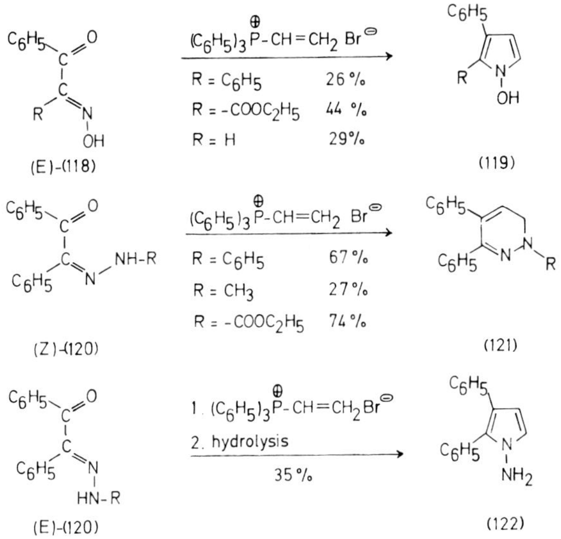

5 Ylides in Heterocyclic Synthesis

Scheme 65

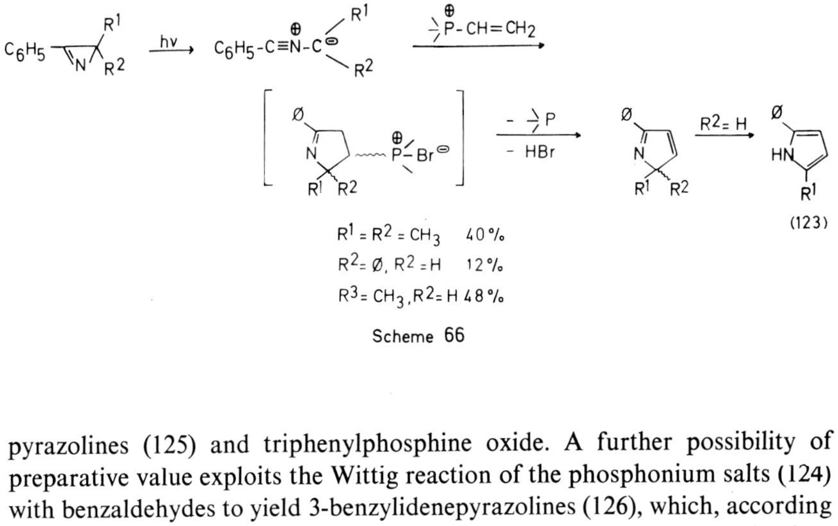

Scheme 66

pyrazolines (125) and triphenylphosphine oxide. A further possibility of preparative value exploits the Wittig reaction of the phosphonium salts (124) with benzaldehydes to yield 3-benzylidenepyrazolines (126), which, according to their substitution pattern, can be converted into pyrazole derivatives (127, 128). An occasional side-reaction, not shown in Scheme 67, is Michael addition of (124) to the triphenylvinylphosphonium bromide taken as starting material.[83]

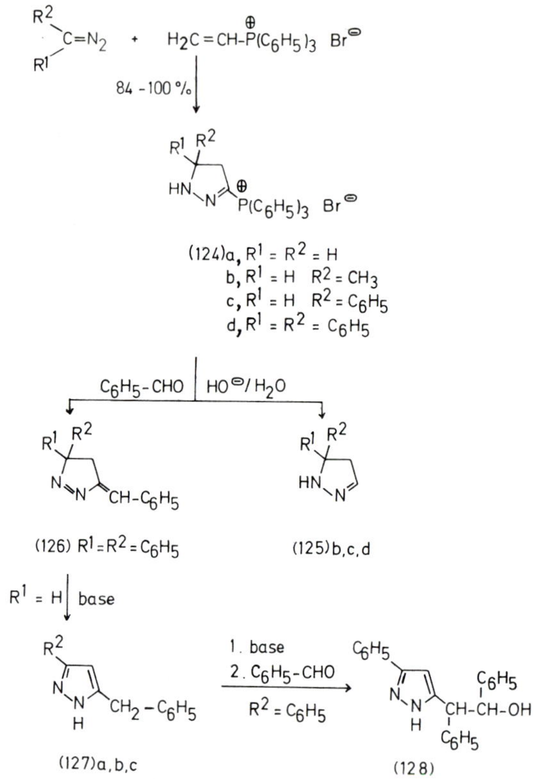

Scheme 67

3.2. Syntheses using 3-Oxoalk-1-enylphosphonium Salts (2-Acylvinylphosphonium Salts)

3-Oxoalk-1-enylphosphonium salts (2-acylvinylphosphonium salts) (129) are readily prepared by treating 2-chlorovinyl ketones with triphenylphosphine.[84] The presence of an additional functional group (carbonyl group) makes compounds of type (129) versatile starting materials for the synthesis of various heterocyclic systems. However, the positively charged phosphorus exerts the overriding influence on the reactivity of this class of compound, in which an "umpolung" of the acylenyl group with respect to nucleophiles is observed, resulting in "inverse" Michael additions.

3.2.1. 1,2,3-Triazoles

The reaction of 2-acylvinyltriphenylphosphonium bromides (129) with azide ion leads to the formation of 4-acyl-1,2,3-triazoles (130) and triphenyl-

5 Ylides in Heterocyclic Synthesis

phosphine (Scheme 68).[85] As primary product a 2-azido-3-oxoalkylidenephosphorane may be assumed, which cyclizes to a triazole betaine. The latter collapses in a Hofmann-type reaction to yield the final products. An analogous reaction has been recorded[86] for 1-acylalk-1-enylphosphonium salts (1-acylvinylphosphonium salts) (131), which on treatment with azide ion undergo cyclization to 4-acyl-1,2,3-triazoles (132), although in only moderate yields (Scheme 69).

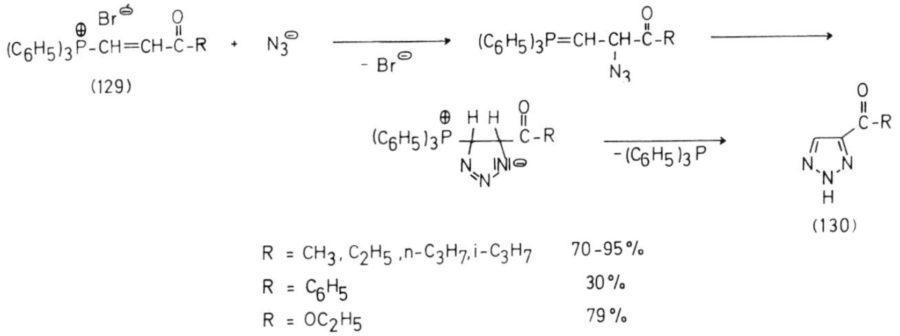

$R = CH_3, C_2H_5, n-C_3H_7, i-C_3H_7$ 70-95%
$R = C_6H_5$ 30%
$R = OC_2H_5$ 79%

Scheme 68

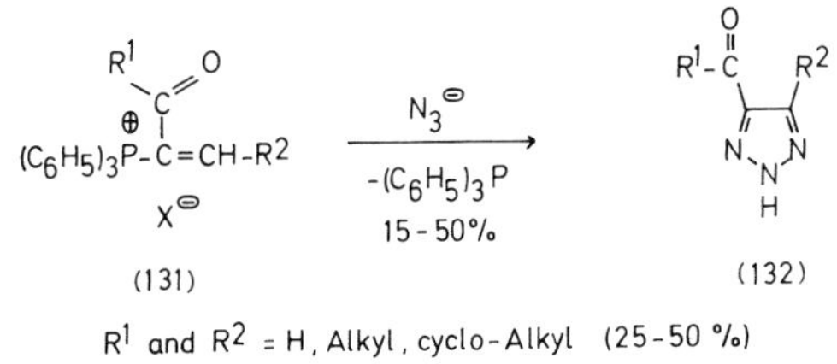

R^1 and R^2 = H, Alkyl, cyclo-Alkyl (25-50 %)

Scheme 69

3.2.2. Pyrazoles

In the reaction of 2-acylvinylphosphonium salts (129) with diazo compounds[87] the electron-withdrawing effect of the carbonyl group appears to compete with that of the positive phosphorus, with the result that addition in both directions can sometimes be observed. It is occasionally possible to influence the direction of addition of the diazo component by varying the solvent; in such cases, choice of a non-absolute reaction medium causes the positive phosphorus to direct the nucleophilic carbon atom of the diazo component to the α-carbon atom relative to the carbonyl group. In no case has it proved possible to isolate the primary pyrazoline adduct; the pyrazole betaines formed from the adducts can be cleaved by alkaline hydrolysis into the corresponding pyrazoles (133, 134) and triphenylphosphine oxide (Scheme 70).[87] In this sequence it is evident that a second molecule of the diazo

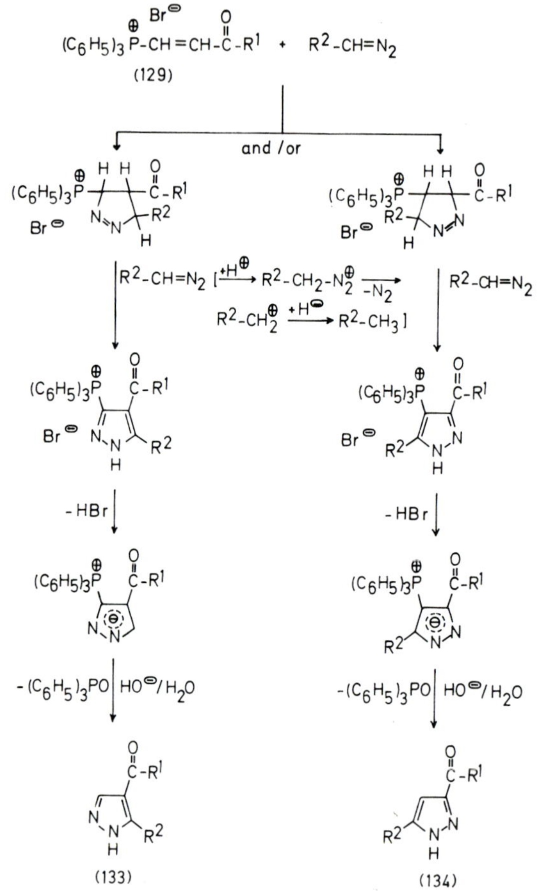

Scheme 70

component functions as a deprotonating agent, the resulting diazonium ion losing nitrogen to yield a carbenium ion which in turn, as hydride acceptor, causes aromatization of the pyrazolinyl system.

3.2.3. 1,3-Thiazoles

2-Acylvinylphosphonium salts (129) react with bidentate nucleophiles such as thioamides or thioureas as shown in Scheme 71. The resulting 5-thiazolyl-methyltriphenylphosphonium salts (135) undergo alkaline hydrolysis to the corresponding 1,3-thiazole (136) and triphenylphosphine oxide.[88] Alternatively, deprotonation of (135) followed by acetylation yields a phosphorane from which a 1,2-diketone (137) can be obtained by oxidative cleavage with potassium permanganate.[89]

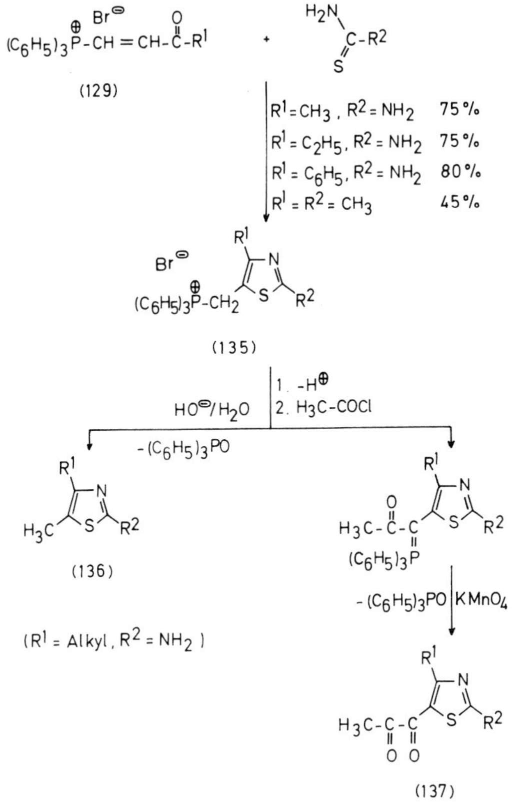

Scheme 71

In a recent patent[90] a series of aminothiazole heterocycles (135a) were prepared in yields of 50–80% according to Scheme 71. The ready nucleophilic displacement of the phosphonio group was exploited in a subsequent step (Scheme 72).

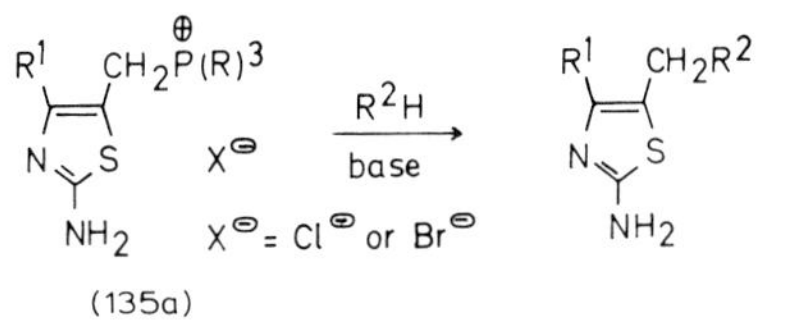

R^1 = H or Alkyl, R = Alkyl resp. C_6H_5

R^2 = CH_3O-, C_2H_5O-, n-C_3H_7O-, n-C_4H_9O-, i-C_4H_9O-, NH_2-, C_2H_5S-

Scheme 72

3.2.4. Pyrroles and imidazoles

The reaction of 2-acylvinylphosphonium salts (129) with β-enaminocarbonyl compounds (3-aminoalk-2-enoates, 2-aminoalk-1-enyl ketones) as bidentate nucleophilic components represents a versatile synthesis of pyrrole derivatives (Scheme 73).[91] The 3-pyrrolylmethyltriphenylphosphonium bromides (138) formed in the reaction shown in Scheme 73 can serve as starting materials for

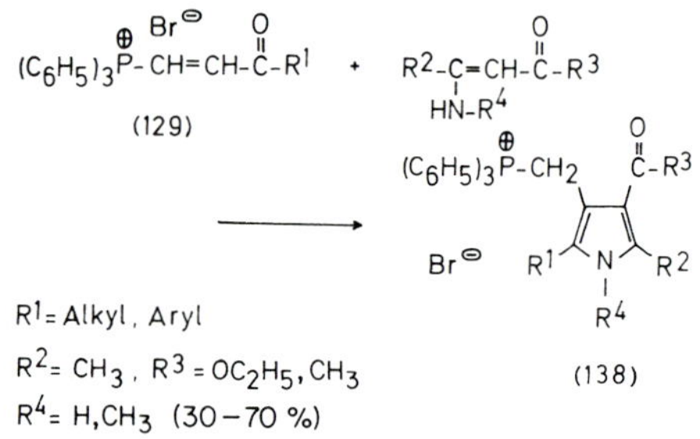

Scheme 73

the synthesis of more complex pyrrole-based structures (140) as shown in Scheme 74, since addition of alkali causes Hofmann-type elimination with formation of a reactive azafulvene intermediate (139). If a nucleophile $(R^5)^-$ is added, addition to the azafulvene with regeneration of the pyrrole system (140) occurs.[91]

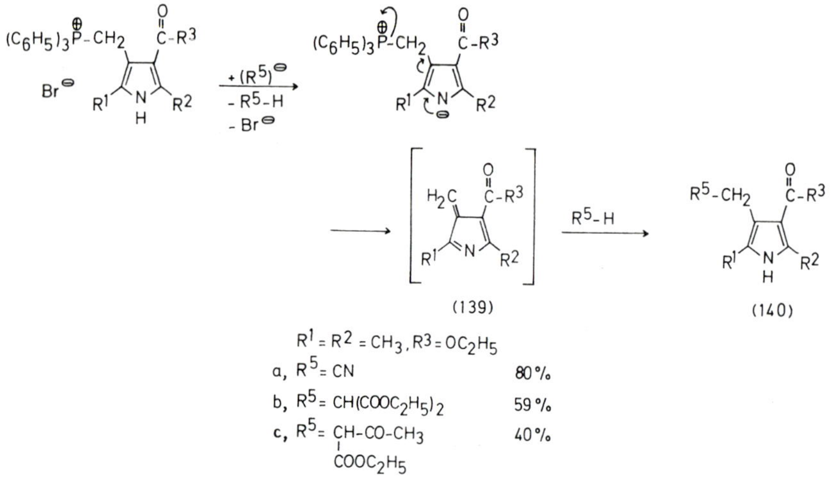

Scheme 74

5 Ylides in Heterocyclic Synthesis

4-Imidazolylmethylphosphonium salts (141) (42–91%) can be obtained from a reaction analogous to that of Scheme 73, in which amidines in place of enaminocarbonyl compounds are treated with 2-acylvinylphosphonium salts (129) (Scheme 75).[91] These phosphonium salts can likewise be subjected to

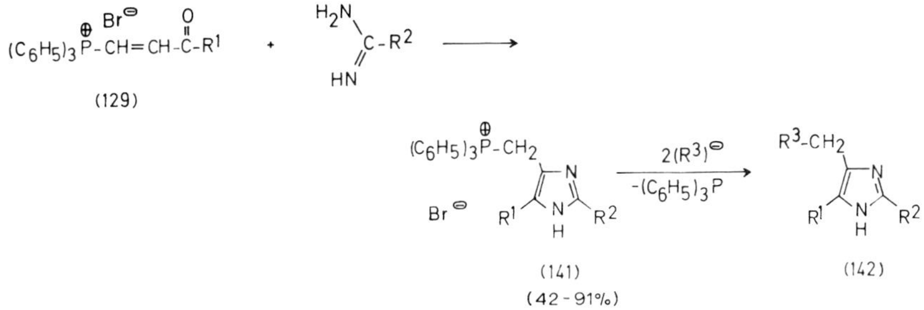

Scheme 75

Hofmann elimination, affording a new imidazole derivative as final product. A series of imidazole derivatives prepared by this method are the subject of a recent patent.[92]

3.2.5. Heterocyclic-fused imidazoles

The reaction outlined in Scheme 75 should in principle be applicable to all N-heterocycles that contain the structural element $-N=C-NH_2$ of the amidines. As expected, 2-aminopyridine reacts with the phosphonium bromide (129a) to form 2-methyl-3-triphenylphosphoniomethylimidazo-[1,2-a]pyridine bromide (143) (Scheme 76). This phosphonium salt offers

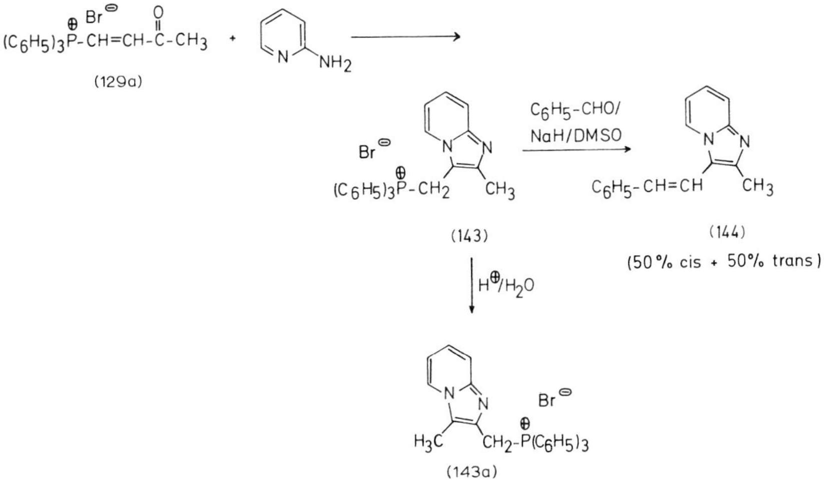

Scheme 76

interesting synthetic possibilities, exemplified by the Wittig reaction with benzaldehyde to yield 2-methyl-3-styrylimidazo[1,2-*a*]pyridine (144).[93] Heating (143) in protic solvents causes a most unusual rearrangement to (143a).[94] Such a Dimroth rearrangement, never previously observed in imidazopyridines with an unsubstituted pyridine ring, is presumably dependent on the presence of the phosphonio substituent.

Analogous products are obtained when amino-pyrimidines and -purines are used instead of 2-aminopyridine (Schemes 77, 78); the imidazopyrimidinyl phosphonium salt (149) is obtained as product of the reaction under kinetic

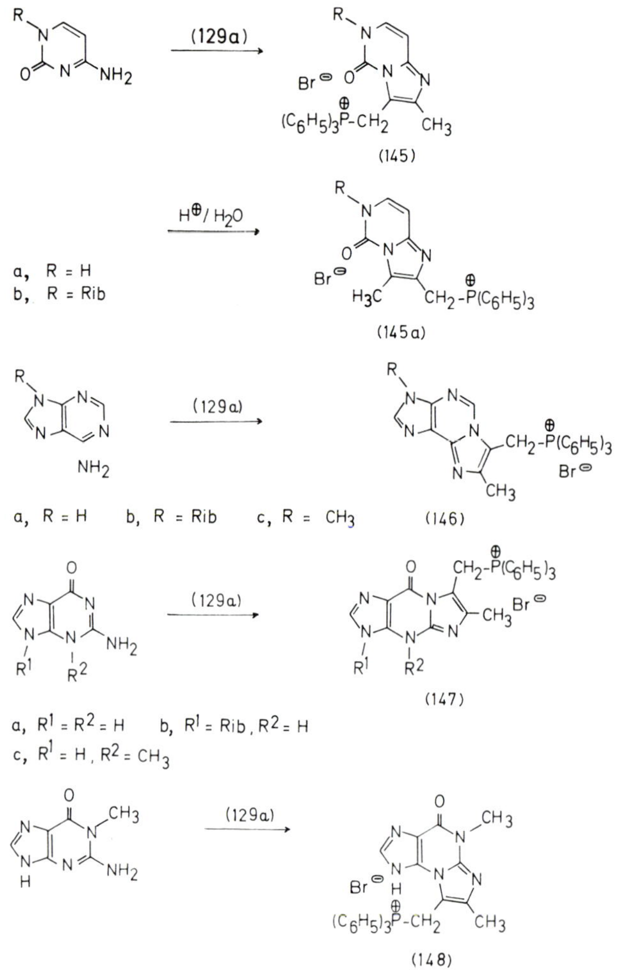

Scheme 77

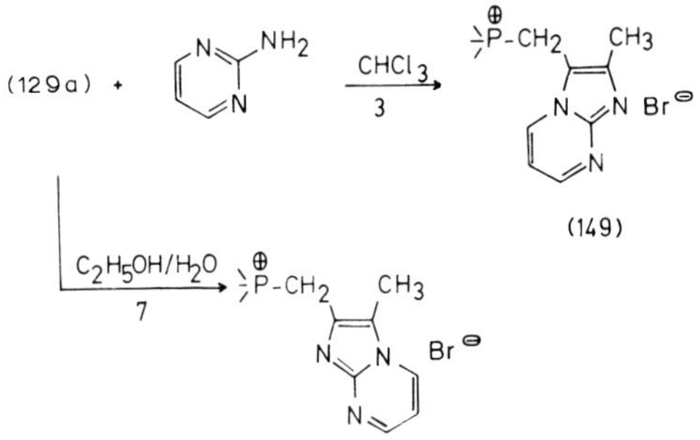

5 Ylides in Heterocyclic Synthesis

Scheme 78

control, while thermodynamic control (Dimroth rearrangement) gives rise to the imidazopyrimidinyl phosphonium salt (150).[94]

Many compounds of biological significance, among them cytosine,[93] adenine,[93,95] 3-methyladenine,[95] guanine,[96] 1-methylguanine,[89] 3-methylguanine,[96] and the nucleosides cytidine,[95] adenosine,[95] and guanosine[95] have been modified in a similar way by treating them with 2-acylvinyltriphenylphosphonium bromide (129) (Schemes 77, 79, 80). Of the products, compound (155) is structurally related to several natural transfer-RNA base units.[97] Compound (145) undergoes transformation to (145a) in aqueous medium, a reaction analogous to the conversion of (143) into (143a).

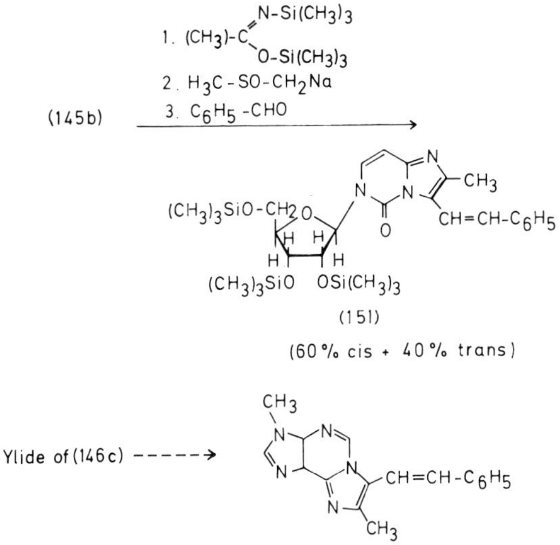

Scheme 79

The synthesis[96] of (151) shown in Scheme 79 illustrates that the Wittig reaction is applicable to nucleosides, provided that the primary adduct obtained with 2-acylvinyltriphenylphosphonium bromide is first protected by silylating the hydroxy groups of the furanose, followed by olefination under the conditions worked out by Corey. Similarly, carbonyl olefination of benzaldehyde with the phosphorane derived from (146c) yields (152) as a mixture of *cis–trans* isomers.

Further interesting examples of structural modification of phosphoniomethyl-N-heterocycles with a mobile N–H proton are shown in Scheme 80.[95] It is highly probable that a conjugated exomethylene intermediate is formed by Hofmann degradation, as earlier outlined in Scheme 74. The reaction is also applicable to (147a), which yields (155) on treatment with the carbanion of malonic ester.[96]

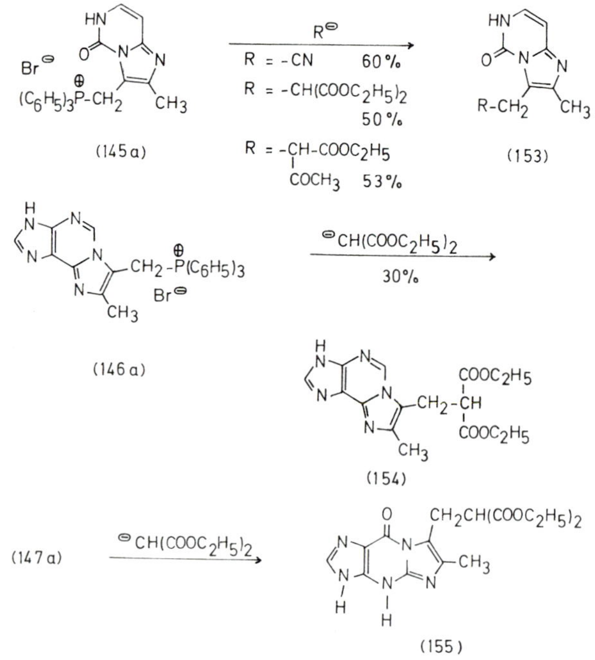

Scheme 80

The reaction of triphenyl-2-cyanovinylphosphonium bromide (156) with 2-aminopyrimidines or 2-aminopyridine respectively yields the expected 2-amino-3-imidazo[1,2-*a*]pyrimidinylmethyltriphenylphosphonium bromides (157a and b) and 2-amino-3-imidazo[1,2-*a*]pyridinylmethyltriphenylphosphonium bromide (158).[98]

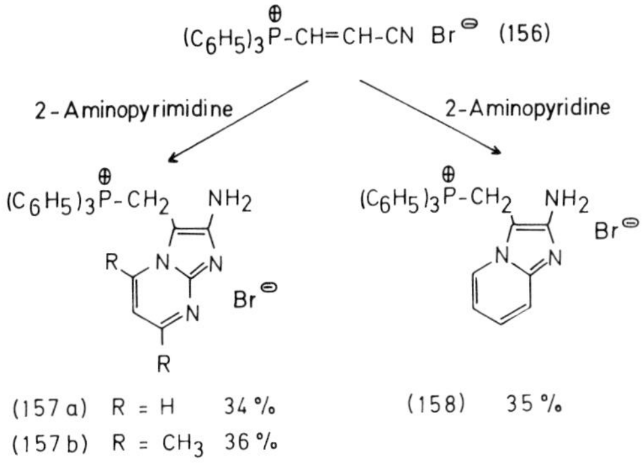

5 Ylides in Heterocyclic Synthesis

$(C_6H_5)_3\overset{\oplus}{P}-CH=CH-CN\ Br^{\ominus}$ (156)

2-Aminopyrimidine ← → 2-Aminopyridine

(157a) R = H   34%
(157b) R = CH$_3$ 36%

(158)   35%

Scheme 81

Notes added in proof

Vinylphosphonium salts with substituents in the α-position give with nucleophiles type RCOA–Y⁻ reaction patterns[100] which are very similar to those presented in Schemes 54, 55, and 60. The yields are generally lower, and anomalous reaction products are observed in some cases in comparison with the unsubstituted phosphonium salts.

(Prop-2-ynyl)triphenylphosphonium bromide cited in Scheme 63 and triphenyl-(2-phenylethynyl)phosphonium bromide have been used by Schweizer for excellent syntheses of a number of heterocycles, e.g. 2-methylbenzimidazole, 2-methylbenzoxazole, 2-methyl(phenyl)benzothiazole, and 2-phenylquinazolin-4-one.[101] Most of them are formed by a rarely observed extrusion of methylenetriphenylphosphorane. (For a survey of this principle see ref. 101.) The same propynylphosphonium salts on reacting with aziridines yield (2-aziridinyl-1-alkenyl)phosphonium salts, which are very useful synthons for the preparation of 2-methyl- and 2-phenyl-2-pyrrolin-3-yltriphenylphosphonium bromide and some substituted (2H-1,3,4,7-tetrahydroazepin-2-ylidenemethyl)triphenylphosphonium bromides and (1H-2,5-dihydroazepin-6-yl)triphenylphosphonium bromides.[102]

An interesting thiophen synthesis has been achieved on reacting dithioesters with alkylidenephosphoranes.[103] 2,5-Dihydrothiophens prepared according to Scheme 58 were dehydrogenated by chloranil to thiophens too.[104]

A synthesis of dihydrofurans and dihydropyrans closely related to that shown in Scheme 57 has been realized by the intramolecular Wittig reaction of ω-acyloxy-propylidene- and -butylidene-triphenylphosphorane respectively.[105]

In addition to the reactivity of vinylphosphonium salts the behaviour of vinyl-1,2-bis-triphenylphosphonium dibromide towards the anion of pyrrole-2-aldehyde has been investigated. Among some other products, pyrrolo[1,2-g]-5-azacyclo[3,2,2]azine is formed in 10% yield.[106]

It may be noted that work recently presented[107] as a novel method is sufficiently considered in Scheme 75.

4. References

1. H. J. Bestmann and R. Zimmermann, *Fortschr. Chem. Forsch.*, 1971, **20**, 1; H. J. Bestmann, *Angew. Chem.*, 1965, **77**, 651, 850; *Angew. Chem. Internat. Ed.*, 1965, **4**, 645, 830; *Neuere Methoden der praparativen organischen Chemie*, Band V, Verlag Chemie, Weinheim, 1967, S. 1; H. J. Bestmann and R. Zimmermann, *Chem. Ztg.*, 1972, **96**, 649.
2. F. Ramirez, *Synthesis*, 1974, 90.
3. J. I. G. Cadogan, *Quart. Rev.*, 1968, **22**, 222; *Synthesis*, 1969, 11.
4. (a) H. J. Bestmann, *Angew. Chem.*, 1977, **89**, 361. (b) K. P. C. Vollhardt, *Synthesis*, 1977, 765.
5. H. J. Bestmann and E. Kranz, *Angew. Chem. Internat. Ed.*, 1967, **6**, 81; *Chem. Ber.*, 1969, **102**, 1802.
6. H. J. Bestmann, H. Häberlein, H. Wagner, and O. Kratzer, *Chem. Ber.*, 1966, **99**, 2848.
7. (a) H. J. Bestmann, R. Armsen, and H. Wagner, *Chem. Ber.*, 1969, **102**, 2259. (b) H. J. Bestmann and O. Kratzer, *Chem. Ber.*, 1963, **96**, 1899. (c) H. J. Bestmann, L. Kisielowski, and W. Distler, *Angew. Chem. Internat. Ed.*, 1976, **15**, 298.
8. (a) P. J. Garratt and K. P. C. Vollhardt, *J. Amer. Chem. Soc.*, 1972, **94**, 7087. (b) P. J. Garratt and D. N. Nicolaides, *J. Chem. Soc., Chem. Comm.*, 1970, 109; 1972, 1014. (c) H. Pommer, *Angew. Chem. Internat. Ed.*, 1977, **16**, 423.
9. P. J. Garratt, A. B. Holmes, F. Sondheimer, and K. P. C. Vollhardt, *J. Amer. Chem. Soc.*, 1970, **92**, 4492.
10. H. J. Bestmann and H. J. Lang, *Tetrahedron Lett.*, 1969, 2101.
11. W. W. Sullivan, D. Ullmann, and H. Shechter, *Tetrahedron Lett.*, 1969, 457.
12. M. von Strandtmann, M. P. Cohen, C. Puchalski, and J. Shavel, *J. Org. Chem.*, 1968, **33**, 4306.
13. A. K. Sorensen and N. A. Klitgard, *Acta Chem. Scand.*, 1970, **24**, 343.
14. H. J. Bestmann, H. J. Lang, and W. Distler, *Angew. Chem.*, 1972, **84**, 65; *Angew. Chem. Internat. Ed.*, 1972, **11**, 59.
15. C. Botteghi, G. Caccia, and S. Gladiali, *Synth. Comm.*, 1976, **6**(8), 549.
16. G. A. Harvey, *J. Org. Chem.*, 1966, **31**, 1587.
17. G. L'abbé and H. J. Bestmann, *Tetrahedron Lett.*, 1969, 63.
18. G. L'abbé, P. Ykman, and G. Smets, *Bull. Soc. Chim. Belg.*, 1969, **78**, 147.
19. P. Ykman, G. L'abbé, and G. Smets, *Tetrahedron Lett.*, 1970, 5225.
20. G. L'abbé, P. Ykman, and G. Smets, *Tetrahedron*, 1969, **25**, 5421.
21. P. Ykman, G. L'abbé, and G. Smets, *Tetrahedron*, 1971, **27**, 5623.
22. E. Zbiral and J. Stroh, *Monatsh. Chem.*, 1969, **100**, 1438.
23. P. Ykman, G. L'abbé, and G. Smets, *Tetrahedron*, 1971, **27**, 845.
24. P. Ykman, G. L'abbé, and G. Smets, *Tetrahedron*, 1973, **29**, 195.
25. P. Ykman, G. Mathys, and G. L'abbé, *J. Org. Chem.*, 1972, **37**, 3213.
26. W. Schörkhuber and E. Zbiral, unpublished results.
27. J. Wulff and R. Huisgen, *Chem. Ber.*, 1969, **102**, 1841.

28. R. Fusco and P. Dalla Croce, *Chim. Ind.* (Milan), 1970, **52**, 45.
29. P. Dalla Croce, *Ann. Chim.* (Italy), 1973, **63**, 867, 895.
30. P. Dalla Croce and D. Pocar, *J. Chem. Soc., Perkin I*, 1976, 619.
31. (a) F. Texier and R. Carrie, *Tetrahedron Lett.*, 1971, 4163; (b) M. Vaultier, R. Danion-Bougot, D. Danion, J. Hamelin, and R. Carrie, *Compt. Rend.* (C), 1973, 1041; (c) *Bull. Soc. Chim. France*, 1976, 1537.
32. H. J. Bestmann and B. Arnason, *Chem. Ber.*, 1962, **95**, 1513.
33. H. J. Bestmann and R. Kunstmann, *Chem. Ber.*, 1969, **102**, 1816.
34. M. Dupré and H. Strzelecka, *Compt. Rend.* (C), 1972, **274**, 1091.
35. H. Strzelecka and M. Simalty-Siemiatycki, *Compt. Rend.*, 1965, **260**, 3989.
36. M. Simalty, H. Strzelecka, and M. Dupré, *Compt. Rend.*, 1967, **265**, 1284.
37. M. Simalty, H. Strzelecka, and M. Dupré, *Compt. Rend.*, 1968, **266**, 1306.
38. H. Strzelecka, M. Dupré, and M. Simalty, *Tetrahedron Lett.*, 1971, 617.
39. H. Strzelecka, M. Simalty-Siemiatycki, and C. Prevost, *Compt. Rend.*, 1962, **254**, 696.
40. H. Strzelecka, *Compt. Rend.*, 1962, **255**, 731; 1962, **257**, 926.
41. H. Strzelecka, M. Simalty, and C. Prevost, *Compt. Rend.*, 1964, **258**, 6167.
42. H. Strzelecka, *Ann. Chim.* (Paris), [14] 1966, **1**, 201.
43. E. Zbiral and H. Hengstberger, *Annalen*, 1969, **721**, 121.
44. E. Zbiral and J. Stroh, *Annalen*, 1969, **725**, 29.
45. E. Zbiral and J. Stroh, *Annalen*, 1969, **727**, 231.
46. C. C. Walker and H. Shechter, *J. Amer. Chem. Soc.*, 1968, **90**, 5626.
47. E. Zbiral, E. Bauer, and J. Stroh, *Monatsh. Chem.*, 1971, **102**, 168.
48. E. Zbiral and E. Bauer, *Phosphorus*, 1972, **2**, 35.
49. E. Zbiral and A. Wolloch, *Tetrahedron*, 1976, **32**, 1289.
50. G. W. Brown, R. C. Cookson, and I. D. R. Stevens, *Tetrahedron Lett.*, 1964, 1263.
51. H. Plieninger and D. vor der Brück, *Tetrahedron Lett.*, 1968, 4371.
52. R. Appel and P. Volz, *Chem. Ber.*, 1975, **108**, 623.
53. E. Zbiral, *Tetrahedron Lett.*, 1966, 2005.
54. R. Appel and G. Siegmund, *Z. Anorg. Chem.*, 1968, **363**, 183.
55. E. E. Schweizer, C. S. Kim, C. S. Labaw, and W. P. Murray, *Chem. Comm.*, 1973, 7.
56. (a) S. A. Foster, L. J. Leyshon, and D. G. Saunders, *Chem. Comm.*, 1973, 29. (b) L. J. Leyshon and D. G. Saunders, *ibid.*, 1971, 1608.
57. M. Pailer and E. Haslinger, *Monatsh. Chem.*, 1970, **101**, 508.
58. A. Hassner and J. E. Galle, *J. Amer. Chem. Soc.*, 1970, **92**, 3733.
59. I. Okada, K. Ichimura, and R. Sudo, *Bull. Chem. Soc. Japan*, 1970, **43**, 1185.
60. P. T. Keough and M. Grayson, *J. Org. Chem.*, 1964, **29**, 631.
61. E. E. Schweizer and K. K. Light, *J. Amer. Chem. Soc.*, 1964, **86**, 2963; *J. Org. Chem.*, 1966, **31**, 870.
62. E. E. Schweizer, *J. Amer. Chem. Soc.*, 1964, **86**, 2744.
63. M. J. Kabachnik, T. Y. Medved, Y. M. Polikarpov, and K. S. Yudena, *Izv. Akad. Nauk SSSR, Otd. Khim. Nauk*, 1962, **9**, 1584.
64. E. E. Schweizer and J. G. Liehr, *J. Org. Chem.*, 1968, **33**, 583.
65. E. E. Schweizer *et al.*, *J. Org. Chem.*, 1970, **35**, 601.
66. M. E. Garst and Th. A. Spencer, *J. Org. Chem.*, 1974, **39**, 584.
67. W. G. Dauben and D. J. Hart, *Tetrahedron Lett.*, 1975, 4353.
68. J. M. McIntosh, H. B. Goodbrand, and G. M. Masse, *J. Org. Chem.*, 1974, **39**, 202; J. M. McIntosh and R. S. Steevensz, *Canad. J. Chem.*, 1973, **52**, 1974.

69. J. M. McIntosh and R. S. Steevensz, *Canad. J. Chem.*, 1977, **55**, 2442.
70. J. M. McIntosh and F. P. Seguin, *Canad. J. Chem.*, 1975, **53**, 3526.
71. J. M. McIntosh and G. M. Masse, *J. Org. Chem.*, 1975, **40**, 1294.
72. J. M. McIntosh and H. Khalil, *Canad. J. Chem.*, 1976, **54**, 1923.
73. E. E. Schweizer, J. G. Liehr, and D. J. Monaco, *J. Org. Chem.*, 1968, **33**, 2416.
74. E. E. Schweizer, C. J. Berninger, and J. G. Thompson, *J. Org. Chem.*, 1968, **33**, 336.
75. E. E. Schweizer et al., *J. Org. Chem.*, 1969, **34**, 207.
76. E. E. Schweizer and L. D. Smucker, *J. Org. Chem.*, 1967, **31**, 3146.
77. E. E. Schweizer, S. De Voe Goff, and W. P. Murray, *J. Org. Chem.*, 1977, **42**, 200.
78. E. E. Schweizer and S. De Voe Goff, *J. Org. Chem.*, 1975, **40**, 144.
79. E. E. Schweizer and C. M. Kopay, *J. Org. Chem.*, 1972, **37**, 1561.
80. N. Gakis, H. Heimgartner, and H. Schmid, *Helv. Chim. Acta*, 1974, **57**, 1403.
81. E. E. Schweizer, C. S. Kim, and R. A. Jones, *Chem. Comm.*, 1970, 39.
82. E. E. Schweizer and C. S. Kim, *J. Org. Chem.*, 1971, **36**, 4033; E. E. Schweizer and C. S. Labaw, *ibid.*, 1973, **38**, 3069.
83. E. E. Schweizer and C. S. Kim, *J. Org. Chem.*, 1971, **36**, 4041.
84. E. Zbiral and E. Werner, *Annalen*, 1967, **707**, 130.
85. M. Rasberger and E. Zbiral, *Monatsh. Chem.*, 1969, **100**, 64.
86. E. Zbiral, M. Rasberger, and H. Hengstberger, *Annalen*, 1969, **725**, 22.
87. E. Zbiral and E. Bauer, *Tetrahedron*, 1972, **28**, 4189.
88. E. Zbiral, *Tetrahedron Lett.*, 1970, 5107.
89. E. Hugl, Dissertation, Universität Wien, 1972.
90. Smith Kline Corporation, Belg. Pat. 847,739 (1976).
91. E. Zbiral and E. Hugl, *Phosphorus*, 1972, **2**, 29.
92. Smith Kline Corporation, Belg. Pat. 847,738 (1976).
93. E. Zbiral and E. Hugl, *Tetrahedron Lett.*, 1972, 439.
94. C. Ivancsics and E. Zbiral, *Annalen*, 1975, 1934.
95. E. Hugl, G. Schultz, and E. Zbiral, *Annalen*, 1973, 278.
96. C. Ivancsics and E. Zbiral, *Monatsh. Chem.*, 1975, **106**, 417.
97. U. G. Khorana, *Proc. Nat. Acad. Sci. USA*, 1967, **57**, 751; B. S. Dudock, G. Katz, E. K. Taylor, and R. W. Holley, *ibid.*, 1969, **62**, 941; K. Nakanishi et al., *Nature*, 1971, **234**, 107; M. Funamizu, A. Teranara, A. Feinberg, and K. Nakanishi, *J. Amer. Chem. Soc.*, 1971, **93**, 6706; H. Kasai and M. Goto, *Tetrahedron Lett.*, 1971, **29**, 2725.
98. C. Ivancsics and E. Zbiral, *Monatsh. Chem.*, 1975, **106**, 839.
99. E. Öhler and U. Schmidt, *Chem. Ber.*, 1975, **108**, 2907.
100. E. E. Schweizer, A. T. Wehman, and D. M. Nycz, *J. Org. Chem.*, 1973, **38**, 1583.
101. E. E. Schweizer and S. De Voe Goff, *J. Org. Chem.*, 1978, **43**, 2972.
102. E. E. Schweizer and M. A. Calagno, *J. Org. Chem.*, 1978, **43**, 4207.
103. H. J. Bestmann and W. Schaper, *Tetrahedron Lett.*, 1979, 243.
104. J. M. McIntosh and H. Khalil, *Canad. J. Chem.*, 1975, **53**, 209.
105. A. Hercouet and M. Le Corre, *Tetrahedron Lett.*, 1979, 5.
106. W. Flitsch and E. R. Gesing, *Tetrahedron Lett.*, 1976, 1997.
107. R. L. Webb, C. S. Labaw, and G. R. Wellman, Abstracts of Papers, ACS/CSJ Chemical Congr. Honolulu, Hawaii, April 1–6, 1979, Org. Chemistry, ORGN, Nr. 162.

6
Deoxygenation via Phosphorus(III) Reagents
1: Heterocyclic Syntheses using Aromatic Nitro Compounds

J. I. G. CADOGAN

1. Introduction, 269
2. Insertion into adjacent aromatic C–H bonds: formation of carbazoles and heterocyclic analogues, 272
3. Insertion into adjacent olefinic and imino C–H bonds: synthesis of indoles and related compounds, 273
4. Addition to adjacent conjugated C=C double bonds: synthesis of indoles and related compounds, 275
5. Insertion into adjacent saturated aliphatic C–H bonds: synthesis of dihydroindoles and related compounds, 276
6. Cyclization at adjacent imino nitrogen (C=N) and azo nitrogen (N=N): synthesis of indazoles, triazoles, and related compounds, 278
7. Interaction with adjacent carbonyl groups, 281
8. Interaction with adjacent, but not conjugated, aromatic rings: synthesis of phenothiazines and related compounds, 286
9. Ring enlargement reactions: formation of azepines and related compounds, 287
10. Incorporation of phosphorus into the heterocyclic ring: formation of amino(oxy)- and amino(thiyl)-phosphoranes, 288
11. Some conundrums, 290
12. References, 292

1. Introduction

Reductive cyclization of aromatic nitro compounds by phosphorus(III) reagents as a general route to nitrogen-containing heterocycles was discovered in 1962.[1] In general the nitro compound is allowed to boil under reflux, overnight, under nitrogen in a solvent, e.g. t-butylbenzene or isopropylbenzene, with 2 equivalents of the phosphorus compound, usually triethyl phosphite. Many of the earlier published applications of the method involved the use of a large excess of the phosphite, in the absence of solvent. We now know that this sometimes leads to lower yields. In such cases it is possible that repetition, now, using a solvent would lead to better conversions.

There is strong evidence in some cases[2] that the reaction proceeds via a nitrene, although in others a nitrene precursor, e.g. (1), is more likely[3-5] (Scheme 1). For simplicity, most mechanisms in this chapter are discussed in

$$(RO)_3P + ArNO_2 \longrightarrow (RO)_3\overset{+}{P}\text{-}O\text{-}\overset{|}{\underset{\overset{|}{O^-}}{N}}Ar \longrightarrow (RO)_3PO + ArNO$$

$$(RO)_3P + ArNO \longrightarrow (RO)_3\overset{+}{P}\text{-}O\text{-}\bar{N}Ar \longrightarrow (RO)_3PO + ArN\text{:}$$

Overall

$$2(RO)_3P + ArNO_2 \longrightarrow 2(RO)_3PO + ArN\text{:}$$

Scheme 1

terms of nitrenes, although this does not imply that the mechanism has been established in each case. The possible participation of the corresponding nitroso compound, as an intermediate, is even less well established. Indeed there is no direct evidence available on this point, because aromatic nitroso compounds react with trialkyl phosphites at 0°, i.e. some 10^6 times faster than the corresponding nitro compound. There is good evidence that the first, and slow, step of the doxygenation involves nucleophilic attack of the phosphorus(III) reagent on the nitro group,[2,6] thus paralleling the deoxygenation of the nitroso compound.[7a] The range of reactivity of the reagent varies widely (Table 1).[6] The most reactive phosphorus compounds are $(Me_2N)_3P$, $EtOPPh_2$, $(EtO)_2PMe$, and the amidites. Of these only the first is readily available, although the second is easily synthesized from readily available materials. Against the first reagent is its noxious character, and the carcinogeneity of the oxidation product, and against the second is the additional preparative difficulty of separating the resulting ethyl diphenylphosphonate from the desired heterocyclic product. The results suggest that ethyl N,N'-diethylphosphorodiamidite [$EtOP(NHEt)_2$] would be a particularly reactive reagent. The cyclic phosphites offer little in terms of reactivity and are not readily prepared in a pure state. Of the acyclic phosphites, triethyl phosphite is less noxious and more hydrolytically stable than trimethyl phosphite, and the resulting triethyl phosphate is more easily removed by distillation than triisopropyl phosphate. These factors, combined with availability and cheapness, are in favour of triethyl phosphite as a general reagent for the deoxygenation of aromatic nitro compounds.

The method is particularly suitable for synthesis of certain five-, six, and seven-membered heterocyclic rings containing the nitrogen atom originally in the nitro group. In general, internuclear cyclizations proceed to give five-membered N-heterocyclic rings, and even when the final product is six-

Table 1. Rates of deoxygenation of 2-nitrobenzylideneaniline in various tervalent phosphorus reagents (mole excess reagent)

Reagent	Temp. (°C)	$10^5 k_1$ (s^{-1})	$t_{1/2}$ (min)
$(Pr^iO)_3P$	105	9·7	120
$(MeO)_3P$	105	10·0	116
$(MeO)_3P$	91·5	2·3	505
$(MeO)_3P$*	91·5	2·65	630
$(EtO)_3P$	105	13·1	89
$(EtO)_3P$	91·5	3·4	337
$(EtO)_3P$*	91·5	3·4	338
$(EtO)_2PNC_5H_{10}$	105	41·2	28
$EtOP(NEt_2)_2$	105	80	20
	40	7·6	153
$EtOPPh_2$	30	6·2	188
$(EtO)_2PMe$	40	14·0	83
⌈NH–POEt–NH⌉ (cyclic)	40	46·5	25
⌈O–POEt–O⌉ (cyclic)*	91·5	0·5	2224
⌈O–PNEt$_2$–O⌉ (cyclic)	91·5	3·1	378
⌈O–PNEt$_2$–O⌉ (cyclic)*	91·5	2·9	396
$(Me_2N)_3P$	91·5	Reaction complete in < 20 min	

* Rate of appearance of 2-phenylindazole.

membered the initial reaction is often via the more accessible five-membered intermediate.

The main modes of cyclization occur formally by insertion of the nitrogen moiety into adjacent aromatic, aliphatic, or alicyclic C–H bonds, by regiospecific addition to adjacent C=C, by combination with adjacent aromatic or non-aromatic nitrogen to give N–N bonds, by interaction with adjacent carbonyl groups, or by ring expansion.

Almost invariably the reactions also produce easily separable phosphorus-containing by-products such as phosphorimidates and their phosphoramidic products of thermolysis and hydrolysis. Occasionally some reduction of starting nitro compound to the corresponding amine occurs; this is considered to be indicative of nitrene participation (Scheme 2).[2]

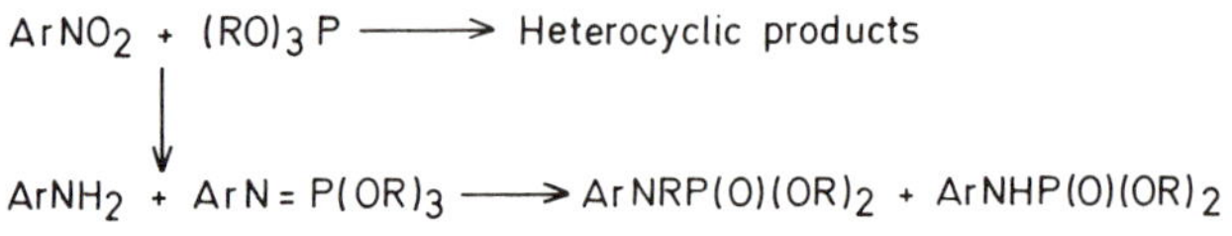

$$ArNO_2 + (RO)_3P \longrightarrow \text{Heterocyclic products}$$

$$\downarrow$$

$$ArNH_2 + ArN=P(OR)_3 \longrightarrow ArNRP(O)(OR)_2 + ArNHP(O)(OR)_2$$

Scheme 2

The other main product of the reaction is the oxidized phosphorus compound. When this is trimethyl or triethyl phosphate, which are alkylating agents, amino or hydroxy functions present may be alkylated to small extents, leading to loss of product.

It is necessary to refer to the deoxygenation of aromatic nitroso compounds. As mentioned already these reactions have the great advantage that they proceed rapidly at room temperature and below, but there have been very few applications in general heterocyclic synthesis. They may be useful in unusual cases, an example of which is the ready conversion of 2,4,6-tri-t-butylnitrosobenzene into 3,5-dimethyl-5,7-di-t-butyl-2,3-dihydroindole (80%) by triethyl phosphite at 20°, whereas the corresponding nitro compound is untouched at 150°.[7b]

2. Insertion into Adjacent Aromatic C–H Bonds: Formation of Carbazoles and Heterocyclic Analogues

This was the first example of the reductive cyclization;[1] 2-nitro-biaryls having a free 2'-position give carbazoles[8,9] in moderate to excellent yield (35–83%) (Scheme 3). Similarly are obtained the related bridged carbazoles, 8,10-dihydrothiepino[3,4,5,6-*def*]carbazole (1; X = S), its dioxide (1; X = SO$_2$), its oxepino and azepino analogues (1; X = O, NR), and diethyl 8,10-dihydro-cyclohepta[1,2,3,4-*def*]carbazole-9,9-dicarboxylate (1; X = C(CO$_2$Et)$_2$).[10] Also formed by this route are the indolo[3,2-*b*]indole (2), the novel indolo-[2,3-*b*]indole (3),[11] 4*H*-furo[3,2-*b*]indole (4),[12] and 10*H*-[1]benzothieno[3,2-*b*]-indoles (5).[13] An attempt to prepare the isomer (6) of (5) failed,[13] possibly as a result of the unfavourable *peri* proton interaction shown in (7). The reaction has been used in a synthesis of harman (8; R = Me, R^1 = H), a β-carboline derivative,[14] the first of many extensions of the reaction to the synthesis of natural products by Kametani and his coworkers.[15] It is noteworthy that 1-(2-nitrophenyl)naphthalene is reduced to 7*H*-benzo[*c*]carbazole (9) rather than the six-membered heterocycle (10).[8]

6 Heterocycles from Nitro-compounds

Note: In many instances in this chapter the product, only, of the phosphorus (III)- induced cyclisation will be shown. In these cases the superimposed line indicates the position of the starting nitro group, thus

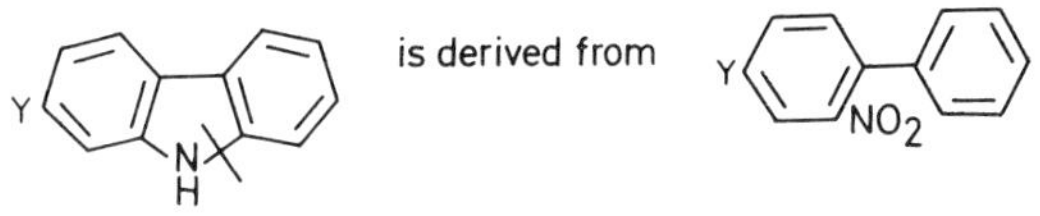

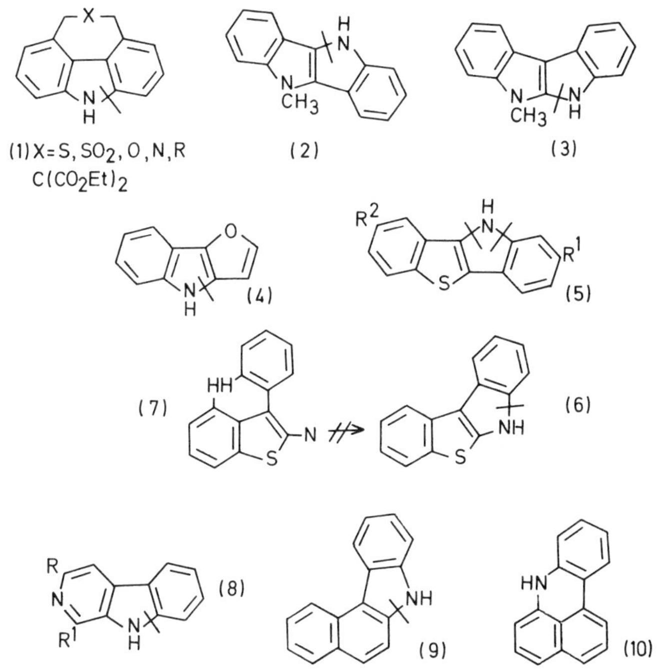

Scheme 3: Carbazoles

3. Insertion into Adjacent Olefinic and Imino C–H Bonds: Synthesis of Indoles and Related Compounds

The first example,[1,8] that of the conversion of *cis*- and *trans*-2-nitrostilbene into 2-phenylindole (58–85%),[9] has been elaborated in many ways (Scheme 4). Thus, α-(2-nitrophenyl)-β-(2-chlorophenyl)acrylic acid gives 2-(2-chlorophenyl)-3-ethoxycarbonylindole[16] (46%), indicating the occurrence of esterification, presumably by triethyl phosphate, during the reaction.

The reductive cyclization of 2-(2-nitrostyryl)pyridine, which presents two possible points of ring closure, also gives rise to an indole (11) rather than

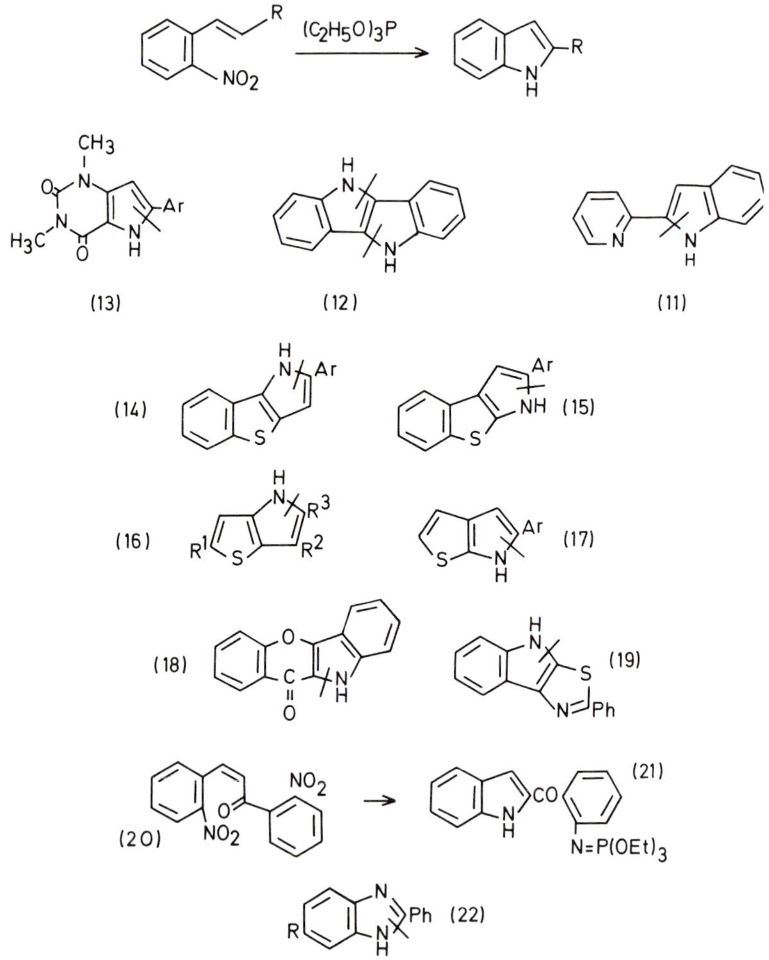

Scheme 4

proceeding via reaction at the electron-rich nitrogen to give a diazepine.[16] Under similar conditions, 2-nitrostyrene and 2,2'-dinitrostilbene give small yields (1–2%) only of indole and indolo[3,2-b]indole (12), respectively, while the reaction of β-nitrostyrene gives no indole, and 2-nitrocinnamic acid is reduced and esterified to give a low yield of indole-2-carboxylic acid ester.[8] There is evidence that the o-nitro-esters themselves give much higher yields of indoles (58%).[17] The biologically interesting pyrrolo[3,2-d]pyrimidine (13) has been prepared, in low yield, from the corresponding 5-nitro-6-styrylpyrimidine derivatives thermally and by u.v. irradiation in the presence of triethyl phosphite.[18] In the latter case, however, some warming occurred, so the products may be arising by a thermal process. The indole synthesis has been extended[19] to include 2-alkyl- (CH_3, C_2H_5; 50–60% yields) and 2-acyl- (CH_3CO, C_6H_5CO; 16% yields) -indoles. Minor by-products of mechanistic interest are also formed. Attempts to improve the yield in the latter case by converting the acyl group into its cyclic acetal had little effect, although this route has been used to synthesize 2-(2-quinuclidinyl)indole.[19c]

The reaction provides a convenient entry into the synthesis of otherwise relatively inaccessible fused heterocyclic systems. Thus, acceptable yields (25–75%) of 2-aryl-1*H*-[1]benzothieno[3,2-*b*]pyrroles (14),[20] the isomeric [2,3-*b*] derivatives (15),[20] thieno[3,2-*b*]pyrroles (16)[21–23] (but not the unsubstituted derivative[22]) and the isomeric [2,3-*b*] derivatives (17),[21] 2-methyl-11*H*-[1]benzopyrano[3,2-*b*]indol-11-one (18),[24] and 2-phenylthiazolo[5,4-*b*]-indole (19) (41%)[25] have been produced from the corresponding nitro compounds (Scheme 4). It is of interest that, when presented with the opportunity, as in (20), of indole formation or reaction with adjacent carbonyl (see Section 7.1), the indole wins, (21) being formed.[26]

Closely related to this general synthesis of indoles is the conversion of *N*-benzylidene-2-nitroanilines into 2-phenylbenzimidazoles (Scheme 4).[27]

4. Addition to Adjacent Conjugated C=C Bonds: Synthesis of Indoles and Related Compounds

An example of this class of reaction is the conversion of 2-phenyl-1-(2-nitrophenyl)propene into the rearranged indole, 2-methyl-3-phenylindole (23), suggesting the intermediacy of an electrophilic species, possibly a nitrene (Scheme 5).[28] In the corresponding reaction of β,β-disubstituted

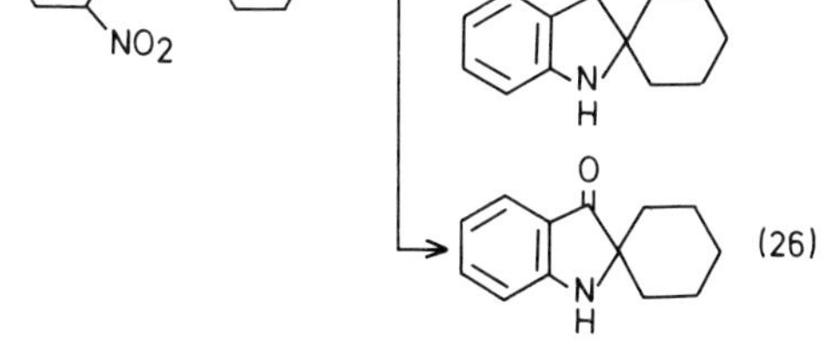

Scheme 5

2-nitrostyrenes, indoles are again the major products. Thus, cyclohexylidene-(2-nitrophenyl)methane undergoes ring closure with rearrangement to give 5,6,7,8,9,10-hexahydrocyclohepta[b]indole (24) (35%) together with lower yields of the bi-indolyl (25) (24%) and the spiro-indolinone (26) (8%) shown in Scheme 5. The cyclopentylidene analogue similarly gives 1,2,3,4-tetrahydrocarbazole (15%) while 2-nitro-β,β-dimethylstyrene gives 2,3-dimethylindole (33%).[28] A completely satisfactory mechanism for the formation of the spiro derivative (25) is still awaited.

Kametani and his coworkers attempted to exploit this reaction in a synthesis of rutaecarpine (27).[29] Unfortunately the "wrong" substituent migrated to give the isomeric pseudorutaecarpine (28) (6%) (Scheme 6).

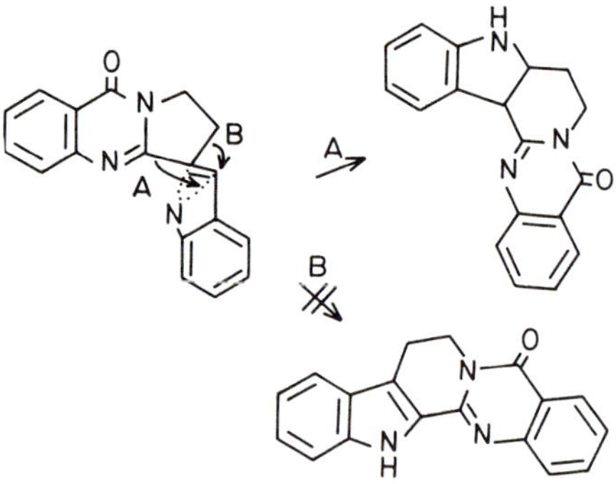

Scheme 6

5. Insertion into Adjacent Saturated Aliphatic C–H Bonds: Synthesis of Dihydroindoles and Related Compounds

Low yields (< 10%) of indolines can be prepared by phosphorus(III) reduction of 2-nitroalkylbenzenes (where alkyl > ethyl).[19b] This reaction is of more mechanistic than synthetic interest however, because small amounts of tetrahydroquinolines are also formed when the alkyl group is propyl or larger (Scheme 7). These results, and those of Feuer and Smolinsky[30] on cyclization of optically active 2-nitro-(2-methylbutyl)benzene provide good circumstantial evidence for nitrene participation.[31] Nevertheless an adaptation to provide spiroindolinones from α-(o-nitrobenzoyl)butyrolactone appears to be useful.[19d]

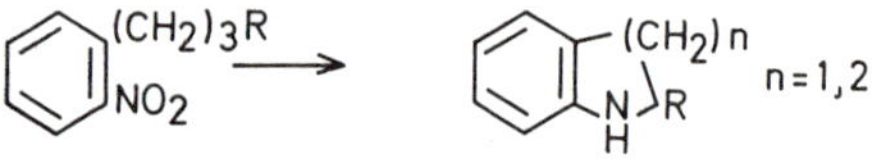

Scheme 7

6 Heterocycles from Nitro-compounds

Of more practical value are insertion reactions[32,33] leading, for example, to benz- and naphth-imidazoles (28),(29) and imidazoquinolines (30) in 23–77% yield (Scheme 8), formed by spontaneous oxidation of the precursing

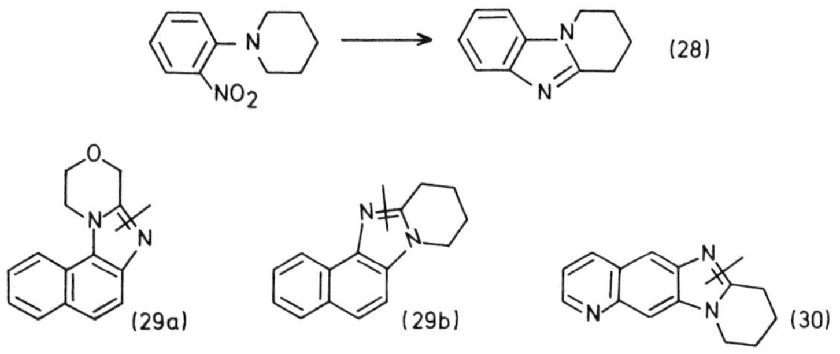

Scheme 8

2,3-dihydrobenzimidazoles. It is noteworthy that preferential deoxygenation of the 2-nitro group in the 2,4-dinitro compounds (31) occurs to give the stable nitro-2,3-dihydrobenzimidazoles (32) (50–60%);[33] this is attributed to interaction between the trialkyl phosphite and the tertiary amino group, the geometry of the resulting intermediate (33) favouring reduction of the *o*-nitro group (Scheme 9).

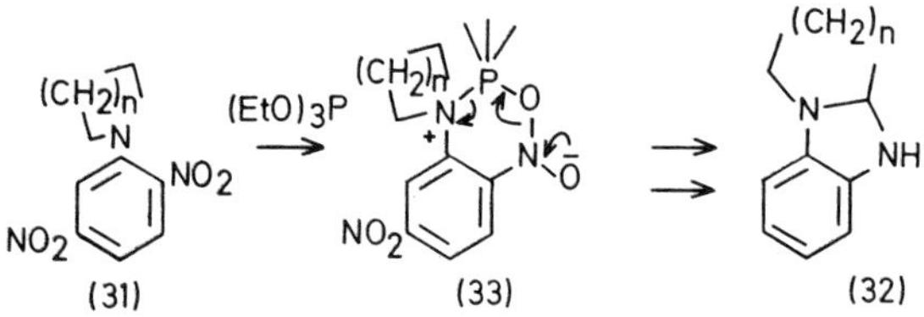

Scheme 9

In *N*-methyl-2-nitrodiphenylamines (34), where the nitrene is presented with the possibilities of insertion into the adjacent aromatic ring to give phenazines or into the *N*-alkyl group to give the five-membered benzimidazole (35), the latter is favoured (36% vs. 12%) (Scheme 10).[34] Also formed is a fused seven-membered azepinobenzimidazole whose genesis is more conveniently discussed in Section 9.

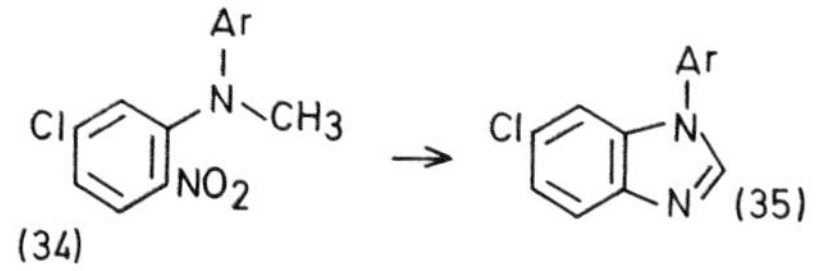

Scheme 10

The related insertion into adjacent o-alkyl side-chains in compounds such as 2-nitro-2',4',6'-trimethylbiphenyl (36) is of limited applicability though it does furnish a route to 8,10-dimethylphenanthridine (37) (14%)[2] (Scheme 11).

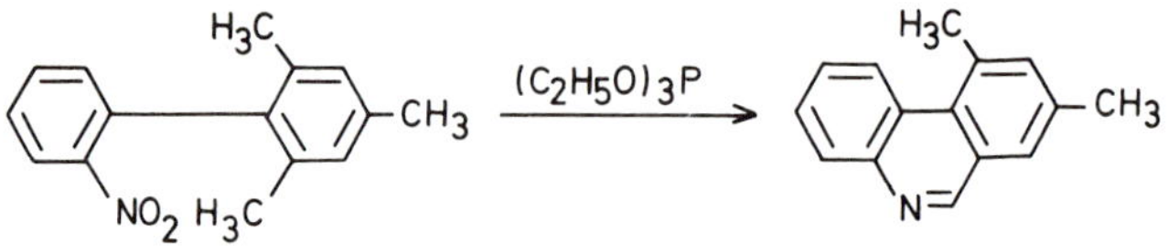

$$(36) \xrightarrow{(C_2H_5O)_3P} (37)$$

Scheme 11

6. Cyclization at Adjacent Imino Nitrogen (C=N) and Azo Nitrogen (N=N): Synthesis of Indazoles, Triazoles, and Related Compounds

The prototype of this reaction is the formation of pyrido[1,2-b]indazole (38) (>90%) by phosphorus(III) cyclization of 2-(2-nitrophenyl)pyridine[8] (Scheme 12). Here the nitrene or its precursor prefers to attack the electron-rich pyridine nitrogen atom rather than react by C–H insertion at the alternative *ortho* position. With only one exception, to which we return below, this is the pattern of such phosphite-induced cyclizations, and this has led to arguments in favour of the participation of singlet nitrenes in these reactions.[35]

Directly analogous cyclizations which have been achieved include (Scheme 12) the formation of the following fused polycyclic heterocycles in 25–95% yields: a wide variety of 2-substituted indazoles (39) from 2-nitrobenzylidene amines,[36,8] thieno[3,2-c]pyrazoles (40),[23] 2-aryl-[1]benzothieno[3,2-c]pyrazoles (41)[37a] and their [2,3-c] isomers (42),[37a] benzothiazolo[3,2-b]indazole (43) (66%),[35b,38] and the corresponding benzoxazolo[3,2-b]indazole (44) (60%)[38] in which ring closure at nitrogen is preferred to that at sulphur or oxygen. Similarly are obtained 11-alkyl-11H-benzimidazolo[1,2-b]indazole (45) (83%)[35b] and 2,2'-bi-2H-indazolyl (46) (22%)[8] from 2-nitrobenzaldazine. Closely related to the latter reaction is the formation of 3-[N-methyl(indazol-2-yl)amino]cyclohex-2-enone (47) (82%) from the hydrazone (48).[39]

The formation of ethyl 3-phenylthiazolo[3,4-b]indazole-1-carboxylate (49; R = CO$_2$Et) (24%) from the nitro compound (50; R = CO$_2$Et)[25] is of interest because the unsubstituted analogue of (49; R = H) provides the only example of preferential CH insertion to give the indolo derivative (51) rather than reaction of the nitrene at the available ring nitrogen atom to give (49; R = H). This is attributed to thermodynamic, rather than kinetic, control.

Also of some synthetic potential are the conversions in low yield of the readily available 2-nitrophenyl 2-pyrimidyl sulphide into pyrimido[1,2-b]-

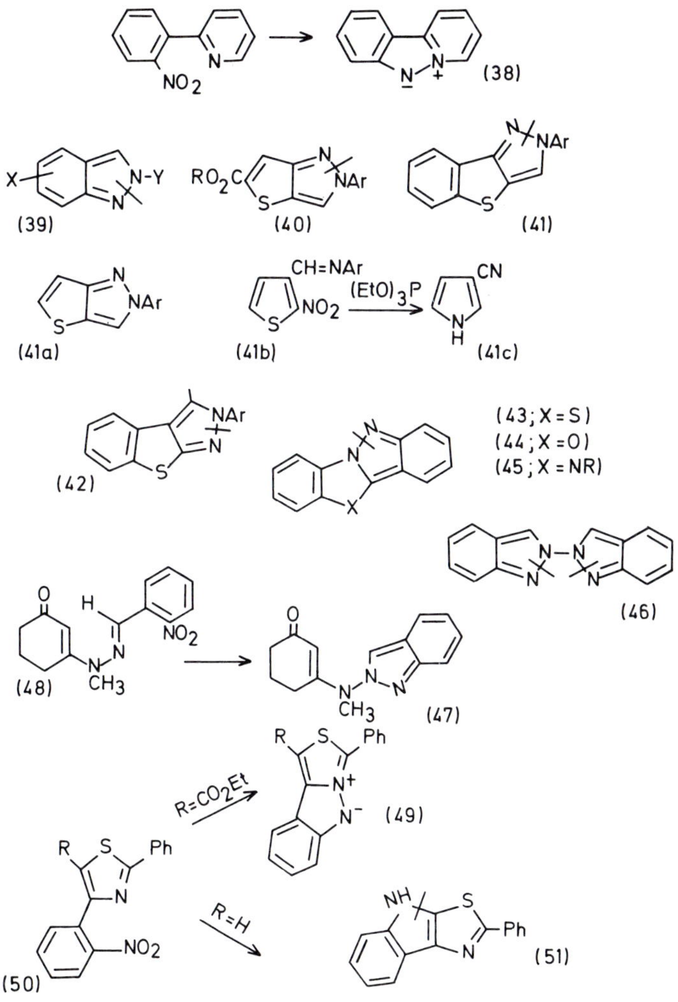

Scheme 12

indazole (52), the 2-pyridyl analogue giving pyrido[1,2-b]indazole (38).[40] Both deoxygenation and desulphurization as in Scheme 13 are suggested.

A by-product in the formation of benzothieno[2,3-c]pyrazole (42) from the corresponding 3-nitrobenzothiophen-2-anil is benzo[b]thiophen-3-carbonitrile,[37a] formed via a ring-opening–ring-closure sequence. Following this it has been shown[37b] that, whereas 3-nitrothiophen-2-anils give the expected 2-arylthieno[3,2-c]pyrazoles (41a), the isomeric 2-nitrothiophen anil (41b) gave 1-phenylpyrrole-3-carbonitrile (41c) (55%), a novel ring transformation of a thiophen into a pyrrole.

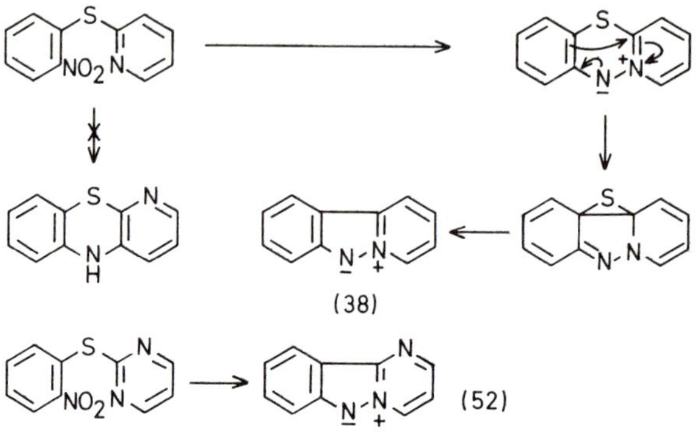

Scheme 13

The parent reaction in the synthesis of triazoles is the conversion of 2-nitroazoarenes into the corresponding 2-aryl-2H-benzotriazoles (53) (30–75%)[36a,8] (Scheme 14), hydroxy derivatives undergoing ethylation as well as cyclization. The variations on this theme are summarized in Scheme 14 and include the syntheses of the pyrazolo[1,2-a]benzotriazole (54) (15–65%)[41,35b]

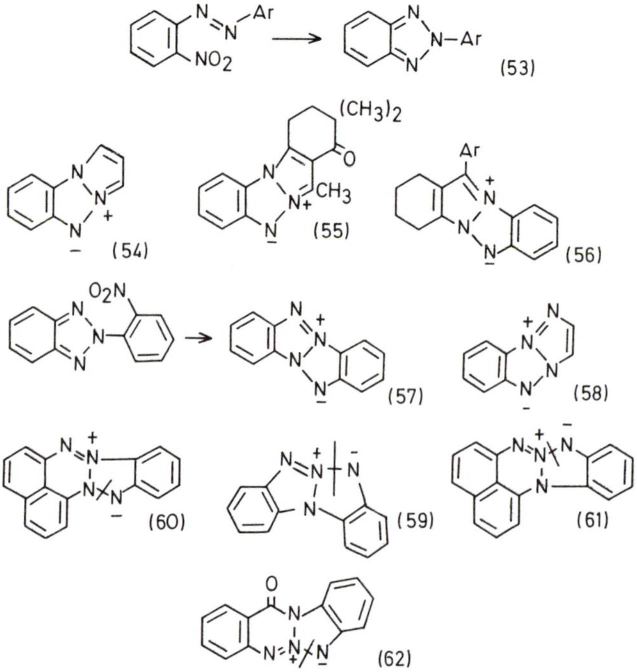

Scheme 14

and its analogue (55) (37%),[39] 12-aryl-1,2,3,4-tetrahydro-6H-indazolo[2,1-a]-benzotriazole (56) (36%),[39] dibenzo[b,f]-1,3a,4,6a-tetraazapentalenes (57) in good yields from either 2,2'-dinitroazobenzene[8] or 2H-2-o-nitrophenylbenzotriazoles[42] (Scheme 14), the benzotetraazapentalenes (58) and (59),[42] benzotriazolo[2,1-a]naphtho[1,8-d,e]triazine (60) (46%)[43] and its [1,2-a] isomer (61),[44] and 13-oxobenzotriazolo[2,1-b]benzo[1,2-e]triazine (62) (55%).[45]

Finally, a most interesting case of concomitant deoxygenation of o-nitro and o-carbonyl groups has been reported. Thus reductive cyclization of 2-benzoyl-2'-nitroazobenzene leads to the indazolotriazole (63) (57%)[46a] (Scheme 15), probably via a concerted pathway. These reactions follow related conversions of 2-acylazo compounds,[46b] e.g. 2-arylazobenzophenone into 2-aryl-3-phenylindazole, and benzoylpyrazoles into pyrazolopyrazoles.

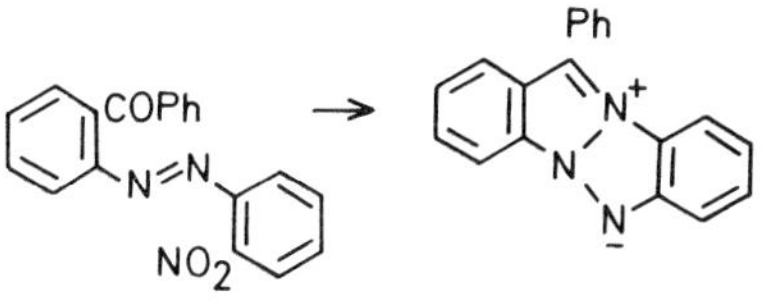

Scheme 15

7. Interaction with Adjacent Carbonyl Groups

7.1. Reaction with o-Carbonyl Groups

The analogy with reductive cyclizations of 2-nitrobenzylidene imines (Scheme 12) is obvious. o-Acylnitrobenzenes readily give anthranils (1,2-benzoxazoles), e.g. (64), (65) (Scheme 16), in good to moderate yields.[47,27,48,49] Anthranils, e.g. (65), are in turn useful precursors, via nitrenes produced on thermolysis, of acridones such as 1-chloro-1-methylimidazo[2,1-b]quinazolin-5-one (66) (63%).[49]

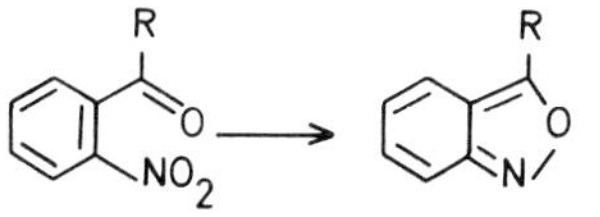

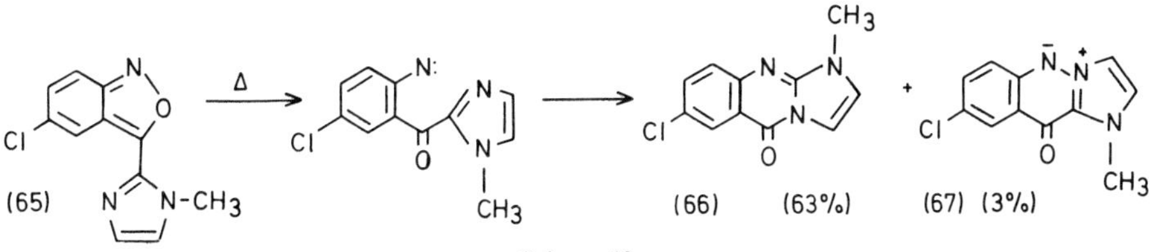

Scheme 16

7.2. Reaction with o-α,β-Unsaturated Carbonyl Groups

We have already seen that mono-β-acyl- or β-carboxy-2-nitrostyrenes, e.g. (20), react normally to give 2-acylindoles, albeit in moderate yield (Scheme 4). However, when the second β position is blocked by another group a series of interesting reactions occur, some of which pose unanswered mechanistic questions.

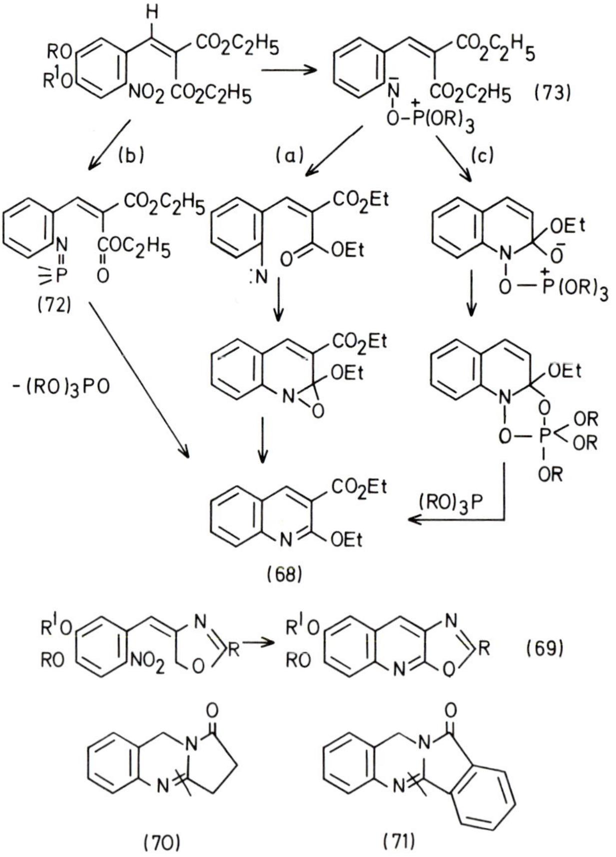

Scheme 17

Thus quinolines (68) (53–68%)[17,50] and oxazolo[5,4-b]quinolines (69) (37–68%)[17,50,51] are formed as shown in Scheme 17. Similarly are formed the quinazoline derivatives (70)[51b] and (71).[51c] The mechanism of this quinoline-forming reaction has not been established. Suggested routes include addition of a nitrene across the carbonyl double bond, followed by deoxygenation [route (a)], via the iminophosphorane (72) [route (b)], or via the nitrene precursor (73) [route (c)] which your reviewer favours by analogy with Saunders' results,[52] discussed below and in Chapter 5. Even this may be an oversimplification because quinolines (74) are also produced in surprisingly high yields from benzylidenecyanoacetates (75; R = CO_2Et) in which the *o*-nitrophenyl- and carbonyl groups are *trans*. Kametani and his coworkers[53] suggest that triethyl phosphite induced *trans–cis* isomerization has occurred, possibly via reversible addition to the double bond, though a thorough controlled mechanistic study has not been carried out (Scheme 18).

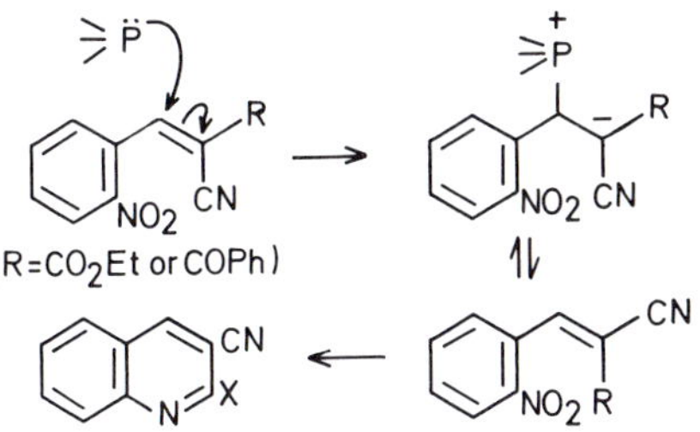

Scheme 18

Perhaps surprisingly, in view of these observations, the corresponding benzoylacetonitriles (75; R = COPh) give low yields of the quinolines (74; X = Ph),[53] possibly reflecting steric hindrance to the larger benzoyl groups. Ethyl α-phenylcinnamates, in which the nitrophenyl and carbonyl groups are *trans*, similarly give quinolines.[54] The sensitivity of the deoxygenation reaction to stereochemistry and substituents is apparent from Scheme 19; the γ-lactone (76), in which the carbonyl and nitrophenyl groups are *cis*, gives 3,4-dihydro-5-methoxy[1,3]oxazino[3,4-*a*]indol-1-one (77; R^3 = MeO) (45%)[55a] whereas the *trans* analogue (78) gives only a trace of the corresponding indolone (77; R^3 = H) and other minor products in a low accountance reaction[55b] (Scheme 19). Of considerable mechanistic interest in the latter reaction, however, is the isolation, for the first time from such reactions, of an azırıne, 1-hydroxymethyl-4,5-dimethoxy-7*H*-azirino[1,2-*a*]indole-7a-carboxylic acid γ-lactone (79), suggesting nitrene or nitrenoid cycloaddition to the adjacent double bond in the isomer (80), of (78) postulated as an intermediate.

Direct evidence for attack of the phosphorus(III) reagent on the α,β double bond comes from the isolation of the dihydropyrazolo[3,4-*b*]quinolin-4-ylphosphonates (81) (7–25%)[56] (Scheme 20) from the Δ^2-pyrazolin-5-ones (82). Trace amounts of the spiro compound (83), pointing to a mechanism via (84), and the 2*H*-pyridazino[4,5-*b*]indole (85), analogous to (77) (Scheme 19), are also formed.

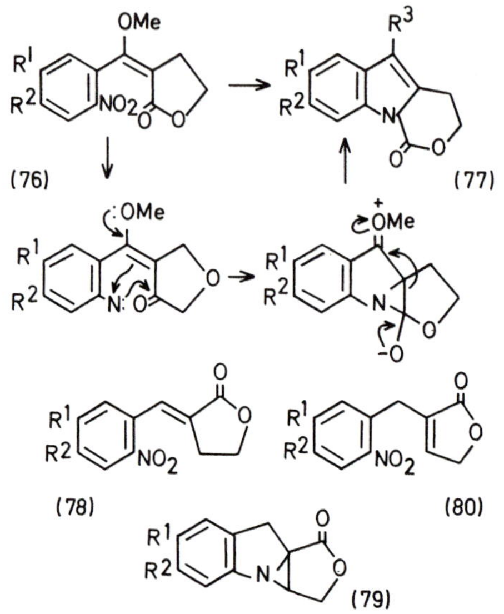

Scheme 19

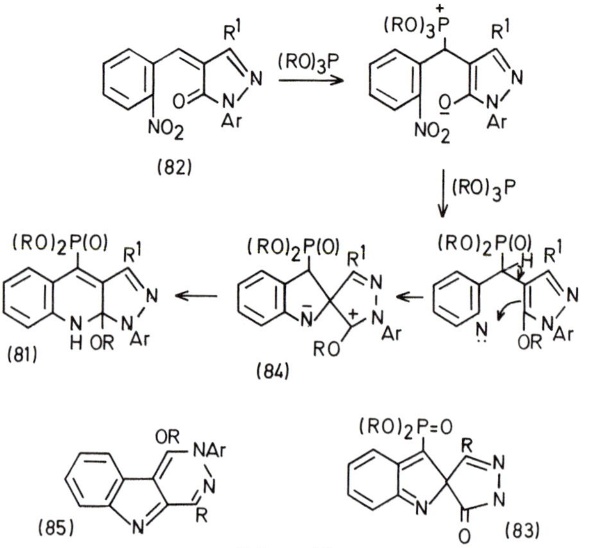

Scheme 20

An aromatic analogue of these reactions involves the phosphorus(III) deoxygenation of 2-nitro-2'-alkoxycarbonyl-biaryls, e.g. (86) (Scheme 21),[57] to give β-carbolines (87) formed by C–C insertion, and benzonaphthyridines (88) in about 20% yields.

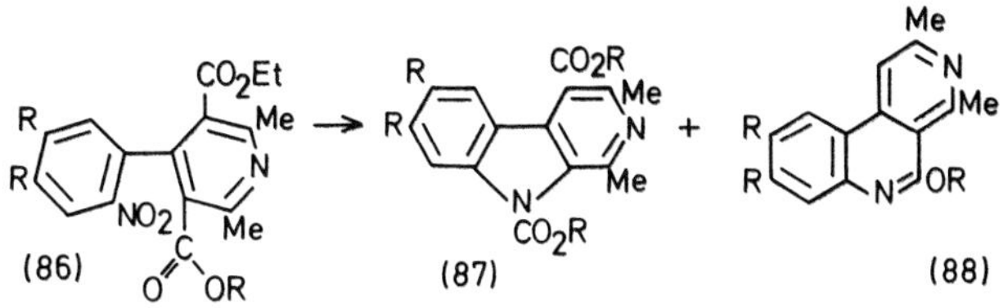

6 Heterocycles from Nitro-compounds   285

Scheme 21

7.3. Reaction with Adjacent, but not Conjugated, Carbonyl Groups

The first example appears to be the formation of methyl 4-methoxytriazolo-[3,4-a]quinoxaline-3-carboxylate (89) (30%) from the o-nitrophenyltriazole (90) when the more nucleophilic tributylphosphine is used instead of triethyl phosphite (Scheme 22). The reaction of related 4-(2-nitroaryl)-3,5-diethoxycarbonyl-1,4-dihydropyridines with triethyl phosphite has also been studied; the low yields of products make the reaction of little use synthetically, but it is of interest mechanistically.[58]

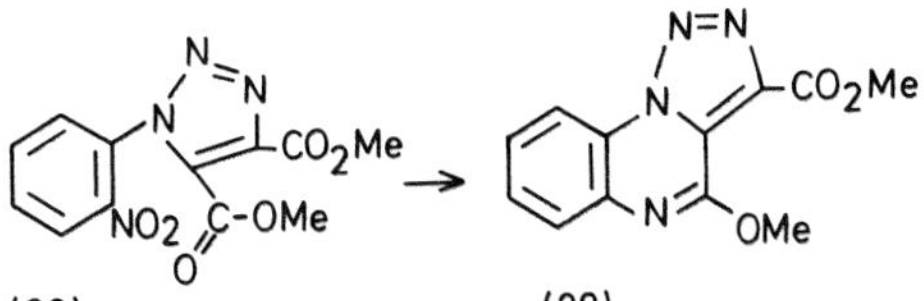

Scheme 22

The good conversion of o-nitrophenyl benzoate into 1,3-benzoxazole (68%) (Scheme 23)[52a] unfortunately is not paralleled with other o-nitrophenyl esters.[52b] Better results are obtained using the azido derivatives (see Chapter 5). There is good evidence that iminophosphoranes [(RO)$_3$P=NAr] and nitrenes are not intermediates in the former reaction,[52b] a species of the type (91) being favoured.

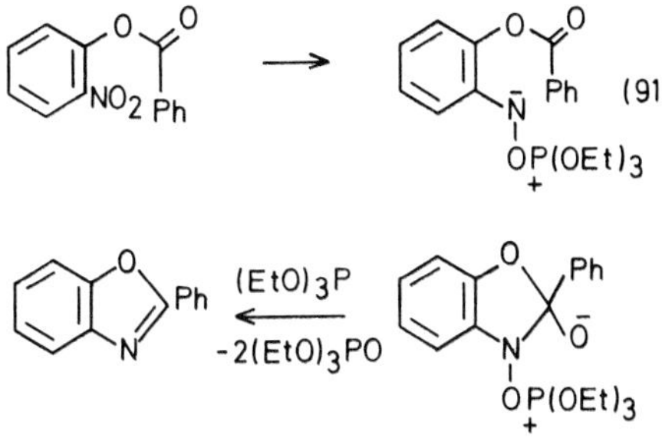

Scheme 23

A possible mechanism is suggested in Scheme 23 in which the intermediacy and fragmentation of a phosphorane (Chapter 11) cannot be ruled out. Similar mechanisms and intermediates must also be considered for most of the reactions described in the preceding sections.

8. Interaction with Adjacent, but not Conjugated, Aromatic Rings: Synthesis of Phenothiazines and Related Compounds

These reactions include deoxygenations of the *o*-nitroaryl aryl ethers and sulphides, *o*-nitrodiarylamines, and *o*-nitrodiphenylmethane (Scheme 24). In practice the expected cyclization to give the six-membered heterocycles (92) occurs in surprisingly few instances.[59]

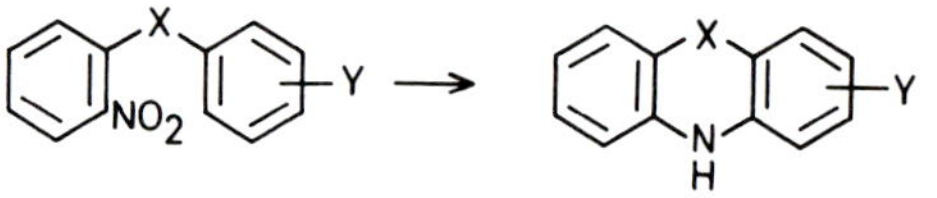

$X = S, O, CH_2, CHPh, NCOCH_3$

Scheme 24

Cyclization of a wide range of 2-nitroaryl 4-substituted-aryl sulphides (93) gives the corresponding 2-substituted phenothiazines (94) in good to excellent yields (Scheme 25).[60] 2-Substituted analogues similarly give 1-substituted phenothiazines.[60d]

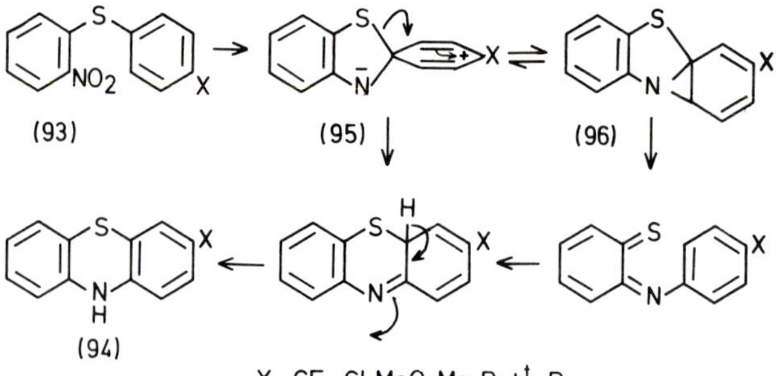

$X = CF_3, Cl, MeO, Me, Bu^t, D$

Scheme 25

The reaction is remarkable because the 3-substituted isomer which, at first sight, might have been expected to be formed by direct insertion into the available *ortho* C–H bond is not formed. Reaction via a five-membered spirodienyl intermediate (95) followed by sigmatropic shift and rearomatization is likely, although reaction via the strained azanorcaradiene (96) is also a

6 Heterocycles from Nitro-compounds

possibility, in which case an allowed $(\sigma^{2s} + \pi^{2s} + \sigma^{2s})$ ring opening followed by Cope-type cyclization can be written.

Blocking both *ortho* positions in the receiving aromatic rings leads to appreciable yields of phosphorus-containing products, to which we return in Section 10, but significant amounts of new heterocyclic systems are also obtainable.[60d] Thus, 2,6-dimethylphenyl 2-nitrophenyl sulphide gives 5,11-dihydro-4-methyldibenzo[b,e][1,4]thiazepine (97) (11%) (Scheme 26),[60d] while the corresponding 2,6-diethoxycarbonyl derivative gives 1,4a-diethoxycarbonyl-4aH-phenothiazine (98) (50%),[60d] providing direct evidence for the non-aromatic intermediate in the general reaction. The corresponding 2-nitroaryl aryl ethers give negligible yields of phenoxazines and high yields of phosphoranes (see Section 10). The corresponding 2-nitrodiarylamines give clean products only when the nitrogen atom is acylated;[34,61] thus *N*-acetyl-2-nitro-2'-methylthiodiphenylamine gives, via rearrangement, 1-methylthio-5-acetyl-5,10-dihydrophenazine (99) (22%) as the main product (Scheme 26).[61]

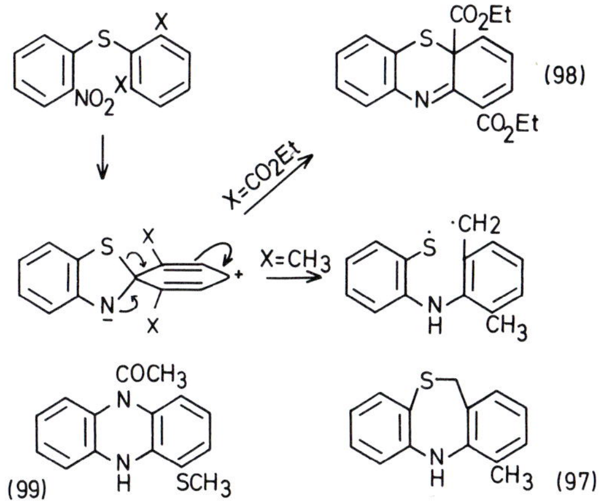

Scheme 26

9. Ring Enlargement Reactions: Formation of Azepines and Related Compounds

The reagents diethyl methylphosphonite [(EtO)$_2$PMe] and ethyl diphenylphosphinite [Ph$_2$POEt] are reactive enough to deoxygenate simple nitrobenzene derivatives in boiling diethylamine to give acceptable yields (20–85%) of 2-diethylamino-3H-azepines (100) (Scheme 27).[62] Hexaethylphosphorus triamide under autoclave conditions,[63] and triethyl phosphite under photolysis,[64] have also been used. Reactions carried out in an excess of triethyl phosphite and the absence of diethylamine lead to low yields (*ca* 20%) of diethyl 3*H*-azepin-7-ylphosphonates (101).[62,65] Reaction as in Scheme 27 is assumed.

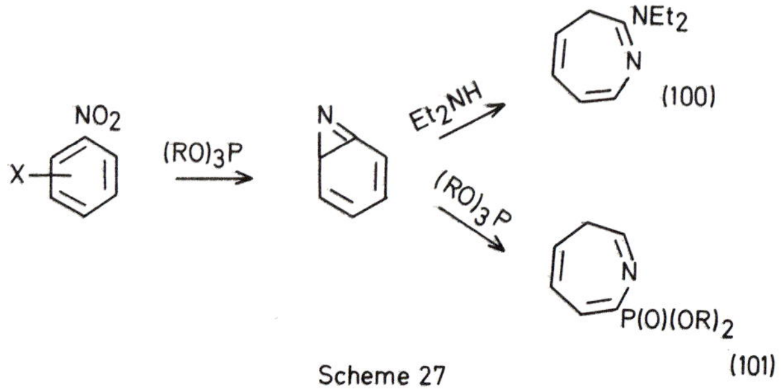

Scheme 27

In a few instances involving bicyclic aromatic nitro compounds ring expansion of the ring other than that carrying the nitro group can occur. Thus 2-nitrophenyl-2,4,6-trimethylphenylmethane with triethyl phosphite gives 6,8,10-trimethyl-10*H*-azepino[1,2-*a*]indole (102) (10%)[66a] (Scheme 28). 2-Chloro-8-methyl-6*H*-azepino[1,2-*a*]benzimidazole (103) (11%) was similarly produced from *N*-acetyl-4-chloro-4'-methyl-2-nitrodiphenylamine[34] (Scheme 28). Interestingly, the related *o*-nitrophenyldithienylmethanes give thieno-[3,2-*b*]quinolines by direct insertion.[75]

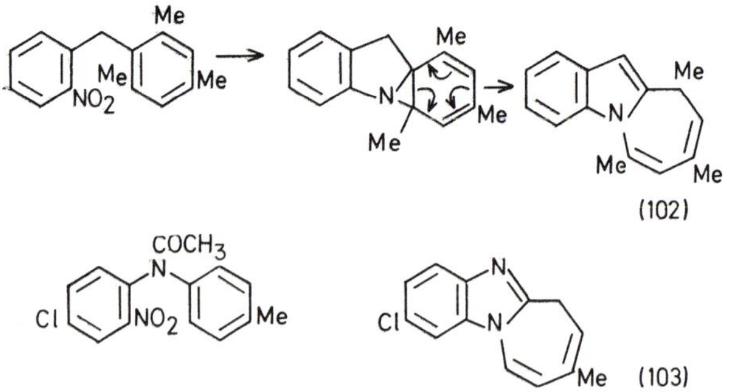

Scheme 28

10. Incorporation of Phosphorus into the Heterocyclic Ring: Formation of Amino(oxy)- and Amino(thiyl)-phosphoranes

We have already seen that whereas 2-nitroaryl aryl sulphides give N–S-heterocycles the corresponding ethers do not give O–N-heterocycles. In the latter case phosphoranes (104; X = O) [3-aryl-2,3-dihydro-1,3,2-benzoxaza-phosph(v)oles] are formed in very high yield (up to 95%) presumably by capture of an intermediate iminoquinone species (105; X = O) by the

6 Heterocycles from Nitro-compounds

phosphorus(III) reagent (Scheme 29).[66] Stable amino(thiyl)phosphoranes (104; X = S, $R^1 = R^2$ = Ph, R^3 = MeO) and (106) are formed in significant yield only when the *ortho* positions in the receiving aromatic ring are blocked, e.g. (106).[67] In the case of (104; $R^1 = R^2 = R^3$ = EtO, X = S) the phosphorane is too unstable to be isolated; instead isomeric open-chain phosphoramidates are formed.[40]

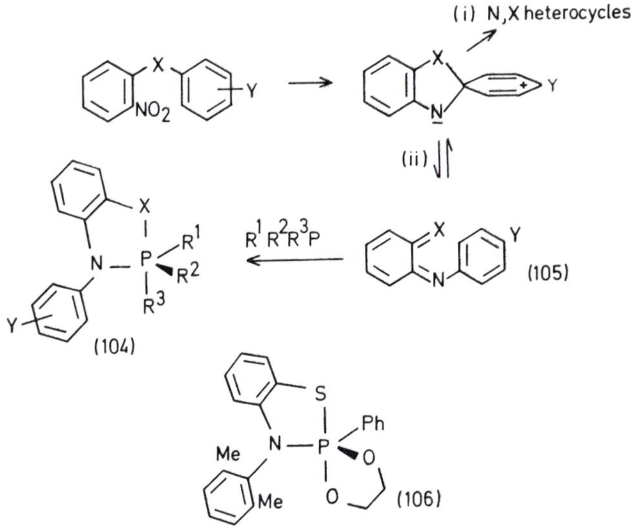

Scheme 29

It is of interest that another class of amino(oxy)phosphoranes, the stable 3,4-diaryl-4,5-dihydro-2-oxo-1,2,5-oxazaphosph(v)oles (107), are formed at room temperature by Michael addition of phosphorus(III) esters to (*E*)-1,2-diaryl-1-nitroethenes (Scheme 30),[68] thermolysis of which in the presence of more phosphorus(III) reagent leads to indoles, thus suggesting that the direct high-temperature conversion of the nitroalkenes into indoles proceeds via (107).[8] Deoxygenation of nitroalkenes provides one of the few phosphorus(III) reductions of aliphatic nitro compounds that is of synthetic value.

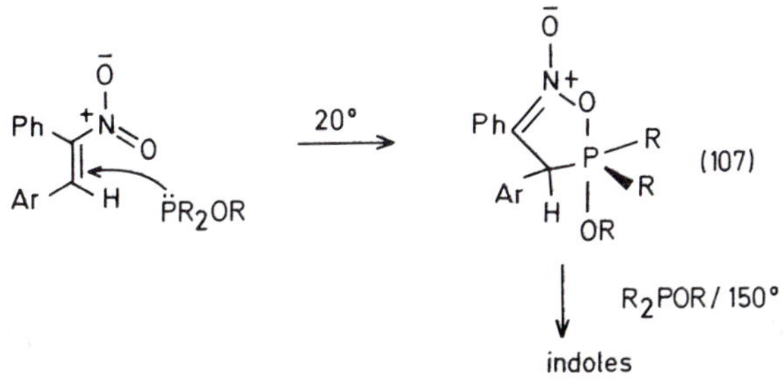

Scheme 30

11. Some Conundrums

Not all reactions of phosphorus(III) reagents with aromatic nitro compounds are easily explained. Three good examples are as follows: a possible explanation of the conversion of 6-nitrolaudanosine (108) into 5,6-dihydro-2,3,9,10-tetramethylbenzo[a]carbazole (109) (39%)[69] involves quaternization by triethyl phosphate, thus giving the diethyl phosphate anion which in turn induces an elimination as shown in Scheme 31. 1,2-Dihydro-2-methyl-1-(2-nitrobenzyl)isoquinoline behaves similarly but the related 6'-nitropapaverine (110) gave a low yield of product[70a] later identified[70b] as azaberbinone (111).

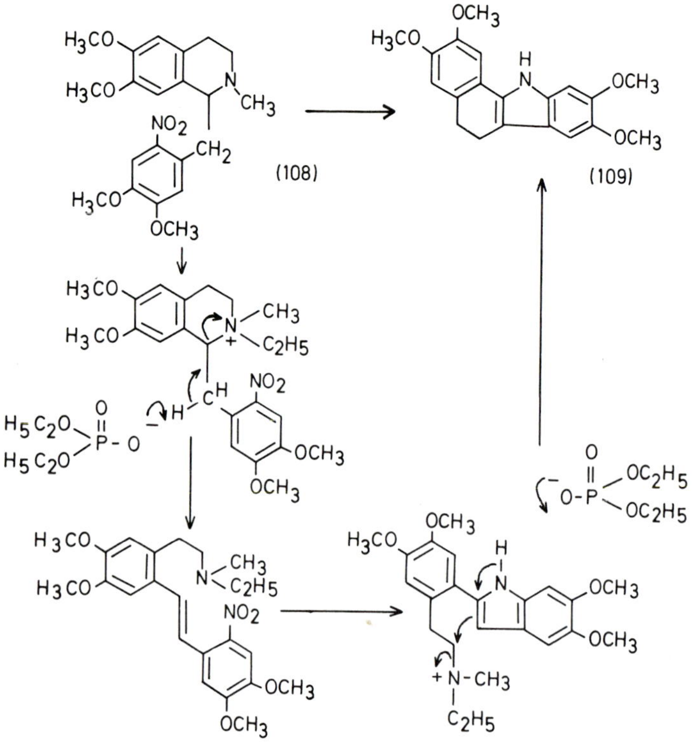

Scheme 31

This suggests that the first step is dehydration to give the anthranil which then isomerizes via the ring-opened acylnitrene, which is a precedented reaction,[59,71] to give the observed product. In accord with this the anthranil (112) has been shown to react under these conditions to give azaberbinone.[70b]

More puzzling is the conversion of α,α-di-(2-furyl)-o-nitrotoluenes (113) into furo[3,2-c]carbazolylphosphonates (114; R = Me, Et, Bu^t) in about 28%

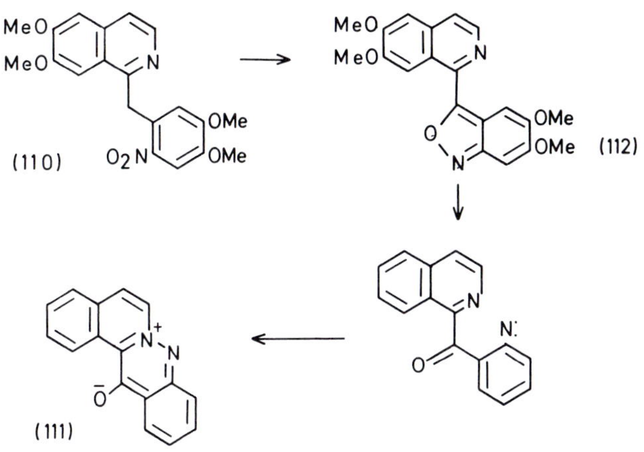

6 Heterocycles from Nitro-compounds

Scheme 32

yield[72] (Scheme 33). It is known that reactions with nitrenes can lead to the fission of furans[73] to give an unsaturated ketone. Such an intermediate is (114a) which could add phosphite at the arrowed position, which would eventually be position 5 in the furocarbazole, but no evidence on the mechanism beyond the identification of acetaldehyde is available.

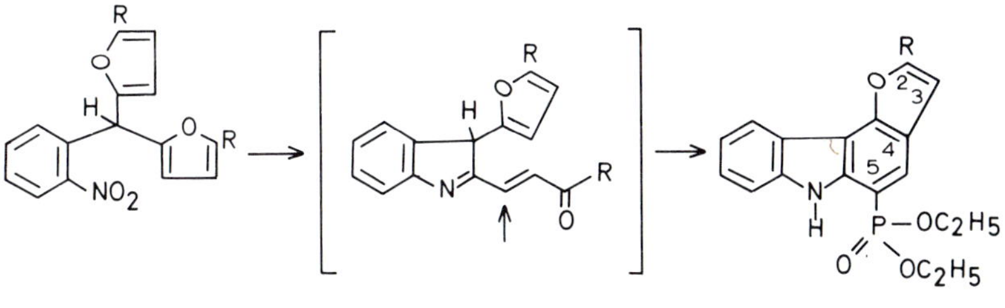

Scheme 33

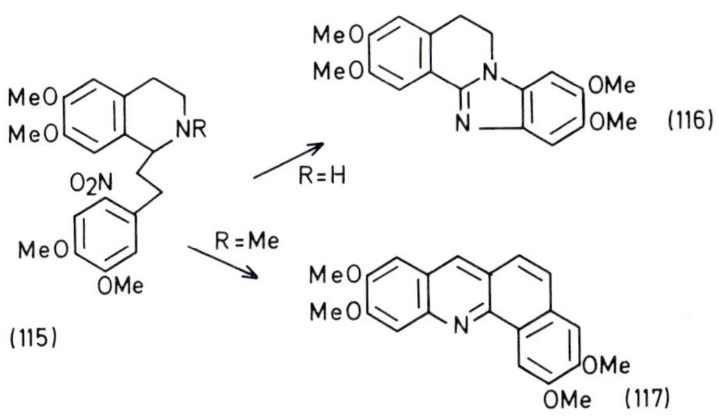

Scheme 34

Finally, the reader is invited to write mechanisms for the conversion of the 1,2,3,4-tetrahydro-1-(2-nitrophenethyl)isoquinolines (115; R = H) and (115; R = Me) into 5,6-dihydro-2,3,9,10-tetramethoxybenzimidazo[2,1-*a*]isoquinoline (116) (12%), which has two carbons less, and 2,3,9,10-tetramethoxybenz-[*c*]acridine (117) (50%), respectively.[74]

12. References

1. J. I. G. Cadogan and M. Cameron-Wood, *Proc. Chem. Soc.*, 1962, 361.
2. J. I. G. Cadogan and M. J. Todd, *J. Chem. Soc. (C)*, 1969, 2808.
3. J. I. G. Cadogan and S. Kulik, *J. Chem. Soc. (C)*, 1971, 2621.
4. P. K. Brooke, R. B. Herbert, and F. G. Holliman, *Tetrahedron Lett.*, 1973, 761.
5. (a) D. G. Saunders, *J. Chem. Soc., Chem. Comm.*, 1969, 680. (b) L. J. Leyshon and D. G. Saunders, *ibid.*, 1971, 1608.
6. M. A. Armour, J. I. G. Cadogan, and D. S. B. Grace, *J. Chem. Soc., Perkin II*, 1975, 1185.
7. (a) J. I. G. Cadogan and A. Cooper, *J. Chem. Soc. (B)*, 1969, 883. (b) L. R. C. Barclay, P. G. Khazanie, K. A. H. Adams, and E. Reid, *Canad. J. Chem.*, 1977, **55**, 3273.
8. J. I. G. Cadogan, M. Cameron-Wood, R. K. Mackie, and R. J. G. Searle, *J. Chem. Soc.*, 1965, 4831.
9. I. Puskas and E. K. Fields, *J. Org. Chem.*, 1968, 33, 4237.
10. D. E. Ames, K. J. Hansen, and N. D. Griffiths, *J. Chem. Soc., Perkin I*, 1973, 2818.
11. A. H. Jackson, D. N. Johnston, and P. V. R. Shannon, *J. Chem. Soc., Chem. Comm.*, 1975, 911.
12. A. Tanaka, K. Yakushijin, and S. Yoshina, *J. Heterocyclic Chem.*, 1977, 975.
13. K. E. Chippendale, B. Iddon, and H. Suschitzky, *J. Chem. Soc., Perkin I*, 1971, 2023.
14. T. Kametani, K. Ogasawara, and T. Yamanaka, *J. Chem. Soc. (C)*, 1968, 1006.
15. T. Kametani, F. F. Ebetino, T. Yamanaka, and K. Nyu, *Heterocycles*, 1974, 2, 209.
16. J. I. G. Cadogan and M. J. Todd, unpublished.
17. T. Kametani, K. Nyu, T. Yamanaka, H. Yagi, and K. Ogasawara, *Tetrahedron Lett.*, 1969, 1027.
18. E. E. Taylor and E. C. Garcia, *J. Org. Chem.*, 1965, 30, 655.
19. (a) R. J. Sundberg, *J. Org. Chem.*, 1965, 30, 3604. (b) *J. Amer. Chem. Soc.*, 1966, **88**, 3781. (c) *J. Org. Chem.*, 1968, 33, 487. (d) T. Kametani, F. F. Ebetino, and K. Fukomoto, *Tetrahedron Lett.*, 1973, 5229.
20. K. E. Chippendale, B. Iddon, and H. Suschitzky, *J. Chem. Soc., Perkin I*, 1973, 125.
21. K. Srinivasan, K. G. Srinivasan, and K. K. Balasubramanian, *Synthesis*, 1973, 313.
22. S. Gronowitz and I. Ander, *Acta Chem. Scand.*, 1975, B29, 513.
23. V. M. Colburn, B. Iddon, and H. Suschitzky, *J. Chem. Soc., Perkin I*, 1977, 2436.

24. F. M. Dean, C. Patampongse, and V. Podimuang, *J. Chem. Soc., Perkin I*, 1974, 583.
25. K. T. Potts and J. L. Marshall, *J. Org. Chem.*, 1976, **41**, 129.
26. T. Kametani, T. Yamanaka, K. Nyu, and S. Takano, *J. Pharm. Soc. Japan*, 1971, **91**, 1033.
27. J. I. G. Cadogan, R. Marshall, D. M. Smith, and M. J. Todd, *J. Chem. Soc. (C)*, 1970, 2441.
28. R. J. Sundberg and T. Yamazaki, *J. Org. Chem.*, 1967, **32**, 290.
29. T. Kametani, T. Yamanaka, and K. Nyu, *J. Heterocyclic Chem.*, 1972, **9**, 1281.
30. B. I. Feuer and G. Smolinsky, *J. Amer. Chem. Soc.*, 1964, **86**, 3085.
31. J. I. G. Cadogan, *Quart. Rev.*, 1968, **22**, 222.
32. H. Suschitzky and M. E. Sutton, *J. Chem. Soc. (C)*, 1968, 3058.
33. R. Garner, G. V. Garner, and H. Suschitzky, *J. Chem. Soc. (C)*, 1970, 825.
34. R. G. R. Bacon and S. D. Hamilton, *J. Chem. Soc., Perkin I*, 1974, 1975.
35. I. M. Robbie, O. Meth-Cohn, and H. Suschitzky, *Tetrahedron Lett.*, (a) 1976, 925; (b) 1976, 929.
36. (a) J. I. G. Cadogan and R. J. G. Searle, *Chem. and Ind.*, 1963, 1282. (b) J. I. G. Cadogan and R. K. Mackie, *Org. Synth.*, 1968, **48**, 113. (c) T. J. Schwan, C. S. Davies, and L. J. Honkomp, U.S. Pat. 4,002,652 (1977) (*Chem. Abs.*, 1977, **86**, 155647r). (d) M. A. Armour, J. I. G. Cadogan, and D. S. B. Grace, *J. Chem. Soc., Perkin II*, 1975, 1185.
37. (a) K. E. Chippendale, B. Iddon, and H. Suschitzky, *J. Chem. Soc., Perkin I*, 1973, 129. (b) V. M. Colburn, B. Iddon, H. Suschitsky, and P. T. Gallagher, *J. Chem., Soc., Chem. Comm.*, 1978, 453.
38. J. I. G. Cadogan and K. Itoh, unpublished.
39. A. J. Nunn and F. J. Rowell, *J. Chem. Soc., Perkin I*, 1975, 629.
40. J. I. G. Cadogan and B. S. Tait, *J. Chem. Soc., Perkin I*, 1975, 2396.
41. Y. Y. Hung and B. M. Lynch, *J. Heterocyclic Chem.*, 1965, **2**, 218.
42. J. C. Kauer and R. A. Carboni, *J. Amer. Chem. Soc.*, 1967, **89**, 2633.
43. H. Sieper, *Tetrahedron Lett.*, 1967, 1987.
44. H. Sieper and P. Tavs, *Ann. Chem.*, 1967, **704**, 161.
45. A. W. Murray and K. Vaughan, *Chem. Comm.*, 1967, 1282.
46. J. H. Lee, A. Matsumoto, M. Yoshida, and O. Simamura, (a) *Chem. Lett.*, 1974, 951; (b) *Bull. Chem. Soc. Japan*, 1974, **47**, 1039.
47. J. I. G. Cadogan, R. K. Mackie, and M. J. Todd., *Chem. Comm.*, 1966, 491.
48. A. U. Rahman and A. J. Boulton, *Tetrahedron*, 1966, Supp. 7, 49.
49. R. Y. Ning, J. F. Blount, P. D. Madan, and R. I. Fryer, *J. Org. Chem.*, 1977, **42**, 1791.
50. T. Kametani, K. Nyu, T. Yamanaka, H. Yagi, and K. Ogasawara, *Chem. Pharm. Bull.* (Japan), 1969, **17**, 2093.
51. (a) T. Kametani, T. Yamanaka, and K. Ogasawara, *J. Chem. Soc. (C)*, 1969, 385. (b) T. Kametani, K. Nyu, and T. Yamanaka, *J. Pharm. Soc. Japan*, 1972, **92**, 1184. (c) *Idem, J. Heterocyclic Chem.*, 1971, **8**, 1071.
52. (a) D. G. Saunders, *Chem. Comm.*, 1969, 680. (b) L. Leyshon and D. G. Saunders, *ibid.*, 1971, 1608.
53. T. Kametani, K. Nyu, and T. Yamanaka, *Chem. Pharm. Bull.* (Japan) 1971, **19**, 1321.
54. T. Kametani, K. Nyu, and T. Yamanaka, *J. Pharm. Soc., Japan*, 1972, **92**, 1180.

55. (a) T. Kametani, F. F. Ebetino, and K. Fukumoto, *J. Chem. Soc., Perkin I*, 1974, 861. (b) *Idem, Tetrahedron*, 1975, **31**, 1241.
56. T. Nishiwaki, G. Fukuhara, and T. Takahashi, *J. Chem. Soc., Perkin I*, 1973, 1606.
57. T. Kametani, T. Yamanaka, and K. Ogasawara, *Chem. Comm.*, 1968, 996; *J. Chem. Soc. (C)*, 1969, 138.
58. T. Kametani, T. Yamanaka, and K. Ogasawara, *J. Chem. Soc. (C)*, 1969, 1616.
59. J. I. G. Cadogan, *Acc. Chem. Res.*, 1972, **5**, 303.
60. (a) J. I. G. Cadogan, R. K. Mackie, and M. J. Todd, *Chem. Comm.*, 1966, 491. (b) J. I. G. Cadogan, S. Kulik, and M. J. Todd, *ibid.*, 1968, 736. (c) J. I. G. Cadogan, S. Kulik, C. Thomson, and M. J. Todd, *J. Chem. Soc. (C)*, 1970, 2437. (d) J. I. G. Cadogan and S. Kulik, *ibid.*, 1971, 2621.
61. Y. Maki, T. Hosokami, and M. Suzuki, *Tetrahedron Lett.*, 1971, 3509.
62. J. I. G. Cadogan and M. J. Todd, *Chem. Comm.*, 1967, 178; *J. Chem. Soc. (C)*, 1969, 2808; T. de Boer, J. I. G. Cadogan, H. M. McWilliam, and A. G. Rowley, *J. Chem. Soc., Perkin II*, 1975, 554.
63. F. R. Atherton and R. W. Lambert, *J. Chem. Soc., Perkin I*, 1973, 1079.
64. R. J. Sundberg, B. P. Das, and R. W. Smith, *J. Amer. Chem. Soc.*, 1969, **91**, 658.
65. J. I. G. Cadogan, R. K. Mackie, and M. J. Todd, *Chem. Comm.*, 1968, 736; J. I. G. Cadogan, D. J. Sears, D. M. Smith, and M. J. Todd, *J. Chem. Soc. (C)*, 1969, 2813; J. I. G. Cadogan and R. K. Mackie, *ibid.*, 1969, 2819.
66. (a) J. I. G. Cadogan, D. S. B. Grace, P. K. K. Lim, and B. S. Tait, *Chem. Comm.*, 1972, 520. (b) *Idem, J. Chem. Soc., Perkin I*, 1975, 2376.
67. J. I. G. Cadogan and N. J. Tweddle, *J. Chem. Soc., Perkin I*, 1979, 1728.
68. J. I. G. Cadogan, R. A. North, and A. G. Rowley, *J. Chem. Res. (S)*, 1978, 1; *(M)*, 1978, 0178–0188; J. I. G. Cadogan and J. Hay, unpublished.
69. T. Kametani, T. Yamanaka, and K. Ogasawara, *J. Org. Chem.*, 1968, **33**, 4446.
70. (a) T. Kametani, T. Yamanaka, K. Ogasawara, and K. Fukumoto, *J. Chem. Soc. (C)*, 1970, 380. (b) M. P. Cava, M. J. Mitchell, and D. T. Hill, *Chem. Comm.*, 1970, 1601.
71. R. Y. Ning, W. Y. Chen, and L. H. Sternbach, *J. Heterocyclic Chem.*, 1974, **11**, 125.
72. G. Jones and W. H. McKinley, *Tetrahedron Lett.*, 1977, 2457.
73. K. Hafner and W. Kaiser, *Tetrahedron Lett.*, 1964, 2185; D. W. Jones, *J. Chem. Soc., Perkin I*, 1972, 225, 2728.
74. T. Kametani, Y. Fujimoto, and M. Mizushima, *Heterocycles*, 1975, **3**, 619.
75. G. Jones, C. Keates, I. Klado, and P. Radley, *Tetrahedron Lett.*, 1979, 1445.

7
Deoxygenations using Phosphorus(III) Reagents
1: General Functional Group Conversions

A. G. ROWLEY

1. Direct deoxygenation by tervalent phosphorus reagents, 295
2. Indirect deoxygenation by tervalent organophosphorus reagents, 327
3. References, 345

1. Direct Deoxygenation by Tervalent Phosphorus Reagents

1.1. Removal of Oxygen attached to Nitrogen

1.1.1. *Deoxygenation of heterocyclic N-oxides*

The conversion of a heterocyclic *N*-oxide into the parent amine is a necessary sequel to the use of the *N*-oxide to obtain substitution patterns in the heterocycle that are not accessible from the amine directly.[1] Phosphines, phosphites, and phosphorus halides, particularly phosphorus trichloride, have all been used to deoxygenate *N*-oxides. Although the last reagent is not an organophosphorus species it will be dealt with here because of its great importance as a reagent for carrying out the reaction and because it occupies a key role in the conceptual and historical development of the method.

The ease of deoxygenation of *N*-oxides by phosphorus trichloride is very variable; the general pattern of reactivity is that reaction is retarded by electron withdrawal from the reacting centre in the heterocyclic system and facilitated by electron supply to the *N*-oxide function. For example, Emerson and Rees[2] found, in a kinetic study, that whereas 3-nitropyridine-1-oxide was deoxygenated at a rate so fast as to defy measurement, even at 25°C, the 4-nitro isomer was amongst the least reactive of *N*-oxides.

The reaction is inhibited by the presence of free hydrogen chloride in the reagents, which presumably serves to protonate the *N*-oxide oxygen.[2] As a consequence the reaction rate is increased in the presence of a stronger base than the *N*-oxide. The weakly nucleophilic 2,6-lutidine has been shown to be suitable in this role,[2] although there has been little interest in its use in synthetic applications of the reaction. No doubt the anticipated difficulty of separating the product amine from 2,6-lutidine is a factor here.

The deoxygenation fails with 3,4-benzocinnoline-5-oxide and also with the structurally similar azoxybenzene,[2] possibly owing to the ability of both materials to form dimers, which will lack a free N-oxide function. This may not be the sole reason, however, since the monomeric 3,4-benzocinnoline-5,6-dioxide is also inert to reaction with phosphorus trichloride.[2]

The main complication likely to be encountered in carrying out the deoxygenation of heterocyclic N-oxides with phosphorus trichloride is the tendency for nitro and hydroxy substituents to be replaced by chlorine. This has been noted particularly in the case of 4-nitro- and 4-hydroxy-quinoline-1-oxides.[3] The former gives 4-chloroquinoline as the only product of reaction with phosphorus trichloride, and the latter N-oxide gives roughly equal quantities of 4-chloro- and 4-hydroxy-quinoline. This problem does not seem to arise in simpler systems such as the pyridine series.

Other functionality present in the N-oxide which might be expected to react with phosphorus trichloride will do so. Hence carboxylate is converted into acid chloride,[4] for example. Complications of this type are best circumvented by a change in reagent to a phosphine or to a phosphite.

Substitution of the halogen atoms in phosphorus trichloride by organic groups generally leads to reagents that are less reactive. The ease of reduction of pyridine-N-oxide by phosphorus(III) reagents decreases in the order shown in Scheme 1.[5]

$$PCl_3 > PhPCl_2 > Ph_2PCl \gg (PhO)_3P$$
$$> (EtO)_3P \gg Ph_3P > Bu_3^nP > Et_2PPh$$

Scheme 1

Together with the observation quoted earlier,[2] that the reaction is retarded by electron withdrawal from the N-oxide function, the order of reactivity shown in Scheme 1 implies a mechanism involving electrophilic attack by the phosphorus reagent at the N-oxide.

In spite of the generally observed order of reactivity, however, it has been repeatedly found that triethyl phosphite can be used to carry out deoxygenations that are not possible with phosphorus trichloride. For example 3,4-benzocinnoline-5,6-dioxide and azoxybenzene are both deoxygenated in greater than 80% yield by triethyl phosphite.[6] The ineffectiveness of the more electrophilic phosphorus halide in this deoxygenation and the success achieved with the more nucleophilic phosphite may imply a different mechanism for deoxygenations by phosphites, as compared to phosphorus trichloride, in some cases at least.

7 Conversions via Deoxygenation

As in the deoxygenations of heterocyclic N-oxides by phosphorus trichloride, the triethyl phosphite reactions may, although only rarely, be complicated by the displacement of labile substituents, this time by nucleophilic attack by phosphorus. Hence from 2-nitropyridine-1-oxide, diethyl 2-pyridylphosphonate (35%) is obtained. The 4-nitro isomer of the N-oxide is, however, apparently insufficiently activated to nucleophilic attack to undergo the substitution reaction.[7]

Triethyl phosphite is also reported to convert heterocyclic N-oxides into the corresponding amines by a free-radical chain reaction which can be initiated by solvent-derived peroxides.[8] This reaction does not seem to have any advantage as a synthetic method and has not been applied as such.

Triphenylphosphine is widely used to deoxygenate heterocyclic N-oxides.[9] The reaction appears to be similar in scope and in yield to deoxygenation with phosphorus trichloride although more severe reaction conditions are required with the phosphine, and this may lead to complications. For example, reaction with triphenylphosphine fails to produce the desired product from 4-nitropyridine-1-oxide, the starting material being decomposed presumably as a result of attack of the phosphine at the nitro group under the relatively severe reaction conditions. Deoxygenation under photolytic conditions is a recent useful development.[159]

1.1.2. Selective monodeoxygenation of heterocyclic di-N-oxides

Most of the information available on this topic is derived from work on quinoxaline-1,4-dioxides. Phosphorus trichloride converts quinoxaline-2-carboxy-N-methylanilide-1,4-dioxide (1a) (Scheme 2) into the 1-monoxide in 80% yield and the second oxygen is unaffected.[10] Similarly methyl and ethyl quinoxalin-2-ylcarbamate-1,2-dioxide (1b) undergo selective deoxygenation to the 1-oxide.[11]

(1a) R^1 = CON(CH$_3$)(Ph) ; R^2 = H
(1b) R^1 = NHCO$_2$CH$_3$ or NHCO$_2$C$_2$H$_5$; R^2 = H
(1c) R^1 = CO$_2$CH$_3$; R^2 = CH$_3$
(1d) R^1 = COCH$_3$; R^2 = CH$_3$

(1)   Scheme 2

The origin of the selectivity may be steric hindrance to attack at the position adjacent to the substituent.[10] Consistent with this view is the loss of regiospecificity on deoxygenation of 2,3-disubstituted quinoxaline dioxides by phosphorus trichloride. For example, 2-methoxycarbonyl (1c) and 2-acetyl-3-methylquinoxaline-1,4-dioxide (1d) both give mixtures of the two possible mono-N-oxides together with substantial yields of the fully deoxygenated parent quinoxalines.[12] Selectivity is restored, however, when trimethyl phosphite in boiling propanol is used as the reagent,[12] although, owing to transesterification, there must now be some question as to the precise nature of the phosphorus ester which is the deoxygenating agent. Under these conditions (1c) and (1d) are selectively deoxygenated at the N-oxide function adjacent to the electron-withdrawing 2-substituent giving the 4-oxide in quantitative yield. The authors of this work suggest that it is indeed the electronic character of the substituent that determines the regioselectivity of the process, and in support of this view they report that 2-trifluoromethyl-3-methylquinoxaline-1,4-dioxide undergoes deoxygenation specifically at the 1-oxygen.

Quinoxaline di-N-oxides that lack an electron-withdrawing substituent adjacent to an N-oxide function do not react with trimethyl phosphite under these conditions.[12]

The full scope of this reaction has not yet been explored but it is known to be less straightforward if the electron-withdrawing substituent is itself reducible. Hence quinoxaline-1,4-dioxide-2-aldehyde (1; $R^1 = CHO$, $R^2 = H$) is reduced to the corresponding carbinol (1; $R^1 = CH_2OH$, $R^2 = H$), the N-oxide functions remaining intact.[12] The expected quinoxaline-1-oxide-3-aldehyde can be obtained, however, by treatment of the carbinol with acid.[12]

1.1.3. Deoxygenation of furoxans and furazans

The deoxygenation of furoxans, as compared to the deoxygenation of other heterocyclic N-oxides, involves the complication that the reaction may in some circumstances proceed beyond the formation of a furazan, the latter being first ring-opened and then deoxygenated to a cyano compound.

Thus, whereas triethyl phosphite at 150–160°C converts diphenylfuroxan into the corresponding furazan in quantitative yield,[13] triphenyl phosphite at 270°C gives an 87% yield of benzonitrile.[14]

In practice the reaction is easily controlled as desired because, except in some special cases discussed below, the conditions required for the furoxan deoxygenation are insufficiently severe to achieve the cleavage of the furazan.

Grundmann[13] has studied the effect of substituents in the furoxan (2) on the ease of deoxygenation by a variety of phosphines and phosphites.

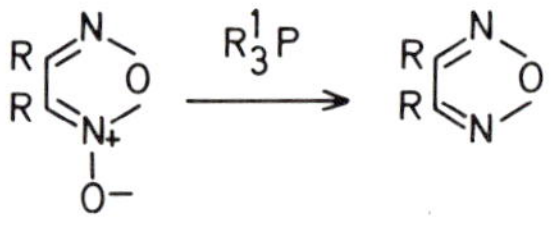

Scheme 3

The pattern that emerges is that the reaction is most facile when R is electron-withdrawing, and more difficult to achieve in the cases where R is an electron-supplying or an aryl group.

7 Conversions via Deoxygenation

There is no clear conclusion as to whether any particular phosphine or phosphite is the ideal reagent except to note that tributylphosphine is peculiarly effective in the deoxygenation of diphenylfuroxan, giving the furazan in 82% yield.

Simple fused benzofuroxans are also very readily reduced; 5-chlorobenzofurazan, for example, is formed in 60% yield on reaction of the corresponding furoxan with triethyl phosphite for 0·5 h in boiling ethanol.[15] Fusion to a more complex aromatic system, however, makes the furoxan less easily reducible, and phenanthrofuroxan is recovered unchanged after 7 h at 112°C with trimethyl phosphite.[16] The reaction can, however, be carried through in quantitative yield with the higher boiling triethyl phosphite at 158°C.[16]

If the furoxan is strained by fusion to a 5-membered ring it becomes impossible to stop the reaction at the furazan stage, and even under relatively mild conditions the reaction proceeds straight through to give a dicyano compound[16] (Scheme 4).

The evidence as to the mechanism of this reaction[16] favours a scheme whereby an unstable furazan, formed as a transient intermediate, opens spontaneously to its nitrile–nitrile oxide tautomer (Scheme 4), which is then

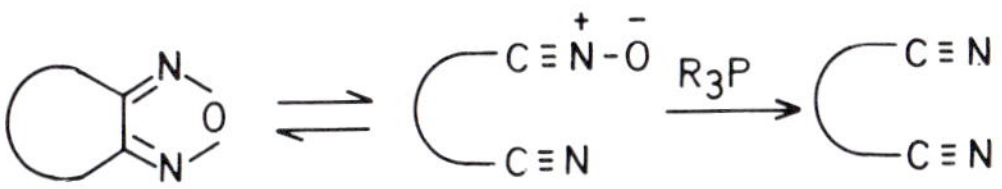

Scheme 4

further deoxygenated. Furazans fused to 5-membered rings are unknown as stable compounds.

Examples of furoxans that are known to react with trialkyl phosphites in this way, giving dinitriles directly under mild conditions, are the bornenofuroxan (3) and acenaphtho[1,2-c]furazan-1-oxide (4).[16]

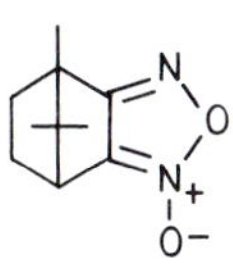

(3) 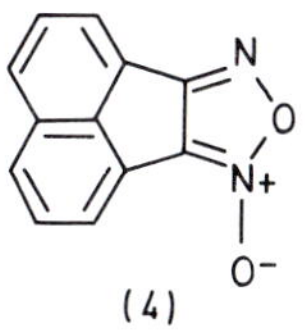   (4)

Scheme 5

As will be apparent from the foregoing discussion, relatively severe conditions are normally required for the ring-opening and deoxygenation of unstrained furazans. The reaction can, however, be carried out at room temperature under ultraviolet irradiation.[17] Thus benzofurazan is converted into (Z,Z)-1,4-dicyanobuta-1,3-diene in 80% yield on irradiation with a high-pressure mercury lamp in the presence of triethyl phosphite. Only small quantities of the other geometrical isomers of the diene are isolated. The authors suggest that the product is not in fact formed stereoselectively but the observed isomer is the most stable under the reaction conditions since its chromophore coincides with that of the benzene used as solvent and it is hence selectively protected from photochemical isomerization.

Naphtho- and phenanthro-furazans are similarly opened and deoxygenated under the same conditions.[17]

1.1.4. Deoxygenation of azoxy compounds

Neither phosphorus trichloride[2] nor triphenylphosphine[18] reacts with azoxybenzene, but both triethyl phosphite[6,19] and triethylphosphine[18] give azobenzene in nearly quantitative yield although elevated temperatures are required.

There has been no systematic attempt to rationalize the observed differing reactivities of phosphorus(III) reagents towards azoxybenzenes although there has been speculation as to the mechanism of the reaction.[19] In the absence of such a complete study it must merely be recorded that a trialkyl phosphite, most conveniently triethyl phosphite, is the reagent of first choice for deoxygenation of azoxy compounds and that temperatures in the region of 150°C are normally required.

1.1.5. Deoxygenation of nitrones

Aldonitrones are reported to be reduced by tertiary phosphines to the expected Schiff bases[18] although details of the experimental conditions and of the scope of the reaction have not been published.

Diphenylnitrone[20] is deoxygenated to a Schiff base on heating at 140–150°C with either triethyl phosphite or tris(dimethylamino)phosphine. A cleaner reaction is apparently obtained with the phosphite.

1.1.6. Deoxygenation of nitrile oxides

The cleavage and deoxygenation of furazans to give nitriles can, as has already been discussed, be regarded as proceeding via deoxygenation of their nitrile–nitrile oxide tautomers but simple nitrile oxides can also be deoxygenated by phosphites.[21]

The choice of phosphite is apparently unimportant from the point of view of carrying out the reaction, so the selection of the reagent can be dictated by the ease of removal of the unreacted phosphite, and of the phosphate formed, from the product. Triethyl or trimethyl phosphite is therefore recommended for the preparation of water-insoluble nitriles, whereas nitriles that can themselves be extracted with water are preferably formed using the water-insoluble triphenyl phosphites as the deoxygenating reagent.[21]

7 Conversions via Deoxygenation

1.2. Removal of Oxygen attached to Sulphur

1.2.1. *Deoxygenation of sulphones and sulphoxides*

Sulphones are generally difficult to reduce [22] and this trait is also observed in their inertness to deoxygenation by organophosphorus reagents. Sulphoxides, however, can be reduced to the corresponding sulphides by a suitable organophosphorus reagent. The best conditions for the reaction and the best choice of reagent depend upon the structure of the sulphoxide.

Triphenylphosphine will convert dimethyl sulphoxide into dimethyl sulphide only under very severe conditions; the reaction is very slow even at 225°C.[23] Triphenylphosphine is in fact the least reactive of the phosphorus(III) reagents studied in this context, and both phosphites and trialkylphosphines are more effective reducing agents for sulphoxides.[23] In contrast to the situation encountered in the deoxygenation of *N*-oxides, there is no easily described electronic effect of the groups on the phosphorus on the facility with which the reaction can be carried out with a particular organophosphorus reagent.

The reaction is, however, acid-catalysed and triphenylphosphine is capable of deoxygenating dimethyl sulphoxide at much lower temperatures than 225°C (*ca.* 100°C) in the presence of a trace of sulphuric acid or, preferably, in glacial acetic acid as solvent.[24] The reaction involves attack of the triphenylphosphine on the conjugate acid of the sulphoxide, and consequently the choice of an effective catalyst depends upon the basicity of the sulphoxide. Hence acetic acid is not effective as a catalyst in the deoxygenation of diaryl sulphoxides by triphenylphosphine, and, whereas sulphuric acid can be used, the Lewis acid boron trifluoride is preferred since it avoids possible complications arising from nuclear sulphonation of reactive substrates. Since it forms a complex with triphenylphosphine oxide, the boron trifluoride must be added in equimolar rather than in catalytic amounts.[24] The reaction is still carried out in glacial acetic acid, usually at the boiling point, and gives rapid conversion. Yields of the diaryl sulphides are good except from the more weakly basic diaryl sulphoxides.

Thus 4,4'-dinitrodiphenyl sulphoxide does not react at all and 4,4'-dichlorodiphenyl sulphoxide gives the corresponding sulphide in only 24% yield. Di-*p*-tolyl sulphide, on the other hand, is formed from the corresponding sulphoxide in 75% yield.

The reaction has also been applied to heterocyclic systems; for example phenoxythiin-10-oxide is deoxygenated by triphenylphosphine in 68% yield.

The presence of the Lewis acid catalyst in the system can lead to complications, since it may complex with the electron pairs of substituent groups on the benzene rings of substituted diaryl sulphoxides and thus reduce the ability of such groups to donate electrons to the π-system. This will cause a drop in the effective basicity and hence a drop in the reactivity of the sulphoxide function.

An effect of this type almost certainly explains the observation that 4,4'-dimethoxydiphenyl sulphoxide is substantially less reactive than would be predicted from the known electron-donating ability of methoxy.

If the reaction of triphenylphosphine and dimethyl or a diaryl sulphoxide is carried out in boiling carbon tetrachloride rather than in acetic acid the corresponding sulphides are obtained in almost quantitative yield, even from 4,4'-dinitrodiphenyl sulphoxide, and no acid catalyst is required.[25] The carbon tetrachloride is actively involved in the reaction and the effective deoxygenating reagent is probably a species such as triphenylphosphine dichloromethylide derived from reaction of the solvent with triphenylphosphine. This reaction appears to be preferable in most instances to the acid-catalysed deoxygenation of diaryl sulphoxides by triphenylphosphine, being independent of the basicity of the substrate and consequently of wider applicability; in addition higher yields are generally obtained.

The applicability of the deoxygenations discussed so far in this section to aliphatic sulphoxides has not been studied except in the instances referred to with dimethyl sulphoxide. There seems no reason to suppose, however, that dimethyl sulphoxide is unique amongst the dialkyl compounds.

There has recently, however, been published a report[26] that a mixture of triphenylphosphine, iodine, and sodium iodide can effect the deoxygenation of a wide spectrum of sulphoxides to sulphides in virtually quantitative yield. The reaction is carried out in acetonitrile at reflux.

If the early promise shown by this reaction is continued it seems destined to become the method of choice for the conversion of sulphoxides into sulphides. The active deoxygenating reagent is believed to be triphenylphosphine diiodide generated *in situ*.

Dialkyl, cycloalkyl, and aralkyl sulphoxides are deoxygenated in a matter of minutes whereas reaction times of 1 hour are required for the deoxygenation of diaryl sulphoxides.

No systematic study of substituent effects in the diaryl sulphoxides is presented but diphenyl and di-*p*-tolyl sulphoxide are both of equal reactivity.[26]

Triphenyl phosphite deoxygenates methyl phenyl sulphoxide to give thioanisole.[27] The reaction is slow, taking 2–3 days at 110°C, but the yield is quantitative. This reaction is apparently general for aryl methyl sulphoxides and is facilitated by electron-supplying substituents in the aryl group and retarded by electron withdrawal. This indicates a process involving electrophilic attack at the sulphoxide. In accord with this suggestion tris-(4-chlorophenyl) phosphite reacts more rapidly than does triphenyl phosphite itself.

Triaryl phosphites cannot be successfully replaced by trialkyl phosphites in this reaction. Trimethyl and triethyl phosphite do not reduce the sulphoxide but rather the latter acts as a catalyst for rearrangement of the phosphites to the isomeric phosphonates.[27]

2-Phenoxy-1,3,2-benzodioxaphosph(III)ole[28] (5a) and the closely related 2-chloro-1,3,2-benzodioxaphosph(III)ole[29] (5b) are both capable of deoxygenating sulphoxides under very mild conditions.

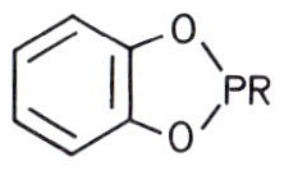

(5a) R = PhO
(5b) R = Cl
Scheme 6

The 2-phenoxyphosphole (5a) is used in carbon tetrachloride solution. The solvent apparently takes no part in the reaction in this case. The reaction is catalysed by a trace of iodine and is complete in less than 5 minutes at room temperature with sulphoxides that are soluble in carbon tetrachloride. With less soluble sulphoxides heating to 60–70°C for 0·25–0·5 h may be necessary to complete the reaction, which gives yields of sulphides in the region of 90% for all reported cases. These encompass dialkyl, diaryl, dibenzyl, arylbenzyl, and heterocyclic sulphoxides.

The 2-chlorophosphole (5b) has also been shown[29] to give excellent yields of sulphides from representatives of these same classes of sulphoxides, and it has the advantage over the 2-phenoxy compound that it can be used in benzene solution, a better solvent than carbon tetrachloride for most sulphoxides, and hence heating is generally unnecessary. The iodine catalyst can also be dispensed with but pyridine is added in equimolar amount. One additional advantage of the use of the 2-chlorophosphole is that, since (5a) must be prepared via (5b), the use of the latter directly represents an increased convenience at the stage of preparation of the reagent.

1.2.2. *Deoxygenation of other oxidized sulphur compounds*

Although, as has already been mentioned, sulphones are not deoxygenated by organophosphorus reagents, some simple derivatives of aryl sulphonic acids are known to be converted into thiols by triphenyl- or triethyl-phosphine.[30] For example, both benzene- and *p*-toluene-sulphonyl halides give the corresponding thiophenols in a rapid exothermic reaction with the phosphine at room temperature, followed by aqueous work-up. The yield is *ca.* 50% and there are several other products, notably the diaryl sulphide, *ca.* 35%.

Benzenesulphinic acid is deoxygenated by triphenylphosphine in virtually quantitative yield to give thiophenol.[30]

The deoxygenation of thiolsulphinates has been investigated.[31] The reaction proceeds in quantitative yield with triphenylphosphine. It has been applied to the preparation of both diaryl and dialkyl sulphides, although rather more severe conditions and longer reaction times are required for conversion of the dialkyl sulphinates than are necessary in the diaryl cases.

Esters of sulphenic acids on reaction with phosphorus(III) reagents undergo exclusive desulphurization to give ethers; none of the deoxygenation product, the sulphide, is observed.[32]

Sulphenic acids themselves, very few stable examples of which are known, have not been deoxygenated as such, but the reactions of trimethyl phosphite with penicillin sulphoxides may proceed via deoxygenation of the ring-opened sulphenic acid tautomer to give, in the first instance, a thiol.[33]

1.3. Deoxygenation of Carbonyl Compounds

1.3.1. *Principles*

Reaction of a carbonyl compound with a tervalent organophosphorus reagent appears to occur by attack at the carbonyl oxygen, although it has been argued in the past that nucleophilic attack at the positive end of the carbonyl dipole may occur.[35] Suffice it to say that in the many cases where the adducts of carbonyl compounds and phosphorus(III) reagents are isolable they invariably have phosphorus–oxygen bonds. If initial attack *is* at carbon, therefore, rearrangement must follow.

The final product of reaction of a phosphorus(III) reagent with a carbonyl compound is often a stable 2:1 adduct (6), a dioxaphospholan, but it is normally supposed that the dipolar 1:1 adduct (7), which is not isolable, is formed as an intermediate.

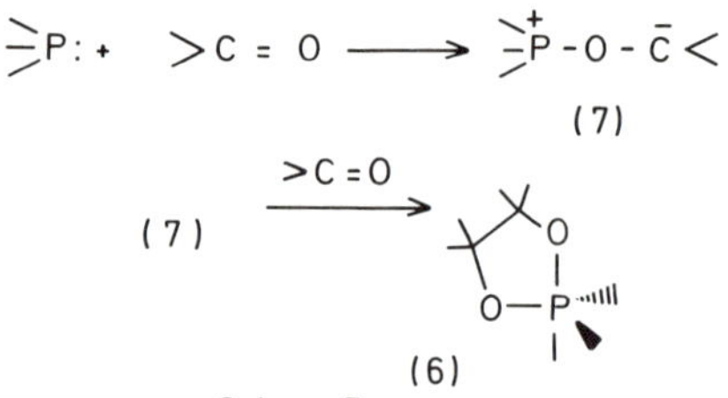

Scheme 7

In many cases the dioxaphospholan is isolable, and these compounds have assumed great theoretical importance in organophosphorus chemistry.[36]

There are, however, many synthetically useful reactions of carbonyl compounds with organophosphorus reagents in which either the 2:1 adducts are not detectable, although their intermediacy is often assumed, or the severity of the reaction conditions causes the immediate decomposition of (6) to useful, non-phosphorus-containing, products.

Although the intermediacy of the 1:1 adduct (7) is only normally inferred the relative reactivities of different carbonyl compounds with phosphorus(III) species can be rationalized on the basis of the expected stability of the dipolar adducts of the type of (7) which they may be supposed to form. What this means in practice is that the only carbonyl compounds that are likely to react with phosphorus(III) reagents are those that, after formation of (7), can provide features capable of stabilizing the carbanion centre.[37] Consequently simple aliphatic ketones and aldehydes do not normally react, nor do aromatic aldehydes and ketones except when they carry electron-withdrawing substituents.

Reaction of benzophenone itself with triisopropyl phosphite[38] appears to be an exception to this rule, but phosphonates are the principal products of the reaction and no synthetically useful yields of non-phosphorus-containing compounds are formed.

Those ketones where the carbanion centre, when formed, is incorporated in a cyclopentadiene system, e.g. cyclopentadienone and fluorenone, are particularly reactive, as are compounds such as phthalic anhydride that provide special features capable of stabilizing a negative charge on carbon.

Although the normal further reaction of the 1:1 adduct is with a second molecule of the carbonyl compound to form the dioxaphospholan,[37] there are sometimes products that may arise directly from the 1:1 adduct itself.

One can also write a mechanism for release of a carbene from (7) (Scheme 8), but there is little experimental confirmation from trapping experiments of the intermediacy of carbenes in the reactions of phosphorus(III) reagents with carbonyl compounds.

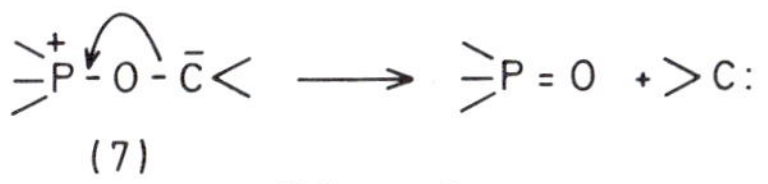

$$\overset{+}{\underset{\diagup}{\supset}}\text{P}-\overset{\frown}{\text{O}}-\bar{\text{C}}{<} \longrightarrow \underset{\diagup}{\supset}\text{P}=\text{O} + {>}\text{C}{:}$$
(7)

Scheme 8

In some cases reactions occur that can be interpreted as Wittig type processes involving ylides (8). Such ylides can be supposed to be formed by the trapping of a carbene by the organophosphorus reagent[39,40] (Scheme 9), and it can be argued that the efficiency of such trapping might preclude the detection of carbenes by their interception with *added* carbene traps.

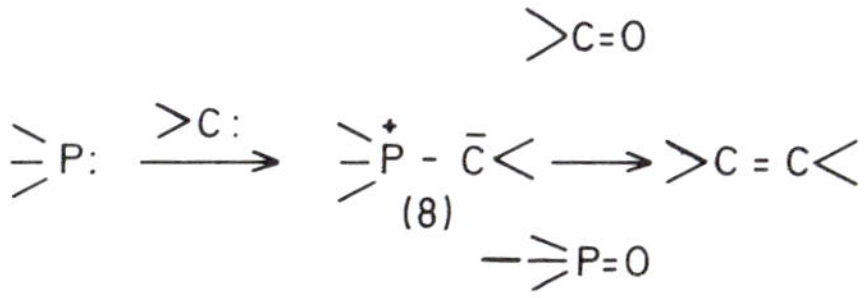

$$\underset{\diagup}{\supset}\text{P:} \xrightarrow{>\text{C:}} \underset{\diagup}{\overset{+}{\supset}}\text{P}-\bar{\text{C}}{<} \xrightarrow[-\underset{\diagup}{\supset}\text{P}=\text{O}]{>\text{C}=\text{O}} {>}\text{C}=\text{C}{<}$$
(8)

Scheme 9

The decomposition of the dioxaphospholans, the 2:1 adducts (6), can be understood in terms of two possible routes (Scheme 10). Either a molecule of the oxidized phosphorus reagent is extruded in a unimolecular process, leaving an alkene epoxide, or, alternatively, a bimolecular reaction occurs between the dioxaphospholan and a further molecule of the phosphorus reagent resulting in an alkene as product. The alkene may also arise via deoxygenation of a first formed, but undetected, epoxide.

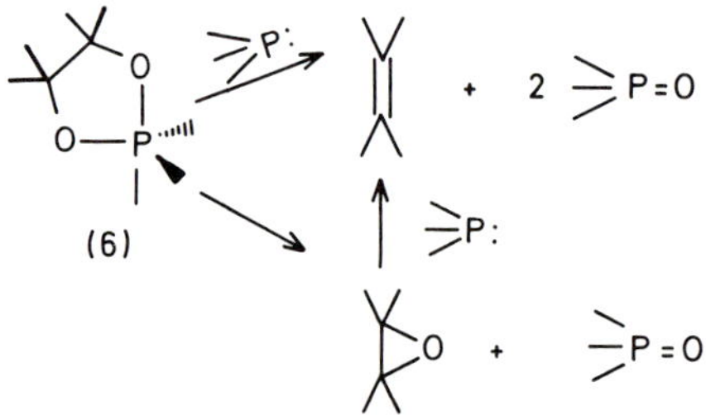

Scheme 10

The precise details and the nature of the products from these reactions are, as will be appreciated from the discussion of the specific cases that follows, very dependent upon the structure of the carbonyl compound under study and upon the nature of the phosphorus(III) reagent employed.

Much of the work on reactions of phosphorus(III) species with carbonyl compounds has been directed at the preparation of dioxaphospholans, and this approach has not always lent itself to the accumulation of data ideally suited to throw light on the scope of such reactions as routes to non-phosphorus-containing products.

It should also be emphasized at this point that the foregoing outline of the possible pathways of reaction of carbonyl compounds with phosphorus(III) reagents is in no way intended to be a critical review of the evidence on this subject, but is merely intended to serve as a framework on which to build the discussion of specific cases that follows. More importantly it is hoped that, together with these specific examples, the foregoing discussion will provide a background against which the potential user may be able to assess the possible outcomes and feasibility of untried reactions of this type.

1.3.2. *Synthesis of epoxides from aromatic aldehydes*

Reaction of some aromatic aldehydes with tris(dialkylamino)phosphines, at room temperature, gives good yields of stilbene oxides.[25,41] Formally two molecules of the aldehyde condense with the elimination of only one of the carbonyl oxygen atoms.

The reaction may involve the intermediacy of a dioxaphospholan but with tris(dimethylamino)phosphine, the most common reagent for carrying out the conversion, no such intermediate is detectable. With some cyclic aminophosphines, however, at low temperature, Ramirez et al.[34] have isolated stable 2:1 adducts with nitrobenzaldehydes and show that these adducts give dinitrostilbene oxides on warming.

Only benzaldehydes with electron-withdrawing substituents undergo this reaction, and an *ortho* disposition of the substituent seems to be particularly advantageous. Yields of the stilbene oxides are of the order of 80–95%. This substituent effect on the reaction is consistent with the requirement, outlined in the introduction, for the formation of an adduct between the carbonyl compound and the phosphorus(III) reagent, in which the carbanion centre can be stabilized.

The stereochemical course of the reaction is a classic pattern of reactivity versus selectivity. A very reactive aldehyde such as 2-chlorobenzaldehyde, on reaction with tris(dimethylamino)phosphine, gives an 80% yield of 2,2'-dichlorostilbene oxide, with no apparent stereoselectivity; equal amounts of the two geometrical isomers of the epoxide are present. With the less reactive 3-bromobenzaldehyde, however, the yield of the corresponding stilbene oxide is only *ca.* 50% and the product is a 3:1 mixture of (E)- and (Z)-isomers respectively.

This exemplifies the general pattern, that any factor reducing the reactivity of the aldehyde leads to increased stereoselectivity, which generally favours formation of the presumably more stable (E)-stilbene oxide. Apart from the observation already made as regards the relationship between the reactivity of the aldehyde and the presence of an electron-withdrawing group, it is reported that the reactivity of the aldehyde is lowered, and hence the predominance of the (E)-epoxide in the products is enhanced, by bulky *ortho* and *meta* substituents,[35] and by carrying out the reaction at low temperature, although this resort is only practicable when an acceptable reaction rate is still maintained on cooling.[34,35]

The method is not entirely confined to conversion of benzaldehydes. 1,2-Di-(2-pyridyl)- and 1,2-di-(2-thienyl)-ethylene oxide have also been obtained from the appropriate heterocyclic aldehydes, although the thienyl compound in particular is not formed in especially good yield.[35,41] These two examples, however, illustrate the value of this method in giving access to disubstituted ethylene oxides with oxidation-sensitive substituents. Such materials are not obtainable by the normal route, from the alkene, on account of the strong oxidants necessarily employed.

The differing reactivities of aldehydes in this reaction enable some "mixed" 1,2-disubstituted ethylene oxides to be prepared. Hence the highly reactive 2-chlorobenzaldehyde and an excess of 3,4-methylenedioxybenzaldehyde give, on reaction with tris(dimethylamino)phosphine, the product (9); 1-(2-chlorophenyl)-2-(2-furyl)ethylene oxide (10) has been prepared in a similar manner.[35]

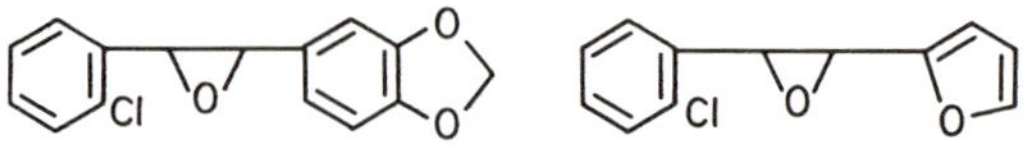

(9)   (10)

Scheme 11

A reaction that complements the tris(dialkylamino)phosphine-induced condensation of substituted benzaldehydes, in that it converts benzaldehyde itself into stilbene oxide, has been reported.[42] The deoxygenating agent is the anion of diphenylphosphine oxide generated *in situ* by sodium hydride. With excess of benzaldehyde at 180°C, stilbene oxide is obtained in *ca.* 90% yield as an equimolar mixture of the (*E*)- and (*Z*)-isomers. Since the reaction conditions are rather severe, if an excess of benzaldehyde is not employed the epoxide is further deoxygenated and (*E*)-stilbene is the sole product; (*Z*)-stilbene is isomerized to the more stable (*E*)-isomer under the reaction conditions.

1.3.3. *Synthesis of epoxides from other aroyl derivatives*

Although the reaction of benzaldehydes discussed in the preceding section is the most extensively studied epoxide synthesis of this type, aroyl derivatives, other than aldehydes, have been used in the preparation of particular epoxides.

Benzoyl cyanide reacts with triethyl phosphite in boiling benzene to give a mixture of equal amounts of the (*E*)- and (*Z*)-isomers of 1,2-dicyanostilbene oxide in *ca.* 50% yield.[43] Mukaiyama and coworkers have carried out a similar conversion[44] under milder conditions and isolated a phosphorane intermediate which was thermolysed in a separate step to give one geometrical isomer of 1,2-dicyanostilbene oxide. This is claimed to be the (*E*)-isomer, although this assignment has been questioned.[43]

An intramolecular version of the reaction of benzoyl cyanides with phosphorus(III) reagents has been carried out[45] to prepare 9,10-dicyanophenanthrene-9,10-oxide (12) from 2,2'-dicyanoformylbiphenyl (11) by reaction with triethyl phosphite.

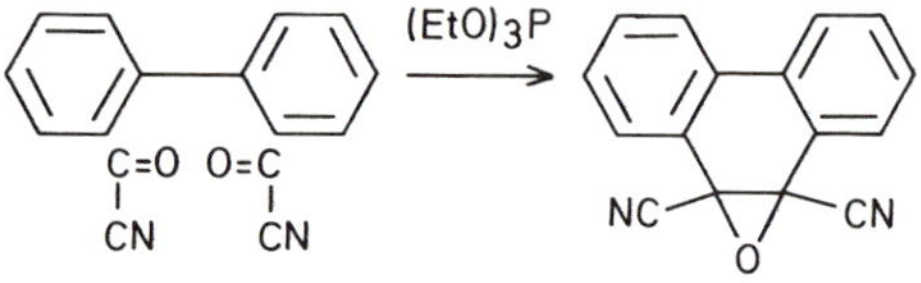

(11)   (12)

Scheme 12

The methyl esters of arylglyoxylic acids also react with triethyl phosphite or, preferably, with tris(dimethylamino)phosphine, to give 1,2-dimethoxycarbonylstilbene oxides.[46] The yields are variable depending on the substituents in the aryl group, and the reaction gives only the (Z)-isomer product. The general pattern of dependence on the substituents in the aryl group is that yields are increased by electron withdrawal, particularly from the *para* position, and decreased by electron supply. Indeed a *p*-methyl substituent suppresses the yield of the stilbene oxide to a preparatively unattractive 2%.

It is interesting to note, at this point, that in the reactions of aroyl derivatives discussed in this section an electron-withdrawing substituent in the aryl moiety, although helpful, is not essential. This is to be contrasted with the absolute requirement for such a substituent in the case of reactions of benzaldehydes. Presumably the presence of a cyano or ester group directly attached to the carbonyl carbon of the aroyl moiety provides sufficient electron withdrawal to activate the carbonyl to attack by a phosphorus(III) reagent.

1.3.4. Intramolecular condensations of aryl aldehydes: synthesis of aromatic epoxides

In 1964 Newman and Blum[47] found that an intramolecular version of the tris(dimethylamino)phosphine-induced condensation of benzaldehydes, to give stilbene oxides, could be carried out. Thus 2,2'-diformylbiphenyl reacts in a few minutes at room temperature to give phenanthrene-9,10-oxide in 89% yield. The reaction is at least as facile as the intermolecular version and, in contrast to the latter, no electron-withdrawing group in the aldehyde is required apart from the mutual effect of the two formyl functions.

2-(2-Formylphenyl)naphthalene-3-aldehyde (13) undergoes a similar reaction but heating to 55°C is required. There is considerable reduction in the yield of epoxide if a 4-methyl substituent is present in the anthracene nucleus (14), and the reaction appeared to fail completely with 1,4-dimethyl substitution (15). This is not a real substituent effect, however, since it has been shown that in the case of reaction of (15) the failure to observe a product arises not from poor reactivity of the starting material but from the instability of the product, and with care the expected 7,12-dimethylbenz[*a*]anthracene-5,6-oxide (16; $R^1 = R^2 = Me$) can be obtained in 75% yield.[48]

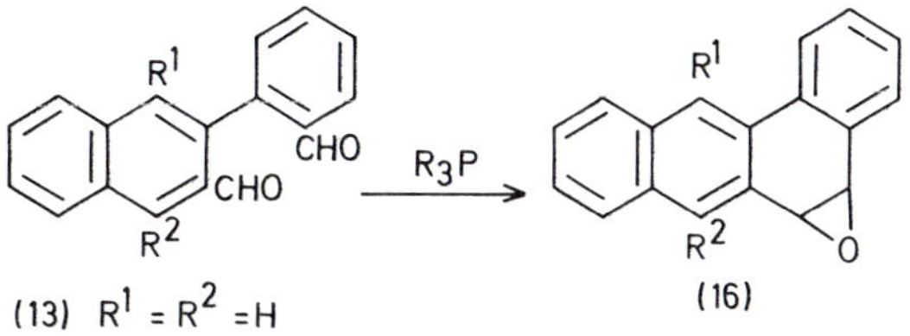

(13) $R^1 = R^2 = H$
(14) $R^1 = H; R^2 = CH_3$
(15) $R^1 = R^2 = CH_3$

Scheme 13

This reaction has now been applied to the synthesis of a variety of K-region aromatic oxides,[48,49,50] compounds of great importance because of their possible role in the carcinogenic activity of polycyclic aromatics. Since the dialdehyde starting materials are often obtained by selective oxidative cleavage of the parent hydrocarbon[48] at the K-region double bond, the sequence as a whole can be used to provide what amounts to a regiospecific epoxidation of the K-double bond.

The cyclization of a dialdehyde to an epoxide appears to have been mainly investigated for the preparation of 6-membered rings. Newman and Blum[47] in their original work, however, attempted to cyclize o-phthalaldehyde to benzocyclobutene epoxide but obtained only an intermolecular condensation product, a stilbene oxide. The strain in the product hoped for is apparently too formidable for the intramolecular process to compete.

1.3.5. *Synthesis of alkenes by reductive dimerization of carbonyl compounds*

It has already been established that simple aldehydes and ketones do not react with phosphines and phosphites unless the carbonyl compounds are activated by the presence of an electron-withdrawing group. At 200°C, however, benzaldehyde and 4-chlorobenzaldehyde are converted, in good yield, into stilbene and 4,4'-dichlorostilbene, respectively, on reaction with the anion of diphenylphosphine oxide.[51] The reaction probably involves the intermediacy of the stilbene oxide which is the product isolated under milder conditions.[42] The reaction has been extended to include 1-naphthaldehyde and anthracene-9-aldehyde as well as the ketone benzophenone, and yields of substituted ethylenes are good.[51]

Benzophenone is also reported to give tetraphenylethylene on reaction both with triisopropyl phosphite[38] and with tris(dimethylamino)phosphine[52] but not in preparatively useful yield.

In contrast, those ketones in which the carbonyl carbon is capable of stabilizing a carbanion react readily with phosphorus(III) reagents and undergo reductive dimerization.[53]

Thus fluorenone is converted into bifluorenylidene in 40% yield by tributylphosphine at 155–165°C, and tetracyclone gives octaphenylfulvalene (28%) at 100°C with the same reagent.[53]

Bifluorenylidene is also obtained from fluorenone on treatment with tris(dimethylamino)phosphine at 20°C,[54] in a reaction interpreted by Ramirez[54] as proceeding via the 1:1 adduct of the phosphine and the carbonyl compound. The intermediacy of fluorenylidene in this reaction has been inferred but attempts to trap such a carbene have been unsuccessful.[53]

Reaction of fluorenone with triethyl[53] or triisopropyl phosphite[38] does not give bifluorenylidene; 9-biphenylenephenanthrone (17) (74%) is obtained instead.

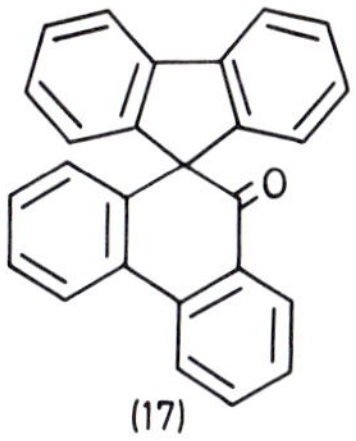

7 Conversions via Deoxygenation

(17)

Scheme 14

This product (17) is believed[54] to arise by pinacol–pinacolone rearrangement of a dioxaphospholan, the 2:1 adduct of fluorenone and the phosphite. Thus, with fluorenone, differing products are obtained from reaction with phosphines as opposed to that with phosphites because the former react via 1:1 adducts and the latter via the 2:1 adducts, the dioxaphospholans.

A ring expansion also occurs on reaction of acenaphthaquinone with trimethyl phosphite,[55] the product being the lactone (18) in 90% yield.

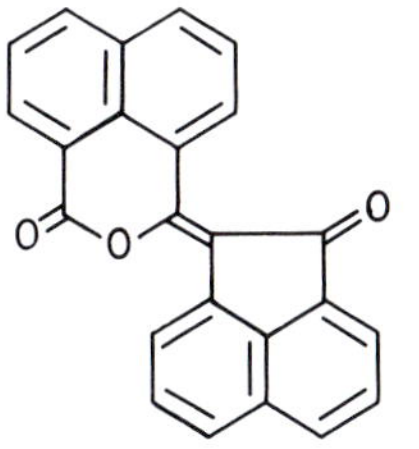

(18)

Scheme 15

In this case a definite intermediate 2:1 adduct of the phosphite and the quinone is isolated and decomposed in a separate step to give the product. In contrast, with 2-N-pyrrolidino-1,3-dimethyl-1,3,2-diazaphospholan as the deoxygenating reagent, acenaphthaquinone gives a 1:1 adduct, which may decompose to a carbene, but from which no synthetically useful products are forthcoming.[56]

Quinones in general do not give useful non-phosphorus-containing products on treatment with phosphorus(III) reagents. The general pattern that is observed is that *p*-quinones are converted into the alkyl ethers of *p*-quinol phosphate esters and *o*-quinones give stable dioxaphospholans.[57]

1.3.6. Reductive dimerization of dicarboxylic acid anhydrides

Phthalic anhydride reacts with trialkyl phosphites to give *trans*-3,3'-biphthalyl in up to 70% yield.[58] The reaction is claimed to proceed via a 1:1 adduct of the phosphite and the anhydride (Scheme 16). Such an adduct would be stabilized by delocalization of the carbanion centre (Scheme 16).

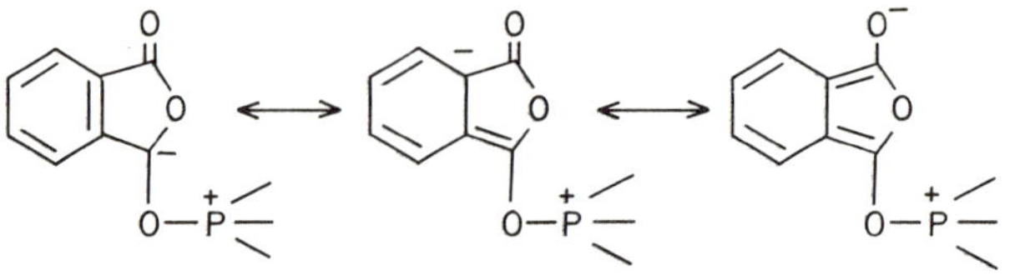

Scheme 16

If this view is correct, electron-withdrawing substituents in the phthalic anhydride should facilitate the reaction, and indeed the conversion of tetrachlorophthalic anhydride into trans-3,3'-octachlorobiphthalyl is much faster than the corresponding reaction of the unsubstituted anhydride although the yield of biphthalyl is lower (40%)[58] from the tetrachloro compound.

The best reagents for carrying out these reactions of phthalic anhydrides are triethyl and tributyl phosphite. This choice represents a balance of two factors. Firstly, the phosphite must be of high enough boiling point to allow an adequate reaction temperature to be used. Secondly, it appears that some component of the reaction mixture induces the rearrangement of trialkyl phosphites to dialkyl alkylphosphonates, and this reaction is more rapid with phosphites derived from the higher alcohols. Triethyl and tributyl phosphites do rearrange but only slowly, and if they are used in moderate excess efficient conversion of the anhydride is possible.

The reaction has been extended to the conversion of 2-thiophthalic anhydride[59,60] (19; X = S) into trans-3,3'-(2,2'-dithio)biphthalyl (21), but an attempt to deoxygenate N-substituted phthalimides was unsuccessful.[59] In fact, the thiophthalic anhydride is more reactive than the oxygen analogue and gives the biphthalyl in 82% yield.

Ramirez's original suggestion,[58] that the phosphite reagent attacks the phthalic anhydrides at the carbonyl oxygen, as shown in Scheme 17, has not necessarily been supported by other workers and attack at the carbonyl carbon has been postulated.[59] The route shown in Scheme 16 has also been quite widely suggested[59,60] but the intermediacy of a carbene in this reaction has never been conclusively demonstrated.

Reaction of 2-thiophthalic anhydride with tris(dimethylamino)phosphine, in contrast to its reaction with trialkyl phosphites, gives not trans-3,3'-(2,2'-dithio)biphthalyl (21; X = S) but the "mixed" product trans-3,3'-(2-thio)-biphthalyl (22) and, predominantly, the ring-expanded δ-lactone-δ-thio-lactone (23).[60] A tentative mechanism has been postulated for this unexpected conversion.[60]

Maleic anhydride is structurally related to phthalic anhydride but, with triphenylphosphine at least, reacts only by addition at the double bond to give triphenylphosphoranylidene succinic anhydride.[61] Substituted maleic

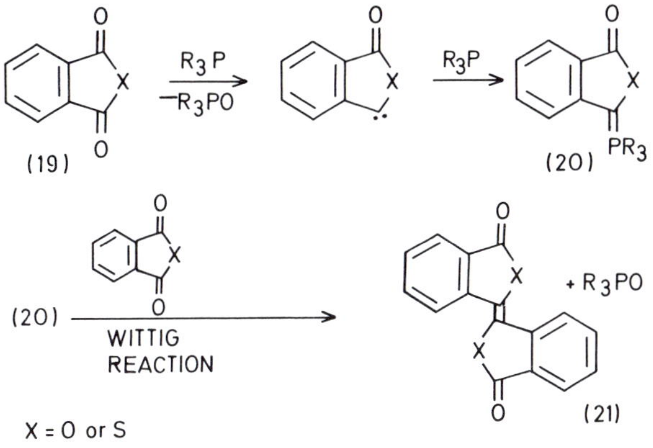

7 Conversions via Deoxygenation

Scheme 17

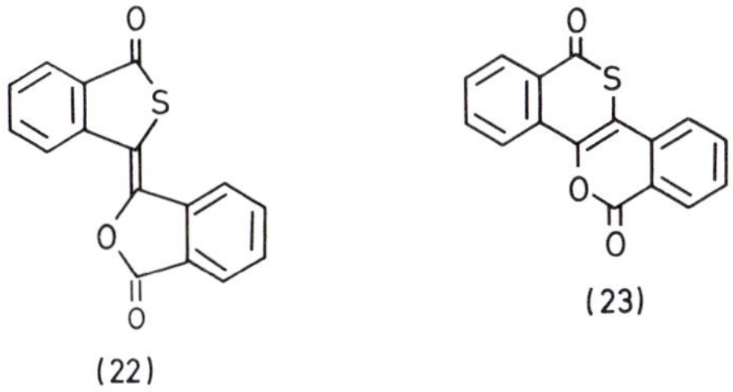

X = O or S

Scheme 18

anhydrides, however, react with triethyl phosphite to give the expected bifurandiones.[61]

Diphenylmaleic anhydride (24; R = Ph) gives (25; R = Ph) in 50% yield but with dimethylmaleic anhydride (24; R = Me) and with tetrahydrophthalic anhydride (24; R–R = –(CH$_2$)$_4$–) the yield of bifurandione is less than 20%. Dichloromaleic anhydride is reported[61] to react violently with triethyl phosphite but the products do not include bifurandiones.

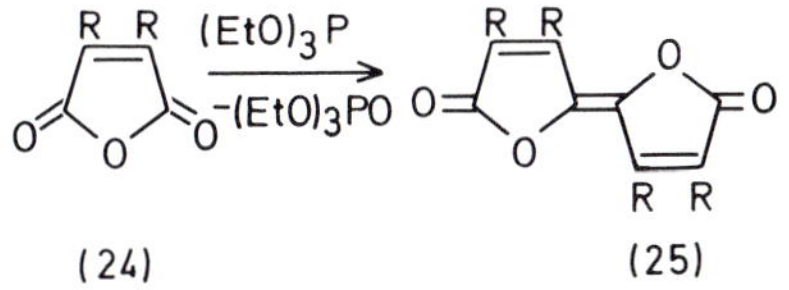

Scheme 19

The reductive co-condensation of diphenylmaleic anhydride and dimethylmaleic anhydride has also been carried out and an unsymmetrical bifurandione obtained in 22% yield.[61]

1.3.7. Miscellaneous reductive condensations of carbonyl compounds

The dithiocarbonate (26) is converted into $\Delta^{2,2'}$-bis-(4,5-dicyano-1,3-dithiolidene) (27) in quantitative yield by triethyl phosphite in boiling benzene.[62] No reaction is observed between (26) and either triphenyl- or tributyl-phosphine.

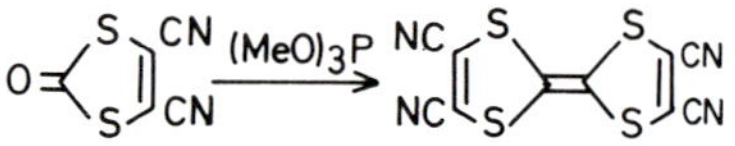

Scheme 20

As is the case with the reductive condensation of aroyl compounds to give epoxides, so with the formation of alkenes, an ester function can provide the electron withdrawal necessary to activate an adjacent carbonyl function to reaction with a phosphorus(III) reagent. Thus methyl pyruvate gives 2,4-dimethoxycarbonylbut-2-ene (38%) on reaction with tributylphosphine.[63] Both the (E)- and (Z)-isomers of the product are formed.

The relative reactivities of a series of α-keto-esters with phosphorus(III) reagents have been reported.[63] Mesoxalates are more reactive than glyoxylates which are in turn more reactive than pyruvates. Details of the non-phosphorus-containing products from these reactions are not given, however, except in the case of methyl pyruvate noted above. The reported order of reactivity of organophosphorus reagents with these compounds is shown in Scheme 21.

$$[(CH_3)_2N]_3P \geqslant (n\text{-}C_4H_9)_3P > (PhO)_3P > Ph_3P$$

Scheme 21

1.3.8. Conversion of aryl trifluoromethyl ketones into alkenes

On the basis of the statement, which has been frequently stressed in the foregoing discussion, that those carbonyl compounds carrying electron-withdrawing groups are particularly reactive towards phosphorus(III) reagents, it might be a reasonable assumption that aryl trifluoromethyl ketones would have high reactivity with tervalent organophosphorus species.

This is found to be the case[64] but the products, although alkenes, are not the symmetrical diarylbis(trifluoromethyl)alkenes that might be expected.

Trifluoromethyl phenyl ketone (28a) and trifluoromethyl p-tolyl ketone (28b) react with tributylphosphine to give a mixture of the (E)- and (Z)-isomers of the alkene (29). The (E)-isomer predominates.[64]

The reaction also occurs with triethylphosphine and may have some generality. A possible mechanism has been suggested[64] although it is not entirely in accord with the experimental results.

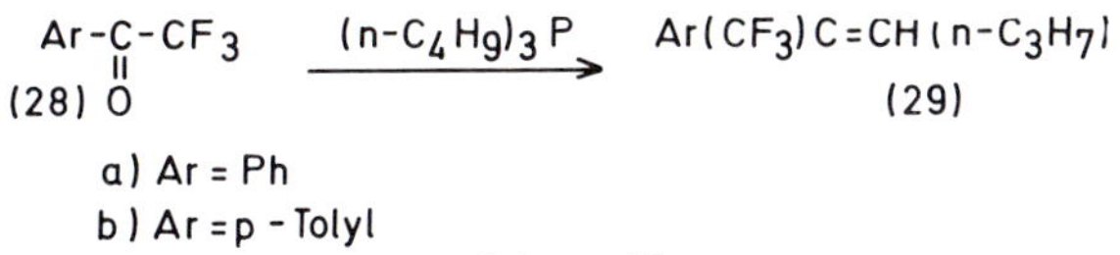

$$\text{Ar-}\underset{\underset{O}{\|}}{C}\text{-CF}_3 \xrightarrow{(n\text{-C}_4\text{H}_9)_3\text{P}} \text{Ar(CF}_3)\text{C=CH}(n\text{-C}_3\text{H}_7)$$
(28)   (29)

a) Ar = Ph
b) Ar = p-Tolyl

Scheme 22

Triphenylphosphine undergoes no reaction with trifluoromethyl phenyl ketone.[64]

1.3.9. *Deoxygenation of isocyanates*

Triethyl phosphite deoxygenates isocyanates to the corresponding isocyanides at 180°C. Conversion of starting material is not normally complete, even on prolonged reaction, and yields are often poor. Aryl isocyanates in particular give yields of isocyanide of 20% and below. Hexyl isocyanide is formed in 57% yield from hexyl isocyanate, and this must be counted as one of the most efficient of these reactions.[19]

If, rather than a phosphite, 2-phenyl-3-methyl-1,3,2-oxazaphospholan[65] is used as the reagent, the deoxygenation of aryl isocyanates can be carried out at ambient temperature and the yields are acceptable. For example, 4-chlorophenyl isocyanide and *o*-tolyl isocyanide are obtained in 70 and 51% yields respectively from the corresponding isocyanates.

Aliphatic isocyanates are also deoxygenated by 2-phenyl-3-methyl-1,3,2-oxazaphospholan in good yield.[65] Cyclohexyl isocyanide and ethyl isocyanide have been obtained in 74 and 65% yield respectively by this method.

1.3.10. *Deoxygenation of 1,2-dicarbonyl compounds: synthesis of alkynes*

1,2-Dicarbonyl compounds, on reaction with phosphorus(III) reagents, undergo a simple 1:1 addition to give the most stable examples known of dioxaphospholan systems[66] (Chapter 11), and severe treatment is normally required to induce the phosphorus heterocycle to react further, with a second molecule of the deoxygenating reagent, and thus undergo conversion into an acetylene (see Scheme 23).

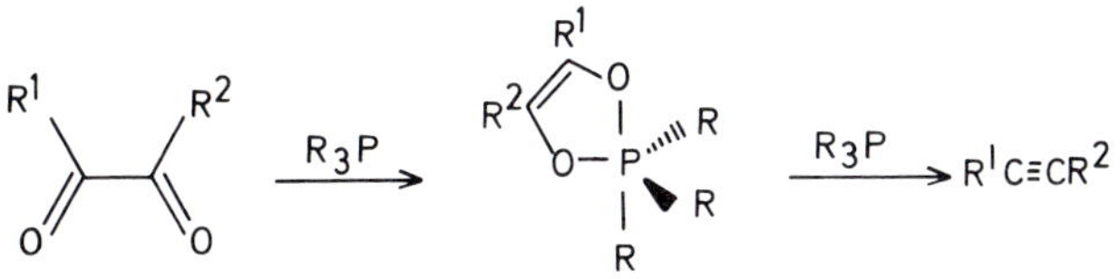

Scheme 23

Thus reaction at 215°C with triethyl phosphite is required to convert benzil into diphenylacetylene (60%).[67] The reaction has been extended to 4,4'-disubstituted benzils but yields of acetylenes are low, ranging from 25–30%

with 4,4′-dimethyl- and 4,4′-dimethoxy-benzil up to 43% from 4,4′-dichlorobenzil. 1-Phenylpropane-1,2-dione (acetylbenzoyl) gives methylphenylacetylene (41%) under the same reaction conditions, but the reaction cannot be successfully applied to the synthesis of acetylenes that have only alkyl substituents.[67]

As an alternative to the "one pot" processes so far described, the dioxaphospholan intermediates can be isolated, by moderating the reaction conditions, and then caused to react with more phosphorus reagent in a second, separate step.[67] Conversions of the phospholans into acetylenes are excellent but whether the two-step process is advantageous overall will depend upon the efficiency with which the phospholans can be obtained, and information on this point is not recorded.[67]

The reaction of triethyl phosphite with 1,2-dicarbonyl compounds has recently been extended to the preparation of bis-diphenylacetylenes from starting materials with two phenylglyoxalyl functions.[68]

1.3.11. *Other syntheses of alkynes*

Diphenylacetylene can be prepared by reductive condensation of diphenylketene in the presence of triethyl phosphite; the yield is 65%.[19]

Trippett and Walker[69] have reported the conversion of 1,2-diphenyl-2-chloroethan-1-one (30) into diphenylacetylene in 90% yield on treatment with triphenylphosphine. The authors suggest that the reaction proceeds via attack of the phosphine on the oxygen of the enol of (30) but point out that this is an unusual reaction of α-halogeno-ketones with phosphines; quaternization of the phosphine by displacement of halogen is more typical. It is not therefore likely that reductive eliminations of this type from α-halogenoketones will provide a general route to acetylenes.

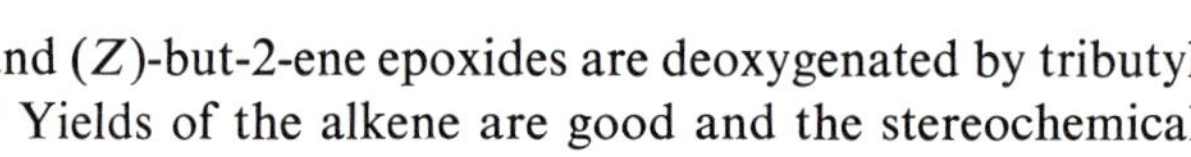

$$\text{PhCH-CPh} \rightleftharpoons \text{PhC=CPh} \xrightarrow{\text{Ph}_3\text{P}} \text{PhC}\equiv\text{CPh}$$
$$\quad |\quad\;\|\qquad\qquad\;\;|\quad\;|$$
$$\;\text{Cl}\quad\text{O}\qquad\qquad\;\;\text{Cl}\;\;\text{OH}$$
$$(30)$$

Scheme 24

1.4. Deoxygenation of Epoxides

Both (*E*)- and (*Z*)-but-2-ene epoxides are deoxygenated by tributylphosphine at 150°C.[70] Yields of the alkene are good and the stereochemical course of the reaction has been studied. The general pattern is one of inversion of stereochemistry, (*E*)-epoxides giving (*Z*)-alkenes and vice versa.

These stereochemical consequences of the reaction have been rationalized in terms of the intermediacy of a betaine (31), which undergoes rotation followed by *cis*-elimination of phosphine oxide (Scheme 25).

The reaction is not, however, entirely stereospecific, and (*E*)-but-2-ene

7 Conversions via Deoxygenation

(Scheme 25, showing structure (31))

epoxide gives a mixture of (E)- and (Z)-alkenes containing some 20% of the (E)-isomer. Similarly the product from the (Z)-epoxide is not solely the (E)-alkene but contains ca. 20% of the (E)-isomer.

Phenylglycidic esters are deoxygenated in good yield by triphenylphosphine[71,72,73] and by tributylphosphine[72,73] to give esters of cinnamic acid. The reaction can be carried out in boiling ethanol but is slow, taking at least 20 h. The neat materials heated at 125–170°C react in 0·5 h. The inclusion of hydroquinone in the reaction mixture is said[71] to be advantageous although not apparently essential. The role of the hydroquinone may be a dual one, that of acid catalyst and of polymerization inhibitor.

The stereochemical consequences of the deoxygenation of glycidates are complicated by the fact that (Z)-cinnamates, but not the (E)-isomers, are isomerized by both tributyl- and triphenyl-phosphines, but different rates of isomerization with the two phosphines are observed.[72,73] Tributylphosphine is the better catalyst for the isomerization. Deoxygenation of both (E)- and (Z)-glycidates leads therefore only to (E)-cinnamates if tributylphosphine is used as the reagent.

Triphenylphosphine gives almost entirely (97%) (E)-cinnamate from (Z)-glycidates, as would be expected on the basis of a mechanism of the type shown in Scheme 25, but the (E)-glycidate gives an almost equimolar mixture of (E)- and (Z)-cinnamates. This presumably reflects partial isomerization of the initially formed all-(Z) product.

Styrene oxide,[71] ethyl dimethylglycidate,[71] and stilbene oxides[73] have all been deoxygenated by phosphines. In the case of the stilbene oxides the stereochemical situation is not clear owing to isomerization of (Z)- but not of (E)-stilbene by the phosphines. Ethylene and propylene oxides are quantitatively deoxygenated to the corresponding alkenes by triethyl phosphite at 150°C under pressure.[74] The generality of this reaction, when applied to other epoxides, is in doubt in view of the report[75] that a poor yield of alkene is obtained from but-2-ene epoxide and triethyl phosphite at 200°C.

1.5. Deoxygenation of other Oxygen Heterocycles

3,3,4,4-Tetramethyl-1,2-dioxetan is converted into tetramethylethylene oxide in overall quantitative yield by a two-step process[76] involving treatment with triphenylphosphine at 6°C to give the product via a dioxaphospholan.

A similar two-step process,[77] this time via a betaine intermediate, has been used to reduce some aromatic endoxides to the parent systems.

1.6. Deoxygenation of Peroxy Compounds

Although the details of the mechanisms of reactions of peroxy compounds with tervalent organophosphorus reagents are in many cases still poorly delineated, the products of the synthetically useful deoxygenation reactions are generally predictable.

However, a cautious note must be sounded since peroxy compounds are typically thermally unstable, and reactions carried out at elevated temperatures may give products of homolysis of the peroxy compound without involvement of the phosphorus reagent.

The picture is further complicated by the fact that the initial homolysis products themselves can often react with the organophosphorus reagent. The phosphorus(III) deoxygenation of alkoxy radicals derived from dialkyl peroxides is well known.[78]

Thus there is always the possibility of a plurality of mechanism in reactions between peroxides and tervalent organophosphorus reagents, and one must read the older literature, in particular, with some caution.

For example, although Horner[79] originally reported that reaction of di-t-butyl peroxide with triphenylphosphine gave di-t-butyl ether, it was subsequently shown[80] that, at the temperature used (110–120°C) for the reaction the "product" was in fact a complex mixture derived from t-butoxyl and t-butyl radicals. The latter are presumably derived by deoxygenation of the former.

1.6.1. Deoxygenation of hydroperoxides

Hydroperoxides react rapidly and exothermically with tervalent organophosphorus reagents, even at room temperature, to give an alcohol.[79] Since the reaction is so facile there is little danger of a competing homolysis of the peroxide and the yields of alcohol are typically quantitative.

The organophosphorus reagents that have been most used for the deoxygenation of hydroperoxides are triphenylphosphine,[79,81] triphenyl phosphite,[82] and triethyl phosphite,[83] and the reaction appears to be equally facile with any of these.

The reaction solvent chosen seems unimportant, and a range of solvent polarities from aqueous ethanol through to pentane have been used with equal success.[81]

The deoxygenation is reported[84] to be catalysed by acids and by molybdenum and vanadium ions, at least in the case where the reagent is triphenyl-

7 Conversions via Deoxygenation   319

phosphine, but since the process is so facile it is not anticipated that a catalyst will normally be required in preparative applications of the reactions.

The stereochemical course of the reaction has been examined and the configuration at the carbon carrying the hydroperoxy function shown to be retained. Thus *trans*-9-decalyl hydroperoxide is converted stereospecifically into *trans*-decalol by triphenylphosphine,[81] and the same reagent gives 1-phenylethanol of retained configuration from optically active 1-phenylethyl hydroperoxide.[85]

The mechanism of the reaction cannot be regarded as fully established but evidence is recorded[81,84] supporting initial attack at the hydroxy oxygen of the hydroperoxide. Homolysis of the peroxide is not involved.

The main synthetic use of phosphorus(III) deoxygenation of hydroperoxides is as part of a hydroxylation sequence in which a hydrocarbon is first autoxidized and the resulting hydroperoxide is then converted into an alcohol by the phosphorus(III) reagent.[86] Both the autoxidation and the deoxygenation can be carried out at low temperatures, under which conditions phosphines and phosphites do not react directly with oxygen. Hence the hydroxylation can be carried out as a "one pot" process.[86] Thus 17α-hydroxypregnan-20-ones have been prepared by passing oxygen into a solution of the parent ketone containing a slight excess of trimethyl phosphite, at $-25°C$.[86]

The scope of these reactions is apparently limited only by the ease of the autoxidation step, but such "limitations" provide useful regiospecificity.

Only tertiary hydrogens that are α to a ketonic function are attacked, not those α to an ester carbonyl. The relative difficulty of enolization of the ester may be a factor here. There is also a steric effect; for example, again in the pregnan-20-one series, the presence of a 16α-methyl group prevents the introduction of a 17α-hydroxy function by an autoxidation-deoxygenation route.[86]

Another interesting application of this type of hydroxylation is in the preparation of α-hydroxy-diazo compounds (34) from hydrazones (32)[87] (Scheme 26). The reaction appears to be general and proceeds in good to moderate yield with both aldehyde and ketone alkylhydrazones.

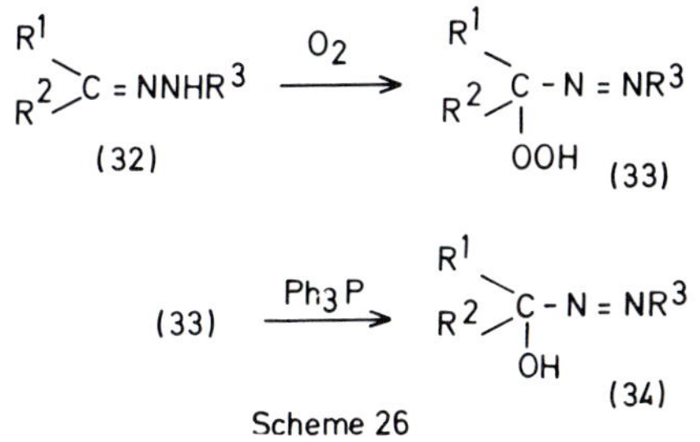

$$R^2R^1C=NNHR^3 \xrightarrow{O_2} R^2R^1C(OOH)-N=NR^3$$
(32) → (33)

$$(33) \xrightarrow{Ph_3P} R^2R^1C(OH)-N=NR^3$$
(34)

Scheme 26

Deoxygenation by triphenylphosphine has recently been used to prepare the previously unknown α-cyclopropylidene acetaldehydes (36) from the peroxyhemiacetals (35).[88] The hydroperoxides (35) are derived by treatment of vinylcyclopropane derivatives with singlet oxygen.

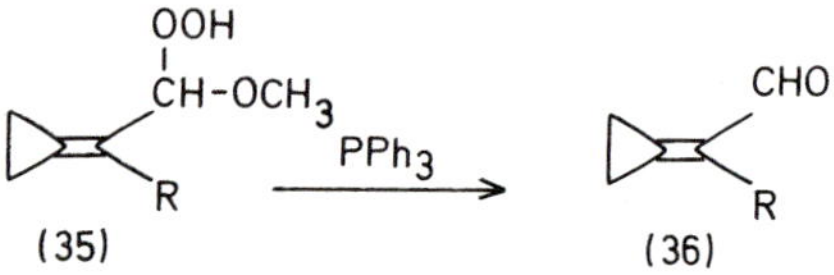

Scheme 27

1.6.2. Deoxygenation of dialkyl peroxides

Simple dialkyl peroxides react only very slowly with phosphorus(III) reagents at room temperature.[79] If the reaction is carried out in the presence of water the product is an alcohol, but under dry conditions the expected dialkyl ether is formed.[89] Thus, after 7 days in wet benzene with tributylphosphine, dihexyl peroxide gave a mixture of hexanol and dihexyl ether. The ratio of the two products depends upon the quantity of water in the solvent, and on deliberately adding an excess of water only the alcohol is formed.[89] As a result of the participation of water, in the presence of less than the stoichiometric amount of water, the reaction exhibits a variation in products with time. Alcohol is formed initially and then, when the water is consumed, ether becomes the major product. This type of situation may pertain if undried solvents are used[89] for the reaction.

Denney et al.[90] reported a similar time variation of products in the reaction of tributylphosphine with diethyl peroxide in the absence of solvent. Ethanol and diethyl ether were ultimately formed in the ratio 69:31 respectively, but again the alcohol was the product characteristic of the early stages of the reaction and the ether was formed only at high conversions of starting materials. The participation of water was not invoked in this case, and a mechanism was suggested (Scheme 28) where S_N2 attack of the phosphine on the peroxide competes with a biphilic process[90,91] leading to a phosphorane

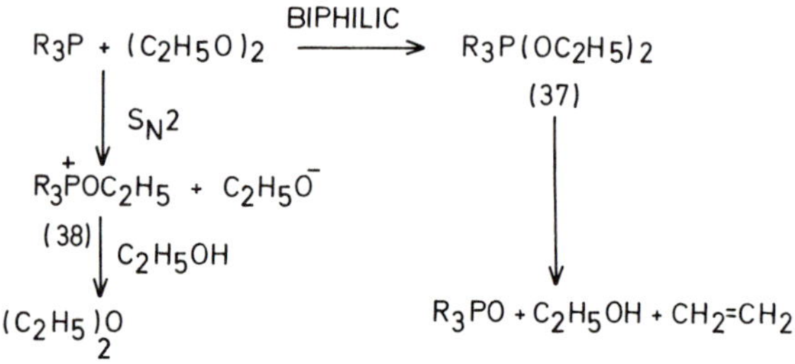

Scheme 28

(37). Direct unimolecular fragmentation of the phosphorane then leads to tributylphosphine oxide, ethylene, and ethanol.

The initial product of the S_N2 reaction is the quasiphosphonium ion (38) and ethoxide ion. Nucleophilic attack of ethoxide on the ethanol formed by fragmentation of (37) is then postulated to be the route of formation of diethyl ether. On this picture, therefore, high ethanol levels, which will be characteristic of the later stages of the reaction, will favour ether formation, not only as a result of the availability of ethanol as a reagent, but also as a result of the generally more polar conditions prevailing.

It is not appropriate to speculate here as to whether one should try to unify these two pieces of work and, for example, ascribe ethanol formation in the early stages of Denney's reaction to hydrolysis of a species analogous to (37), but it certainly seems that reaction of dialkyl peroxide with tributylphosphine is a reaction of some complexity and that predictable results might best be achieved by following the work of Holtz et al.[89] and working in dry benzene when it is desired to obtain ethers and in aqueous acetone when the preparation of alcohols is required.

The generality of the reaction is in some doubt; dialkyl peroxides are, as has been noted, of rather low reactivity with phosphorus(III) reagents, and di-t-butyl peroxide for example is reported to be completely inert to tributylphosphine at temperatures below the decomposition temperature of the peroxide. The cyclic peroxide 1,2-dioxan, however, reacts with the same reagent over 6 days at room temperature, to give butane-1,4-diol, in aqueous solvents. In benzene[89] a mixture of tetrahydrofuran and butane-1,4-diol is formed.

Tributylphosphine is the reagent most studied for the preparative deoxygenation of dialkyl peroxides and is of relatively high reactivity. Triphenylphosphine and phenyldiisopropylphosphine have also been used but the former is reported to have only one twenty-fifth of the reactivity of tributylphosphine in this context.[89]

Phosphites have also been studied but interest has centred largely around their reactions with dialkyl peroxides at low temperatures when pentaalkoxyphosphoranes are the major products.[91,92] The order of reactivity of phosphites with diethyl peroxide is triisopropyl phosphite < triethyl phosphite < 6-membered cyclic phosphites < 5-membered cyclic phosphites. This is the reverse of the order of reactivity observed in reactions involving the quaternization of the phosphite,[92] and Denney[91] has cited this as evidence against an ionic mechanism for the reaction between dialkyl peroxides and phosphorus(III) reagents and as being supportive of a biphilic process with no net polarity.

The scope of the deoxygenation of dialkyl peroxides has recently been extended[93] to include the preparation of silyl alkyl ethers by deoxygenation of an alkyl trimethylsilyl peroxide with triphenylphosphine. The alkyl group used was alicyclic but there is no reason to suppose that the reaction is restricted to such systems.

1.6.3. Deoxygenation of endocyclic peroxides

Endocyclic peroxides are much more easily deoxygenated by tervalent organophosphorus reagents than are their acyclic counterparts.

Horner and Jurgeleit[79] carried out the first reaction of this type when they deoxygenated ascaridole (39) with triphenylphosphine. The original assignment of the structure of the product was subsequently shown to be in error and the correct structure to be that of the 3,4-epoxymenthene derivative (40).[94]

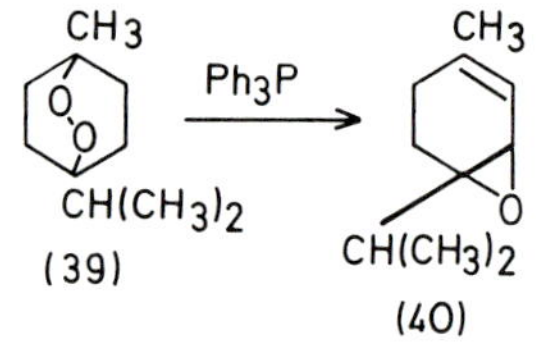

Scheme 29

This conversion into an exocyclic epoxide on reaction with triphenylphosphine is the typical reaction of a phosphorus(III) reagent with an endocyclic peroxide.[95,96,97] For example,[97] the endoperoxide (41) is deoxygenated by triphenyl phosphite at low temperature to give the benzene di-epoxide (42). The two epoxy functions are mutually *trans*.

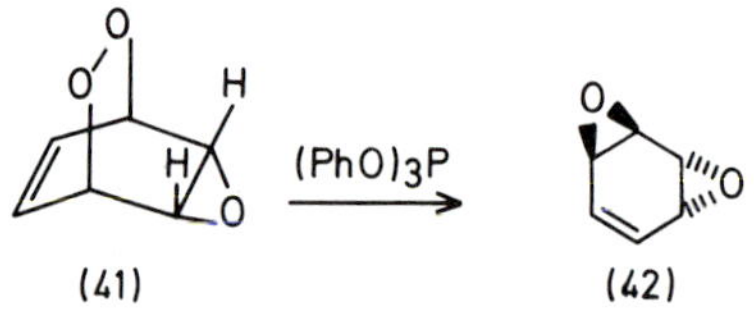

Scheme 30

This type of stereospecific conversion of the endoperoxide into epoxide is typical. Thus *trans*-3,4-dibromo-9,10-dioxatricyclo[4.2.2.0^{2,5}]dec-7-ene (43), on treatment with triphenylphosphine in benzene at 10–20°C, is converted into (44)[95] where the stereochemical relationship between the epoxy group and the junction between the 4- and 5-membered rings is the same as that between the ring junction and the endoperoxy function in the starting material (43).

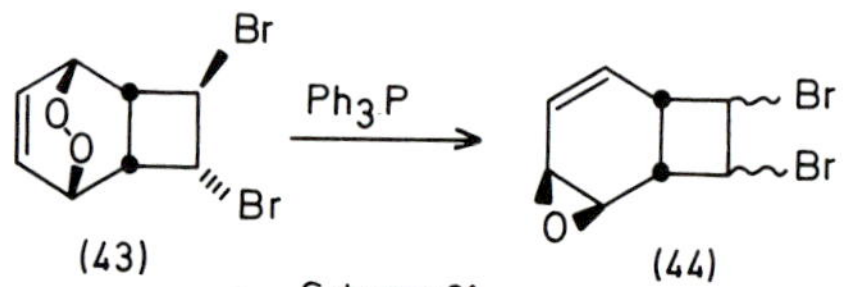

Scheme 31

7 Conversions via Deoxygenation   323

The stereochemical relationship between the bromine atoms and the rest of the molecule has been lost, indicating a lack of selectivity by the reagent as to which of the endoperoxy oxygen atoms is removed. This lack of regioselectivity is typical, so in an unsymmetrical situation two products will always result. An additional example is provided by the deoxygenation of 2,3-homotropone-4,7-epidioxide (45).[96] The observed products are the monocyclic conjugated dienedione (48) and only one of the expected epoxides (46), but the dienedione (48) is almost certainly derived from the first formed alternative epoxide (47) by rearrangement and ring expansion.

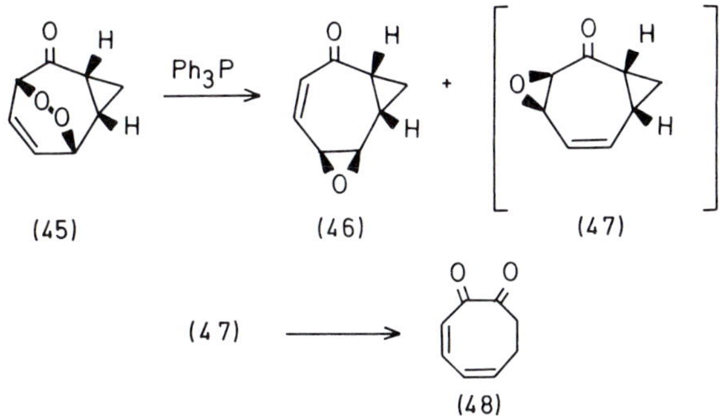

Scheme 32

1.6.4. Deoxygenation of diaroyl peroxides

Diaroyl peroxides react with phosphines[79,98,99] and with triethyl phosphite[100,101] at room temperature to give carboxylic anhydrides. The reaction is highly exothermic and careful temperature control is necessary if thermal decomposition of the peroxide and formation of the great complexity of products, other than the anhydride, is to be avoided.[100] The reaction is more easily controlled if carried out in a solvent, and yields of acid anhydrides are typically quantitative.[99]

Denney and coworkers[99] have studied the mechanism of the reaction between some diaroyl peroxides and tributylphosphine in detail and have suggested a process (Scheme 33) involving ion-pairs, on the basis of oxygen labelling studies.

In the case of mixed diaroyl peroxides (49) the distribution of reaction pathways via the two possible ion-pairs (50a) and (50b) is controlled by the nature of the substituents R^1 and R^2. The phosphorus reagent attacks the most electron-deficient of the two peroxide oxygen atoms, and hence, for example, in the case of reaction of 4-nitrobenzoyl benzoyl peroxide (49; $R^1 = NO_2$, $R^2 = H$) the ion-pair (50a; $R^1 = NO_2$, $R^2 = H$) is involved. In the case of an essentially electroneutral substituent such as, for example, when R^1 = phenyl and R^2 = hydrogen (Scheme 33) it can be shown, by

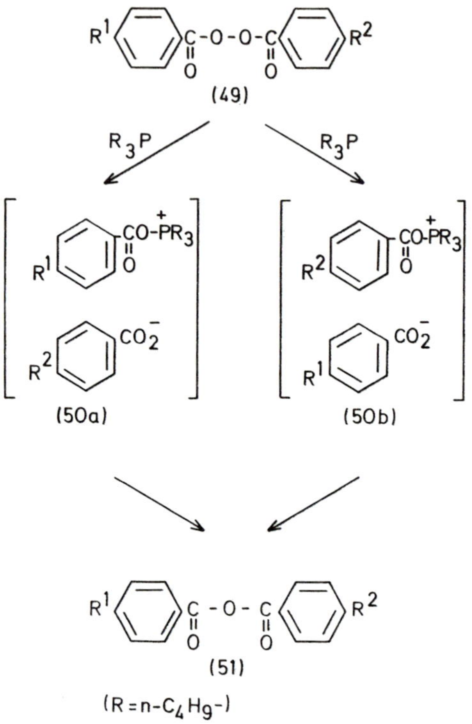

Scheme 33

labelling studies, that both ion-pairs (50a) and (50b) participate. This does not, however, lead to a mixture of products; only 4-phenylbenzoyl benzoyl peroxide is observed and no 4,4'-diphenyl or unsubstituted benzoyl peroxide is formed. This indicates that the ion-pairs (50a) and (50b) are "tight" and this seems to be the case in solvents ranging from halogenomethanes to mixtures of chloroform and ether.

A similar conclusion was drawn by Cadogan and coworkers[100] on the basis of the observation that only the monosubstituted anhydride was formed when a series of monosubstituted benzoyl benzoyl peroxides were allowed to react with triethyl phosphite.

1.6.5. *Deoxygenation of peroxy-esters and peroxy-lactones*

t-Butyl perbenzoate is reduced to t-butyl benzoate by triphenyl- or tributyl-phosphine.[102] Some benzoic acid is formed as a side-product but it is not clear whether this is an artifact of the work-up procedure.

Oxygen-18 labelling studies indicate that the attack by the phosphorus reagent occurs at the peroxide oxygen that is directly attached to the carbonyl

7 Conversions via Deoxygenation

group. This is the more electropositive of the two oxygens and also, particularly in the case of t-butyl esters, is the more sterically accessible.

Tributylphosphine in chloroform reduces *trans*-9-decalyl perbenzoate stereospecifically to the *trans*-benzoate.[102] This is the only information available on the stereochemistry of this type of reaction but, particularly since such a view is in accord with all other known cases of deoxygenations of peroxy compounds, it is probably safe to suggest that deoxygenation of peresters in general will proceed with retention of configuration at the carbinol carbon atom in the alcohol moiety.

β-Alkyl-β-phenyl-β-peroxypropiolactones (52) react with triphenylphosphine in hexane at room temperature.[103] Only small quantities of the expected reduction product, the β-alkyl-β-phenyl-β-propiolactone (54), are formed and the major products are an α-alkylstyrene and an aryl alkyl ketone (Scheme 34) in proportions that vary with the reaction conditions as discussed below.

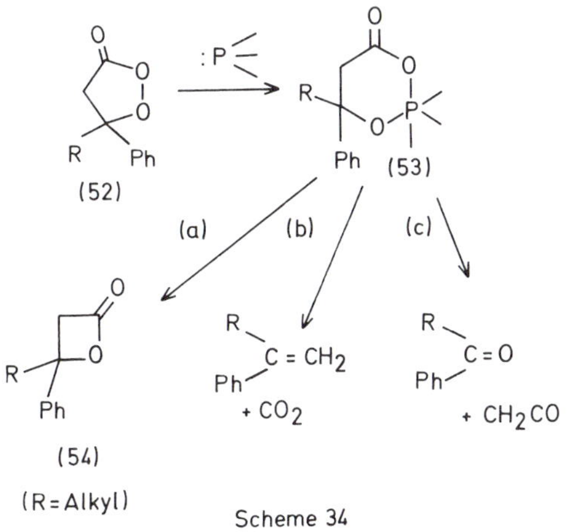

Scheme 34

The reaction is postulated[103] to proceed by the initial formation of the dioxaphosphorinan (53) which then breaks down by one of the routes (a), (b), or (c) shown in Scheme 34. Styrene formation is favoured by high solvent polarity and is enhanced greatly if R is a small group such as methyl as compared to isopropyl. Studies of electronic effects on the reaction show that, whereas the rate of the reaction is increased by raising the nucleophilicity of the phosphorus reagent, the ratio of the products from a particular peroxylactone (52; R = Me) is rather insensitive to changes in the phosphorus reagent.

Another reaction in which simple deoxygenation is in competition with a deoxygenation-fragmentation pathway is that of triphenylphosphine with dibutylmalonyl peroxide (55).[104] Dibutylmalonic anhydride (57) is formed in

55–60% yield, and the remainder of the starting material is accounted for as dibutyl ketone. The reaction probably involves the intermediacy of the dioxaphosphorinan (56).

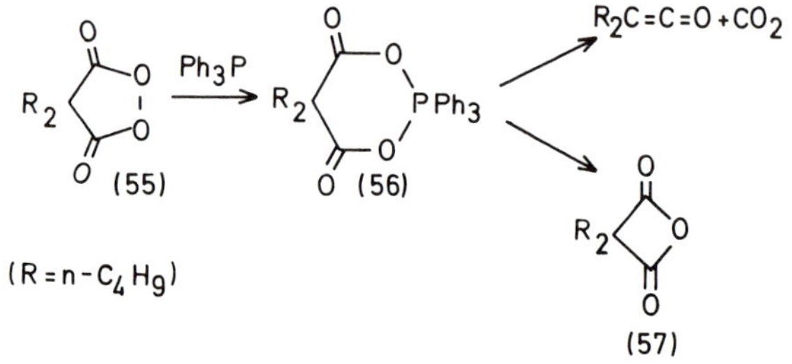

Scheme 35

1.6.6. Deoxygenation of peroxydicarbonates

Adam and Rois[105] have studied the reaction of diisopropyl peroxydicarbonates (58) with phosphorus(III) reagents at room temperature. Here again the reaction probably proceeds via a phosphorus(V) species which can undergo extrusion of the oxidized phosphorus reagent to give a pyrocarbonate (60), or which can form the simple dialkyl carbonate (61) by loss of the oxidized phosphorus reagent together with carbon dioxide. The mechanism suggested

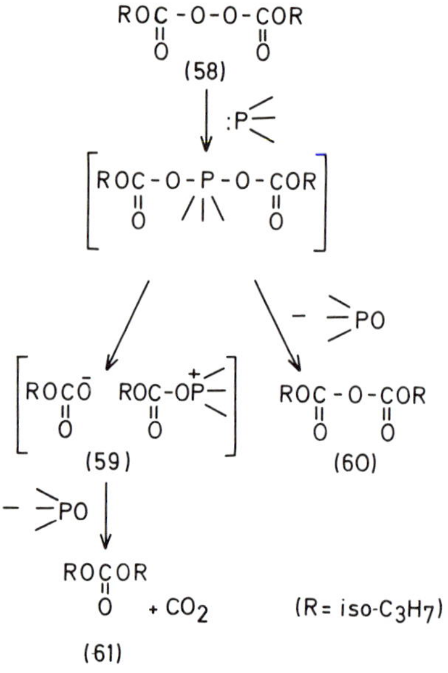

Scheme 36

7 Conversions via Deoxygenation

for this reaction (Scheme 36) is consistent with the observation that carbonate formation predominates in polar solvents with triphenyl phosphite as the deoxygenating reagent, whereas non-polar solvents favour the pyrocarbonate formation, particularly if the reagent is a phosphite. The ions (59) postulated to be intermediates in the formation of the carbonate are free even under non-polar conditions, since, on reaction of a mixture of diisopropyl peroxydicarbonate (58; R = i-C_3H_7) and di-s-butyl peroxydicarbonate (58; R = s-C_4H_9), all three possible carbonates are formed in statistical relative amounts.

2. Indirect Deoxygenation by Tervalent Organophosphorus Reagents

The deoxygenation reactions discussed so far in this chapter have largely been concerned with conversions where direct use is made of the deoxygenating capacity of an organophosphorus reagent to obtain a product that is derived simply by removal of oxygen from the starting material. The remaining section of this chapter will deal with some reagents that are not quite so straightforward in application and in which the deoxygenating capacity of the phosphorus(III) species is used as only one component of the reagent system.

2.1. Use of the Dialkyl Azodicarboxylate–Triphenylphosphine Reagent (DEADC–TPP)

Many condensations involving the elimination of the elements of water can be brought about by the combination of a reducing agent, which removes the oxygen, and an oxidant which will accept the two hydrogen atoms. Reactions brought about in this way, which have the advantage over conventionally catalysed condensations of proceeding under neutral, non-polar, conditions, have recently been reviewed[106] with particular reference to their uses in the synthesis of polypeptides, nucleotides, and macrolides.

One redox condensation reagent in particular, however, has been applied in a much more general sense. This reagent is the betaine (63) derived by reaction of diethyl azodicarboxylate (DEADC) (62) with triphenylphosphine (TPP).[106,107] Dimethyl azodicarboxylate has occasionally been used instead of the diethyl ester.

One of the most valuable uses of the DEADC–TPP reagent is that it enables alcohols to be used as alkylating agents for the replacement of acid hydrogen atoms by alkyl groups.

The generally accepted mechanism for the reaction[106] is shown in Scheme 37. In essence the alcohol has been activated to nucleophilic substitution by

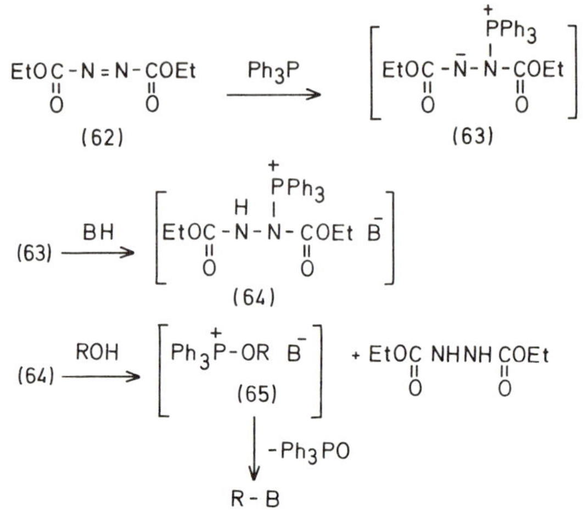

$$\text{EtOC-N=N-COEt} \xrightarrow{Ph_3P} \left[\text{EtOC-}\overset{+PPh_3}{\overset{|}{N}}\text{-}\overset{-}{N}\text{-COEt} \right]$$
$$\quad\overset{\parallel}{O}\quad\overset{\parallel}{O}\qquad\qquad\qquad\overset{\parallel}{O}\qquad\overset{\parallel}{O}$$
$$\qquad(62)\qquad\qquad\qquad\qquad\qquad(63)$$

$$(63) \xrightarrow{BH} \left[\text{EtOC-}\overset{\overset{H}{|}}{N}\text{-}\overset{\overset{+PPh_3}{|}}{N}\text{-COEt } B^- \right]$$
$$\qquad\qquad\qquad\quad\overset{\parallel}{O}\qquad\overset{\parallel}{O}$$
$$\qquad\qquad\qquad\qquad(64)$$

$$(64) \xrightarrow{ROH} \left[Ph_3\overset{+}{P}\text{-OR } B^- \right] + \text{EtOC NHNH COEt}$$
$$\qquad\qquad\qquad(65)\qquad\quad\overset{\parallel}{O}\qquad\overset{\parallel}{O}$$

$$\downarrow -Ph_3PO$$
$$R-B$$

Scheme 37

conversion into the quasiphosphonium salt (65) and the overall result of the reaction is that of a condensation with formal elimination of water. It is important to note that, although the products of DEADC–TPP catalysed condensations are often similar, or identical, to the products obtained from acid or base catalysed condensations, the mechanism of the two reactions is completely different. The DEADC–TPP catalysed reactions are nucleophilic substitutions, and this has a profound effect upon the stereochemistry of these "condensation" reactions and upon their regioselectivity.

The alkylation by alcohols and DEADC–TPP of centres carrying an acidic hydrogen is applicable not only to conventionally acidic substrates such as carboxylic acids and phenols but also to weakly acidic materials such as carboximides and to some 1,2-dicarbonyl compounds.

It has been estimated that a pK_a of less than 11 for the hydrogen to be replaced is required before reaction can occur[108] and thus, for example, the *N*-alkylation of carboxamides is not normally possible by this method.

2.1.1. *Preparation of esters*

Work in this area has been concerned mainly with the preparation of esters of benzoic acid.[109] Reaction of DEADC–TPP, benzoic acid, and an alcohol proceeds smoothly in ether at room temperature, to give benzoates in 80–90% yield. A similar reaction with pentanoic acid[109] was not so successful, yields of pentanoyl esters being less than 50%. The reaction proceeds with inversion of the configuration[110] at the carbinol carbon of the alcohol, which is consistent with the idea[106] that the reaction is properly regarded as an S_N2 type attack by the acid on the activated alcohol and is mechanistically quite distinct from a conventional esterification.

7 Conversions via Deoxygenation

This stereospecificity of the reaction has led to its development as a method for the inversion of hydroxylated chiral centres in steroids[111] by the sequence alcohol → inverted benzoate → inverted alcohol. Alternatively formate or phenylacetyl esters can be employed.

Since the benzoylation is very susceptible to steric hindrance the reaction shows high regioselectivity. For example, at room temperature in tetrahydrofuran as solvent, only hydroxy groups on the β-face of steroids normally react.[111] Thus 5α-cholestan-3β-ol can be converted via its 3α-benzoate into 5α-cholestan-3α-ol. Since the 3α-hydroxy compound will not form a benzoate under these conditions and is consequently not inverted, a mixture of the two epimers at the 3-position can be converted into pure 5α-cholestan-3α-ol by this process. This stereochemical control of regioselectivity also allows cholic acid derivatives to be selectively esterified at the 3-hydroxy function and hence inverted in configuration at C-3, with the configuration of C-7 and C-12 unaffected.

The selectivity of the reaction is lifted to some extent[112] if the solvent is changed to benzene and the reaction temperature raised to the boiling point. Hydroxy groups on the α-face of steroids then react with benzoic acid in the presence of DEADC–TPP.

This method of esterification of steroids, being a nucleophilic substitution, appears to suffer from the usual problems encountered due to the participation of i-steroid intermediates in the nucleophilic substitution reactions of cholesterol. Reaction of this steroid with benzoic acid, DEADC, and TPP gives a complex mixture of products. It is claimed, however, that under suitable reaction conditions, which are not adequately specified, cholesterol reacts conventionally to give the expected 3α-benzoate ester.[111,113]

The major complication with the esterification reaction occurs in the case of keto-steroids when direct reaction of DEADC with the ketone may occur to give an en-hydrazodicarboxylate derivative.[112] Steroids that contain 1,2-*trans*-diaxial diol functions may also give problems, tending to form epoxides on reaction with the phosphorus reagent.[112]

There appears to be no general tendency for DEADC–TPP to induce the elimination of water from steroids, in competition with nucleophilic substitution, but this has been observed to occur in one instance.[112]

The selectivity of the DEADC–TPP catalysed benzoylation of the hydroxy functions in some carbohydrates has been examined.[114] The reaction again proceeds with inversion of configuration of the hydroxy-bearing carbon and shows stereochemical control of regioselectivity. Methyl 3-deoxy-α-D-*arabino*-hexopyranoside (66)[114a] is benzoylated preferentially at the exocyclic 6-hydroxy. The most reactive ring hydroxy is that at the 4-position. The tendency for elimination of water to compete with substitution is much more prevalent in carbohydrates than in steroids, and as much as 23% of the product on attempted benzoylation of (66) is the alkene methyl 4,6-di-*O*-benzoyl-2,3-dideoxy-α-D-*threo*-hex-2-enopyranoside (67). Where the geometry of the intermediate quasiphosphonium salt is favourable it appears that elimination is likely to occur. It is presumed that a 1,2-diaxial arrangement of the quasiphosphonium centre and the leaving proton will be required.[114a]

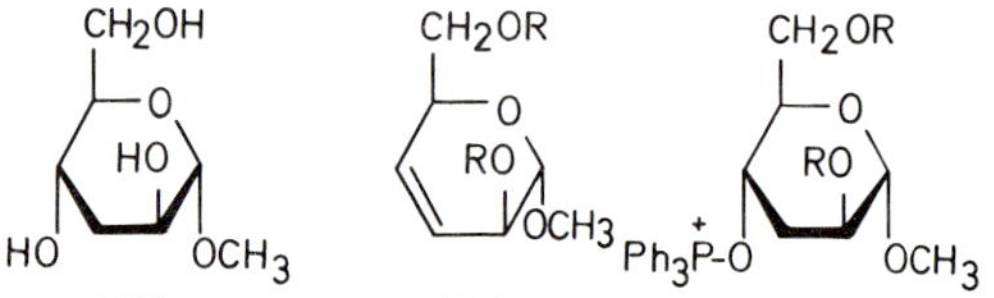

Scheme 38

R = Benzoyl

The carbohydrate residues of the nucleosides uridine and adenosine have also been benzoylated and *p*-nitrobenzoylated by the corresponding acids and DEADC–TPP.[114b] Reaction occurred selectively at the 5'-hydroxy.

One further use of the DEADC–TPP reagent in the preparation of carboxylic esters is worthy of note; 1,3-diacyl derivatives of glycerol can be benzoylated at C-2 by this method[115] without any of the complications normally associated with reactions at this position in such systems. These complications result from 1,2- and/or 3,2-acyl migrations. Again the reaction proceeds with inversion as shown by study of derivatives where C-2 is chiral.

DEADC–TPP has also been used to prepare esters of phosphoric acid.[110,116] The reagent is as selective in the phosphorylation of polyhydric alcohols as it is in their acylation. Hence the sugar residues of the nucleosides thymidine and uridine can be selectively phosphorylated at the 5'-hydroxy by dibenzyl hydrogen phosphite.[116] Subsequent hydrogenolysis removes the benzyl groups leaving the free, specifically 5'-phosphorylated, nucleotides.

2.1.2. *Conversion of alcohols into amines* via *N-alkylphthalimides*

Phthalimide and alcohols condense under the influence of DEADC–TPP to give *N*-alkylphthalimides which can then be cleaved to amines.[117,118] The method therefore provides a reasonably direct means of converting alcohols into amines. The reaction is carried out in tetrahydrofuran at room temperature and gives an amine of inverted configuration relative to the starting alcohol. Yields of the *N*-alkylphthalimides are generally better than 80%.[117]

The reaction shows good functional selectivity; the hydroxy groups of allyl alcohol, 2-chloroethanol, and (±)-ethyl lactate were replaced by a phthalimide moiety without affecting the other functionality in the molecule.[118]

The reaction of ethyl lactate with phthalimide and DEADC–TPP provides an example of the first step in what may serve as a general method for the

7 Conversions via Deoxygenation

conversion of the esters of α-hydroxy-acids into α-amino-acid esters of inverted configuration relative to the starting hydroxy-acid. One complication that may arise, however, is as a result of a tendency for α-hydroxy-acids with acidic β-hydrogens to undergo elimination (Scheme 39), presumably via an intermediate quasiphosphonium salt such as (70). Thus diethyl malate (69) gives only diethyl fumarate on treatment with DEADC–TPP and phthalimide and the phthalimide is not involved in the reaction.[118]

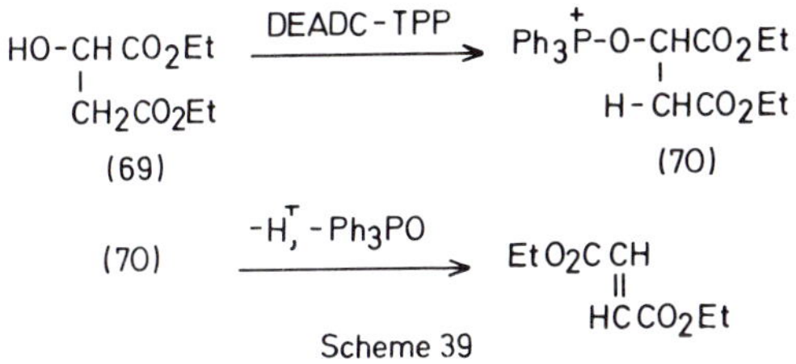

$$HO-CHCO_2Et \atop CH_2CO_2Et \quad (69) \quad \xrightarrow{DEADC-TPP} \quad Ph_3\overset{+}{P}-O-CHCO_2Et \atop H-CHCO_2Et \quad (70)$$

$$(70) \quad \xrightarrow{-H^+, -Ph_3PO} \quad EtO_2C\,CH \atop \underset{HCCO_2Et}{\|}$$

Scheme 39

The replacement of a hydroxy group by phthalimide in the presence of DEADC–TPP has also been applied to the conversion of sugars with one "isolated," unprotected, hydroxy group into the corresponding deoxy-phthalimido derivatives and hence into amino-sugars.[119] If the hydroxy group undergoing reaction is at the anomeric centre then the product is a mixture of anomers,[120] but it has been presumed, although not demonstrated, that the expected complete inversion occurs on reaction of hydroxy groups at other positions.[119]

2.1.3. Other condensations at nitrogen

Condensations on to acidic nitrogen centres other than that of phthalimide have been carried out with phosphorus(III) reagents and DEADC. N-Alkyl-succinimides can be prepared from succinimides and alcohols,[117] and sugar–saccharin conjugates have also been obtained in this way.[121]

Nucleosides can be prepared by condensation of sugars at the 9-position of 6-chloropurine.[122] In this case diphenylmethylphosphine replaces triphenyl-phosphine as the reducing component of the reagent system. The stereochemical integrity at the anomeric centre of the sugar is largely lost although there is a tendency towards net inversion.

Amides, even p-nitrobenzamides, are not sufficiently acidic to react with DEADC–TPP,[108] but N-acylcarbamates,[108] in which the N–H group is flanked on both sides by carbonyl functions, are smoothly alkylated by alcohols in the presence of the DEADC–TPP reagent.

2.1.4. Alkylation of N-hydroxy groups

N-Alkoxy-phthalimides and -succinimides can be prepared by alkylation of the corresponding N-hydroxy-imides with an alcohol and DEADC–TPP.[123]

The reaction has been used for the preparation of O-phthalimido derivatives from sugars with one unprotected hydroxy group.[124]

Unfortunately the potentially useful O-alkylation of arylhydroxamic acids cannot be achieved in this way since a competing Lössen rearrangement intervenes, apparently catalysed by DEADC–TPP.[125] The product of the reaction is either an O-(N-arylcarbamoyl)hydroxamate or, as a result of a further rearrangement, a diarylurea. Thus, although unsuccessful as a means of alkylating hydroxamic acids, the method provides a means of rearranging unacylated hydroxamic acids at room temperature under neutral conditions and in aprotic solvents.

2.1.5. Synthesis of ethers

Phenols are smoothly converted into alkyl aryl ethers on reaction with alcohols in the presence of DEADC–TPP.[126,127]

The reaction is quite general in application and the aprotic conditions mean that even t-butanol can be successfully used as the alkanol component.[126] Naphthols cannot replace phenols.

The reaction has been applied in the steroid field; 5α-cholestan-3β-ol is converted into 3α-phenoxy-5α-cholestane on reaction with phenol;[127] this stereochemical course of the reaction, inversion of the alcohol, is the same as that observed in benzoylation with benzoic acid and DEADC–TPP. Cholesterol, however, forms a phenyl ether with retention of configuration, presumably via an i-steroid intermediate.

The reaction cannot be simply applied to the preparation of glycosides. 2,3:5,6-Di-O-isopropylidene-α-D-mannofuranose does not condense with alcohols in the presence of DEADC–TPP unless a mercuric halide is present.[128] The mechanism of this reaction is clearly different from the alkylations so far discussed and may involve a glycosyl halide as an intermediate. The stereochemical course of the mercuric halide catalysed reaction is also different; although not entirely stereospecific it goes with predominant retention at the carbohydrate anomeric centre.

Quinazolin-4(3H)-ones (71) have been converted into enol ethers (72) by alkylation with a variety of alkanols in tetrahydrofuran in the presence of DEADC–TPP.[127] Uracil has also been alkylated, at both oxygens, by

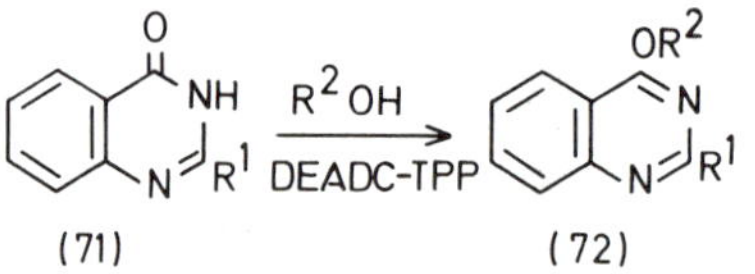

$R^1 = {}^i Pr$ or $pMeO C_6H_4$

Scheme 40

7 Conversions via Deoxygenation

cholesterol under similar conditions although, because of solubility problems, the reaction was carried out with hexamethylphosphoramide as solvent.[127]

In perhaps surprising contrast to these reactions of (71) and of uracil it is reported[129] that the 3-phenyltoxoflavin (73) is alkylated not on oxygen but at nitrogen, to give a 1-alkyl-3-methyl-6-phenyl-7-azalumazine (74) on reaction with DEADC–TPP and an alkanol in dioxan.

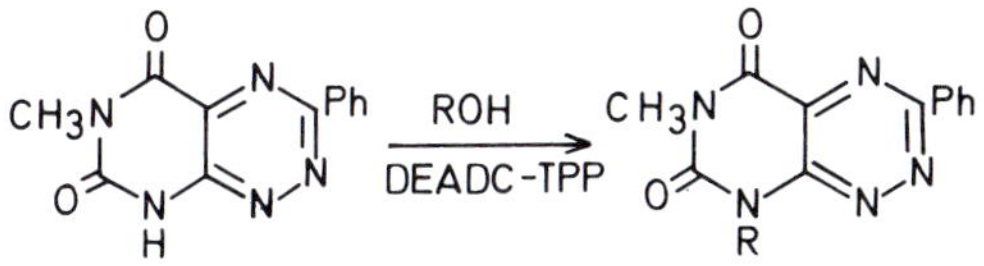

$$(73) \xrightarrow[\text{DEADC-TPP}]{\text{ROH}} (74)$$

Scheme 41

2.1.6. Synthesis of anhydrides

DEADC–TPP does not appear to have application to the self-condensation of carboxylic acids to give anhydrides; indeed the absence of such a reaction is necessary to the successful application of the reagent in the acylation of alcohols by carboxylic acids. A combination of tributyl- or triphenylphosphine as reductant and 1,2-dibenzoylethylene as hydrogen acceptor has been successfully used, however, and converts both aryl and alkyl carboxylic acids into anhydrides in good yield.[130] The method is also applicable to the self-condensation of phosphates to give pyrophosphates.[130]

2.1.7. Alkylation of active hydrogen compounds

The alkylation of a series of 1,2-dicarbonyl compounds by propanol in the presence of DEADC–TPP has been studied.[108] Ethyl cyanoacetate ($pK_a > 9$) and malononitrile (pK_a 11·2) are C-alkylated in the expected manner to give ethyl 2-cyanopentanoate (52%) and 1,1-dicyanobutane (51%) respectively. Ethyl acetoacetate (pK_a 10·7) gives a mixture of C- and O-alkylation products with O-alkylation predominating by a factor of 3. Diethyl malonate (pK_a 13·3) is not alkylated under the same conditions and it is suggested that this is due to its low acidity.[108]

The stereochemical course of the condensation of ethyl cyanoacetate with (R)-octan-2-ol has been investigated[108] and although the absolute stereochemistry of the product, ethyl 2-cyano-3-methylnonanoate, has not been assigned it is reported to be optically active and the tentative suggestion has been advanced that the reaction proceeds with inversion at the alcohol.

2.1.8. Other nucleophilic substitutions

The mode of action of the DEADC–TPP reagent can, as should now be clear, be summed up in its ability to activate alcohols to nucleophilic substitution by converting them into quasiphosphonium salts and thus providing triphenylphosphine oxide as a leaving group. It is largely an accident of the history of the development of the reaction that, until recently, the nucleophiles employed have been such as to result in reactions that are equivalent to acid or base catalysed condensations. As a result of recent work, however, it is becoming clear that DEADC–TPP can be used in conjunction with a wide spectrum of anions to bring about nucleophilic substitution with inversion.

The hydroxy group of 2-substituted cyclohexanes and the 3β-hydroxy function in a variety of cholesterol and cholestane derivatives have been replaced by azide, thiocyanate, iodide, bromide, nitrite, and nitrate by reaction with the relevant anionic nucleophile and DEADC–TPP.[131] The reactions are carried out in benzene or benzene–acetonitrile mixtures at 20°C, and a convenient form of the nucleophile has been found to be as a triphenylmethylphosphonium salt prepared from triphenylmethylphosphonium bromide by ion exchange.

The reactions all go with inversion of configuration to give 3α-substituted products, even with cholesterol; there is presumably no participation by an i-steroid intermediate in this case.

An interesting observation reported by the same workers[131] is that diethyl malate, on reaction with DEADC–TPP and hydrazoic acid, gives diethyl 2-azidosuccinate in 74% yield. This is in contrast to the situation already quoted[118] when an attempt to condense a malate with phthalimide gave only the elimination product, an ester of fumaric acid. The change in solvent from tetrahydrofuran to benzene may be of significance here.

An ester can also serve as a source of nucleophile; thus tosylates and sulphates can be prepared from alcohols and methyl tosylate and dimethyl sulphate respectively in the presence of DEADC–TPP. Similarly halogen nucleophiles can be derived from alkyl halides rather than from inorganic salts.[131]

2.1.9. Transesterification of carboxylic esters

It has already been recorded that diethyl malonate is not C-alkylated by propanol in the presence of DEADC–TPP. Attempts to methylate the same substrate[132] with methanol at 0°C similarly did not give C-alkylation, but a rapid ester interchange occurred giving principally dimethyl malonate with traces of the mixed ester. The reaction appears to be generally applicable to

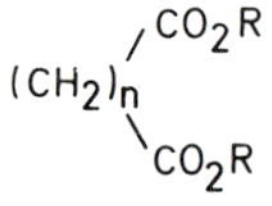

$$(CH_2)_n \begin{matrix} CO_2R \\ CO_2R \end{matrix}$$

(75)

Scheme 42

7 Conversions via Deoxygenation

the transesterification of the diesters of dicarboxylic acids of the type of (75). The mutual inductive effects of the ester functions are important and reactivity falls along the sequence $n = 0 > 1 > 2 > 3$. The reaction is most successful for the conversion of higher esters into their lower homologues, particularly into methyl esters.

Monocarboxylic acid esters do react but only slowly, and in this case the reaction is almost wholly restricted to the preparation of methyl esters.

Only a tentative mechanism has been suggested for this reaction, involving the participation of the betaine, from reaction of DEADC with TPP, as a bidentate catalyst.[132]

2.1.10. *Intramolecular dehydrations*

The DEADC–TPP reagent is, in principle, capable of the intramolecular removal of the elements of water and the creation of unsaturation. The occasional occurrence of such a process as an unwanted side-reaction in some attempted condensations has already been referred to.[112,114,118]

There have, however, been few attempts to put the reaction to use. A notable exception[133] is provided by the conversion of formamides of the type of (76) into isocyanides by reaction with DEADC–TPP in tetrahydrofuran.

$$\text{RNH-CHO} \xrightarrow{\text{DEADC - TPP}} \text{R-}\overset{+}{\text{N}}\equiv\overset{-}{\text{C}}$$
(76)

Scheme 43

The reaction has only been shown to be successful when R is cyclohexyl, benzyl, phenyl, 2,6-dimethylphenyl, and 2,4,6-trimethylphenyl, and fails completely when R is a simple alkyl group or 4-nitrophenyl. In spite of the lack of generality of the reaction it has the advantage, over many other isocyanide syntheses, of avoiding exposure of the product to base, and the method may give access to some isocyanides that are racemized or decomposed by such exposure.

β-Hydroxy-carboxylic acids undergo decarboxylative dehydration to alkenes in the presence of DEADC–TPP.[134] The full scope of the reaction has not been investigated but a great diversity of starting materials have been employed successfully. The stereochemical course of the reaction is at present unknown.

2.2. Use of the Ozone–Triphenyl Phosphite Adduct

Reaction of ozone with phosphites and phosphines at $-70°C$ generally results in their oxidation to phosphates and phosphine oxides respectively.[135] In the case of triaryl phosphites, however, a 1:1 adduct is formed between the phosphite and ozone which decomposes to the phosphate and oxygen only on warming above −35°C.[135] It is now known that the ability to form such an adduct is not restricted to triaryl phosphites but that other phosphites can undergo the reaction. This point is discussed later. The present section will deal only with the triphenyl phosphite–ozone adduct which is the most extensively studied and hence the most synthetically useful of these compounds.

In his initial study Thompson[135] showed that the adduct of ozone and triphenyl phosphite could be used to oxidize sulphides to sulphones and ultimately to sulphoxides; phosphorus(III) compounds were also oxidized, including, importantly, the oxidation of triphenyl phosphite to triphenyl phosphate. Hence the adduct between the former and ozone is formed efficiently only if the experimental conditions are arranged such that free phosphite is never present in substantial excess over the ozone.

Since triphenyl phosphate is always formed in quantitative yield in oxidations by the triphenyl phosphite–ozone adduct it was suggested, at an early stage, that the effective oxidizing agent might be oxygen, the deoxygenation product of ozone, which would, because of spin conservation, be in the singlet state.[136]

A great deal of argument and investigative work has surrounded this point as to whether the triphenyl phosphite–ozone adduct can genuinely be considered to be a singlet oxygen source. From a synthetic point of view the relevance of this argument is as to whether the adduct can usefully replace conventional singlet oxygen generation. At present the conclusions can be summarized as follows.

(a) In most cases thermal decomposition of the triphenyl phosphite–ozone adduct in the presence of singlet oxygen acceptors leads to products identical in type to those obtained with authentic singlet oxygen sources. Differences that do occur are generally associated with different selectivities, both as between acceptors and different regio- and stereo-selectivities within any given acceptor.[137]

(b) There are some reactions between the triphenyl phosphite–ozone adduct and singlet oxygen acceptors that occur at temperatures substantially below the decomposition temperature of the adduct and must thus be regarded as direct oxygen transfers, not involving free singlet oxygen. In many cases, however, these reactions also show products and product distributions very close to those obtained with authentic singlet oxygen.[138]

Thus, for synthetic purposes the use of the triphenyl phosphite–ozone adduct can largely be regarded as an alternative to singlet oxygen generation of a conventional type, although selectivity, particularly stereo- and regio-selectivity, may not parallel precisely that of "authentic" singlet oxygen.

There are three important synthetically useful reactions of singlet oxygen,[139] and it is convenient to classify the reactions of the triphenyl phosphite–ozone adduct into the same categories: 1,4 Diels–Alder type addition to dienes;

7 Conversions via Deoxygenation

1,2-addition to olefins leading, via oxetans, to carbonyl compounds; and "ene" reactions giving allylic hydroperoxides.

2.2.1. *1,4-Addition to dienes*

1,4-Addition to dienes is normally carried out by decomposing the triphenyl phosphite–ozone adduct, in dichloromethane or chloroform, at temperatures above $-35°C$ in the presence of the diene.[140] The reaction appears to be quite general in scope and the products are either an endoperoxide itself or the decomposition products of such an intermediate. Hence cyclohexa-1,3-diene gives 5,6-dioxabicyclo[2.2.2]oct-2-ene (77) in 67% yield and tetraphenylcyclopentadienone is converted into (Z)-1,2-dibenzoylstilbene (78) in 37% yield.[140]

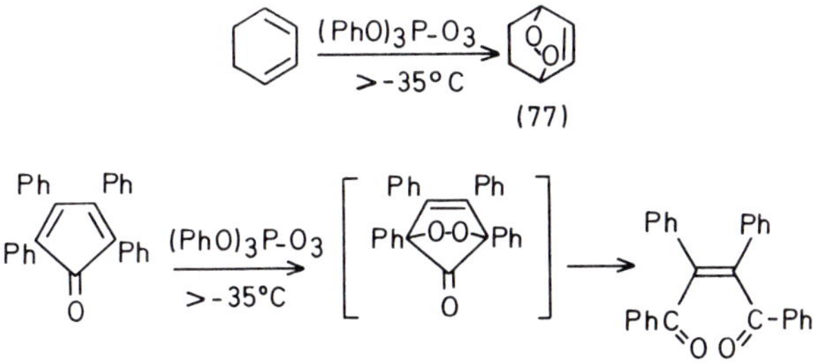

Scheme 44

The reaction extends generally to endoperoxide formation in aromatic systems; hence, 9,10-diphenylanthracene is converted, in 77% yield, into its endoperoxide by reaction with the triphenyl phosphite–ozone adduct at temperatures above $-35°C$.[140]

There have been some interesting applications of the endoperoxide forming reaction to synthetic problems.

Benzene monoxide (79) is converted, by the triphenyl phosphite–ozone adduct, ultimately into *trans*-benzene trioxide (81) (*trans*-3,6,9-trioxatetracyclo[6.1.0.0^{2,4}.0^{5,7}]nonane).[141] The reaction involves the 1,4-endoperoxide (80) as an intermediate, which is rearranged thermally at 45°C to the desired product.

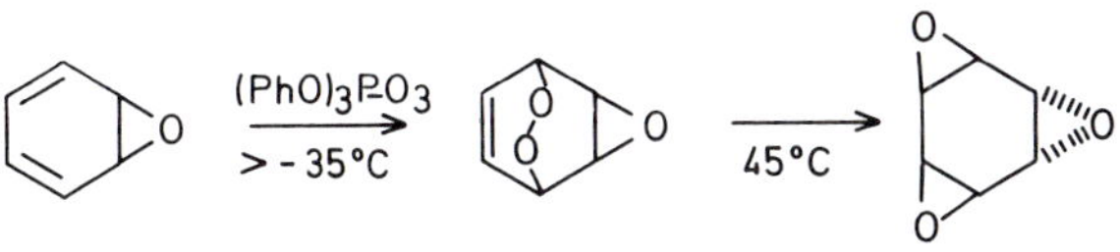

Scheme 45

Barnett and Needham[142] have developed the use of 9-anthroxy for protecting the hydroxy group of alcohols. The protecting group is removed by a two-step process, the first step of which is conversion of the anthracene residue into its endoperoxide by treatment with the triphenyl phosphite–ozone adduct. The endoperoxide is then hydrogenolysed regenerating the alcohol.

The reaction seems to be generally applicable, and satisfactory methods of preparing the anthroxy ethers have been developed.[142] In conclusion it should be recorded that the reactions described in this section are all equally capable of being carried out with singlet oxygen from a conventional source.

2.2.2. The "ene" reaction: formation of allylic hydroperoxides

The "ene" reaction by which alkenes are converted into allylic hydroperoxides is characteristic of singlet oxygen[139] and can also be carried out with the triphenyl phosphite–ozone adduct;[138,143] hence[137] tetramethylethylene is converted readily into the hydroperoxide (82). The reaction occurs both at higher

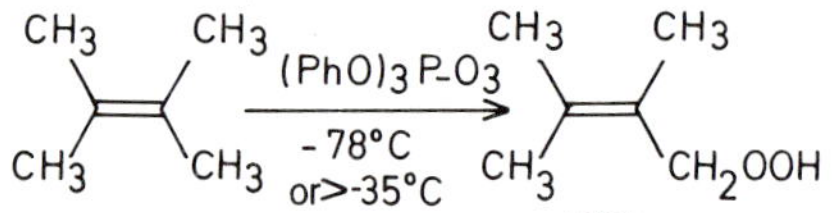

$$(CH_3)_2C=C(CH_3)_2 \xrightarrow[\text{or} > -35°C]{(PhO)_3P\cdot O_3, -78°C} (CH_3)_2C=C(CH_3)(CH_2OOH)$$
(82)

Scheme 46

temperatures ($> -35°C$) and at $-78°C$ at which temperature the adduct is indefinitely stable, and it is this latter observation that led Bartlett[138,143] to postulate that the triphenyl phosphite–ozone adduct was capable of the direct transfer of oxygen to an acceptor without the formation of free singlet oxygen. From the point of view of the synthetic use of the reaction in this case, however, this is of little import since the product is the same for either mechanism. This identity of products as between the two routes is preserved in most such cases, even when there is the possibility of a mixture of products. Thus the hydroperoxides (83) and (84) are formed in essentially the same ratio $(1:1)$[137] from 2-methylbut-2-ene irrespective of whether the reaction is carried out above the decomposition temperature of the triphenyl phosphite–ozone adduct or at $-78°C$, and moreover the product ratio is identical to

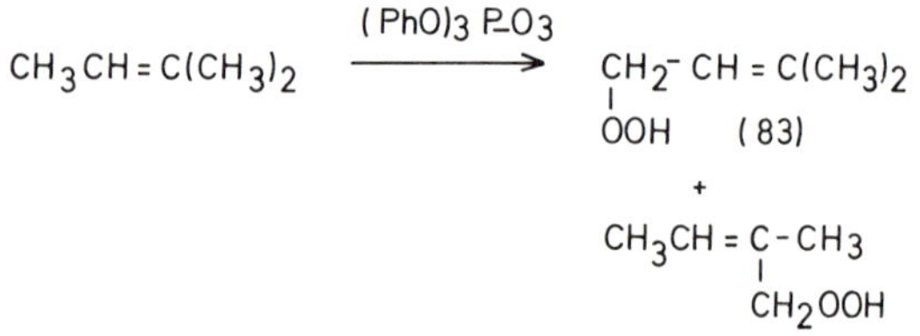

$$CH_3CH=C(CH_3)_2 \xrightarrow{(PhO)_3P\cdot O_3} \underset{OOH}{CH_2-CH=C(CH_3)_2} \quad (83)$$

$$+$$

$$CH_3CH=\underset{CH_2OOH}{C-CH_3}$$
(84)

Scheme 47

7 Conversions via Deoxygenation

that obtained with singlet oxygen from a photochemical source.[137] Thus in this particular reaction it seems that, for reactions of simple olefins at least, the regioselectivity of the triphenyl phosphite–ozone adduct is essentially the same as that of singlet oxygen even at temperatures too low to allow the adduct to act as a singlet oxygen source.

The situation may be more complex, however, in more sterically constrained systems. Hence Sam and Sutherland[144] found that, whereas germacrene (85) reacted only at one endocyclic double bond to give hydroperoxide (86) (50%) with the triphenyl phosphite–ozone adduct, with "authentic" singlet oxygen the products were (87) and (88) with (87) predominating by an amount

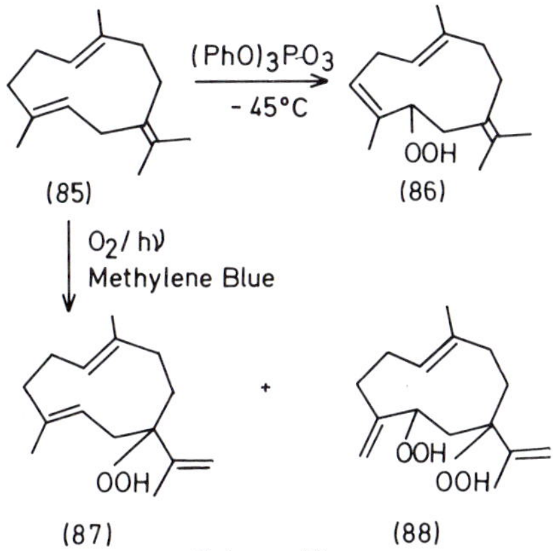

Scheme 48

indicating that the isopropylidene system was nine times more reactive than the endocyclic double bond. Thus the preferred position of attack with the two reagents is different, and even where attack does occur at the same site the allylic system formed is different.

Since Sam and Sutherland's reaction was carried out at $-45°C$, 10 degrees lower than the decomposition temperature of the triphenyl phosphite–ozone adduct, it can be presumed to be a direct oxygen transfer. The bulky adduct will have a more serious stereochemical constraint upon its approach to the germacrene than will singlet oxygen and thus its regioselectivity and stereoselectivity will, not surprisingly, be different.

Although the generality of this effect is unexploited there is some reason to suppose that at temperatures below $-35°C$, when direct oxygen transfer to the acceptor occurs, the use of the triphenyl phosphite–ozone adduct provides a means of selective attack, both regio- and stereo-selective, at the least hindered site in a polyfunctional system.

Another area in which singlet oxygen and direct oxygen transfer from the triphenyl phosphite adduct to an acceptor show different characteristics is when there is direct competition between two substrates, one of which is capable of hydroperoxide formation and the second of which reacts to form only an endoperoxide. Thus if a mixture of 2,5-dimethylfuran and tetramethylethylene[138,143] is treated with the triphenyl phosphite–ozone adduct at $-78°C$ the ethylene reacts, to give an allylic hydroperoxide, some ten times faster than the furan, which gives the 2,5-endoperoxide. This order of reactivity is completely reversed on competition of the same two substrates for singlet oxygen from Methylene Blue sensitized photolysis of oxygen. The generality of this observation is untested but potentially it provides the basis of methods for the selective formation of either allylic hydroperoxides or endoperoxides in systems capable of reaction in either mode.

2.2.3. *1,2-Cycloadditions: formation of dioxetans*

1,2-Addition to alkenes is another very characteristic reaction of singlet oxygen;[139] the reaction may give an isolable dioxetan[139,145] or, more commonly, the isolated products are carbonyl compounds from complete oxidative cleavage of the alkene double bond[138,139] and the intermediacy of a dioxetan is only inferred.

The triphenyl phosphite–ozone adduct has been little exploited for carrying out this reaction, but one particular example is worthy of note since it provides probably the clearest example of a case where different results can be achieved with the triphenyl phosphite–ozone adduct as compared with singlet oxygen from a photochemical source.

Thus Schaap and Bartlett[145] showed that reaction of either (E)- or (Z)-1,2-diethoxyethylene with the adduct at $-78°C$ gave the same mixture of *cis*- and *trans*-diethoxydioxetans (89) and (90), irrespective of the stereochemistry of the starting material.

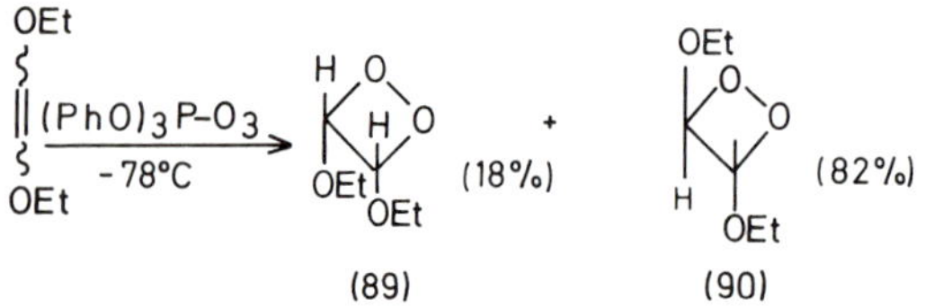

Scheme 49

It was previously shown[146] that reaction of "authentic" singlet oxygen with either isomer of 1,2-diethoxyethylene is entirely stereospecific, the (Z)-isomer giving only *cis*-dioxetan whereas the (E)-isomer leads specifically to the dioxetan with the ethoxy groups *trans*.

The reaction of either of the diethoxyethylenes with the triphenyl phosphite–ozone adduct is presumed to involve intermediates of the type of (91) and (92) in which free rotation, leading to the more stable *trans*-dioxetan, is possible.

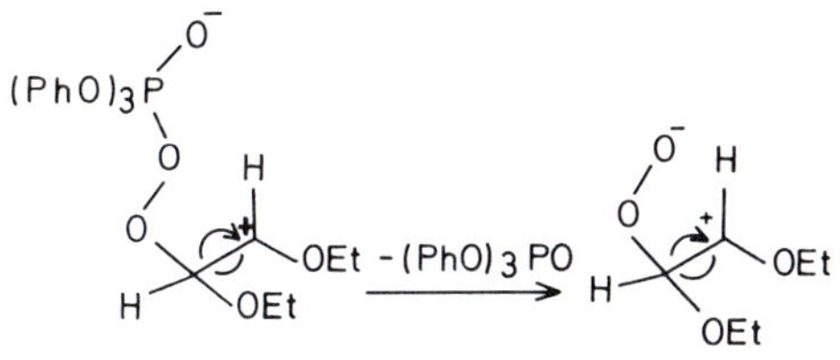

7 Conversions via Deoxygenation

Scheme 50

This reaction appears to be another example of a case where, although singlet oxygen and the triphenyl phosphite adduct give the same types of product, their stereoselectivity is different. Again it is not known how general the effect is but it points, again, to the possibility that there may be scope for the development of synthetically useful sequences where the use of singlet oxygen or the triphenyl phosphite–ozone adduct may provide complementary selectivities. This appears most likely to be found when the adduct is used as a direct oxygen transfer reagent at temperatures below its decomposition point ($< -35°C$).

2.2.4. The ozone–triphenyl phosphite adduct as a simple oxidant

In addition to giving the singlet oxygen type reactions discussed above the adduct can also be employed, at $-78°C$, as a simple oxygen transfer oxidant. As already mentioned, Thompson oxidized phosphorus(III) compounds and simple sulphides during his original study[135] of the reagent. Disulphides can also be oxidized by the adduct.[147] The reaction is selective in the sense that, initially at least, the thiolsulphinate (93) is formed with a preference of 10:1 as compared to thiolsulphonate (94) although prolonged reaction (5–6 days)

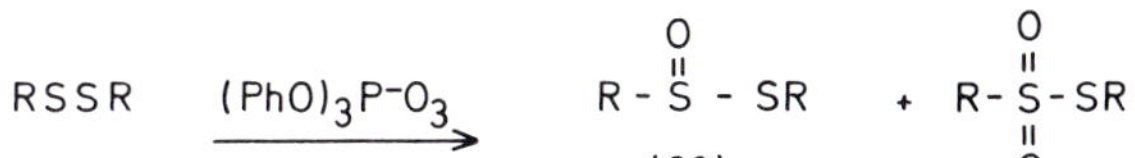

Scheme 51

reverses this ratio of products. Diethyl and dimethyl sulphides, for example, give the corresponding thiolsulphinates in ca. 50% yield. Even di-t-butyl sulphide undergoes oxidation to the thiolsulphonate with the triphenyl phosphite–ozone adduct in 48% yield, a reaction that cannot be accomplished with conventional oxidants even of the power of peracetic acid.

Bestmann et al.[148] have reported an interesting and useful application of the adduct as an alternative to autoxidation, for the conversion of phosphorus ylides, of the type of (95), into alkenes. The reaction apparently involves the

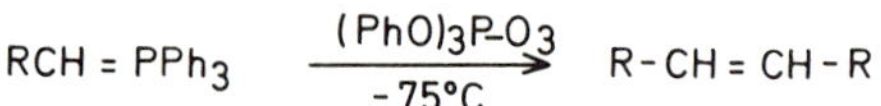

$$\text{RCH} = \text{PPh}_3 \xrightarrow[-75^\circ\text{C}]{(\text{PhO})_3\text{P-O}_3} \text{R-CH} = \text{CH-R}$$
(95)

Scheme 52

conversion of one molecule of the ylide into an aldehyde, by the oxidant. The aldehyde then undergoes a conventional Wittig reaction with a second molecule of ylide. In the case of ylides of the type of (96) the reaction stops at the first stage and ketones are obtained. The reaction is apparently generally

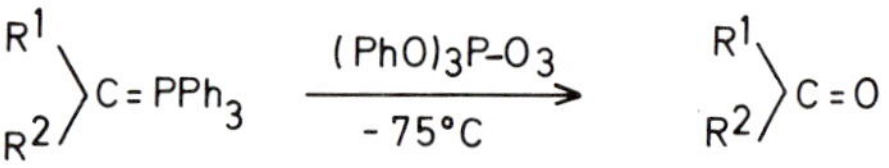

$$\begin{array}{c} R^1 \\ R^2 \end{array}\!\!\!\!\!C = \text{PPh}_3 \xrightarrow[-75^\circ\text{C}]{(\text{PhO})_3\text{P-O}_3} \begin{array}{c} R^1 \\ R^2 \end{array}\!\!\!\!\!C = O$$
(96)

Scheme 53

applicable and yields are consistently better than those obtained from autoxidation. Additional advantages over the autoxidation routes are the mild conditions (-75°C) required and the greater control over the amount of oxidant used. Perhaps the most important advantage over autoxidation, however, is that "stable" acyl ylides, which are not autoxidized, can be converted into 1,4-diacylethylenes and ketones by this new method.

2.2.5. Other ozone–phosphite adducts

The attraction of having a phosphite–ozone adduct that could be stored at room temperature and then heated to provide a source of singlet oxygen on demand has stimulated the examination of reactions of a wide range of phosphites with ozone.[149–151] The ozone adducts of the cage compounds 4-ethyl-2,6,7-trioxa-1-phosphabicyclo[2.2.2]octane[149] (97) and 1-phospha-2,8,9-trioxaadamantane[150] (98) are so far the most stable such compounds that have been used as singlet oxygen sources.

(97)   (98)

Scheme 54

The ozone adduct of (97) has a half-life of 76·2 min at 10°C and has successfully replaced the triphenyl phosphite–ozone adduct in its reactions with tetraphenylcyclopentadiene and 9,10-diphenylanthracene.[149] A particular advantage claimed for the use of the ozone adduct of (97) is that, since the by-product, phosphate, 4-ethyl-2,6,7-trioxa-1-phosphabicyclo[2.2.2]octane-1-oxide, is insoluble in solvents such as benzene, its removal from the reaction products is very easy.

7 Conversions via Deoxygenation

The ozone adduct of (98), which has a half-life of 25 min at 11°C, has also been shown to be able to replace the triphenyl phosphite–ozone adduct in a wide spectrum of its reactions, and to give broadly similar yields. This new adduct is water-soluble and therefore capable of providing a source of singlet oxygen in aqueous systems.

A phosphite–ozone adduct that is indefinitely stable at room temperature is not yet known, and a fundamental study of the stability of a series of such adducts, derived from different phosphites, by Stephenson and McClure[151] suggests that such a material may not be readily attainable. These workers confirmed the particular stability of the ozone adducts of caged phosphites and of certain cyclic phosphites, whose decomposition is retarded because pseudorotation at phosphorus is prevented, thus precluding attainment of the ideal geometry for extrusion of oxygen. All phosphite–ozone adducts in which pseudorotation is prevented have broadly similar stabilities and additional stability cannot be conferred by further increasing stereochemical constraints.

2.3 Other Reactions involving Ozone and Phosphorus(III) Reagents

The reactions discussed here are distinct from those discussed in the previous section in that here the phosphorus(III) reagent acts as the reductive component of what is otherwise a conventional ozonolysis reaction.

The full spectrum of simple alkenes has been shown to be cleaved to the expected ozonolysis products by treatment with ozone at $-40°C$ and below, followed by reduction, still at low temperature, by one of the lower trialkyl phosphites.[152] Methanol, or methanol plus a halogenomethane, is the solvent of choice for the reaction, and the choice of phosphite need be dictated only by the necessity to have a by-product phosphate with physical properties that facilitate its removal from the carbonyl products. The reaction is best documented with trimethyl phosphite, and yields are good and similar to those obtained with conventional reducing agents.

The reaction works equally well in the ozonolysis of substrates other than simple alkenes. Hence indene and phenanthrene are cleaved to the expected products, phenylacetylene is oxidized to phenylglyoxal, and butyne-1,4-diol diacetate is converted into 1,4-diacetoxybutane-1,2-dione (65%), the latter yield being almost twice that reported for conventional ozonolysis.

There is thus no reported case in which the use of trimethyl phosphite as reducing agent is significantly inferior to more conventional reducing methods.

Since the reaction is carried out at low temperatures ($< -40°C$) the work-up of an ozonolysis with a trialkyl phosphite can be particularly valuable when working with sensitive polyfunctional systems.

The conversion of the tricyclic ketone (99) into the dialdehyde (100) is an important step in Woodward's classic steroid synthesis,[153] and was originally carried out by a complex multi-step route. It can now be performed in satisfactory (>50%) yield by low-temperature ozonolysis (−60°C) followed by trimethyl phosphite reduction. Conventional ozonolysis work-up is quite unsatisfactory, giving roughly half the yield of the improved method.[152]

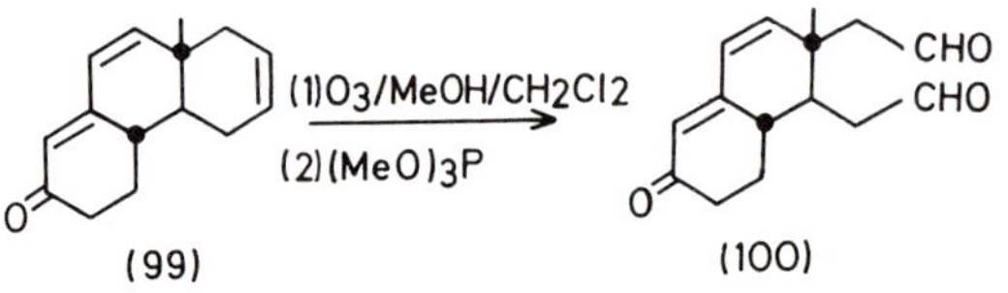

Scheme 55

A particularly elegant application of this mild ozonolysis work-up is the conversion of the 9-octalin-2-ones (101) into the dihydroazulenes (bicyclo-[5.3.0]dec-1(7)-ene-6,8-diones) (103) in high yield.[154] The intermediate triones (102) were not observed since they spontaneously undergo transannular condensation with loss of water.

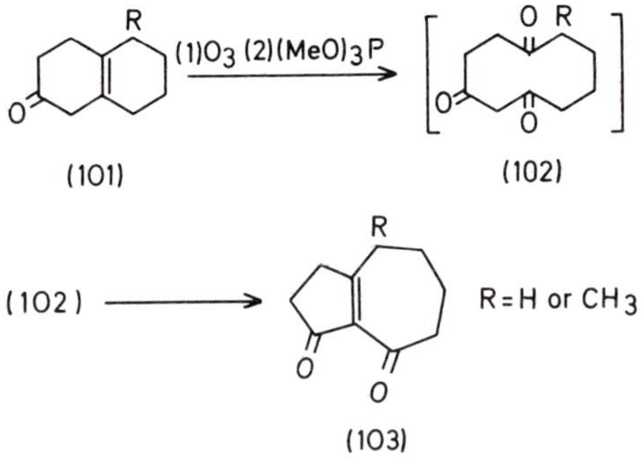

Scheme 56

Phosphorus(III) compounds other than phosphites have been used to reduce the ozonolysis intermediates but their synthetic utility is not so extensively documented. Hence Lorenz[155] has reduced ozonides with triphenylphosphine at room temperature. The reaction takes about 3 days but is sufficiently quantitative to form the basis of an estimation of ozonides. Although best performed in an alcoholic solvent, the reduction can be carried out in heptane provided that it is catalysed by small quantities of phenol.

The future possibility of a more preparatively useful development of this reaction may stem from the report that the time required for the reaction can be reduced to 60 min by carrying out the reduction in a sealed tube at 100°C.[156]

Triphenylphosphine can be used as reductant in the ozonolysis of acetylenes to α,β-diketones.[157] Hence both hex-3-yne-2,5-diol (104; $R^1 = R^2 = Me$, $X = H$) and its diacetate (104; $R^1 = R^2 = Me$, $X = CH_3CO$) are converted in >50% yield into 2,5-dihydroxyhexane-3,4-dione (105; $R^1 = R^2 = Me$, $X = H$) and 2,5-diacetoxyhexane-3,4-dione (105; $R^1 = R^2 = Me$, $X = CH_3CO$) respectively.

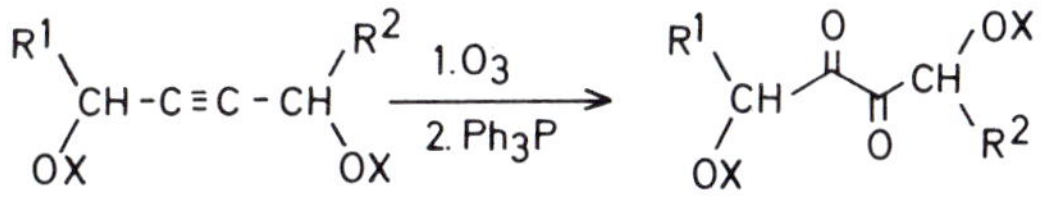

$$\underset{(104)}{\overset{R^1}{\underset{OX}{>}}CH-C\equiv C-CH\overset{R^2}{\underset{OX}{<}}} \xrightarrow[2.\ Ph_3P]{1.\ O_3} \underset{(105)}{\overset{R^1}{\underset{OX}{>}}CH\overset{O}{\underset{}{-C-}}\overset{}{\underset{O}{-C-}}CH\overset{OX}{\underset{R^2}{<}}}$$

Scheme 57

The reaction appears to be general when both R^1 and R^2 are alkyl groups, but if one of the hydroxy or acetoxy functions is primary (R^1 or $R^2 = H$) yields are considerably depressed owing to spontaneous fragmentation of the intermediate products from reaction with ozone, prior to reduction by the phosphine. Thus 1,4-diacetoxypent-2-yne gives only *ca.* 33% of dione on ozonolysis followed by triphenyl phosphite reduction, and the corresponding free diol does not give a dione in preparatively useful yield at all.

Tris(dimethylamino)phosphine has also been used as the reducing component of an ozonolysis in a steroid synthesis and shows some promise as a particularly mild reagent,[158] being effective at −60°C, and possibly rivals the phosphites in this respect although the generality of its application is unexplored.

3. References

J. I. G. Cadogan and R. K. Mackie, "Tervalent Phosphorus Compounds in Organic Synthesis," *Quart. Rev.*, 1974, **3**, 87.

T. Mukaiyama, "Oxidation-Reduction Condensation," *Angew. Chem. Internat. Ed.*, 1976, **15**, 94.

L. Horner and H. Hoffmann, "Preparative and Analytical Importance of Phosphorus and Related Compounds," *Newer Methods of Preparative Organic Chemistry* (ed. W. Foerst), Academic Press, New York and London, 1963, Vol. II, p. 163.

1. A. R. Katritzky and J. M. Lagowski, *Chemistry of the Heterocyclic N-oxides*, Academic Press, London, 1971.
2. T. R. Emerson and C. W. Rees, *J. Chem. Soc.*, 1964, 2319.
3. E. Ochai, *J. Org. Chem.*, 1953, **18**, 534.
4. M. E. Baumgarten and J. E. Dirks, *J. Org. Chem.*, 1958, **23**, 900.
5. F. Ramirez and A. M. Aguair, Amer. Chem. Soc., Abstracts of 134th Meeting, 1958, p. 42N.
6. J. I. G. Cadogan, M. Cameron-Wood, R. K. Mackie, and R. J. G. Searle, *J. Chem. Soc.*, 1965, 4831.

7. J. I. G. Cadogan, D. J. Sears, and D. M. Smith, *J. Chem. Soc. (C)*, 1969, 1314.
8. T. R. Emerson and C. W. Rees, *J. Chem. Soc.*, 1962, 1917.
9. E. Howard and W. F. Olszewski, *J. Amer. Chem. Soc.*, 1959, **81**, 1483.
10. M. S. Habib and C. W. Rees, *J. Chem. Soc.*, 1960, 3386.
11. R. A. Burrell, J. M. Cox, and E. G. Savins, *J. Chem. Soc., Perkin I*, 1973, 2707.
12. J. P. Dirlam and J. W. McFarland, *J. Org. Chem.*, 1977, **42**, 1360.
13. C. Grundmann, *Chem. Ber.*, 1964, **97**, 575.
14. S. M. Katzman and J. Moffat, *J. Org. Chem.*, 1972, **37**, 1842.
15. A. J. Boulton, A. C. G. Gray, and A. R. Katritzky, *J. Chem. Soc.*, 1965, 5958.
16. J. Ackrell, M. Altaf-ur-Rahman, A. J. Boulton, and R. C. Brown, *J. Chem. Soc., Perkin I*, 1972, 1587.
17. T. Mukai and M. Nitta, *J. Chem. Soc., Chem. Comm.*, 1970, 1192.
18. L. Horner and H. Hoffmann, in *Newer Methods of Preparative Organic Chemistry* (ed. W. Foerst), Academic Press, New York and London, 1963, Vol. II.
19. T. Mukaiyama, H. Nambu, and M. Okamoto, *J. Org. Chem.*, 1962, **27**, 3651.
20. B. A. Arbusov, E. N. Dianova, V. S. Vinogradova, and A. F. Lisin, *Izv. Akad. Nauk S.S.S.R., Ser. Khim.*, 1975, 695 (*Chem. Abs.*, 1975, **83**, 28335).
21. C. Grundmann and H. D. Frommeld, *J. Org. Chem.*, 1965, **30**, 2077.
22. W. E. Truce, T. C. Klingler, and W. W. Brand, in *Organic Chemistry of Sulphur* (ed. S. Oae), Plenum, New York, 1977.
23. E. H. Amonoo-Neizer, S. K. Ray, R. A. Shaw, and B. C. Smith, *J. Chem. Soc.*, 1965, 4296.
24. H. Szmant and O. Cox, *J. Org. Chem.*, 1966, **31**, 1595.
25. J. P. A. Costrillon and H. H. Szmant, *J. Org. Chem.*, 1965, **30**, 1338.
26. G. A. Olah, B. G. B. Gupta, and S. C. Narang, *Synthesis*, 1978, 137.
27. S. Oae, A. Nakanishi, and S. Kozuka, *Tetrahedron*, 1972, **44**, 549.
28. M. Dreux, Y. Leroux, and P. Savignac, *Synthesis*, 1974, 506.
29. D. W. Chasar and T. W. Pratt, *Synthesis*, 1976, 262.
30. L. Horner and H. Nickel, *Annalen*, 1955, **597**, 20.
31. J. F. Carson and F. F. Wong, *J. Org. Chem.*, 1961, **26**, 1467.
32. D. H. R. Barton, G. Page, and D. A. Widdowson, *J. Chem. Soc., Chem. Comm.*, 1970, 1466.
33. R. D. G. Cooper and F. L. Jose, *J. Amer. Chem. Soc.*, 1970, **92**, 2575; L. D. Hatfield, J. Fisher, F. L. Jose, and R. D. G. Cooper, *Tetrahedron Lett.*, 1970, 4897.
34. F. Ramirez, A. G. Gulati, and C. P. Smith, *J. Org. Chem.*, 1968, **33**, 13.
35. V. Mark, *J. Amer. Chem. Soc.*, 1963, **85**, 1884.
36. F. Ramirez, *Accounts Chem. Res.*, 1968, **1**, 168.
37. F. Ramirez, S. B. Bhatai, and C. P. Smith, *Tetrahedron*, 1967, **23**, 2067, and references therein.
38. A. C. Poshkus and J. E. Herweh, *J. Org. Chem.*, 1964, **29**, 2567.
39. P. J. Bunyan and J. I. G. Cadogan, *J. Chem. Soc.*, 1963, 42.
40. C. W. Bird and D. Y. Yong, *J. Chem. Soc., Chem. Comm.*, 1969, 932.
41. V. Mark, *Org. Synth.*, 1966, **46**, 31.
42. W. M. Horspool, S. T. McNeilly, J. A. Miller, and I. M. Young, *J. Chem. Soc., Perkin I*, 1972, 1113.

43. J. H. Boyer and R. Selvarajan, *J. Org. Chem.*, 1970, **35**, 1229.
44. T. Mukaiyama, I. Kuwajima, and K. Ohno, *Bull. Chem. Soc. Japan*, 1965, **38**, 1954.
45. G. W. Griffin, S. K. Satra, N. E. Brightwell, K. Ishikawa, and N. S. Bhacca, *Tetrahedron Lett.*, 1976, 1239.
46. G. W. Griffin, D. M. Gibson, and K. Ishikawa, *J. Chem. Soc., Chem. Comm.*, 1975, 595.
47. M. S. Newman and S. Blum, *J. Amer. Chem. Soc.*, 1964, **86**, 5598.
48. R. G. Harvey, S. H. Goh, and C. Cortez, *J. Amer. Chem. Soc.*, 1975, **97**, 3468; S. C. Agarwal and B. L. Van Duuren, *J. Org. Chem.*, 1977, **42**, 2730.
49. R. M. Moriarty, P. Dansette, and D. M. Jerina, *Tetrahedron Lett.*, 1975, 2557.
50. S. C. Agarwal and B. L. Van Duuren, *J. Org. Chem.*, 1975, **40**, 2307.
51. L. Horner, P. Beck, and V. G. Toscano, *Chem. Ber.*, 1961, **94**, 1323.
52. M. J. Gallacher and I. D. Jenkins, *J. Chem. Soc. (C)*, 1969, 2605.
53. I. J. Borowitz, M. Anschel, and P. D. Readio, *J. Org. Chem.*, 1971, **36**, 553.
54. F. Ramirez, A. S. Gulati, and C. P. Smith, *J. Amer. Chem. Soc.*, 1967, **89**, 6283.
55. F. Ramirez and M. Ramanathan, *J. Org. Chem.*, 1961, **26**, 3041.
56. F. Ramirez, A. W. Patwardhan, H. J. Kugler, and C. P. Smith, *Tetrahedron*, 1968, **24**, 2275.
57. F. Ramirez, S. B. Bhatia, A. V. Patwardhan, E. H. Chen, and C. P. Smith, *J. Org. Chem.*, 1968, **33**, 20.
58. F. Ramirez, H. Yamanaka, and O. H. Basedow, *J. Amer. Chem. Soc.*, 1961, **83**, 173.
59. C. W. Bird and D. Y. Wong, *J. Chem. Soc., Chem. Comm.*, 1969, 932.
60. J. H. Markgraf, C. I. Heller, and N. L. Avery, *J. Org. Chem.*, 1970, **35**, 1588.
61. C. W. Bird and D. Y. Wong, *Tetrahedron*, 1975, **31**, 31, and references therein.
62. M. G. Miles, J. D. Wilson, D. J. Dahm, and J. H. Wagenknecht, *J. Chem. Soc. Chem. Comm.*, 1974, 751; M. G. Miles, J. S. Wager, J. D. Wilson, and A. R. Siedle, *J. Org. Chem.*, 1975, **40**, 2577.
63. A. N. Pudovik, I. V. Gur'yanova, V. P. Kakurina, and N. P. Anoshina, *J. Gen. Chem. (U.S.S.R.)*, 1971, **41**, 1237.
64. D. J. Burton, F. E. Herkes, and K. J. Klabunde, *J. Amer. Chem. Soc.*, 1966, **88**, 5042.
65. T. Mukaiyama and Y. Yakota, *Bull. Chem. Soc. Japan*, 1965, **38**, 858.
66. F. Ramirez, R. B. Mitra, and N. B. Desai, *J. Amer. Chem. Soc.*, 1960, **82**, 2651; F. Ramirez and N. B. Desai, *ibid.*, p. 2652.
67. T. Mukaiyama, H. Nambu, and T. Kumamoto, *J. Org. Chem.*, 1964, **29**, 2243.
68. J. A. Harvey and M. A. Ogliarso, *Org. Preparations and Procedures Int.*, 1976, **8**, 37.
69. S. Trippett and D. M. Walker, *J. Chem. Soc.*, 1960, 2976.
70. M. J. Boskin and D. B. Denney, *Chem. and Ind.*, 1959, 330.
71. G. Wittig and W. Haag, *Chem. Ber.*, 1955, **88**, 1654.
72. D. E. Bissing and A. J. Speziale, *J. Amer. Chem. Soc.*, 1963, **85**, 3878.
73. D. E. Bissing and A. J. Speziale, *J. Amer. Chem. Soc.*, 1965, **87**, 2683.
74. C. B. Scott, *J. Org. Chem.*, 1957, **22**, 1118.
75. N. P. Neureiter and F. G. Bordwell, *J. Amer. Chem. Soc.*, 1959, **81**, 578.

76. R. Appel and R. Kleinstuck, *Chem. Ber.*, 1974, **107**, 5; P. D. Bartlett, A. L. Baumstark, and M. E. Landis, *J. Amer. Chem. Soc.*, 1973, **95**, 6486.
77. C. D. Weis, *J. Org. Chem.*, 1962, **27**, 3520.
78. C. Walling and R. Rabinowitz, *J. Amer. Chem. Soc.*, 1959, **81**, 1243; P. J. Krusic, W. Mahler, and J. K. Kochi, *ibid.*, 1972, **94**, 6033; A. G. Davies, D. Griller, and B. P. Roberts, *J. Chem. Soc., Perkin II*, 1972, 993.
79. L. Horner and W. Jurgeleit, *Annalen*, 1955, **591**, 138.
80. C. Walling, O. H. Basedow, and E. S. Savos, *J. Amer. Chem. Soc.*, 1960, **82**, 2181.
81. D. B. Denney, W. F. Goodyear, and B. Goldstein, *J. Amer. Chem. Soc.*, 1960, **82**, 1393.
82. K. J. Humphris and G. Scott, *J. Chem. Soc., Perkin II*, 1973, 826, 831.
83. M. S. Kharasch, R. A. Mosher, and I. S. Bengelsdorf, *J. Org. Chem.*, 1960, **25**, 1000.
84. R. Hiatt and C. McColeman, *Canad. J. Chem.*, 1971, **49**, 1712; R. Hiatt, R. J. Smythe, and C. McColeman, *ibid.*, p. 1707.
85. A. Davies and R. Feld, *J. Chem. Soc.*, 1958, 4637.
86. J. N. Gardner, F. E. Carlon, and O. Gnoj, *J. Org. Chem.*, 1968, **33**, 3294.
87. M. Schulz, U. Missol, and H. Bohm, *J. Prakt. Chem.*, 1974, **316**, 47.
88. G. Rousseau, P. Le Perchec, and J. M. Conia, *Tetrahedron Lett.*, 1977, 2517.
89. H. D. Holtz, P. W. Solomon, and J. E. Mahan, *J. Org. Chem.*, 1973, **38**, 3175.
90. D. B. Denney, H. M. Relles, and A. K. Tsolis, *J. Amer. Chem. Soc.*, 1964, **86**, 4487.
91. D. B. Denney, D. Z. Denney, C. D. Hall, and K. L. Marsi, *J. Amer. Chem. Soc.*, 1972, **94**, 245.
92. D. B. Denney and D. H. Jones, *J. Amer. Chem. Soc.*, 1969, **91**, 5821, and references therein.
93. C. W. Jefford and C. G. Rimbault, *Tetrahedron Lett.*, 1977, 2375.
94. G. O. Pierson and O. A. Runquist, *J. Org. Chem.*, 1969, **34**, 3654.
95. Y. Ito, M. Oda, and Y. K. Kitahana, *Tetrahedron Lett.*, 1976, 839.
96. Y. Ito, M. Oda, and Y. K. Kitahana, *Tetrahedron Lett.*, 1975, 239.
97. C. H. Foster and G. A. Berchtold, *J. Org. Chem.*, 1975, **40**, 3743.
98. F. Challenger and V. K. Wilson, *J. Chem. Soc.*, 1927, 209.
99. M. A. Greenbaum, D. B. Denney, and A. K. Hoffmann, *J. Amer. Chem. Soc.*, 1956, **78**, 2563; D. B. Denney and M. A. Greenbaum, *ibid.*, 1957, **79**, 979.
100. P. J. Bunyan, A. J. Burn, and J. I. G. Cadogan, *J. Chem. Soc.*, 1963, 1527.
101. D. I. Brydon and J. I. G. Cadogan, *J. Chem. Soc.*, (*C*), 1968, 819.
102. D. B. Denney, W. F. Goodyear, and B. Goldstein, *J. Amer. Chem. Soc.*, 1961, **83**, 1726.
103. W. Adam, R. J. Ramirez, and S. C. Tsai, *J. Amer. Chem. Soc.*, 1969, **91**, 1254.
104. W. Adam and J. W. Diehl, *J. Chem. Soc., Chem. Comm.*, 1972, 797.
105. W. Adam and A. Rois, *J. Org. Chem.*, 1971, **36**, 407.
106. T. Mukaiyama, *Angew. Chem. Internat. Ed.*, 1976, **15**, 94.
107. E. Brunn and R. Huisgen, *Angew. Chem. Internat. Ed.*, 1969, **8**, 513.
108. M. Wada and O. Mitsunobu, *Tetrahedron Lett.*, 1972, 1279.
109. O. Mitsunobu and M. Yamada, *Bull. Chem. Soc. Japan*, 1967, **40**, 2380.
110. O. Mitsunobu and M. Eguchi, *Bull. Chem. Soc. Japan*, 1971, **44**, 3427.
111. A. K. Bose, B. Lal, W. A. Hoffman, and M. S. Manhas, *Tetrahedron Lett.*, 1973, 1619.

7 Conversions via Deoxygenation

112. H. Loibner and E. Zbiral, *Helv. Chim. Acta*, 1977, **60**, 417.
113. R. Aneja, A. P. Davies, and J. A. Knaggs, *Tetrahedron Lett.*, 1975, 1033.
114. (a) G. Alfredsson and P. J. Garegg, *Acta Chem. Scand.*, 1973, **27**, 724. (b) S. Shimokawa, J. Kimura, and O. Mitsunobu, *Bull. Chem. Soc., Japan*, 1976, **49**, 3357.
115. R. Aneja, A. P. Davies, A. Harkes, and J. A. Knaggs, *J. Chem. Soc., Chem. Comm.*, 1974, 963.
116. O. Mitsunobu and J. Kimura, *J. Amer. Chem. Soc.*, 1969, **91**, 6510; O. Mitsunobu, M. Yamada, and T. Mukaiyama, *Bull. Chem. Soc. Japan*, 1967, **40**, 935.
117. O. Mitsunobu, M. Wada, and T. Sano, *J. Amer. Chem. Soc.*, 1972, **94**, 679.
118. M. Wada, T. Sano, and O. Mitsunobu, *Bull. Chem. Soc. Japan*, 1973, **46**, 2833.
119. A. Zamojski, W. A. Szarek, and J. K. N. Jones, *Carbohydrate Res.*, 1972, **23**, 460.
120. J. Jurezak, G. Grynkiewiez, and A. Zamojski, *Carbohydrate Res.*, 1975, **39**, 147.
121. W. A. Szarek, C. Depew, and J. K. N. Jones, *J. Heterocyclic Chem.*, 1976, **13**, 1131.
122. W. A. Szarek, C. Depew, H. C. Jarell, and J. K. N. Jones, *J. Chem. Soc., Chem. Comm.*, 1975, 648.
123. E. Grochowski and J. Jurezak, *Synthesis*, 1977, 277.
124. E. Grochowski and J. Jurezak, *Carbohydrate Res.*, 1976, **50**, C15.
125. S. Bittner, S. Grinberg, and I. Kartoon, *Tetrahedron Lett.*, 1974, 1965.
126. S. Bittner and Y. Assaf, *Chem. and Ind.*, 1975, 281.
127. M. S. Manhas, W. H. Hoffman, B. Lal, and A. K. Bose, *J. Chem. Soc., Perkin I*, 1975, 461.
128. W. A. Szarek, H. C. Jarrell, and J. K. N. Jones, *Carbohydrate Res.*, 1977, **57**, C13.
129. F. Yoneda and N. Nagamatsu, *Tetrahedron Lett.*, 1973, 1577.
130. I. Kuwajima and T. Mukaiyama, *J. Org. Chem.*, 1964, **29**, 1385.
131. H. Loibner and E. Zbiral, *Helv. Chim. Acta*, 1976, **59**, 2100.
132. S. Bittner, Z. Barneis, and S. Felix, *Tetrahedron Lett.*, 1975, 3871.
133. B. Bejer, E. von Hirrichs, and I. Ugi, *Angew. Chem. Internat. Ed.*, 1972, **11**, 929.
134. J. Mulzer and G. Bruntrup, *Angew. Chem. Internat. Ed.*, 1977, **16**, 255.
135. Q. E. Thompson, *J. Amer. Chem. Soc.*, 1961, **83**, 845.
136. E. J. Corey and W. C. Taylor, *J. Amer. Chem. Soc.*, 1964, **86**, 3880.
137. R. W. Murray, J. W.-P. Lin, and M. L. Kaplan, *Ann. New York Acad. Sci.*, 1970, 171.
138. P. D. Bartlett, G. D. Mendenhall, and A. P. Schaap, *Ann. New York Acad. Sci.*, 1970, **171**, 79.
139. D. R. Kearns, *Chem. Rev.*, 1971, **71**, 395.
140. R. W. Murray and M. L. Kaplan, *J. Amer. Chem. Soc.*, 1969, **91**, 5358.
141. C. H. Foster and G. A. Berchtold, *J. Amer. Chem. Soc.*, 1972, **94**, 7939.
142. W. E. Barnett and L. L. Needham, *J. Org. Chem.*, 1971, **36**, 4134.
143. P. D. Bartlett and G. D. Mendenhall, *J. Amer. Chem. Soc.*, 1970, **92**, 210.
144. T. W. Sam and J. K. Sutherland, *J. Chem. Soc., Chem. Comm.*, 1972, 424.
145. A. P. Schaap and P. D. Bartlett, *J. Amer. Chem. Soc.*, 1970, **92**, 6055.
146. P. D. Bartlett and A. P. Schaap, *J. Amer. Chem. Soc.*, 1970, **92**, 322.

147. R. W. Murray, R. D. Smetana, and E. Block, *Tetrahedron Lett.*, 1971, 299.
148. H. J. Bestmann, L. Kisielowski, and W. Distler, *Angew. Chem. Internat. Ed.*, 1976, **15**, 298.
149. M. E. Brennan, *J. Chem. Soc., Chem. Comm.*, 1970, 956.
150. A. P. Schaap, K. Kees, and A. L. Thayer, *J. Org. Chem.*, 1975, **40**, 1185.
151. L. M. Stephenson and D. E. McClure, *J. Amer. Chem. Soc.*, 1973, **95**, 3074.
152. W. S. Knowles and Q. E. Thompson, *J. Org. Chem.*, 1960, **25**, 1031.
153. R. B. Woodward, F. Sondheimer, D. Taub, K. Heusler, and W. M. McLamone, *J. Amer. Chem. Soc.*, 1952, **74**, 4223.
154. R. A. Kretchmer and W. M. Schafer, *J. Org. Chem.*, 1973, **38**, 95.
155. O. Lorenz, *Analyt. Chem.*, 1965, **37**, 101.
156. O. Lorenz and C. R. Parks, *J. Org. Chem.*, 1965, **30**, 1976.
157. L. De, B. Maurer, and G. Ohloff, *Helv. Chim. Acta*, 1973, **56**, 1882.
158. A. Furlenmeier, A. Furst, A. Langemann, G. Waldvogel, P. Hocks, U. Kerb, and R. Wiechert, *Helv. Chim. Acta*, 1967, **50**, 2387.
159. C. Kaneko, M. Yamamori, A. Yamamoto, and R. Hayashi, *Tetrahedron Lett.*, 1978, 2799; C. Kaneko, A. Yamamoto, and M. Gomi, *Heterocycles*, 1979, **12**, 227.

8
Desulphurization via Phosphorus(III) Reagents: General Functional Group Conversions

R. K. MACKIE

1. Reactions with thiirans: synthesis of olefins, 351
2. Vinylogous amides and β-diketones from thioimidates and related compounds, 354
3. Alkenes from thionocarbonates and related reactions, 354
4. Alkenes from 1,3-oxathiolan-5-ones and azosulphides, 359
5. Desulphurization of sulphides and thiols, 360
6. Reactions with sulphoxides, sulphenic acids, and sulphenates, 363
7. Desulphurization of di- and poly-sulphides, 365
8. Formation of pyridyl thioesters and related synthons, 373
9. Condensation reactions, including formation of amides, phosphorylation, and application in nucleoside synthesis, 374
10. Reactions involving other sulphenyl compounds, 376
11. Reactions with carbon disulphide, thiobenzophenone, and dicyanosulphonium ylides, 379
12. Miscellaneous reactions, 380
13. References, 382

The size and polarizability of tervalent phosphorus enable it to interact more effectively with the vacant orbitals on sulphur than it can with those on first-row elements oxygen and nitrogen. Thus, phosphines and phosphites react with elemental sulphur in air to give phosphine sulphides[1] and thionophosphates[2] via a Menshutkin reaction involving successive nucleophilic displacements of phosphorus on sulphur.[3] This is the parent reaction which adumbrates many of synthetic significance.

1. Reactions with Thiirans: Synthesis of Olefins

Thiirans react with tervalent phosphorus compounds under much milder conditions than do oxirans, and they are smoothly converted into olefins.[4]

The reaction is completely stereospecific,[5] *meso*-2,3-dimethylthiiran giving (Z)-but-2-ene and (±)-2,3-dimethylthiiran giving (E)-but-2-ene, and in each case with less than 1% of the other isomer being formed. The mechanism is thought to involve a simple displacement on sulphur as shown in Scheme 1.[6]

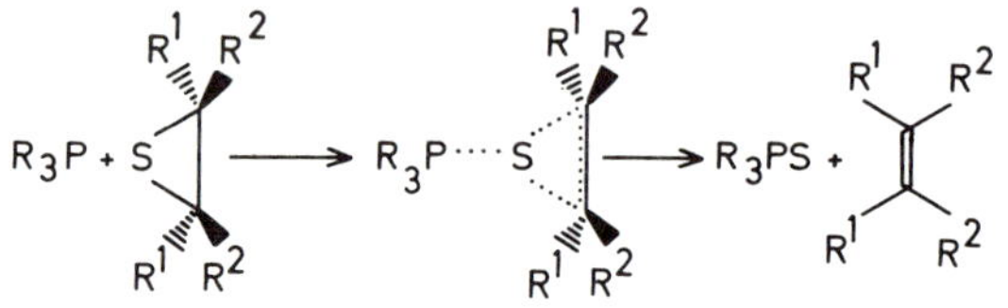

Scheme 1

Until recently the reaction was of little preparative significance because thiirans were often prepared from olefins, although the synthesis of some thermochromic alkenes had been described.[7] However, in the past few years new syntheses of thiirans have been reported,[8] and these have led to a new synthesis of olefins.[9] Although the intermediate thiirans have in some cases

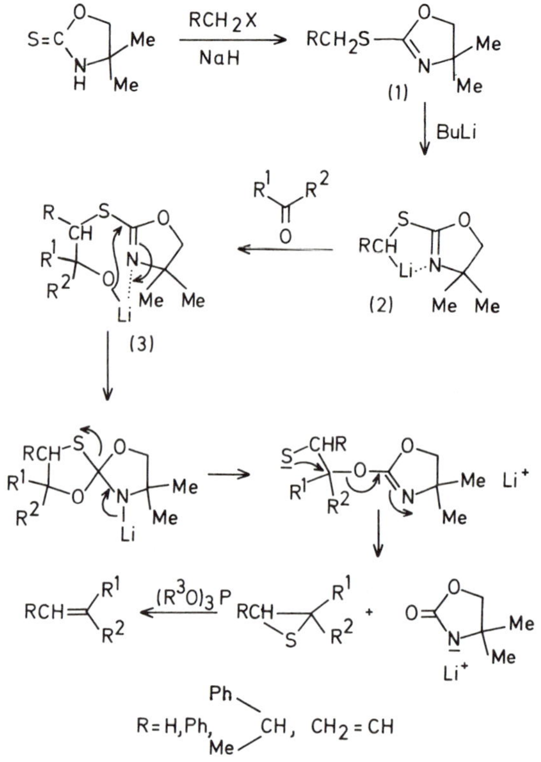

$R = H, Ph, \underset{Me}{\overset{Ph}{>}}CH, CH_2=CH$

Scheme 2

8 Conversions via Desulphurization

been isolated, it is by no means certain that this is necessary and, in the case of vinyl thiirans, which extrude sulphur, it is undesirable (Scheme 2). Table 1 shows the olefins formed from the carbonyl compounds indicated. Where (E)–(Z) isomerism is possible there is a predominance of the (E)-isomer.

Table 1. Synthesis of olefins from carbonyl compounds

Carbonyl compound	Olefin (E:Z ratio)
Heptanal	oct-1-ene
Formylcyclohexane	cyclohexylethylene
Cyclooctanone	methylenecyclooctane
2-Decalone	2-methylenedecalin
1-Formylcyclohexene	1-vinylcyclohexene
Cinnamaldehyde	1-phenylbuta-1,3-diene (96:4)
1-Decalone	1-methylenedecalin
Hexane-2,5-dione	2,5-dimethylhexa-1,5-diene
2-Methylcyclohexanone	1-methyl-2-methylenecyclohexane
Benzaldehyde	stilbene (86:14)
Heptanal	1-phenyloct-1-ene (60:40)
Cinnamaldehyde	1,4-diphenylbuta-1,3-diene (96:4)
Cyclooctanone	benzylidenecyclooctane
Benzophenone	1,1,2-triphenylpropene
Cyclohexyl phenyl ketone	1-cyclohexyl-1,2-diphenylpropene (91:9)
Benzaldehyde	1-phenylbuta-1,3-diene (96:4)
Heptanal	deca-1,3-diene

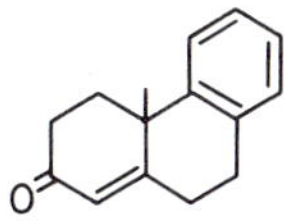

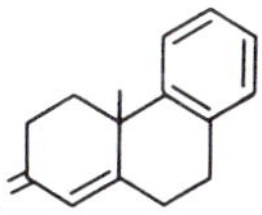

This is the result of minimizing steric effects in the intermediate (3). Using the chiral oxazole (4) asymmetric synthesis can be achieved; for example, optically active (5) is formed from racemic 2-methylcyclohexanone.[9] Some oxazoles form anions (2; R = CO_2Et) which are too stable to react with aldehydes or ketones, and the methylene group of S-ethyloxazoles (1; R = Me) is not of sufficient acidity to form a lithium salt.

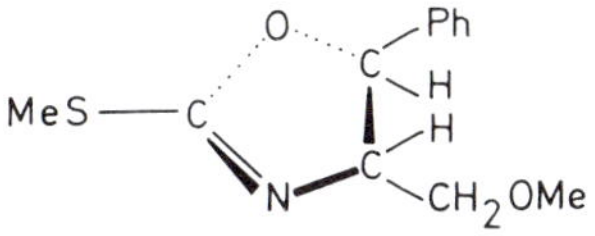

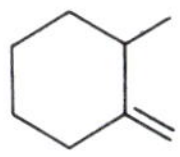

2. Vinylogous Amides and β-Diketones from Thioimidates and Related Compounds

Eschenmoser and his coworkers have developed a coupling reaction as a key step in the synthesis of corrins. The development from the original concept

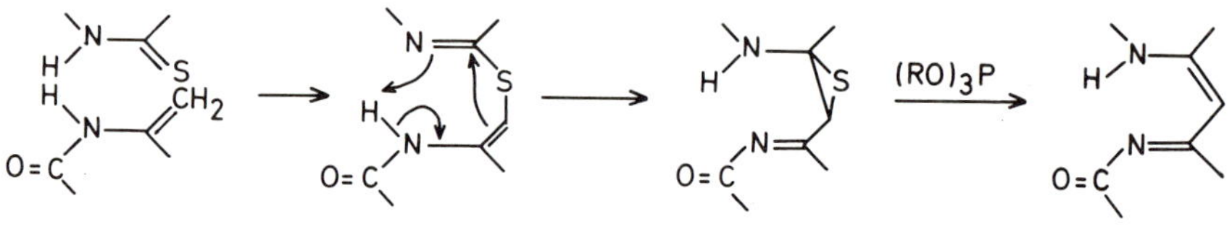

Scheme 3

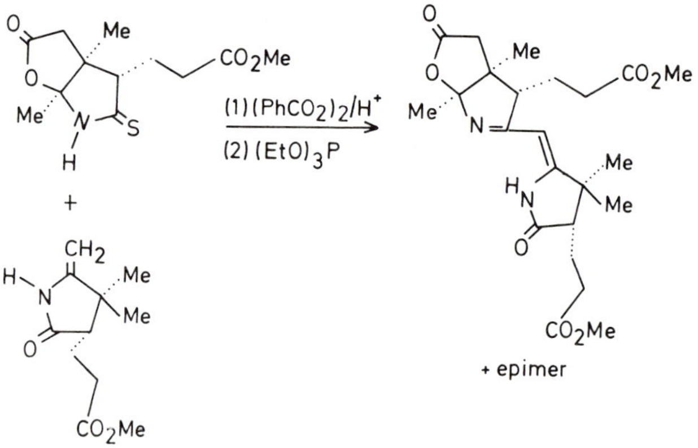

Scheme 4

(Scheme 3) to the preparative process (Scheme 4) has been described in detail.[10] Problems have been encountered in the oxidative coupling reaction but, where these can be solved, for example by reaction of a thioamide with an iodinated enamide, the overall reaction is successful.[11] The method has also been applied to the synthesis of vinylogous amides and of β-diketones,[12] representative examples of which are in Scheme 5. A report indicating that tervalent phosphorus compounds may not be the most satisfactory desulphurizing agents in the coupling reaction has appeared.[13]

3. Alkenes from Thionocarbonates and Related Reactions

cis-Elimination from an intermediate thionocarbonate permits the conversion of a *vic*-diol into an olefin with retention of stereochemistry. This reaction, developed by Corey and Winter, is a most valuable stereospecific olefin-

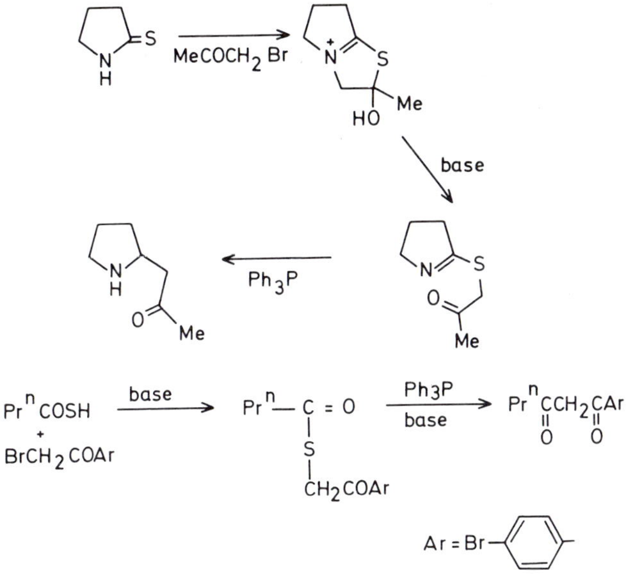

8 Conversions via Desulphurization

Scheme 5

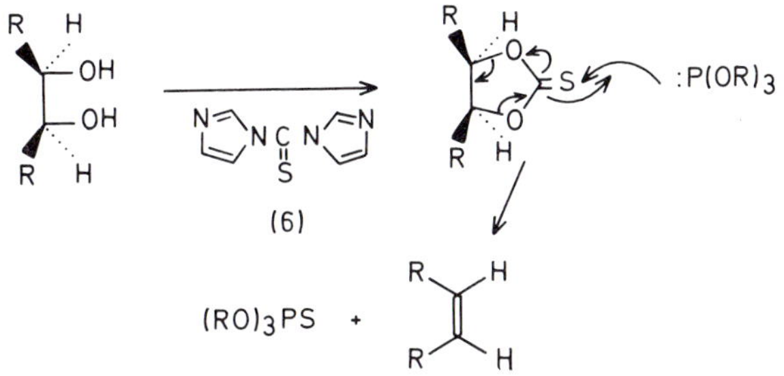

Scheme 6

forming reaction[14,15] (Scheme 6). The diol is converted into the thionocarbonate using thiocarbonyldiimidazole (6), and desulphurization is most often effected using trimethyl phosphite. It appears that the reaction may involve an intermediate carbene since a by-product (8) formed in the reaction from the glucofuranose derivative (7) can be accounted for by postulating an intermediate carbene (9) (Scheme 7).[16] The reaction results in (Z)-alkenes being formed from *meso* or *erythro* diols and (E)-alkenes from racemic or *threo* diols both with a very high degree of specificity. The versatility of the reaction is demonstrated by the range of examples in which the reaction may be employed: synthesis of unsaturated sugars,[16,17] cyclobutenes,[18,19] optically active *trans*-cyclooctene,[20] twistene,[21] and *cis*-cinnamic acid.[22] Very highly substituted diols do not react with (6) but the thionocarbonate may be formed

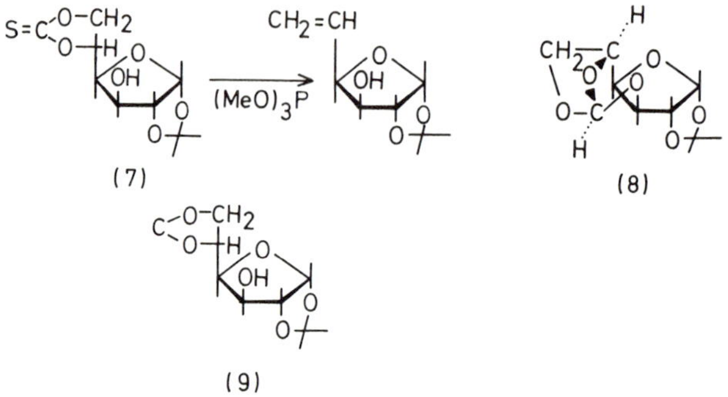

Scheme 7

by reaction of the di-anion of the diol with carbon disulphide followed by reaction with methyl iodide. The olefin is obtained from the thionocarbonate in the usual way.[23]

Unsaturated thionocarbonates can be used in two ways. In one, internal acetylenes can be prepared, but the authors suggest that synthesis from the α-diketone dihydrazone is the preferred method[24] (Scheme 8). The other involves the use of vinylene thionocarbonate (10) in the Diels–Alder reaction as an equivalent of acetylene[25] (Scheme 9).

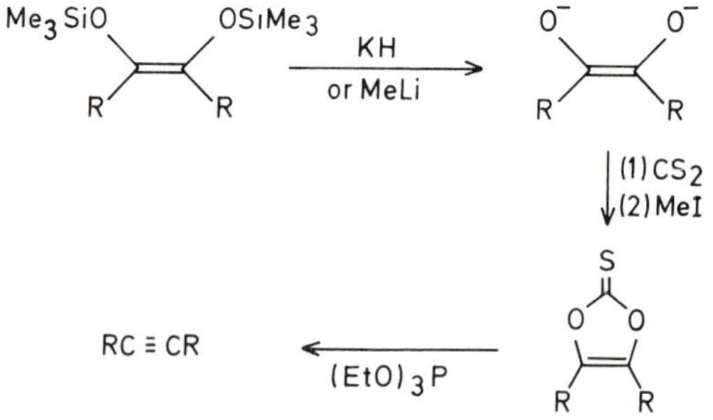

Scheme 8

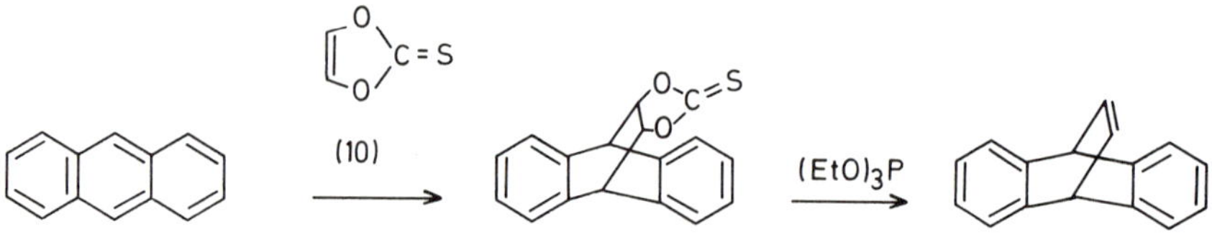

Scheme 9

By use of 1,2-trithiocarbonates, (Z)-alkenes can be isomerized to (E)-alkenes. Two routes are outlined in Scheme 10.[15] *trans*-Cyclohexene 1,2-trithiocarbonate gives (11) and *o*-phenylene thionocarbonate gives (12).[26]

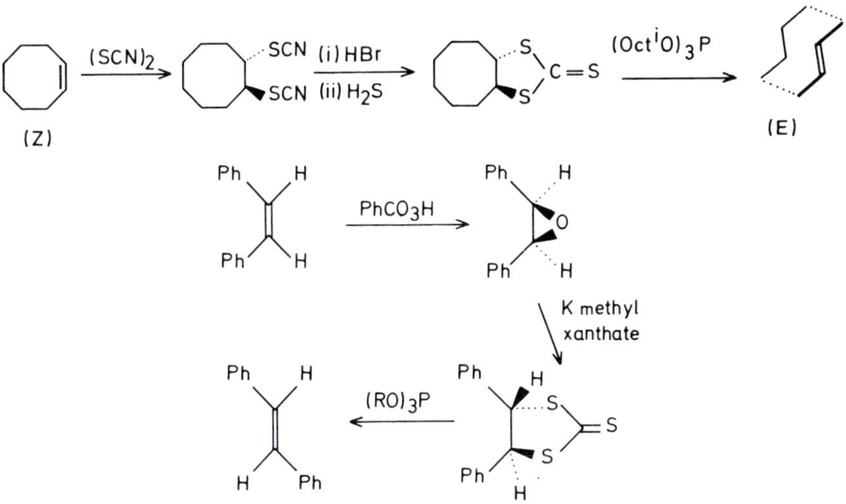

8 Conversions via Desulphurization

Scheme 10

cis-Cyclohexene 1,2-trithiocarbonate, on the other hand, gives the expected elimination product, cyclohexene, on reaction with triethyl phosphite.[27]

Charge transfer complexes of tetrathio-ethylenes and -fulvalenes and their selenium analogues have important electrical properties. Tetrathioethylenes have been prepared by the action of triethyl phosphite on acyclic trithiocarbonates,[28] and the synthesis of the fulvalenes from thio- and selenocarbonates using phosphines and phosphites has recently been reviewed.[29] It appears that selenone derivatives (13a) are more reactive than thione derivatives (13b).[30] However, the 1,2-trithiocarbonate (14a) gives a lower yield of 2,2'-bis-(4,5-dicyano-1,3-dithiolidene) than does the 1,2-dithiocarbonate (14b).[31] A more extensive investigation of the reactions of (14a) with triphenylphosphine shows that (16) and (17) are formed at lower temperatures

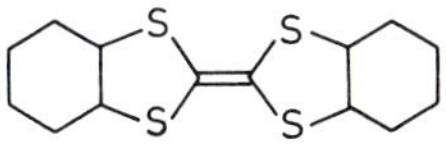

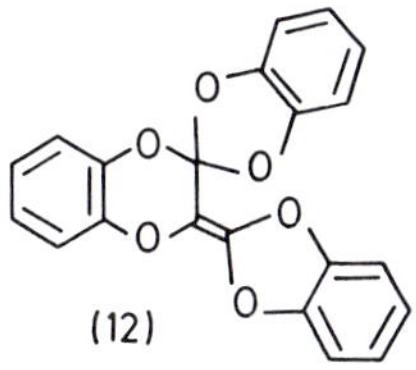

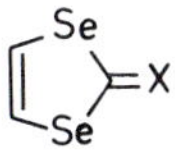

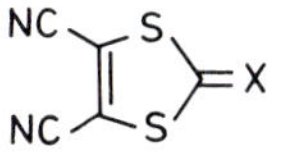

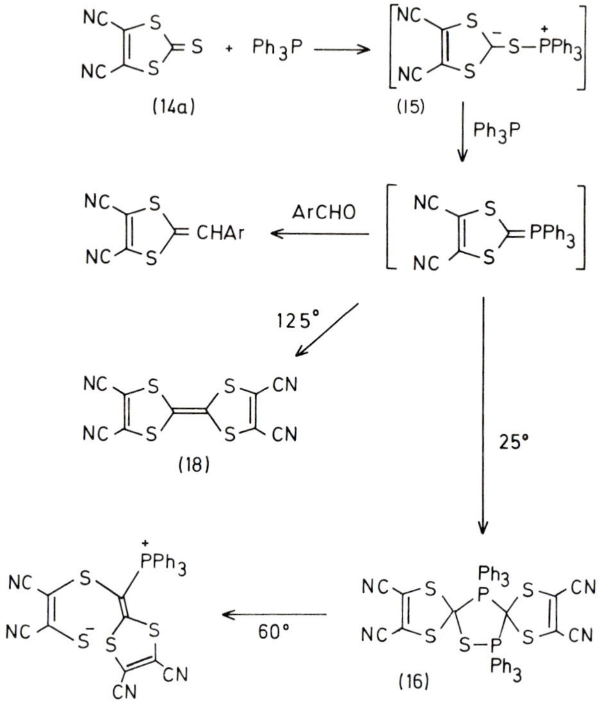

Scheme 11

and that the expected fulvalene (18) is only formed at 125°. In the presence of aromatic aldehydes a Wittig reaction is observed and the authors believe that the dipolar intermediate (15) rather than a carbene is involved[32] (Scheme 11).

A similar conclusion had previously been reached in an investigation of the reaction of o-phenylene trithiocarbonate with phosphites.[33] Further work on this reaction has shown that tetrathiofulvalene formation is greatly affected by the functional groups attached to the trithiocarbonate.[34]

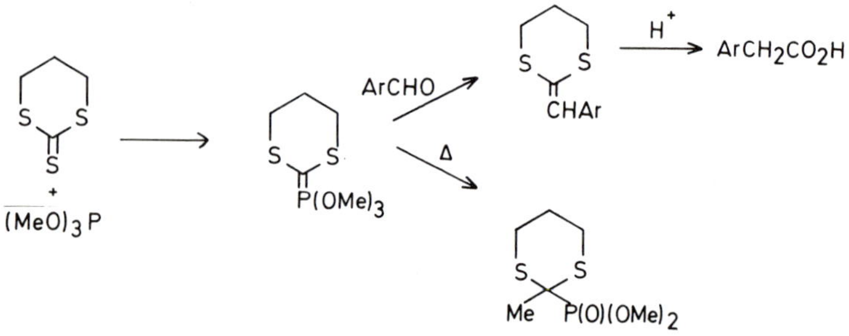

Scheme 12

8 Conversions via Desulphurization

Ylides are formed by reaction of phosphites with 1,3-trithiocarbonates, and these undergo Wittig reactions giving ketene thioketals which on hydrolysis yield substituted acetic acids[35] (Scheme 12).

4. Alkenes from 1,3-Oxathiolan-5-ones and Azosulphides

In 1970 Barton enunciated the principles of an olefin synthesis involving a two-fold extrusion process[36] (Scheme 13). Two successful applications using

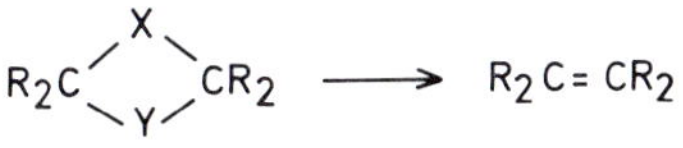

$$R_2C\underset{Y}{\overset{X}{\diagup\diagdown}}CR_2 \longrightarrow R_2C=CR_2$$

Scheme 13

tervalent phosphorus compounds to initiate the extrusion process have been reported. In the first, 1,3-oxathiolan-5-ones, prepared from thiobenzilic acid and either ketones or benzaldehyde, are treated with tris(dimethylamino)-phosphine at high temperature (Scheme 14). Sterically hindered olefins may

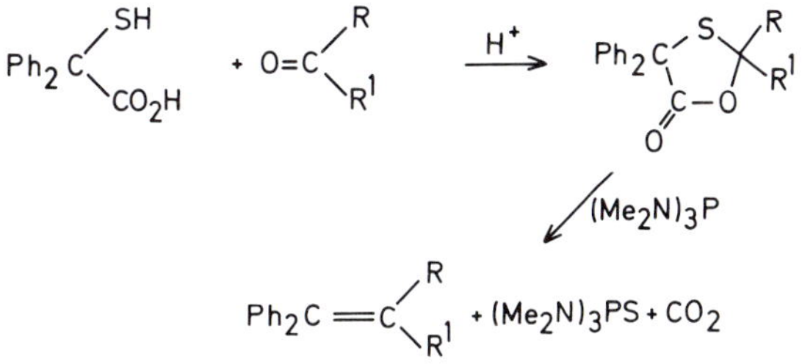

Scheme 14

be prepared but the scope of the reaction is limited by the need to have an aryl or other conjugated residue to facilitate the loss of carbon dioxide. The second method does not suffer from this disadvantage and is therefore of wider applicability. In this method, elimination of nitrogen accompanies extrusion of sulphur from azosulphides (Scheme 15). No olefin is formed

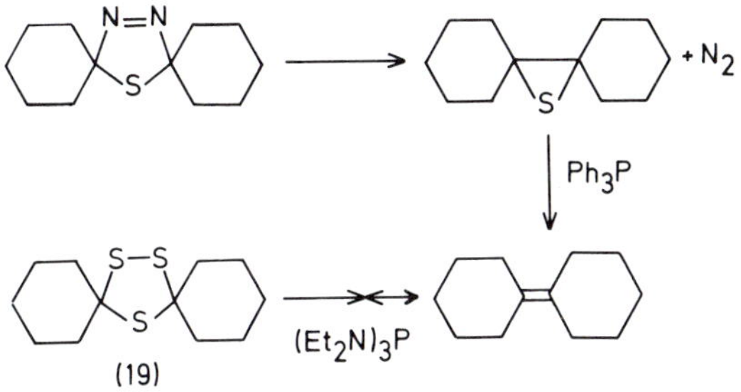

Scheme 15

from the corresponding trithia compound (19) on treatment with tervalent phosphorus compounds. The synthesis of symmetrical azosulphides is accomplished by reaction of a ketone azine with hydrogen sulphide followed by oxidation of the resulting tetrahydrothiadiazole with lead tetraacetate[37] or with diethyl azodicarboxylate.[38] Extrusion of sulphur is effected by using either triphenylphosphine or triethyl phosphite. The stereochemistry of the intermediate thiirans and of the olefins formed has been investigated.[39] Unsymmetrical and very hindered olefins can be obtained by forming the azosulphide from a thione and either diphenyldiazomethane or di-t-butyldiazomethane.[40] Two methods that do not involve the preparation of diazomethanes have also been described (Scheme 16).

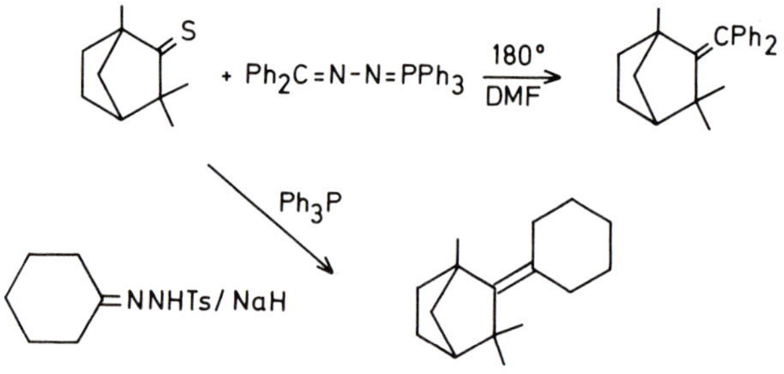

Scheme 16

5. Desulphurization of Sulphides and Thiols

Under ionic conditions simple thiols and monosulphides do not react with tervalent phosphorus compounds. However, under the influence of ultraviolet light or in the presence of a radical initiator (e.g. AIBN), alkanethiols are quantitatively converted into alkanes.[41] The reaction involves a thiophosphoranyl radical (20) as shown in Scheme 17.[42,43]

$$R^1SH \xrightarrow[\text{initiator}]{\text{u.v. or}} R^1S^\bullet$$

$$R^1S^\bullet + R_3P \longrightarrow R_3\overset{\bullet}{P}SR^1 \longrightarrow R_3PS + R^{1\bullet}$$
$$\hspace{5cm}(20)$$
$$R^{1\bullet} + R^1SH \longrightarrow R^1H + R^1S^\bullet$$

Scheme 17

Corey and Block[44] investigated the decomposition of thioethers with phosphines on irradiation with ultraviolet light. With the cyclic sulphide (21) and tributylphosphine or trioctyl phosphite a mixture of cyclohexane, cyclohexene, and hexa-1,5-diene is formed; the findings are rationalized in Scheme

8 Conversions via Desulphurization

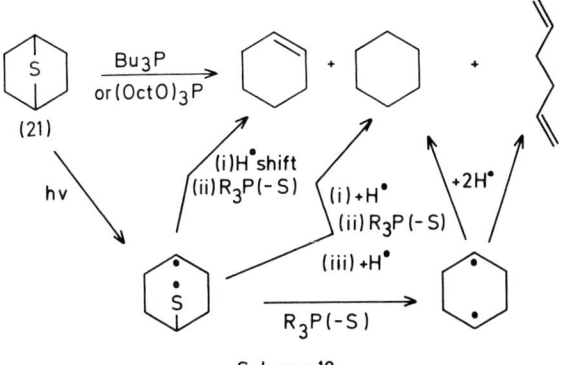

Scheme 18

18. Use of the more thiophilic tris(dimethylamino)phosphine leads to an increase in the proportion of hexa-1,5-diene formed. With larger cyclic compounds bicyclic hydrocarbons result rather than acyclic dienes. Diallyl sulphide gives hexa-1,5-diene and allyl methallyl sulphide a statistical distribution of the three possible dienes, hexa-1,5-diene, 2-methylhexa-1,5-diene, and 2,5-dimethylhexa-1,5-diene. Irradiation of a mixture of trioctyl phosphite and dibenzyl sulphide gives a 60% yield of bibenzyl. This led Bruhin and Jenny[45] and Boekelheide et al.[46] to devise a new synthesis of cyclophanes (Scheme 19). This method is now widely used for the synthesis of 2,2-[47] and 3,3-cyclophanes.[48]

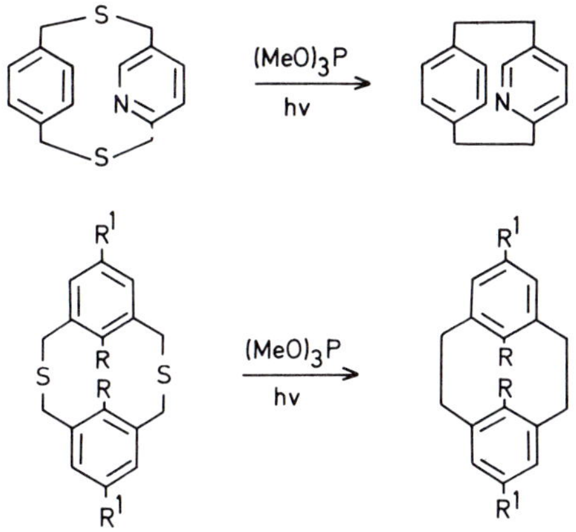

Scheme 19

In more complex molecules, some desulphurizations of sulphides have been reported to take place without the use of radical initiators or of ultraviolet light. A smooth desulphurization of a 5-mercaptocorrin–zinc complex has

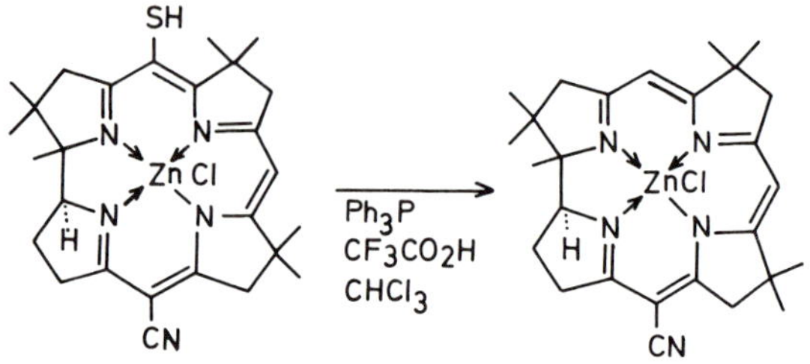

Scheme 20

been reported[49] (Scheme 20) and, at 200° in trichlorobenzene, triphenylphosphine converts N-methylthiaphlorin into N-methylcorrole.[50] Thiaphlorin itself thermally extrudes sulphur but yields are higher in the presence of triphenylphosphine. A method for the desulphurization of S-phenylacyl derivatives of 6-thiopurine and 2- or 4-thiopyrimidine nucleosides involves triphenylphosphine and potassium t-butoxide. This contributes to a valuable synthesis of C-substituted nucleosides.[51] Syntheses of stilbenes[52] from α-chlorodibenzyl sulphides and of diarylacetylenes[53] from α,α-dichlorodibenzyl sulphides have been reported. In neither case is the mechanism known but in both cases the order of addition of reagents (triphenylphosphine followed by potassium t-butoxide) is critical. Although syntheses of symmetrical molecules only have been reported, there appears to be no reason why the reaction cannot be extended to unsymmetrical stilbenes and diarylacetylenes (Scheme 21).

Scheme 21

There is some evidence that $6H$-1,3,4-thiadiazines can be converted into a valence isomer (22). They can be desulphurized to pyrazoles although in some cases a thiol may also be formed and in others a small amount of disulphide.[54] In both cases the by-product can be converted into the desired pyrazoles by reduction (Scheme 22).

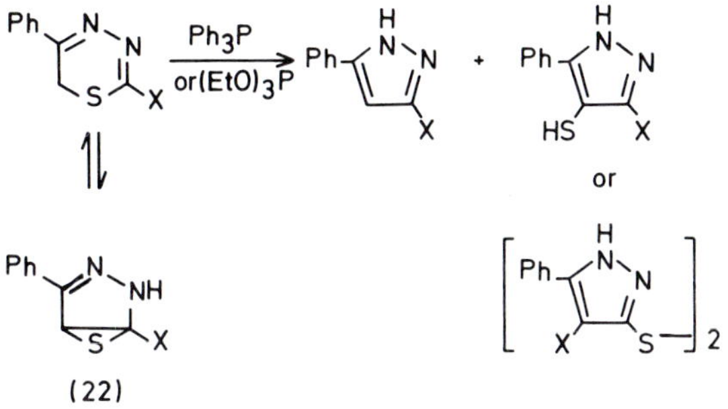

8 Conversions via Desulphurization

Scheme 22

A reported desulphurization of 3-chlorothietan may involve a thiiran intermediate or a displacement of chloride ions followed by a displacement of the phosphine sulphide by chloride ions[55] (Scheme 23). It may be significant that both thietan and 3-hydroxythietan react more slowly to give unidentified products.

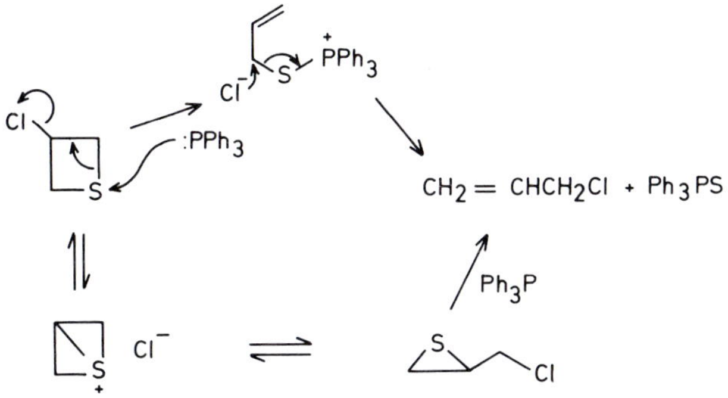

Scheme 23

6. Reactions with Sulphoxides, Sulphenic Acids, and Sulphenates

Deoxygenation of simple sulphoxides has been achieved using a variety of reagents.[56] Phosphorus trichloride appears to be most suitable for aromatic sulphoxides.[57] The use of triphenylphosphine with iodine has been described in Chapter 10 and another method is described later in this chapter (Section 11).

Deoxygenation of penicillin sulphoxides has also been reported[58] but, in this case, an examination of the products indicates that deoxygenation of the tautomeric sulphenic acid (23) is involved and the thiol so formed condenses with the amido side-chain (Scheme 24). Triethyl phosphite has been used to

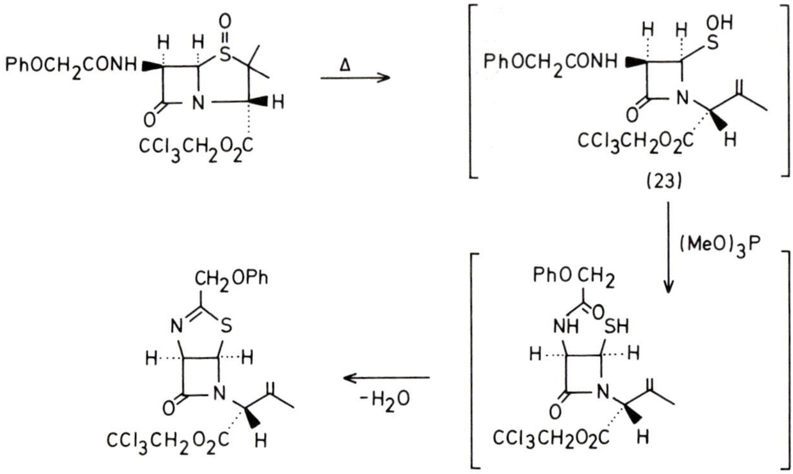

Scheme 24

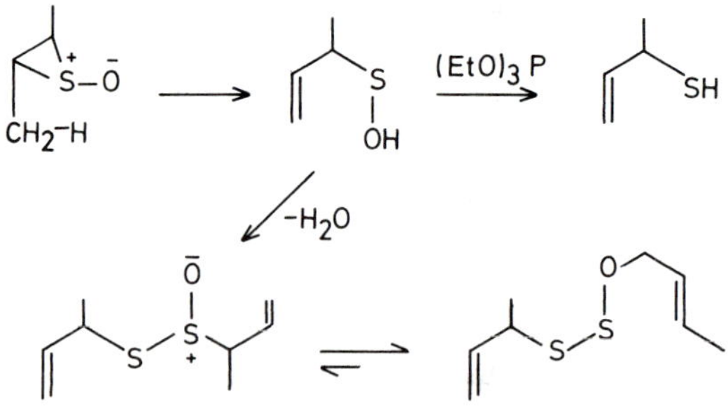

Scheme 25

trap a sulphenic acid postulated to be an intermediate in the sigmatropic rearrangement of episulphoxides[59] (Scheme 25).

Alkyl sulphenates are desulphurized by tri-n-butylphosphine to the corresponding ethers but triphenylphosphine and triphenyl phosphite are relatively ineffective.[60]

Aryl allyl sulphoxides readily rearrange to the allyl sulphenate, and treatment with trimethyl phosphite gives dimethyl allyl phosphate and a methyl sulphide as the major products. However, if the reaction is carried out in methanol the formation of trimethyl phosphate and allyl alcohol results[61] (Scheme 26), and in cyclic systems a predominance of the less stable alcohol is reported. Since aryl allyl sulphoxides can be alkylated,[62] a stereoselective synthesis of olefins is possible. For example, the synthesis shown in Scheme 27 produces (±)-nuciferal with less than 5% of the (Z)-isomer as impurity.[63] Reactions via sulphenates with trimethyl phosphite have been used to trans-

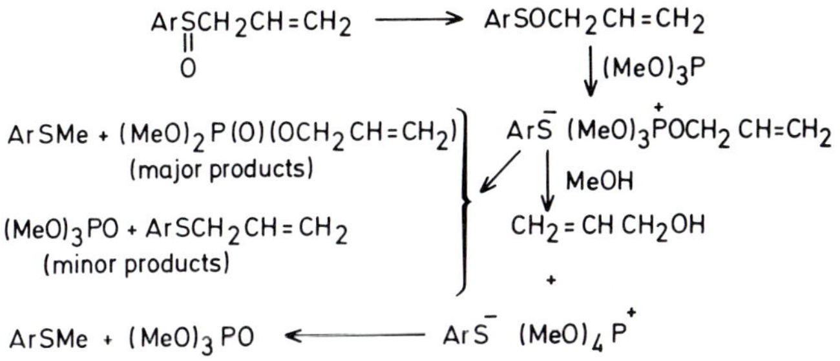

8 Conversions via Desulphurization

Scheme 26

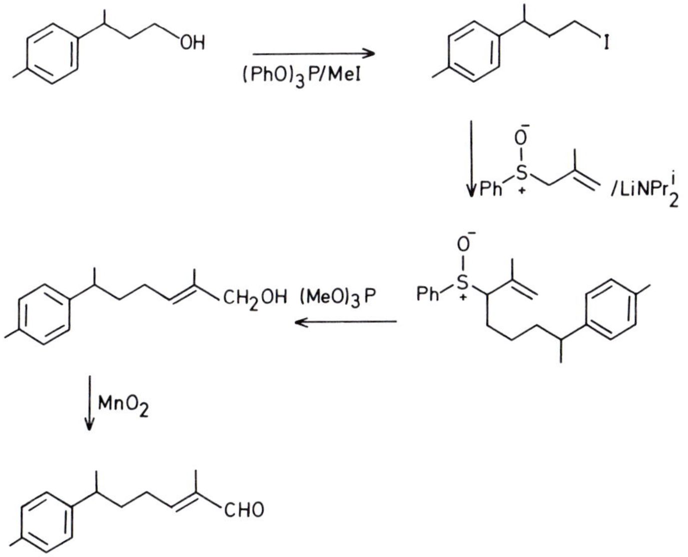

Scheme 27

form the more easily synthesized (13Z)-15β-prostaglandin into the naturally occurring (13E)-15α-isomer[64] (Scheme 28), and also to convert a (14E)-13-sulphide into prostaglandin E.[65]

7. Desulphurization of Di- and Poly-sulphides

The thermal reaction of trialkyl phosphites with diethyl disulphide gives Arbusov products[66] (Scheme 29), and an analogous reaction with diphenyl disulphide has been reported,[67] but no reaction takes place if triphenyl phosphite is used. Indeed, this reagent has been used to prepare diphenyl disulphide from benzenesulphenyl chloride. With ultraviolet irradiation or

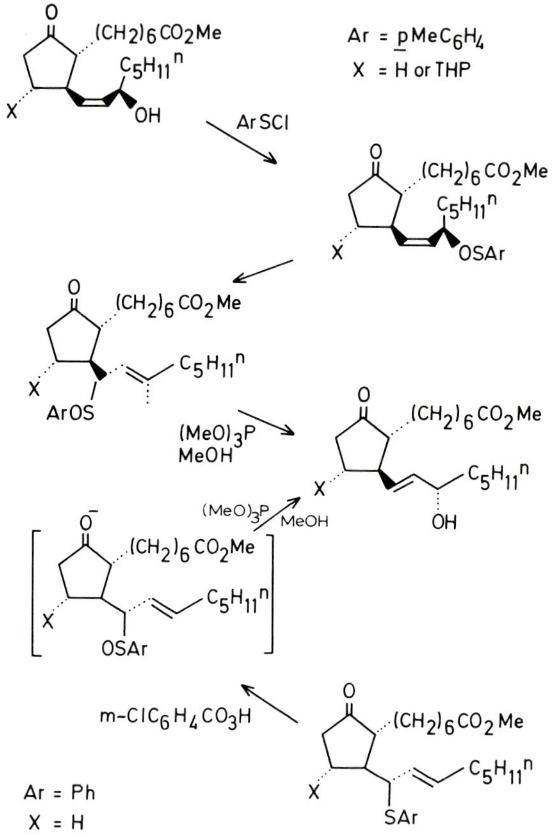

Scheme 28

$$RSSR + (R^1O)_3P \longrightarrow (R^1O)_3\overset{+}{P}SR \; \bar{S}R \longrightarrow RSR^1 + (R^1O)_2P(O)SR$$

$$PhSCl + (PhO)_3P \longrightarrow PhSSPh + (PhO)_3PCl_2$$

Scheme 29

in the presence of an initiator such as di-t-butyl peroxide, however, triethyl phosphite and diethyl disulphide react to give diethyl sulphide and triethyl phosphorothionate[42] by a free-radical process (Scheme 30).

$$EtSSEt \xrightarrow[\text{initiator}]{} 2EtS^\bullet$$

$$EtS^\bullet + (EtO)_3P \longrightarrow Et\overset{\bullet}{S}P(OEt)_3 \longrightarrow Et^\bullet + (EtO)_3PS$$

$$EtSSEt + Et^\bullet \longrightarrow EtSEt + EtS^\bullet$$

Scheme 30

8 Conversions via Desulphurization

The radical reaction leads to an overall desulphurization of the disulphide, and the close similarity to the reaction of phosphites with thiols should be noted. By carrying out the reaction of triethyl phosphite with dibutyl disulphide under radical conditions and in the presence of a high pressure of carbon monoxide, it is possible to cause the alkyl radicals to combine with carbon monoxide to give acyl radicals which then attack the disulphide. A high yield of S-butyl thiopentanoate results[43] (Scheme 31). Competitive

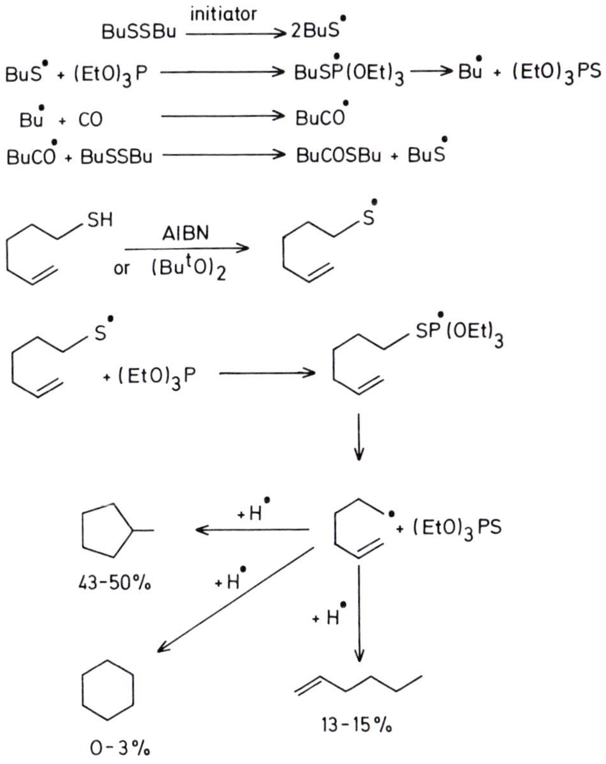

$$BuSSBu \xrightarrow{\text{initiator}} 2BuS^\bullet$$

$$BuS^\bullet + (EtO)_3P \longrightarrow BuS\overset{\bullet}{P}(OEt)_3 \longrightarrow Bu^\bullet + (EtO)_3PS$$

$$Bu^\bullet + CO \longrightarrow BuCO^\bullet$$

$$BuCO^\bullet + BuSSBu \longrightarrow BuCOSBu + BuS^\bullet$$

Scheme 31

reactions involving styrene or methyl acrylate are less successful but a yield of up to 50% of methylcyclopentane can be obtained from hex-5-enethiol.[68] The order of reactivity towards thiyl radicals is reported to be $Bu_3P > (EtO)_3P > Ph_3P > (PhO)_3P$. The ionic Arbusov reaction is very rapid for sulphides having α-alkoxycarbonyl groups,[69] and in a series of unsymmetrical disulphides studied only one of the two possible sets of products was obtained, that involving the sulphide formed from the more stable thiolate anion.[70] At temperatures over 170° the S-aryl thiophosphates formed by reaction of diaryl disulphides and trialkyl phosphites decompose to give sulphides[71] (Scheme 32).

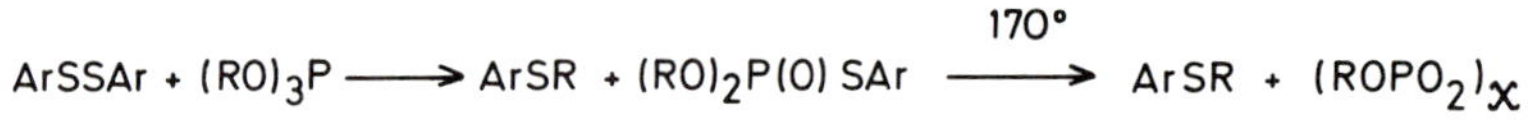

$$\text{ArSSAr} + (RO)_3P \longrightarrow \text{ArSR} + (RO)_2P(O)\text{SAr} \xrightarrow{170°} \text{ArSR} + (ROPO_2)_x$$

Scheme 32

Desulphurizations of a tetrathiin and of a trithiole with triethyl phosphite have been reported[72] (Scheme 33).

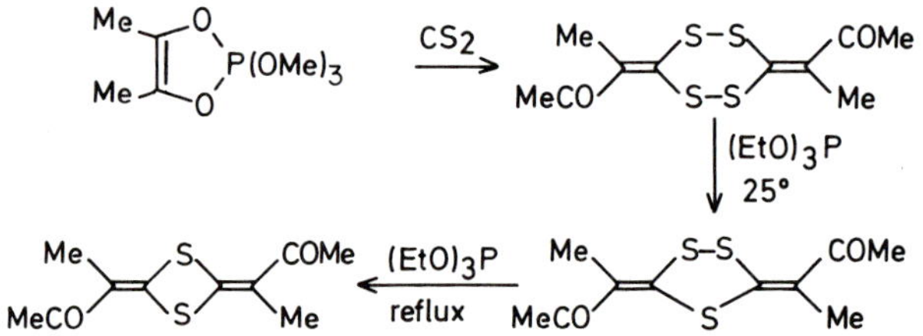

Scheme 33

The desulphurization of acyl disulphides with triphenylphosphine has been known for many years[73] and it has been shown that dialkyl, diaryl, and aralkyl disulphides are not desulphurized by triphenylphosphine in boiling benzene.[74,75] Triphenylphosphine-induced desulphurization of acyl disulphides has been used in the synthesis of thietan-2,4-diones.[76] Disulphides having carbonyl groups on the α-carbon atoms, for example (24)[77] and (25),[78] are readily desulphurized by triphenylphosphine. Where the disulphide links chiral carbon atoms, inversion of configuration is observed. Sulphur chelates may also be desulphurized.[79]

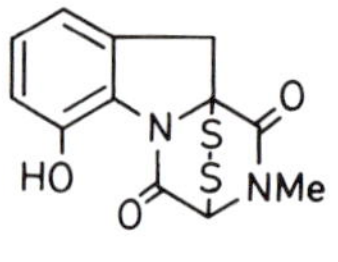

(24) 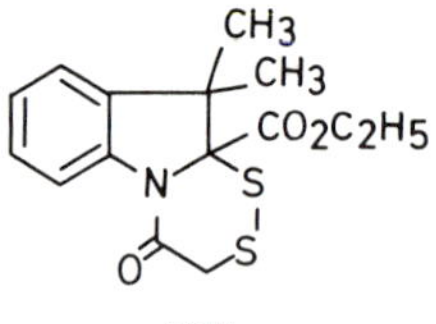   (25)

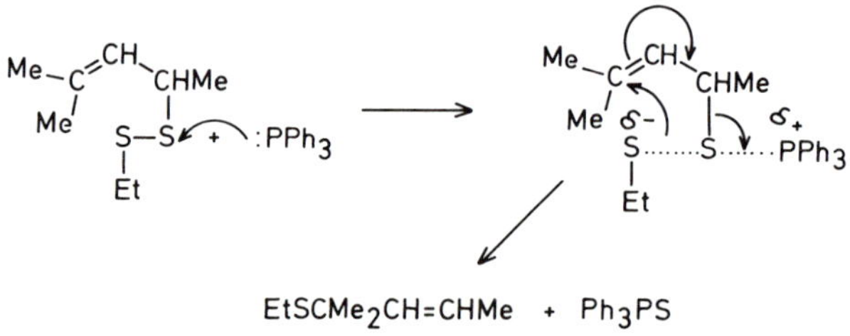

$$\text{EtSCMe}_2\text{CH=CHMe} + \text{Ph}_3\text{PS}$$

Scheme 34

8 Conversions via Desulphurization

The ease with which diallyl disulphides are desulphurized by triphenylphosphine was first observed by Challenger and Greenwood.[80] The reaction involves an allylic rearrangement[81,82] and an $S_N i$ mechanism was suggested (Scheme 34) but Höfle and Baldwin[83] prefer a mechanism involving isomerization to a thiosulphoxide followed by desulphurization (Scheme 35). The

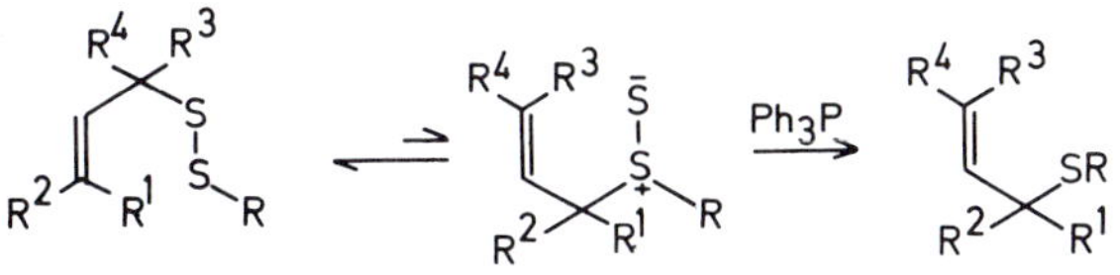

Scheme 35

reaction has been used in an elegant synthesis of squalene from difarnesyl disulphide.[84] Cyclic phosphoramidites (26) desulphurize allyl disulphides but, in addition, they react with diaryl disulphides[85] (Scheme 36).

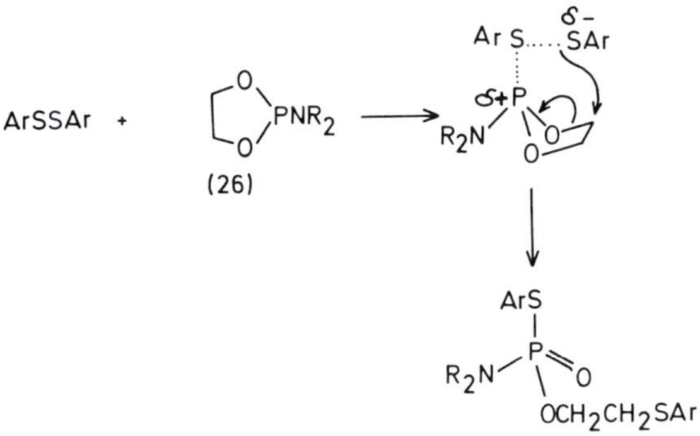

Scheme 36

Tris(diethylamino)phosphine is a very effective desulphurizing agent,[86] converting dialkyl and aralkyl disulphides into the corresponding sulphides. The mechanism does not appear to involve either ion pairs or quinquecovalent intermediates since all three possible sulphides are formed from unsymmetrical disulphides such as 4-bromodibenzyl disulphide. However, this cannot be true in all cases since cyclic phosphoranes have been isolated from the reaction of certain cyclic phosphines and phosphites with the cyclic

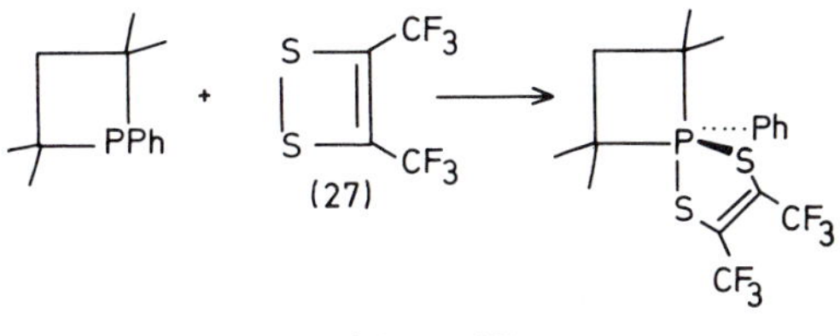

Scheme 37

disulphide (27)[87] (Scheme 37). Phosphoranes having phosphorus–sulphur bonds appear to be less stable than oxyphosphoranes; that derived by reaction of (27) with trimethyl phosphite decomposes to a thiophosphate (Scheme 38) but from reaction of the disulphide with triphenylphosphine only triphenylphosphine disulphide has been isolated.

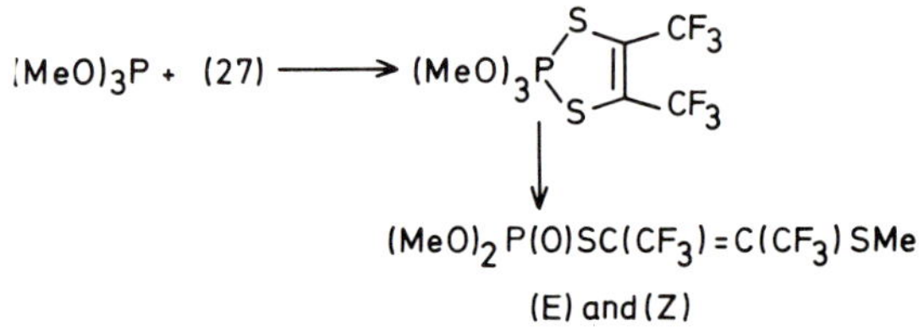

$$(MeO)_3P + (27) \longrightarrow (MeO)_3P\begin{pmatrix} S & CF_3 \\ S & CF_3 \end{pmatrix}$$

$$\downarrow$$

$$(MeO)_2 P(O)SC(CF_3)=C(CF_3)SMe$$

(E) and (Z)

Scheme 38

The formation of the *trans*-tetrahydrothiophen-2,5-dicarboxylate (28) by reaction of the *cis*-disulphide with tris(diethylamino)phosphine[75] implies displacement of phosphine sulphide by a thiolate anion on a quasiphosphonium salt (Scheme 39). Such phosphonium salts have been observed by

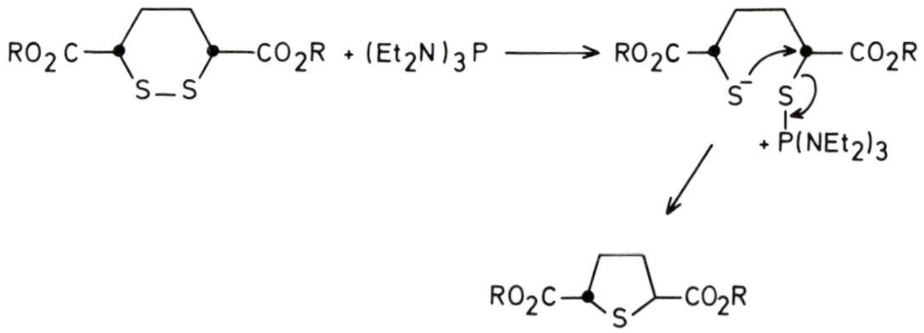

Scheme 39

^{31}P n.m.r. spectroscopy while that from di-2-benzothiazolyl disulphide can be isolated. It decomposes, on heating, to the sulphide but di-2-pyridyl disulphide gives a phosphonium salt which is thermally stable. Derivatives of α-lipoic acid can be desulphurized and L-cystine gives L-lanthionine. However, some exceptions to the normal course of the reaction have been reported[88] (Scheme 40).

Alkyl tetrasulphides are readily desulphurized by triphenylphosphine to disulphides.[81] In the reaction of triphenylphosphine with the trisulphides sporidesmin E and thiodehydrogliotoxin, the central sulphur atom is eliminated[77] but desulphurization of dibenzyl trisulphide proceeds by two mechanisms.[89] Using triphenylphosphine, 90% of the reaction is at the central sulphur and 10% at the terminal sulphur atoms; using tris(dimethylamino)-

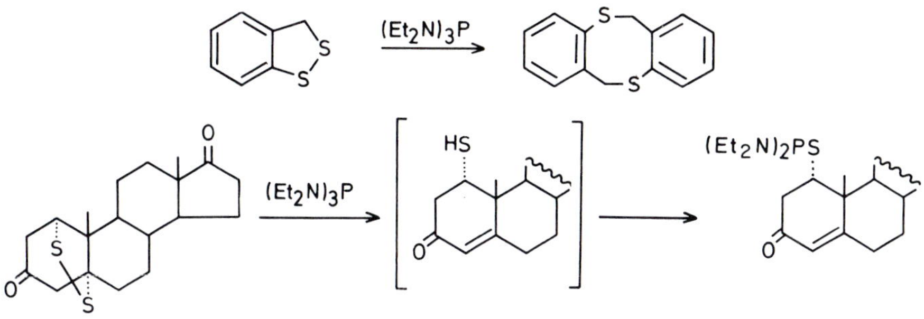

8 Conversions via Desulphurization

Scheme 40

phosphine, 96% of the reaction is at the terminal sulphur atoms and 4% at the central sulphur. Tributylphosphine reacts with less regiospecificity.

Benzyl hydrodisulphide reacts with either triethyl- or triphenyl-phosphine to give toluene, dibenzyl disulphide, hydrogen sulphide, and phosphine sulphide.[90] Triphenylmethyl hydrodisulphide, on the other hand, gives a mixture of phosphine sulphide and triphenylmethanethiol. The latter indicates that reaction takes place at the sulphydryl sulphur and in the former case reaction takes place at the sulphenyl sulphur, although subsequent details of the mechanism are uncertain (Scheme 41). The products formed from

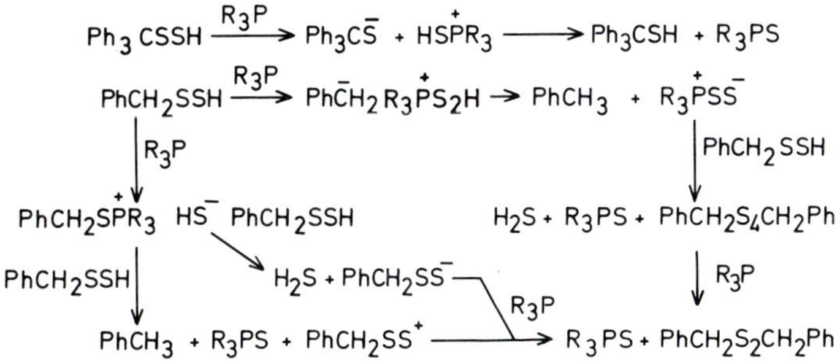

$$Ph_3CSSH \xrightarrow{R_3P} Ph_3C\bar{S} + HS\overset{+}{P}R_3 \longrightarrow Ph_3CSH + R_3PS$$

$$PhCH_2SSH \xrightarrow{R_3P} PhCH_2\bar{R}_3\overset{+}{P}S_2H \rightarrow PhCH_3 + R_3\overset{+}{P}S\bar{S}$$

$$\downarrow R_3P \qquad\qquad\qquad\qquad \downarrow PhCH_2SSH$$

$$PhCH_2\overset{+}{S}PR_3 \; HS^- \; PhCH_2SSH \qquad H_2S + R_3PS + PhCH_2S_4CH_2Ph$$

$$PhCH_2SSH \downarrow \qquad\searrow H_2S + PhCH_2SS^- \qquad\qquad \downarrow R_3P$$

$$PhCH_3 + R_3PS + PhCH_2SS^+ \xrightarrow{R_3P} R_3PS + PhCH_2S_2CH_2Ph$$

Scheme 41

benzhydryl hydrodisulphide suggest that reaction takes place at both sulphur atoms. Benzyl hydrodisulphide with triethyl phosphite gives products arising from reaction at the sulphenyl sulphur, but reactions of benzhydryl, and triphenylmethyl hydrodisulphides with either triethyl phosphite or triphenyl phosphite give mixtures of the thiol and the thionophosphate[91] in each case.

Although diphenyl disulphide is not desulphurized by triphenylphosphine under anhydrous conditions, it is reduced to thiophenol if water is present.[74] Triphenylphosphine reacts rapidly with diaryl disulphides in aqueous methanol to give a quantitative yield of thiol.[92] The reaction is catalysed by strong acids. Dialkyl disulphides, on the other hand, react only slowly but both alkyl and aryl disulphides are rapidly reduced by tributylphosphine in aqueous methanol.[93] While both tris(hydroxymethyl)- and tris(diethylamino)-phosphine reduce disulphides, formation of by-products lessens the value of these reagents, but tributylphosphine in aqueous propanol is a specific reagent for the quantitative reduction of disulphide linkages in wool keratin.[94] If required, alkylation of the thiol groups formed in such reactions can be carried out *in situ*.[95] Similarly reaction of insulin with tributylphosphine in the presence of ethylenimine leads to complete reduction and aminoethylation.[96] A further use of tributylphosphine is in the stabilization of aqueous solutions of cysteine and glutathione.[97] The phosphine reduces disulphides produced by oxidation, excess of phosphine being subsequently removed from the solution by extraction with a dilute solution of sulphur in chloroform. Water-soluble phosphines such as tris(hydroxymethyl)phosphine have been used to activate papain, a sulphydryl-dependent enzyme,[98] and to reduce human gamma-globulin.[99] *Care should be exercised in handling tris(hydroxymethyl)phosphine; violent explosion on distillation may occur.*

Triphenylphosphine in aqueous dioxan has been reported to be useful in the *in situ* preparation of easily oxidizable thiols since the by-product, triphenylphosphine oxide, is sufficiently inert not to interfere with a subsequent reaction in, for example, the synthesis of unsymmetrical disulphides[100] (Scheme 42). Reaction of diphenyl disulphide with triphenylphosphine in

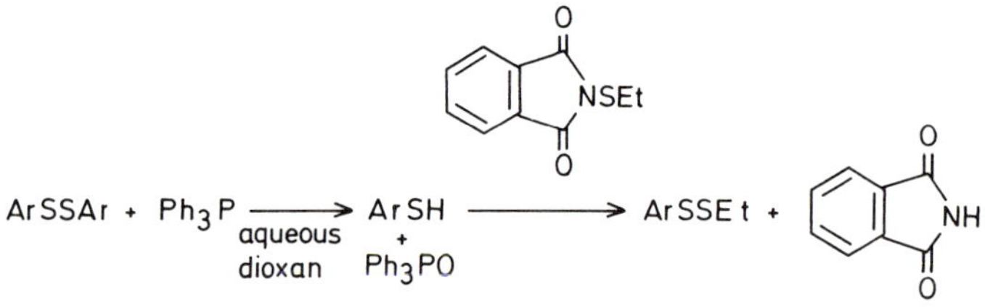

Scheme 42

methanol results in the formation of thioanisole, thiophenol, and triphenylphosphine oxide.[101] The conversion of an alcohol into a *p*-tolyl thioether using *p*-tolyl disulphide and tributylphosphine is one step in the synthesis of (*R*)- and (*S*)-muscone which involves a cyclofragmentation of an epoxy-

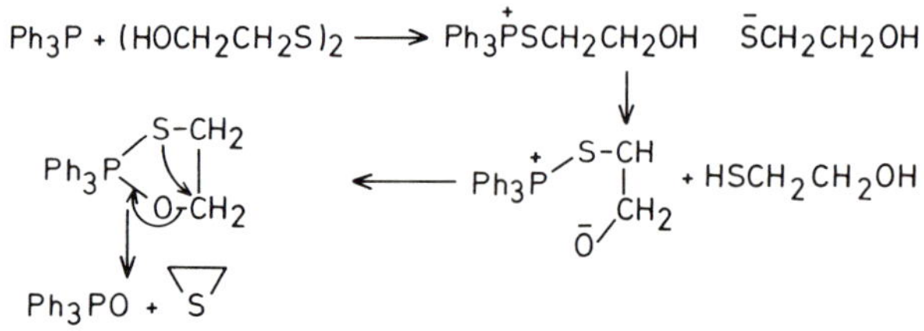

Scheme 43

sulphone derived from the thioether.[102] Triphenylphosphine reacts with bis(hydroxyethyl) disulphide to form triphenylphosphine oxide, 2-mercaptoethanol, and thiiran[103] (Scheme 43).

8. Formation of Pyridyl Thioesters and Related Synthons

The formation of thioanisole in the reaction of diphenyl disulphide with triphenylphosphine in methanol suggests that S-phenyl thioesters would be formed if the reaction were carried out in the presence of carboxylic acids. This has been shown to be the case,[104] the only complication being that α,β-unsaturated acids, in addition to being converted into the thioester, undergo a Michael addition of thiophenol.

2-Pyridyl thioesters derived from di-2-pyridyl disulphide are of considerable interest as synthetic intermediates. For example, they react with Grignard reagents to give ketones[105] and are rapidly converted into esters under very mild conditions[106] (Scheme 44). Corey and his coworkers have used dipyridyl disulphide and triphenylphosphine to convert hydroxy-acids into lactones by a "double activation" process[107] (Scheme 45). The method is of significance in the synthesis of macrocyclic lactones up to a ring size of 21 and of macrolides.[108] The syntheses of these compounds are the subject of two recent

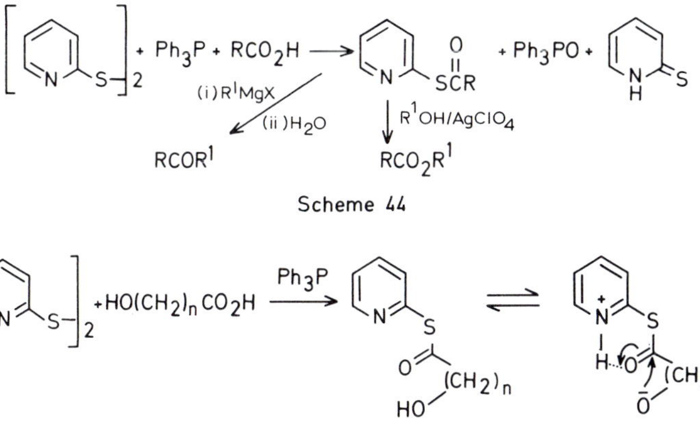

Scheme 44

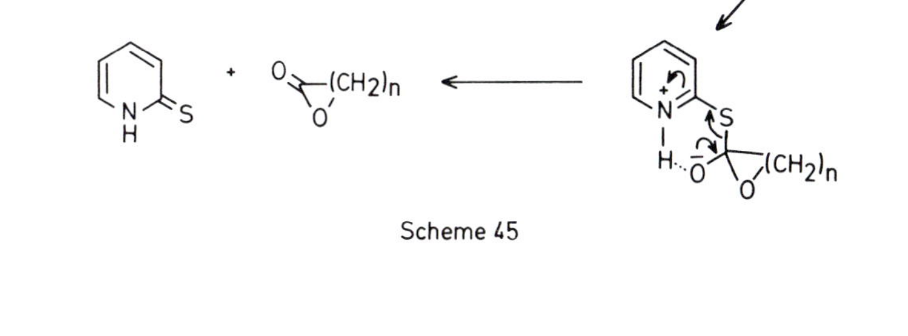

Scheme 45

reviews.[109] An attempt to find alternatives to dipyridyl disulphide for the cyclization employed bis-N-methyl-2-imidazolyl disulphide, but this reagent is acylated at nitrogen as well as at sulphur.[110] However, acylation at sulphur is enhanced by a t-butyl group on the carbon adjacent to nitrogen; thus bis-N-methyl-4-t-butyl-2-imidazolyl disulphide and its N-isopropyl analogue are superior to dipyridyl disulphide.

9. Condensation Reactions, Including Formation of Amides, Phosphorylation, and Application in Nucleoside Synthesis

Mukaiyama has made an extensive study of "oxidation–reduction" condensation reactions, the basis of which is outlined in Scheme 46. Several reviews

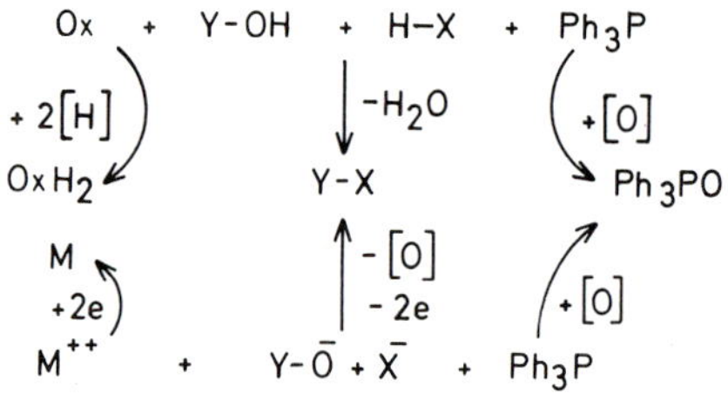

Scheme 46

describing the procedure have appeared.[111] Condensations have been reported involving the use of tervalent phosphorus compounds with mercuric salts,[112] dibenzoylethylene,[113] diethyl azodicarboxylate,[114] and disulphides[115] as oxidants. In the formation of amides using triphenylphosphine and an aryl disulphide, it is necessary to minimize the formation of thiol ester by use of a mercaptan scavenger. Cupric chloride is used as an example in Scheme 47 but

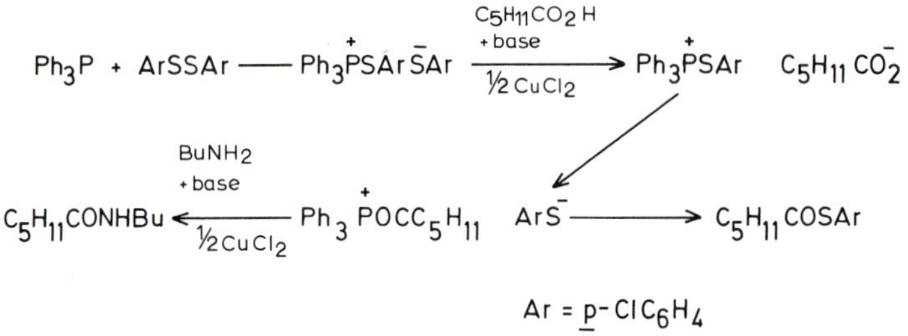

Scheme 47

8 Conversions via Desulphurization

soft metal ions (e.g. Ag^+, Hg^+, Hg^{2+}, or Cd^{2+}) are most effective. The use of di-2-pyridyl disulphide eliminates the need for a scavenger since pyridine thione is formed by reduction of the disulphide and it is not sufficiently nucleophilic to compete with the amide-forming reaction. Applications of the condensation to peptide synthesis will be described in Chapter 12.

Triphenylphosphine and di-2-pyridyl disulphide can be used in phosphorylation of alcohols,[116] amines,[116] monophosphate esters,[116] and thiols.[117] However, in some cases a symmetrical pyrophosphate may also be formed but active phosphorylating agents such as phosphoramidates derived from imidazole and morpholine, and pyridyl phosphates derived from 2-hydroxypyridine, have been successfully prepared in this way (Scheme 48).[118]

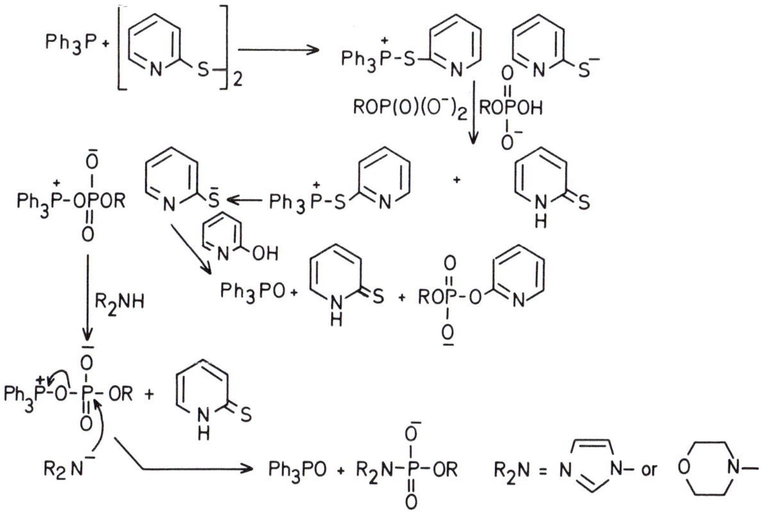

Scheme 48

Synthesis of phosphoramidates, phosphates, and pyrophosphates using triphenylphosphine and di-2-pyridyl disulphide are especially useful in nucleotide chemistry. In particular, it is found that the intermediate (29) is stable. Therefore, reaction with water, amines, and monoesters of phosphoric acid can be carried out without the formation of a symmetrical pyrophosphate as a by-product[119] (Scheme 49). Nucleoside 2',3'- and 3',5'-cyclic phosphates can be formed by reaction of a phosphate with triphenylphosphine and dipyridyl disulphide.

Several syntheses of di- and tri-nucleotides[120] and a synthesis of coenzyme A have been reported in which each condensation was carried out using triphenylphosphine and dipyridyl disulphide.[121]

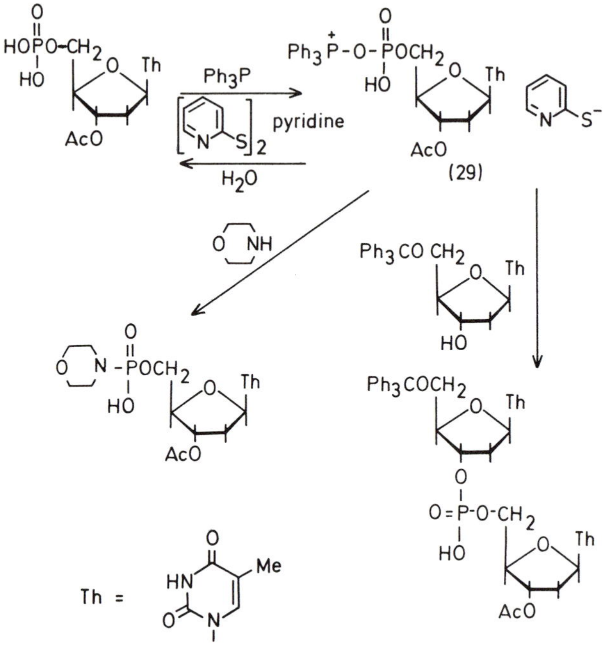

Scheme 49

10. Reactions Involving other Sulphenyl Compounds

10.1. Sulphenyl Chlorides

Trialkyl phosphites react with sulphenyl chlorides to give the expected Arbusov product[122] but in the presence of added base benzenesulphenyl chloride reacts with triethyl phosphite to give thiophenetole.[123] The phosphonium salt formed on reaction of an arylsulphenyl chloride with triphenylphosphine reacts with sodium benzoate to give an S-aryl thiobenzoate (Scheme 50).

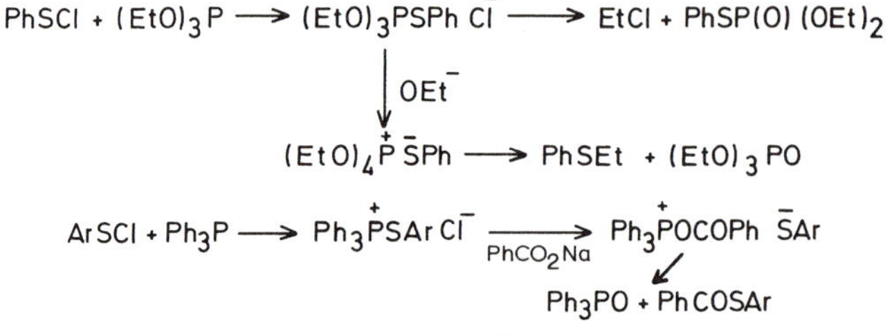

Scheme 50

8 Conversions via Desulphurization

10.2. Compounds Containing Sulphur–Nitrogen Bonds

Sulphenamides do not react with triethyl phosphite but reaction takes place between triphenylphosphine and a sulphenamide in the presence of carboxylate ions. The products are amides but, in order to minimize by-products formed by reaction of the thiolate anions, copper carboxylates are used[124] (Scheme 51). The metal ion also serves to activate the sulphur–nitrogen bond. This reaction has been applied to the synthesis of peptides.

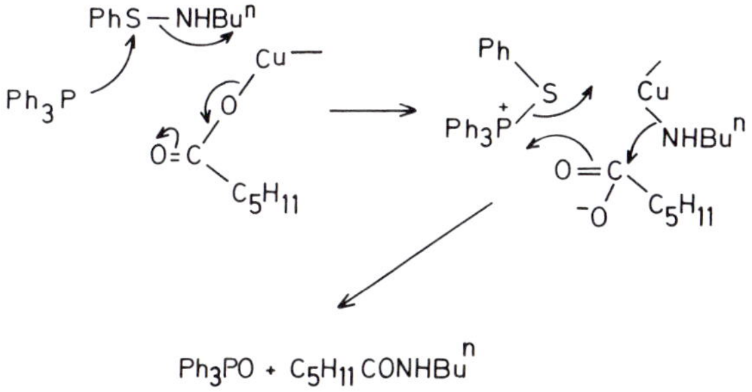

Scheme 51

Sulphenamides obtained by reaction of potassium phthalimide with sulphenyl chlorides are desulphurized by tris(dimethylamino)phosphine[125] (Scheme 52). This provides an alternative to the Gabriel synthesis for some branched alkyl amines, such as isopropyl, for which a direct synthesis of the N-alkylphthalimide is unsuccessful. However, elimination takes place with cyclohexyl and t-butyl groups.

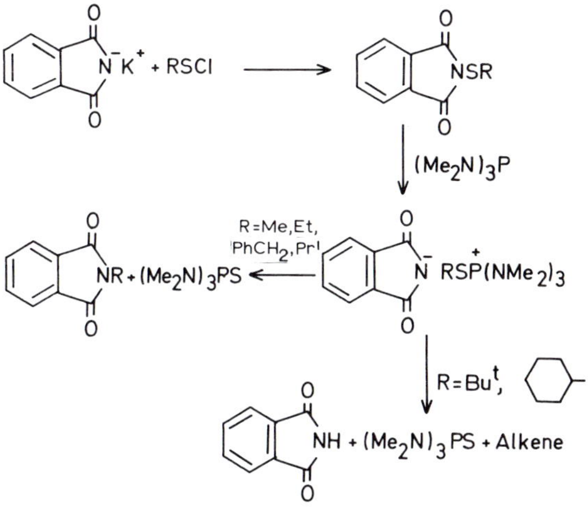

Scheme 52

Thioamides[126] and thioureas[127] form complexes with ethyl azodicarboxylate. The complexes are desulphurized by triphenylphosphine giving ketenimines and carbodiimides (Scheme 53). The mechanism of the oxidation of

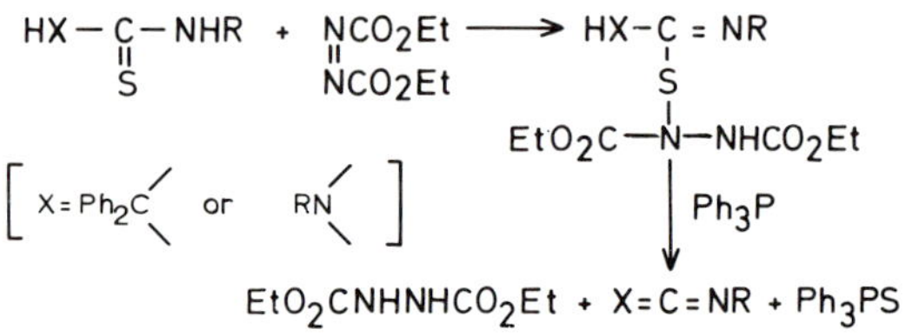

Scheme 53

thiols to disulphides using diethyl azodicarboxylate and triphenylphosphine is unclear but it appears to involve a charge transfer complex.[128]

Heterocumulenes have been prepared by desulphurization of 1,2,4-dithiazoles, 1,2,4-thiadiazoles, and isothiazoles[129] (Scheme 54).

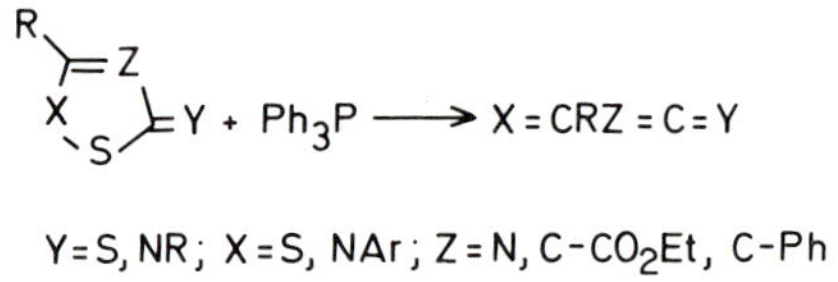

Y = S, NR; X = S, NAr; Z = N, C-CO$_2$Et, C-Ph

Scheme 54

10.3. Thiolsulphinates and Sulphenylthiolsulphinates

Trialkyl phosphites react with alkyl thiolsulphinate esters to give Arbusov products[130] but with aryl thiolsulphinates deoxygenation to the disulphide which undergoes a further Arbusov reaction is also observed. Triphenylphosphine is reported to deoxygenate thiolsulphinates[131] while tris(dimethylamino)phosphine desulphurizes alkyl thiolsulphinate esters.[132] In open-chain compounds the sulphone is the major product but in cyclic systems the cyclic sulphinate ester predominates (Scheme 55). With aryl esters the salts (32) do not react further and they are also formed in the desulphurization of aryl sulphenylthiolsulphinates. Triphenylphosphine reacts with sulphenylthiol-

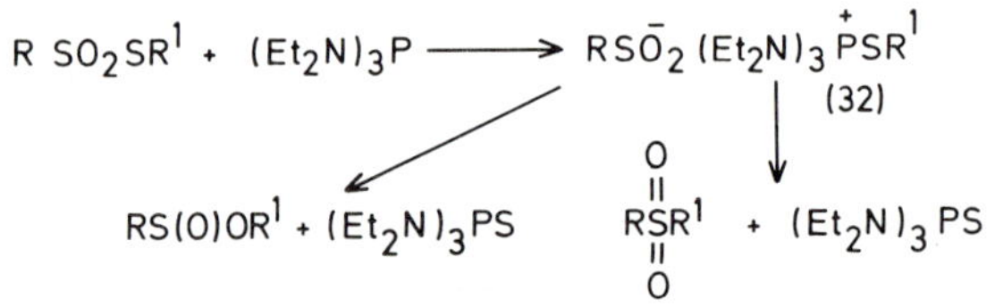

Scheme 55

sulphinates to give disulphides and the reaction is thought to involve desulphurization to a thiolsulphonate followed by deoxygenation.

10.4. Miscellaneous Sulphenyl Compounds

Normal alkyl chlorides can be prepared from the corresponding thiol by reaction of the alkyl chlorocarbonyl disulphide with triphenylphosphine[133] (Scheme 56). Cyclohexanethiol is converted into cyclohexene by this method.

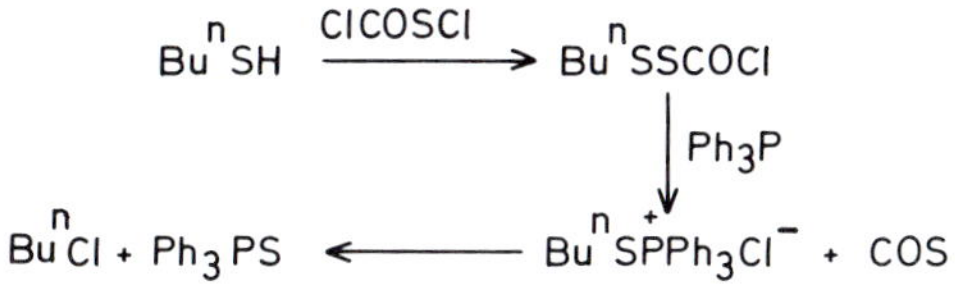

$$Bu^nSH \xrightarrow{ClCOSCl} Bu^nSSCOCl$$
$$\downarrow Ph_3P$$
$$Bu^nCl + Ph_3PS \leftarrow Bu^nS\overset{+}{P}Ph_3Cl^- + COS$$

Scheme 56

Alkyl thiocyanates undergo an Arbusov reaction with trialkyl phosphites[134] leading to nitriles based on the alkyl group of the phosphite: a potentially useful homologation (Scheme 57). No report of a reaction of a thiocyanate with a phosphine has been reported but a reaction involving selenocyanates will be discussed in Section 12.

$$(RO)_3P + R^1SCN \longrightarrow (RO)_3\overset{+}{P}SR^1 \ \overset{-}{CN} \longrightarrow (RO)_2P(O)SR^1 + RCN$$

Scheme 57

11. Reactions with Carbon Disulphide, Thiobenzophenone, and Dicyanosulphonium Ylides

The complex formed by triethylphosphine and carbon disulphide was reported more than one-hundred years ago.[135] Although reports of complexes of trialkylphosphines have appeared since that time,[136] little is known of their chemistry. Recently, however, interest has been shown in the reaction of acetylenes with the red crystalline complex formed from tributylphosphine and carbon disulphide. With an acetylene having one electron-withdrawing group, a tetrathiafulvalene is formed[137] (Scheme 58) but if the reaction is carried out in the presence of an aromatic aldehyde a very rapid Wittig reaction takes place. Indeed, if methyl propiolate is added slowly to a solution containing the aldehyde and tributylphosphine in carbon disulphide at −30°, the product precipitates as the ester is added. Phenylacetylene reacts slowly at room temperature while acetylene requires a temperature of over 100° before the rate of reaction is acceptable. Reaction of the complex with dimethyl

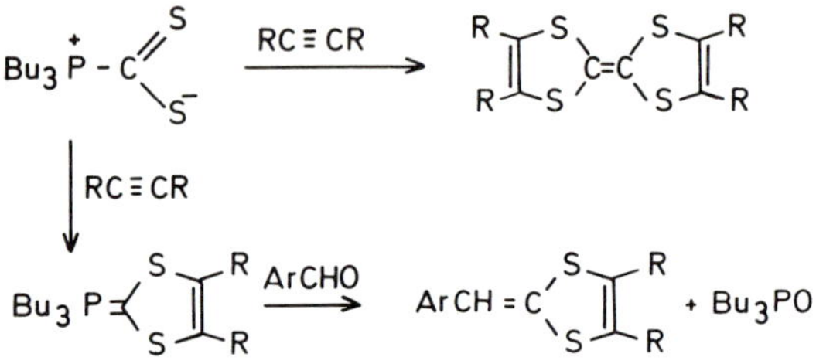

Scheme 58

maleate in methanol results in the formation of dimethyl succinate, tributylphosphine oxide, dimethyl ether, and carbon disulphide. The mechanism of tetrathiafulvalene formation is still uncertain.[138]

Reaction of the complex with propiolic acid or with acetylenedicarboxylic acid gives dipolar compounds which on hydrolysis give dithioles.[139]

Thiobenzophenone polymerizes in the presence of tributylphosphine at room temperature, but at 100° tetraphenylethane and tetraphenylethylene are formed. A similar reaction takes place with thiofluorenone at lower temperatures. The mechanism has not been elucidated but a carbene and an episulphide may be involved.[140] The reaction of thiobenzophenone with triethyl phosphite gives a complex mixture of products including tetraphenylethylene along with triethyl phosphorothionate.[141]

Dicyanosulphonium ylides react with triphenylphosphine to give alkylmalondinitriles.[142]

12. Miscellaneous Reactions

Primary and secondary alcohols are smoothly converted into selenides by reaction with tributylphosphine and phenyl or o-nitrophenyl selenocyanate.[143] A suggested mechanism is shown in Scheme 59. Treatment of the selenides

Scheme 59

formed from primary alcohols with hydrogen peroxide in tetrahydrofuran converts them into olefins. This sequence has been used in the synthesis of β-elemenone (32)[144] (Scheme 60).

Deoxygenation of sulphoxides with tervalent phosphorus compounds has

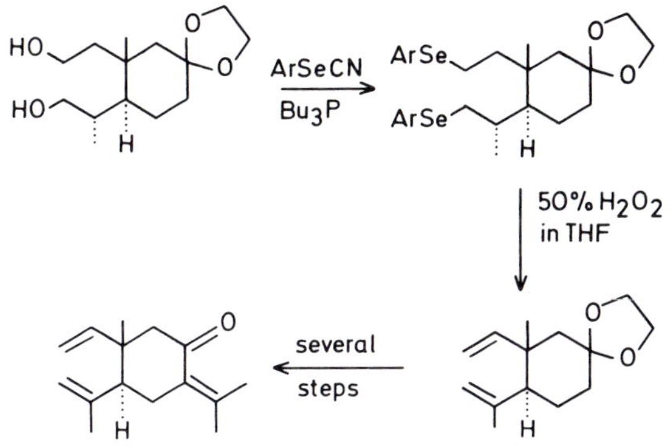

8 Conversions via Desulphurization

Scheme 60

been discussed in this chapter (Section 6) and with phosphine–halogen compounds in Chapter 10.

A report has appeared of a method for the deoxygenation of sulphoxides using O,O-diethyl hydrogen phosphoroselenoate[145] (Scheme 61). The corresponding telluroate as its sodium salt converts 1,2-epoxides into terminal olefins without double-bond migration[146] (Scheme 62). Only a catalytic amount of tellurium is required and the reaction is carried out by adding sodium diethyl phosphite to a mixture of tellurium and a solution of the epoxide. Non-terminal epoxides react slowly and limonene diepoxide can be converted into 1,2-epoxy-p-menth-8-ene.

Scheme 61

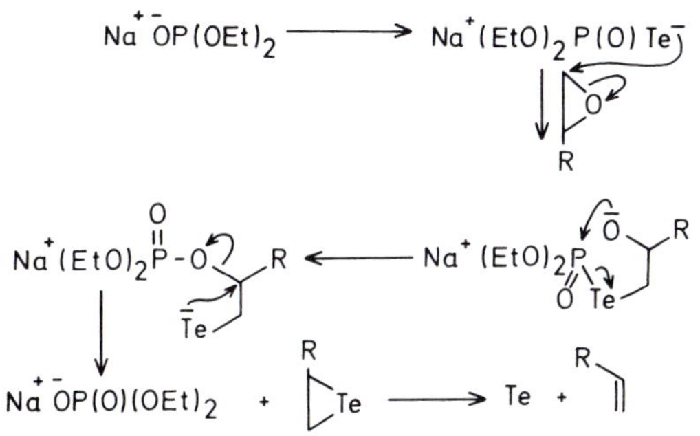

Scheme 62

13. References

1. D. Seyferth and D. E. Welch, *J. Organometal. Chem.*, 1964, **2**, 1.
2. F. W. Hoffmann and T. R. Moore, *J. Amer. Chem. Soc.*, 1958, **80**, 1150; W. Strecker and R. Spitaler, *Ber. Deut. Chem. Ges.*, 1926, **59**, 1772.
3. P. D. Bartlett and G. Meguerian, *J. Amer. Chem. Soc.*, 1956, **78**, 3710; P. D. Bartlett, E. F. Cox, and R. E. Davis, *ibid.*, 1961, **83**, 103.
4. R. D. Schuetz and R. L. Jacobs, *J. Org. Chem.*, 1958, **23**, 1799; 1961, **26**, 3467; R. E. Davis, *ibid.*, 1958, **23**, 1767.
5. N. P. Neureiter and F. G. Bordwell, *J. Amer. Chem. Soc.*, 1959, **81**, 578.
6. M. J. Boskin and D. B. Denney, *Chem. and Ind.*, 1959, 330.
7. M. M. Sidky, M. R. Mahran, and L. S. Boulos, *J. Prakt. Chem.*, 1970, **312**, 51.
8. K. Hirai, H. Matsuda, and Y. Kishida, *Chem. and Pharm. Bull.* (Japan), 1972, **20**, 2067; C. R. Johnson, A. Nakanishi, N. Nakanishi, and K. Tanaka, *Tetrahedron Lett.*, 1975, 2865.
9. A. I. Meyers and M. E. Ford, *J. Org. Chem.*, 1976, **41**, 1735.
10. A. Eschenmoser, *Quart. Rev.*, 1970, **24**, 366.
11. E. Götschi, W. Hunkeler, H.-J. Wild, P. Schneider, W. Fuhrer, J. Gleason, and A. Eschenmoser, *Angew. Chem. Internat. Ed.*, 1973, **12**, 910; E. Götschi and A. Eschenmoser, *ibid.*, p. 912.
12. M. Roth, P. Dubs, E. Götschi, and A. Eschenmoser, *Helv. Chim. Acta*, 1971, **54**, 710.
13. H. Singh, S. S. Narula, and C. S. Gandhi, *Tetrahedron Lett.*, 1977, 3747.
14. E. J. Corey and R. A. E. Winter, *J. Amer. Chem. Soc.*, 1963, **85**, 2677.
15. E. J. Corey, F. A. Carey, and R. A. E. Winter, *J. Amer. Chem. Soc.*, 1965, **87**, 934.
16. D. Horton and C. G. Tindall, *J. Org. Chem.*, 1970, **35**, 3558.
17. W. V. Ruyle, T. Y. Shen, and A. A. Patchett, *J. Org. Chem.*, 1965, **30**, 4353; A. H. Haines, *Carbohydrate Res.*, 1972, **21**, 99; T. L. Nagabhushan, *Canad. J. Chem.*, 1970, **48**, 383.
18. W. Hartmann, H.-M. Fischler, and H.-G. Heine, *Tetrahedron Lett.*, 1972, 853.
19. L. A. Paquette and J. C. Philips, *Tetrahedron Lett.*, 1967, 4645; W. Hartmann, L. Schrader, and D. Wendisch, *Chem. Ber.*, 1973, **106**, 1076.
20. E. J. Corey and J. I. Shulman, *Tetrahedron Lett.*, 1968, 3655.
21. M. Tichy and J. Sicher, *Tetrahedron Lett.*, 1969, 4609.
22. C. Sandris, *Tetrahedron*, 1968, **24**, 3589.
23. L. A. Paquette, I. Itoh, and W. B. Farnham, *J. Amer. Chem. Soc.*, 1975, **97**, 7280.
24. D. P. Bauer and R. S. Macomber, *J. Org. Chem.*, 1976, **41**, 2640.
25. W. K. Anderson and R. H. Dewey, *J. Amer. Chem. Soc.*, 1973, **95**, 7161.
26. R. Hull and R. Farrand, *Chem. Comm.*, 1965, 164.
27. R. Okazaki, K. Okawa, and N. Inamoto, *Chem. Comm.*, 1971, 843.
28. W. Adams and J.-C. Liu, *J. Chem. Soc., Chem. Comm.*, 1972, 73.
29. M. Narita and C. U. Pittman, Jr., *Synthesis*, 1976, 489.
30. E. M. Engler and V. V. Patel, *J. Chem. Soc., Chem. Comm.*, 1975, 671.
31. M. G. Miles, J. D. Wilson, D. J. Dahm, and J. H. Wagenknecht, *J. Chem. Soc., Chem. Comm.*, 1974, 751.

32. M. G. Miles, J. S. Wager, J. D. Wilson, and A. R. Siedle, *J. Org. Chem.*, 1975, **40**, 2577.
33. G. Scherowsky and J. Weiland, *Chem. Ber.*, 1974, **107**, 3155.
34. C. U. Pittman, Jr., M. Narita, and Y. F. Liang, *J. Org. Chem.*, 1976, **41**, 2855.
35. E. J. Corey and G. Märkl, *Tetrahedron Lett.*, 1967, 3201.
36. D. H. R. Barton and B. J. Willis, *Chem. Comm.*, 1970, 1225; D. H. R. Barton, E. H. Smith, and B. J. Willis, *ibid.*, p. 1226. D. H. R. Barton and B. J. Willis, *J. Chem. Soc., Perkin I*, 1972, 305.
37. H. Sauter, H. G. Horster, and H. Prinzbach, *Angew. Chem. Internat. Ed.*, 1973, **12**, 991; A. P. Schaap and G. R. Falcr, *J. Org. Chem.*, 1973, **38**, 3061; L. K. Bee, J. Beeby, J. W. Everett, and P. J. Garratt, *ibid.*, 1975, **40**, 2212.
38. G. Seitz and H. Hoffmann, *Synthesis*, 1977, 201.
39. R. M. Kellogg, M. Noteboom, and J. K. Kaiser, *Tetrahedron*, 1976, **32**, 1641.
40. D. H. R. Barton, F. S. Guziec, Jr., and I. Shahak, *J. Chem. Soc., Perkin I*, 1974, 1794.
41. F. W. Hoffmann, R. J. Ess, T. C. Simmons, and R. S. Hanzel, *J. Amer. Chem. Soc.*, 1956, **78**, 6414; C. Walling and R. Rabinowitz, *ibid.*, 1957, **79**, 5326.
42. C. Walling and R. Rabinowitz, *J. Amer. Chem. Soc.*, 1959, **81**, 1243.
43. C. Walling, O. H. Basedow, and E. S. Savas, *J. Amer. Chem. Soc.*, 1960, **82**, 2181.
44. E. J. Corey and E. Block, *J. Org. Chem.*, 1969, **34**, 1233.
45. J. Bruhin and W. Jenny, *Tetrahedron Lett.*, 1973, 1215.
46. V. Boekelheide, I. D. Reingold, and M. Tuttle, *J. Chem. Soc., Chem. Comm.*, 1973, 406.
47. M. Brink, *Synthesis*, 1975, 807; H. A. Staab and W. Rebafka, *Chem. Ber.*, 1977, **110**, 3333.
48. M. W. Haenel, A. Flatow, V. Taglieber, and H. A. Staab, *Tetrahedron Lett.*, 1977, 1733; M. Matsumoto, T. Otsubo, Y. Sakata, and S. Misumi, *ibid.*, p. 4425.
49. A. Fischli and A. Eschenmoser, *Angew. Chem. Internat. Ed.*, 1967, **6**, 866.
50. M. J. Broadhurst, R. Grigg, and A. W. Johnson, *J. Chem. Soc., Perkin I*, 1972, 1124.
51. H. Vorbrüggen and K. Krolikiewicz, *Angew. Chem. Internat. Ed.*, 1976, **15**, 689.
52. R. H. Mitchell, *Tetrahedron Lett.*, 1973, 4395.
53. R. H. Mitchell, *J. Chem. Soc., Chem. Comm.*, 1973, 955.
54. W. D. Pfeiffer and E. Bulka, *Synthesis*, 1977, 485.
55. D. C. Dittmer and S. M. Kotin, *J. Org. Chem.*, 1967, **32**, 2009.
56. E. H. Amonoo-Neizer, S. K. Ray, R. A. Shaw, and B. C. Smith, *J. Chem. Soc.*, 1965, 4296; S. Oae, A. Nakanishi, and S. Kozuka, *Tetrahedron*, 1972, **28**, 549.
57. I. Granoth, A. Kalir, and Z. Pelah, *J. Chem. Soc. (C)*, 1969, 2424.
58. R. D. G. Cooper and F. L. José, *J. Amer. Chem. Soc.*, 1970, **92**, 2575.
59. J. E. Baldwin, G. Höfle, and S. C. Choi, *J. Amer. Chem. Soc.*, 1971, **93**, 2810.
60. D. H. R. Barton, G. Page, and D. A. Widdowson, *Chem. Comm.*, 1970, 1466.
61. D. A. Evans and G. C. Andrews, *J. Amer. Chem. Soc.*, 1972, **94**, 3672.
62. D. A. Evans, G. C. Andrews, and C. L. Sims, *J. Amer. Chem. Soc.*, 1971, **93**, 4956; D. A. Evans, G. C. Andrews, T. T. Fujimoto, and D. Wells, *Tetrahedron Lett.*, 1973, 1385.

63. D. A. Evans, G. C. Andrews, T. T. Fujimoto, and D. Wells, *Tetrahedron Lett.*, 1973, 1389.
64. J. G. Miller, W. Kurz, K. G. Untch, and G. Stork, *J. Amer. Chem. Soc.*, 1974, **96**, 6774.
65. K. Kondo, T. Umemoto, Y. Takahatake, and D. Tunemoto, *Tetrahedron Lett.*, 1977, 113.
66. H. I. Jacobson, R. G. Harvey, and E. V. Jensen, *J. Amer. Chem. Soc.*, 1955, **77**, 6064.
67. A. C. Poshkus and J. E. Herweh, *J. Amer. Chem. Soc.*, 1957, **79**, 4245.
68. C. Walling and M. S. Pearson, *J. Amer. Chem. Soc.*, 1964, **86**, 2262.
69. D. E. Ailman, *J. Org. Chem.*, 1965, **30**, 1074.
70. R. G. Harvey, H. I. Jacobson, and E. V. Jensen, *J. Amer. Chem. Soc.*, 1963, **85**, 1618.
71. K. Pilgram and F. Korte, *Tetrahedron*, 1965, **21**, 203.
72. A. J. Kirby, *Tetrahedron*, 1966, **22**, 3001.
73. A. Schönberg, *Ber. Deut. Chem. Ges.*, 1935, **68**, 163.
74. A. Schönberg and M. Z. Barakat, *J. Chem. Soc.*, 1949, 892.
75. D. N. Harpp and J. G. Gleason, *J. Amer. Chem. Soc.*, 1971, **93**, 2437.
76. J. H. Schauble and J. D. Williams, *J. Org. Chem.*, 1972, **37**, 2514.
77. S. Safe and A. Taylor, *Chem. Comm.*, 1969, 1466; *J. Chem. Soc. (C)*, 1971, 1189.
78. H. C. J. Ottenheym, T. F. Spande, and B. Witkop, *J. Amer. Chem. Soc.*, 1973, **95**, 1989.
79. J. P. Fackler, Jr., J. A. Fetchin, and J. A. Smith, *J. Amer. Chem. Soc.*, 1970, **92**, 2910.
80. F. Challenger and D. Greenwood, *J. Chem. Soc.*, 1950, 26.
81. C. G. Moore and B. R. Trego, *Tetrahedron*, 1962, **18**, 205.
82. M. B. Evans, G. M. C. Higgins, B. Saville, and A. A. Watson, *J. Chem. Soc.*, 1962, 5045.
83. G. Höfle and J. E. Baldwin, *J. Amer. Chem. Soc.*, 1971, **93**, 6307.
84. G. M. Blackburn, W. D. Ollis, C. Smith, and I. O. Sutherland, *Chem. Comm.*, 1969, 99.
85. K. Pilgram, D. D. Philips, and F. Korte, *J. Org. Chem.*, 1964, **29**, 1844.
86. D. N. Harpp, J. G. Gleason, and J. P. Snyder, *J. Amer. Chem. Soc.*, 1968, **90**, 4181.
87. N. J. De'Ath and D. B. Denney, *J. Chem. Soc., Chem. Comm.*, 1972, 395.
88. D. N. Harpp and J. G. Gleason, *J. Org. Chem.*, 1970, **35**, 3259.
89. D. N. Harpp and D. K. Ash, *Chem. Comm.*, 1970, 811.
90. J. Tsurugi, T. Nakabayashi, and T. Ishihara, *J. Org. Chem.*, 1965, **30**, 2707; T. Nakabayashi, S. Kawamura, T. Kitao, and J. Tsurugi, *ibid.*, 1966, **31**, 861.
91. T. Nakabayashi, J. Tsurugi, S. Kawamura, T. Kitao, M. Ui, and M. Nose, *J. Org. Chem.*, 1966, **31**, 4174.
92. R. E. Humphrey and J. M. Hawkins, *Analyt. Chem.*, 1964, **36**, 1812.
93. R. E. Humphrey and J. L. Potter, *Analyt. Chem.*, 1965, **37**, 164.
94. B. J. Sweetman and J. A. Maclaren, *Austral. J. Chem.*, 1966, **19**, 2347.
95. J. A. Maclaren and B. J. Sweetman, *Austral. J. Chem.*, 1966, **19**, 2355; J. A. Maclaren, *Text. Res. J.*, 1971, **41**, 713.
96. U. Th. Rüegg and J. Rudinger, *Israel J. Chem.*, 1974, **12**, 391.
97. A. Kirkpatrick and J. A. Maclaren, *Analyt. Biochem.*, 1973, **56**, 137.
98. D. M. Kirschenbaum, E. Bakker, L. Thiel, and S. Plotch, *Fed. Proc.*, 1967, **26**, 837.

99. M. E. Levison, A. S. Josephson, and D. M. Kirschenbaum, *Experientia*, 1969, **25**, 126.
100. L. E. Overman, J. Smoot, and J. D. Overman, *Synthesis*, 1974, 59.
101. R. S. Davidson, *J. Chem. Soc. (C)*, 1967, 2131.
102. Q. Branca and A. Fischli, *Helv. Chim. Acta*, 1977, **60**, 925.
103. M. Grayson and C. E. Farley, *Chem. Comm.*, 1967, 831.
104. T. Endo, S. Ikenaga, and T. Mukaiyama, *Bull. Chem. Soc. Japan*, 1970, **43**, 2632.
105. T. Mukaiyama, M. Araki, and H. Takei, *J. Amer. Chem. Soc.*, 1973, **95**, 4763.
106. H. Gerlach and A. Thalmann, *Helv. Chim. Acta*, 1974, **57**, 2661.
107. E. J. Corey and K. C. Nicolaou, *J. Amer. Chem. Soc.*, 1974, **96**, 5614; E. J. Corey, K. C. Nicolaou, and L. S. Melvin, Jr., *ibid.*, 1975, **97**, 653, 654; E. J. Corey, D. J. Brunelle, and P. J. Stork, *Tetrahedron Lett.*, 1976, 3405.
108. E. J. Corey, P. Ulrich, and J. M. Fitzpatrick, *J. Amer. Chem. Soc.*, 1976, **98**, 222; H. Gerlach, K. Oertle, A. Thalman, and S. Servi, *Helv. Chim. Acta*, 1975, **58**, 2036; U. Schmidt, J. Gombos, E. Haslinger, and H. Zak, *Chem. Ber.*, 1976, **109**, 2628.
109. K. C. Nicolaou, *Tetrahedron*, 1977, **33**, 683; T. G. Back, *ibid.*, p. 3041.
110. E. J. Corey and D. J. Brunelle, *Tetrahedron Lett.*, 1976, 3409.
111. T. Mukaiyama and H. Takei, *Topics Phosphorus Chem.*, 1976, **8**, 587; T. Mukaiyama, *Angew. Chem. Internat. Ed.*, 1976, **15**, 94.
112. T. Mukaiyama, H. Nambu, and I. Kuwajima, *J. Org. Chem.*, 1963, **28**, 917.
113. I. Kuwajima and T. Mukaiyama, *J. Org. Chem.*, 1964, **29**, 1385.
114. O. Mitsunobu and M. Eguchi, *Bull. Chem. Soc. Japan*, 1971, **44**, 3427.
115. R. Matsueda, H. Maruyama, M. Ueki, and T. Mukaiyama, *Bull. Chem. Soc. Japan*, 1971, **44**, 1373.
116. T. Mukaiyama and M. Hashimoto, *Bull. Chem. Soc. Japan*, 1971, **44**, 196.
117. T. Mukaiyama and M. Hashimoto, *Tetrahedron Lett.*, 1971, 2425.
118. T. Mukaiyama and M. Hashimoto, *Bull. Chem. Soc. Japan*, 1971, **44**, 2284.
119. T. Mukaiyama and M. Hashimoto, *J. Amer. Chem. Soc.*, 1972, **94**, 8528.
120. T. Hata, I. Nakagawa, and N. Takebayashi, *Tetrahedron Lett.*, 1972, 2931; M. Hashimoto and T. Mukaiyama, *Chem. Lett.*, 1973, 513; T. Hata, I. Nakagawa, and Y. Nakada, *Tetrahedron Lett.*, 1975, 467.
121. M. Hashimoto and T. Mukaiyama, *Chem. Lett.*, 1972, 595.
122. D. C. Morrison, *J. Amer. Chem. Soc.*, 1955, **77**, 181.
123. T. Mukaiyama and M. Ueki, *Tetrahedron Lett.*, 1967, 3429.
124. T. Mukaiyama, M. Ueki, H. Maruyama, and R. Matsueda, *J. Amer. Chem. Soc.*, 1968, **90**, 4490; M. Ueki, H. Maruyama, and T. Mukaiyama, *Bull. Chem. Soc. Japan*, 1971, **44**, 1108.
125. D. N. Harpp and B. A. Orwig, *Tetrahedron Lett.*, 1970, 2691.
126. O. Mitsunobu, K. Kato, and F. Kakase, *Tetrahedron Lett.*, 1969, 2473.
127. O. Mitsunobu, K. Kato, and M. Wada, *Bull. Chem. Soc. Japan*, 1971, **44**, 1362.
128. K. Kato and O. Mitsunobu, *J. Org. Chem.*, 1970, **35**, 4227.
129. J. Goerdeler, J. Haag, C. Lindner, and R. Losch, *Chem. Ber.*, 1974, **107**, 502.
130. J. Michalski, T. Modro, and J. Wieczorkowski, *J. Chem. Soc.*, 1960, 1665.
131. L. Horner and H. Nickel, *Annalen*, 1955, **597**, 20.
132. D. N. Harpp, J. G. Gleason, and D. K. Ash, *J. Org. Chem.*, 1971, **36**, 322.
133. D. L. J. Clive and C. V. Denyer, *J. Chem. Soc., Chem. Comm.*, 1972, 773.
134. W. A. Sheppard, *J. Org. Chem.*, 1961, **26**, 1460.
135. A. W. Hofmann, *Annalen*, 1861, Supp. **1**, 1.

136. A. Hantzsch and H. Hibbert, *Ber. Deut. Chem. Ges.*, 1907, **40**, 1508; W. C. Davies and W. J. Jones, *J. Chem. Soc.*, 1929, 33.
137. H. D. Hartzler, *J. Amer. Chem. Soc.*, 1971, **93**, 4961.
138. L. R. Melby, H. D. Hartzler, and W. A. Sheppard, *J. Org. Chem.*, 1974, **39**, 2456.
139. C. U. Pittman, Jr., and M. Narita, *J. Chem. Soc., Chem. Comm.*, 1975, 960.
140. Y. Ogata, M. Yamashita, and M. Mizutani, *Bull. Chem. Soc. Japan*, 1976, **49**, 1721.
141. Y. Ogata, M. Yamashita, and M. Mizutani, *Tetrahedron*, 1974, **30**, 3709.
142. K. Friedrich and J. Rieser, *Synthesis*, 1970, 479.
143. P. A. Grieco, S. Gilman, and M. Nishizawa, *J. Org. Chem.*, 1976, **41**, 1485.
144. G. Majetich, P. A. Grieco, and M. Nishizawa, *J. Org. Chem.*, 1977, **42**, 2327.
145. D. L. J. Clive, W. A. Kiel, S. M. Menchen, and C. K. Wong, *J. Chem. Soc., Chem. Comm.*, 1977, 657.
146. D. L. J. Clive and S. M. Menchen, *J. Chem. Soc., Chem. Comm.*, 1977, 658.

9
Functional Group Conversions using Phosphorus(III) Reagents and Polyhalogenoalkanes

ROLF APPEL AND MECHTHILD HALSTENBERG

1. Introduction, 387
2. Initial steps of the reaction between a tertiary phosphine and a tetrahalogenomethane, 388
3. Basis of the preparative application of the reagent R_3P–CCl_4 in multicomponent reactions, 391
4. The three-component reaction: triphenylphosphine–CCl_4–water, 394
5. Chlorinations, 400
6. Dehydrations, 408
7. Reactions providing P–N linkage, 411
8. Four-component reactions, 414
9. The reagent R_3P–C_2Cl_6, 416
10. Phosphorus(III) esters–CCl_4, 417
11. Other reagent combinations, 422
12. Summary and outlook, 423
13. References, 424

1. Introduction

Since the first synthesis of phosphoramidates by means of carbon tetrachloride, reported by Atherton, Openshaw, and Todd,[1] tervalent phosphorus compounds have found increasing application in combination with polyhalogenoalkanes as most versatile reagents, e.g. for halogenation of O-functional compounds, for dehydration and P–N linking reactions.[2,3] To understand these apparently very different reactions, the processes in the two-component system will have to be closely examined. The highly reactive intermediates occurring in this system are a key factor in the understanding of the preparative use of multicomponent reactions.

Of the tervalent phosphorus derivatives, tertiary phosphines, tris(dimethylamino)phosphine, and phosphites are used; of the polyhalogenoalkanes, tetrachloromethane and hexachloroethane are favoured. Combinations of these reagents will be considered in turn in this chapter.

2. Initial Steps of the Reaction between a Tertiary Phosphine and a Tetrahalogenomethane

The reaction of a phosphine with a tetrahalogenomethane follows an ionic route.[4] It is initiated by the polarizing action of the permanently dipolar phosphine on the symmetrical but readily polarizable tetrahalogenomethane. In accord with Scheme 1 it leads to heterolytic bond cleavage by a direct interaction between the phosphorus atom and a halogen atom. The amount of charge transfer in the dipolar associate (1) then depends on the substituents carried by the phosphorus atom. With strongly basic amino groups this tendency is so great that in the case of tris(dimethylamino)phosphine the

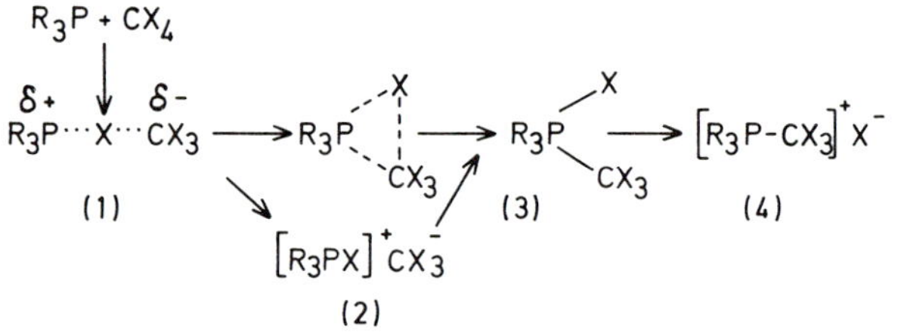

$$R_3P + CX_4$$

$$\overset{\delta+}{R_3P} \cdots X \cdots \overset{\delta-}{CX_3} \longrightarrow R_3P\overset{X}{\underset{CX_3}{\cdots}} \longrightarrow R_3P\overset{X}{\underset{CX_3}{\diagup}} \longrightarrow [R_3P-CX_3]^+ X^-$$

$$(1) \qquad\qquad\qquad \searrow \qquad (3) \qquad\qquad (4)$$

$$[R_3PX]^+ CX_3^-$$

$$(2)$$

Scheme 1

primary complex already exists largely as an ion pair (2). With more weakly basic phosphines the charge transfer to the CX_4 molecule is much less pronounced. The ion pair (2) with its trihalogenomethanide ion has not yet been detected, so in the reaction of triorganylphosphines with CX_4 the ion pair (2) has only the subordinate importance of a canonical limiting form of a bonding state. In most cases, however, the dipolar associate (1) will form, by way of a quinquevalent phosphorane (3), the (trichloromethyl)phosphonium chloride (4) which has been detected and isolated from the systems [$(CH_3)_2N]_3$-P–CCl_4[5] and Ph_3P–CCl_4.[6,7]

2.1. Triphenylphosphine–CCl_4

In the system Ph_3P–CCl_4, however, the (trichloromethyl)phosphonium chloride (4) can only be regarded as a short-lived and highly reactive intermediate which, independently of the molar proportions of the reactants, reacts with further phosphine to give [chloro(triphenylphosphoranediyl)-methyl]triphenylphosphonium chloride (8a) and dichlorotriphenylphosphorane (6a). According to Scheme 2 the overall reaction thus proceeds by way of the (dichloromethylene)phosphorane (5)[8–10] which reacts with a third molecule of phosphine under the autocatalytic influence of (6) to give via the intermediate (7) the stable end-product (8). In the absence of (6) the ylide (5)

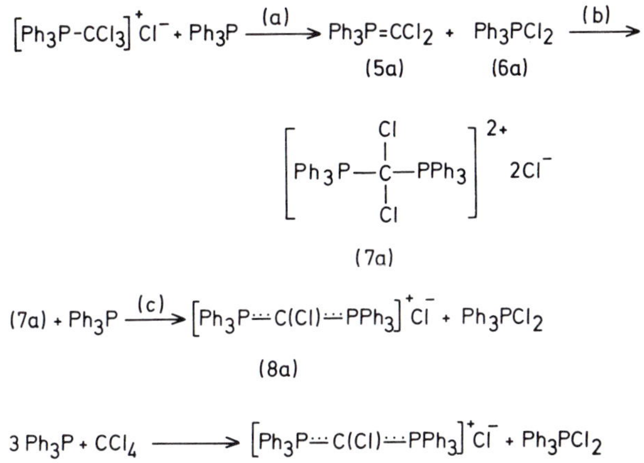

$$[Ph_3P-CCl_3]^+ Cl^- + Ph_3P \xrightarrow{(a)} Ph_3P=CCl_2 + Ph_3PCl_2 \xrightarrow{(b)}$$
$$\text{(5a)} \quad \text{(6a)}$$

$$\left[Ph_3P-\underset{\underset{Cl}{|}}{\overset{\overset{Cl}{|}}{C}}-PPh_3 \right]^{2+} 2Cl^-$$
$$\text{(7a)}$$

$$(7a) + Ph_3P \xrightarrow{(c)} [Ph_3P\cdots C(Cl)\cdots PPh_3]^+ Cl^- + Ph_3PCl_2$$
$$\text{(8a)}$$

$$3\, Ph_3P + CCl_4 \longrightarrow [Ph_3P\cdots C(Cl)\cdots PPh_3]^+ Cl^- + Ph_3PCl_2$$

Scheme 2

and triphenylphosphine do not undergo any reaction. The existence of the intermediate bisphosphonium cation (7) was demonstrated by conversion into the corresponding hexachloroantimonate.[9]

Very remarkable is the enormous solvent-dependence of the reaction rate. The reaction is fastest in acetonitrile and slowest in pure excess CCl_4. Comparative kinetic studies indicate that charge separation is the rate-determining step, which depends markedly on the solvation properties of the solvent.[6]

In apolar inert solvents such as benzene the ylide (5) can partly react with dichlorophosphorane re-forming (trichloromethyl)phosphonium chloride (4) and triphenylphosphine, i.e. at least this step is an equilibrium.[11] Observations in the system Ph_3P–CCl_4–H_2O suggest that (trichloromethyl)phosphonium chloride, too, may reversibly decompose to give Ph_3P and CCl_4.

2.2. Aliphatic Phosphines–CCl_4

Investigations of a series of trialkyl- and dialkylphenyl-phosphines (trimethyl-, triethyl-, tri-n-propyl-, tri-n-butyl-, dimethylphenyl-, and diethylphenyl-) showed that under the proper reaction conditions these react with CCl_4 in the same way as triphenylphosphine.[12,13] According to Scheme 2 they give mainly [chloro(triorganylphosphoranediyl)methyl]triorganylphosphonium chloride and dichlorophosphorane. This uniform course of the reaction, however, is observed only if the reaction is carried out at low temperatures and in a very dilute solution in solvents that do not solve the dichlorophosphorane.

Even though in the case of aliphatic phosphines the corresponding (trichloromethyl)phosphonium chlorides have not yet been detected as significant intermediates, their transient existence can be deduced from the subsequent reactions. Dechlorination by a second molecule of phosphine gives the ylide (dichloromethylene)phosphorane (5) and the dichlorophosphorane (6) which, as in the system Ph_3P–CCl_4, may be detected by Wittig or Horner reactions with benzaldehyde respectively. Up to this stage the reaction hence follows the same path as in the system Ph_3P–CCl_4. The further course of the reaction depends on the solubility of the dichlorophosphorane and the proton activity of the α-hydrogen atom in the alkyl chain.

Dependent on the solvent the dichlorophosphorane may exist in the covalent form or in the ionic form R_3PCl^+ Cl^-, in which the α-hydrogen atom becomes so easily movable that the ylide can effect HCl transfer as shown in Scheme 3 [reaction (b)]. The subsequent products of the unstable P-chloroalkylidenephosphorane (10) are either the dimer (11)[14] [reaction (c)] or the rearranged α-chlorophosphine (12)[15] [reaction (d)]. Both reaction paths were observed in several cases.

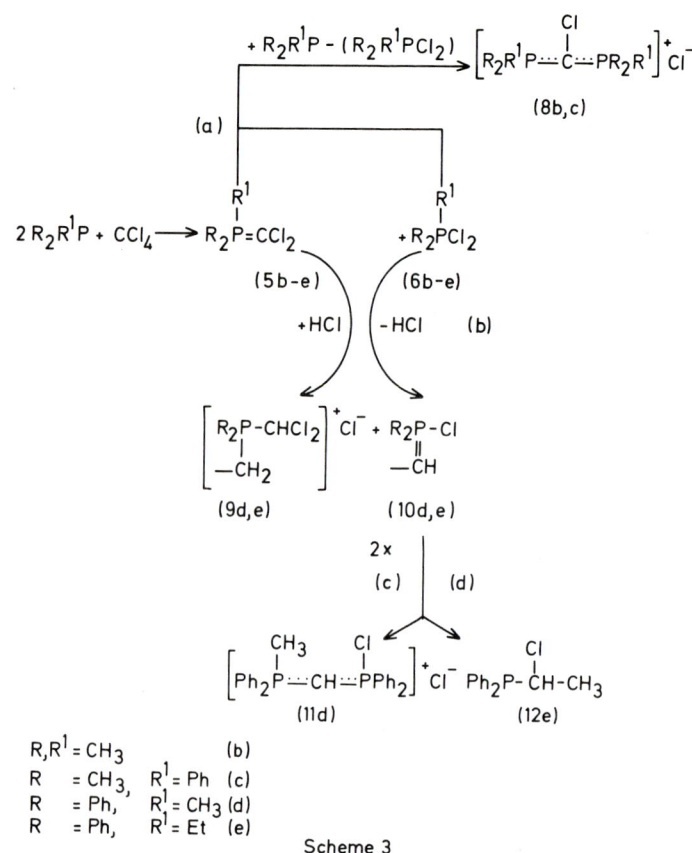

Scheme 3

3. Basis of the Preparative Application of the Reagent R_3P-CCl_4 in Multicomponent Reactions

In all preparatively useful reactions in which the R_3P-CCl_4 reagent is treated with a proton-active substrate to effect chlorination or dehydration, the active agents responsible for phosphorylating* the substrate are to be sought among the intermediates shown in Scheme 1. From this point of view it is advisable to bring the phosphine and CCl_4 into reaction in the molar ratio of 1:1. Stoichiometry exceeding this molar ratio favours the subsequent reactions of the two-component system according to Scheme 2, which lead to the compound (8) which is inactive as a phosphorylating agent. Because of these very rapid reactions it is advisable to mix the two components only in the presence of the substrate. This can be done, for example, by adding the easily measurable CCl_4 to a premixed solution of phosphine and substrate. The reaction time can be considerably influenced by proper choice of the solvent; it increases in the order acetonitrile, dichloromethane, benzene; mixtures of polar solvents with benzene or other inert apolar solvents are equally suitable. Even with a large excess of CCl_4, which acts at the same time as a solvent, satisfactory results are obtained in multicomponent reactions. This does not contradict the above-mentioned unreactivity of the system Ph_3P-CCl_4,[16] since in this case the polar substrate functions at first as a solvating agent and thus initiates the reactions shown in Scheme 1. All the reactive intermediates are very susceptible to hydrolysis, making it often necessary to use carefully dried solvents.

In general the reagent triphenylphosphine-CCl_4 is to be preferred for multicomponent reactions, since triphenylphosphine is an odourless, readily available, and conveniently handled solid. More important, the corresponding dichlorophosphorane (6a) carries no active α-hydrogen atom, so the subsequent reactions described in Section 2.2 are prevented.

3.1. Mechanism of the Three-Component Reaction

Most of the preparatively useful chlorination, dehydration, and P-N linking reactions are multicomponent reactions in which the reagent R_3P-CCl_4 is treated with a proton-active nucleophile (often referred to below as substrate). It is characteristic of all these reactions that one hydrogen atom of the nucleophile is transformed into chloroform. According to this result the first investigators of the chlorination of alcohols took the ion pair (2) to be the

---

* Here and below, "phosphorylation" is often used to mean in general the formation of a phosphorus derivative, rather than its usual meaning of the introduction of a phosphoryl group.

active agent, its cation phosphorylating the substrate and the trichlorocarbanion giving chloroform:[17-27]

$$R_3PCl^+ CCl_3^- + ROH \rightarrow R_3POR^+Cl^- + HCCl_3 \qquad (1)$$

In all reactions with Ph_3P-CCl_4, however, the yield of chloroform is much lower than calculated. Furthermore the (chloromethyl)phosphonium chloride $Ph_3PCH_2Cl^+Cl^-$ always appears as by-product, which does not agree with equation (1).

According to recent kinetic and stereochemical investigations two different mechanisms are involved in all reactions of the reagent Ph_3P-CCl_4 with a substrate.[28-30] As shown in Scheme 4 the chloroform is formed by reaction

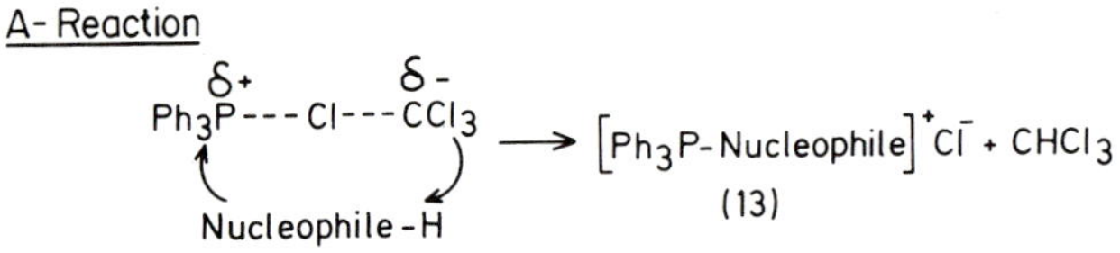

A-Reaction

$$\underset{\text{Nucleophile-H}}{Ph_3\overset{\delta+}{P}---Cl---\overset{\delta-}{CCl_3}} \rightarrow [Ph_3P\text{-Nucleophile}]^+Cl^- + CHCl_3$$
$$(13)$$

Scheme 4

with the dipolar associate. This is shown by parallel experiments with (trichloromethyl)phosphonium chloride (4a) whose reactivity towards several amines and alcohols differs from that of the combination phosphine–CCl_4.[6] Stereochemical investigations[30] gave no positive support for a reaction via the chlorophosphonium trichloromethanide ion pair (2) which should be highly reactive owing to the extremely unstable trichloro-carbanion, and should instantaneously and nearly quantitatively afford chloroform.

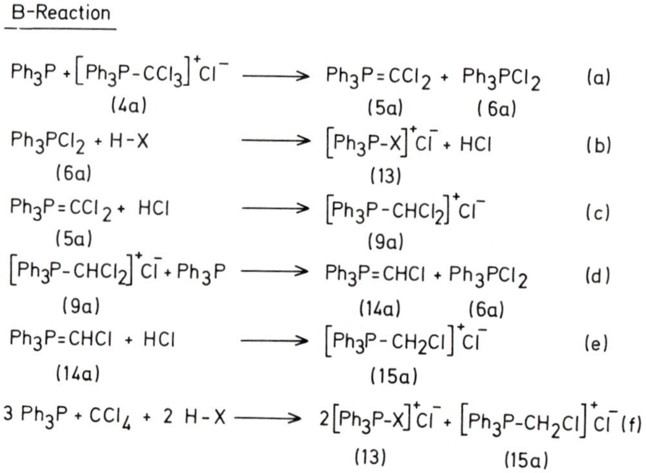

B-Reaction

$Ph_3P + [Ph_3P-CCl_3]^+Cl^- \longrightarrow Ph_3P=CCl_2 + Ph_3PCl_2$ (a)
  (4a)   (5a)   (6a)

$Ph_3PCl_2 + H-X \longrightarrow [Ph_3P-X]^+Cl^- + HCl$ (b)
  (6a)   (13)

$Ph_3P=CCl_2 + HCl \longrightarrow [Ph_3P-CHCl_2]^+Cl^-$ (c)
  (5a)   (9a)

$[Ph_3P-CHCl_2]^+Cl^- + Ph_3P \longrightarrow Ph_3P=CHCl + Ph_3PCl_2$ (d)
  (9a)   (14a)   (6a)

$Ph_3P=CHCl + HCl \longrightarrow [Ph_3P-CH_2Cl]^+Cl^-$ (e)
  (14a)   (15a)

$3 Ph_3P + CCl_4 + 2 H-X \longrightarrow 2[Ph_3P-X]^+Cl^- + [Ph_3P-CH_2Cl]^+Cl^-$ (f)
                                         (13)   (15a)

Scheme 5

9 Conversions via R₃P-Halogenoalkanes

In reality the attack at the dipolar associate and the elimination of chloroform is so slow that the subsequent reaction of the two-component system can compete with it. As mentioned before, the dichlorophosphorane (6) is an excellent phosphorylating agent, reacting with most nucleophiles to give the same end-products as the combination triphenylphosphine–CCl₄, while according to Scheme 5 [reaction (c)] the ylide (dichloromethylene)phosphorane (5) acts as an HCl acceptor. In this way the (dichloromethyl)phosphonium chloride (9) is formed which reacts with further phosphine to give the ylide (chloromethylene)phosphorane (14) and the dichlorophosphorane. Renewed reaction of (14) leads to the stable (chloromethyl)phosphonium chloride (15) which is a side-product of all three-component reactions. The salt (13) composed of the phosphine and the nucleophile is stable when XH is amine, but in other cases it decomposes in the characteristic manner by elimination of phosphine oxide to give the desired chlorination or dehydration products, as shown by the example of chloroformamidines (Scheme 6).

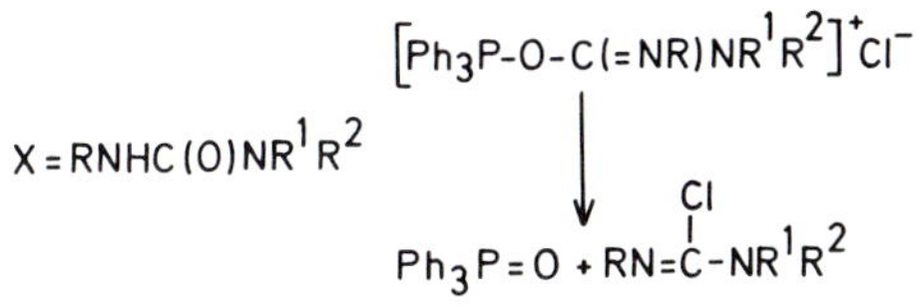

$$X = RNHC(O)NR^1R^2$$

$$[Ph_3P-O-C(=NR)NR^1R^2]^+Cl^-$$

$$\downarrow$$

$$Ph_3P=O + RN=\overset{Cl}{\underset{|}{C}}-NR^1R^2$$

Scheme 6

The reaction routes A and B can both be detected in varying amounts in all preparatively applied three-component reactions, if the phosphine component is triphenylphosphine. In the system Ph₃P–CBr₄–proton-active nucleophile, too, it has been shown by the synthesis of hydrazonyl halides (16), which is analogous to the imide chloride formation, that the chloroform-producing A reaction mainly entails attack at the dipolar associate according to Scheme 4.[6,28] Comparative experiments with the corresponding bromine salt gave only poor yields.[31]

In all reactions of the reagent Ph₃P–CCl₄ with a third component the phosphorylations are effected to a considerable degree by the dichlorophosphorane formed as an intermediate. This explains why most of the reactions treated in Sections 5–8 can also be effected directly by means of Ph₃PCl₂ or, even more conveniently, by the reagent R₃P–C₂Cl₆ (see also Section 9). For preparative application it is important that in the reaction route B one-third of the phosphine used is withdrawn as Ph₃PCH₂Cl⁺Cl⁻ from the phosphorylating reaction. This requires a small correction to the 1:1 stoichiometry derived in Section 3; on the assumption that each of the two reaction routes

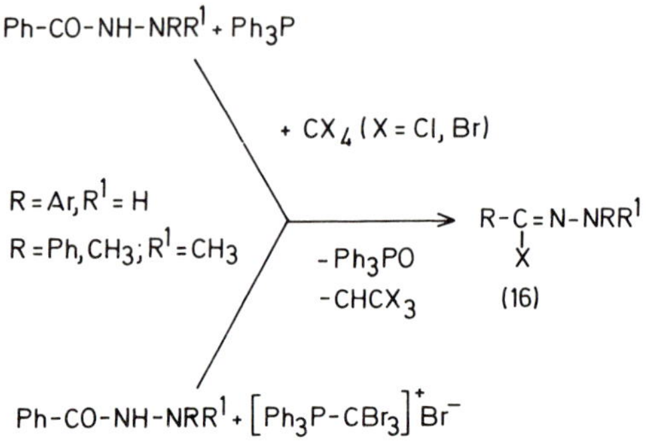

$$\text{Ph-CO-NH-NRR}^1 + \text{Ph}_3\text{P}$$

Scheme 7

$$R = Ar, R^1 = H$$
$$R = Ph, CH_3; R^1 = CH_3$$

$$+ CX_4 \ (X = Cl, Br)$$

$$-Ph_3PO$$
$$-CHCX_3$$

$$R-C=N-NRR^1$$
$$\quad \ \ \downarrow$$
$$\quad \ \ X$$
$$\quad \quad (16)$$

$$\text{Ph-CO-NH-NRR}^1 + [\text{Ph}_3\text{P-CBr}_3]^+ \text{Br}^-$$

A and B accounts for half of the total reaction, which is in fact often found, a molar ratio of 1·2:1 is advisable.[28] There are, however, reactions that scarcely produce chloroform, indicating that in these reactions only the intermediates of path B and the third component are involved. Examples are the treatment of Ph_3P–CCl_4 with enolizable ketones[29] or with water.

The three-component reaction triphenylphosphine–CCl_4–water will be treated in detail in the following section because it provides an easy route to (dichloromethyl)- (9a) and (monochloromethyl)-triphenylphosphonium chloride (15a), both being important starting materials for new organophosphorus reagents, such as (chloromethylene)phosphoranes and carbodiphosphoranes, now obtainable as pure substances.

4. The Three-Component Reaction: Triphenylphosphine–CCl_4–Water

On dissolving triphenylphosphine and CCl_4 in a mixture of acetonitrile and benzene the equilibrium, given by Scheme 8, is observed. If water is added to this system in the molar proportion indicated by the equation (a), pure (dichloromethyl)phosphonium (9a) is obtained. Here the dichlorophosphorane is at first hydrolysed to oxide and hydrogen chloride which is trapped in the next step by the basic ylide, giving the salt.

The (dichloromethyl)phosphonium chloride (9a) displays the same reactions with tertiary phosphines as the (trichloromethyl)phosphonium salt (4a), and a corresponding scheme can be formulated. In addition to the ylide (14a) and dichlorophosphorane, the bisphosphonium salt (17a) is present in equilibrium; this is dechlorinated by one more phosphine molecule to give the methine-bridged salt (11a), thus leaving the cycle. With water the system also

9 Conversions via R₃P-Halogenoalkanes

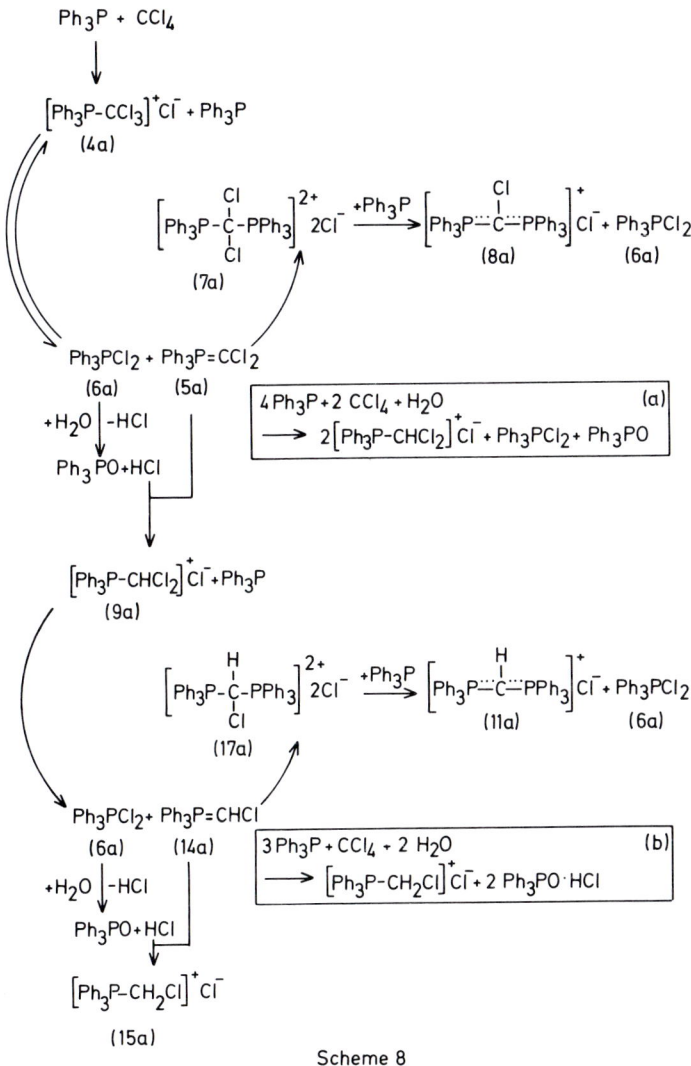

Scheme 8

reacts in such a way that the hydrogen chloride formed by hydrolysis of the dichlorophosphorane adds to the (chloromethylene)phosphorane (14a) to give the (monochloromethyl)phosphonium chloride (15a). As a result of the link between these two cycles it is possible to obtain, by addition of the correct amount of water, either the (dichloromethyl)- (9a) or the (monochloromethyl)-phosphonium chloride (15a), according to the equation (b), in high yield and purity (Scheme 8).[7]

The reaction with water is so effective that it is not only suitable for the preparation of these salts but may also be used for very precise water determinations in organic solvents. This new method is based on reaction rate determinations by means of conductivity measurements (see Fig. 1), based on the protons formed by hydrolysis of the dichlorophosphorane. Along the branch A of the curve the conductivity continuously increases due to the hydrogen chloride formed by hydrolysis of (6a). At the maximum the water is totally consumed and the conductivity now decreases along branch B since

A $2 Ph_3P + CCl_4 \longrightarrow Ph_3P=CCl_2 + Ph_3PCl_2$

$Ph_3PCl_2 + H_2O \longrightarrow Ph_3PO + 2 HCl$

$Ph_3P=CCl_2 + 2 HCl \longrightarrow [Ph_3PCHCl_2]^+ Cl^- + HCl$

B $2 Ph_3P + CCl_4 \longrightarrow Ph_3P=CCl_2 + Ph_3PCl_2$

$Ph_3P=CCl_2 + HCl \longrightarrow [Ph_3PCHCl_2]^+ Cl^-$

C $3 Ph_3P + CCl_4 \longrightarrow \left[Ph_3P\!-\!\underset{\underset{Cl}{|}}{C}\!-\!PPh_3 \right]^+ Cl^-$
$+ Ph_3PCl_2$

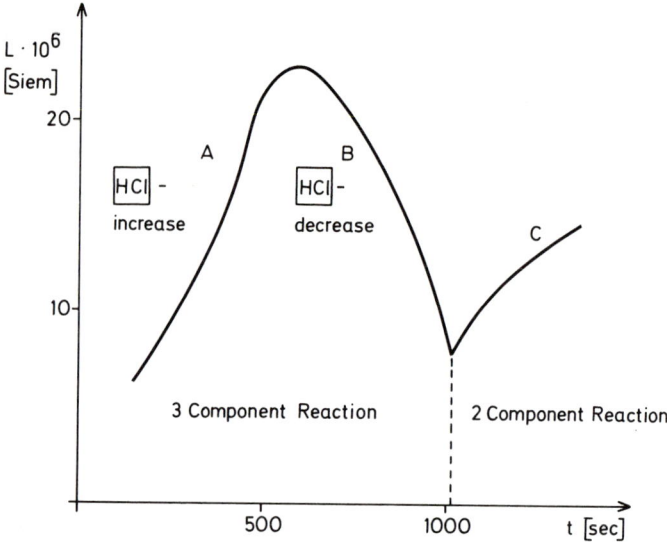

Figure 1

the hydrogen chloride is removed by the ylide (5a) giving the salt. At the minimum all hydrogen chloride is used up and a renewed increase in conductivity results which is caused by the just commencing formation of the chloromethine-bridged phosphonium chloride (8a).

This method allows the determination of very small amounts of water in a large number of organic solvents, with the exception of strong amines, down to a concentration of 10^{-3} mole %. The system is thus probably equal to the Karl Fischer reagent if not better.[32]

4.1. Chloromethylenephosphoranes and Hexaorganylcarbodiphosphoranes from Intermediates of the Reactions Ph_3P-CCl_4 and $Ph_3P-CCl_4-H_2O$

The various (chloromethyl)triphenylphosphonium chlorides formed as intermediates in the systems Ph_3P-CCl_4 and $Ph_3P-CCl_4-H_2O$ proved to be excellent and easily available starting materials for the two (chloromethylene)-phosphoranes (5a) and (14a). The (dichloromethylene)phosphorane (5a) is obtained by dechlorination of the (trichloromethyl)phosphonium chloride (4a) with tris(dimethylamino)phosphine:[8,9]

$$Ph_3PCCl_3{}^+Cl^- + P[N(CH_3)_2]_3 \rightarrow Ph_3P{=}CCl_2 + Cl_2P[N(CH_3)_2]_3 \quad (2)$$
(4a)   (5a)

The missing member in the series of chlorinated methylenephosphoranes, i.e. the ylide (monochloromethylene)phosphorane (14a), can be obtained by transylidation starting from (chloromethyl)phosphonium chloride (15a). This salt is a much stronger Brönsted acid than methylphosphonium chloride, so the strongly basic methylenephosphorane can easily dehydrochlorinate it to the desired (chloromethylene)phosphorane:[33]

$$Ph_3PCH_2Cl^+Cl^- + Ph_3P{=}CH_2 \rightarrow Ph_3P{=}CHCl + Ph_3PCH_3{}^+Cl^- \quad (3)$$
(15a)   (14a)

Of considerably greater preparative importance is probably the easy access to the class of carbodiphosphoranes by means of the system tertiary phosphine–CCl_4. The first member was the hexaphenylcarbodiphosphorane (18a), first isolated by Ramirez in 1961, which can be much more conveniently obtained in any quantity from the end-product of the reaction of triphenylphosphine–CCl_4, the salt (8a), by dechlorination with tris(dimethylamino)-phosphine.[10]

$$\left[Ph_3P{\cdots}\overset{Cl}{\underset{|}{C}}{\cdots}PPh_3\right]^+ Cl^- + P[N(CH_3)_2]_3 \longrightarrow Ph_3P{=}C{=}PPh_3 + Cl_2P[N(CH_3)_2]_3$$
         (8a)   (18a)

Scheme 9

Application of this simple synthesis to the alkyl substituted salts has not yet been achieved. The alkyl compounds are nearly insoluble in benzene or toluene, so some acetonitrile or dichloromethane has to be added. The alkylcarbodiphosphoranes which are in fact formed at first are so strongly basic that they immediately abstract protons. Hence only the stable methine-bridged salt (11) can be isolated. However, a whole series of asymmetrical

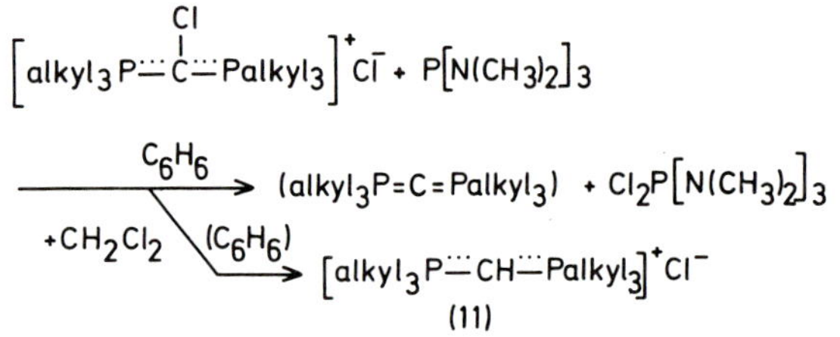

$$[alkyl_3P\overset{Cl}{\overset{|}{=}}C\overset{\cdot\cdot}{=}Palkyl_3]^+Cl^- + P[N(CH_3)_2]_3$$

$$\xrightarrow[+CH_2Cl_2]{C_6H_6} (alkyl_3P=C=Palkyl_3) + Cl_2P[N(CH_3)_2]_3$$

$$\xrightarrow{(C_6H_6)} [alkyl_3P\overset{\cdot\cdot}{=}CH\overset{\cdot\cdot}{=}Palkyl_3]^+Cl^-$$
$$(11)$$

Scheme 10

phenylalkyl-substituted carbodiphosphoranes can be obtained. The starting material is the (dichloromethyl)phosphonium chloride (9a); this reacts with aliphatic phosphines to give the CH-bridged salts (11) which can be deprotonated by butyllithium to the corresponding carbodiphosphoranes (18).[34]

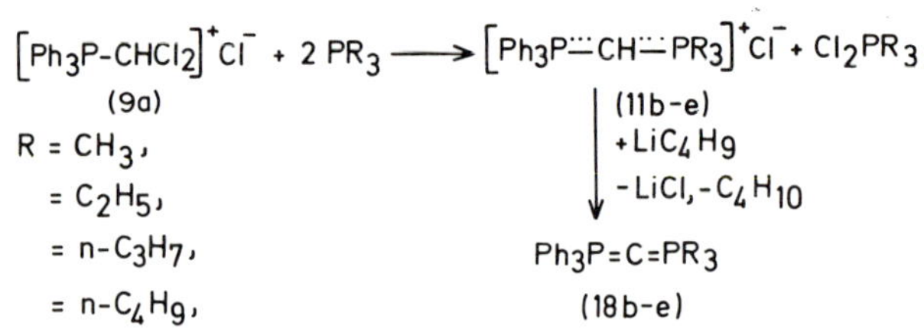

$$[Ph_3P-CHCl_2]^+Cl^- + 2\ PR_3 \longrightarrow [Ph_3P\overset{\cdot\cdot}{=}CH\overset{\cdot\cdot}{=}PR_3]^+Cl^- + Cl_2PR_3$$

(9a)   (11b-e)

R = CH$_3$,   +LiC$_4$H$_9$

 = C$_2$H$_5$,   −LiCl, −C$_4$H$_{10}$

 = n-C$_3$H$_7$,

 = n-C$_4$H$_9$,   Ph$_3$P=C=PR$_3$

           (18b-e)

Scheme 11

Compared with hexaphenylcarbodiphosphorane the semi-alkyl-substituted bis-ylides show increased reactivity. They are not only stronger bases but also thermally unstable. Gentle heating already causes hydrogen migration from the C-atom of the alkyl chain to the bridge C-atom accompanied by migration of a phenyl group to give the phosphine-substituted alkylidenephosphorane (19). The structure follows from a Wittig reaction with benzaldehyde affording phosphine, substituted styrene, and phosphine oxide. Addition of hydrogen chloride yields the salt (20) which can be reduced by LiAlH$_4$ to give the mixed substituted bisphosphino-methane (Scheme 12).[34]

9 Conversions via R_3P-Halogenoalkanes

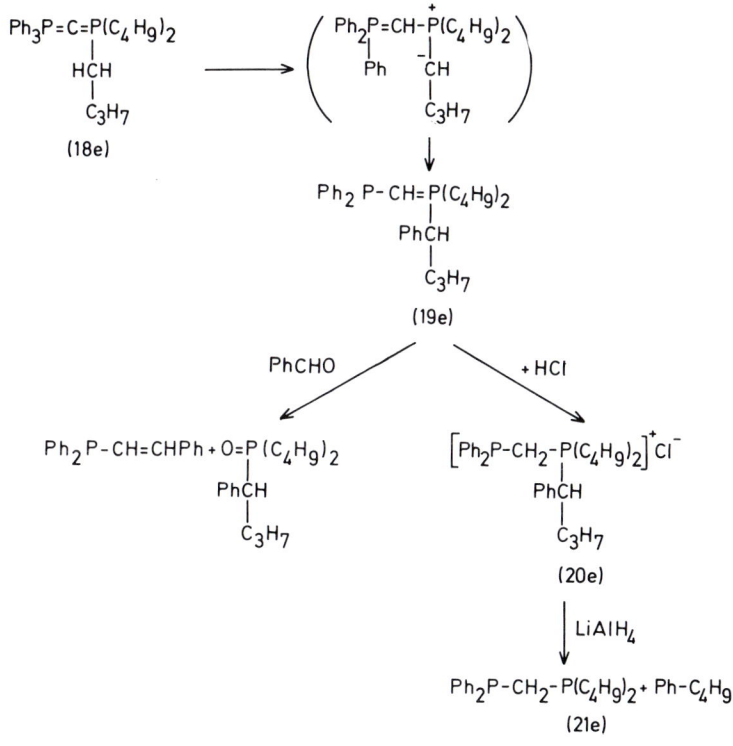

Scheme 12

The preparative significance of carbodiphosphoranes lies especially in two main ranges of application. On the one hand they are extremely strong bases which often can replace organometallic reagents in reactions such as deprotonation or dehalogenation. On the other hand they offer an easy route to the versatile class of phosphacumulene ylides which recently have been reviewed in detail by Bestmann.[35]

Furthermore, specifically substituted carbodiphosphoranes, such as chloropentaphenylcarbodiphosphorane, open up a gateway to hitherto unknown cyclic phosphorus–carbon compounds. Thus, to take one example, the reaction of (trichloromethyl)phosphonium chloride (4a) with chlorodiphenylphosphine affords the PCP-bridged salt (8f) which can be dechlorinated by tris(dimethylamino)phosphine to give the thermally labile chloropentaphenylcarbodiphosphorane (18f). Gentle heating causes smooth cycloaddition yielding the diphosphacyclobutadiene (22f).[36,14,37]

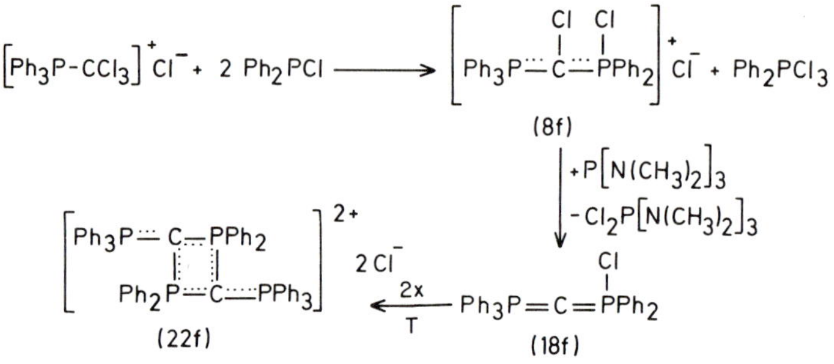

$$[Ph_3P\text{-}CCl_3]^+ Cl^- + 2\, Ph_2PCl \longrightarrow \left[Ph_3P\overset{Cl}{\underset{|}{\text{---}}}C\overset{Cl}{\underset{|}{\text{---}}}PPh_2\right]^+ Cl^- + Ph_2PCl_3$$

(8f)

$$\left[\begin{array}{c}Ph_3P\text{---}C\text{---}PPh_2\\ Ph_2P\text{---}C\text{---}PPh_3\end{array}\right]^{2+} 2\,Cl^- \xleftarrow[T]{2x} Ph_3P=C=\overset{Cl}{\underset{|}{P}}Ph_2$$

$$\Big\updownarrow \begin{array}{l}+P[N(CH_3)_2]_3\\ -Cl_2P[N(CH_3)_2]_3\end{array}$$

(22f)   (18f)

Scheme 13

5. Chlorinations

5.1. Chlorinations of O-Functional Compounds

5.1.1. *Alcohols and thiols*

The predominant reaction of tertiary phosphines and tetrachloromethane with alcohols[17–27] or thiols[23,38–40] occurs in the sense of equation (4), yielding the salt (13) and chloroform:

$$R_3P + CCl_4 + R^1XH \rightarrow R_3PXR^{1+}Cl^- + CHCl_3 \quad (4)$$

(13) X = O, S

The phosphonium chlorides (13), in which R = Ph and R^1 = alkyl, decompose in a rapid Arbusov rearrangement to the phosphine oxide or sulphide and alkyl chloride:

$$R_3PXR^{1+}Cl^- \rightarrow R_3P{=}X + R^1Cl \quad (5)$$

Thus the reagent is suitable for the conversion of alcohols into the corresponding alkyl chlorides under very mild conditions,[41–49] although in the case of secondary alcohols the formation of olefins can be favoured by slight variation of the reaction conditions.[50] The chlorination usually is so mild that it can be used to convert, for example, the hydroxy groups of 5-spiro[3.4]-octen-7-ol,[51] glycerides,[52] carbohydrates,[18,19,52–54] lincomycins,[55] or geraniol[56] into the corresponding chloro derivatives; the mild halogenation of OH groups in ribose bound in nucleosides has been achieved too.[57] The stereochemistry of this reaction has been clarified by experiments with optically active alcohols;[23–35,51,58] it is largely stereoselective, with inversion. Regio- and stereo-selective chlorinations of alkyl alcohols with the analogous combination Ph_3P–hexachloroacetone have also been reported.[59]

9 Conversions via R_3P-Halogenoalkanes

The phosphonium ion from tris(dimethylamino)phosphine is considerably more stable, so it can be isolated as the perchlorate or hexafluorophosphate.[21,26] Even secondary alcohols[60] and propane-1,3-diols[61] form stable alkoxy-tris-(dimethylamino)phosphonium salts, while α-glycols immediately condense to give epoxides or spiro-phosphoranes (subsequent deamination).[62] The isolated salts are suitable for the synthesis of alkyl aryl ethers or thioethers.[63]

Treatment with alkali-metal halides or pseudohalides enforces the Arbusov rearrangement, and covalent halogen or pseudohalogen compounds are obtained,[58,64] e.g.:

$$[(CH_3)_2N]_3POR^+ClO_4^- + NaN_3 \rightarrow [(CH_3)_2N]_3PO + RN_3 + NaClO_4 \quad (6)$$

Primary and secondary alcohols can similarly be converted into the corresponding carbonitriles; these are formed when the alcohols are heated with

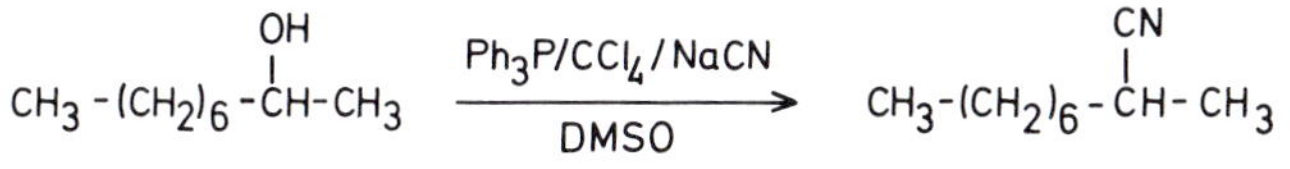

$$CH_3-(CH_2)_6-\underset{\underset{OH}{|}}{CH}-CH_3 \xrightarrow[DMSO]{Ph_3P/CCl_4/NaCN} CH_3-(CH_2)_6-\underset{\underset{CN}{|}}{CH}-CH_3$$

Scheme 14

Ph_3P-CCl_4 and NaCN in dimethyl sulphoxide (DMSO):[65] Analogously the iodination of alcohols can be achieved by a four-component reaction:[66]

$$ROH \xrightarrow{Ph_3P-CCl_4-NaI} RI \quad (7)$$

The aryloxy- and arylthio-phosphonium salts (13; R^1 = aryl) are markedly more stable than the alkoxy- and alkylthio-phosphonium salts, and when R = Ph they can be isolated as chlorides[40] or hexachloroantimonates.[39] The phenoxy- and phenylthio-triphenylphosphonium chlorides are extremely sensitive to hydrolysis and decompose immediately even in air to HCl, triphenylphosphine oxide, and phenol or thiophenol; when they are treated with an alcohol the corresponding alkyl halide is formed together with the phosphine oxide and phenol or thiophenol. Alkanethiols decompose these salts analogously with formation of the phosphine sulphide.[40]

As in all such three-component reactions, the side-reaction (B reaction) according to equation (8) occurs alongside the main reaction [A reaction, equation (4)].

$$3\,Ph_3P + CCl_4 + 2\,R^1XH \rightarrow 2\,Ph_3PXR^{1+}Cl^- + Ph_3PCH_2Cl^+Cl^- \quad (8)$$

$$(13)\ X = O, S$$

5.1.2. Aldehydes, ketones, and epoxides

The systems tris(dimethylamino)phosphine–CCl_4 and triphenylphosphine–CCl_4 behave very differently with aldehydes. Owing to the differing charge transfer discussed in Section 2, only in the case of tris(dimethylamino)-phosphine do significant amounts of the ion pair (2) and the trichloromethanide ion appear. The overall reaction expressed in equation (d), Scheme 15, is based on a multistep mechanism [equations (a)–(c)] in which the first step is nucleophilic attack of the CCl_3^- ion on the carbonyl carbon.[67-70]

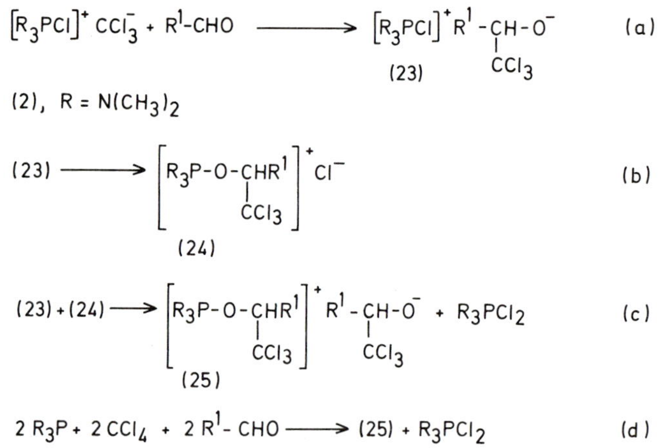

$$[R_3PCl]^+ \, CCl_3^- + R^1\text{-CHO} \longrightarrow [R_3PCl]^+ \, R^1\text{-}\underset{\underset{CCl_3}{|}}{CH}\text{-}O^- \quad \text{(a)}$$

(2), R = N(CH$_3$)$_2$ \hspace{3cm} (23)

$$(23) \longrightarrow \left[R_3P\text{-}O\text{-}\underset{\underset{CCl_3}{|}}{CHR^1} \right]^+ Cl^- \quad \text{(b)}$$

(24)

$$(23) + (24) \longrightarrow \left[R_3P\text{-}O\text{-}\underset{\underset{CCl_3}{|}}{CHR^1} \right]^+ R^1\text{-}\underset{\underset{CCl_3}{|}}{CH}\text{-}O^- + R_3PCl_2 \quad \text{(c)}$$

(25)

$$2\,R_3P + 2\,CCl_4 + 2\,R^1\text{-CHO} \longrightarrow (25) + R_3PCl_2 \quad \text{(d)}$$

Scheme 15

On hydrolysis, the alkoxyphosphonium salt (26) affords α-substituted trichloroethanols.[68-70] If, however, the starting materials are used in the ratio phosphine/CCl_4/aldehyde = 2:1:1, then 1,1-dichloroalkenes are obtained in accordance with Scheme 16.

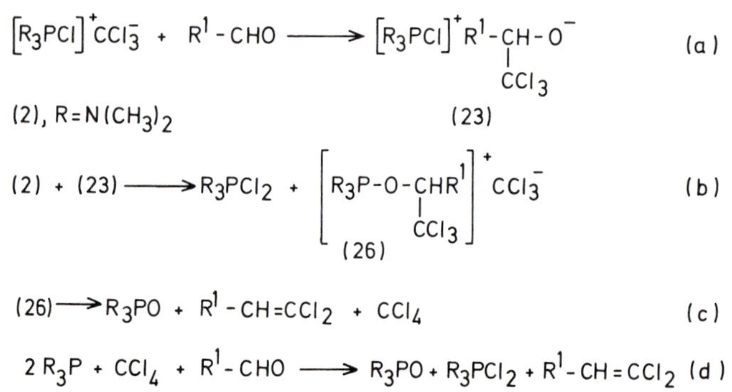

$$[R_3PCl]^+ \, CCl_3^- + R^1\text{-CHO} \longrightarrow [R_3PCl]^+ \, R^1\text{-}\underset{\underset{CCl_3}{|}}{CH}\text{-}O^- \quad \text{(a)}$$

(2), R = N(CH$_3$)$_2$ \hspace{3cm} (23)

$$(2) + (23) \longrightarrow R_3PCl_2 + \left[R_3P\text{-}O\text{-}\underset{\underset{CCl_3}{|}}{CHR^1} \right]^+ CCl_3^- \quad \text{(b)}$$

(26)

$$(26) \longrightarrow R_3PO + R^1\text{-CH=}CCl_2 + CCl_4 \quad \text{(c)}$$

$$2\,R_3P + CCl_4 + R^1\text{-CHO} \longrightarrow R_3PO + R_3PCl_2 + R^1\text{-CH=}CCl_2 \quad \text{(d)}$$

Scheme 16

By analogy with the system triphenylphosphine–CCl_4 the (dichloromethylene)phosphorane (5) containing $N(CH_3)_2$ instead of Ph might be expected to be formed and to react with the carbonyl compound by a Wittig-type reaction; this, however, is excluded because the (trichloromethyl)phosphonium chloride (4; $R = N(CH_3)_2$) (Scheme 1) formed as intermediate is not dechlorinated by excess phosphine as in the case of (4a) where $R^1 = Ph$.[5] The ylide mechanism is also ruled out by the fact that not only aldehydes but also α-(trichloromethyl)-alcohols can be converted into 1,1-dichloroalkenes in good yields.[68] Formation of the dichloroalkenes probably results directly from dechlorination of the alkoxyphosphonium salt (26).

In the reaction of the reagent Ph_3P–CCl_4 with carbonyl compounds none of the intermediates formulated in Scheme 1 plays a significant part. The reaction occurs only by way of the dichlorophosphorane (6) formed according to Scheme 2 and leads to 1,1-dichloroalkenes and dichloromethylene compounds:[4,6,71-74]

$$Ph_3P\!=\!CCl_2 + Ph_3PCl_2 + 2\,O\!=\!C\!< \;\rightarrow\; Cl_2C\!=\!C\!< + Cl_2C\!< + 2\,Ph_3PO$$
(5a)   (6a)   (9)

Enolizable ketones can also afford 1,1-dichloroalkenes with Ph_3P–CCl_4, but the chlorine-substitution product of the enol is formed too (Scheme 17).[75]

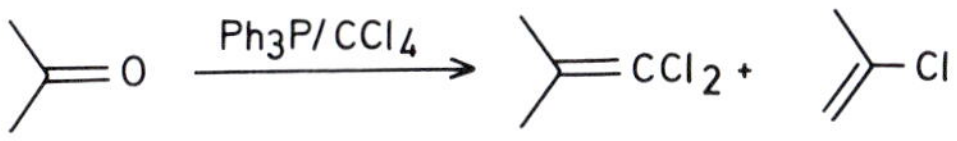

Scheme 17

With five-membered ring ketones the dichloroalkene predominates to a great extent, whereas the reverse is true for six-membered ring ketones. Good yields of β-chloro-α,β-unsaturated ketones are obtained from 1,3-diketones[76] (Scheme 18).

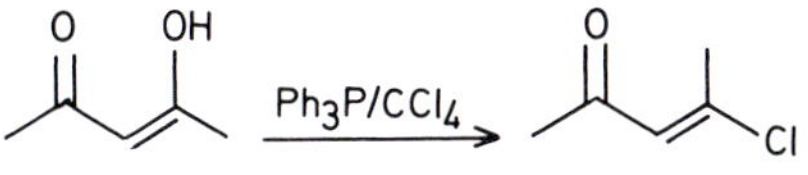

Scheme 18

Epoxides are converted into dichloroalkanes by Ph_3P–CCl_4 (Scheme 19); in the case of cycloalkene oxides this reaction leads stereospecifically to cis-1,2-dichlorocycloalkanes.[77] Presumably this reaction does not occur by

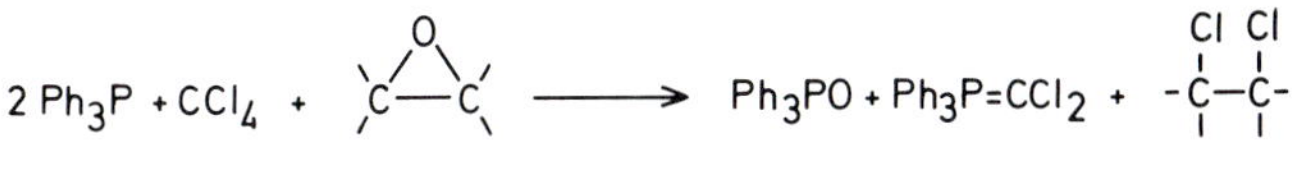

Scheme 19

the postulated ionic mechanism[77] but rather, by analogy with equation (9), through the intermediates (5) and (6), the dichlorophosphorane reacting with the epoxide in the known manner (Scheme 20).[78]

$$Ph_3PCl_2 + R_2C\overset{O}{\underset{}{-}}CR_2 \longrightarrow Ph_3PO + R_2\overset{Cl}{\underset{}{C}}-\overset{Cl}{\underset{}{C}}R_2$$

Scheme 20

5.1.3. Carboxylic acids and oxophosphoric acids

By analogy with the conversion of alcohols into alkyl chlorides, carboxylic acids can easily be converted into acyl chlorides by Ph_3P–CCl_4 [cf. equations (4) and (5); $X = O$, $R^1 = C(O)R$].[79] This synthesis is especially advantageous in the case of substances that are sensitive to acidic conditions because the reaction medium is neutral, in contrast to the usual methods for preparing acid chlorides by means of thionyl chloride, phosphorus pentachloride, oxalyl chloride, or phosgene. This process has also proved suitable for the preparation of acid chlorides from other than carboxylic acids, e.g. from phosphoric esters and phosphinic acids;[80] 50–60% yields of the corresponding acid chlorides were obtained by the reaction of dimethyl- or diphenyl-phosphinic acid or of dibutyl hydrogen phosphate in acetonitrile:

$$Ph_3P + CCl_4 + R_2P(O)OH \rightarrow R_2P(O)Cl + CHCl_3 + Ph_3PO \quad (10)$$
$$R = CH_3, C_6H_5, OC_4H_9\text{-}n$$

Phosphoric monoesters can be converted into phosphorodichloridates $ROP(O)Cl_2$, $R = i\text{-}C_3H_7$ or C_6H_5, in the same way.

In these cases it was shown that chlorination does not occur exclusively according to equation (10), but rather that the B reaction also occurs:

$$3\,Ph_3P + CCl_4 + 2\,R_2P(O)OH \rightarrow$$
$$2\,R_2P(O)Cl + 2\,Ph_3PO + Ph_2PCH_2Cl^+Cl^- \quad (11)$$

Participation of the B reaction in the synthesis of carboxylic acid chlorides has not yet been proved experimentally. Nevertheless, it can be assumed that it does so in that case too, as is usual in the reactions of the Ph_3P–CCl_4 system with proton-active substrates.

Diethyl phosphorochloridate is not obtained according to equation (10), the anhydride being formed instead (Scheme 21).[80] This anhydride is formed

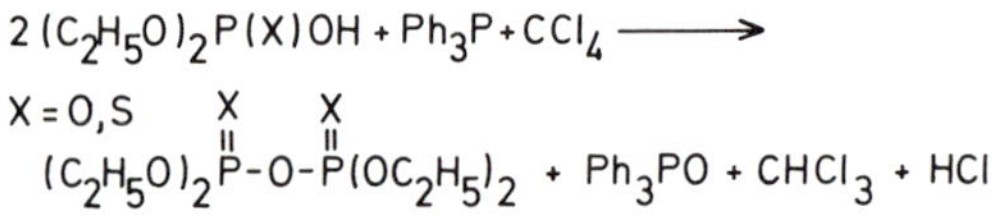

$$2\,(C_2H_5O)_2P(X)OH + Ph_3P + CCl_4 \longrightarrow$$
$$X = O, S \quad$$
$$(C_2H_5O)_2\overset{X}{\underset{}{P}}-O-\overset{X}{\underset{}{P}}(OC_2H_5)_2 + Ph_3PO + CHCl_3 + HCl$$

Scheme 21

not by way of the acid chloride, but directly from the phosphoryloxyphosphonium salt as shown in Scheme 22. Similarly anhydrides are obtainable

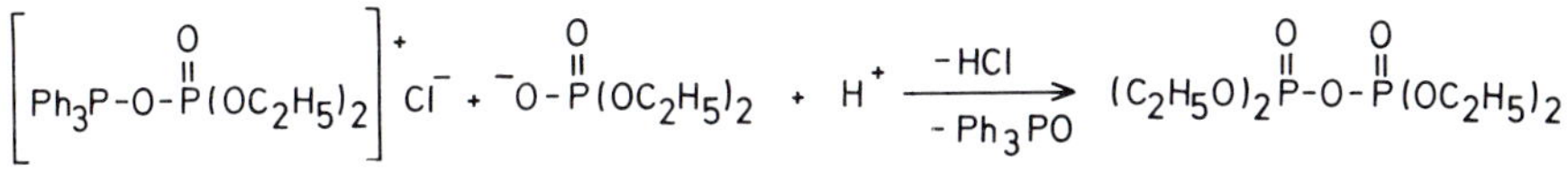

$$\left[Ph_3P-O-\overset{O}{\overset{\|}{P}}(OC_2H_5)_2\right]^+ Cl^- + {}^-O-\overset{O}{\overset{\|}{P}}(OC_2H_5)_2 + H^+ \xrightarrow[-Ph_3PO]{-HCl} (C_2H_5O)_2\overset{O}{\overset{\|}{P}}-O-\overset{O}{\overset{\|}{P}}(OC_2H_5)_2$$

Scheme 22

from phosphoric monoesters with $Ph_3P–CCl_4$, while, in the presence of alcohols, the corresponding diesters are formed.[81]

As a result of the greater stability of the intermediate acyloxyphosphonium chloride (13), the reaction in the three-component system $[(CH_3)_2N]_3P–CCl_4–R^1COOH$ also leads directly to the acid anhydride (Scheme 23).[82]

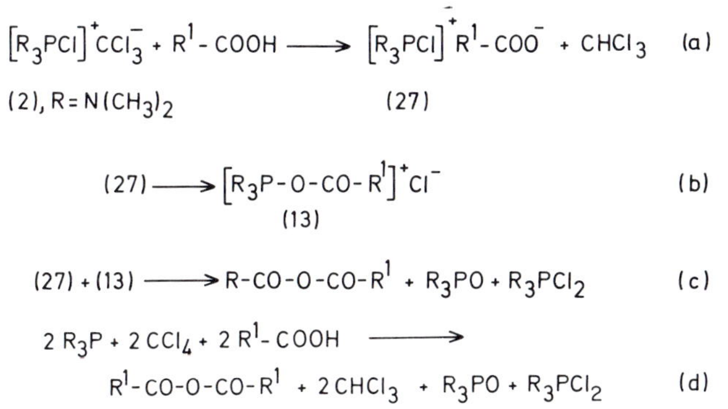

$$[R_3PCl]^+ CCl_3^- + R^1\text{-COOH} \longrightarrow [R_3PCl]^+ R^1\text{-COO}^- + CHCl_3 \quad (a)$$

(2), R = N(CH$_3$)$_2$ \qquad\qquad\qquad (27)

$$(27) \longrightarrow [R_3P\text{-O-CO-}R^1]^+ Cl^- \quad (b)$$
(13)

$$(27) + (13) \longrightarrow R\text{-CO-O-CO-}R^1 + R_3PO + R_3PCl_2 \quad (c)$$

$$2\,R_3P + 2\,CCl_4 + 2\,R^1\text{-COOH} \longrightarrow$$
$$R^1\text{-CO-O-CO-}R^1 + 2\,CHCl_3 + R_3PO + R_3PCl_2 \quad (d)$$

Scheme 23

5.2. Chlorinations of –C(X)NH– Functional Compounds (X = O or S)

5.2.1. Chloroformamidines from trisubstituted ureas

The action of $Ph_3P–CCl_4$ on N,N,N'-trisubstituted ureas gives chloroformamidines in good yields.[83] The very mild reaction conditions allow the synthesis of compounds that are accessible only with difficulty, if at all, by the usual methods, e.g. chloroformamidines containing one or two aryl groups on the disubstituted nitrogen atom. Purely aliphatic chloroformamidines can also be prepared by this method [84] as well as chloroformamidines from tri-N-substituted bis-ureas.[85] Here, as with all –C(O)NH– functional compounds,

$$Ph_3P + CCl_4 + R\text{-NH-CO-}NR^1R^2 \longrightarrow R\text{-N=}\underset{Cl}{\overset{}{C}}\text{-}NR^1R^2 + Ph_3PO + CHCl_3$$

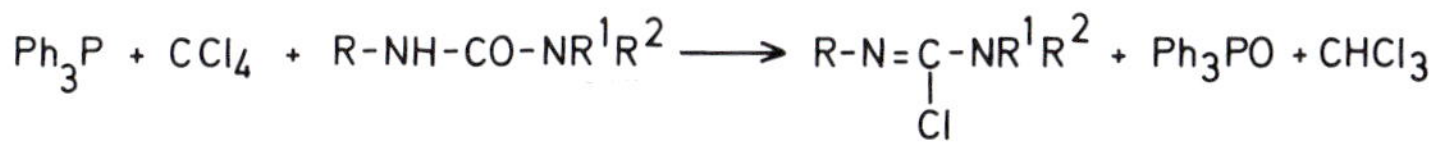

Scheme 24

the phosphorylation occurs on the oxygen atom. The O-phosphorylated salt then decomposes to phosphine oxide and chloroformamidine as in the Arbusov rearrangement (Scheme 25). The B reaction could be shown to take

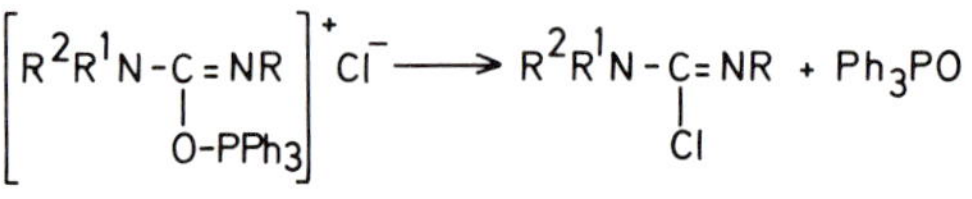

$$\left[R^2R^1N-\underset{O-PPh_3}{C}=NR \right]^+ Cl^- \longrightarrow R^2R^1N-\underset{Cl}{C}=NR + Ph_3PO$$

Scheme 25

part in the chloroformamidine formation too (Scheme 26), the active agent being the intermediate dichlorophosphorane (6).

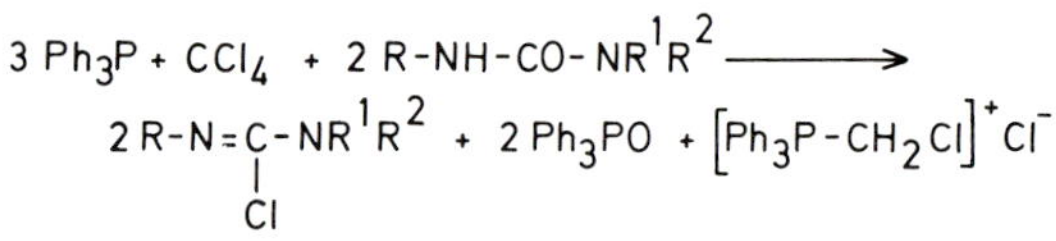

$$3\,Ph_3P + CCl_4 + 2\,R-NH-CO-NR^1R^2 \longrightarrow$$
$$2\,R-N=\underset{Cl}{C}-NR^1R^2 + 2\,Ph_3PO + [Ph_3P-CH_2Cl]^+Cl^-$$

Scheme 26

5.2.2. Imidoyl halides from monosubstituted carboxamides and by Beckmann rearrangement of ketoximes

Imidoyl halides which are important starting materials for the synthesis of amidines and imino-ethers are obtained from N-monosubstituted carboxamides according to Scheme 27.[86]

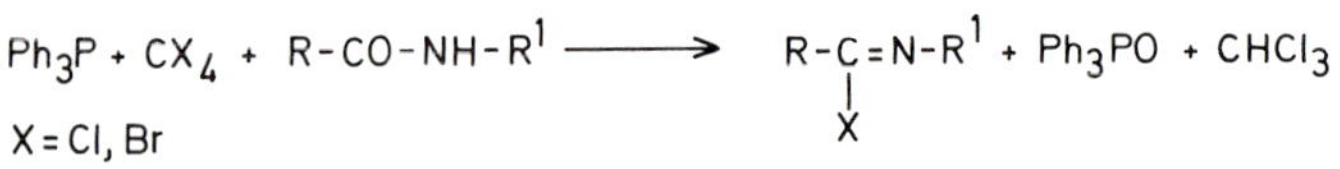

$$Ph_3P + CX_4 + R-CO-NH-R^1 \longrightarrow R-\underset{X}{C}=N-R^1 + Ph_3PO + CHCl_3$$

X = Cl, Br

Scheme 27

Under the mild reaction conditions imidoyl halides hitherto difficult to obtain can be synthesized.[87] As to the participation of the B reaction, the criteria of Section 5.2.1 also apply. Another method of preparation of imidoyl halides by means of the Ph_3P-CCl_4 reagent is the reaction with aromatic or aliphatic ketoximes (Scheme 28).[88]

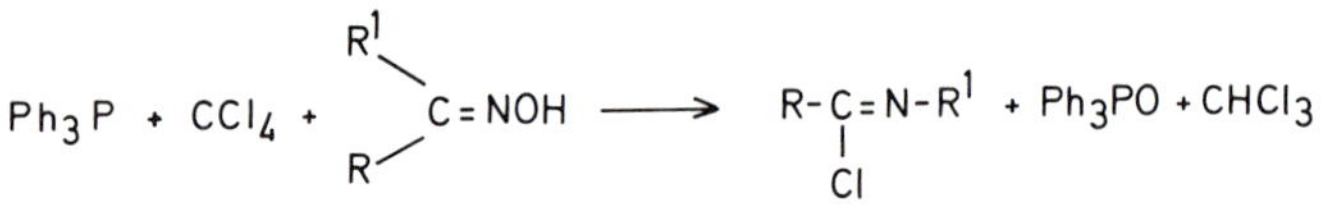

$$Ph_3P + CCl_4 + \underset{R}{\overset{R^1}{>}}C=NOH \longrightarrow R-\underset{Cl}{C}=N-R^1 + Ph_3PO + CHCl_3$$

Scheme 28

This formation of imidoyl halides might proceed either by way of the N-chloro compound (Scheme 29) or, as shown in Scheme 30, by Beckmann rearrangement of the O-phosphorylated ketoxime itself.

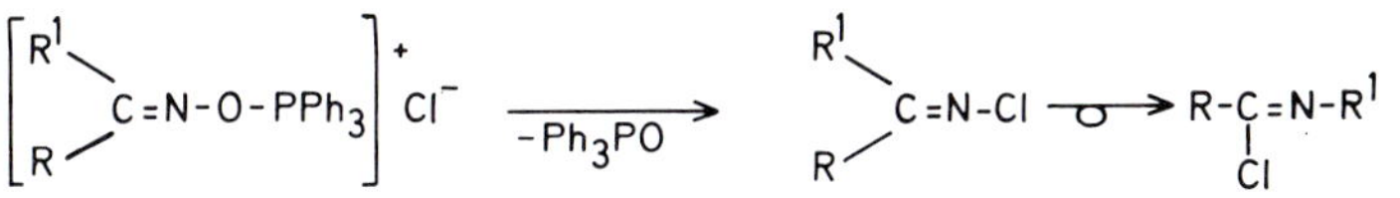

9 Conversions via R_3P-Halogenoalkanes

$$\left[\begin{array}{c}R^1\\ \diagdown\\ C=N-O-PPh_3\\ \diagup\\ R\end{array}\right]^+ Cl^- \xrightarrow{-Ph_3PO} \begin{array}{c}R^1\\ \diagdown\\ C=N-Cl\\ \diagup\\ R\end{array} \xrightarrow{} R-\underset{\underset{Cl}{|}}{C}=N-R^1$$

Scheme 29

$$\left[\begin{array}{c}R^1\\ \diagdown\\ C=N-O-PPh_3\\ \diagup\\ R\end{array}\right]^+ Cl^- \xrightarrow[-Ph_3PO]{\sim R^1} [R-C\equiv N-R^1]^+ Cl^- \longrightarrow R-\underset{\underset{Cl}{|}}{C}=N-R^1$$

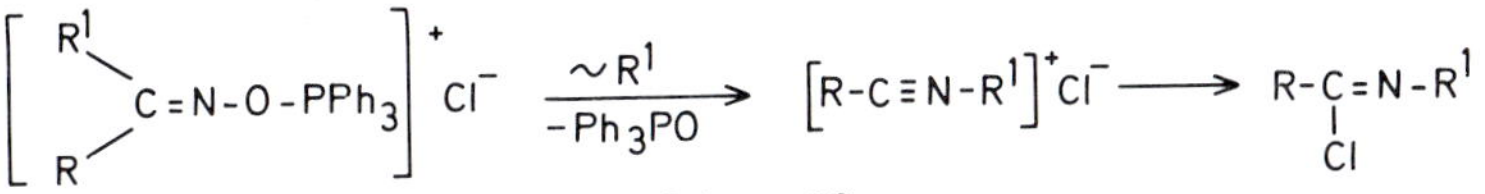

Scheme 30

Since most imidoyl chlorides are markedly sensitive to hydrolysis, chromatographic processes are rarely suitable for the separation of the phosphine oxide; depending on the nature of the groups R and R^1, distillation under high vacuum often leads to complicated decompositions involving loss of hydrogen chloride.[89] In such cases isolation of the imidoyl chloride is abandoned and the product is used directly, or it is hydrolysed to amide which can be readily separated from the phosphine oxide.

By hydrolysis to the amides it has also been shown that di-aliphatic ketoximes give imidoyl chlorides too.[90] The migration tendency of the two groups R and R^1 follows the same rules as for the Beckmann rearrangement by the usual reagents (H_2SO_4, PCl_5, or $COCl_2$). In di-aromatic compounds the unsubstituted phenyl group migrates preferentially and in aromatic aliphatic ketoximes the aromatic group undergoes the shift.[88]

5.2.3. Reaction with carbamoyl halides

No isocyanide dihalides are obtained from carbamoyl chlorides and fluorides with Ph_3P–CCl_4 as might be expected by analogy with other reactions of –CO–NH– compounds, but isocyanates and dihalogenotriphenylphosphorane are formed instead:[91]

$$Ph_3P + CCl_4 + RNHCOCl \rightarrow RN{=}C{=}O + Ph_3PCl_2 + CHCl_3 \quad (12)$$

The reaction of carbamoyl fluoride gives a mixture of dichloro- and difluorotriphenylphosphorane in place of the chlorofluorophosphorane. It is probable that in this reaction the halogen atom and not the oxygen atom attacks the phosphorus of the charge-transfer complex (1) nucleophilically, thus increasing the electron density of the trichloromethyl group to such an extent that it deprotonates the carbamoyl halide (Scheme 31). Gas-chromatographic

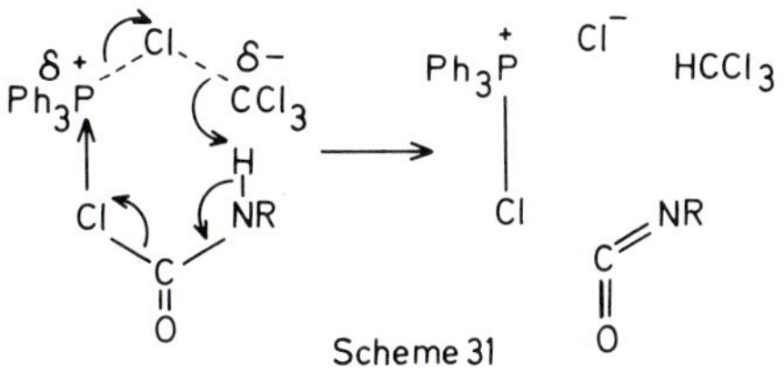

Scheme 31

determinations of the increase in $CHCl_3$ concentration and decrease in that of CCl_4 show that in this case too [in addition to the reaction according to equation (12)] the B reaction [equation (13)] makes a considerable contribution to the overall conversion:

$$3\,Ph_3P + CCl_4 + 2\,RNHCOCl \rightarrow$$
$$2\,RN{=}C{=}O + 2\,Ph_3PCl_2 + Ph_3PCH_2Cl^+Cl^- \quad (13)$$

5.2.4. Reaction with thiocarbamic O-esters

In contrast to N-unsubstituted or monosubstituted carbamic esters which afford stable N-phosphonium salts on treatment with Ph_3P-CCl_4,[92] N-substituted thiocarbamic O-esters form no phosphonium salts but directly give phosphine sulphide, isocyanate, and alkyl chloride (Scheme 32).[93]

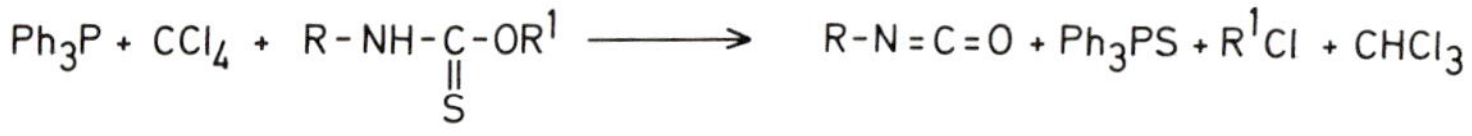

$$Ph_3P + CCl_4 + R-NH-\underset{\underset{S}{\|}}{C}-OR^1 \longrightarrow R-N{=}C{=}O + Ph_3PS + R^1Cl + CHCl_3$$

Scheme 32

5.2.5. N-Phenylchlorothioformimidic esters from thiocarbamic S-esters

Unlike the O-esters, the isomeric S-esters of thiocarbamic acid can be chlorinated by Ph_3P-CCl_4 with retention of the molecular skeleton on the carbonyl group.[93] By this method N-phenyl-chlorothioformimidic esters (pale yellow oils with an unpleasant odour) are obtained in 70–80% yield (Scheme 33). Depending on the nature of the group R^1 the thermal stability

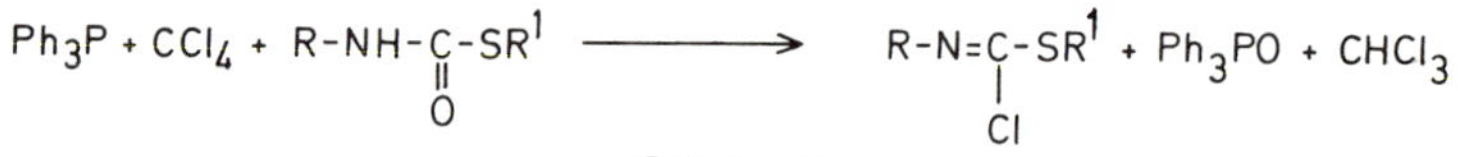

$$Ph_3P + CCl_4 + R-NH-\underset{\underset{O}{\|}}{C}-SR^1 \longrightarrow R-N{=}\underset{\underset{Cl}{|}}{C}-SR^1 + Ph_3PO + CHCl_3$$

Scheme 33

varies greatly. When R^1 = t-butyl, complete decomposition to isothiocyanate and t-butyl chloride takes place during working-up, while decomposition is of minor importance when $R^1 = C_2H_5$, n-C_3H_7, or i-C_3H_7. Measurement of the yield of chloroform indicates that in the case of O- and S-ester formation the B reaction is involved to the extent of approximately 50%.

6. Dehydrations

6.1. Nitriles from Carboxamides and Aldoximes

The ability of the Ph_3P-CCl_4 system to remove water was discovered during investigations of reactions with carboxamides.[94,95] In the presence of tertiary nitrogen bases such as triethylamine, pyridine, etc. nitriles are thus formed according to equation (14) in 80–90% yields.

$$Ph_3P + CCl_4 + RC(X)NH_2 \xrightarrow[-(Et_3NH)Cl]{+Et_3N} RC{\equiv}N + Ph_3PX + CHCl_3 \quad (14)$$
$$X = O, S$$

Thiocarboxamides react in the same way with Ph_3P-CCl_4. This nitrile synthesis is suitable for aliphatic as well as for aromatic carboxamides and can be applied to the preparation of such nitriles as 4,8-dioxoadamantane-2-carbonitrile[95] and 7-chloro-2,3-dihydro-1-methyl-2-oxo-5-phenyl-1H-1,4-benzazepine-3-carbonitrile[96] which are not accessible by the use of the usual acidic dehydrating agents. Again only 50% of the nitrile is formed according to equation (14), while the other half is obtained via the B reaction shown in equation (15). The dehydrating agent here is the dichlorophosphorane (6) formed as intermediate.

$$3\,Ph_3P + CCl_4 + 2\,RC(X)NH_2 \xrightarrow[-2\,(Et_3NH)Cl]{+Et_3N}$$
$$2\,RC{\equiv}N + 2\,Ph_3PX + Ph_3PCH_2Cl^+Cl^- \quad (15)$$

Removal of phosphine oxide or sulphide occurs from the O- or S-phosphorylated salt respectively. This is also the case for the dehydration of aldoximes, since it has been shown by using deuterated acetaldoxime that the hydrogen of the chloroform comes solely from the OH group [equations (16) and (17)].[97]

$$Ph_3P + CCl_4 + DON{=}CHR \rightarrow Ph_3PON{=}CHR^+Cl^- + DCCl_3 \quad (16)$$
$$Ph_3PON{=}CHR^+Cl^- + Et_3N \rightarrow RC{\equiv}N + Ph_3PO + (Et_3NH)Cl \quad (17)$$

6.2. Isocyanides from Monosubstituted Formamides

Removal of water from N-substituted formamides can also be achieved with considerable success by the Ph_3P-CCl_4 reagent. According to equation (18) alkyl and aryl isocyanides are obtained in yields of 90%,[98] although so far it has been tested on only a few examples.

$$Ph_3P + CCl_4 + RNHCHO \xrightarrow[-(Et_3NH)Cl]{+Et_3N} RNC + Ph_3PO + CHCl_3 \quad (18)$$

6.3. Carbodiimides from N,N'-Disubstituted Ureas

The synthesis of carbodiimides from N,N'-disubstituted ureas or thioureas proceeds equally smoothly and with similarly high yields:[99]

$$Ph_3P + CCl_4 + RNHC(X)NHR \xrightarrow[-(Et_3NH)Cl]{+Et_3N}$$

X = O, S
$$RN{=}C{=}NR + Ph_3PX + CHCl_3 \quad (19)$$

This reaction is also applicable to the preparation of alkylene-bridged di(carbodiimides).[100]

6.4. Ketene Imines from α-Disubstituted Primary Acid Amides

Dehydration of α-disubstituted primary acid amides by means of the Ph_3P–CCl_4 reagent gives ketene imines in similarly high yields:[101]

$$R_2HCC(O)NHR^1 \xrightarrow{-H_2O} R_2C{=}C{=}NR^1 \quad (20)$$

6.5. C-Ethoxycarbonyl-C-imidoylketene Imines from (Aminoalkylene)-malonamic Esters

The (aminoalkylene)malonamic esters obtained by the reaction of secondary enamines with isocyanates can be regarded as vinylogous ureas and, accordingly, react in the same way as N,N′-disubstituted ureas themselves to give the very labile (ethoxycarbonyl)imidoylketene imines.[102]

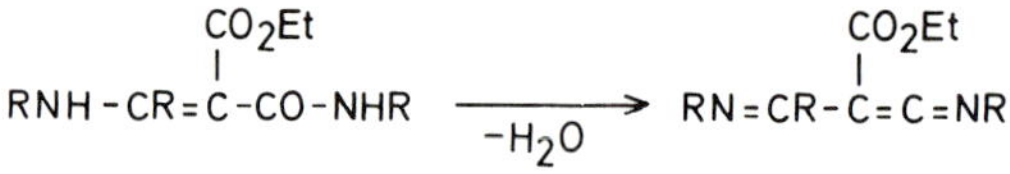

$$\begin{array}{c} CO_2Et \\ | \\ RNH-CR{=}C-CO-NHR \end{array} \xrightarrow{-H_2O} \begin{array}{c} CO_2Et \\ | \\ RN{=}CR-C{=}C{=}NR \end{array}$$

Scheme 34

6.6. Nitrogen Heterocycles from N-Substituted Amino-Alcohols

In contrast to the usual methods of preparing aziridines by way of β-halogenoamines or β-amino-sulphonic acids, the action of Ph_3P–CCl_4 on N-substituted amino-alcohols provides a single-step aziridine synthesis under mild conditions and in good yields (Scheme 35).[103] The ring-closure from the alkoxyphos-

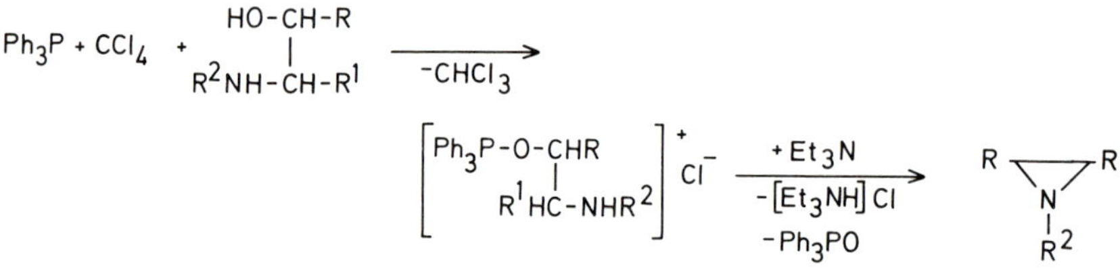

$$Ph_3P + CCl_4 + \begin{array}{c} HO-CH-R \\ | \\ R^2NH-CH-R^1 \end{array} \xrightarrow{-CHCl_3} \left[\begin{array}{c} Ph_3P-O-CHR \\ | \\ R^1HC-NHR^2 \end{array}\right]^+ Cl^- \xrightarrow[-Ph_3PO]{+Et_3N \atop -[Et_3NH]Cl} \begin{array}{c} R\triangle R^1 \\ N \\ | \\ R^2 \end{array}$$

Scheme 35

phonium salt occurs with inversion as in similar synthesis by means of Ph_3PBr_2.[104] Amino-alcohols not substituted on the nitrogen undergo P–N linkage to a predominant extent, and this procedure is thus unsuitable for the synthesis of aziridines not substituted on nitrogen. α-Amino-δ-hydroxy-alkanes react analogously with Ph_3P–CCl_4 to give N-substituted pyrrolidines.[105]

6.7. Removal of Water from Phosphoramidates and Phosphinamides

In most cases P–N linkage is observed when phosphoramides or their diesters are treated with Ph_3P–CCl_4 (cf. Section 7.2). An exception is the reaction of diethyl phosphoramidate or diphenylphosphinamide with Ph_3P–CCl_4 which are dehydrated by this reagent to give cyclotri-λ^5-phosphazenes;[106] formation of the cyclophosphazenes presumably occurs via a four-membered ring as shown in Scheme 36. These reactions by no means exhaust the possible

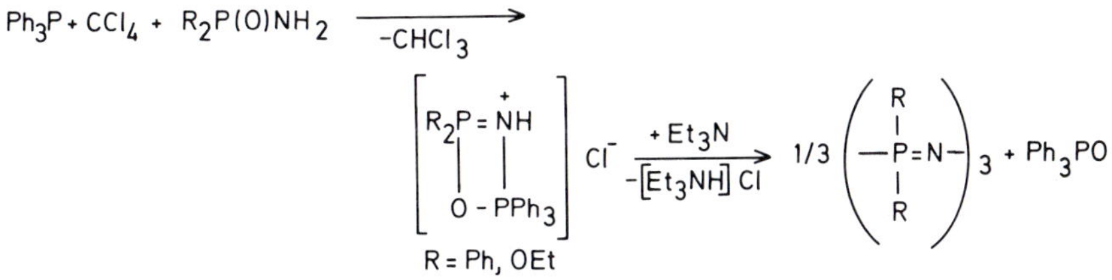

$$Ph_3P + CCl_4 + R_2P(O)NH_2 \xrightarrow{-CHCl_3} \begin{bmatrix} R_2\overset{+}{P}=NH \\ | \quad | \\ O-PPh_3 \end{bmatrix} Cl^- \xrightarrow[-[Et_3NH]Cl]{+Et_3N} 1/3 \begin{pmatrix} R \\ | \\ -P=N- \\ | \\ R \end{pmatrix}_3 + Ph_3PO$$

R = Ph, OEt

Scheme 36

examples of the use of intramolecular elimination of water. This is apparent from the recently reported 1,1-dehydration of N-p-tolylhydroxylamine,[107] which affords mainly p-azotoluene, and the dehydration of N-acylphenyl-glycine esters which proceeds with concomitant cyclization to give the 5-alkoxyoxazole.[108]

6.8. Intermolecular Removal of Water

In Section 5.1.3 examples of intermolecular removal of water have been discussed, for instance the formation of carboxylic anhydrides and phosphoric diesters or phosphinic anhydrides. Intermolecular dehydrations occur also during esterification and amide formation which will be treated in great detail in Section 8 in connection with four-component reactions.

7. Reactions Providing P–N Linkage

The action of the tertiary phosphine–CCl_4 system on ammonia and its derivatives resembles the reaction with alcohols only in the first step. But, in contrast to the alkoxyphosphonium salts, aminophosphonium chlorides do not readily rearrange, and this procedure is therefore particularly suitable for the formation of P–N bonds under mild conditions and in high yield. The two-component system reacts also with amides and imides of sulphur and phosphorus affording P–N linkage. These P–N linking reactions have the advantages of a smaller salt content in the product than after the reaction of phosphorus–halogen compounds with amines, and of less danger in the handling of the starting materials than in the chloramination or azide method. Chiral phosphines lose their optical activity on treatment with CCl_4 and amine, indicating that the reaction involves a pentacoordinate phosphorane intermediate.[30] In all such reactions with the Ph_3P-CCl_4 reagent the B reaction occurs besides the main reaction by route A that generates chloroform. However, the contribution of route B to the overall reaction is insignificant and the formulation of this side-reaction is therefore neglected in the discussion of the examples below.

7.1. Aminophosphonium Chlorides and Iminophosphoranes

Aminophosphonium chlorides are obtained in good yields by treatment of ammonia, primary, or secondary amines with a tertiary phosphine and CCl_4 [equation (21)], and they can readily be deprotonated to iminophosphoranes by suitable bases.[109,110]

$$R_3P + CCl_4 + R^1R^2NH \rightarrow R_3PNR^1R^{2+}Cl^- + CHCl_3$$
$$R^1, R^2 = H, Alkyl, Aryl, N(CH_3)_2 \tag{21}$$

The reaction temperature depends on the reactivity of the phosphines; with aliphatic phosphines, which react violently with CCl_4 at room temperature, the reactions are effected at $-50°C$,[111] but in other cases best between 20 and 60 °C.[109] Recently this method of P–N linkage has also been used for the phosphorylation of amino-sugars available only with difficulty.[112]

On application of this reaction to ditertiary bisphosphines with amines, bis(aminophosphonium) dichlorides are obtained which can be deprotonated by bases to give bis(iminophosphoranes) of the type (I).[113] Bis(iminophosphoranes) containing the molecular skeleton (II) can be obtained by the reaction of diamines with CCl_4 and tertiary phosphines with subsequent

$$-N=P{\diagdown \atop \diagup}{\left[{\matrix{| \cr C \cr |}}\right]}_n{\diagup \atop \diagdown}P=N- \qquad {\diagup \atop \diagdown}P=N-{\left[{\matrix{| \cr C \cr |}}\right]}_n-N=P{\diagdown \atop \diagup}$$

(I)   (II)

deprotonation.[113] Hydrazine behaves like a diamine towards Ph_3P-CCl_4; by twofold P–N linkage N,N'-bis(triphenylphosphonio)hydrazine chloride is formed, which is dehydrochlorinated already by liquid NH_3 to the bis-(phosphoranediyl)hydrazine (28).[114,115]

$$Ph_3P=N-N=PPh_3 \quad (28)$$

Differently substituted chlorophosphines can be converted into cyclotri- and cyclotetra-λ^5-phosphazenes by directed cyclocondensation with NH_3–CCl_4.[116] By separation of the diastereoisomeric 1-aminodiphosphazen-(1)-yl chlorides (29) this process could be worked out to a stereoselective synthesis of cyclotriphosphazenes.[117]

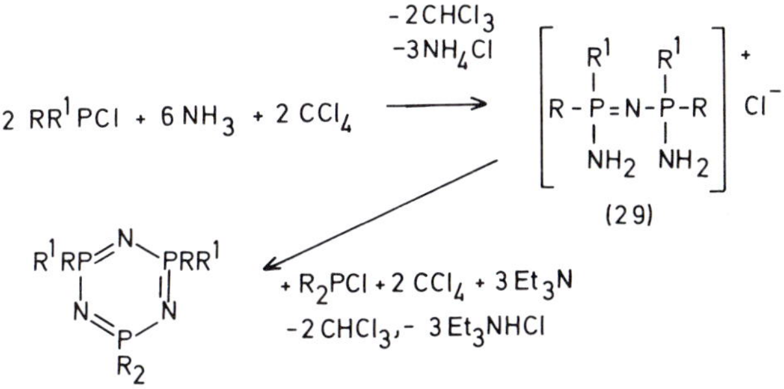

Scheme 37

Only brief reference can be made here to other possibilities of directed cyclocondensation of either diphosphines or difunctional amino compounds with CCl_4 to give diazadiphosphorines[118] or triazadiphosphorines.[119] Other cyclic compounds containing the atomic sequence –S–N–P– can also be obtained by this method.[118]

7.2. N-(Triphenylphosphoranediyl)amides of Sulphuric Acid, Sulphonic Acids, Alkyl Sulphates, and Fluorosulphamides, and N-(Triphenylphosphoranediyl)-phosphoramidic Diesters and -phosphinamides

Unlike carboxamides, the amides of sulphonic acids or alkyl sulphates, N,N-dialkylsulphamic acids, and fluorosulphamic acid react with Ph_3P–CCl_4 not with dehydration but with P–N linkage.[120,121] The primary products formed thereby are N-substituted aminophosphonium chlorides, which are converted, in some cases spontaneously, and in others only on the addition of a base, into the corresponding iminotriphenylphosphoranes, e.g.:

$$Ph_3P + CCl_4 + H_2NSO_2R \xrightarrow[-(Et_3NH)Cl]{+Et_3N} Ph_3P=NSO_2R + CHCl_3 \quad (22)$$

$$R = \text{Alkyl, Aryl, N(Alkyl)}_2\text{, OAlkyl, F}$$

The reaction of dialkyl phosphoramidates or phosphinamide with triphenylphosphine–CCl$_4$ takes a similar course (Scheme 38),[122] but in analogous

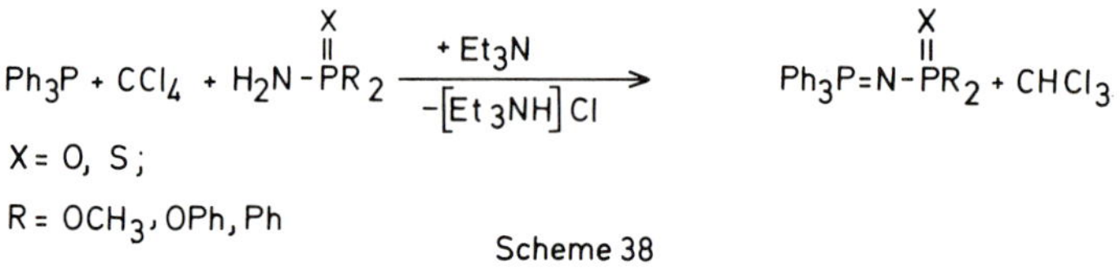

$$Ph_3P + CCl_4 + H_2N-\overset{\overset{X}{\|}}{P}R_2 \xrightarrow[-[Et_3NH]Cl]{+Et_3N} Ph_3P=N-\overset{\overset{X}{\|}}{P}R_2 + CHCl_3$$

X = O, S;
R = OCH$_3$, OPh, Ph

Scheme 38

experiments with diethyl phosphoramidate and diphenylphosphinamide removal of water occurred with the formation of cyclo-λ^5-phosphazenes (cf. Section 6.7).[106]

8. Four-Component Reactions

Not only intramolecular but also intermolecular dehydrations can be effected by the tertiary phosphine–CCl$_4$ reagent, as has been illustrated above by the formation of acid anhydrides (Section 5.1.3). Further examples are the condensations of the phosphine–CCl$_4$–carboxylic acid system to which an alcohol or an amine has been added as a fourth component.

8.1. Esterification

Ester formation by means of Ph$_3$P–CCl$_4$ as a condensing agent has been investigated for the esterification of ethanol and butanol with chloroacetic, butyric, succinic, or benzoic acid. High yields are obtained only when the alcohol is added as the last component after a pre-reaction of the three-component system (ca. 2 h). Under the conditions of a true four-component system, i.e. if the alcohol and acid are added simultaneously to the condensing agent, the appreciably poorer yields depend on the strength of the acid. The acyloxyphosphonium salt assumed as an intermediate may be attacked by the

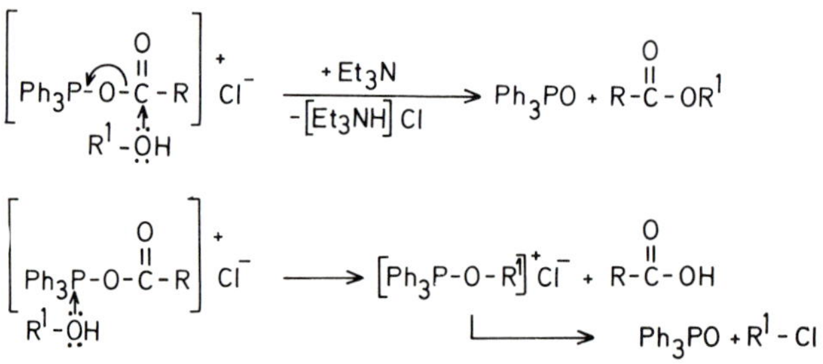

$$\left[Ph_3\overset{\curvearrowleft}{P}-O-\overset{\overset{O}{\|}}{\underset{\uparrow}{C}}-R\right]^+ Cl^- \xrightarrow[-[Et_3NH]Cl]{+Et_3N} Ph_3PO + R-\overset{\overset{O}{\|}}{C}-OR^1$$
$$R^1-\overset{..}{O}H$$

$$\left[Ph_3\overset{\curvearrowleft}{P}-O-\overset{\overset{O}{\|}}{\underset{\uparrow}{C}}-R\right]^+ Cl^- \longrightarrow [Ph_3P-O-R^1]^+Cl^- + R-\overset{\overset{O}{\|}}{C}-OH$$
$$R^1-\overset{..}{O}H \qquad\qquad\qquad \longrightarrow Ph_3PO + R^1-Cl$$

Scheme 39

alcohol either on the phosphorus or on the carbonyl carbon; the more electrophilic the central carbon atom, i.e. the stronger the carboxylic acid, the more is ester formation favoured.

8.2. Acid Amides and Peptides

Amides are obtained in high yields by condensation of a carboxylic acid and an amine with the $Ph_3P–CCl_4$ reagent:[123]

$$Ph_3P + CCl_4 + RCOOH + 2HNR^1R^2 \rightarrow$$
$$RC(O)NR^1R^2 + Ph_3PO + CHCl_3 + (R^1R^2NH_2)Cl \quad (23)$$

In an analogous four-component reaction N-protected amino-acids and amino-acid esters can be linked to form peptides.[124–130] Racemization is observed only in very few cases and can largely be suppressed by addition of 1-hydroxybenzotriazole.[129,130] The reaction conditions are those of a true four-component reaction, but surprisingly there is no P–N linkage with the amine component, or conversion of the alcoholic group in serine, threonine, and tyrosine, or of the amide group in glutamine.[128]

The formation of lactams from substituted acetic acids and ketenimines with $Ph_3P–CBr_4–Et_3N$ is to be considered as a further example of a four-component reaction.[131]

8.3. Esters, Amidic Esters, and Amides of Phosphinic and Phosphoric Acids

The esterification of mono- and di-esters of phosphoric acid and of phosphinic acids can also be achieved by means of $Ph_3P–CCl_4$. Good to very good yields are obtained by the reaction effected as a single-stage process according to equation (24) if additional triethylamine is added to bind the hydrogen chloride released.

$$Ph_3P + CCl_4 + R_2P(O)OH + R^1OH \xrightarrow[-(Et_3NH)Cl]{+Et_3N}$$
$$R_2P(O)OR^1 + Ph_3PO + CHCl_3 \quad (24)$$

$R = CH_3$, OAlkyl; $R^1 =$ Alkyl

The same applies to the synthesis of phosphinamides and phosphoramidic esters, which can be obtained from phosphinic acids or the mono- or di-esters of phosphoric acid with ammonia or primary or secondary amines together with $Ph_3P–CCl_4$ as condensing agent;[132] presumably the reaction proceeds by way of the O-phosphorylated salt and not of the acid chloride or the anhydride as an intermediate:

$$Ph_3POP(O)R_2{}^+Cl^- + 2\,HNR^1{}_2 \xrightarrow[-(R^1{}_2NH_2)Cl]{} Ph_3PO + R^1{}_2NP(O)R_2 \quad (25)$$

9. The Reagent $R_3P-C_2Cl_6$

As discussed in detail in Section 3.1, the intermediate dichlorophosphorane represents an active agent in the application of the system R_3P-CCl_4. Many of the reactions reported in Sections 5–8 thus can also be effected by dichlorophosphorane or by the more easily prepared and therefore more often applied R_3PBr_2 (cf. Chapter 10). Dichlorophosphorane has been used for chlorination[132–135] and dehydration reactions[132] as well as for P–N linkage.[136] For a long time the range of synthetic application has been limited by the laborious preparation with elemental chlorine and the impurity of the product obtained by this procedure.

Significant progress has now been made by application of the combination $R_3P-C_2Cl_6$ to these reactions, the effective reaction principle being the smooth and quantitative formation of dichlorophosphorane:

$$R_3P + C_2Cl_6 \rightarrow R_3PCl_2 + C_2Cl_4 \qquad (26)$$

The advantages of the method are the simple procedure and the easy handling of the conveniently measurable and (compared with elemental chlorine) mild and harmless halogenating agent C_2Cl_6. The tetrachloroethylene formed along with the dichlorophosphorane is easily removed together with the solvent by distillation. Furthermore the dichlorophosphorane is obtained analytically pure and thus can be used for subsequent reactions without further working-up. Many reactions are not affected by tetrachloroethylene and can therefore be achieved in a single-step synthesis, for example the preparation of aminophosphonium chlorides from phosphines, C_2Cl_6, and amines[137] or the synthesis of the (trimethylsilyl)iminophosphoranes (30) according to equation (27) which is analogous to the preparation of (30) from

$$R_3P + C_2Cl_6 + HN[Si(CH_3)_3]_2 \xrightarrow[-ClSi(CH_3)_3]{-C_2Cl_4}$$

$$[R_3PNHSi(CH_3)_3]^+Cl^- \xrightarrow[2\times]{-ClSi(CH_3)_3} R_3P{=}NSi(CH_3)_3 + (R_3PNH_2)^+Cl^-$$
$$\qquad\qquad\qquad\qquad\qquad\qquad (30) \qquad\qquad\qquad (27)$$

dibromotriorganylphosphorane and hexamethyldisilazane.[138]

For the synthesis of peptides the combination tris(dimethylamino)phosphine–hexachloroethane turned out to be especially helpful. Good results, with respect to the yields and degree of racemization, absolutely comparable or even superior to those of the DCCI method, are obtained when the aminophosphine is first converted into the dichlorophosphorane by C_2Cl_6 and the cooled solution is then mixed with the amino-acid or peptide component. According to the Anderson test, addition of 1-hydroxybenzotriazole largely suppresses racemization. ^{31}P n.m.r. spectroscopic studies show that the

9 Conversions via R₃P-Halogenoalkanes

reaction proceeds via the (benzotriazolyloxy)tris(dimethylamino)phosphonium chloride (31) isolated by Castro,[139,140] which reacts with the carboxy component to give the activated ester (32). The laborious and expensive isolation

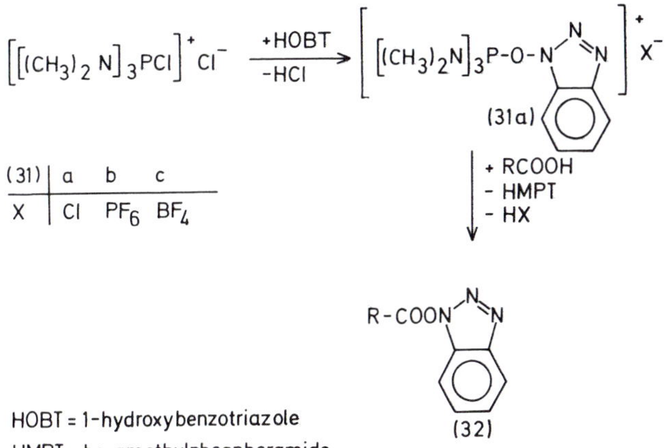

$$[[(CH_3)_2N]_3PCl]^+ Cl^- \xrightarrow[-HCl]{+HOBT} [[(CH_3)_2N]_3P\text{-}O\text{-}Bt]^+ X^-$$

(31a)

(31)	a	b	c
X	Cl	PF₆	BF₄

↓ + RCOOH
  − HMPT
  − HX

R−COO−Bt

(32)

HOBT = 1-hydroxybenzotriazole
HMPT = hexamethylphosphoramide

Scheme 40

and stabilization of (31a) in the form of (31b)[139,140] or (31c)[141] can thus be avoided.

In addition to high yields, little racemization, good availability of the reagents required, and easy separation from by-products, an advantage of this method is the avoidance of protecting functional groups because those of serine, threonine, tyrosine, asparagine, and glutamine have been shown not to undergo any reaction. Also the lower sensitivity to hydrolysis, compared with Ph₃P–C₂Cl₆, has proved advantageous because careful drying of the solvents becomes unnecessary.[142]

10. Phosphorus(III) Esters–CCl₄

The Michaelis–Arbusov reaction of phosphites with alkyl halides has been known for a long time and is the most widely used method of forming P–C bonds. Trying to extend this reaction, Kamai[143] was the first to treat phosphites with polyhalogenomethanes. The two-component reactions in most cases indeed yielded the expected products while in the presence of a nucleophilic third component P–C linkage was prevented.

In contrast to the system tertiary phosphine–CX₄ the compounds treated in this section are mainly used as starting materials for the synthesis of various classes of organophosphorus compounds of which some have been claimed to be useful as fire retardants, plasticizers, and agricultural chemicals,[144-148] while mere conversion of organic substrates without insertion of phosphorus by these reagents, e.g. for the synthesis of peptides,[124] is an exception.

Included in this section are reactions of CX_4 with compounds of the type $ROPR^1_2$ with R = alkyl or aryl and R^1 = alkyl, aryl, OR, or NR_2.

10.1. Two-Component Reactions

10.1.1. *Reaction products and conditions*

Esters of tervalent phosphorus acids react with carbon tetrahalides rather uniformly under elimination of R–Hal to give (trihalogenomethyl)phosphoryl compounds (Scheme 41).

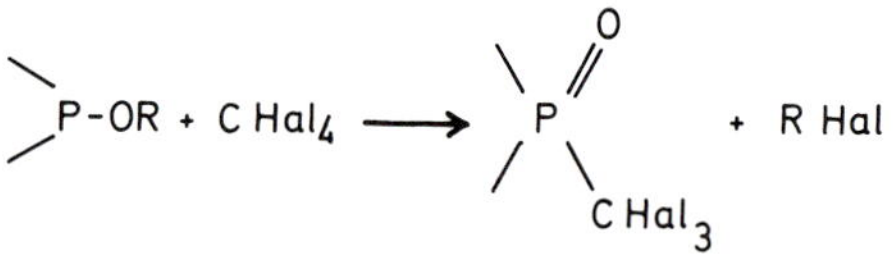

$$\text{>P-OR} + \text{C Hal}_4 \longrightarrow \text{>P(=O)-C Hal}_3 + \text{R Hal}$$

Scheme 41

Phosphinous acid esters are thus transformed into phosphine oxides by CCl_4:[149-154]

$$R_2POR + CCl_4 \rightarrow R_2P(O)CCl_3 + RCl \qquad (28)$$

phosphinates are obtained from phosphonites (with CCl_4[149,153,155-157] or CCl_3Br[155]):

$$RP(OR)_2 + CX_3Y \rightarrow R(RO)P(O)CX_3 + RY \qquad (29)$$
$$X = Cl;\ Y = Cl,\ Br$$

or directly starting from dichlorophosphine and ethylene oxide in CCl_4 (Scheme 42):[151]

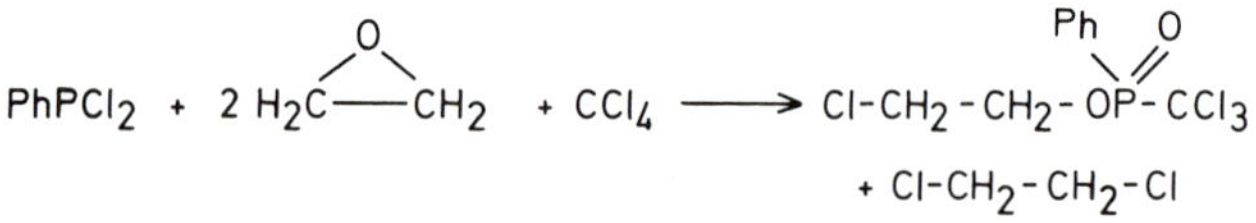

$$\text{PhPCl}_2 + 2\ H_2C\text{—}CH_2 + CCl_4 \longrightarrow Cl\text{-}CH_2\text{-}CH_2\text{-}OP(Ph)(=O)\text{-}CCl_3$$
$$+ Cl\text{-}CH_2\text{-}CH_2\text{-}Cl$$

Scheme 42

and phosphites give phosphonates with CCl_4,[143,156-171] CCl_3Br,[151,159,162,165,169] $CFBr_3$,[172] and CF_2Br_2:[172]

$$P(OR)_3 + CXYZ_2 \rightarrow (RO)_2P(O)CYZ_2 + RX \qquad (30)$$
$$X,Y,Z = Cl,\ Cl,\ Cl;\ Cl,\ F,\ Cl;\ Br,\ Cl,\ Cl;\ Br,\ F,\ Br;\ Br,\ Br,\ F$$

From mixed trialkyl phosphites the alkyl groups are preferentially removed in the order $CH_3 > C_2H_5 > i\text{-}C_3H_7$.[161] An exception is the reaction with CBr_4, which in the case of aryl or mixed aryl alkyl phosphites gives phosphorobromidates:[173]

$$PhOP(OR)_2 \xrightarrow{CBr_4} PhO(RO)P(O)Br \tag{31}$$
$$R = Ph, \text{alkyl}$$

Alkyl esters yield only undistillable mixtures with CBr_4 besides alkyl bromide.[155,159,173] With CCl_4 the formation of phosphorochloridates has been observed only as a side-reaction in some cases, e.g. with $(PhO)_2POEt$,[174] $(MeO)_3P$,[159] and tetraethyl diphosphite:[175]

$$(EtO)_2POP(OEt)_2 \xrightarrow{CCl_4} (EtO)_2P(O)Cl + (EtO)_2P(O)CCl_3 + (EtO)_2PCl \tag{32}$$

Ester amides, however, have been reported to give higher amounts of phosphoryl chloride, and with ester diamides this becomes the main product:[176,177]

$$(R_2N)_2POR^1 + CCl_4 \rightarrow (R_2N)_2P(O)Cl + R^1CCl_3 \tag{33}$$

The reaction conditions vary greatly with the reactivity of the halogenomethane as well as the phosphorus compound. Thus, CBr_4 reacts much more vigorously than do CCl_4, CBr_2F_2, $CBrF_3$, and $CFCl_3$, while CF_3I and $CFCl_3$ have been reported to be inert under normal reaction conditions.[178] Dialkyl esters of aryl phosphorus acids react with CCl_4 vigorously at room temperature,[149,155] trialkyl phosphites require several hours' boiling to react completely,[156] while triphenyl phosphite does not react even on boiling for several hours at 160 °C.[149] An investigation of the reactivities of phosphites by Kamai and Kharrasova[179] gave the following order:

$$(i\text{-}PrO)_3P > (EtO)_3P > (MeO)_3P > (EtO)_2POPh > (PhO)_2POEt \gg (PhO)_3P$$

showing that the reactivity increases with increasing electron-donating power of the substituents. The same increase in the nucleophilicity of phosphorus becomes evident in the order:

$$(R_2N)_2P(OR) > (R_2N)P(OR)_2 > P(OR)_3$$

While trialkyl phosphites show no detectable reaction with CCl_4 at room temperature and dialkoxy compounds react only slowly, $(Me_2N)_2POMe$ reacts even at 0°C almost explosively with carbon tetrachloride.[177]

10.1.2. Mechanism

At first the reactions were thought to be radical-chain processes[161–163,179,180] because they were reported to be light- and peroxide-catalysed.[162,163,181] The presumed peroxide catalysis, however, was later shown to be an oxidation reaction giving trialkyl phosphate, and the formation of trichloromethylphosphonate was not initiated thereby.[182,183]

It is now believed that, especially in the case of bromotrichloromethane, radical and ionic mechanisms occur, while the reaction of CCl_4 with phosphites in the dark and in the absence of free-radical catalysts follows a purely ionic route.[163] The reaction then is initiated by a nucleophilic attack of phosphorus at halogen (Scheme 43), giving a chlorophosphonium trichloromethanide,

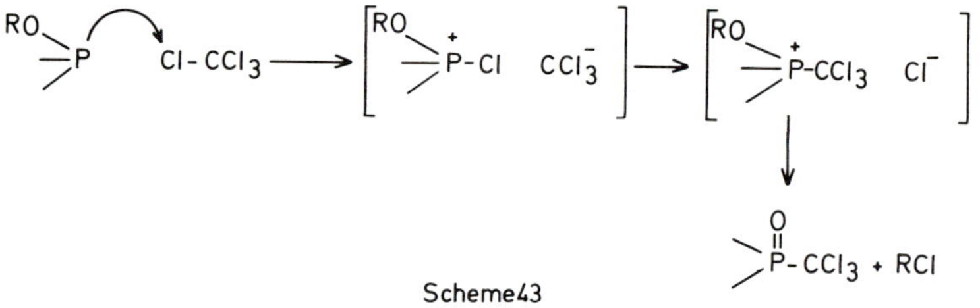

$$RO\!\!>\!\!P \quad Cl\text{-}CCl_3 \longrightarrow \left[RO\!\!>\!\!\overset{+}{P}\text{-}Cl \quad CCl_3^- \right] \longrightarrow \left[RO\!\!>\!\!\overset{+}{P}\text{-}CCl_3 \quad Cl^- \right]$$

$$\downarrow$$

$$>\!\!\overset{\overset{O}{\|}}{P}\text{-}CCl_3 + RCl$$

Scheme 43

which in the absence of alcohols or thiols rearranges to the corresponding (trichloromethyl)phosphonium chloride. The latter easily eliminates alkyl chloride to give the phosphoryl compound. Though the existence of the intermediates chlorophosphonium and (trichloromethyl)phosphonium salt has not yet been proved, this route seems very reasonable not only by analogy with the systems $R_3P\text{–}CX_4$ and $(R_2N)_3P\text{–}CX_4$ but also because of the findings in the three-component system phosphite–CX_4–nucleophile, since the action of phosphites on CCl_4 in the presence of alcohols, thiols, or other nucleophiles gives no PC-coupling, but formation of $HCCl_3$ and phosphates instead (cf. Section 10.2).

10.2. Three-Component Reactions

10.2.1. OH Compounds

The interaction of trialkyl phosphites with CCl_4,[17,38,53,169,184,185] $CCl_3\text{-}Br$,[38,185,186] or CBr_4[184] in the presence of alcohols results in the formation of haloform and phosphate:

$$XCY_3 + (RO)_3P \longrightarrow (RO)_3PX^+CY_3^- \xrightarrow[-CHY_3]{+ROH}$$

$$(RO)_4P^+X^- \longrightarrow (RO)_3PO + RX \quad (34)$$

X, Y = Cl, Cl; Br, Br; Br, Cl

Trihalogenomethylphosphonates are merely side products. A uniform course of the reaction, however, is observed only if the alkyl groups of alcohol and phosphite are the same. Otherwise the intermediate tetraalkoxyphosphonium salts can on the one hand eliminate different alkyl halides and on the other hand a ligand exchange is possible at this stage as is shown by the formation of $(EtO)_3PO$ from $(PhO)_3P$ and $EtOH.^{38}$ These side-reactions do not occur if phenols are used instead of alcohols. Thus dialkyl aryl phosphates are obtained from $(MeO)_3P$ or $(EtO)_3P$ with $CCl_3Br.^{186,187}$ $(PhO)_3P$ does not react with CCl_4 and phenols; with CBr_4 a reaction occurs at elevated temperature but only decomposition is observed.[3] Similar reactions occur with phosphonites or when phosphinites,[188] prepared *in situ* by alcoholysis of chlorophosphines, are allowed to react with CCl_4 (Scheme 44).

$$R_2PCl + R^1OH + B \longrightarrow R_2POR^1 + HCl \cdot B$$

$$R_2POR^1 + CCl_4 + R^1OH \longrightarrow [R_2P(OR^1)_2]^+Cl^- + CHCl_3 \longrightarrow R_2\overset{O}{\underset{\|}{P}}\text{-}OR^1 + R^1Cl$$

Scheme 44

10.2.2. SH Compounds

The reaction between trialkyl phosphites, thiols[163,165,169,189] or thiophenols,[190] and CCl_4 or $CCl_3Br^{165,169}$ analogously leads to O,O,S-trialkyl or O,O-dialkyl-S-aryl phosphorothioates:

$$(RO)_3P + R^1SH \xrightarrow[-CHCl_3]{+CCl_3Hal} (RO)_2P(O)(SR^1) + R\text{-}Hal \quad (35)$$

$Hal = Cl, Br$

The trihalogenomethyl phosphonate is again merely a by-product. Some O,O,O-trialkyl phosphorothionates are also produced, probably resulting from free-radical reactions between thiol and trialkyl phosphite.[188] The reaction here appears to be ionic when carried out in hexane while in the presence of azobisisobutyronitrile radical intermediates are involved.[165,190]

10.2.3. NH Compounds

Treatment of trialkyl phosphites with tetrahalogenomethane in the presence of secondary amines results in phosphorylation of the amine:[144,145,147]

$$(RO)_3P + HNR^1_2 + CX_4 \rightarrow (RO)_2P(O)NR^1_2 + CHX_3 + RX \quad (36)$$

while in the case of N-aryl-phosphoramidites dehydrohalogenation by excess secondary amine gives imino-phosphoramidates:[191]

$$(RO)_2PNHAr + 2R^1_2NH + CCl_4 \rightarrow$$
$$(RO)_2P(=NAr)NR^1_2 + CHCl_3 + R^1_2NH_2{}^+Cl^- \quad (37)$$

11. Other Reagent Combinations

In the halogenophosphines the nucleophilicity of the phosphorus is so greatly repressed that no reaction occurs with tetrachloromethane.[192] If, however, a primary or secondary amine is added to the chlorophosphine–CCl_4 system, then replacement of halogen is followed at once by a series of CCl_4 oxidation and P–N linkage reactions. Di-, tri- and tetra-aminophosphonium salts are thus obtained starting from, respectively, chlorodiorganylphosphines, dichloroorganylphosphines, or phosphorus trichloride.[193]

The action of amines on diphosphines and CCl_4 yields diaminodiorganylphosphonium chlorides,[194] and analogously triaminoorganylphosphonium chlorides are the stable products from the three-component cyclophosphine–CCl_4–amine system.[195]

While two component reactions of white phosphorus, which also contains three-coordinate phosphorus atoms, with CF_3I[196] [equation (38)], CCl_4[197] [equation (39)], and CCl_3Br[198] (Scheme 45) have been known for some time and have successfully been applied to several other polyhalogenated alkanes,[199]

$$P_4 + CF_3I \xrightarrow{200-220°C} CF_3PI_2 + (CF_3)_2PI + (CF_3)_3P + P_xI_y \quad (38)$$

$$P_w + CCl_4 \xrightarrow[104\ h]{157°C} Cl_3CPCl_2 + PCl_3 + P_{red} \quad (39)$$

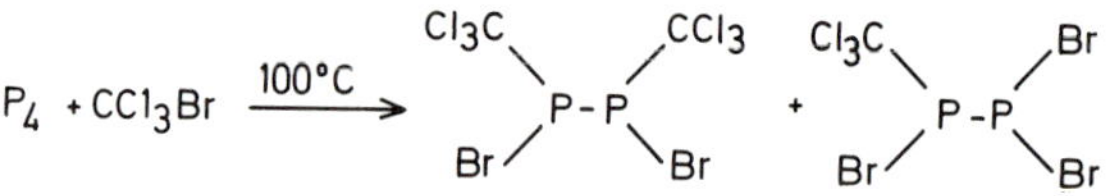

$$P_4 + CCl_3Br \xrightarrow{100°C} \underset{Br}{\overset{Cl_3C}{>}}P-P\underset{Br}{\overset{CCl_3}{<}} + \underset{Br}{\overset{Cl_3C}{>}}P-P\underset{Br}{\overset{Br}{<}}$$

Scheme 45

three-component reactions of white phosphorus–CCl_4 with secondary amine[195,200] or alkoxide[201] have only recently been reported.

The reaction with amines directly yields the peralkylated tetraaminophosphonium salts, and trialkyl phosphites are obtained in high yields by conversion of white phosphorus with alkoxides in alcohol-containing tetrachloromethane. Phosphites, phosphates, or alkylphosphonic acid alkyl esters can be obtained by the action of sodium alkoxide on Ph_3P–CCl_4.[202]

The conversion of dialkyl hydrogen phosphites into phosphoramidates by CCl_4 and amines:

$$(RO)_2POH + CCl_4 + R^1_2NH \rightarrow (RO)_2P(O)NR^1_2 + CHCl_3 + HCl \quad (40)$$

first reported by Atherton, Openshaw, and Todd[1] has been further investigated and not only been applied to a great variety of amines,[203–215] guanidine derivatives,[216] biguanides,[217,218] and amino acids,[219,220] but also used for

9 Conversions via R_3P-Halogenoalkanes

the phosphorylation of alcohols[221-223] and H_2O.[221] Similarly phosphorochloridates can be obtained from dialkyl hydrogen phosphites and CCl_4.[204,224]

Some substituted alkyl phosphines[225,226] and phosphonites[227] have been treated with CCl_4, and numerous reactions of very different polyhalogenated compounds with tervalent phosphorus compounds have also been reported which, owing to their specific character, cannot be mentioned here.

Treatment of ketones with CBr_4,[228-230] $CBrCl_3$,[231,232] CF_2Br_2,[233] $CFBr_3$,[234] or $CFCl_3$[235] and Ph_3P results in dihalogenomethylation of the carbonyl compound.

12. Summary and Outlook

The results obtained up to now show already that tervalent phosphorus compounds in combination with polyhalogenomethanes are simple and useful reagents for the synthesis of many important classes of compounds and can smoothly effect chlorinations and dehydrations even with very sensitive substrates. The reaction conditions can be suited to a large extent to the demands made by the various donor strengths of the substituents chosen for attachment to the phosphorus. Thus many reactions that can be carried out with Ph_3P–CCl_4 in the range 20–60°C require cooling to −60 to −20°C if they are effected with $[(CH_3)_2N]_3P$–CCl_4. Furthermore, the halogen in the tetrahalogenomethane may be changed (as shown in some examples above) so that, for example, bromine derivatives analogous to the chloro compounds can be obtained by using CBr_4.

While tervalent phosphorus acid derivatives are mainly used for the preparation of organophosphorus compounds suitable as flame retardants, corrosion inhibitors, detergents, lubricants or additives to these, agricultural chemicals, or pharmaceuticals, the system tervalent phosphine–tetrahalogenomethane is of more general importance as a mild and selective halogenating or dehydrating agent, and it can be expected that a controlled introduction of the R_3P–CCl_4 reagent will find its greatest successes in this field. Whether this system, which is primarily a laboratory reagent, will also become suitable for industrial application is probably a matter of economics. In this connection it remains to be found out whether recovery and reactivation of the phosphine oxide is worthwhile; possibilities for this are offered by conversion of the oxide by phosgene[236] into the dichlorophosphorane, which itself is an excellent chlorinating and dehydrating agent,[132-135] or by further treatment of the latter with elemental phosphorus to yield, in particular, phosphorus trichloride and triphenylphosphine.[237]

Chlorinations and dehydrations on polymers bearing R_2P groups appear to hold considerable promise for the future. Resins[238] suitable for this purpose can be obtained, for example, from polystyrene crosslinked with divinylbenzene by bromination and subsequent reaction with sodium or lithium diphenylphosphide. Apart from the chlorination of alcohols[239,240] the modified procedure is also suitable for numerous intra- and inter-molecular eliminations of water such as the synthesis of imide halides,[241] nitriles,[241] or peptides.[242,243] A particular advantage of this method is that the solutions of the reaction products are obtained free from phosphine oxide and the $R_2P(O)$ groups of the resin present after the reaction can easily be regenerated. In addition to reduction with trichlorosilane[239,244] they can be transformed by phosgene into the dichlorophosphorane resin[236,245,246] which can likewise be used for many of the reactions (Scheme 46). Effecting the same reactions with

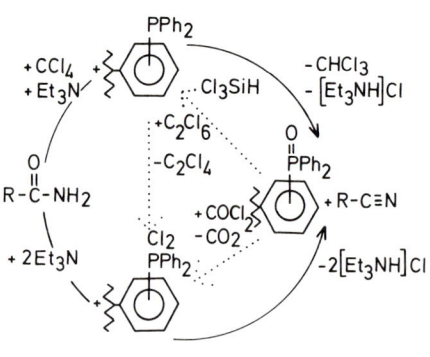

Scheme 46

the combination polymer-supported phosphine–C_2Cl_6 has turned out to be even more advantageous. With this reagent the synthesis of imide halides,[247] nitriles,[247] isocyanides,[247] carbodiimides,[247] or peptides[248,249] occurs only via the dichlorophosphorane, that is to say, no participation of the B-reaction and hence no formation of (chloromethyl)phosphonium salt as by-product is observed, so the polymer can quantitatively be converted into the dichlorophosphorane resin after the reaction. The resin thus can be used repeatedly without loss and the method has been worked out to a continuous process with high efficiency.[249]

13. References

Key reviews for further reading:
B. Miller, *Topics Phosphorus Chem.*, 1965, **2**, 133.
P. A. Chopard, *Chimia*, 1966, **20**, 420.
J. I. G. Cadogan and R. K. Mackie, *Chem. Soc. Rev.*, 1974, **3**, 87.

9 Conversions via R₃P-Halogenoalkanes

H. Teichmann, *Z. Chem.*, 1974, **14**, 216.
R. Appel, *Angew. Chem. Internat. Ed.*, 1975, **14**, 801.
1. F. R. Atherton, H. T. Openshaw, and A. R. Todd, *J. Chem. Soc.*, 1945, 660.
2. J. I. G. Cadogan and R. K. Mackie, *Chem. Soc. Rev.*, 1974, **3**, 87.
3. H. Teichmann, *Z. Chem.*, 1974, **14**, 216.
4. R. Rabinowitz and R. Marcus, *J. Amer. Chem. Soc.*, 1962, **84**, 1312.
5. W. Reid and H. Appel, *Annalen*, 1964, **679**, 51.
6. R. Appel, F. Knoll, W. Michel, W. Morbach, H.-D. Wihler, and H. Veltmann, *Chem. Ber.*, 1976, **109**, 58.
7. R. Appel and W. Morbach, *Synthesis*, 1977, 699.
8. R. Appel, F. Knoll, and H. Veltmann, *Angew. Chem. Internat. Ed.*, 1976, **15**, 315.
9. R. Appel and H. Veltmann, *Tetrahedron Lett.*, 1977, 399.
10. R. Appel, F. Knoll, H.-F.Schöler, and H.-D. Wihler, *Angew. Chem. Internat. Ed.*, 1976, **15**, 701.
11. H. Veltmann, Dissertation, University of Bonn, 1977.
12. R. Appel, R. Milker, and I. Ruppert, *Chem. Ber.*, 1977, **110**, 2385.
13. R. Appel and H.-F. Schöler, *Chem. Ber.*, 1978, **111**, 2056.
14. R. Appel and H.-D.Wihler, *Chem. Ber.*, 1978, **111**, 2054.
15. R. Appel and M. Huppertz, reported at the Arbusov-Jubilee Congress, Kiev, U.S.S.R., December 1977. *Z. Anorg. Allg. Chem.*, in press.
16. A. J. Speziale and K. W. Ratts, *J. Amer. Chem. Soc.*, 1962, **84**, 857.
17. I. M. Downie, J. B. Holmes, and J. B. Lee, *Chem. and Ind.*, 1966, 900.
18. J. B. Lee and T. J. Nolan, *Canad. J. Chem.*, 1966, **44**, 1331.
19. J. B. Lee and I. M. Downie, *Tetrahedron*, 1967, **23**, 359.
20. J. Hooz and S. S. H. Gilani, *Canad. J. Chem.*, 1968, **46**, 86.
21. I. M. Downie, J. B. Lee, and M. F. S. Matough, *Chem. Comm.*, 1968, 1350.
22. B. Castro and C. Selve, *Bull. Soc. Chim. France*, 1971, 4368.
23. R. G. Weiss and E. I. Snyder, *Chem. Comm.*, 1968, 1358.
24. R. G. Weiss and E. I. Snyder, *J. Org. Chem.*, 1970, **35**, 1627.
25. E. I. Snyder, *J. Amer. Chem. Soc.*, 1969, **91**, 5118.
26. B. Castro and C. Selve, *Bull. Soc. Chim. France*, 1971, 2296.
27. R. Boigegrain, B. Castro, and C. Selve, *Tetrahedron Lett.*, 1975, 2529.
28. R. Appel and K. Warning, *Chem. Ber.*, 1975, **108**, 6.
29. I. Tömösközi, L. Gruber, and L. Radics, *Tetrahedron Lett.*, 1975, 2437.
30. R. Appel and K. Warning, *Phosphorus*, 1974, **4**, 29.
31. P. Wolkoff, *Canad. J. Chem.*, 1975, **53**, 1333.
32. R. Appel, W. Michel, and W. Morbach, *Z. Naturforsch.*, 1978, **336**, 1205.
33. R. Appel and W. Morbach, *Angew. Chem. Internat. Ed.*, 1977, **16**, 180.
34. R. Appel and G. Erbelding, *Tetrahedron Lett.*, 1978, 2689.
35. H. J. Bestmann, *Angew. Chem. Internat. Ed.*, 1977, **16**, 349.
36. R. Appel and H.-D. Wihler, *Angew. Chem. Internat. Ed.*, 1977, **16**, 402.
37. J. Weiss, personal communication.
38. A. J. Burn and J. I. G. Cadogan, *J. Chem. Soc.*, 1963, 5788
39. H. Teichmann and W. Gerhard, *Z. Chem.*, 1974, **14**, 233.
40. R. Appel, K. Warning, and K. D. Ziehn, *Annalen*, 1975, 406.
41. D. Brett, I. M. Downie, J. B. Lee, and M. F. S. Matough, *Chem. and Ind.*, 1969, 1017.
42. E. H. Axelrod, G. M. Milne, and E. E. v. Tamelen, *J. Amer. Chem. Soc.*, 1970, **92**, 2139.

43. R. G. Weiss and E. J. Snyder, *J. Org. Chem.*, 1971, **36**, 403.
44. E. I. Snyder, *J. Org. Chem.*, 1972, **37**, 1466.
45. G. Stork, M. E. Jung, E. Colvin, and Y. Noel, *J. Amer. Chem. Soc.*, 1974, **96**, 3684.
46. R. Aneja, A. P. Davies, and J. A. Knaggs, *Tetrahedron Lett.*, 1974, 67.
47. C. Georgoulis and G. Ville, *Bull. Soc. Chim. France*, 1975, 607.
48. H. L. Goering and S. L. Trenbeath, *J. Amer. Chem. Soc.*, 1976, **98**, 5016.
49. P. J. Kocienski, G. Cernigliaro, and G. Feldstein, *J. Org. Chem.*, 1977, **42**, 353.
50. R. Appel and H.-D. Wihler, *Chem. Ber.*, 1976, **109**, 3446.
51. R. D. Miller, M. Schneider, and D. L. Dolce, *J. Amer. Chem. Soc.*, 1973, **95**, 8469.
52. R. Aneja, A. P. Davies, and J. A. Knaggs, *J. Chem. Soc., Chem. Comm.*, 1973, 110.
53. J. B. Lee and T. J. Nolan, *Tetrahedron*, 1967, **23**, 2789.
54. B. Castro, Y. Chapleur, and B. Gross, *Bull. Soc. Chim. France*, 1973, 3034.
55. R. D. Birkenmeyer (Upjohn Co.), U.S. 3,475,407 (1969) (*Chem. Abs.*, 1970, **7**, 25835w).
56. L. B. Hunt, D. F. MacSweeny, and R. Ramage, *Tetrahedron*, 1971, **27**, 1491.
57. J. P. H. Verheyden and J. G. Moffatt, *J. Org. Chem.*, 1972, **37**, 2289.
58. A. W. Friederang and D. S. Tarbell, *J. Org. Chem.*, 1968, **33**, 3797.
59. R. M. Magid, O. S. Fruchey, and W. L. Johnson, *Tetrahedron Lett.*, 1977, 2999.
60. B. Castro, Y. Chapleur, and B. Gross, *Tetrahedron Lett.*, 1974, 2313.
61. B. Castro and C. Selve, *Bull. Soc. Chim. France*, 1974, 3004.
62. R. Boigegrain and B. Castro, *Tetrahedron*, 1976, **32**, 1283.
63. I. M. Downie, H. Heaney, and G. Kemp, *Angew. Chem. Internat. Ed.*, 1975, **14**, 370.
64. B. Castro, Y. Chapleur, B. Gross, and C. Selve, *Tetrahedron Lett.*, 1972, 5001.
65. D. Brett, I. M. Downie, and J. B. Lee, *J. Org. Chem.*, 1967, **32**, 855.
66. S. Miyano, H. Ushiyama, and H. Hashimato, *Nippon Kagaku Kaishi*, 1977, 138.
67. G. Lavielle, J. C. Combret, and J. Villiéras, *Bull. Soc. Chim. France*, 1971, 2042.
68. J. C. Combret, J. Villiéras, and G. Lavielle, *Tetrahedron Lett.*, 1971, 1035.
69. B. Castro, R. Burgada, G. Lavielle, and J. Villiéras, *Compt. Rend. (C)*, 1969, **268**, 1067.
70. B. Castro, R. Burgada, G. Lavielle, and J. Villiéras, *Bull. Soc. Chim. France*, 1969, 2770.
71. D. J. Burton and J. R. Greenwald, *Tetrahedron Lett.*, 1967, 1535.
72. R. L. Soulen, D. B. Clifford, F. F. Crim, and J. A. Johnston, *J. Org. Chem.*, 1971, **36**, 3386.
73. R. L. Soulen, S. C. Carlson, and F. Lang, *J. Org. Chem.*, 1973, **38**, 479.
74. C. Gadreau and A. Foucaud, *Bull. Soc. Chim. France*, 1976, 2068.
75. N. S. Isaacs and D. Kirkpatrick, *J. Chem. Soc., Chem. Comm.*, 1972, 443.
76. L. Gruber, I. Tömösközi, and L. Radics, *Synthesis*, 1975, 708.
77. N. S. Isaacs and D. Kirkpatrick, *Tetrahedron Lett.*, 1972, 3869.
78. A. N. Thakore, P. Pope, and A. C. Oehlschlager, *Tetrahedron*, 1971, **27**, 2617.
79. J. B. Lee, *J. Amer. Chem. Soc.*, 1966, **88**, 3440.

80. R. Appel and H. Einig, *Z. Anorg. Allg. Chem.*, 1975, **414**, 236.
81. H. Kaye and L. Todd, *J. Chem. Soc. (C)*, 1967, **15**, 1420.
82. B. Castro and J. B. Dormoy, *Bull. Soc. Chim. France*, 1971, 3034.
83. R. Appel, K. D. Ziehn, and K. Warning, *Chem. Ber.*, 1973, **106**, 2093.
84. R. Appel, K. Warning, and K. D. Ziehn, *Chem. Ber.*, 1974, **107**, 698.
85. W. Ried, N. Kothe, R. Schweitzer, and A. Höhle, *Chem. Ber.*, 1976, **109**, 2921.
86. R. Appel, K. Warning, and K. D. Ziehn, *Chem. Ber.*, 1973, **106**, 3450.
87. R. Appel, K. D. Ziehn, and K. Warning (Bayer A. G.), Ger. Offen. 2,333,110 (1975) (*Chem. Abs.*, 1975, **82**, 156284z).
88. R. Appel and K. Warning, *Chem. Ber.*, 1975, **108**, 1437.
89. J. v. Braun, F. Jostes, and A. Heymons, *Ber. Deut. Chem. Ges.*, 1927, **60**, 92.
90. R. M. Waters, N. Wakabayaski, and E. S. Fields, *Org. Prep. Proced. Int.*, 1974, **6**, 53.
91. R. Appel, K. Warning, K. D. Ziehn, and A. Gilak, *Chem. Ber.*, 1974, **107**, 2671.
92. R. Appel and K. Giesen, unpublished results.
93. R. Appel and K. Giesen, *Chem. Ber.*, 1976, **109**, 810.
94. E. Yamato and S. Sugasawa, *Tetrahedron Lett.*, 1970, 4383.
95. R. Appel, R. Kleinstück, and K. D. Ziehn, *Chem. Ber.*, 1971, **104**, 1030.
96. R. Jaunin and W. Arnold, *Helv. Chim. Acta*, 1973, **56**, 2569.
97. R. Appel, R. Kleinstück, and K. D. Ziehn, *Chem. Ber.*, 1971, **104**, 2025.
98. R. Appel, R. Kleinstück, and K. D. Ziehn, *Angew. Chem. Internat. Ed.*, 1971, **10**, 132.
99. R. Appel, R. Kleinstück, and K. D. Ziehn, *Chem. Ber.*, 1971, **104**, 1335.
100. E. L. Lawton (Monsanto Co.), U.S. 3,972,933 (1976) (*Chem. Abs.*, 1977, **86**, 72231z).
101. H. Teichmann and A. Thai, *Z. Chem.*, 1977, **17**, 93.
102. J. Goerdeler and C. Lindner, *Tetrahedron Lett.*, 1972, 1519.
103. R. Appel and R. Kleinstück, *Chem. Ber.*, 1974, **107**, 5.
104. J. Okada, K. Ichimura, and R. Sudo, *Bull. Chem. Soc. Japan*, 1970, **43**, 1185.
105. R. Appel and K. Friemann, unpublished results.
106. R. Appel and H. Einig, *Chem. Ber.*, 1975, **108**, 914.
107. T. Ohashi and R. Appel, *Bull. Chem. Soc. Japan*, 1975, **48**, 1667.
108. N. Engel, B. Kübel, and W. Steglich, *Angew. Chem. Internat. Ed.*, 1977, **16**, 394.
109. R. Appel, R. Kleinstück, K. D. Ziehn, and F. Knoll, *Chem. Ber.*, 1970, **103**, 3631.
110. I. N. Zhumovova and A. P. Martynyuk, *Zhur. Obshch. Khim.*, 1974, **44**, 82.
111. B. Ross and K. P. Reetz, *Chem. Ber.*, 1974, **107**, 2720.
112. I. Pinker, J. Kovaćs, and A. Messmer, *Carbohydrate Res.*, 1977, **53**, 117.
113. R. Appel, B. Blaser, R. Kleinstück, and K. D. Ziehn, *Chem. Ber.*, 1971, **104**, 1847.
114. R. Appel, B. Blaser, and G. Siegemund, *Z. Anorg. Allg. Chem.*, 1968, **363**, 176.
115. R. Appel and P. Volz, *Chem. Ber.*, 1975, **108**, 623.
116. R. Appel and G. Saleh, *Chem. Ber.*, 1973, **106**, 3455.
117. H. Friemann, Dissertation, University of Bonn, 1977.
118. R. Appel, R. Kleinstück, and K. D. Ziehn, *Chem. Ber.*, 1972, **105**, 2476.
119. R. Appel and G. Saleh, *Annalen*, 1972, **766**, 98.

120. R. Appel, R. Kleinstück, and K. D. Ziehn, *Chem. Ber.*, 1971, **104**, 2250.
121. R. Appel and H. Einig, *Z. Naturforsch.*, 1975, **30b**, 134.
122. H. Einig, Dissertation, University of Bonn, 1974.
123. L. C. Barstow and V. J. Haruby, *J. Org. Chem.*, 1971, **36**, 1305.
124. S. Yamada and Y. Takeuchi, *Tetrahedron Lett.*, 1971, 3595.
125. Th. Wieland and A. Seeliger, *Chem. Ber.*, 1971, **104**, 3992.
126. Y. Takeuchi and S. Yamada, *Chem. Pharm. Bull.* (Japan), 1974, **22**, 832.
127. Y. Takeuchi and S. Yamada, *Chem. Pharm. Bull.* (Japan), 1974, **22**, 841.
128. R. Appel, G. Bäumer, and W. Strüver, *Chem. Ber.*, 1975, **108**, 2680.
129. R. Appel, G. Bäumer, and W. Strüver, *Chem. Ber.*, 1976, **109**, 801.
130. M. S. Manhas, S. G. Amin, B. Ram, and A. K. Bose, *Synthesis*, 1976, 689.
131. R. Appel and H. Einig, *Z. Anorg. Allg. Chem.*, 1975, **414**, 241.
132. L. Horner, H. Oediger, and H. Hoffmann, *Annalen*, 1959, **626**, 26.
133. G. A. Wiley, R. L. Hershkowitz, B. M. Rein, and B. C. Chung, *J. Amer. Chem. Soc.*, 1964, **86**, 964.
134. G. A. Wiley, B. M. Rein, and R. L. Hershkowitz, *Tetrahedron Lett.*, 1964, 2509.
135. L. Kaplan, *J. Org. Chem.*, 1966, **31**, 3454.
136. L. Horner and H. Oediger, *Annalen*, 1959, **627**, 142.
137. R. Appel and H.-F. Schöler, *Chem. Ber.*, 1977, **110**, 2382.
138. H. Nöth, L. Meinel, and H. Madersteig, *Angew. Chem. Internat. Ed.*, 1965, **4**, 709.
139. B. Castro, J. R. Dormoy, G. Evin, and C. Selve, *Tetrahedron Lett.*, 1975, 1219.
140. B. Castro, J. R. Dormoy, B. Dourtoglou, G. Evin, C. Selve, and J. G. Ziegler, *Synthesis*, 1976, 751.
141. I. J. Galpin, P. F. Gordon, R. Ramage, and W. D. Thorpe, *Tetrahedron*, 1976, **32**, 2417.
142. R. Appel and L. Willms, *Chem. Ber.*, 1979, **112**, 1057, 1064.
143. G. Kamai and L. P. Egorova, *Zhur. Obshch. Khim.*, 1946, **16**, 1521 (*Chem. Abs.* 1947, **41**, 5439g).
144. G. F. D'Alelio, U.S. 3,985,835 (1968) (*Chem. Abs.*, 1977, **86**, 56146v).
145. R. P. Napier (Mobil Oil Corp.), U.S. 4,046,774 (*Chem. Abs.*, 1977, **87**, 201310w).
146. S. L. Giolito (Stauffer Chemical Co.), U.S. Publ. Pat. Appl. B 483,606 (1971) (*Chem. Abs.*, 1976, **84**, 122903w).
147. G. F. D'Alelio, U.S. 3,886,237 (1968) (*Chem. Abs.*, 1975, **83**, 59046m).
148. Pure Chemicals Ltd., Fr. 1,395,595 (1963) (*Chem. Abs.*, 1965, **63**, 5839a).
149. G. Kamai, *Dokl. Akad. Nauk SSSR*, 1949, **66**, 389 (*Chem. Abs.*, 1950, **44**, 127c).
150. F. M. Kharrasova and G. Kamai, *Zhur. Obshch. Khim.*, 1964, **34**, 2195 (*Chem. Abs.*, 1964, **61**, 10705f).
151. G. Kamai, F. M. Kharrasova, G. I. Rakhimova, and R. B. Sultanova, *Zhur. Obshch. Khim.*, 1969, **39**, 625 (*Chem. Abs.*, 1969, **71**, 50104y).
152. A. E. Arbusov and K. V. Nikonorov, *Zhur. Obshch. Khim.*, 1949, **18**, 2008 (*Chem. Abs.*, 1949, **43**, 3801i).
153. F. M. Kharrasova, G. I. Rakhimova, T. V. Zykova, and R. A. Salakhutdinov, *Zhur. Obshch. Khim.*, 1973, **43**, 1930 (*Chem. Abs.*, 1974, **80**, 3584f).
154. F. M. Kharrasova, V. D. Efimova, R. A. Salakhutdinov, and T. V. Zykova, *Zhur. Obshch. Khim.*, 1976, **46**, 2237 (*Chem. Abs.*, 1977, **86**, 16751g).
155. G. Kamai, *Zhur. Obshch. Khim.*, 1948, **18**, 443 (*Chem. Abs.*, 1948, **42**, 7723d).

156. S. L. Khisamova and G. Kamai, *Zhur. Obshch. Khim.*, 1950, **20**, 1162 (*Chem. Abs.*, 1951, **45**, 1531d).
157. G. Kamai, *Dokl. Akad. Sci. URSS*, 1947, **55**, 219 (*Chem. Abs.*, 1947, **41**, 5863y).
158. G. M. Kosolapoff, *J. Amer. Chem. Soc.*, 1947, **69**, 1002.
159. G. Kamai, *Dokl. Akad. Nauk SSSR*, 1951, **79**, 795 (*Chem. Abs.*, 1952, **46**, 6081f).
160. M. S. Kharasch and J. S. Bengelsdorf, *J. Org. Chem.*, 1955, **20**, 1356.
161. G. Kamai and F. M. Kharrasova, *Zhur. Obshch. Khim.*, 1957, **27**, 953 (*Chem. Abs.*, 1958, **52**, 3666a).
162. C. E. Griffin, *Chem. and Ind.*, 1958, 415.
163. J. I. G. Cadogan and W. R. Foster, *J. Chem. Soc.*, 1961, 3071.
164. J. I. G. Cadogan and J. T. Sharp, *Tetrahedron Lett.*, 1966, 2733.
165. R. E. Atkinson, J. I. G. Cadogan, and J. T. Sharp, *J. Chem. Soc.* (*B*), 1969, 138.
166. N. V. de Bataafsche Petroleum Maarschappij, Brit. Pat. 692,261 (1953) (*Chem. Abs.*, 1954, **48**, 10052).
167. F. M. Kharrasova, T. V. Zykova, R. A. Salakhutdinov, V. D. Efimova, and R. D. Shafigullina, *Zhur. Obshch. Khim.*, 1974, **44**, 2419 (*Chem. Abs.*, 1975, **82**, 73097q).
168. F. M. Kharrasova, V. D. Efimova, L. V. Nesterov, A. A. Martinov, and R. A. Salakhutdinov, *Zhur. Obshch. Khim.*, 1978, **48**, 229.
169. P. J. Bunyan and J. I. G. Cadogan, *J. Chem. Soc.*, 1962, 2953.
170. S. Nakasato and K. Higuchi, *J. Amer. Oil Chem. Soc.*, 1970, **47**, 283.
171. K. D. Berlin and N. K. Roy, *J. Chem. Eng. Data*, 1970, **15**, 579.
172. D. J. Burton and R. M. Flynn, *J. Fluorine Chem.*, 1977, **10**, 329.
173. G. Kamai and F. M. Kharrasova, *Tr. Kazan. Khim. Tekhnol. Inst.*, 1957, **23**, 122 (*Chem. Abs.*, 1958, **52**, 9980).
174. G. Kamai and F. M. Kharrasova, *Tr. Kazan. Khim. Tekhnol. Inst.*, 1957, **23**, 127 (*Chem. Abs.*, 1958, **52**, 9946).
175. G. Kamai, *Dokl. Akad. Nauk SSSR*, 1950, **70**, 233 (*Chem. Abs.*, 1951, **45**, 5611f).
176. V. S. Abramov and N. A. Iljina, *Zhur. Obshch. Khim.*, 1971, **41**, 100 (*Chem. Abs.*, 1963, **58**, 11394).
177. J. H. Hargis and W. D. Alley, *J. Amer. Chem. Soc.*, 1974, **96**, 5927.
178. A. F. Isbell, *U.S. Dep. Commer. Off. Tech. Serv.*, 1961, AD 266 675 (*Chem. Abs.*, 1963, **51**, 11394).
179. G. Kamai and F. M. Kharrasova, *Izv. Vyssh. Uchebn. Zaved. Khim. Khim. Tekhnol.*, 1961, **4**, 229 (*Chem. Abs.*, 1961, **55**, 21762).
180. C. Walling and R. Rabinowitz, *J. Amer. Chem. Soc.*, 1959, **81**, 1243.
181. C. E. Griffin, Amer. Chem. Soc., Abstracts of 135th Meeting, 1959, p. 690.
182. A. J. Burn, J. I. G. Cadogan, and P. Bunyan, *J. Chem. Soc.*, 1963, 1527.
183. A. J. Burn, J. I. G. Cadogan, and P. Bunyan, *J. Chem., Soc.*, 1964, 4369.
184. P. C. Crofts and J. M. Downie, *J. Chem. Soc.*, 1963, 2559.
185. A. J. Burn and J. I. G. Cadogan, *Chem. and Ind.*, 1963, 736.
186. Ciba Sa., Belg. 748,289 (1970).
187. R. P. Napier and S. T. D. Gough, *Org. Prep. Proced. Int.*, 1971, **3**, 117 (*Chem. Abs.*, 1971, **75**, 62644).
188. R. Appel and U. Warning, *Chem. Ber.*, 1976, **109**, 805.
189. J. I. G. Cadogan, *Adv. Free-Radical Chem.*, 1967, **2**, 203.

190. L. L. Murdock and T. L. Hopkins, *J. Org. Chem.*, 1968, **33**, 907.
191. E. E. Nifant'ev, G. F. Bebikh, and T. P. Sakodynskaja, *Zhur. Obshch. Khim.*, 1972, **41**, 2011 (*Chem. Abs.*, 1972, **76**, 59093e).
192. U. Warning, Diplomarbeit, Universität Bonn, 1975.
193. H. Teichmann, W. Gerhard, and W. Kochmann, Ger. (East) 105,242 (1973) (*Chem. Abs.*, 1975, **82**, 43042a).
194. R. Appel and R. Milker, *Chem. Ber.*, 1975, **108**, 2349.
195. R. Appel and R. Milker, *Z. Anorg. Allg. Chem.*, 1975, **417**, 161.
196. F. W. Bennett, H. J. Emeléus, and R. N. Haszeldine, *J. Chem. Soc.*, 1953, 1565.
197. D. Perner and A. Henglein, *Z. Naturforsch.*, 1962, **17b**, 703.
198. P. L. Airey, *Z. Naturforsch.*, 1969, **24b**, 1393.
199. L. Maier, *Fortschr. Chem. Forsch.*, 1971, **19**, 1 (see references therein).
200. R. Appel and W. Eichenhofer, to be published.
201. C. Brown, R. F. Hudson, G. A. Warten, and H. Coates, *J. Chem. Soc., Chem. Comm.*, 1968, 7.
202. H. A. Lehmann, H. Schadow, H. Richter, R. Kurze, and M. Oertel (VEB Stickstoffwerk, Piesteritz), Ger. Offen. 2,643,282 (1977) (*Chem. Abs.*, 1977, **87**, 184678c).
203. F. R. Atherton and A. R. Todd, *J. Chem. Soc.*, 1947, 674.
204. G. M. Steinberg, *J. Org. Chem.*, 1950, **15**, 637.
205. V. Ettel and M. Zbirovsky, *Coll. Czech. Chem. Comm.*, 1956, **21**, 1607 (*Chem. Abs.*, 1957, **51**, 11643b).
206. L. N. Nikolenko and E. V. Degterev, *Zhur. Obshch. Khim.*, 1967, **37**, 1350 (*Chem. Abs.*, 1968, **68**, 59490s).
207. R. S. Edmundson, *J. Chem. Soc.* (*C*), 1971, 3614.
208. D. Redmore (Petrolite Corp.), U.S. 3,591,330 (1969) (*Chem. Abs.*, 1971, **75**, 119 870x).
209. Petrolite Corp., U.S. 4,048,264 (1977).
210. E. E. Nifant'ev and A. I. Zavalisina, *Zhur. Obshch. Khim.*, 1967, **37**, 1854 (*Chem. Abs.*, 1968, **68**, 29491a).
211. E. S. Rodionov, E. F. Bugerenko, V. D. Sheludyakov, A. J. Popov, V. F. Mironov, and N. V. Mironova, U.S.S.R. 415,267 (1972) (*Chem. Abs.*, 1974, **80**, 146 289z).
212. A. O. M. Okorodudu (Mobil Oil Corp.), U.S. 3,986,967 (1975) (*Chem. Abs.*, 1977, **86**, 192 352m).
213. V. D. Warner, D. B. Mirth, A. S. Dey, S. S. Turesky, and B. Soloway, *J. Medicin. Chem.*, 1973, **16**, 1185.
214. M. A. Pudovik and A. N. Pudovik, *Zhur. Obshch. Khim.*, 1973, **43**, 2144 (*Chem. Abs.*, 1974, **80**, 48105w).
215. A. Zwierzack and J. Brylikowska-Piotrowich, *Angew. Chem. Internat. Ed.*, 1977, **16**, 107.
216. J. Anatol, H. M. Vidalenc, and G. P. M. Loiseau (Ugine Kuhlmann), Ger. Offen. 2,130,304 (1971) (*Chem. Abs.*, 1972, **76**, 59638m).
217. J. Anatol (Etablissements Kuhlmann), Fr. 1,461,378 (1965) (*Chem. Abs.*, 1967, **67**, 43 446a).
218. J. Anatol, H. Vidalenc, and G. P. M. Loiseau (Ugine Kuhlmann), Ger. Offen. 2,130,303 (1971) (*Chem. Abs.*, 1972, **76**, 59745u).
219. N. P. Grechkin and L. K. Nikonorova, *Izv. Akad. Nauk SSSR*, 1969, **5**, 1180 (*Chem. Abs.*, 1969, **71**, 38230d).

220. D. A. Predvoditelev, V. B. Kvantrishvili, and E. E. Nifant'ev, *Zhur. Org. Khim.*, 1976, **12**, 38 (*Chem. Abs.*, 1976, **84**, 121038m).
221. E. E. Nifant'ev, V. S. Blagoveshchenskii, A. M. Sokurenko, and L. S. Skyarskii, *Zhur. Obshch. Khim.*, 1974, **44**, 108 (*Chem. Abs.*, 1974, **80**, 96089m)
222. Kumisi Chemical Ind. Co. Ltd., JA 47 16,469 (1972).
223. C. S. Wang and M. J. Mintz, *J. Heterocyclic Chem.*, 1976, **13**, 891.
224. T. Gajda and A. Zwierzack, *Synthesis*, 1976, **4**, 243.
225. Farbwerke Hoechst AG, NL 7 404,820 (1973).
226. G. Kamai, R. K. Valetdinov, and R. K. Imagilov, *Zhur. Obshch. Khim.*, 1969, **34**, 379 (*Chem. Abs.*, 1969, **70**, 115 227s).
227. O. J. Kolodyazhnyi, L. A. Repina, and Y. G. Golobov, *Zhur. Obshch. Khim.*, 1974, **44**, 951 (*Chem. Abs.*, 1974, **81**, 13607a).
228. F. Ramirez, N. B. Desai, and N. McKelvie, *J. Amer. Chem. Soc.*, 1962, **84**, 1745.
229. E. J. Corey and P. L. Fuchs, *Tetrahedron Lett.*, 1972, 3769.
230. G. H. Posner, G. L. Loomis, and H. Sawaya, *Tetrahedron Lett.*, 1975, 1373.
231. B. A. Clement and R. L. Soulen, *J. Org. Chem.*, 1976, **41**, 557.
232. W. G. Salmond, *Tetrahedron Lett.*, 1977, 1239.
233. D. G. Naae, H. S. Kesling, and D. J. Burton, *Tetrahedron Lett.*, 1975, 3789.
234. R. W. Vanderhaar, D. J. Burton, and D. G. Naae, *J. Fluorine Chem.*, 1977, **1**, 381.
235. M. J. Hamme and D. J. Burton, *J. Fluorine Chem.*, 1977, **10**, 131.
236. R. Appel and W. Heinzelmann (BASF AG), Ger. 1,192,205 (1965) (*Chem. Abs.*, 1965, **63**, 8405g).
237. G. Wunsch, K. Wintersberger, and H. Geierhaas (BASF AG), Ger. 1,247,310 (1967) (*Chem. Abs.*, 1968, **68**, 105, 363c).
238. W. Heitz and R. Michels, *Angew. Chem. Internat. Ed.*, 1972, **11**, 298.
239. S. L. Regen and D. P. Lee, *J. Org. Chem.*, 1975, **40**, 1669.
240. P. Hodge and G. Richardson, *J. Chem. Soc., Chem. Comm.*, 1975, 622.
241. C. R. Harrison, P. Hodge, and W. J. Rogers, *Synthesis*, 1977, 41
242. R. Appel, W. Strüver, and L. Willms, *Tetrahedron Lett.*, 1976, 905.
243. R. Appel and L. Willms, *J. Chem. Res. (S)*, 1977, 84.
244. W. Heitz and R. Michels, *Annalen*, 1973, 227.
245. H. M. Relles and R. W. Schluenz, *J. Amer. Chem. Soc.*, 1974, **96**, 6469.
246. M. Masaki and N. Nakeya, *Angew. Chem. Internat. Ed.*, 1977, **16**, 552.
247. R. Appel and E. Gabel, to be published.
248. R. Appel and L. Willms, *Chem. Ber.*, 1977, **110**, 3209.
249. R. Appel and H. Janssen, to be published.

10
Functional Group Conversions using Phosphorus(III) Reagents with Halides and Halogens

R. K. MACKIE

1. Reactions of phosphorus(III) reagents with halogens, hydrogen halides, and alkyl halides, 433
2. Reactions of phosphorus(III) reagents at positive halogens, 446
3. Reactions with acyl halides, 459
4. Reactions with aryl halides, 460
5. Summary of functional group conversions, 461
6. References, 462

1. Reactions of Phosphorus(III) Reagents with Halogens, Hydrogen Halides, and Alkyl Halides

1.1. Reactions of Phosphites with Halogens and Hydrogen Halides

The reaction of trialkyl phosphites with halogens which results in the formation of a phosphorohalidate and an alkyl halide has been known for many years[1] but the reaction is not normally used for the preparation of alkyl halides. However, a convenient synthesis of alkyl iodides has been reported and has proved to be one of the more successful methods of carrying out this transformation.[2] It is more satisfactory than an earlier method involving the action of iodine on a diphenyl alkyl phosphite.[3] The success of the method, shown in Scheme 1, is largely due to the ease of removal of the phosphoroiodate (1) by hydrolysis.

Triaryl phosphites form 1:1 adducts with halogens, and these adducts have been used in the synthesis of alkyl[4] and aryl halides.[5] Reactions have been carried out successfully either by using preformed adduct or by adding halogen to a mixture of the alcohol and triaryl phosphite. Reaction of acetylenic and allenic alcohols with triphenyl phosphite dibromide results in low yields of impure products. However, if the reaction is carried out under anhydrous conditions at low temperature in the presence of an equimolar

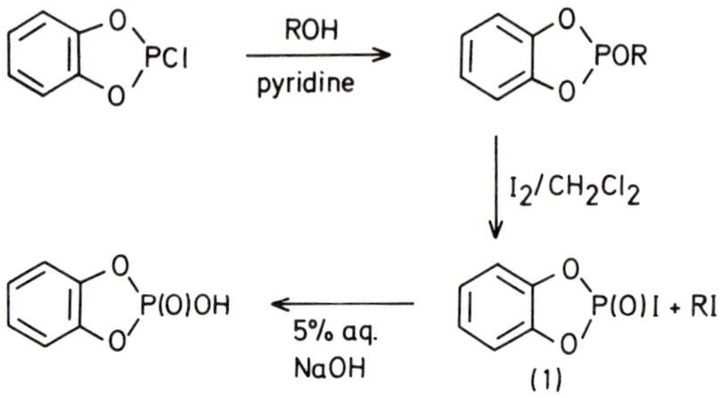

Scheme 1

quantity of pyridine, good yields of pure bromides are obtained.[6] In the case of aryl halides, these are formed by pyrolysis of mixed tetraaryloxyphosphonium salts, $(ArO)_3P^+(OAr')X^-$. Formation of the aryl halide Ar'X is favoured if the aryl group Ar' contains substituents more electron-withdrawing than those in the group Ar. Thus, reactions involving the use of 3 moles of p-t-butylphenol and 1 mole of a second phenol with phosphorus pentachloride are most satisfactory. In this way, good yields of chlorotoluenes, chloronitrobenzenes, dichlorobenzenes, chloronaphthalenes, chloropyridines, and chloropyrimidines may be obtained. Bromides and iodides are formed if the phosphonium chloride is treated with ethyl bromide and ethyl iodide, respectively, before pyrolysis. Much of the early work in this area has been reviewed by Rydon.[7]

The synthesis of optically active halides is the subject of a review[8] which includes the use of hydrogen chloride and of hydrogen bromide in the presence of phosphites, phosphonites, or phosphinites. It is reported that isomerically and optically pure alkyl halides are formed in the reaction of alkyl diphenylphosphinites with hydrogen halides or with halogens[9] (Scheme 2). Bromides and iodides, although optically pure initially, may undergo racemization if longer reaction times are required.

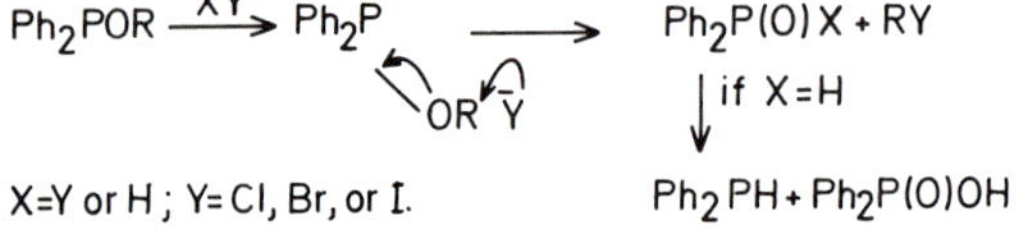

Scheme 2

Triethyl phosphite and pyridine react in the presence of bromine or iodine to give an intermediate (2) which finds use as a condensing agent in the synthesis of esters and of amides[10] (Scheme 3). Phenyl phosphite and diisopropyl phosphite have also been used in this reaction.

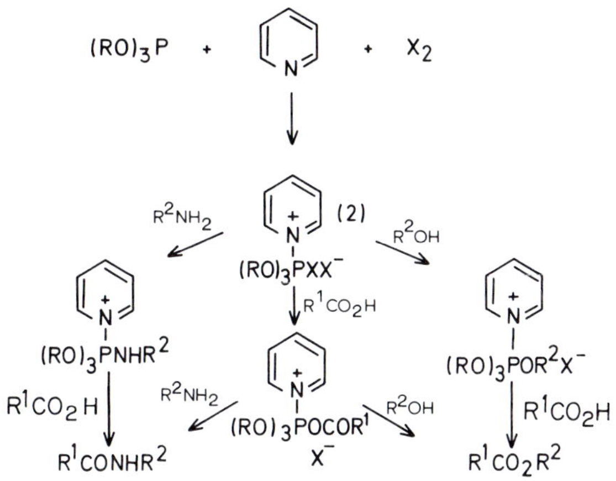

Scheme 3

[Scheme 3 depicting reactions of $(RO)_3P$ with pyridine and X_2, forming $(RO)_3PXX^-$, then branching to products via R^2NH_2, R^2OH, and R^1CO_2H pathways including $(RO)_3PNHR^2$, $(RO)_3POCOR^1 X^-$, $(RO)_3POR^2 X^-$, giving R^1CONHR^2 and $R^1CO_2R^2$]

1.2. Reactions of Phosphites with Alkyl Halides

The Michaelis–Arbusov reaction, involving the reaction of alkyl halides and trialkyl phosphites, has been discussed in the introduction. When triaryl phosphites react with alkyl halides, triaryloxyalkylphosphonium salts are formed.[7,11] These salts have been widely used for the conversion of alcohols into alkyl halides, and reactions are carried out by reaction of the alcohol either with a mixture of triphenyl phosphite and an "active" halide or with the preformed phosphonium salt. The most commonly used "active" halides are methyl iodide, benzyl bromide, and benzyl chloride (Scheme 4).

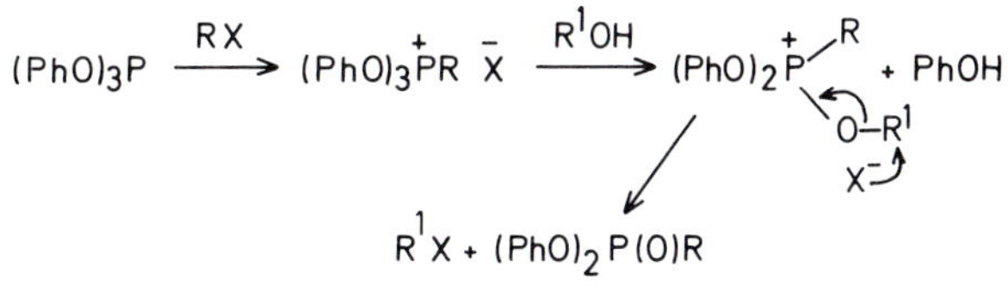

$$(PhO)_3P \xrightarrow{RX} (PhO)_3\overset{+}{P}R\ X^- \xrightarrow{R^1OH} (PhO)_2\overset{+}{P}\underset{X^-}{\overset{R}{\underset{O-R^1}{\diagdown}}} + PhOH$$

$$R^1X + (PhO)_2P(O)R$$

Scheme 4

The most widely used reagent is methyltriphenoxyphosphonium iodide (3), and experimental details of its use in the synthesis of alkyl iodides have been published.[12] The reagent has been widely used in carbohydrate[13–17] and in nucleoside chemistry.[18–22] Reactions are of the S_N2 type, involving inversion of configuration at the reaction centre[11,15] [Scheme 5; (4) → (5)]. Other functional groups are unaffected except the amino group which can be protected by acetylation.[19] Secondary alcohols react more slowly than primary, and retention of configuration may be observed in some iodinations

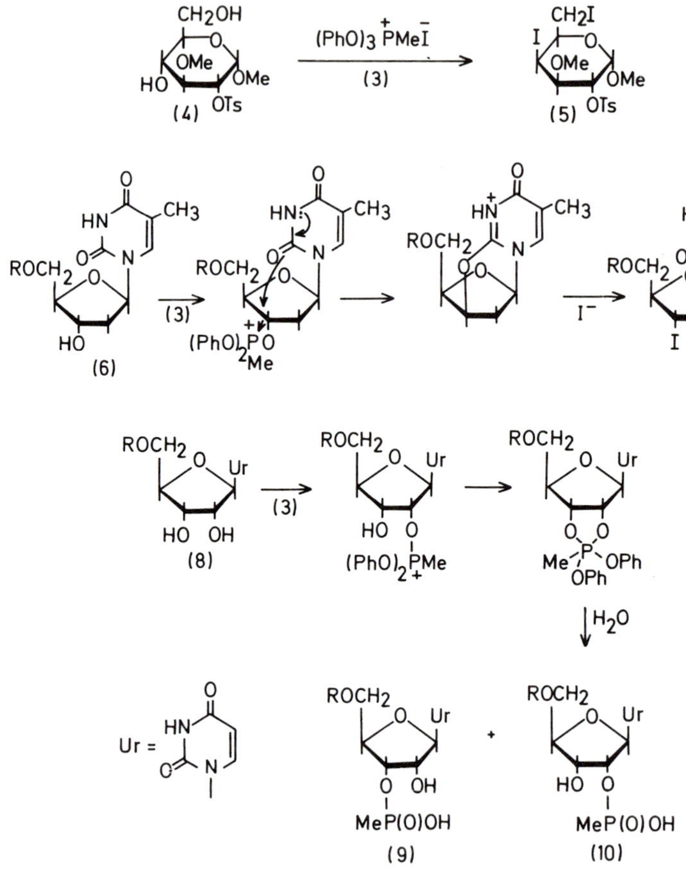

Scheme 5

owing to participation by an adjacent carbonyl group[18,22] [Scheme 5; (6) → (7)]. A chloride ion in a similar situation gives the usual S_N2 reaction. Methyltriphenoxyphosphonium iodide cannot be used to iodinate *vic*-diols owing to an intramolecular reaction which leads to the formation of a mixture of two phosphonates[21] [Scheme 5; (8) → (9) + (10)]. Molecular rearrangements are also observed in reactions involving the 19-hydroxy-3-keto-steroid (11) and the ketal (12)[23] (Scheme 6).

The use of hexamethylphosphoramide (HMPA) as solvent causes the course of the reaction to be modified. Secondary alcohols are readily dehydrated but tertiary alcohols yield alkenes only slowly.[24] The alkenes formed show a predominance of the more stable Saytzeff product with a stereoselective tendency towards the (*E*)-isomer. Reaction with unsaturated alcohols results in the formation of dienes and trienes[25] (Scheme 7). A

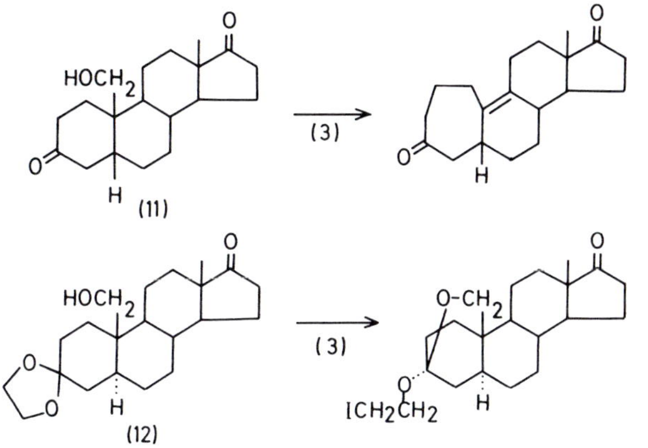

Scheme 6

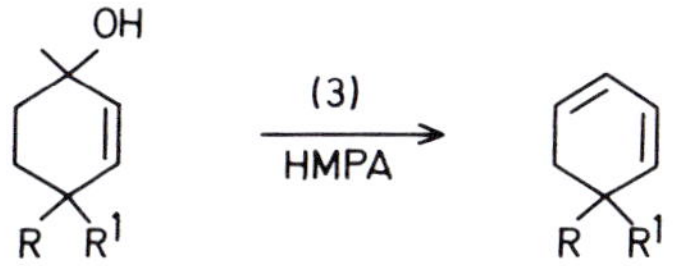

$$\mathrm{MeCH{=}CHCH(OH)CH_2CH{=}CH_2} \xrightarrow[\text{HMPA}]{(3)} \mathrm{MeCH{=}CHCH{=}CHCH{=}CH_2} + \mathrm{CH_2{=}CHCH{=}CHCH_2CH{=}CH_2}$$

Scheme 7

conversion of primary alcohols into hydrocarbons has been reported. This involves the conversion of the alcohol into the iodide followed by reduction with sodium cyanoborohydride, a reagent showing high specificity for reduction of alkyl iodides. Yields are not substantially diminished if the reaction is carried out without isolation of the iodide[26a] (Scheme 8). Recently, methyltriphenoxyphosphonium iodide has been shown to be useful in the stereospecific deoxygenation of epoxides to alkenes, with retention of configuration,[26b] while butyltriphenoxyphosphonium bromide promotes condensations to amides, esters, ureas, and thioureas.[150]

$$\mathrm{RCH_2OH} \xrightarrow[\text{HMPA}]{(3)} \mathrm{RCH_2I} \xrightarrow{\mathrm{NaBH_3CN}} \mathrm{RCH_3}$$

Scheme 8

1.3. Syntheses involving Phosphine Dihalides

The 1:1 adducts of triphenylphosphine and halogens have been widely used in the conversion of alcohols into halides.[27–29] The reaction nearly always follows an S_N2 mechanism and, in the case of chiral alcohols, little or no racemization is observed. The reaction of triphenylphosphine dibromide (13), Ph_3PBr_2, with *endo*-norborneol (14), producing *exo*-norbornyl bromide (15), is an example of a reaction taking the normal course, but that with the *exo*-isomer (16) produces a mixture comprising the *endo*-bromide (17), the *exo*-bromide (15), and nortricycline (18) (Scheme 9).

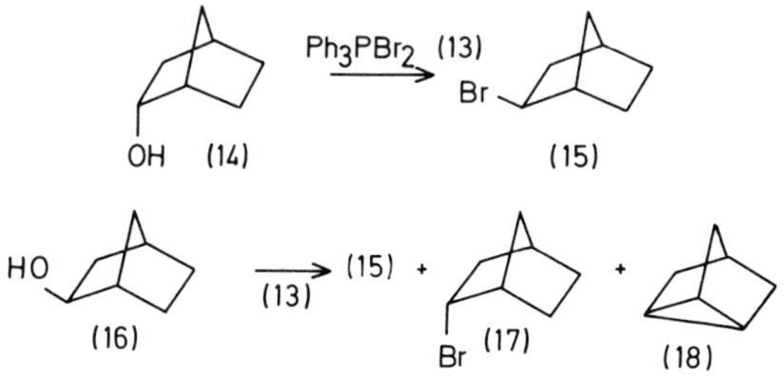

Scheme 9

Stable phosphonium salts (19) and (20), isolated from reaction of triphenylphosphine dibromide with 2,2,2-triphenylethanol and 7-norbornanol respectively, provide evidence for the mechanism suggested by Wiley et al.[29] (Scheme 10). Reactions which can only be carried out in low yield using PBr_3

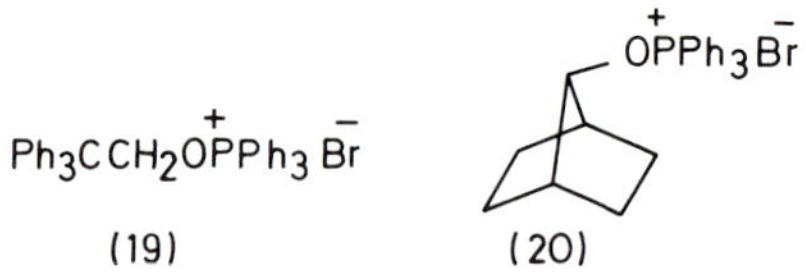

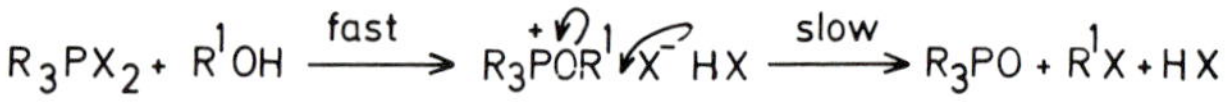

Scheme 10

are often successful. For example, α,α'-dibromoalkynes are prepared in high yield and with little racemization by using triphenylphosphine dibromide in acetonitrile.[30] It is suggested that low racemization is the result of strong association between triphenylphosphine oxide and hydrogen bromide in this solvent.

The use of dimethylformamide (DMF) as solvent causes a mechanistic change (Scheme 11).[31] Primary[32,33] and more reactive secondary alcohols[34] are converted into bromides under these conditions. Less reactive secondary alcohols may be converted into formates[32] or dehydration[34-36] may take place. The latter may be accompanied by formylation while the former leads to a selective oxidation of primary alcohols (Scheme 12). Triphenylphosphine dihalides have been successfully used in carbohydrate chemistry,[15,37] but rearrangements may occur in certain isopropylidene derivatives[15] (Scheme 13).

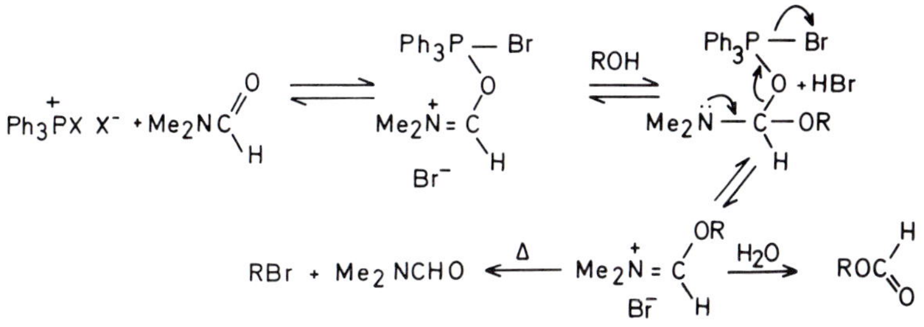

10 Conversions via R_3P-Halides, Halogens

[Scheme 11]

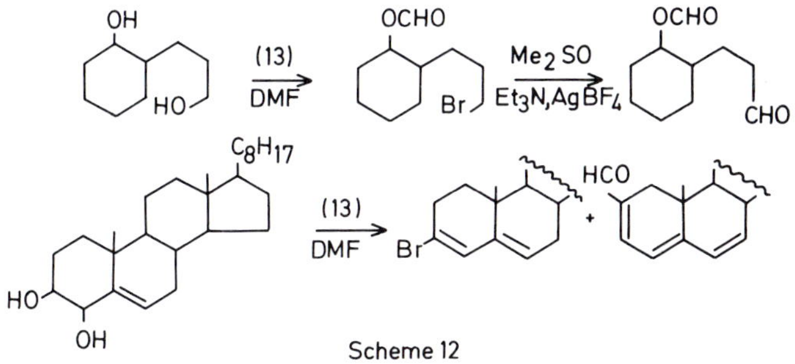

[Scheme 12]

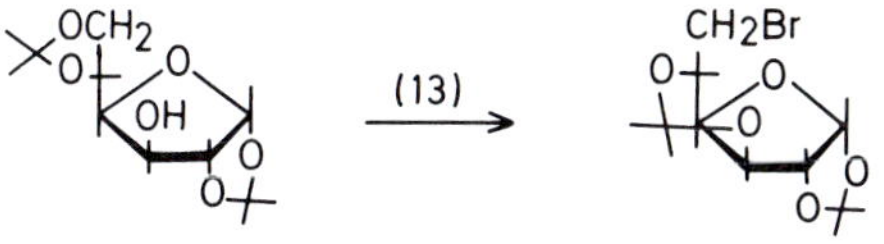

[Scheme 13]

Triphenylphosphine and bromoform react with alcohols to give bromides,[38] thus providing an alternative way of carrying out this transformation which does not appear to have been widely exploited (Scheme 14). Bridgehead

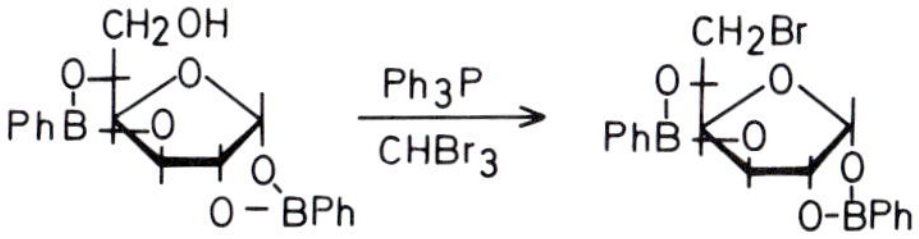

[Scheme 14]

chlorides can be prepared from the alcohol either by means of triphenylphosphine and chlorine or by treatment of the hypochlorite derived from the alcohol with triphenylphosphine[39] (Scheme 15).

In an investigation of the relative effectiveness of various reagents which might be used to halogenate nucleoside derivatives, Verheyden and Moffatt[22] find that for iodination, methyltriphenoxyphosphonium iodide (3) and

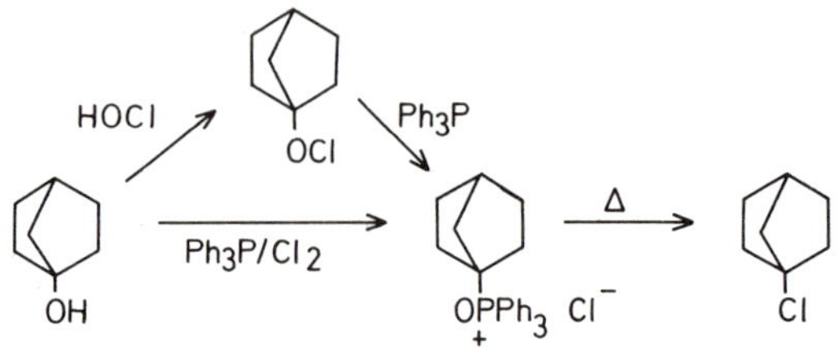

Scheme 15

triphenylphosphine diiodide in dimethylformamide are satisfactory but carbon tetraiodide with triphenylphosphine in dimethylformamide and N-iodosuccinimide with triphenylphosphine are poor; for bromination, side-reactions occur when using carbon tetrabromide with triphenylphosphine in dimethylformamide, dimethylacetamide being a preferable solvent; for chlorination, although benzyltriphenoxyphosphonium chloride gives satisfactory results, the reagent is a syrup that is difficult to store and carbon tetrachloride with triphenylphosphine is preferred. They also point out that the unsuitability of alkyltriphenoxyphosphonium salts for reactions with *vic*-diols does not hold for either phosphine dihalide or carbon tetrahalide reagents. Triphenylphosphine dibromide and bromoform with triphenylphosphine are reported to behave anomalously with nucleosides, and a virtually quantitative yield of the parent base is obtained. For example, adenine is quantitatively formed by reaction of adenosine with either triphenylphosphine dibromide or triphenylphosphine and bromoform in dimethylformamide at room temperature for 18 hours.[40a]

Before leaving the conversion of alcohols into halides it is noteworthy that triphenylphosphine with the pseudohalogen thiocyanogen similarly converts alcohols into thiocyanates.[40b]

Benzoins are converted in almost quantitative yield into benzils by triphenylphosphine dibromide.[41] 2-Aminoethanol derivatives are converted into aziridines[42] and 3-aminopropan-1-ol derivatives give azetidines.[43] In the former case, piperazines may also be formed and, indeed, from 2-N-benzyl- and 2-N-cyclohexyl-aminoethanol only the N,N'-disubstituted piperazines are formed.

Phenols are converted into aryl halides[28,44] and, in contrast to the triaryl phosphite reaction[5] described earlier, the reaction appears to be quite general and has also been applied to hydroxy-pyridines and -quinolines. Diphenyltrichlorophosphorane, Ph_2PCl_3, is reported to be a slightly more powerful reagent than triphenylphosphine dichloride.[45]

Horner and his coworkers[27] demonstrated that, with a suitable choice of triphenylphosphine dihalide, cyclohexanone could be converted into 1-chlorocyclohexene, benzaldehyde into α,α-dichlorotoluene, benzoic acid into benzoyl chloride or benzoyl bromide, and benzamide or benzaldoxime into benzonitrile. The reaction with benzaldehyde has been extended, as shown in Scheme 16, to provide a convenient method of deuterating aromatic aldehydes.[46]

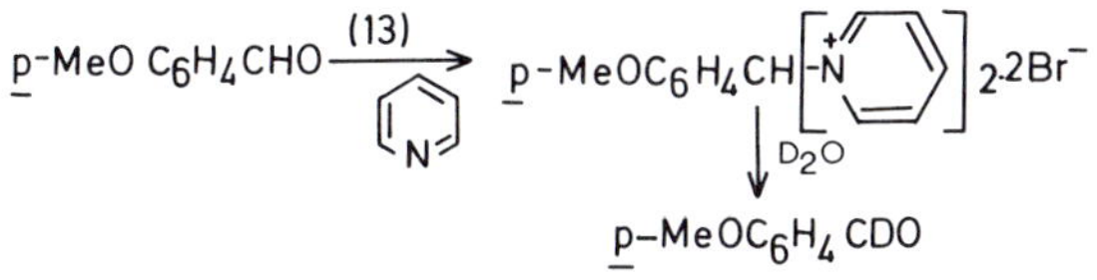

Scheme 16

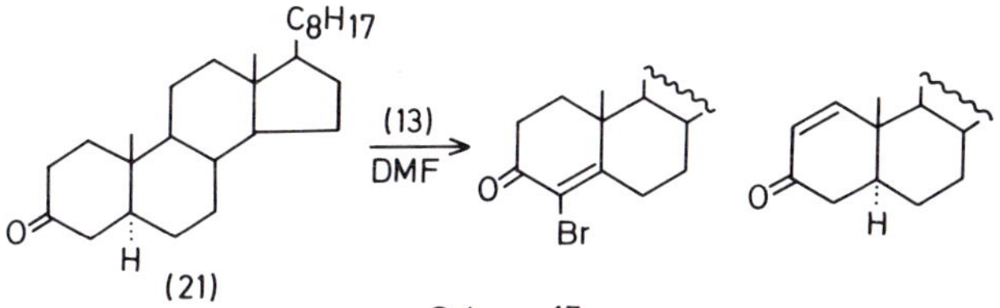

Scheme 17

In contrast to the simple reaction with cyclohexanone, cholestanone (21) reacts to give a mixture of unsaturated ketones[34] (Scheme 17). Cyclic β-diketones are converted into β-bromo(or chloro)-α,β-unsaturated ketones using triphenylphosphine dibromide (or dichloride) in the presence of triethylamine, but with triphenylphosphine diiodide the reaction stops at the phosphonium salt stage (22) and the reaction can only be completed by removal of the solvent and heating the residue to 165° (Scheme 18).

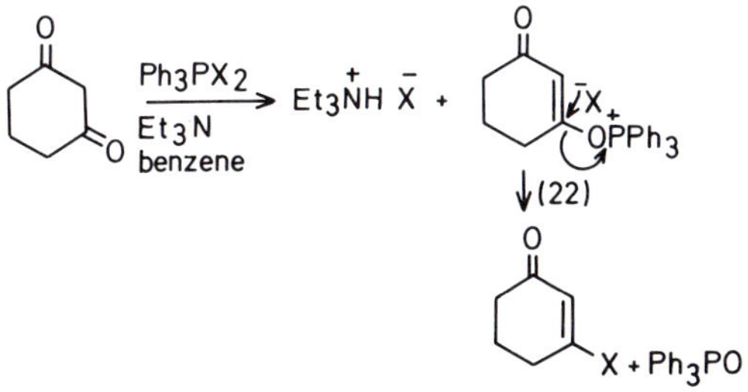

Scheme 18

A wide range of acyl bromides has been prepared by the action of triphenylphosphine dibromides on acids.[47] Triphenylphosphine dihalides catalyse the Beckmann rearrangement of cycloalkanone oximes to amides.[48] However, aryl ketoximes that have one α-hydrogen react with triphenylphosphine dibromide to give ketene imines (23)[49] and, although the same product is formed from secondary amides,[50] the oxime is the preferred starting material, especially for thermally unstable imines such as phenylketene-*N*-phenylimine, owing to the lower reaction temperature required (Scheme 19). Vinylketene imines are useful intermediates[51] in cycloaddition reactions leading to cyclohexene derivatives, as shown in Scheme 20. The imines are prepared by reaction of suitably substituted amides.

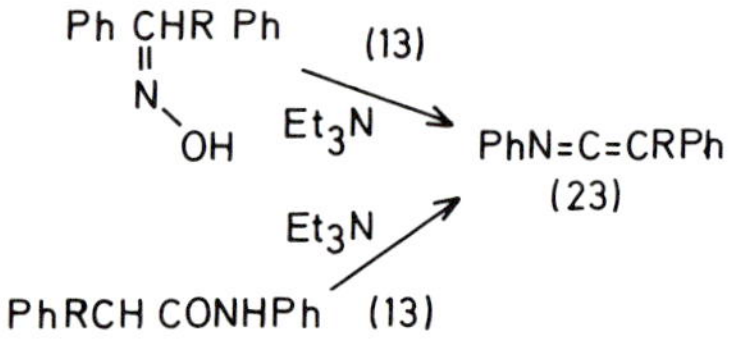

$$\begin{array}{c} \text{Ph }\underset{\underset{\text{OH}}{\overset{\|}{\text{N}}}}{\text{CHR}}\text{ Ph} \quad (13) \\ \searrow \text{Et}_3\text{N} \\ \qquad \qquad \qquad \text{PhN=C=CRPh} \\ \nearrow \text{Et}_3\text{N} \quad (23) \\ \text{PhRCH CONHPh} \quad (13) \end{array}$$

Scheme 19

$$\underset{\text{Me}}{\overset{\text{CH}_2=\text{CH}}{\diagdown}}\text{CH CONH-C}_6\text{H}_4\text{Me-}\underline{p} \xrightarrow[\text{Et}_3\text{N}]{(13)} \underset{\text{Me}}{\overset{\text{CH}_2=\text{CH}}{\diagdown}}\text{C=C=NC}_6\text{H}_4\text{Me-}\underline{p}$$

$$\downarrow \text{PhCH=C(CN)}_2$$

[Cyclohexene adducts: CH$_3$ substituted cyclohexene with (CN)$_2$, Ph, H substituents $\xleftarrow{\text{H}_3\overset{+}{\text{O}}}$ NC$_6$H$_4$Me-$\underline{p}$ imine cyclohexene with (CN)$_2$, Ph, H substituents]

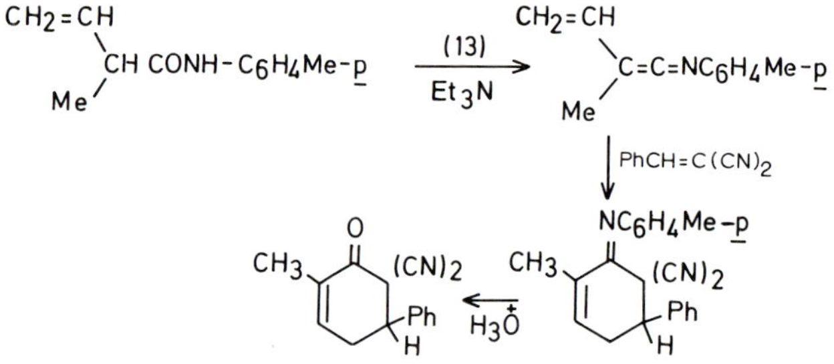

Scheme 20

Dimethylformamide and triphenylphosphine dibromide can be used in a Vilsmeier type aldehyde synthesis (Scheme 21). Aldehydes prepared by this method include indole-3-aldehyde, thiophen-2-aldehyde, diphenylacrolein, cinnamaldehyde, and *p*-dimethylaminobenzaldehyde. Phenylacetylene is converted into β-bromocinnamaldehyde. *N*,*N*'-Disubstituted ureas are converted into carbodiimides, and *N*-substituted formamides into isocyanates, using triphenylphosphine dibromide in the presence of base.[50]

$$\text{RH} \xrightarrow[\text{Me}_2\text{NCHO}]{(13)} \text{RCHO}$$

Scheme 21

Esters can be directly converted into acyl halides using a number of reagents (Scheme 22).[52, 53] If the group R is a halogenated alkyl group, e.g. CF$_3$ or CH$_2$Cl, the esters are cleaved by (13) and (24) but not by (25). If the group R is non-halogenated, (25) is the preferred reagent. This reagent also diminishes the possibility of further reaction between the acyl halide and the phosphine dihalide. Phthalide is readily cleaved by triphenylphosphine dichloride but the

10 Conversions via R_3P-Halides, Halogens

$$Ph_3PCl_2 \quad Ph_3PBr_2 \quad Ph_3\overset{+}{P}Cl \ BF_3Cl^-$$
$$(24) \quad\quad (13) \quad\quad\quad (25)$$

$$RCO_2R^1 + Ph_3PX_2 \longrightarrow RCOX + R^1X + Ph_3PO$$
$$RCO_2R^1 + Ph_3\overset{+}{P}Cl \ BF_3Cl^- \longrightarrow RCOCl + R^1Cl + BF_3 + Ph_3PO$$

Scheme 22

reaction of alicyclic lactones with triphenylphosphine dibromides gives only low yields of ω-bromoacyl bromides owing to polymerization and tar formation.[54] Ethers are readily cleaved by triphenylphosphine dibromide[55,56] (Scheme 23), the order of reactivity being primary greater than secondary.

$$ROR + Ph_3PBr_2 \rightleftharpoons R_2\overset{+}{O}\underset{Br}{P}Ph_3 \ Br^- \longrightarrow Ph_3\overset{+}{P}OR \ Br^- + RBr$$
$$\downarrow$$
$$Ph_3PO + RBr$$

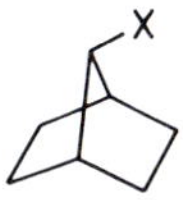

(26) a; X = Br
      b; X = OBu^t

Scheme 23

Tertiary alkyl ethers give olefins. The reaction has been utilized to prepare the otherwise difficult to synthesize 7-bromonorbornane (26a) from 7-t-butoxynorbornane (26b).[57] Alcohols protected as the tetrahydropyranyl ether are readily converted into bromides,[58] and addition of sodium cyanide to the reaction mixture results in the formation of nitriles.[59] In the case of aryl ethers, an aryloxyphosphonium bromide is formed[56] and the aryl bromide may be obtained by pyrolysis of the salt above 230°.

Two groups of workers have examined the phosphine dihalide induced ring opening of epoxides. In the earlier report,[60] the authors find that cyclohexene oxide gives both *cis*- and *trans*-1,2-dihalogenocyclohexanes, the proportion of *trans* increasing from triphenylphosphine dichloride to triphenylphosphine dibromide and tris(dimethylamino)phosphine dichloride. They compare these reactions with that of carbon tetrachloride with triphenylphosphine, from which they obtained a 1:1 mixture of *cis* and *trans* isomers, in contrast to the claim of Isaacs and Kirkpatrick[61] that only a trace of the *trans* isomer is formed. A later report is concerned with the interconversion of (*E*)- and (*Z*)-alkenes by conversion of their epoxides into dibromides followed by debromination using zinc.[62] They find the sequence shown in Scheme 24 to be highly stereospecific for the conversion of (*Z*9)-pentacosene into the (*E*)-isomer, but changing to methylene chloride in the bromination step results in a 1:1

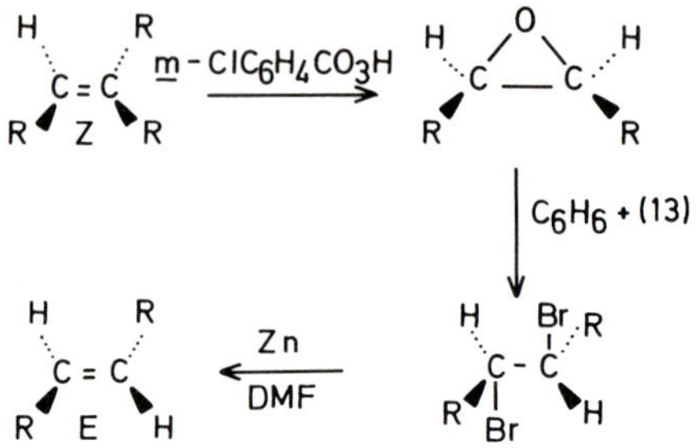

Scheme 24

mixture of (*E*)- and (*Z*)-isomers. The analogous conversion of (*E*) into (*Z*) is less stereospecific and a more satisfactory sequence involves treatment of the epoxide with hydrochloric acid before addition of triphenylphosphine dibromide in methylene chloride solution; this results in the formation of a *threo* bromochloride. Phosphine dihalides are better than lithium diphenylphosphide because ester functions have to be protected in the latter case.

DeWit and Wynberg[63] report that triphenylphosphine dibromide cleavage of endoxides followed by dehydrobromination is a valuable method of aromatizing these compounds (Scheme 25).

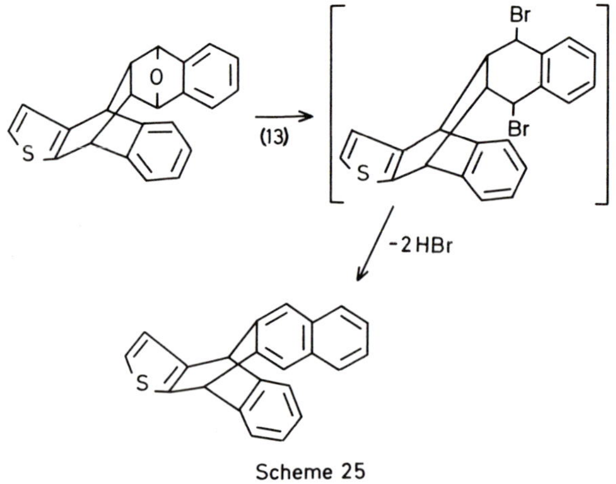

Scheme 25

Primary amines react with triphenylphosphine dibromide in the presence of triethylamine to give a phosphonium bromide (27).[64] This has been used as the first stage in the formation of iminophosphoranes which is discussed in Chapters 1 and 5, and in a synthesis of secondary amines[65] (Scheme 26). The main limitation is that the phosphonium salt (28) is formed when the group R^1 is methyl or ethyl. With higher alkyl halides elimination takes place. *N,N,N'*-Trisubstituted hydrazines can be made from *N,N*-disubstituted hydrazines.[66]

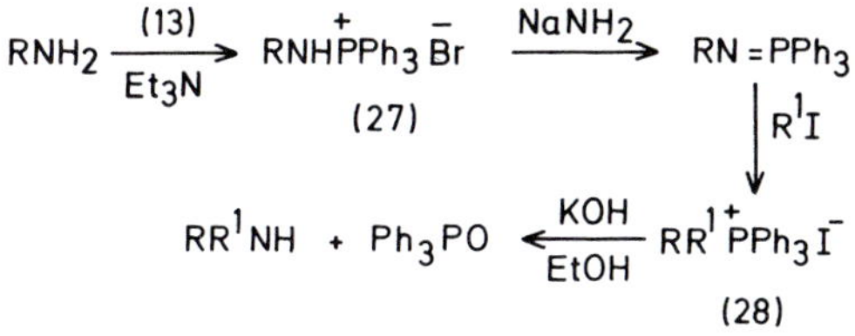

Scheme 26

$$RNH_2 \xrightarrow[Et_3N]{(13)} R\overset{+}{N}HPPh_3 \, \overset{-}{Br} \xrightarrow{NaNH_2} RN=PPh_3$$
$$(27) \qquad\qquad \downarrow R^1I$$
$$RR^1NH + Ph_3PO \xleftarrow[EtOH]{KOH} RR^1\overset{+}{P}Ph_3 \, \overset{-}{I}$$
$$(28)$$

Stabilized ylides can be prepared by the action of triphenylphosphine dichloride in the presence of triethylamine on compounds containing reactive methylene groups[67] (Scheme 27). However, phenylnitromethane gives benzonitrile under these conditions.

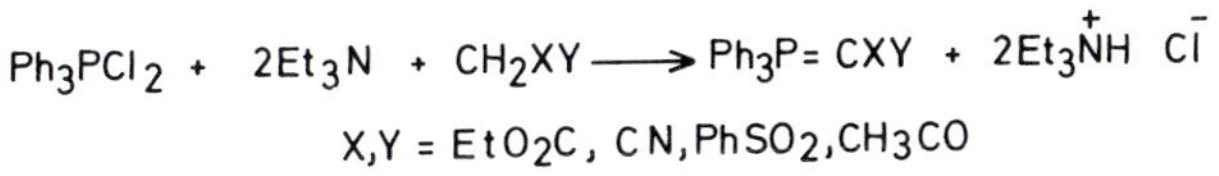

$$Ph_3PCl_2 + 2Et_3N + CH_2XY \longrightarrow Ph_3P=CXY + 2Et_3\overset{+}{N}H \, \overset{-}{Cl}$$
$$X,Y = EtO_2C, \, CN, PhSO_2, CH_3CO$$

Scheme 27

Sulphoxides are slowly deoxygenated by triphenylphosphine,[68] but deoxygenation occurs very rapidly using triphenylphosphine in the presence of iodine with added sodium iodide[69] (Scheme 28). The suggested mechanism is in agreement with one put forward earlier[70a] but the ease with which this reaction takes place casts doubt on the mechanism suggested for the deoxygenation of sulphoxides by 2-phenoxy-1,3,2-benzodioxaphosphole (29) in carbon tetrachloride,[70b] particularly because the latter reaction is catalysed by iodine.

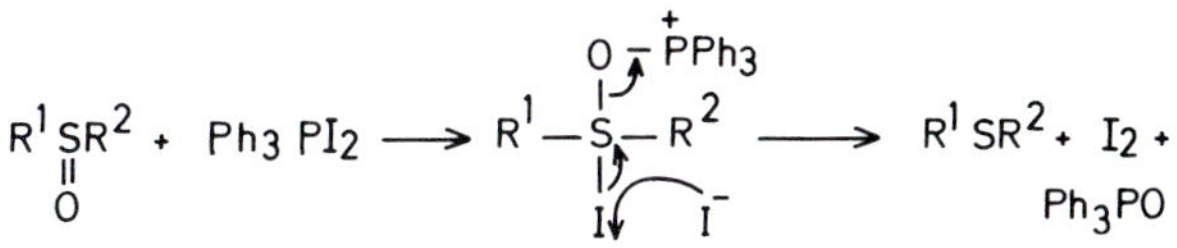

$$R^1\underset{O}{\underset{\|}{S}}R^2 + Ph_3PI_2 \longrightarrow R^1-\underset{I\overset{\frown}{\,\,\,\,}I^-}{\overset{O-\overset{+}{P}Ph_3}{\underset{|}{S}}}-R^2 \longrightarrow R^1SR^2 + I_2 + Ph_3PO$$

Scheme 28

Two solid-phase chlorination reagents (30) and (31) have been used. They perform transformations similar to those reported for triphenylphosphine dichloride.[71] The spent reagent can be regenerated using phosgene, and in some cases reactions are carried out in the presence of phosgene to regenerate the reagent *in situ*.

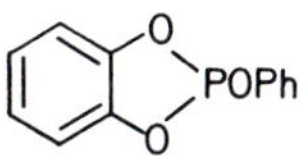

(29) benzodioxaphosphole-POPh

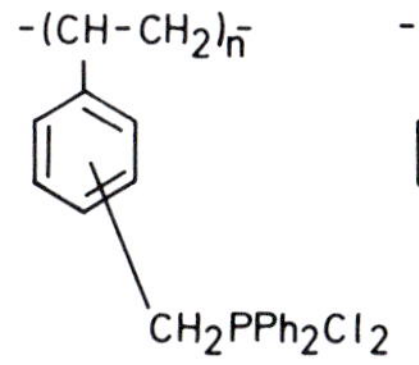

(30) -(CH-CH$_2$)$_n$- phenyl-CH$_2$PPh$_2$Cl$_2$

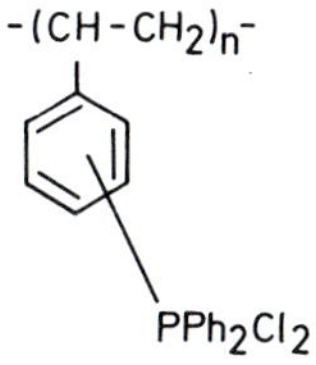

(31) -(CH-CH$_2$)$_n$- phenyl-PPh$_2$Cl$_2$

There is only one report of fluorination using triphenylphosphine difluoride,[72a] and it appears that, since more drastic conditions are required, side-reactions are more prevalent. For example, butane-1,4-diol gives tetrahydrofuran while cyclohexene is formed from cyclohexanol.

Finally we note that sulphenyl chlorides behave as pseudohalogens in reactions with triphenylphosphine to give episulphides.[72b]

1.4. Reactions involving Triphenylphosphine and Hydrobromic Acid

Triphenylphosphine in the presence of hydrobromic acid is a useful reagent for hydrolysis of esters and for cleavage of aryl alkyl ethers in non-aqueous medium to give phenols. From dialkyl ethers, olefins are formed, and the reagent can hydrolyse ketals and convert β-keto-esters into ketones[73] (Scheme 29).

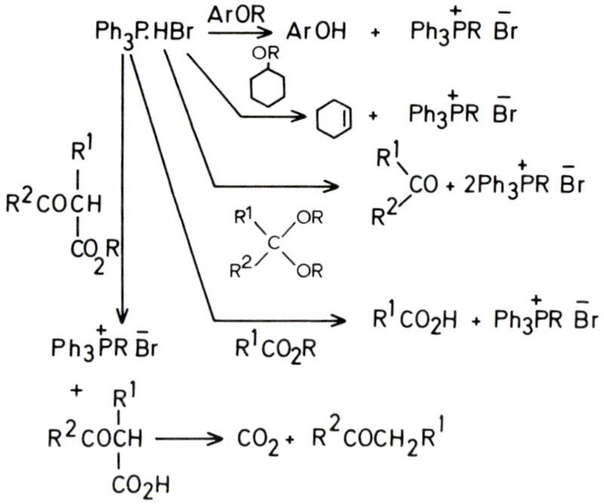

Scheme 29

2. Reactions of Phosphorus(III) Reagents at Positive Halogens

2.1. With Compounds containing Activated Carbon–Halogen Bonds

The Perkow reaction involving the action of a trialkyl phosphite on an α-halogenocarbonyl compound has been mentioned in the introduction, and it is now commonly accepted that the mechanism involves initial attack of phosphite at the carbonyl carbon atom. The reaction has little preparative significance as far as functional group transformations are concerned.

However, the reaction of trialkyl- and triaryl-phosphines with α-halogenocarbonyl compounds is of greater significance. Two possible phosphonium salts may be formed in this reaction, the ketophosphonium salt (32) or the

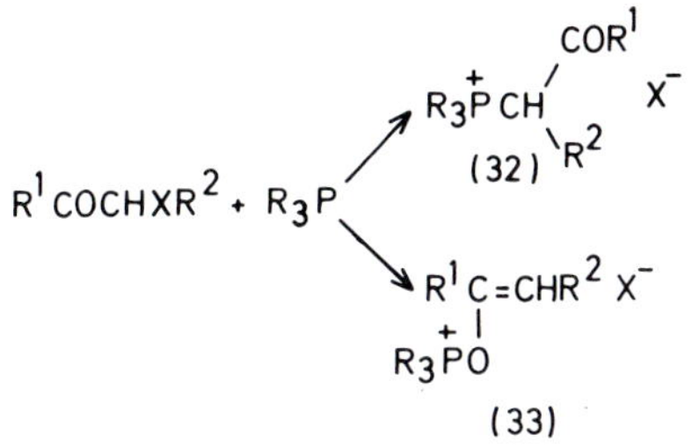

10 Conversions via R₃P-Halides, Halogens

$$R^1COCHXR^2 + R_3P \begin{array}{c} \nearrow R_3\overset{+}{P}CH\overset{COR^1}{\underset{R^2}{\diagdown}} \quad X^- \\ (32) \\ \searrow R^1C=CHR^2 \ X^- \\ \overset{+}{\underset{R_3PO}{|}} \\ (33) \end{array}$$

Scheme 30

enolphosphonium salt (33) (Scheme 30). The former is observed for the reaction of α-chlorobenzyl phenyl ketone with methylphenyl-n-propylphosphine, PhP(Me)Pr^n, and involves a simple S_N2 displacement of the chloride ion,[74] but in the latter, observed in the reaction of α-bromobenzyl phenyl ketone, initial attack is at the halogen (Scheme 31). In the presence of

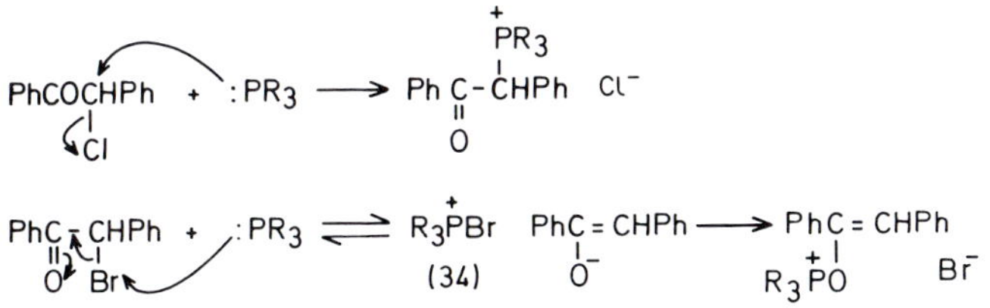

$$PhCO\underset{\underset{Cl}{|}}{C}HPh + :PR_3 \longrightarrow Ph\underset{\underset{O}{||}}{C}-\overset{\overset{+}{PR_3}}{C}HPh \ Cl^-$$

$$Ph\underset{\underset{O}{||}}{C}\underset{Br}{\overset{\curvearrowleft}{C}}HPh + :PR_3 \rightleftharpoons \underset{(34)}{R_3\overset{+}{P}Br} \quad PhC=CHPh \longrightarrow PhC=CHPh \\ \underset{O^-}{|} \quad \overset{+}{\underset{R_3PO}{|}} \quad Br^-$$

Scheme 31

compounds with active hydrogens (e.g. diethyl malonate or dimedone) or with alcohols, bromophosphonium ions (34)[75] are intercepted but only alcohols react with enol phosphonium salts.[76] In both cases dehalogenation of the halogeno compound results.

In inert solvents phosphine dihalides are often the only product formed.[77] 2-Bromocyclohexanone forms both α- (35) and β- (35a) ketophosphonium salts,[77,78] the latter by an elimination–addition reaction, and α-bromoisobutyrophenone gives only the β-ketophosphonium salt (36) (Scheme 32).[79] In

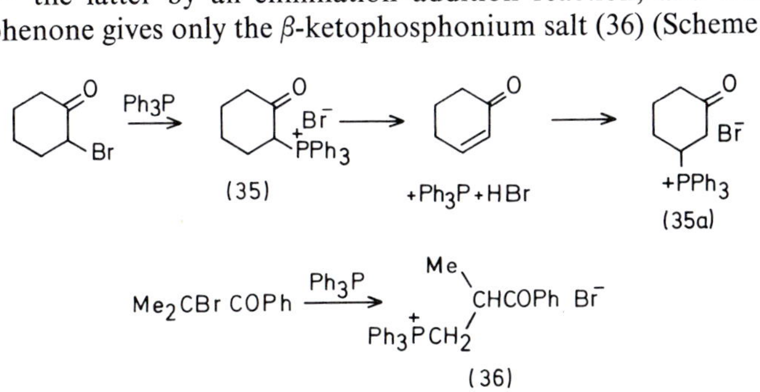

$$Me_2CBrCOPh \xrightarrow{Ph_3P} \underset{Ph_3\overset{+}{P}CH_2}{\overset{Me}{\diagdown}CHCOPh} \ Br^-$$
$$(36)$$

Scheme 32

all cases dehalogenation takes place in methanol. 2-Bromodimedone reacts with triphenylphosphine to give 3-bromo-5,5-dimethylcyclohex-2-enone[78,80] (Scheme 33).

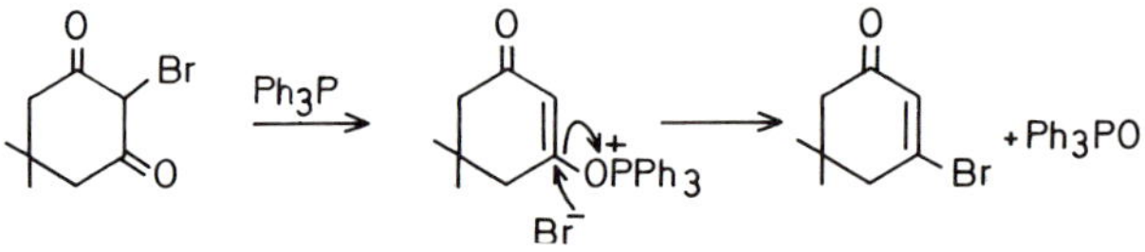
Br
O
Ph3P
OPPh3
Br
+Ph3PO

Scheme 33

If the carbanion formed by dehalogenation has three stabilizing groups attached to it, the chloro compound will react with triphenylphosphine to give a stable enolphosphonium salt[81,82] (Scheme 34). These give α-halogeno-

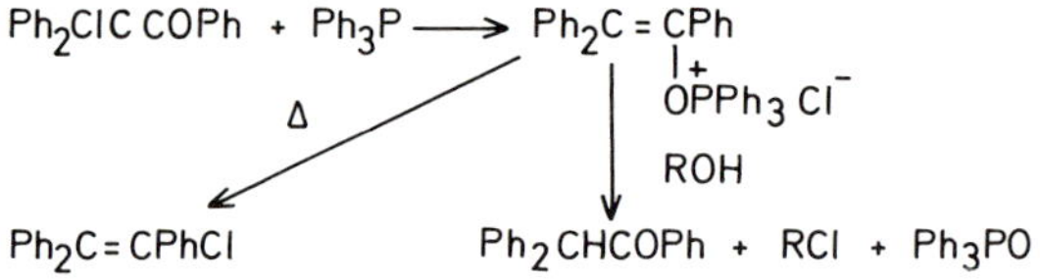
Ph2ClC COPh + Ph3P → Ph2C = CPh
OPPh3 Cl
Δ
ROH
Ph2C = CPhCl
Ph2CHCOPh + RCl + Ph3PO

Scheme 34

olefins on heating, and alkyl halides on reaction with alcohols. Diphenylacetylene has been reported to be the major product of the reaction between α-chlorobenzyl phenyl ketone[83] and triphenylphosphine but later work failed to confirm this.[84]

α-Chloroacetic acid, like other ω-chloro-acids, gives a phosphonium salt which on treatment with sodium bicarbonate gives a betaine. The betaines decompose on heating in ways characteristic of the number of methylene groups between the phosphonium cation and the carboxylate anion[85] (Scheme 35). Bromoacetic acid on the other hand gives acetyl bromide.

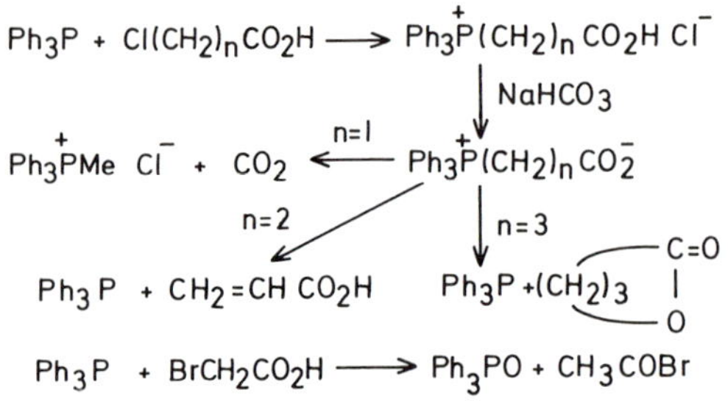
Ph3P + Cl(CH2)nCO2H → Ph3P(CH2)n CO2H Cl
NaHCO3
n=1
Ph3PMe Cl + CO2
Ph3P(CH2)nCO2
n=2
n=3
Ph3P + CH2=CH CO2H
Ph3P +(CH2)3
C=O
O
Ph3P + BrCH2CO2H → Ph3PO + CH3COBr

Scheme 35

The increased stability conferred on carbanions by chlorine atoms leads to the initial attack of triphenylphosphine on such compounds at the halogen atom. For example, polychloroacetamides give products that involve

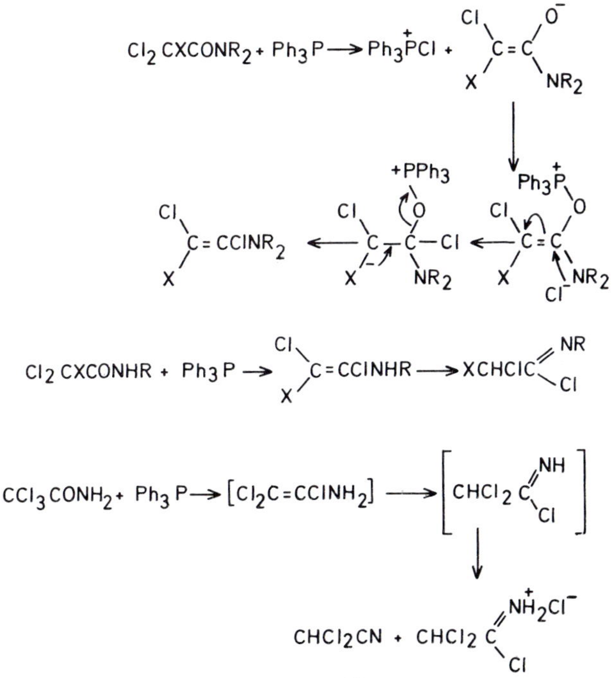

10 Conversions via R_3P-Halides, Halogens

$$Cl_2CXCONR_2 + Ph_3P \longrightarrow Ph_3\overset{+}{P}Cl + \underset{X}{\overset{Cl}{>}}C=C\underset{NR_2}{\overset{O^-}{<}}$$

$$\underset{X}{\overset{Cl}{>}}C=CClNR_2 \longleftarrow \underset{X}{\overset{Cl}{-}}\overset{O}{\underset{NR_2}{C-C-Cl}} \longleftarrow \underset{X}{\overset{Cl}{>}}C=C\underset{Cl^-}{\overset{\overset{+}{P}Ph_3}{<}}\underset{NR_2}{\overset{O}{<}}$$

$$Cl_2CXCONHR + Ph_3P \longrightarrow \underset{X}{\overset{Cl}{>}}C=CClNHR \longrightarrow XCHClC\underset{Cl}{\overset{NR}{\nearrow\!\!\!\!\!=}}$$

$$CCl_3CONH_2 + Ph_3P \longrightarrow [Cl_2C=CClNH_2] \longrightarrow \left[CHCl_2C\underset{Cl}{\overset{NH}{\nearrow\!\!\!\!\!=}}\right]$$

$$\downarrow$$

$$CHCl_2CN + CHCl_2C\underset{Cl}{\overset{\overset{+}{N}H_2Cl^-}{<}}$$

Scheme 36

migration of chlorine[86,87] (Scheme 36). The vinyl amines prepared from tertiary amides react with alcohols to give an alkyl halide and a monodehalogenated amide (Scheme 37).

$$ArCCl_2CONMePh \xrightarrow{Ph_3P} ArCCl=CClNMePh \xrightarrow{PrOH} ArCHClCONMePh + PrCl$$

Scheme 37

Trichloroacetonitrile and alkyl esters of trihalogenoacetic acids in the absence of a proton donor give an ylide and phosphine dihalide. Monodehalogenation occurs in the presence of ethanol.[88] Epoxides or β-halogeno-

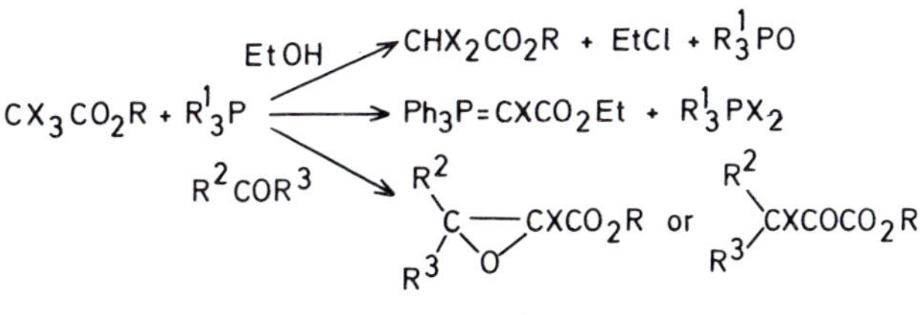

$$CX_3CO_2R + R^1_3P \begin{array}{c} \nearrow \overset{EtOH}{\longrightarrow} CHX_2CO_2R + EtCl + R^1_3PO \\ \longrightarrow Ph_3P=CXCO_2Et + R^1_3PX_2 \\ \searrow \underset{R^2COR^3}{} \underset{R^3}{\overset{R^2}{>}}C\underset{O}{-}CXCO_2R \text{ or } \underset{R^3}{\overset{R^2}{>}}CXCOCO_2R \end{array}$$

Scheme 38

α-keto-esters are formed in the presence of aldehydes and ketones[89] (Scheme 38). These compounds are also formed in the Darzens condensation of dihalogeno-esters with aldehydes and ketones, and the route described here is at least as effective. Dibromoketene, which can be trapped with cyclopentadiene as the bicyclic ketone (37), is formed by the action of triphenylphosphine on trimethylsilyl tribromoacetate.[90]

$$Ph_3\overset{+}{P}CH_2CN\ Br^-$$
(37)   (38)

Triphenylphosphine reacts in inert solvents with α-bromoacetonitrile to give the phosphonium salt (38), and in the presence of excess ethanol the yield of (38) is reduced and some ethyl bromide and acetonitrile are formed.[82,91] This indicates that a proportion of the reaction takes place by attack at bromine. However, only the vinylphosphonium salt (39) is formed on reaction with α-bromo-α-cyanoacetamide (40). This salt has been used in phosphorylation reactions of alcohols, imidazole, and phosphates[92] (Scheme 39). The reagent

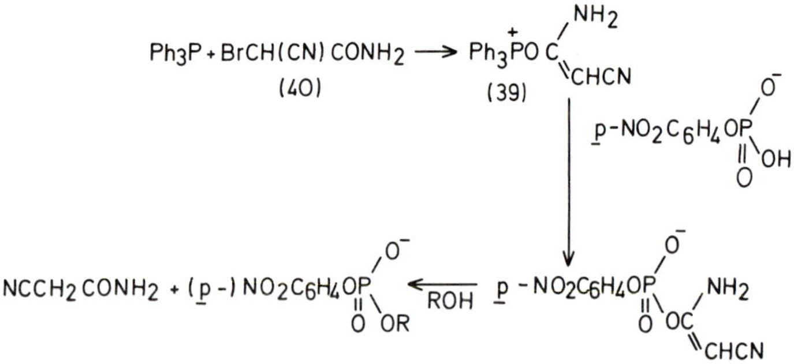

$$Ph_3P + BrCH(CN)CONH_2 \longrightarrow Ph_3\overset{+}{P}OC\overset{NH_2}{\underset{CHCN}{\diagdown}}$$
(40)   (39)

Scheme 39

can be used in conjunction with benzyl alcohol or allyl alcohol to oxidize phosphites to phosphates.[93] Mukaiyama's group have made an extensive investigation into the use of (40) and of α,α-dibromomalonamide (41) in the presence of phosphites to prepare unsymmetrical phosphates and pyrophosphates.[93-95] Some examples are shown in Scheme 40. It should be noted that benzyl phosphites may be used where free hydroxy groups are required in the product. Dialkyl phosphites have also been used in the reaction.

Foucaud and his coworkers have also carried out a series of investigations on reactions of tervalent phosphorus compounds on α-halogeno-α-cyano-esters, amides, and nitriles, and they have synthesized a number of heterocyclic systems.[96] Scheme 41 shows a representative selection of these.

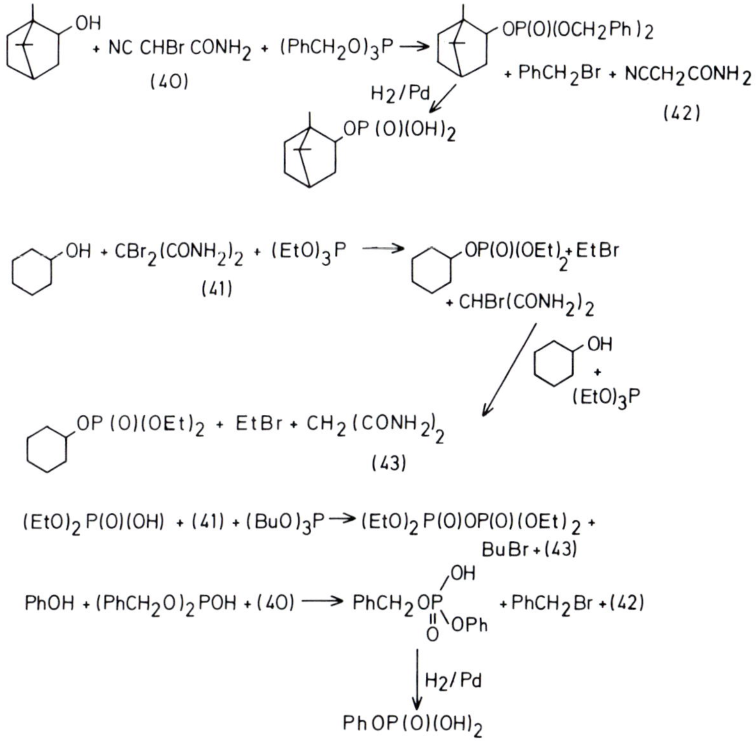

10 Conversions via R₃P-Halides, Halogens

Scheme 40

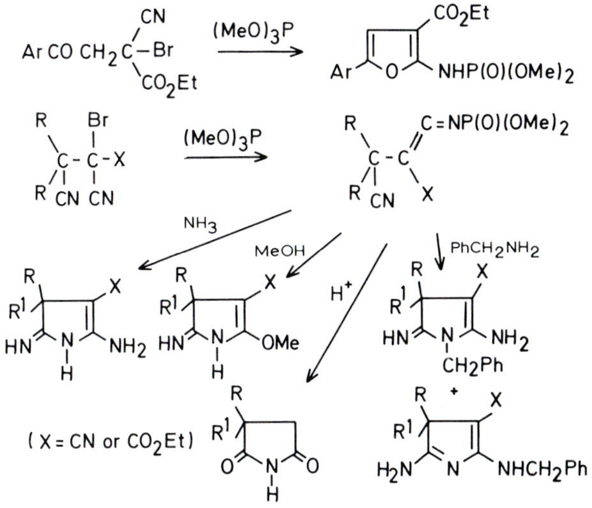

Scheme 41

Reaction of phosphines with an α-bromo-oxime involves attack at the halogen and results in the formation of an imidoyl bromide[49] (Scheme 42) which involves a Beckmann rearrangement. When a trace of base is present, a phosphonium salt (44) is formed which on treatment with base forms azines and amides.

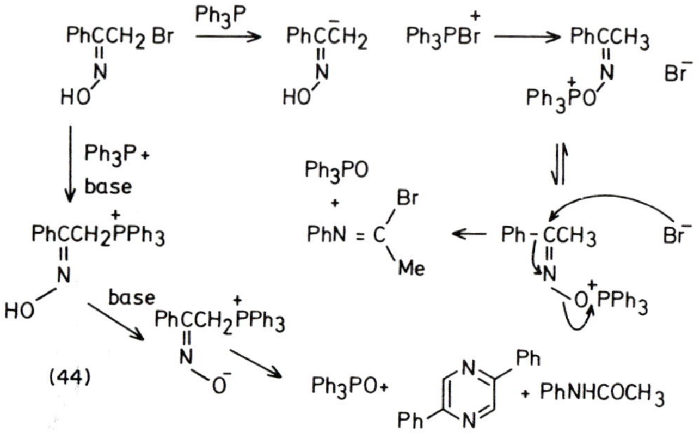

Scheme 42

Stabilized carbanions can also be generated by reaction of phosphines with α-halogeno-sulphones.[97] In addition to the usual reactions of the bromophosphonium ion with water, alcohols, and amines, reaction of the carbanion with formaldehyde is observed (Scheme 43).

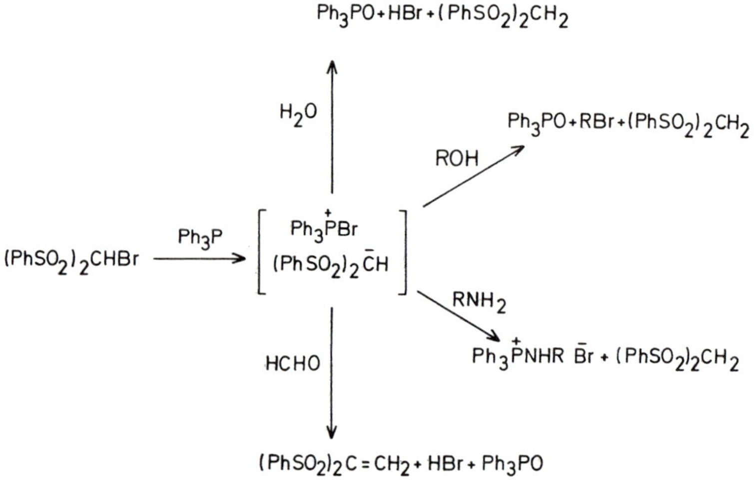

Scheme 43

The reactions of α-halogenobenzyl phenyl sulphones with triphenylphosphine in 90% aqueous dimethylformamide show an unusual order of reactivity ($k_{Cl} < k_I < k_{Br}$) and it is concluded that this results from the strength of the P–Br bond formed rather than from compensation for the strength of the C–Br bond that is broken.[98] Dehalogenation of the corresponding bromides in methanol takes place with retention of configuration.[99]

1,3-Elimination takes place with α,α'-dihalogeno-sulphones[100] to give olefins (Scheme 44) and with α,α'-dihalogeno-sulphoxides[101] to give episulphoxides. In both cases double inversion is observed. Double elimination

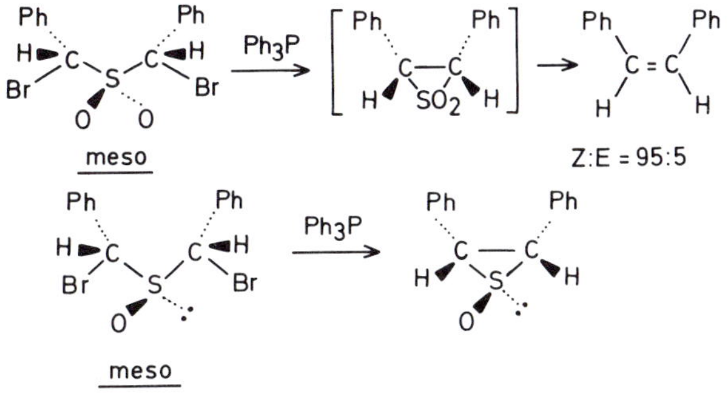

Scheme 44

results when α,α,α',α'-tetrabromo-sulphones (45) react with triphenylphosphine giving an unsaturated cyclic sulphone (46) which on thermolysis would give rise to internal acetylenes.[102] It is of interest that compounds (45) are prepared by an obscure reaction of trialkyl phosphites with bis(tribromomethyl) sulphone which had previously been described in patent literature (Scheme 45).

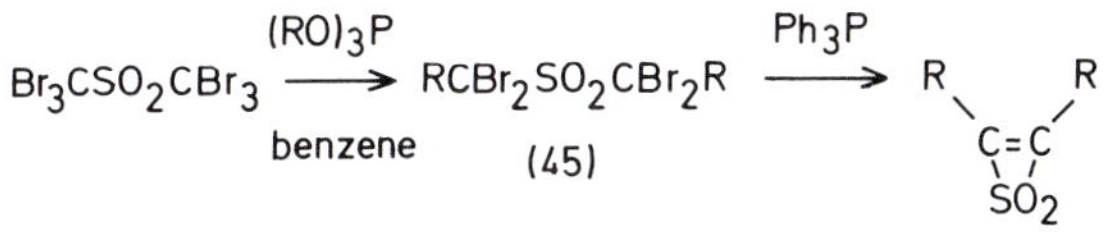

Scheme 45

The special position of α,α'-dihalogeno-sulphones (and -sulphoxides) with respect to the phosphine-induced 1,3-elimination reactions is shown by the fact that, of a wide range of compounds investigated, including α,α'-dibromodibenzyl ketone,[103] 2,4-dibromopentan-3-one, 2,2,4,4-tetrabromopentan-3-one,[102] compounds (47) and (48),[104] and α,α,α'-tribromodibenzyl ketone (49), only (49) gives a good yield of an elimination product. Using a 1:1 mixture of tris(dimethylamino)phosphine and triethylamine, (49) gives diphenylcyclopropenone.[102]

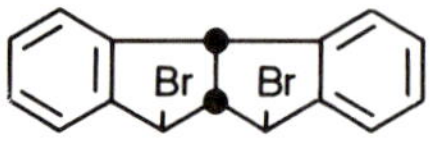

(47) 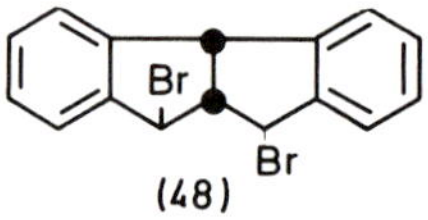 (48)

PhCBr$_2$COCHBrPh

(49)

Acyclic 1-chloro-1-nitroalkanes give Perkow type products with triethyl phosphite.[105] With triphenylphosphine, 1-bromo-1-nitroalkanes give nitriles,[106] and chloroimines are formed from either 1-nitro-[107] or 1-nitroso-1-chlorocycloalkanes[48] (Scheme 46). It should be noted that the phosphonium salt (50) is that produced by reaction of the cycloalkanone oxime with triphenylphosphine dichloride.

$$RCHClNO_2 + 2(EtO)_3P \longrightarrow RCH=NOP(O)(OEt)_2 + (EtO)_3PO + EtCl$$

$$RCHBrNO_2 + Ph_3P \longrightarrow [RCNO] \xrightarrow{Ph_3P} RCN + Ph_3PO$$
$$+ Ph_3PO + HBr$$

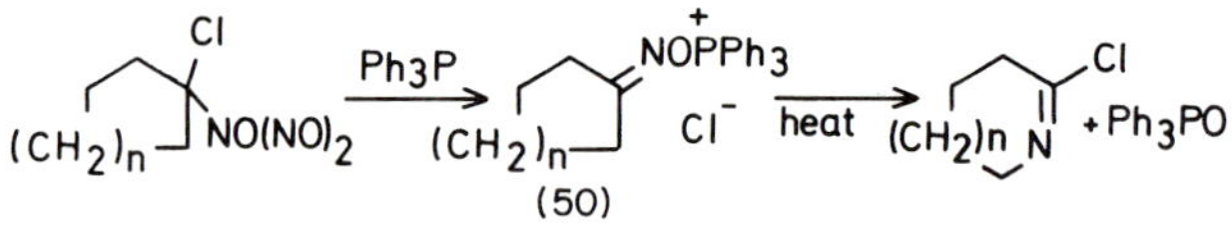

Scheme 46

$$(C_6F_5)_2CHBr \xrightarrow{Ph_3P} (C_6F_5)_2\bar{C}H \xrightarrow{(51)} (C_6F_5)_2CHCH(C_6F_5)_2$$
(51) $\quad\quad$ Ph$_3\overset{+}{P}$Br $\quad\quad$ + Ph$_3$PBr$_2 \quad\quad$ (52)

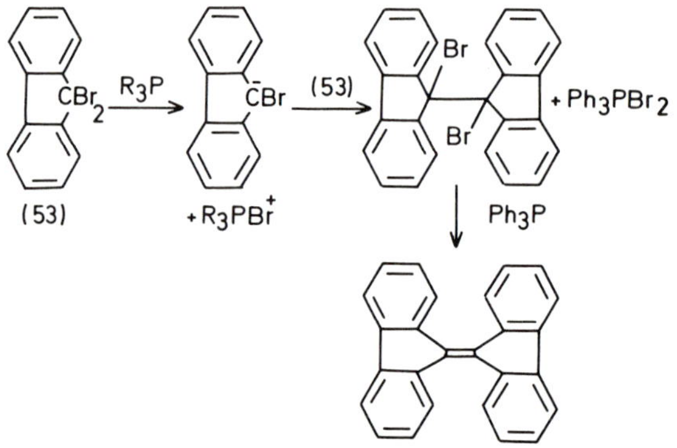

(53)   +R$_3$P$\overset{+}{B}$r   +Ph$_3$PBr$_2$

↓ Ph$_3$P

Scheme 47

Pentafluorophenyl groups stabilize carbanions. Thus triphenylphosphine reacts with (51) to give the tetra(perfluoroaryl)ethane (52)[108] whereas benzhydryl bromide gives a phosphonium salt via an S_N2 mechanism. However, 9,9-dibromofluorene gives difluorenylidene by a mechanism that has been shown not to involve a carbene[109] (Scheme 47).

Certain halogenoacetylenes react rapidly with phosphines.[110] In the case of bromoacetylenes, initial attack is at halogen. In such cases, the presence of an alcohol results in dehalogenation of the bromoacetylene.[111]

2.2. With Compounds containing Nitrogen–Halogen Bonds

Trialkyl phosphites react with N-halogenosuccinimides to give either Arbusov type products (54)[112] or dialkyl phosphorohalidates (55).[113] In either case the mechanism involves initial attack on the halogen atom[114] (Scheme 48).

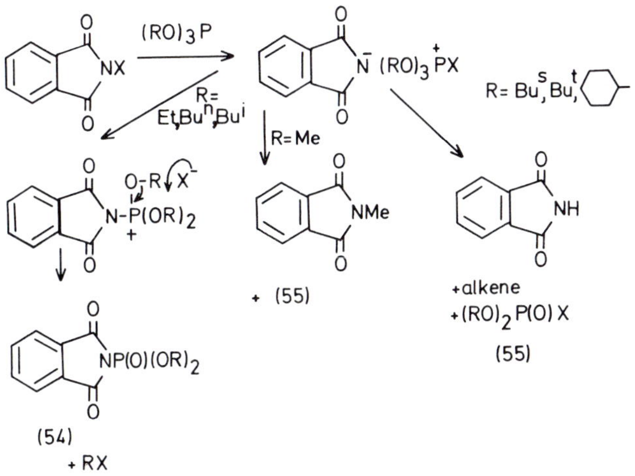

Scheme 48

Alcohols have been phosphorylated using N-bromosuccinimide and benzyl diethyl phosphite. Without added alcohol the Arbusov product (54; R = Et) is formed.[95]

Triphenylphosphine reacts with N-halogenosuccinimides to give a phosphonium salt[114] which decomposes in the presence of an alcohol to give the alkyl halide and triphenylphosphine oxide.[76] The halogenoimide–phosphine combination is often found to be preferable to phosphorus trihalides for halogenating alcohols, but several other mixtures, e.g. triphenylphosphine with N-bromoacetamide and triphenyl phosphite with N,N-dichlorourethane, have been found to be equally satisfactory.[115] In most cases the reactions involve inversion of configuration.[116] In some circumstances, e.g. the iodination of cholesterol,[117] N-iodosuccinimide has proved to be more satisfactory than triphenylphosphine diiodide, but in others, e.g. iodination of some nucleosides,[22] the converse is true. A report that N-bromosuccinimide and triphenylphosphine in the presence of an excess of tetrabutylammonium iodide can iodinate alcohols may indicate that the isolation of N-iodosuccinimide is unnecessary.[118a] This would make the method more attractive and by use of other added nucleophiles widen the scope of functional group transformations that could be carried out with N-bromosuccinimide. In a related reaction sulphenamides with tributylphosphine convert alcohols into thioethers.[118b]

Alkyl phosphites react with N-halogeno-amides to give nitriles and amides, and with N-halogeno-N-alkyl-amides to give N-halogeno-imines.[119]

The use of N-chlorodiisopropylamine with tris(dimethylamino)phosphine (56) is reported to hold advantages over the use of carbon tetrachloride with (56) in the formation of alkoxyphosphonium chlorides (58).[120] For example, reaction at the primary hydroxy group of glucose or mannose derivatives requires a much smaller excess of reagent leading to a reduction in by-products and tars. Further, reactions with allylic alcohols and with secondary alcohols that react slowly with carbon tetrachloride may be carried out in a very much shorter time using N-chlorodiisopropylamine. The difference is due to the greater base strength of the diisopropylamide ion compared with the trichloromethyl anion, thus causing more complete deprotonation of the alcohol to form the ion pair (57). The salts can be isolated as perchlorates (59) which can react with nucleophiles such as amines,[121] potassium bromide,[122] potassium iodide,[122] thiolate anions,[122] and sodium azide.[123] This last results in the formation of azides (60) whose configuration is inverted relative to that of the alcohol (Scheme 49). Tertiary alcohols are not converted into alkoxyphosphonium salts by use of either N-chlorodiisopropylamine or carbon tetrachloride with triphenylphosphine.

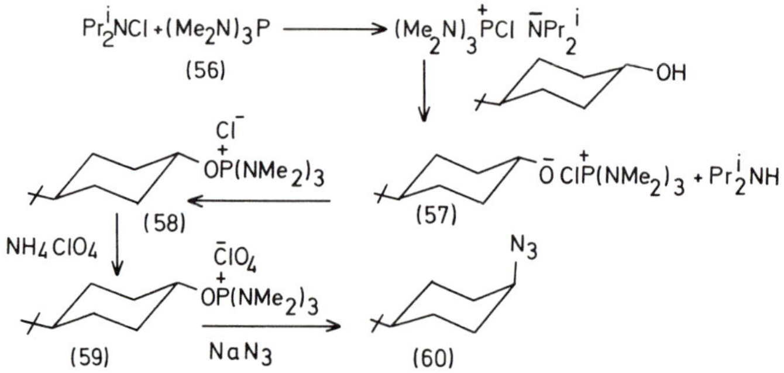

Scheme 49

2.3. Elimination from 1,2-Dihalogeno Compounds

Elimination of bromine from 1,2-dibromoethanes using tervalent phosphorus compounds has been reported to require at least one electron acceptor to be attached to a carbon bearing one of the bromine atoms[124,125] (Scheme 50),

$$PhCOCHBrCHBrCOPh + (RO)_3P \longrightarrow PhCOCH=CHCOPh + RBr + (RO)_2P(O)Br$$
$$(61)$$
$$PhCHBrCH_2Br + (RO)_3P \longrightarrow PhCH=CH_2 + RBr + (RO)_2P(O)Br$$
$$CH_2BrCHBrCN + (RO)_3P \longrightarrow CH_2=CHCN + RBr + (RO)_2P(O)Br$$

Scheme 50

although diethyl vinylphosphonate has been reported to be a product of the reaction of 1,2-dibromoethane with triethyl phosphite.[126] It is probable that the mechanism involves attack at halogen and in the presence of alcohol monodehalogenation takes place (Scheme 51).[124] Debromination of (61) using 1 mole of triethyl phosphite gives a better yield of a purer product than do reagents such as sodium iodide or zinc.[127]

$$ClCH_2CCl_2CN + (RO)_3P \longrightarrow (EtO)_3\overset{+}{P}Cl\ CH_2Cl\overset{-}{C}ClCN \longrightarrow (EtO)_2P(O)Cl$$
$$\downarrow EtOH \quad + EtCl + CH_2=CClCN$$
$$(EtO)_3PO + EtCl + CH_2ClCHClCN$$

Scheme 51

In the debromination of 1,2-dibromocyclohexane, the order of reactivity is $Bu^n_3P > Ph_3P \gg (EtO)_3P$ and the yields of cyclohexene are low.[128] The effect of phenyl groups on the rates of phosphine-induced debromination reactions has been studied and some relevant results are collected in Table 1. Whereas *meso*-1,2-dibromo-1,2-diphenylethane gives *trans*-stilbene, the

Table 1. Debromination of 1,2-dibromoethanes with triphenylphosphine

Compound	Reaction conditions	Reaction time (h)	Alkene yield (%)
meso-PhCHBrCHBrPh	boiling benzene	24	99
(±)-PhCHBrCHBrPh (62)	boiling toluene	24	50
trans-1,2-dibromohydrindane	boiling toluene	21	20
trans-1,2-dibromocyclohexane	boiling toluene	21	10

racemic isomer gives a mixture of *cis*- and *trans*-stilbene. The authors suggest that the *cis* isomer is converted into the *trans* under the influence of triphenylphosphine dibromide formed during the debromination. As evidence for this they show that the presence of excess isopropanol increases the *cis*:*trans* ratio. However, in an extensive investigation, Devlin and Walker[129] show that *erythro* and *threo* dibromides both give rise to *trans* olefins, except in the case of (62) where a 2:1 mixture of *trans*- and *cis*-stilbene results. They show that, under their reaction conditions, *cis*-olefins are not isomerized to *trans*.

Dechlorination of some chloropolyfluoroalkanes has been reported,[130,131] and the phosphite-induced polymerization of 1,4-bis(trichloromethyl)benzene shown in Scheme 52 has been described.[132]

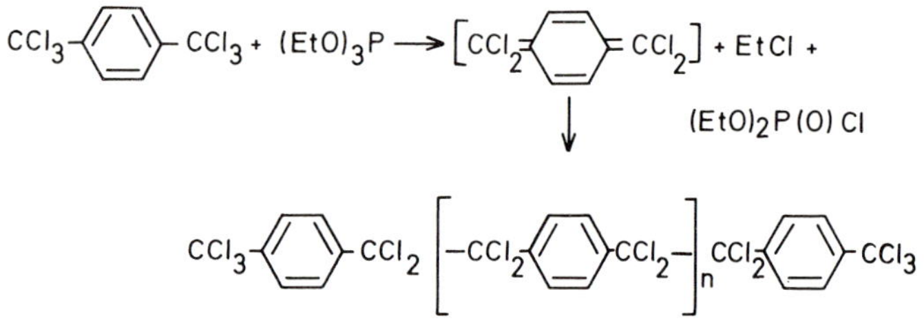

Scheme 52

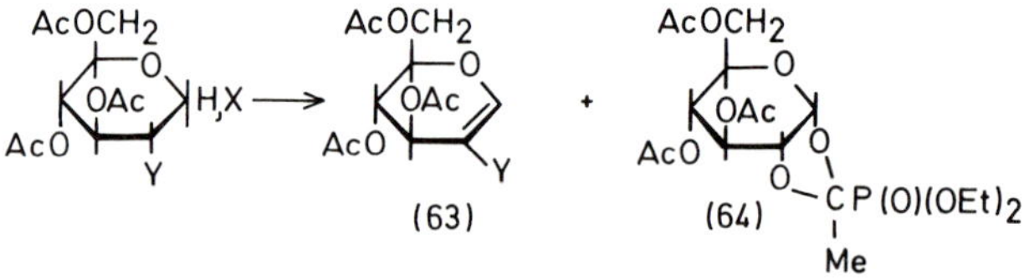

Scheme 53

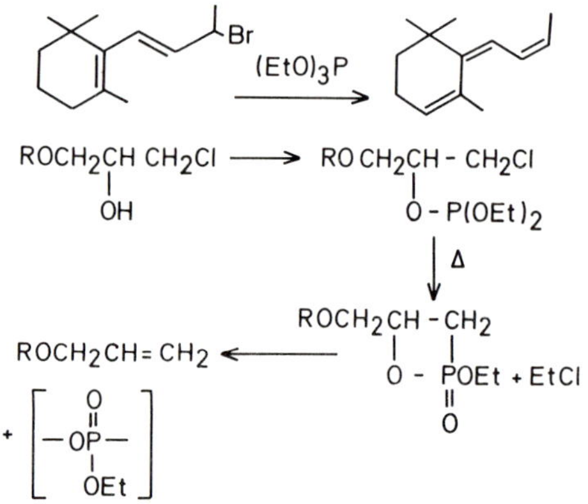

Scheme 54

10 Conversions via R_3P-Halides, Halogens

α-Bromo-ethers react readily with triphenylphosphine. A phosphonium salt is formed from 1,2-dibromo-1-ethoxyethane.[133] α-Bromo-glucose, -galactose, and -mannose derivatives are dehydrobrominated[134] (Scheme 53). Compound (64) is formed only when Y is $-O_2CCH_3$. In all other cases the olefin (63) is the only product.

Dehydrobromination has been observed in allylic bromides,[135] and an elimination of hypochlorous acid has been reported[136] via transesterification (Scheme 54).

3. Reactions with Acyl Halides

The Arbusov reaction of acyl halides and trialkyl phosphites results in the formation of α-ketophosphonate esters which on reduction with sodium borohydride, followed by treatment with a buffer at pH 6, yield aldehydes.[137] This appears to represent a highly selective reduction of acids to aldehydes.

The phosphonium salts formed by reaction of simple acid chlorides with phosphines have little or no synthetic application. However, phthaloyl chloride and dichlorophthalide form (65) when allowed to react with triphenylphosphine (Scheme 55).[138]

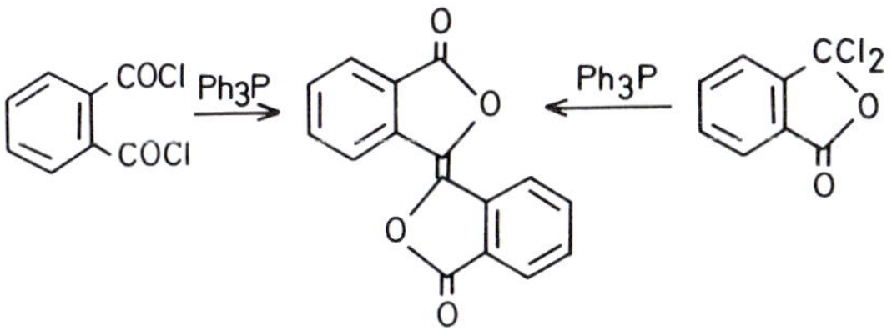

Scheme 55

α-Bromodiphenylacetyl bromide is debrominated by triphenylphosphine, and this represents a very satisfactory route to diphenylketene.[139] However, no dimethylketene is formed from α-bromoisobutyryl bromide while a complex ylide is produced from α-chloroacetyl chloride.[140]

Whereas phosphonium salts result when chloroformate esters react with triphenylphosphine in equimolar quantities, alkyl chlorides are formed in

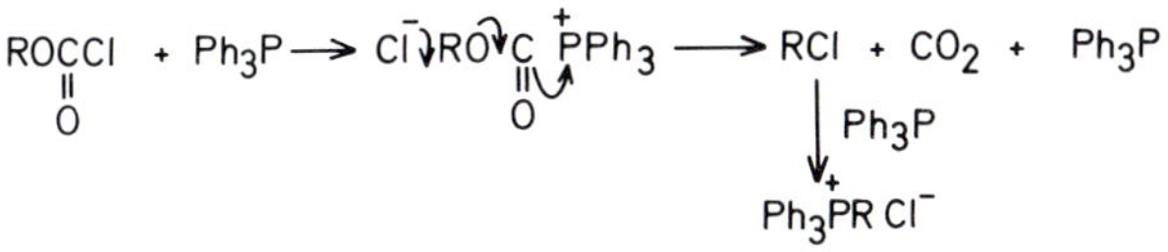

Scheme 56

good yield in the presence of a catalytic amount of phosphine, the reaction taking place with inversion of configuration[141] (Scheme 56).

Acetyl bromide reacts with triphenylphosphine in the presence of hydrogen bromide to give a bisphosphonium salt (66).[142] This reacts with nucleophiles[143] and is a useful intermediate in the synthesis of vinyl ethers[144] (Scheme 57). When the final hydrolysis is carried out in deuterium oxide, a dideuterated vinyl ether is formed.

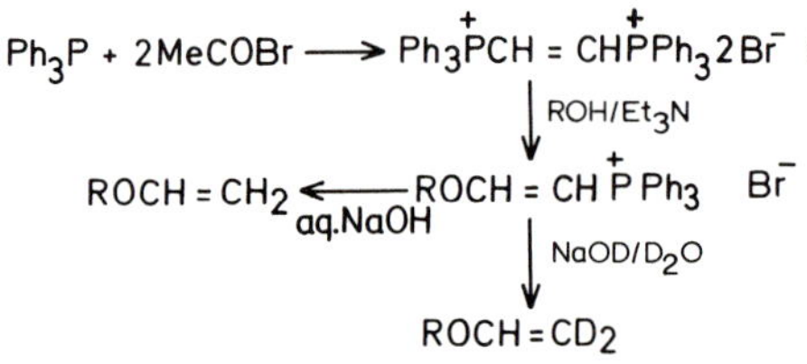

$$Ph_3P + 2MeCOBr \longrightarrow Ph_3\overset{+}{P}CH=CH\overset{+}{P}Ph_3 \, 2Br^- \quad (66)$$

$$\downarrow ROH/Et_3N$$

$$ROCH=CH_2 \xleftarrow{aq.NaOH} ROCH=CH\overset{+}{P}Ph_3 \quad Br^-$$

$$\downarrow NaOD/D_2O$$

$$ROCH=CD_2$$

Scheme 57

4. Reactions with Aryl Halides

Aryl halides do not normally react with trialkyl phosphites unless the reaction is carried out either in the presence of Ni(II) salts[145] or under the influence of ultraviolet radiation.[146] Some very reactive halides (67a) give phosphonates (67b) on reaction with triethyl phosphite. The phosphorus–carbon bond of these phosphonates is cleaved by hydrochloric acid to give an overall dehalogenation to (67c).[147] A similar dehalogenation to give (68c) can be achieved by alkaline cleavage of the phosphonium salt (68b) obtained by reaction of triphenylphosphine on the corresponding chloro compound (68a).[148] These reactions are superior to direct dehalogenations.

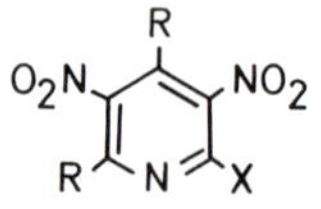

(67) a; X = Cl
    b; X = P(O)(OEt)$_2$
    c; X = H
R = H or Me

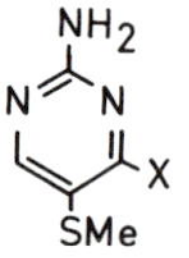

(68) a; X = Cl
    b; X = $\overset{+}{P}Ph_3 \, Cl^-$
    c; X = H

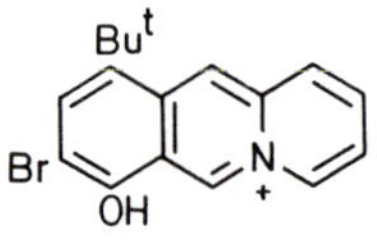

(69)

Unlike other aryl bromides, *o*- and *p*-bromophenols and *o*- and *p*-bromoanilines react with triphenylphosphine at 150°. In the former case, debromination takes place and, in the latter, an aminophosphonium bromide results.[45] An extremely facile debromination is reported to take place when (69) is boiled with triphenylphosphine for 5 minutes in nitromethane.[149]

5. Summary of Functional Group Conversions

Compounds or reaction type	Starting material	Page
Acyl bromides	acids	441
Acyl halides	esters	442
Aldehydes	acids	459
Aldehydes (Vilsmeier reactions)		442
Alkenes	alcohols	436
Alkenes	α,α-dihalogeno-sulphones	453
Alkenes	1,2-dibromoalkanes	457–8
Alkenes (stereospecific)	epoxides	437
Alkyl halides	alcohols	433–4, 435, 437–8, 448–9, 452, 455–6
Alkyl halides (optically active)	phosphites–hydrogen halide	434
Alkyl iodides	alcohols	433–4, 435–7, 440
Alkyl iodides	phosphoroiodates	433
Alkynes	tetrabromo-sulphones	453
Alkynyl bromides		433
Alkylation of hydrazines		444
Amides	acids	434
Amines, secondary	amines, primary	444
Aromatization via endoxide cleavage		444
Aryl halides	phenols	434, 440
Aryl iodides	phenols	434
Azetidines	3-aminopropan-1-ols	440
Azido-carbohydrates	carbohydrates	456
Aziridines	2-aminoethanols	440
Beckmann rearrangements		441, 452
[²H]Benzaldehyde	benzaldehyde	441
Benzils	benzoins	440
Benzonitrile	benzamide or benzaldoxime	441
Benzonitrile	phenylnitromethane	445
Benzoyl halides	benzoic acid	441
β-Bromocinnamaldehyde	phenylacetylene	442
Chlorocyclohexene	cyclohexanone	440
Dehalogenation (RBr → RH)		448, 449, 453, 455, 460
Dibromoketene	trimethylsilyl tribromoacetate	450
α,α-Dichlorotoluene	benzaldehyde	440
Diphenylcyclopropenone	α,α,α'-tribromodibenzyl ketone	453

Compounds or reaction type (Cont.)	Starting material (Cont.)	Page
Diphenylketene	α-bromodiphenylacetyl bromide	460
Episulphoxides	α,α-dihalogeno-sulphones	453
Epoxide fission		437, 443
Episulphides	sulphenyl chlorides	446
Ester hydrolysis		446
Esters	acids	434
Ether cleavage		443, 446
Halogeno-carbohydrates	carbohydrates	435, 456
3-Halogenocyclohex-2-enone	cyclohexane-1,3-dione	441
Halogeno-nucleosides	nucleosides	435, 439
Halogenopyridines	hydroxypyridines	440
Halogenoquinolines	hydroxyquinolines	440
Halogeno-steroids	hydroxy-steroids	436, 456
Ketene imines	aryl ketoximes	441
Ketones	ketals	446
Ketones	β-keto-esters	446
Nitriles	alcohols	443
Nitriles	1-bromo-1-nitroalkanes	454
Perfluorotetraphenylethane	perfluorodiphenylmethane	455
Phosphates	phosphites	450
Phosphorylation of alcohols, imidazole, and phosphates		450
Piperazines	2-N-alkylaminoethanols	440
Pyrrolines synthesis		450
Sulphoxide deoxygenation		445
Thiocyanates	alcohols	440
Thioethers	alcohols	456
Vilsmeier reactions		442
Vinyl ethers	alcohols	460
Vinyl ketene imines	allylamides	442
Ylides, stabilized	active methylene compounds	445

6. References

1. H. McCombie, B. C. Saunders, and G. J. Stacey, *J. Chem. Soc.*, 1945, 380.
2. E. J. Corey and J. E. Anderson, *J. Org. Chem.*, 1967, **32**, 4160; P. A. Grieco and R. S. Finkelhor, *ibid.*, 1973, **38**, 2245.
3. J. P. Forsman and D. Lipkin, *J. Amer. Chem. Soc.*, 1953, **75**, 3145.
4. D. G. Coe, S. R. Landauer, and H. N. Rydon, *J. Chem. Soc.*, 1954, 2281.
5. D. G. Coe, H. N. Rydon, and B. L. Tonge, *J. Chem. Soc.*, 1957, 323.
6. D. K. Black, S. R. Landor, A. N. Patel, and P. F. Whiter, *J. Chem. Soc.* (*C*), 1967, 2260.
7. H. N. Rydon, *Chem. Soc. Special Publ.* No. 8, 1957, **61**.
8. H. R. Hudson, *Synthesis*, 1969, 112.
9. H. R. Hudson, A. R. Qureshi, and D. Ragoonanan, *J. Chem. Soc., Perkin I*, 1972, 1595.

10. N. Yamazaki, F. Higashi, and S. A. Kazaryan, *Synthesis*, 1974, 436.
11. S. R. Landauer and H. N. Rydon, *J. Chem. Soc.*, 1953, 2224.
12. H. N. Rydon, *Org. Synth.*, 1971, **51**, 44.
13. J. B. Lee and M. M. El Sawi, *Chem. and Ind.*, 1960, 839.
14. N. K. Kochetkov and A. I. Usov, *Tetrahedron Lett.*, 1963, 519.
15. N. K. Kochetkov and A. I. Usov, *Tetrahedron*, 1963, **19**, 973.
16. N. K. Kochetkov and A. I. Usov, *Methods Carbohydrate Chem.*, 1972, **6**, 205.
17. K. Kefurt, J. Jary, and Z. Samek, *Chem. Comm.*, 1969, 213.
18. J. P. H. Verheyden and J. G. Moffatt, *J. Amer. Chem. Soc.*, 1964, **86**, 2093.
19. J. P. H. Verheyden and J. G. Moffatt, *J. Org. Chem.*, 1970, **35**, 2319.
20. G. A. R. Johnston, *Austral. J. Chem.*, 1968, **21**, 513.
21. J. P. H. Verheyden and J. G. Moffatt, *J. Org. Chem.*, 1970, **35**, 2868.
22. J. P. H. Verheyden and J. G. Moffatt, *J. Org. Chem.*, 1972, **37**, 2289.
23. E. Santaniello, E. Caspi, W. L. Duax, and C. M. Weeks, *J. Org. Chem.*, 1977, **42**, 482.
24. M. G. Hutchins and C. A. Milewski, *J. Org. Chem.*, 1972, **37**, 4190.
25. C. W. Spangler and T. W. Hartford, *Synthesis*, 1976, 108.
26. (a) R. O. Hutchins, D. Kandasamy, C. A. Maryanoff, D. Masilamani, and B. E. Maryanoff, *J. Org. Chem.*, 1977, **42**, 82. (b) K. Yamada, S. Goto, H. Nagase, Y. Kyotani, and Y. Hirata, *ibid.*, 1978, **43**, 2076.
27. L. Horner, H. Oediger, and H. Hoffmann, *Annalen.*, 1959, **626**, 26.
28. G. A. Wiley, R. L. Hershkowitz, B. M. Rein, and B. C. Chung, *J. Amer. Chem. Soc.*, 1964, **86**, 964.
29. G. A. Wiley, B. M. Rein, and R. L. Hershkowitz, *Tetrahedron Lett.*, 1964, 2509.
30. R. Machinek and W. Lüttke, *Synthesis*, 1975, 255.
31. M. E. Herr and R. A. Johnson, *J. Org. Chem.*, 1972, **37**, 310.
32. R. K. Boeckman, Jr., and B. Granem, *Tetrahedron Lett.*, 1974, 913.
33. R. G. Bergman, *J. Amer. Chem. Soc.*, 1969, **91**, 7405.
34. D. Levy and R. Stevenson, *J. Org. Chem.*, 1965, **30**, 3469.
35. D. Levy and R. Stevenson, *Tetrahedron Lett.*, 1965, 341; *J. Org. Chem.*, 1967, **32**, 1265.
36. T. Dahl, R. Stevenson, and N. S. Bhacca, *J. Org. Chem.*, 1971, **36**, 3243.
37. R. D. Birkenmeyer and F. Kagan, *J. Medicin. Chem.*, 1970, **13**, 616.
38. L. G. Mogel and A. M. Yurkevich, *J. Gen. Chem. (U.S.S.R)*, 1970, **40**, 682.
39. D. B. Denney and R. R. DiLeone, *J. Amer. Chem. Soc.*, 1972, **37**, 2289.
40. (a) S. G. Verenikina, E. G. Chauser, and A. M. Yurkevich, *J. Gen. Chem. (U.S.S.R.)*, 1971, **41**, 1636. (b) Y. Tamura, T. Kawasaki, M. Adachi, M. Tanio, and Y. Kita, *Tetrahedron Lett.*, 1977, 4417.
41. T. L. Ho, *Synthesis*, 1972, 697.
42. I. Okada, K. Ichimura, and R. Sudo, *Bull. Chem. Soc. Japan*, 1970, **43**, 1185.
43. J. P. Freeman and P. J. Mondron, *Synthesis*, 1974, 894.
44. J. P. Schaefer and J. Higgins, *J. Org. Chem.*, 1967, **32**, 1607.
45. H. Hoffmann, L. Horner, H. G. Wippel, and D. Michael. *Chem. Ber.*, 1962, **95**, 523.
46. T. Hase, *Acta Chem. Scand.*, 1970, **24**, 2263.
47. H.-J. Bestmann and L. Mott, *Annalen*, 1966, **693**, 132.
48. M. Ohno and I. Sakai, *Tetrahedron Lett.*, 1965, 4541.
49. M. Masaki, K. Fukui, and M. Ohta, *J. Org. Chem.*, 1967, **32**, 3564.
50. H.-J. Bestmann, J. Lienert, and L. Mott, *Annalen*, 1968, **718**, 24.

51. E. Sonveaux and L. Ghosez, *J. Amer. Chem. Soc.*, 1973, **95**, 5417.
52. D. J. Burton and W. M. Koppes, *J. Chem. Soc., Chem. Comm.*, 1973, 425.
53. A. G. Anderson, Jr., and D. H. Kono, *Tetrahedron Lett.*, 1973, 5121.
54. E. E. Smissman, H. N. Alkaysi, and M. W. Creese, *J. Org. Chem.*, 1975, **40**, 1640.
55. A. G. Anderson and F. J. Freenor, *J. Amer. Chem. Soc.*, 1964, **86**, 5037.
56. A. G. Anderson and F. J. Freenor, *J. Org. Chem.*, 1972, **37**, 626.
57. A. P. Marchand and W. R. Weimar, Jr., *Chem. and Ind.*, 1969, 200.
58. M. Schwarz, J. E. Oliver, and P. E. Sonnet, *J. Org., Chem.*, 1975, **40**, 2410.
59. P. E. Sonnet, *Synth. Comm.*, 1976, **6**, 21.
60. A. N. Thakore, P. Pope, and A. C. Oehlschlager, *Tetrahedron*, 1971, **27**, 2617.
61. N. S. Isaacs and D. Kirkpatrick, *Tetrahedron Lett.*, 1972, 3869.
62. P. E. Sonnet and J. E. Oliver, *J. Org. Chem.*, 1976, **41**, 3279.
63. J. de Wit and H. Wynberg, *Rec. Trav. Chim.*, 1973, **92**, 281.
64. L. Horner and H. Oediger, *Annalen*, 1959, **627**, 142.
65. H. Zimmer and G. Singh, *J. Org. Chem.*, 1963, **28**, 483; H. Zimmer, M. Jayawant, and P. Gutsch, *ibid.*, 1970, **35**, 2826.
66. H. Zimmer and G. Singh, *J. Org. Chem.*, 1964, **29**, 1579.
67. L. Horner and H. Oediger, *Chem., Ber.* 1968, **91**, 437.
68. E. H. Amonoo-Neizer, S. K. Ray, R. A. Shaw, and B. C. Smith, *J. Chem. Soc.*, 1965, 4296.
69. G. A. Olah, B. G. B. Gupta, and S. C. Narang, *Synthesis*, 1978, 137.
70. (a) R. Luckenbach and G. Herweg, *Annalen*, 1976, 2305. (b) M. Dreux, Y. Leroux, and P. Savignac, *Synthesis*, 1974, 506.
71. H. M. Relles and R. W. Schluenz, *J. Amer. Chem. Soc.*, 1974, **96**, 6469.
72. (a) Y. Kobayashi and C. Akashi, *Chem. and Pharm. Bull.* (Japan), 1968, **16**, 1009; Y. Kobayashi, C. Akashi, and K. Morinaga, *ibid.*, p. 1784. (b) J. E. Baldwin and D. P. Hesson, *Chem. Comm.*, 1976, 667.
73. H. J. Bestmann, L. Mott, and J. Lienert, *Annalen*, 1967, **709**, 105.
74. I. J. Borowitz, K. C. Kirby, P. E. Rusek, and E. W. R. Casper, *J. Org. Chem.*, 1971, **36**, 88.
75. I. J. Borowitz and R. Virkhaus, *J. Amer. Chem. Soc.*, 1963, **85**, 2183.
76. S. Trippett, *J. Chem. Soc.*, 1962, 2337.
77. R. F. Hudson and G. Salvadori, *Helv. Chim. Acta*, 1966, **49**, 96.
78. I. J. Borowitz, K. C. Kirby, Jr., and R. Virkhaus, *J. Org., Chem.*, 1966, **31**, 4031.
79. I. J. Borowitz, K. C. Kirby, Jr., P. E. Rusek, and R. Virkhaus, *J. Org. Chem.*, 1968, **33**, 3686.
80. I. J. Borowitz and L. I. Grossman, *Tetrahedron Lett.*, 1962, 471.
81. A. J. Speziale and R. D. Partos, *J. Amer. Chem. Soc.*, 1963, **85**, 3312.
82. R. D. Partos and A. J. Speziale, *J. Amer. Chem. Soc.*, 1965, **87**, 5068.
83. S. Trippett and D. M. Walker, *J. Chem. Soc.*, 1960, 2976.
84. I. J. Borowitz, P. E. Rusek, and R. Virkhaus, *J. Org. Chem.*, 1969, **34**, 1595.
85. D. B. Denney and L. C. Smith, *J. Org. Chem.*, 1962, **27**, 3404.
86. A. J. Speziale and R. Smith, *J. Amer. Chem. Soc.*, 1962, **84**, 1868.
87. A. J. Speziale and L. J. Taylor, *J. Org. Chem.*, 1966, **31**, 2450.
88. D. J. Burton and J. R. Greenwald, *Tetrahedron Lett.*, 1967, 1535.
89. J. Villieras, G. Lavielle, R. Burgada, and B. Castro, *Compt. Rend.* (*C*), 1969, **268**, 1164; J. Villieras, P. Coutrot, and J.-C. Combret, *ibid.*, 1970, **270**, 1250.
90. T. Okada and R. Okawara, *Tetrahedron Lett.*, 1971, 2801.

91. G. P. Schiemenz and H. Engelhard, *Chem. Ber.*, 1961, **94**, 578.
92. F. Cramer and T. Hata, *Annalen*, 1966, **692**, 22.
93. T. Mukaiyama, O. Mitsunobu, and T. Obata, *J. Org. Chem.*, 1965, **30**, 101.
94. T. Hata and T. Mukaiyama, *Bull. Chem. Soc. Japan*, 1962, **35**, 1106; T. Mukaiyama, T. Hata, and K. Tasaka, *J. Org. Chem.*, 1963, **28**, 481; T. Hata and T. Mukaiyama, *Bull. Chem. Soc. Japan*, 1964, **37**, 103; O. Mitsunobu, T. Obata, and T. Mukaiyama, *J. Org. Chem.*, 1965, **30**, 1071.
95. T. Mukaiyama, T. Obata, and O. Mitsunobu, *Bull. Chem. Soc. Japan*, 1965, **38**, 1088.
96. E. Corre and A. Foucaud, *Chem. Comm.*, 1971, 10; A. Foucaud and R. Leblanc, *Tetrahedron Lett.*, 1969, 509; R. Leblanc, E. Corre and A. Foucaud, *Tetrahedron*, 1972, **28**, 4039; D. LeGuern, M. A. LeMoing, G. Morel, and A. Foucaud, *Tetrahedron*, 1977, **33**, 27.
97. H. Hoffmann and H. Förster, *Tetrahedron Lett.*, 1963, 1547.
98. B. B. Jarvis and J. C. Saukaitis, *Tetrahedron Lett.*, 1973, 709.
99. F. G. Bordwell, E. Doomes, and P. W. R. Corfield, *J. Amer. Chem. Soc.*, 1970, **92**, 2581.
100. F. G. Bordwell, B. B. Jarvis, and P. W. R. Corfield, *J. Amer. Chem. Soc.*, 1968, **90**, 5298.
101. B. B. Jarvis, S. D. Dutkey, and H. L. Ammon, *J. Amer. Chem. Soc.*, 1972, **94**, 2136.
102. L. A. Carpino and J. R. Williams, *J. Org. Chem.*, 1974, **39**, 2320.
103. C. J. Devlin and B. J. Walker, *J. Chem. Soc., Perkin I*, 1972, 1249.
104. S. J. Cristol, A. R. Dahl, and W. Y. Lim, *J. Amer. Chem. Soc.*, 1970, **92**, 5670.
105. J. F. Allen, *J. Amer. Chem. Soc.*, 1957, **79**, 3071.
106. S. Trippett and D. M. Walker, *J. Chem. Soc.*, 1960, 2976.
107. M. Ohno and N. Kawabe, *Tetrahedron Lett.*, 1966, 3935.
108. R. Filler and F. P. Avonda, *J. Chem. Soc., Chem. Comm.*, 1972, 943.
109. I. J. Borowitz, P. E. Rusek, Jr., and P. D. Readio, *Phosphorus*, 1971, **1**, 147.
110. S. I. Miller, C. E. Orzech, C. A. Welch, G. R. Ziegler, and J. I. Dickstein, *J. Amer. Chem. Soc.*, 1962, **84**, 2020.
111. J. I. Dickstein and S. I. Miller, *J. Org. Chem.*, 1972, **37**, 2168.
112. E. Graydou, G. Peiffer, A. Guillemonat, and J.-C. Traynard, *Compt. Rend. (C)*, 1972, **275**, 547.
113. D. J. Scharf, *J. Org. Chem.*, 1974, **39**, 922.
114. A. K. Tsolis, W. E. McEwen, and C. A. Van der Werf, *Tetrahedron Lett.*, 1964, 3217.
115. E. E. Schweizer, W. S. Creasy, K. K. Light, and E. T. Shaffer, *J. Org. Chem.*, 1969, **34**, 212; U. Axen, J. L. Thompson, and J. E. Pike, *Chem. Comm.*, 1970, 602.
116. A. K. Bose and B. Lal, *Tetrahedron Lett.*, 1973, 3937.
117. A. V. Bayless and H. Zimmer, *Tetrahedron Lett.*, 1968, 3811.
118. (a) M. M. Ponpipom and S. Hanessian, *Carbohydrate Res.*, 1971, **18**, 342. (b) K. A. Walker, *Tetrahedron Lett.*, 1977, 4475.
119. J. M. Desmarchelier and T. R. Fukuto, *J. Org. Chem.*, 1972, **37**, 4218.
120. B. Castro, Y. Chapleur, and B. Gross, *Tetrahedron Lett.*, 1974, 2313.
121. B. Castro and C. Selve, *Bull. Soc. Chim. France*, 1971, 4368.
122. B. Castro, M. Ly, and C. Selve, *Tetrahedron Lett.*, 1973, 4455.
123. Y. Chapleur, B. Castro, and B. Gross, *Synth. Comm.*, 1977, **7**, 143.

124. K. C. Pande and G. Trampe, *J. Org. Chem.* 1970, **35**, 1169.
125. J. P. Schroeder, L. B. Tew, and V. M. Peters, *J. Org. Chem.*, 1970, **35**, 3181.
126. A. Y. Garner, E. C. Chapin, and P. M. Scanlon, *J. Org. Chem.*, 1959, **24**, 532.
127. S. Dershowitz and S. Proskauer, *J. Org. Chem.*, 1961, **26**, 3595.
128. I. J. Borowitz, D. Weiss, and R. K. Crouch, *J. Org. Chem.*, 1971, **36**, 2377.
129. C. J. Devlin and B. J. Walker, *J. Chem. Soc., Perkin I*, 1972, 1249.
130. P. Johncock, *Synthesis*, 1977, 551.
131. A. E. Platt and B. Tittle, *J. Chem. Soc. (C)*, 1967, 1150.
132. V. W. Gash, *J. Org. Chem.*, 1967, **32**, 2007.
133. J. M. McIntosh and H. B. Goodbrand, *Synthesis*, 1974, 862.
134. H. Paulsen and J. Thiem, *Chem. Ber.*, 1973, **106**, 115, 132.
135. H. Freyschlag, H. Grassner, A. Nürrenbach, H. Pommer, W. Reif, and W. Sarnecki, *Angew. Chem. Internat. Ed.*, 1965, **4**, 287.
136. H. G. Henning, *Z. Chem.*, 1966, **6**, 463.
137. L. Horner and H. Röder, *Chem. Ber.*, 1970, **103**, 2984.
138. H. Kunzek and K. Rühlmann, *J. Organometal. Chem.*, 1972, **42**, 391.
139. S. D. Darling and R. L. Kidwell, *J. Org. Chem.*, 1968, **33**, 3974.
140. P. A. Chopard, *J. Org. Chem.*, 1966, **31**, 107.
141. H. J. Bestmann and K. H. Schnabel, *Annalen*, 1966, **698**, 106.
142. H. Christol, H.-J. Cristau, and J. P. Joubert, *Bull. Soc. Chim. France*, 1974, 1421.
143. H. Christol, H.-J. Cristau, J. P. Joubert, and M. Soleiman, *Compt. Rend. (C)*, 1974, **279**, 167.
144. H. Christol, H.-J. Cristau, and M. Soleiman, *Synthesis*, 1975, 736.
145. P. Tavs, *Chem. Ber.*, 1970, **103**, 2428.
146. R. Obrycki and C. E. Griffin, *J. Org. Chem.*, 1968, **33**, 632.
147. L. Achremowicz, *Synthesis*, 1975, 653.
148. H. Egg, W. Richter, and H. Bretschneider, *Monatsh. Chem.*, 1969, **100**, 518.
149. D. L. Fields and J. B. Miller, *J. Heterocyclic Chem.*, 1970, **7**, 91.
150. N. Yamazaki, M. Yamaguchi, F. Higashi, and H. Katinoki, *Synthesis*, 1979, 355.

11
Pentacoordinate Phosphoranes in Synthesis

K. BURGER

1. Introduction, 467
2. 2,2-Dihydro-1,3,2-dioxaphospholens, 469
3. 2,2-Dihydro-1,3,2-dioxaphospholans, 481
4. 2,2-Dihydro-1,4,2-dioxaphospholans, 488
5. 2,2-Dihydro-1,2-oxaphospholens-4, 488
6. 2,2-Dihydro-1,2-oxaphospholans, 490
7. 3,3-Dihydro-1,3-oxaphospholan-5-ones, 491
8. Trioxaphosphetans [Phosphorus(III)–ozone adducts], 491
9. 2,2-Dihydro-1,3,2-oxathiaphospholans, 492
10. 2,2-Dihydro-1,4,2-oxazaphospholens-4 (Δ^4-1,4,2λ^5-oxazaphospholines), 492
11. 5,5-Dihydro-1,2,5-oxazaphospholens-2 (Δ^2-1,2,5λ^5-oxazaphospholines), 494
12. 2,2-Dihydro-1,3,2-oxazaphospholans (1,3,2λ^5-oxazaphospholidines), 495
13. 5,5-Dihydro-1,2,5-oxazaphospholans (1,2,5λ^5-oxazaphospholidines), 496
14. 2,2-Dihydro-1,3,4,2-oxadiazaphospholens-4 (Δ^4-1,3,4,2λ^5-oxadiazaphospholines), 496
15. 1,1-Dihydrophospholens-3, 497
16. 1,1-Dihydrophospholans, 498
17. 1,1-Dihydrophosphirans, 498
18. Bis-(2,2'-biphenylene)phosphoranes, 499
19. Acyclic phosphoranes, 499
20. Miscellaneous reactions, 501
21. References, 503

1. Introduction

Pentacoordinate phosphoranes are compounds possessing a phosphorus atom to which five ligands are covalently bonded; they are derivatives of the unknown PH_5. In this chapter the synthetic potential only of stable, i.e. isolated, phosphoranes will be reviewed.[1-5] The chemistry of halogenophosphoranes has been discussed several times in great detail[3,6] and therefore will not be included.

Mathematical treatment and experimental measurements indicate that the preferred geometry of the molecular skeleton formed by the five valencies of a phosphorus(v) species is that of a trigonal bipyramid (TBP), slightly favoured energetically over a tetragonal (TB) arrangement,[7,8] having two apical (a) and three equatorial (e) skeletal positions (Scheme 1). However,

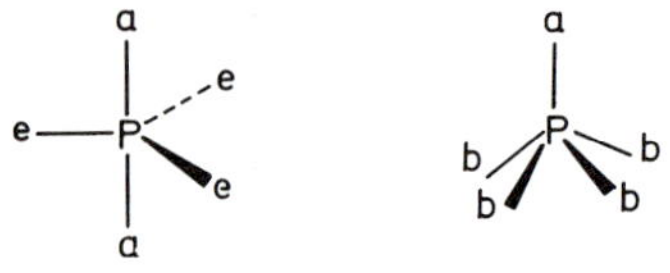

Scheme 1

X-ray analyses have revealed that in certain cases, especially in the cyclospiro phosphorane series, the geometry can be much closer to a square pyramid.[9]

The stability of phosphoranes increases with the number of ligands of relatively high electronegativity, and with decrease of steric bulk in the immediate vicinity of the phosphorus atom. In general large groups will prefer equatorial skeletal positions. Cyclic and spirocyclic phosphoranes exhibit particular stability because the planar or nearly planar ring systems reduce steric crowding around the central atom significantly. The equatorial-apical placement of small ring systems (ring size 4 or 5) and the diequatorial placement of 6- and 7-membered rings represents the energetically favourable situation.[1e,10-12,14,15,19]

Recent studies suggest[12-16] that there is significant participation of the 3d-orbitals in the bonding of the phosphoranes, and that the equatorial positions produce more phosphorus-3d-orbital interaction than the apical positions. This plausibly explains why apical bonds are longer than comparable equatorial bonds. A heteroatom placed apical should consequently be more basic than when in an equatorial position. Furthermore it has been established that in 5-membered cyclic unsaturated oxyphosphoranes the endocyclic apical P–O bond is longer than the exocyclic apical P–O bond. Electrophilic attack on a cyclic oxyphosphorane should therefore occur preferentially at the endocyclic apical position, and in the case of an unsaturated ring system also at the quasi-equatorial carbon atom to produce a maximally stabilized

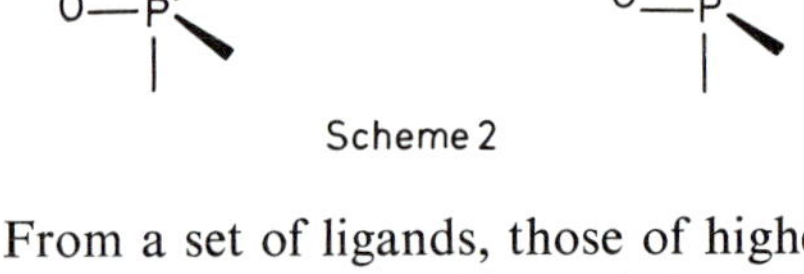

Scheme 2

ionic intermediate. From a set of ligands, those of higher apicophilicity[12,17] will preferentially occupy the apical positions. Apicophilicity often parallels electronegativity, but not always (polarity rule;[18] modified polarity rule[12]).

When phosphorus(v) compounds are transformed into phosphorus(IV) compounds, bond breaking should always occur apically (ionization mechanism).[19] Transformation of a 5-coordinate phosphorus compound into another proceeds via a 6-coordinate transition state or intermediate (substitution mechanism).[19-22] This type of reaction has been shown to be susceptible to base catalysis.

Phosphoranes are able to undergo regular permutational isomerizations (PI), i.e. rapid ligand reorganizations among the equatorial and apical skeletal positions of the TBP without bond rupture[12,14,15,18,23-25] to place the best leaving groups readily in a convenient position for departure.[26]

The majority of phosphoranes known up to now are represented by mono- and spiro-cyclic compounds containing a 1,3,2-dioxaphospholen or a 1,3,2-dioxaphospholan ring system. For nomenclature of phosphoranes see *Chemical Abstracts*.

2. 2,2-Dihydro-1,3,2-dioxaphospholens

2.1. Reaction with Oxygen and Ozone

2,2-Dihydro-1,3,2-dioxaphospholens (1), readily obtained from α-diketones and phosphites, are destroyed by molecular oxygen,[27,28] but these oxidation processes are relatively slow, so storage, preferentially under nitrogen atmosphere, is possible without problems. Oxidation products are phosphate esters, α-diketones, and/or anhydrides.

The reaction of (1) with ozone occurs very readily, apparently by two different mechanisms, producing α-diketones and phosphates, diacyl or diaroyl peroxides and phosphates.[28-30]

2.2. Hydrogenation

Hydrogenation of 2,2-dihydro-1,3,2-dioxaphospholens (1) leads directly to monoketones and phosphates (Scheme 3). 2,2-Dihydro-1,3,2-dioxaphospholans can be recovered unchanged after prolonged treatment under

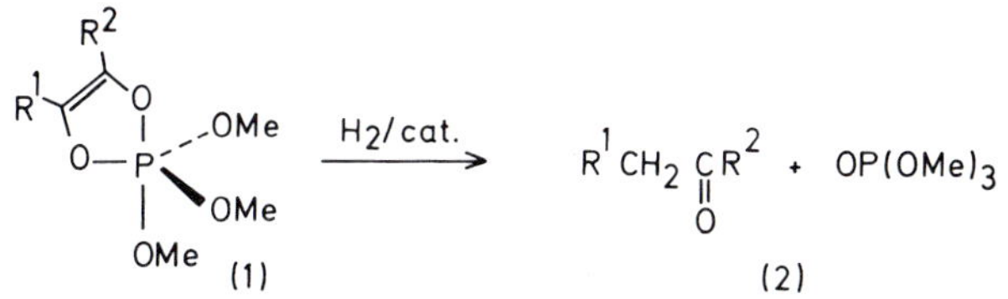

$$R^1\text{-}C(\text{=}O)\text{-}C(R^2)(\text{-}O\text{-})\text{-}O\text{-}P(OMe)_3 \xrightarrow{H_2/cat.} R^1CH_2\underset{O}{\overset{\|}{C}}R^2 + OP(OMe)_3$$

(1)   (2)

Scheme 3

catalytic hydrogenation conditions, therefore they cannot be intermediates. This reaction offers an alternative to standard procedures for converting acyloin condensation products into monoketones, and may be used for ketone transposition.[31]

2.3. Hydrolysis

The vast majority of oxaphosphoranes are more or less sensitive to moisture with only a few exceptions.[32] Hydrolysis may involve different mechanistic pathways depending on structure, substitution pattern, and experimental conditions.

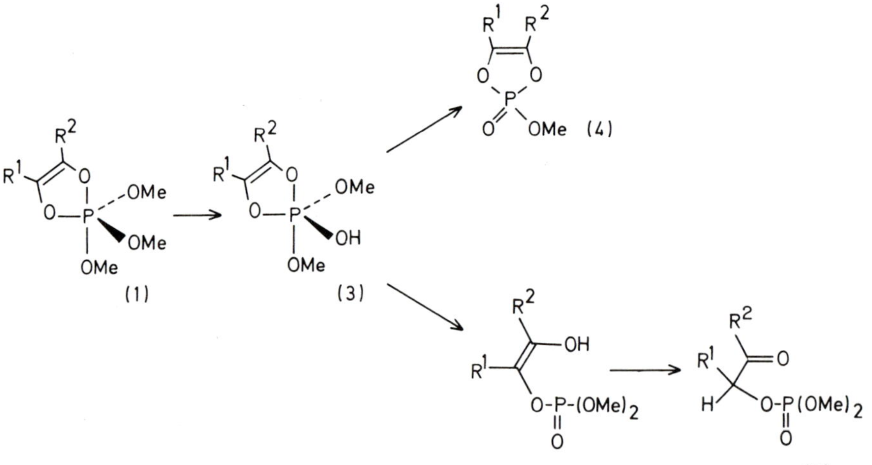

Scheme 4

At low temperature hydrolysis of 2,2-dihydro-1,3,2-dioxaphospholens[33–35] follows the substitution mechanism with preservation of the ring system, when carried out in the presence of 1 equivalent of water. The cyclic hydroxyphosphoranes (3) first formed are transformed into cyclic enediol phosphates (4) by apical loss of an alkoxy group.[33,34] An α-hydroxy-ketone phosphate (5) is the result of rupture of the endocyclic apical P–O bond. At higher temperatures preferential degradation to trialkyl phosphate and enediols occurs. The existence of an equilibrium hydroxyphosphorane (6) $\rightleftharpoons$ phosphate ester (7) in solution has been demonstrated by Ramirez *et al.*[36]

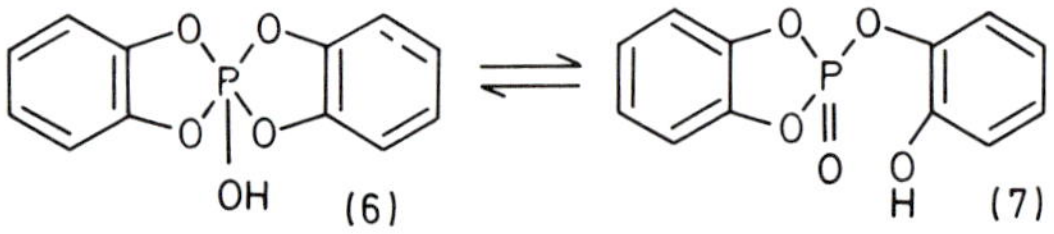

Scheme 5

11 Phosphoranes in Synthesis

Furthermore is has been claimed that cyclic hydroxyphosphoranes have been isolated as stable compounds on hydrolysis of spirocyclic phosphoranes.[37] A hydrolytic pathway similar to the one given above was observed in the 2,2,2-tris(dimethylamino)-2,2-dihydro-1,3,2-dioxaphospholen series.[38]

2.4. Alcoholysis

Alcoholysis of 2,2-dihydro-1,3,2-dioxaphospholens (1) occurs by a mechanism similar to that of hydrolysis.[20-22] The reaction, usually base-catalysed, proceeds via a phosphorus(VI) transition state or intermediate. Attack occurs opposite to an equatorial position; precise and general rules are not yet available with respect to the leaving group placement.[21] Particularly when the reaction is carried out in solvents of low polarity, these ring systems exhibit remarkable stability (Scheme 6).

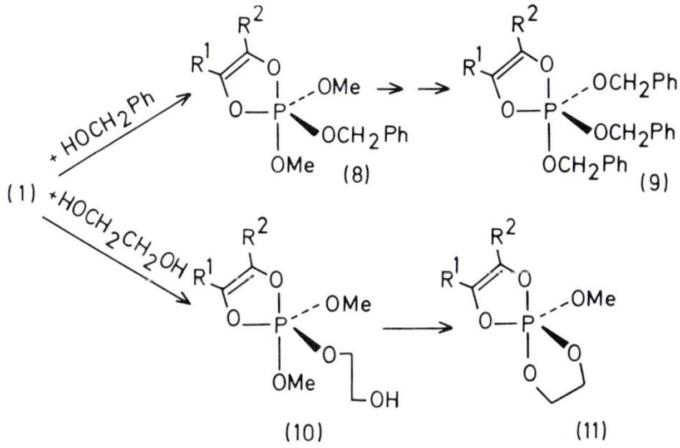

Scheme 6

Successive exchange of three methoxy groups by three benzyloxy groups occurs when (1) is heated with benzyl alcohol at 100°C.[20] Numerous exchange reactions of this type are described[21] including substitution of the $-N(CH_3)_2$ function by methanol.[39]

Two-step alcoholyses and aminolyses with difunctional compounds such as 1,2- and 1,3-diols, 1,2- and 1,3-amino-alcohols, and 1,2- and 1,3-diamines to produce spirocyclic compounds are favourable processes, leading in general to considerable stabilization of the pentacoordinated state.[22,40-42]

Forced conditions may cause a breakdown of the ring system by rupture of the endocyclic apical or equatorial P–O bond, depending on the structure of the phosphorane and the experimental conditions.

2.5. Reactions with Hydrogen Halides

Phosphate esters of α-hydroxy-aldehydes or α-hydroxy-ketones (12) are obtained on reaction of 2,2-dihydro-1,3,2-dioxaphospholens (1) with HCl.[43–45] As minor by-products α-halogeno-aldehydes and α-halogeno-ketones (13) are isolated, respectively. The latter are the only products formed when the 2,2,2-triphenoxy derivatives (14) are the substrates.

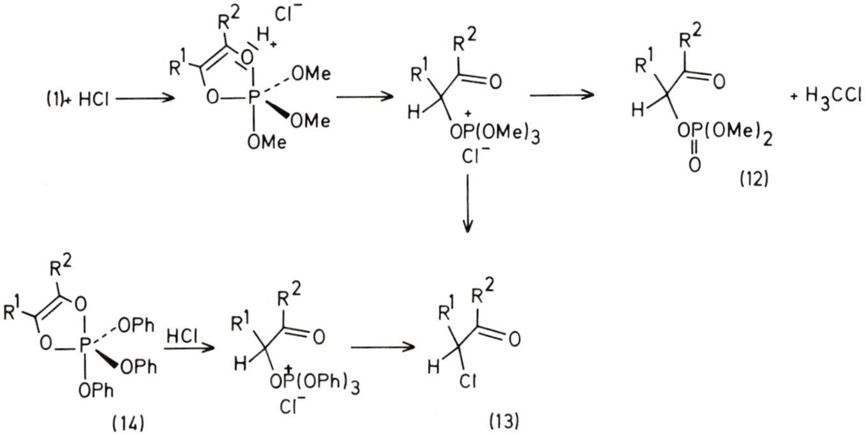

Scheme 7

From unsymmetrically substituted 2,2-dihydro-1,3,2-dioxaphospholens (R^1 = H, R^2 = alkyl, aryl), phosphate esters of α-hydroxy-ketones are produced predominantly. If the ring carries both alkyl and aryl groups, both possible products are formed, with the aryl ketone (R^1 = alkyl, R^2 = aryl) slightly favoured.

2.6. Reactions with Acid Chlorides and Anhydrides

On reaction of unsubstituted 2,2-dihydro-1,3,2-dioxaphospholens (1; R^1 = R^2 = H) (Scheme 8) with acyl chlorides or anhydrides, O-acylation occurs to

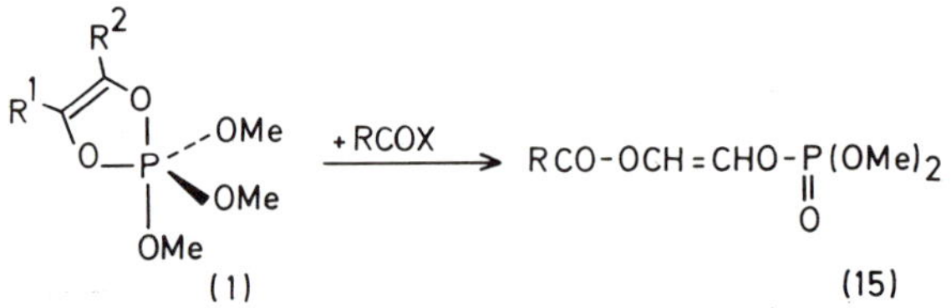

Scheme 8

11 Phosphoranes in Synthesis

produce acylated enediol phosphates (15).[45-47] This contrasts with the *C*-acylation observed with 4,5-disubstituted 2,2-dihydro-1,3,2-dioxaphospholens (1) (Scheme 9) yielding phosphate esters of α-hydroxy-β-diketones of type (16).[48]

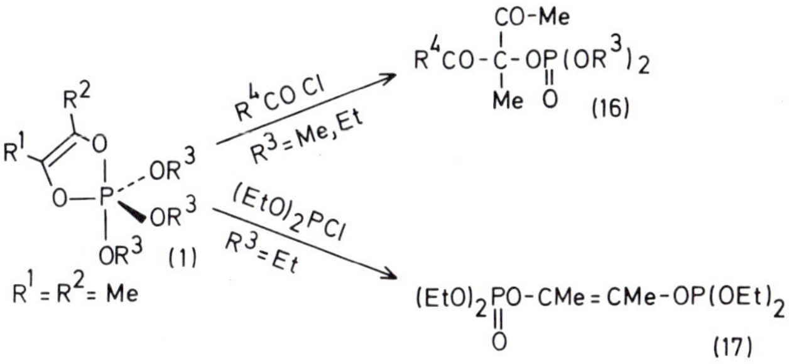

Scheme 9

However, with the biacetyl adduct (1; $R^1 = R^2 = Me$, $R^3 = Et$) (Scheme 9), it has been shown that the attack on carbon or oxygen strongly depends on the electrophile used. From R^4COCl and (1; R^3 = Me or Et) the product (16) corresponds to a *C*-acylation, while from (1; R^3 = Et) and diethyl phosphorochloridite the product (17) indicates attack at the oxygen atom.[49]

Furthermore, the ratio of *C*-acylation to exocyclic *O*-acylation when (1) is treated with acetyl chloride has been shown to depend on the reaction medium[50] (Scheme 10). In the absence of a solvent a 97:3 (*C*:*O*) ratio was obtained, whereas in dichloromethane the ratio was 20:80. Acetyl bromide gives only little *C*-acylation under any conditions and almost exclusive *O*-acylation in acetonitrile.

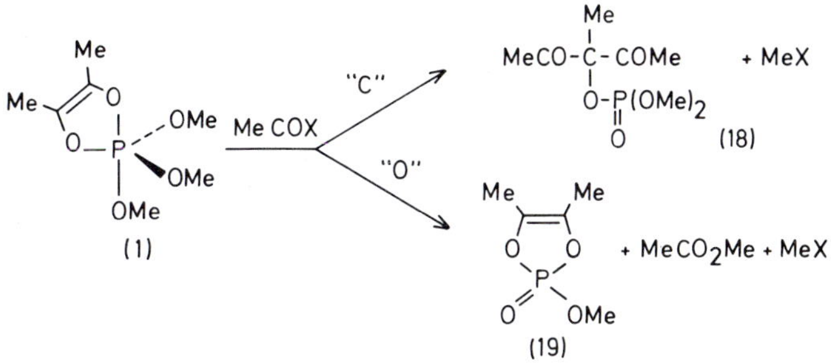

Scheme 10

α-Chloro-β-keto-sulphides (20) have been obtained[51] in an exothermic reaction from (1) and sulphenyl chlorides. An ionic mechanism has been suggested, closely related to the above *C*-acylation.

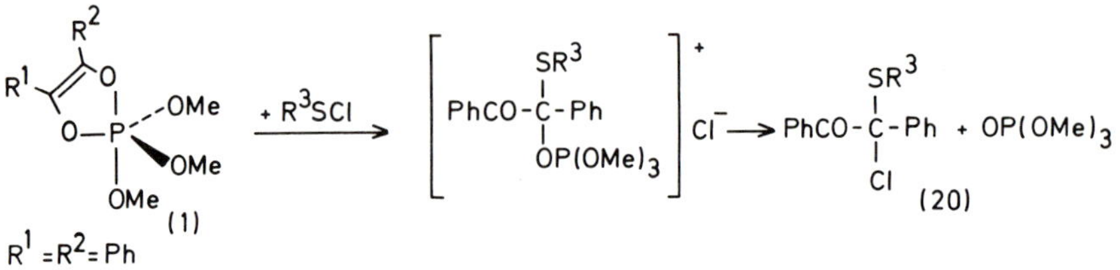

Scheme 11

C-Acylation seems to be the initial step of the reaction of (1) with phosgene, producing the β-keto-α-hydroxy-acid chloride (21),[48] which can be transformed readily into two stereoisomeric 5-membered cyclic acylphosphates (22a/b) on heating.[52] This class of compound is an extraordinarily powerful and selective phosphorylating agent towards primary, secondary, and tertiary alcohols as well as phenols.[53]

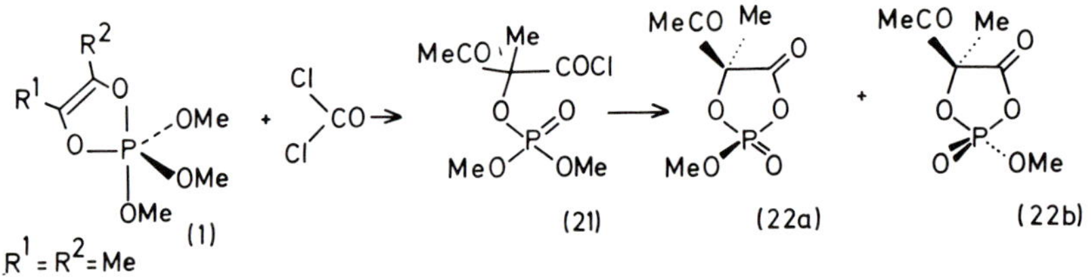

Scheme 12

2.7. Bromination

2,2-Dihydro-1,3,2-dioxaphospholens of type (1) possessing at least one exocyclic alkoxy group react with bromine to give derivatives of α-halogeno-α-hydroxy-ketones (23)[43,54] (Scheme 13). An attack of the electrophile at the

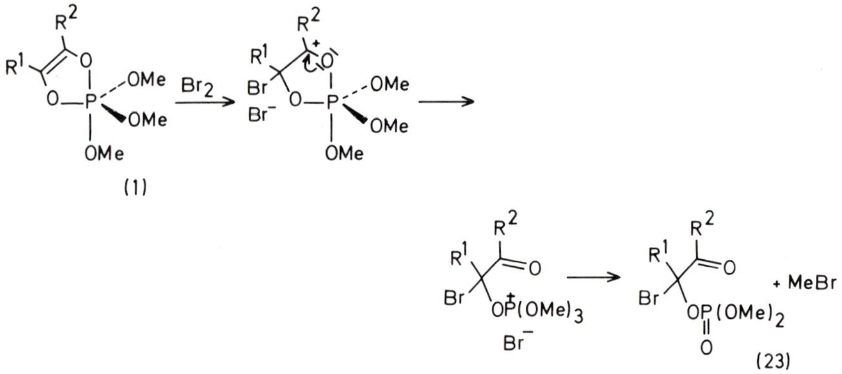

Scheme 13

11 Phosphoranes in Synthesis

quasi-equatorial carbon atom generating a carbonium ion adjacent to the apical, more basic oxygen, with a subsequent P–O bond rupture and stabilization of the resulting phosphonium bromide by an Arbusov-type reaction,[55] seems to be a plausible mechanistic interpretation of the observed experimental facts.

If $R^1 \neq R^2$, the more stable carbonium ion will dictate the course of the reaction. If the alkoxy groups of (1) are replaced by phenoxy groups (Scheme 14), α,α-dibromo-ketones (24) will be the result of the reaction. Bromination

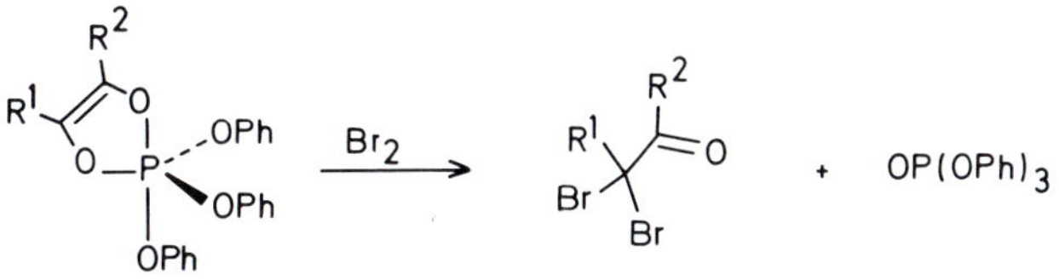

Scheme 14

of the *o*-quinonedioxaphospholen gives the α,α-dibromo-ketone, regardless of the nature of the exocyclic ligands. A similar reaction has been observed when (1) reacted with *N*-bromophthalimide.[56]

2.8. Reactions with Activated Carbon–Carbon Double Bonds

Cyclopropanes (26)[57] are the products of the reaction of (1) with activated double-bond systems such as arylidenemalonodinitriles. The preponderance of stereoisomer (26) is held to be due to the greater stability of the intermediate (25) compared with that of its diastereoisomer.

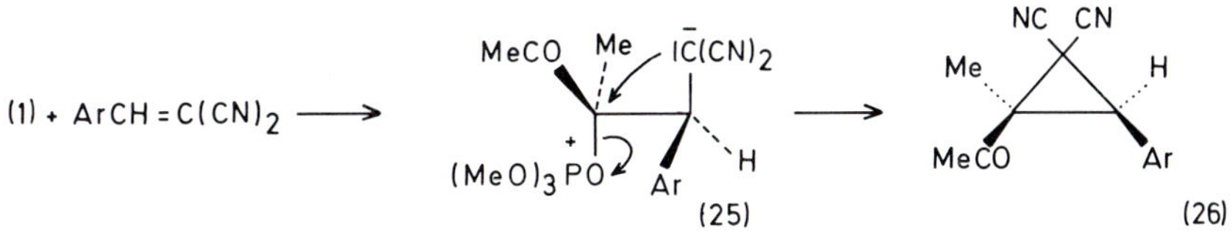

Scheme 15

2.9. Reactions with Carbonyl Compounds

The reaction of tervalent phosphorus reagents with carbonyl compounds assumed much greater synthetic value when Ramirez *et al.*[58–71] discovered that the adducts are capable of reacting further with a variety of carbonyl compounds producing new carbon–carbon bonds under very mild and neutral conditions. For example, the reaction of phosphites with diketones in a 1:2 molar ratio provides a selective pinacolic reduction, the products obtained on hydrolysis being diketols.[58,59,63b] Methyl pyruvate adds to (1) to yield 4-acetyl-4,5-dimethyl-5-methoxycarbonyl-2,2,2-trimethoxy-2,2-dihydro-1,3,2-dioxaphospholan. From the mixture of the two diastereoisomers formed, the *trans* isomer was found to be converted readily into methyl DL-*threo*-α,β-dimethyl-α,β-dihydroxylevulinate via a cyclic phosphate ester. Similarly aldehydes and ketones react with (1) to produce 2,2-dihydro-1,3,2-dioxaphospholans which undergo cleavage at pH 5 in boiling water to give α,β-dihydroxy-ketones.[61,62] With α-keto-esters and certain α-diketones the second step was found to be faster than the first, and in such cases only the 2:1 adducts can be isolated.[72,73] It is assumed that the reaction involves a phosphorus(VI) intermediate (27)[74,75] (Scheme 16). In a subsequent step, rupture of an endocyclic P–O bond with formation of a dipolar species (28) occurs, which still retains a 5-coordinated phosphorus atom, followed by ring-closure to 2,2-dihydro-1,3,2-dioxaphospholans of type (29).

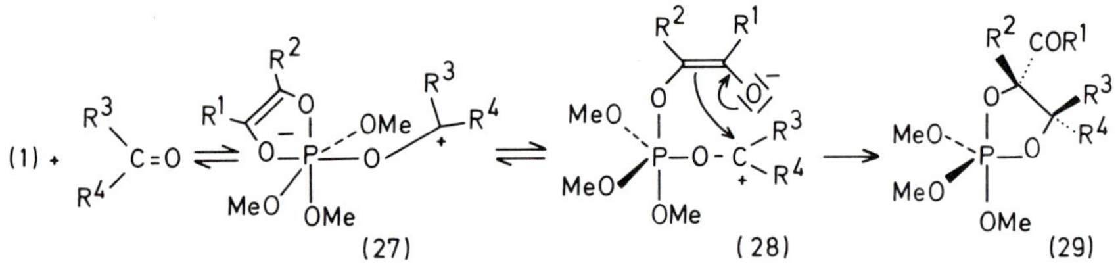

Scheme 16

The suggested O→P coordination mechanism (1) ⇌ (27) parallels the mechanism of the nucleophilic substitution at phosphorus(V)[20–22] and is also consistent with the attack of the 2,2-dihydro-1,3,2-dioxaphospholens occurring at the carbonyl oxygen rather than the carbon–carbon double bond of α,β-unsaturated aldehydes; furthermore the mechanism provides a likely explanation for the higher reactivity of carbonyl than thiocarbonyl compounds. The versatility of this reaction is illustrated by Scheme 17.

2.10. Reactions with Heterocumulenes

Ketene adds to (1) to give stable 2,2-dihydro-1,3,2-dioxaphospholans (30) possessing an exocyclic carbon–carbon double bond[66] (Scheme 18); similarly the reaction of (1) with isocyanates produces 4-imino-2,2-dihydro-1,3,2-dioxaphospholans (31).[66,76–79] The stability of the latter varies considerably, depending on the substitution pattern. For $R^1 = R^2 \neq H$ and R^3 = aryl, stable compounds have been isolated. If compounds (31) are allowed to react further with a second equivalent of an isocyanate, hydantoins (32) are produced.[66,76] Arylsulphonyl isocyanates react entirely in the same way ($R^3 = R^4 = SO_2Ar$).[77]

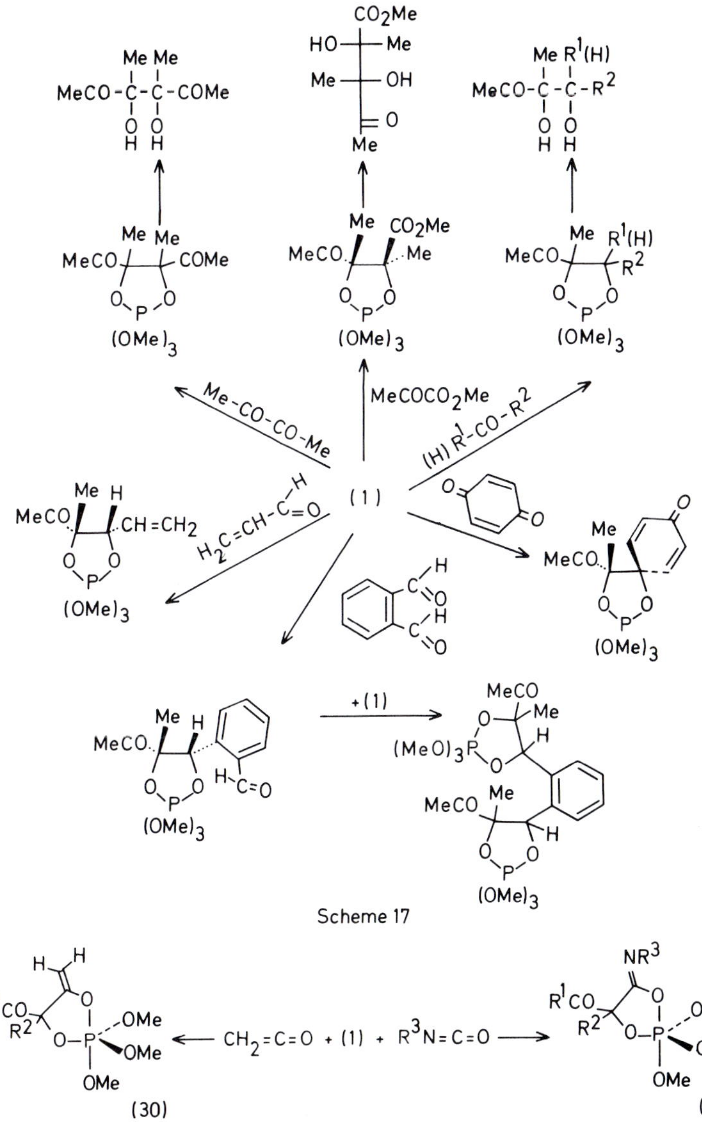

11 Phosphoranes in Synthesis

Scheme 17

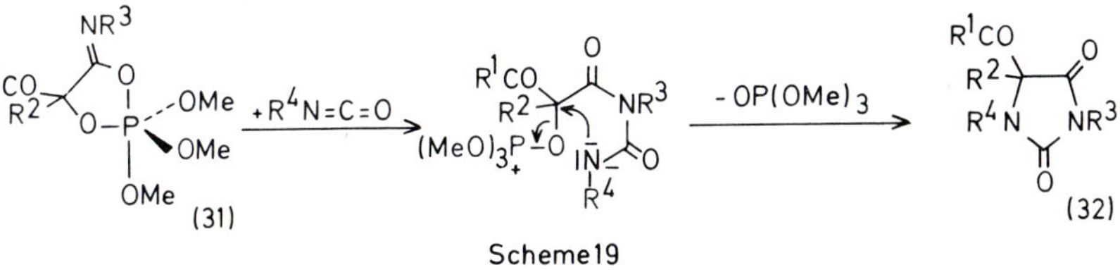

Scheme 18

Scheme 19

With at least one hydrogen present at ring positions 4 and 5, three different types of reaction have been observed (Scheme 20). First, C–H insertion yields carbamoyl-2,2-dihydro-1,3,2-dioxaphospholens (33)[78,80] (i.e. from phenoxycarbonyl isocyanate); (33) on treatment with hydrochloride can readily be transformed into phosphate esters of β-keto-α-hydroxy-amides (34).[78] Second, 2-oxazolin-4-one formation occurs with aroyl isocyanates; compounds (35) are shown to exist in the form of the tautomeric 4-hydroxy-oxazoles (36).

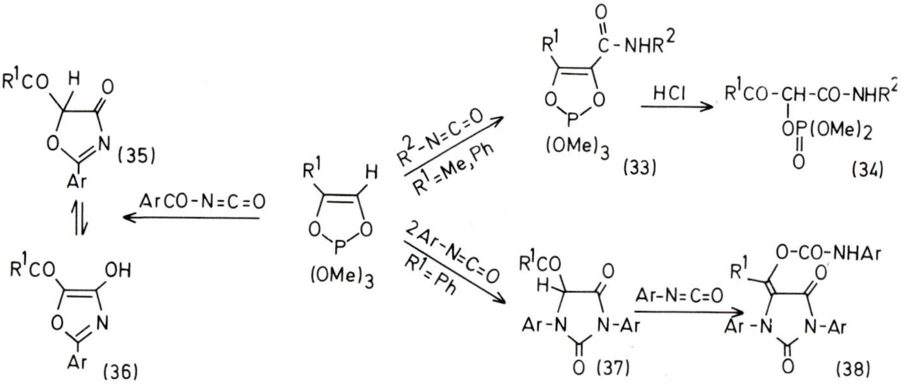

Scheme 20

Third, reaction with aryl isocyanates produces hydantoins (37), which may react further with a third equivalent of an isocyanate to give (38). Acyl, aroyl,[76] alkoxycarbonyl, carbamoyl, and thioaroyl isocyanates[81] react with 2,2-dihydro-1,3,2-dioxaphospholens of type (1) lacking hydrogen atoms at the ring (i.e. $R^1 = R^2 \neq H$), to give 2-oxazolin-4-ones (39) and 2-thiazolin-4-ones (40), respectively; phenoxycarbonyl isocyanate however gives a mixture of 2-oxazolin-4-ones (39) and hydantoins (37). 2-Thiazoline-4,5-diones likewise react with 2-thiazolin-4-one formation (40).[74,81]

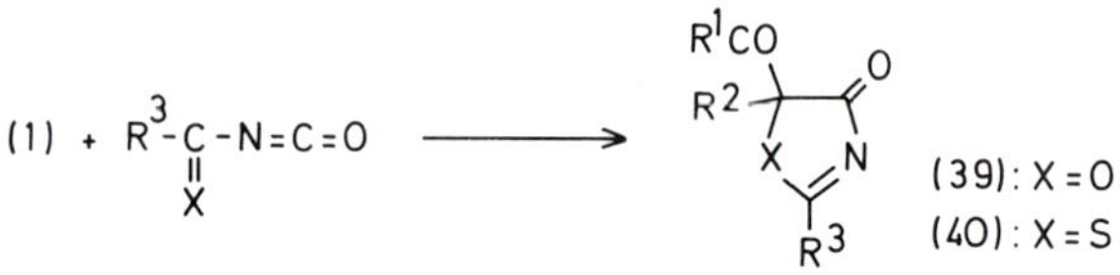

Scheme 21

2,2-Dihydro-1,3,2-dioxaphospholens (1) do not react with aryl isothiocyanates, but acyl, aroyl, and carbamoyl isothiocyanates,[74] being more reactive, give the corresponding 2-oxazoline-4-thiones (41). 2-Oxazolin-4-ones (39), 2-thiazolin-4-ones (40), 2-oxazoline-4-thiones (41), as well as the hydantoins (37), readily accessible via phosphoranes, can be transformed on

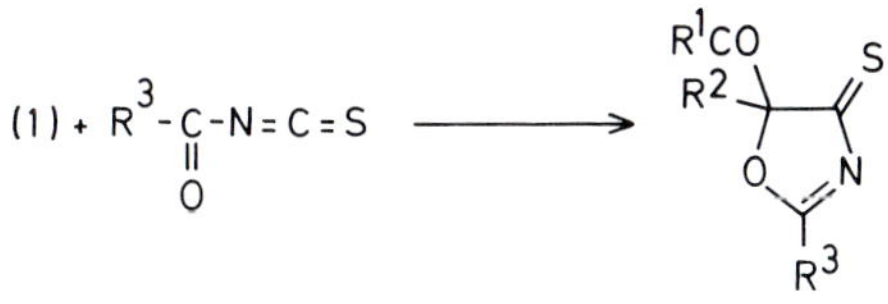

Scheme 22

$$(1) + R^3-\underset{\underset{O}{\|}}{C}-N=C=S \longrightarrow \text{(41)}$$

hydrolysis into α-hydroxy-amides,[77] α-mercapto-amides,[74] α-hydroxy-thio-amides,[74] and α-amino-acids,[82] respectively, possessing an additional carbonyl group in the β-position.

With carbon disulphide (1) produces a 6-membered heterocyclic system (42),[83] while, from the reaction of 2,2,2-trimethoxy-4-phenyl-2,2-dihydro-1,3,2-dioxaphospholen (43) and carbon disulphide under the same experimental conditions, a 1,2,4-trithione (44) was isolated.[79] Compound (42) can be desulphurized in the presence of triethyl phosphite to give a 5-membered and finally a 4-membered ring system.[83]

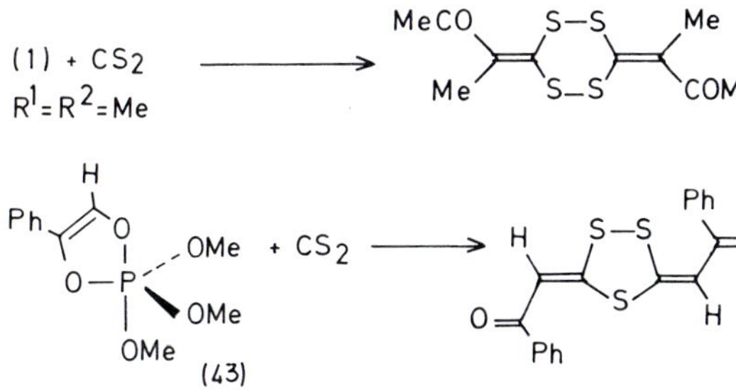

Scheme 23

$$(1) + CS_2,\ R^1 = R^2 = Me \longrightarrow (42)$$

$$(43) + CS_2 \longrightarrow (44)$$

Carbon suboxide, C_3O_2, condenses with (1), in spite of its remarkable tendency to undergo self-polymerization, yielding γ-acyl-β-alkoxy-$\Delta^{\alpha,\beta}$-butenolide phosphonate (45).[84]

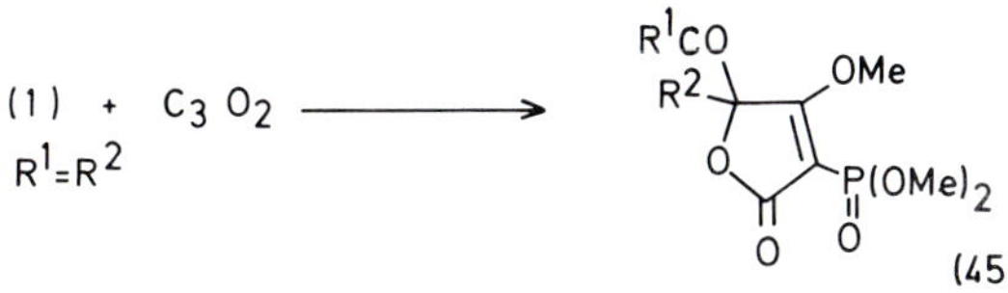

Scheme 24

$$(1) + C_3O_2,\ R^1 = R^2 \longrightarrow (45)$$

2.11. Fragmentation Reactions

Most of the 2,2-dihydro-1,3,2-dioxaphospholens (1) have been shown to be thermally stable under nitrogen up to 120°C.[85] At higher temperatures or on irradiation, two types of fragmentation process can be observed: a [4 + 1] retrograde reaction to give the starting materials,[86,87] then attack of the tervalent phosphorus compound at the carbon atom of one of the carbonyl groups to give a dipolar intermediate which may undergo an Arbusov-type reaction;[88] and/or a [3 + 2] retro process generating a carbene-like species and corresponding O=P< compounds. Quite often both processes compete with each other.

The metal salt catalysed thermolysis of 2,2-dihydro-1,3,2-dioxaphospholens (1; $R^1 = R^2 = Ph$), as shown by Mukaiyama,[89] proceeds via carbenoid species (46), which can be trapped by alcohols, phenyl isocyanate, and dicyclohexylcarbodiimide to give (47), (48), and (49). In the absence of a metal catalyst the dimer (50) and diphenylacetylene (53) are produced.[90]

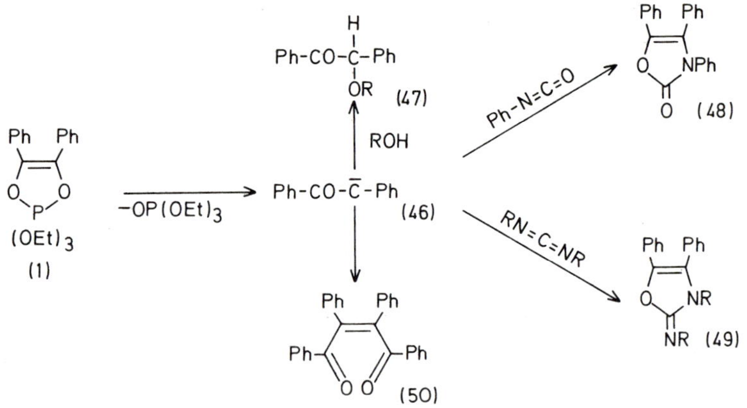
H
Ph-CO-C-Ph
OR
(47)
ROH
Ph Ph
O O
P
(OEt)3
(1)
-OP(OEt)3
Ph-CO-C̄-Ph
(46)
Ph-N=C=O
Ph Ph
O NPh
O (48)
RN=C=NR
Ph Ph
O NR
NR (49)
Ph Ph
Ph- -Ph
O O
(50)
Scheme 25

When the benzil adduct of triethyl phosphite (1; $R^1 = R^2 = Ph$, $R^3 = Et$) is heated to 215°C, in the presence of an excess of the phosphite, only triethyl phosphate and diphenylacetylene (53) are isolated.[90] The intermediate in this reaction is diphenylketene (51) which reacts further on deoxygenation[91] to give the alkyne (53)[92] possibly by rearrangement of the carbene (52) or its precursor.

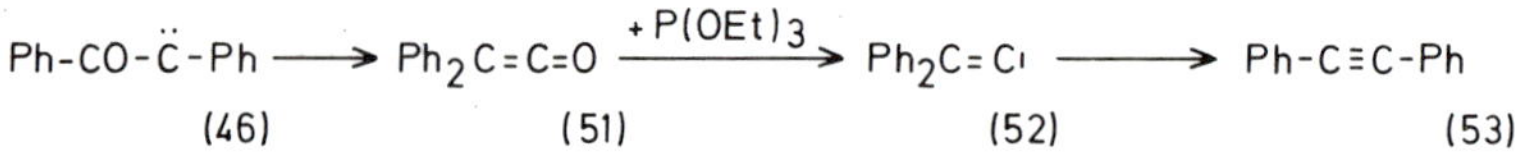
Ph-CO-C̈-Ph
(46)
Ph2C=C=O
(51)
+P(OEt)3
Ph2C=C:
(52)
Ph-C≡C-Ph
(53)
Scheme 26

Bentrude[93] has reported that 2,2-dihydro-1,3,2-dioxaphospholen (1) on photolysis in dry acetone is transformed into 2,2-dihydro-1,3,2-dioxaphospholan (54), the latter being hydrolysed by traces of water. A [2 + 2] cycloaddition process is assumed to occur at the double bond of (1) with formation of an unstable oxetan intermediate which rearranges to give (54).[94]

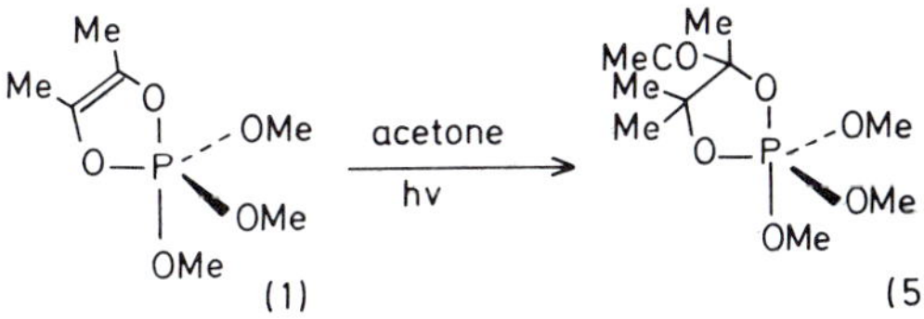

Scheme 27

Neidlein[80,95,96] used the [3 + 2] cycloreversion process which 2,2-dihydro-1,3,2-dioxaphospholens undergo in the presence of certain heterocumulenes for the synthesis of several 5-membered heterocyclic ring systems. 4-Oxazoline-2-thiones (55) were obtained on treating (1) with sulphonyl isothiocyanates,[95] while 2-oxo-Δ^4-1,2,3-oxathiazolines (56) are the products of the reaction of (1) with N-sulphinylsulphonamides.[80,95] The reaction of (57) and N-sulphonyl isocyanate is remarkable; the product has been assigned a β-lactam structure (58).[79]

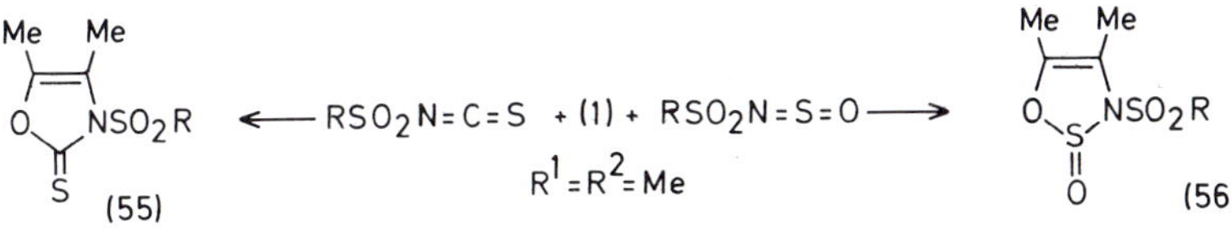

Scheme 28

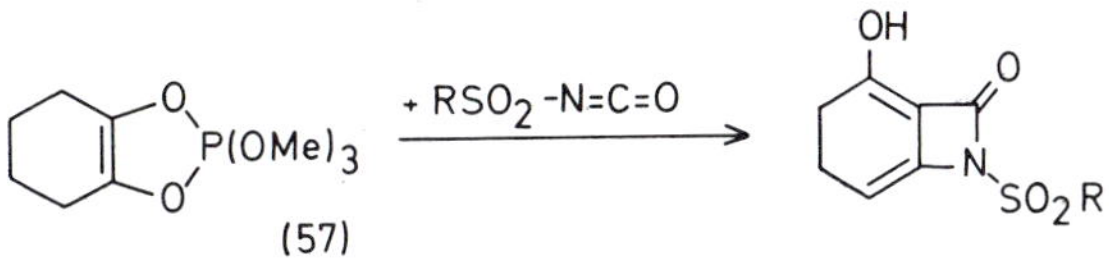

Scheme 29

When (1; $R^1 = R^2 = $ Me) is photolysed in the presence of bromotrichloromethane a 1-trichloromethylated 2-keto-phosphate is formed[97] via a radical route.

3. 2,2-Dihydro-1,3,2-dioxaphospholans

The chemistry of 2,2-dihydro-1,3,2-dioxaphospholans (59) is much less versatile than that of the 2,2-dihydro-1,3,2-dioxaphospholens (1) because the former lack the reactive enediol function. In the majority of the reactions the pentacovalent state of the phosphorus atom is lost.[1g]

3.1. Hydrolysis

Hydrolysis of 2,2-dihydro-1,3,2-dioxaphospholans (59) proceeds by different mechanistic pathways, analogous to those observed for 2,2-dihydro-1,3,2-dioxaphospholens (1), with suitable variations due to structural differences.[98-100] Diol monophosphates (60),[101] 1,2-diols (61),[63b] cyclic phosphates (62), (65),[99,102,103] α,β-dihydroxy-ketones (63),[98] α-hydroxy-β-diketones (64), and 2,3-dihydroxy-1,4-diketones (66)[103] are readily accessible from aldehydes and 1,2-diketones via the corresponding phosphoranes[59,61,62,64,65,68,69,102,103] as shown in Scheme 30.

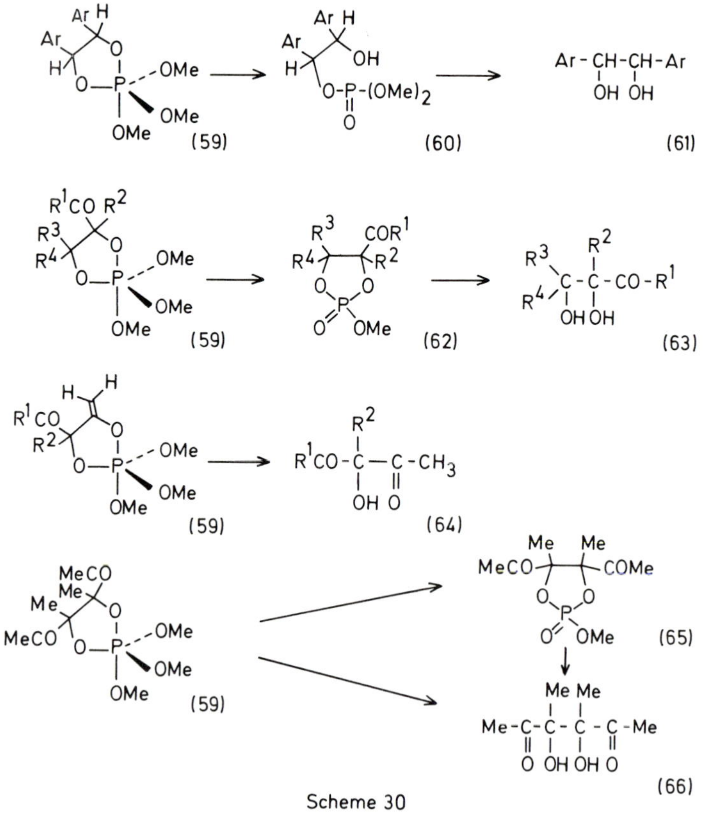

Scheme 30

3.2. Alcoholysis

Mechanistically, alcoholysis parallels hydrolysis and alcoholysis of the 2,2-dihydro-1,3,2-dioxaphospholens (1), i.e. the reaction may occur on acid catalysis by apical bond rupture (ionization mechanism) or in the presence of

11 Phosphoranes in Synthesis

a base (substitution mechanism). It seems to be more difficult to retain the ring system than in the 1,3,2-dioxaphospholen series. Several types of product are accessible as illustrated in Scheme 31.

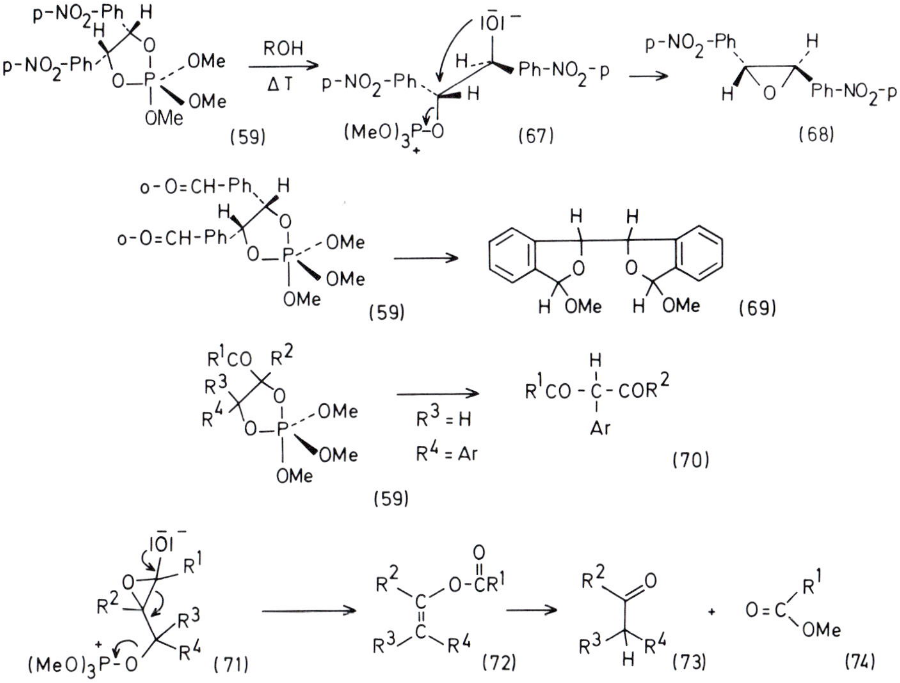

Scheme 31

Thus, if the ring carries one hydrogen atom and an aryl group at C-4 and C-5, epoxide formation is the favoured process (59) → (68). The reaction is stereospecific, and a dipolar species (67) generated by an "ionization step" is presumed to be the intermediate. With an *ortho* aldehyde function present in the aryl groups, cyclic diacetals are formed by reaction of the oxide anion, initially generated by P–O bond rupture, with the *ortho* formyl groups, (59) → (69).

When the 1,3,2-dioxaphospholan ring is equipped with a hydrogen atom and an aryl group at one carbon atom and an acyl group at the other, β-diketones (70) are formed. The overall result of this reaction is a reductive diacylation of an aldehyde by an α-diketone[67] (59) → (70) via a phosphorane.

A reductive monoacylation of an aldehyde is observed if the development of a carbonium ion at one of the carbon atoms of the dioxaphospholan ring is disfavoured by the presence of electron-withdrawing groups.[58,64,65,104] The key step of the reaction is assumed to be the formation of an epoxide or a 3-membered cyclic hemiacetal intermediate (71) which rearranges to give an enol ester (72), the latter being cleaved into (73) and (74) on alcoholysis.

3.3. Bromination

2,2-Dihydro-1,3,2-dioxaphospholans (30) possessing an exocyclic carbon–carbon double bond[66] add bromine readily to give phosphate esters of γ-bromo-α-hydroxy-β-diketones (76).

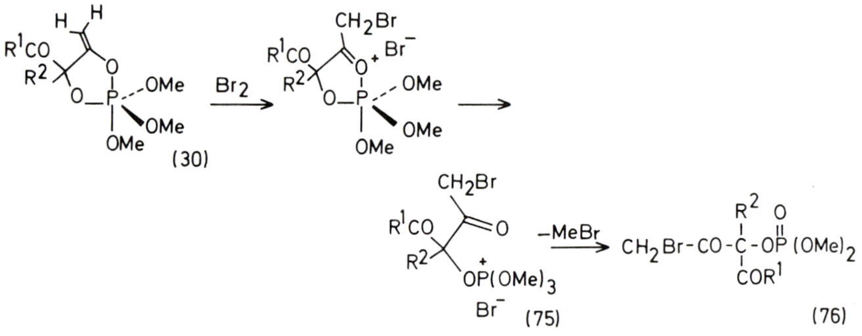

Scheme 32

3.4. Insertion into P–H Bonds

Spirophosphoranes of type (77) containing P–H bonds react with multiple bond systems, such as $>C=O$,[105] $>C=N-$,[105b,106] $>C=C<$,[107] and $-C\equiv C-$,[108] with insertion into the P–H bond to form new P–C bonds (77) → (78).

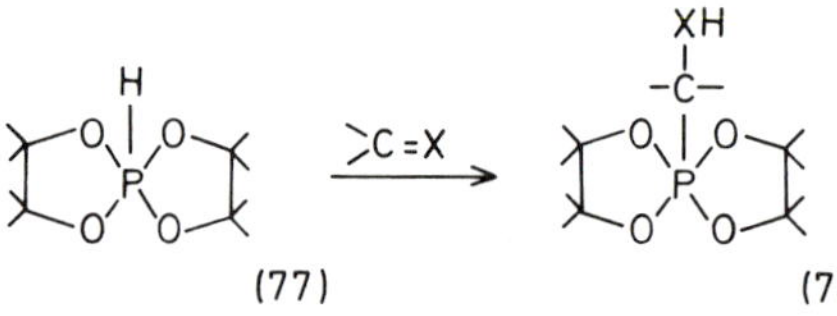

Scheme 33

Metallation of (77) followed by reaction with alkyl halides[109] or benzoyl chloride[110] results likewise in formation of a new P–C bond. The anion (79) involved in this reaction was shown to exist in equilibrium with a phosphorus-(III) species (80) in solution; the latter can be trapped by Me₃SiCl to form (81).

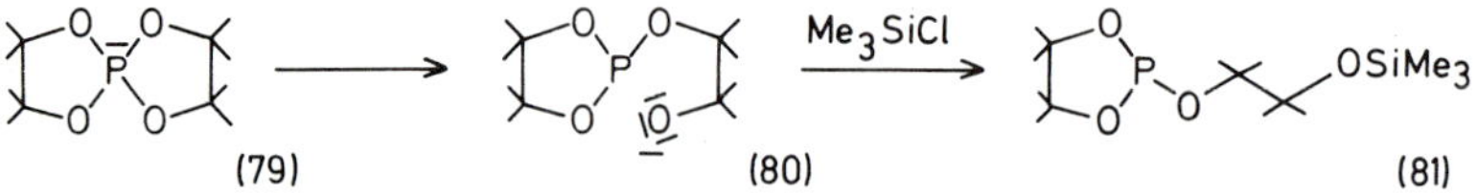

Scheme 34

11 Phosphoranes in Synthesis

The equilibria between tetraoxyphosphoranes[111-113] as well as some N-analogues[114] possessing one P–H bond and their corresponding ring-open phosphites have been investigated intensively.

Phosphoranes of type (77) are potential reducing agents.[115] For instance, in the presence of methanol and a pyrrolidino-enamine, (82) was formed with reduction of the enamine.[116]

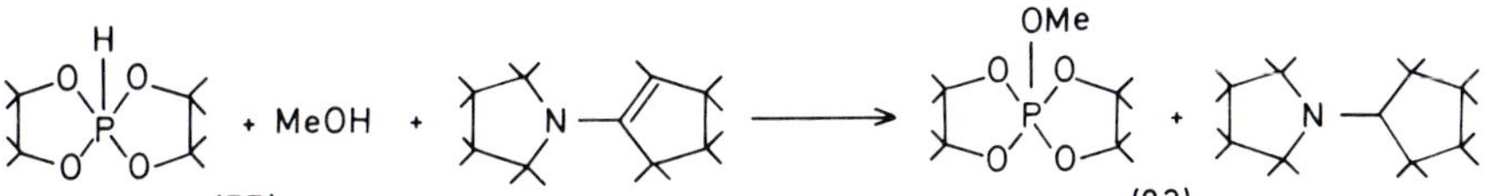

Scheme 35

Bis-(2,2'-biphenylene)hydridophosphorane (83) seems to contain a highly reactive P–H bond. In inert solvents spontaneous radical formation (83) → (84) was observed; when oxygen was excluded dimerization products (85) were isolated. In the presence of strong and bulky nucleophiles proton-abstraction (83) → (86) occurred; in the presence of strong electrophiles hydride-abstraction (83) → (88) took place. The blue-green anion (86) very probably exists in equilibrium with a ring-open species (87), the latter being successfully trapped on protonation.[117]

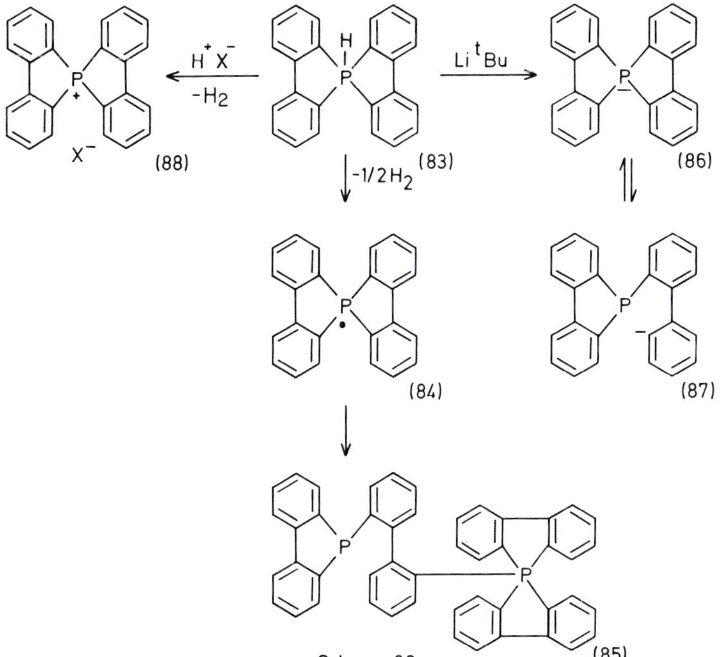

Scheme 36

3.5. Formation of Epoxides

2,2-Dihydro-1,3,2-dioxaphospholans (59) decompose at elevated temperatures to give the corresponding trialkyl phosphates and epoxides.[67,118–122] The reaction seems to be stereospecific, as proved in the case of the decomposition of phosphoranes obtained from a (±)-phosphite(88%)–*meso*-phosphite(12%) mixture and diethyl peroxide giving *cis*-2-butene oxide (85%) and *trans*-2-butene oxide (15%).[120] The formation of a *cis* epoxide from a *trans* phosphorane supports a mechanistic scheme involving a betaine intermediate (89) and back-side attack of the oxygen to displace the phosphate.[122]

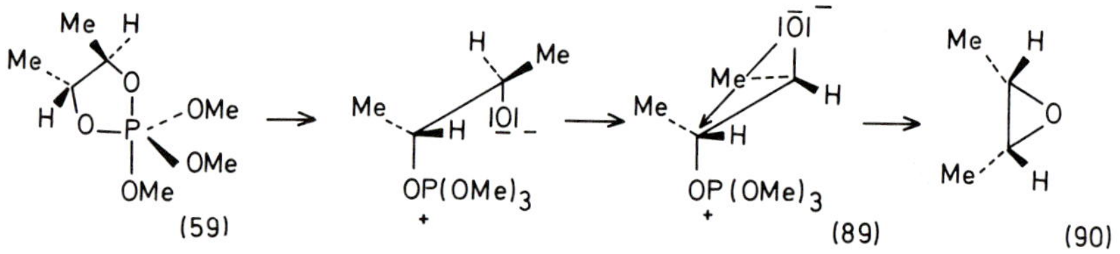

Scheme 37

It has been shown by Denney[41,123] that this type of reaction is of general value in heterocyclic synthesis (see Scheme 72). Epoxides likewise are formed on heating of 2,2-dihydro-1,3,2-dioxaphospholans and -1,4,2-dioxaphospholans respectively, derived from tris(dimethylamino)phosphine[119,124,125] and from the reaction of 1,2-diols and tris(dimethylamino)phosphine in the presence of carbon tetrachloride.[126]

2,2-Dihydro-1,3,2-dioxaphospholans (91) obtained from fluorenones and triethyl phosphite, when heated in acetonitrile[127] give the spiro ketones (92) and oxazolines (93), the former being the only products, besides triethyl phosphate, on heating (91) to 160°C without trapping reagents.[128–130]

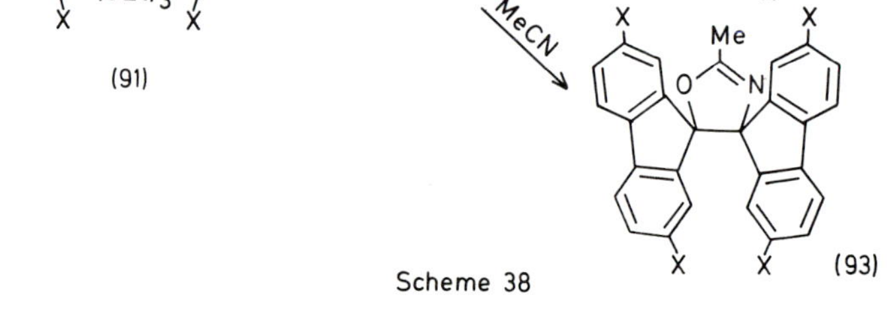

Scheme 38

11 Phosphoranes in Synthesis

3.6. Formation of Carbenes

A variety of 2,2-dihydro-1,3,2-dioxaphospholans of type (59) undergo photocycloelimination of the [5 → 2 + 2 + 1] type,[131] producing carbenes. Irradiation of (59; $R^1 = R^2 =$ Ph, $R^3 =$ Et) in 2-methyl-2-butene affords cis- and trans-1-cyano-1-phenyl-2,2,3-trimethylcyclopropane (94), (95). Photolysis of (91) in trans-2-butene results in the formation of (92), the expected cyclopropane being only the by-product.

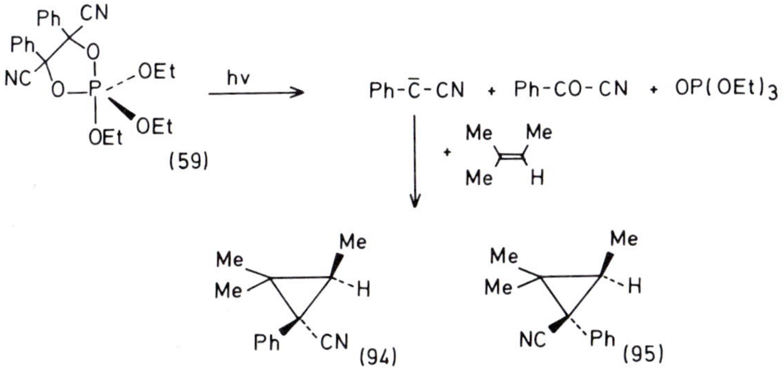

Scheme 39

3.7. The 2,2-Dihydro-1,3,2-dioxaphospholan–2,2-Dihydro-1,2-oxaphosphetan Transformation

Phosphite adducts of hexafluoroacetone exhibit remarkable thermostability and decompose only when heated above 170°C.[132] But certain 2,2-dihydro-1,3,2-dioxaphospholans (59) in which one or more of the exocyclic ligans are α-hydrogen-containing alkyl groups undergo rearrangement on heating to yield a 4-membered ring system retaining a pentacoordinated phosphorus atom. This reaction provides a convenient route to 2,2-dihydro-1,2-oxaphosphetans.[21,133-135] Compounds (96) are of the same type as the intermediates formulated in the Wittig reaction;[135] further heating causes decomposition to yield alkenes and phosphinates.[133]

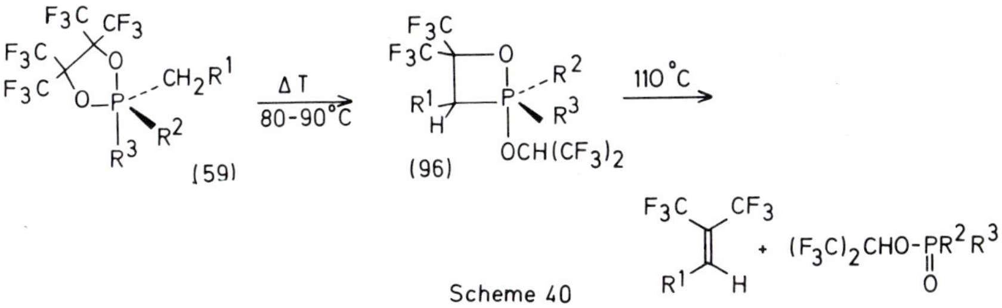

Scheme 40

4. 2,2-Dihydro-1,4,2-dioxaphospholans

4.1. The 2,2-Dihydro-1,4,2-dioxaphospholan–2,2-Dihydro-1,3,2-dioxaphospholan Transformation

2,2-Dihydro-1,4,2-dioxaphospholans (97), being considered the kinetically stable products of the reaction of tervalent phosphorus compounds with hexafluoroacetone and pentafluorobenzaldehyde respectively, rearrange below room temperature to yield 2,2-dihydro-1,3,2-dioxaphospholans of type (98).[134,136]

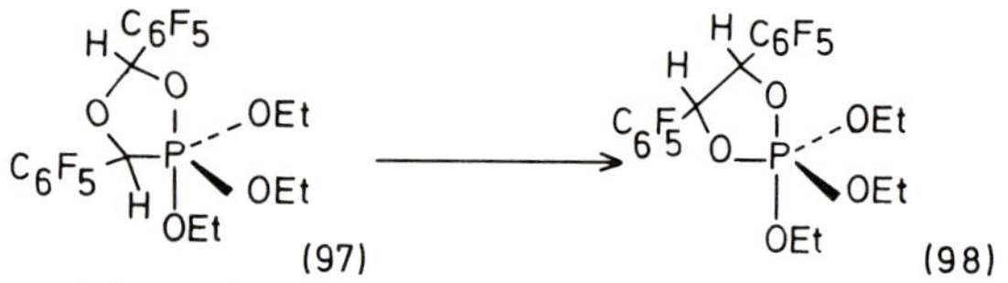

(97) cis/trans isomers (98) cis/trans isomers

Scheme 41

4.2. The 2,2-Dihydro-1,4,2-dioxaphospholan–2,2-Dihydro-1,2-oxaphosphetan Transformation

2,2-Dihydro-1,4,2-dioxaphospholan (99), produced from chlorodimethylphosphine and hexafluoroacetone, rearranges on heating to 100°C to give 2,2-dihydro-1,2-oxaphosphetan (100). A 2,2-dihydro-1,3,2-dioxaphospholan intermediate could not be detected by ^{19}F n.m.r. spectroscopy.[137]

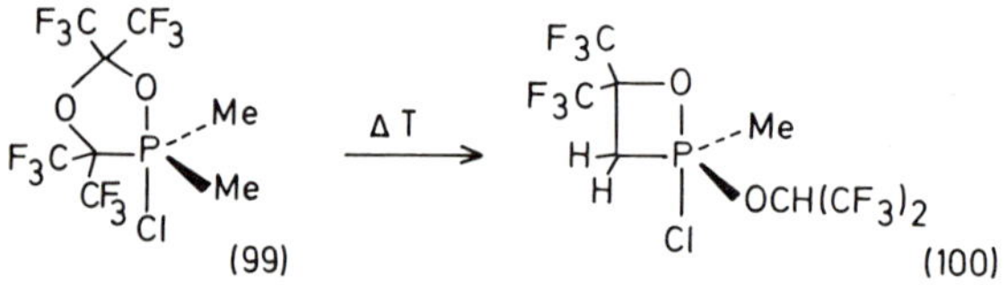

(99) (100)

Scheme 42

5. 2,2-Dihydro-1,2-oxaphospholens-4

5.1. Hydrolysis

Oxaphosphoranes of type (101) and (103) having one endocyclic P–O bond are easily hydrolysed to yield phosphonates (102), (104), and (105), respectively.[138–140]

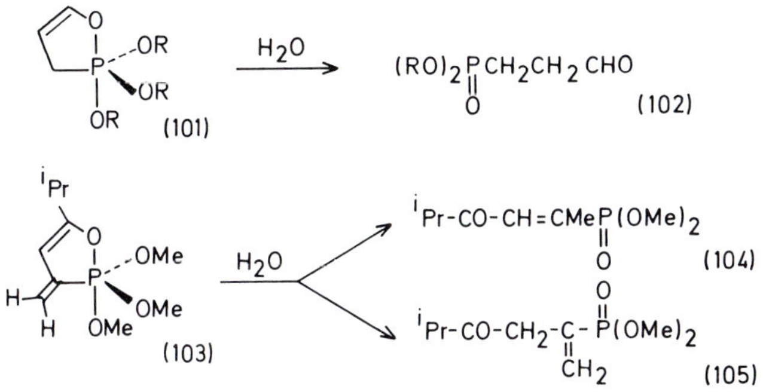

11 Phosphoranes in Synthesis

Scheme 43

5.2. Thermolysis

Phosphoranes (106)[141] at 190°C suffer P–O bond cleavage and subsequent Arbusov rearrangement to give phosphonates (107).

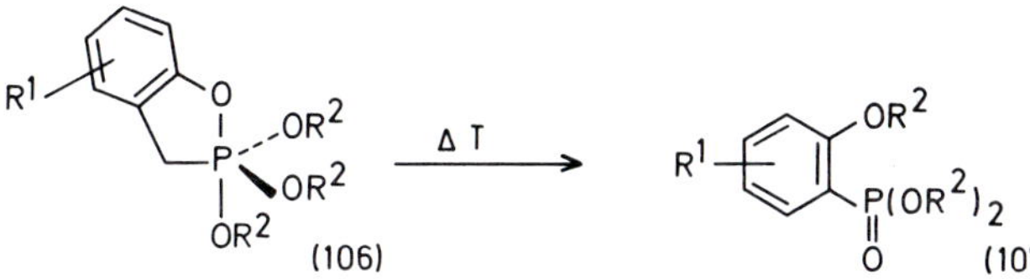

Scheme 44

5.3. The 2,2-Dihydro-1,2-oxaphospholen–2,2-Dihydro-1,2-thiaphospholen Transformation

The formation of phosphonate (109) from P_2S_5 and phosphorane (101) probably involves the intermediacy of a 2,2-dihydro-1,2-thiaphospholen (108),[142] showing that the thio compounds are considerably less stable than their oxygen analogues.

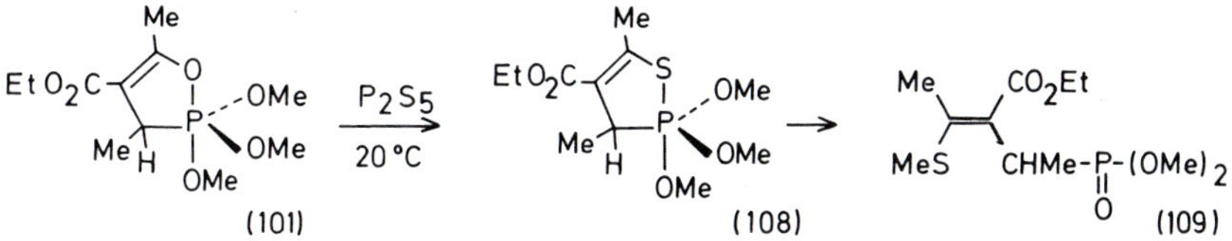

Scheme 45

6. 2,2-Dihydro-1,2-oxaphospholans

6.1. Reactions with Electrophiles

2,2-Dihydro-1,2-oxaphospholan (110), accessible from the reaction of cyclohexene oxide and cyclopropylidenetriphenylphosphorane, suffers ring-opening on treatment with HBr. The phosphonium bromide (111) obtained can be ring-closed again in the presence of bases.[143]

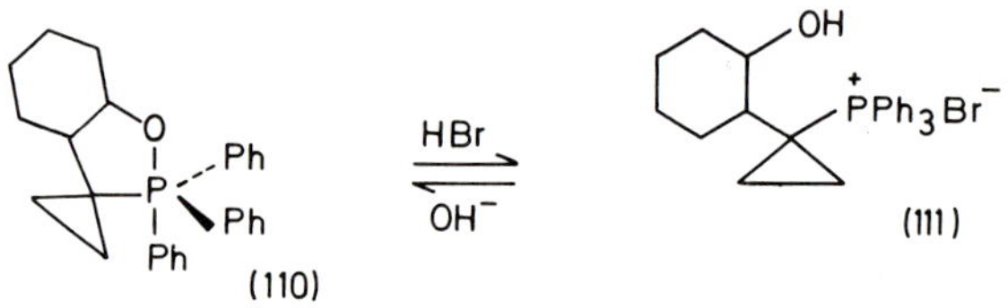

(110) (111)

Scheme 46

6.2. Thermolysis

2,2-Dihydro-1,2-oxaphospholans (112), obtainable as stable compounds from a series of strongly basic phosphine alkylenes[143,144] and epoxides, decompose on heating by P–O bond rupture with loss of triphenylphosphine and subsequent proton transfer (112) → (113).[144] A further reaction pathway was observed on thermolysis of (112),[145,146] namely P–O bond rupture followed

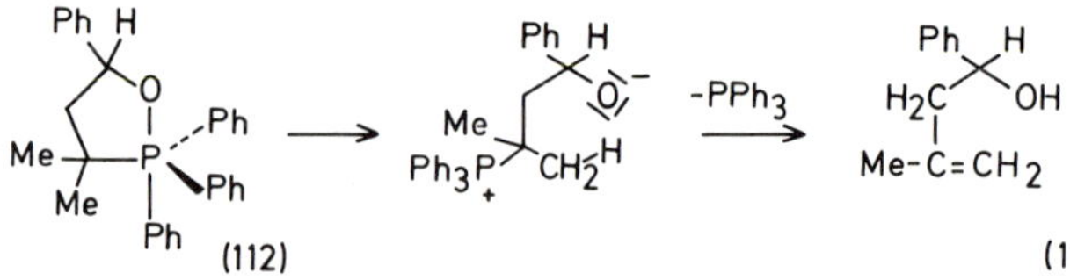

(112) (113)

Scheme 47

by a prototropic shift to give ylides. 2,2-Dihydro-1,2-oxaphospholan (112) at 90°C behaves exactly like an ylide, reacting with aldehydes to give products of a normal Wittig reaction (112) → (114) → (115).[145]

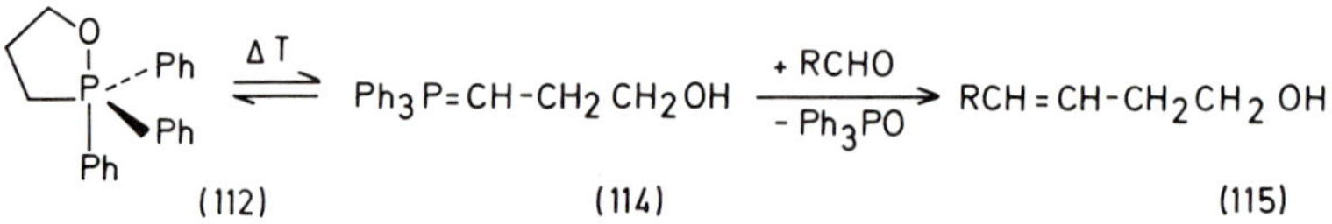

(112) (114) (115)

Scheme 48

7. 3,3-Dihydro-1,3-oxaphospholan-5-ones

The products from the reaction of ketene with phosphites and aminophosphines were assigned a 3,3-dihydro-1,3-oxaphospholan structure (116).[147] Compound (116) on treatment with bromine (116) → (117), carbon disulphide (116) → (118), and carbon dioxide (116) → (119), and on hydrolysis (116) → (120) reacts with P–C bond cleavage as illustrated by Scheme 49. When heated to 60°C the ketene dimer (121) is formed quantitatively with loss of phosphite.

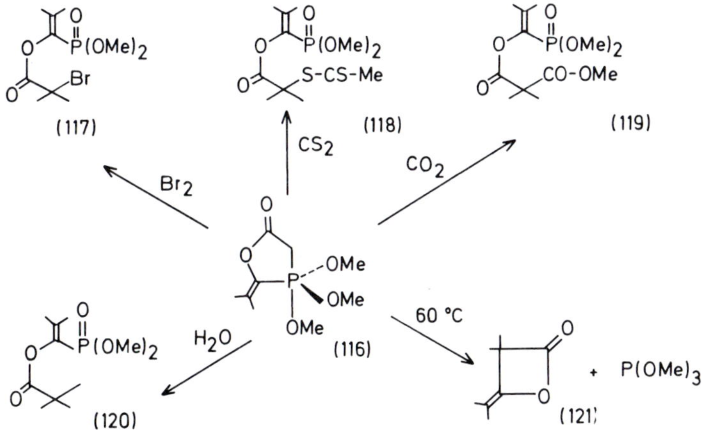

Scheme 49

8. Trioxaphosphetans [Phosphorus(III)–Ozone Adducts]

Triarylphosphines[148] and trialkyl and triaryl phosphites[149] readily react with ozone to give phosphine oxides and phosphate esters respectively. The triphenyl phosphite–ozone 1:1 adduct [(PhO)$_3$PO$_3$] can be isolated at $-70°C$ but decomposes at $-35°C$ to produce triphenyl phosphate and singlet oxygen.[149,150]

It has been suggested that polycyclic ozonides should exhibit increased stability as a result of restricted ligand permutational isomerization.[151,152] 1-Phospha-2,8,9-trioxaadamantane ozonide (122)[153] was found to be a relatively stable adduct, 106 times more stable than (PhO)$_3$PO$_3$ and 1·4 times more stable than 1-ethyl-4-phospha-3,5,8-trioxabicyclo[2.2.0]octane ozonide at $-5°C$. The water-soluble trioxaphosphetan (122) decomposes at 11°C quantitatively to give the phosphate (123) and singlet oxygen with a half-life of 25 min. Trioxaphosphetans represent a class of reagents of considerable synthetic value. These are discussed more fully in Chapter 6.

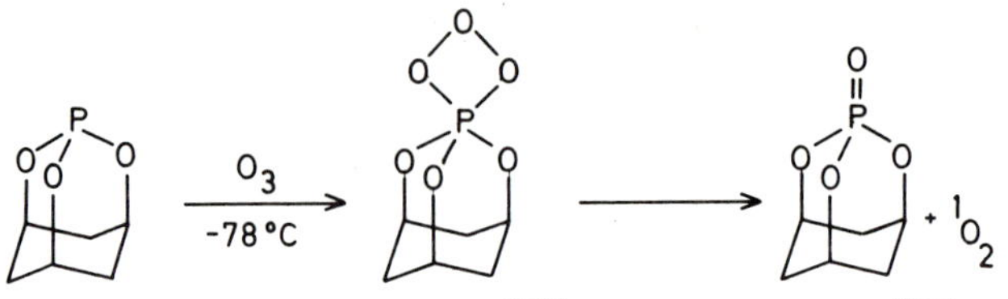

Scheme 50

9. 2,2-Dihydro-1,3,2-oxathiaphospholans

9.1. Formation of Thiirans

A [3 + 2] cycloelimination process is observed[154] when the 1:1 adducts (124), obtained from ethylene phosphonothioites (R = Ph, Bu^t) with biacetyl, benzylideneacetylacetone, and phenanthraquinone, are heated. They readily give thiirans (125) and the corresponding phosphonate or phosphinate esters (126).

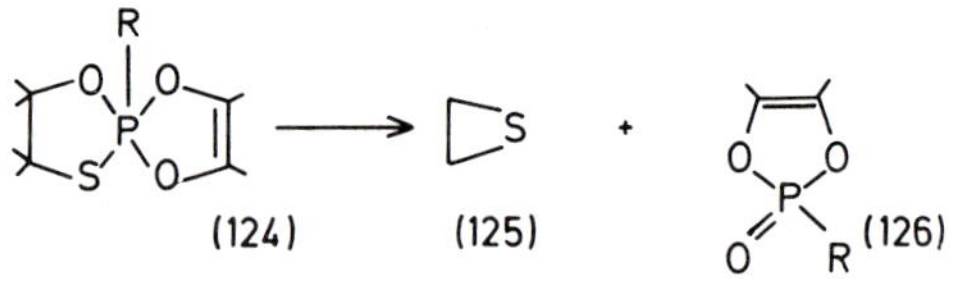

Scheme 51

10. 2,2-Dihydro-1,4,2-oxazaphospholens-4 (Δ^4-1,4,2λ^5-Oxazaphospholines)

10.1. Thermolysis and Photolysis

Δ^4-1,4,2λ^5-Oxazaphospholines (127a),[155,156] readily prepared by treatment of N-(perfluoroisopropylidene)carboxamides[157] with phosphites, on hydrolysis

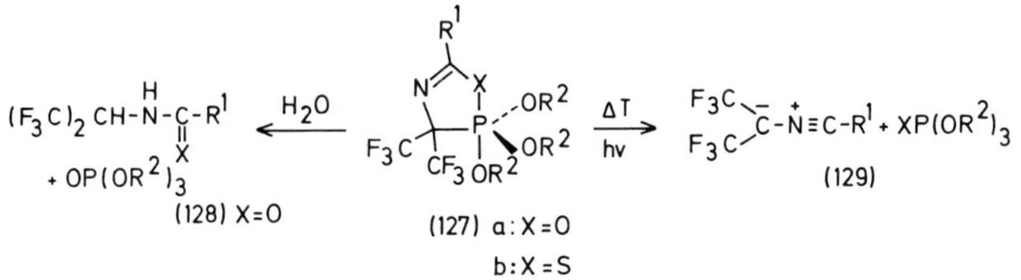

Scheme 52

yield *N*-hexafluoroisopropyl-amides (128).[158] On heating or on photolysis cycloelimination of trialkyl or triaryl phosphates occurs. From trapping reactions with nucleophiles and numerous dipolarophiles it is clearly established that nitrile ylides (129) are the second fragment produced.[158] Δ^4-1,4,2λ^5-Thiazaphospholines (127b)[159] likewise undergo a [3 + 2] cycloreversion reaction producing nitrile ylides. With dipolarophiles such as olefins,[160–162] acetylenes,[160,161,162b,163] azo compounds,[164,165] nitriles,[164] carbonyl compounds,[164] *N*-arenes,[166] nitrosobenzene,[167] isocyanates,[168] and isocyanides,[169,170] various types of 5- and 4-membered heterocyclic systems are conveniently accessible. When photolysis or thermolysis is carried out without

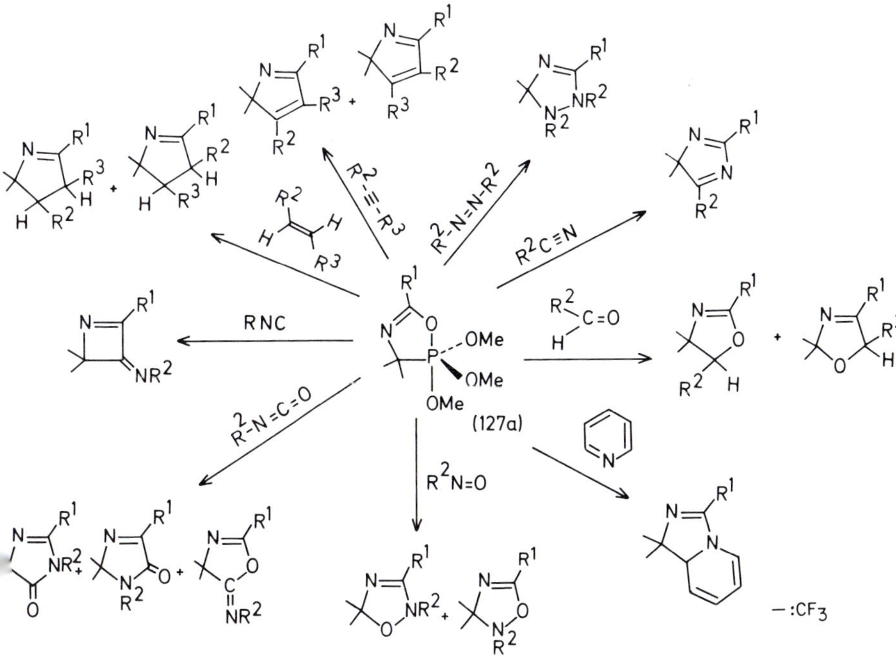

Scheme 53

trapping reagent, the 1,3-dipolar species (129) undergoes dimerization as illustrated in Scheme 54.[171] The initial step seems to be a head-to-head dimerization process in which the carbenoid mesomer of the nitrile ylide[172] may be involved (129) → (130). Partly chlorinated nitrile ylides (133) generated by thermolysis of phosphoranes of type (127) rearrange by a [1,4] chlorine shift to yield *N*-vinylimidoyl chlorides (134)[173] before being trapped by dipolarophiles. Electron-rich dipolarophiles strongly influence the fragmentation process of compounds (127). In the presence of enol ethers,[162] enamines,[174] or electron-rich isocyanides,[174] a [4 + 1] cycloreversion process

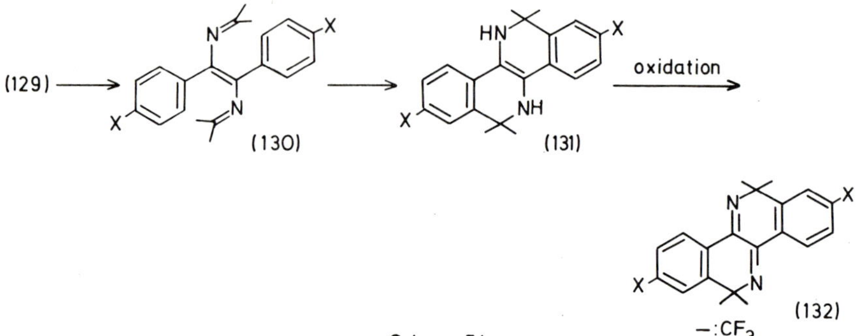

Scheme 54

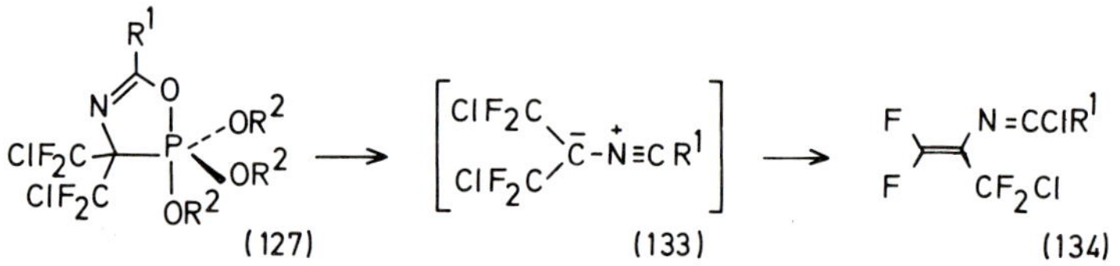

Scheme 55

already successfully competes with the [3 + 2] process, to yield (135) and phosphites. In the presence of ynamines only the [4 + 1] process can be observed;[162b,174] the acylimines (135) thus formed undergo Diels–Alder reactions.[162b]

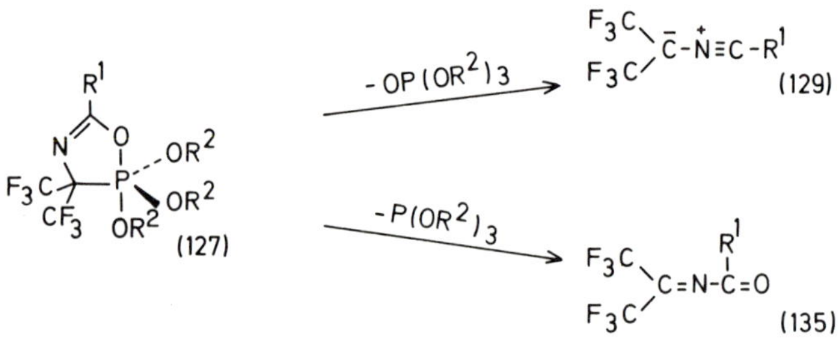

Scheme 56

11. 5,5-Dihydro-1,2,5-oxazaphospholens-2 (Δ^2-1,2,5λ^5-Oxazaphospholines)

11.1. Thermolysis

Δ^2-1,2,5λ^5-Oxazaphospholines (136) obtained by 1,3-dipolar cycloaddition reaction from nitrile oxides and phosphine alkylenes[175–178] decompose on heating. Azirines (137), ketene imines (138), and unsaturated oximes (139), respectively, are the products of this fragmentation process, depending on the substitution pattern of the phosphorus heterocycle (136). The following generalizations have been made,[178] based on the experimental data available.

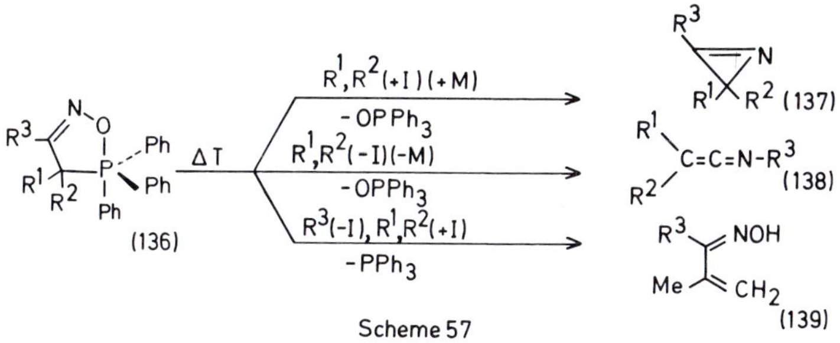

Scheme 57

β-Keto-alkylidenephosphoranes with nitrile oxides give isoxazoles.[179,180] The intermediacy of Δ^2-1,2,5λ^5-oxazaphospholines (136) was shown.[179]

Δ^2-1,2,5λ^5-Oxazaphospholine-N-oxides (140) are stable on boiling in ether but can be converted into phosphonates (141) when heated in the presence of trimethyl phosphite.[181]

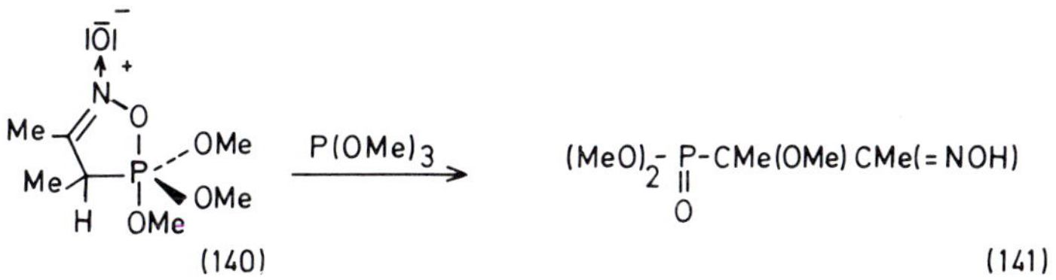

Scheme 58

12. 2,2-Dihydro-1,3,2-oxazaphospholans (1,3,2λ^5-Oxazaphospholidines)

12.1. Thermolysis

1,3,2λ^5-Oxazaphospholidines (142), accessible from iminophosphoranes and epoxides, give aziridines (143) and phosphine oxide in high yields on pyrolysis.[182]

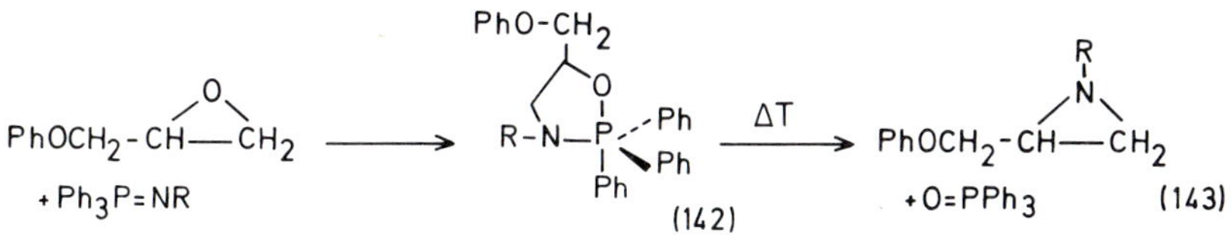

Scheme 59

12.2. Photolysis

Phosphoranes of type (144),[183] structurally closely related to (142), are transformed into carbazoles (146),[184] probably via a diradical intermediate (145), on photolysis.

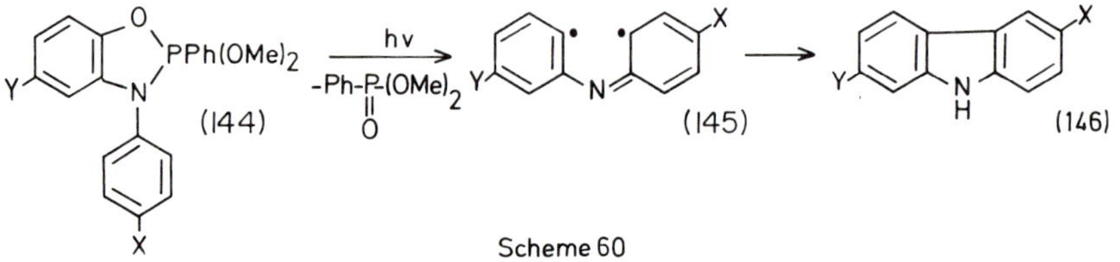

Scheme 60

13. 5,5-Dihydro-1,2,5-oxazaphospholans (1,2,5λ^5-Oxazaphospholidines)

1,2,5λ^5-Oxazaphospholidines of type (147), obtained from phosphine alkylenes and nitrones, formally lose benzyne on heating to give substituted phosphine oxides (148).[185]

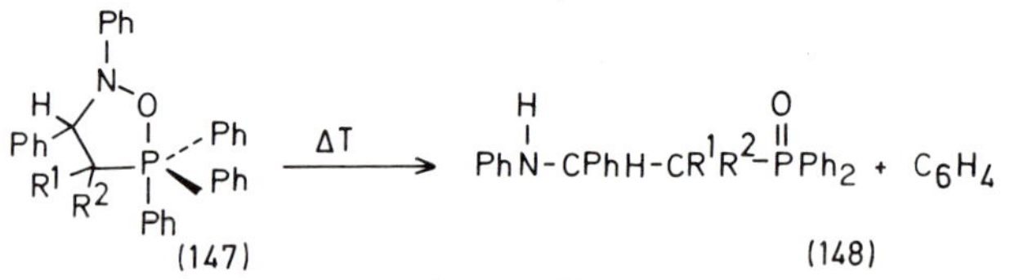

Scheme 61

14. 2,2-Dihydro-1,3,4,2-oxadiazaphospholens-4 (Δ^4-1,3,4,2λ^5-Oxadiazaphospholines)

Δ^4-1,3,4,2λ^5-Oxadiazaphospholines (149),[186,187] readily accessible from azodicarboxylates and triphenyl phosphite, react with aroyl and sulphonyl isocyanates as well as aroyl and sulphonyl isothiocyanates to give 1,2,4-triazolin-3-ones (150a) and 1,2,4-triazoline-3-thiones (150b),[79,96] respectively.

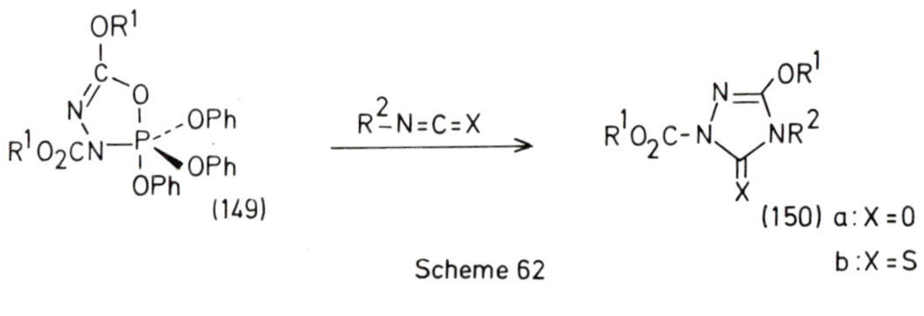

Scheme 62

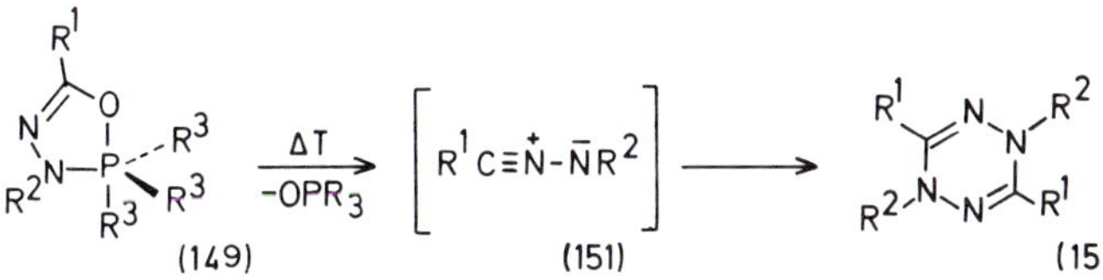

11 Phosphoranes in Synthesis

Scheme 63

In the absence of trapping reagents, compounds of the type (149) were found to decompose on heating by loss of phosphate ester with 1,2,4,5-tetrazine (152) formation. Nitrile imines (151) are the intermediates in this reaction.[188] Protonation, acylation, as well as the attack of PCl_5 was found to occur readily at the equatorial N-3 atom. Subsequent P–N bond rupture after ligand reorganization and nucleophilic displacement of the $>$P=O moiety provides a likely explanation of the experimental facts[188] as summarized in Scheme 64.

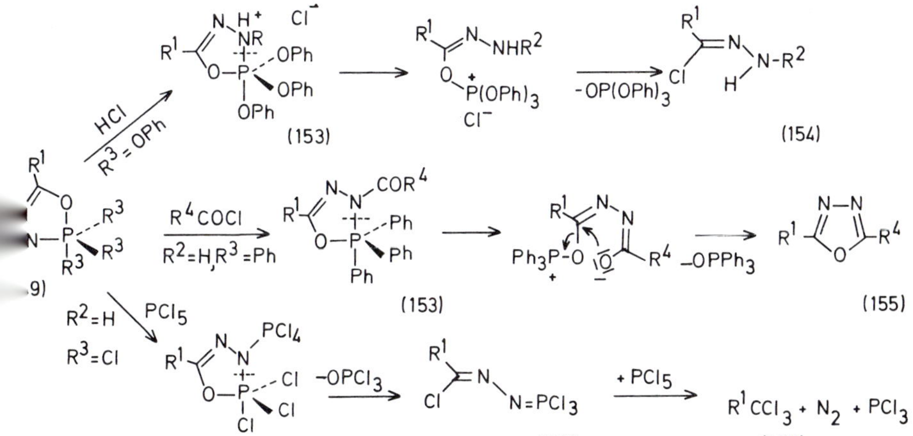

Scheme 64

15. 1,1-Dihydrophospholens-3

Based on orbital symmetry considerations, Hoffmann[189] predicted that concerted reactions of a trigonal bipyramidal species are forbidden for apical-equatorial departure or addition, but allowed for apical-apical and equatorial-equatorial departure and addition, while Howell[190] calculated that the path of lowest energy for the fragmentation ($PH_5 \rightarrow PH_3 + H_2$) involves a non-least-motion departure of an axial and an equatorial ligand.

Fragmentation of 1,1-diethoxy-1,2,5-trimethyl-1,1-dihydrophospholen-3 (158) gives *trans,trans*-hexa-2,4-diene (159).[191] The reaction was shown to be 99% stereospecific and probably concerted. Likewise, the phosphorane (160) fragments stereospecifically to (159) and the corresponding phosphonite.[191] The activation energy ($\Delta G^* \sim 105$ kJ mol^{-1}) was explained in terms of the need to place the phospholen ring diequatorial for the fragmentation process.

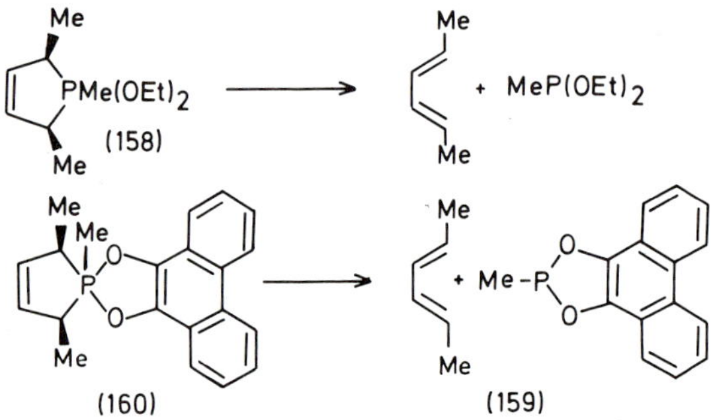

Scheme 65

16. 1,1-Dihydrophospholans

Thermolysis or photolysis of (161) gives triphenylphosphine and (162) which rearranges to give cyclooctatetraene (163),[192] showing that the 6-electron process clearly dominates the 4-electron fragmentation.

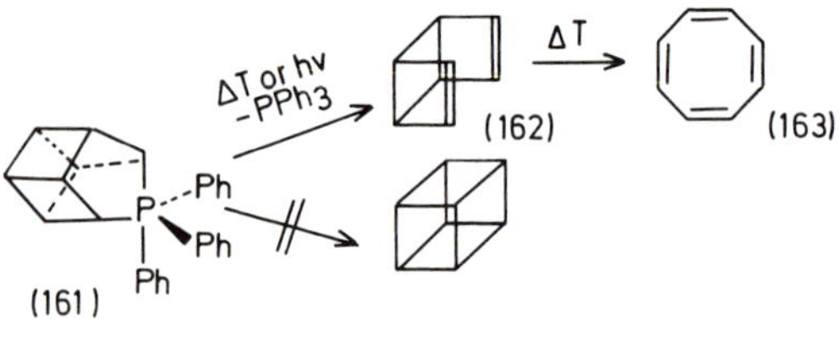

Scheme 66

17. 1,1-Dihydrophosphirans

A highly unstable 1,1-dihydrophosphiran (166) is suggested as the intermediate in the reaction of the phosphiran (164) with the dithieten (165) at $-78°$C to give ethylene and the dithiophosphonite shown in Scheme 67.[193] The reaction is claimed to be stereospecific, *cis*-2,3-dideuterio-1,1-dihydrophosphiran giving *cis*-dideuterioethylene.

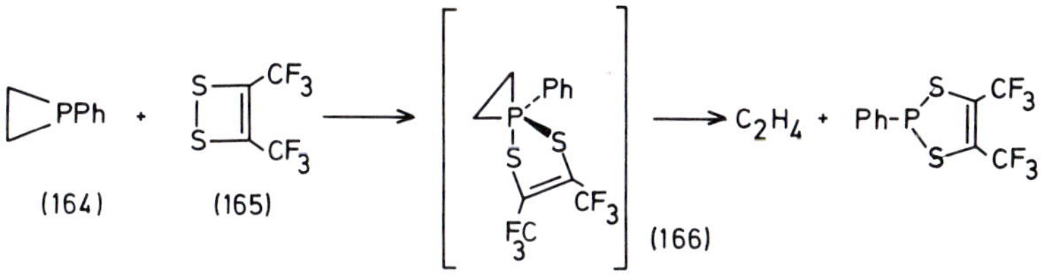

Scheme 67

18. Bis-(2,2'-biphenylene)phosphoranes

Pyrolysis of the bis-(2,2'-biphenylene)phosphorane (167) yields phosphines (168) where the tervalent phosphorus becomes part of a 9-membered ring system, when R is small (R = Me, Ph), but when R is large (R = 8-quinolyl, 9-anthryl), phosphines of type (169) are formed.[194,195]

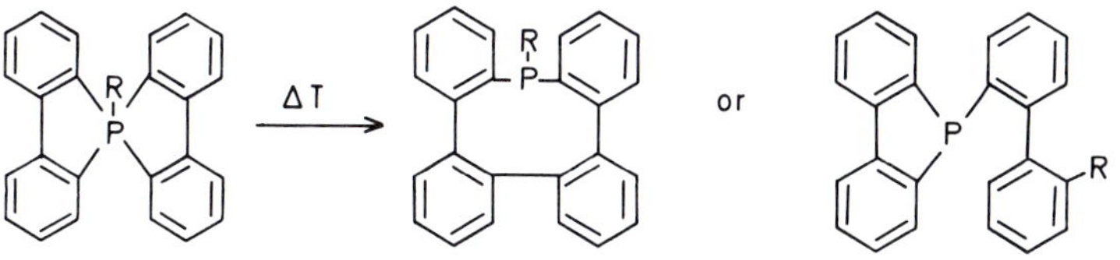

Scheme 68

19. Acyclic Phosphoranes

Pentaalkoxyphosphoranes $P(OR)_5$, the orthoesters of phosphoric acid, readily alkylate any molecule bearing a sufficiently acidic hydrogen, for example carboxylic acids, water, alcohols, and CH-acidic compounds.[196] The latter reaction does not require any acid or base catalyst; it is suggested that dissociation of the phosphorane is the first step of the reaction.[197] Corresponding reactions have been observed with oxyphosphoranes containing

$$(EtO)_5P \rightleftharpoons (EtO)_4P^+ + {}^-OEt$$

Scheme 69

fewer than five alkoxy groups, $R^1{}_nP(OR^2)_{5-n}$. Similarly 2,2-dihydro-1,2-oxaphospholens have been shown to be powerful O-alkylating agents, converting acids and phenols into esters and ethers respectively.[198]

Acyclic phosphoranes containing at least two alkoxy or phenoxy groups are readily susceptible to alcoholysis and aminolysis. With 1,2- and 1,3-diols, cyclic phosphoranes of type (171) are formed. Thus pentaphenoxyphosphorane (170)[199] with catechol in benzene was shown to exchange two phenoxy groups to give (171), while a further exchange in methylene chloride at 25°C[200] produced the spirocyclic phosphorane (172). The reaction of pentaethoxyphosphorane with butane-1,4-diol and pentane-1,5-diol gave tetrahydrofuran and tetrahydropyran respectively. Various heterocyclic systems are readily available by this type of reaction.[41,123] The mechanism has been clearly demonstrated by the formation of the ether (174) from *trans*-cyclohexane-1,4-diol (173). Scheme 72 illustrates the considerable synthetic value of this type of reaction.

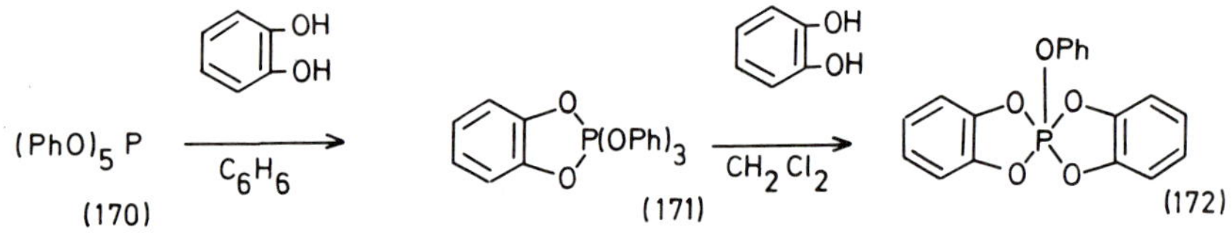

Scheme 70

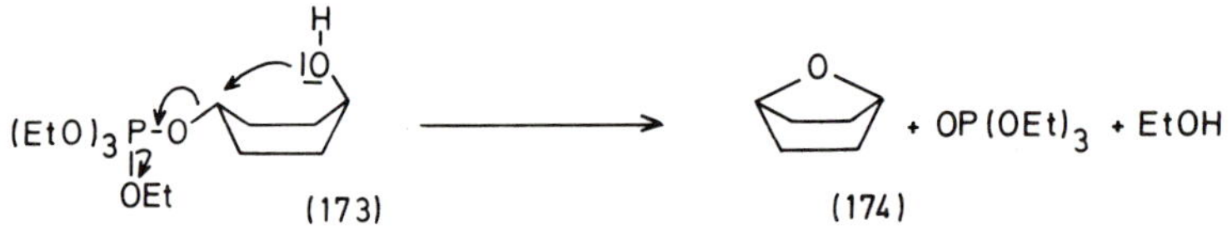

Scheme 71

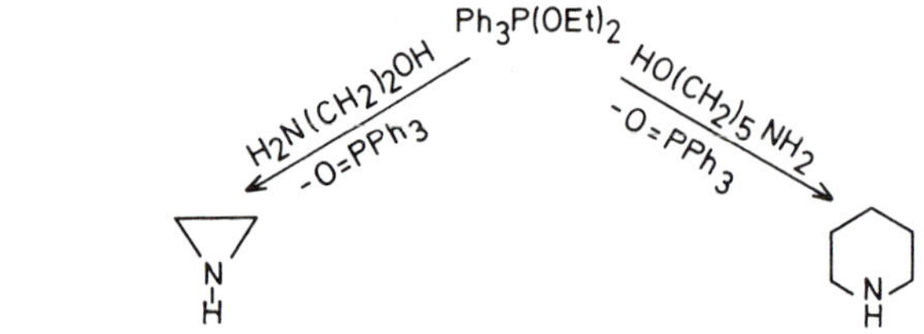

Scheme 72

Intermolecular exchange of ligands occurs slowly between the trimethyl phosphite–biacetyl adduct (1) and pentaphenoxyphosphorane (170) in benzene at 25°C. An oxonium salt (175) is the suggested intermediate.[201]

Compound (170) with HCl reacts to give triphenoxydichlorophosphorane and phenol.[198] On reaction of pentaphenylphosphorane with phenol in pyridine at room temperature tetraphenylphenoxyphosphorane was isolated.[202] On heating to 130°C pentaphenylphosphorane decomposes to give benzene, triphenylphosphine, and 2,2'-biphenylene-phenylphosphine.[203]

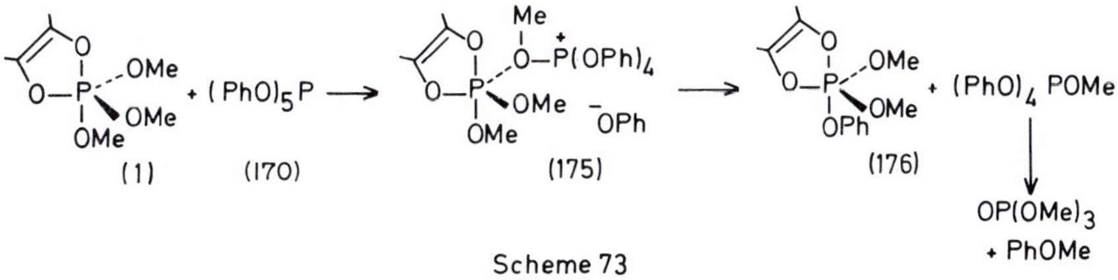

Scheme 73

20. Miscellaneous Reactions

20.1. Phosphoranes as Intermediates in the Michaelis–Arbusov Reaction

The Michaelis–Arbusov reaction is generally formulated with a phosphonium intermediate.[204] But there are indications that phosphoranes have also to be taken into consideration as intermediates in this reaction.[205,206] The reactions of 1-halogeno(pentafluoro)cyclobutenes (177) with trialkyl phosphites are the first examples where phosphoranes (178) have been isolated as stable intermediates in the Michaelis–Arbusov reaction.[205,207]

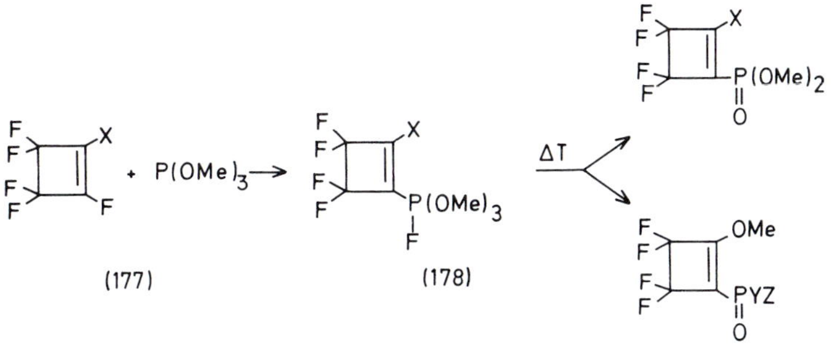

Scheme 74

20.2. Formation of Copolymers

2-Phenyl-1,3,2-dioxaphospholan (179) and electrophilic monomers such as β-propiolactones, derivatives of acrylic acid, vinyl ketones, and 3-hydroxypropane-1-sulphonic acid sultone react on heating to give 1:1 copolymers

(181).[208–210] On mixing equimolar amounts of these compounds at room temperature in ether, phosphoranes of type (182) result in nearly quantitative yield;[211] the latter are transformed into the same type of copolymers (181) when heated to 150°C. Likewise, a mixture of 2-phenoxy-1,3,2-dioxaphospholan (183) and α-keto-acids on heating produces 1:1 copolymers (185). At −20°C (183) and α-keto-acids react to give pentaoxyphosphoranes (186), which again can be transformed into (185) at 120°C. The polymerization is supposed to proceed via zwitterions of type (180) and (184) respectively.[212] Compound (186) on hydrolysis yields α-hydroxy-acids.[213]

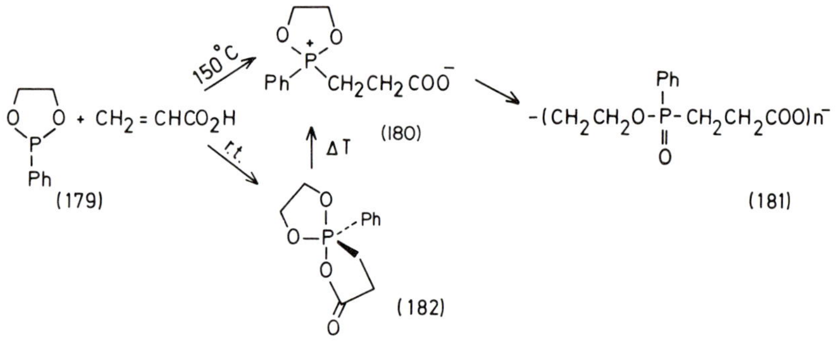

Scheme 75

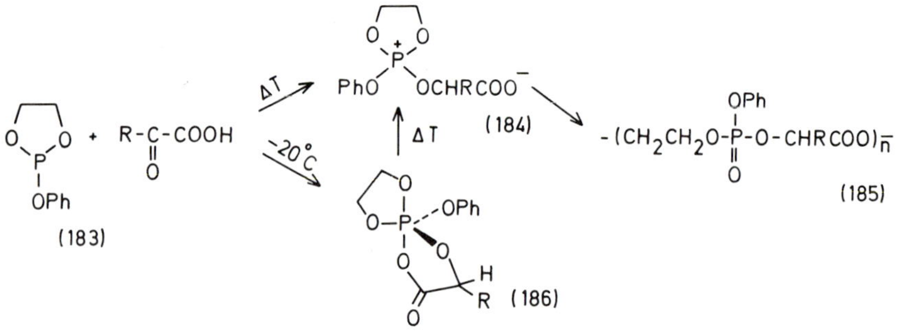

Scheme 76

20.3. Hexacoordination

Access to compounds possessing hexacoordinate phosphorus atoms has become easier recently.[5] One of the early examples of this class of compound was synthesized by base-catalysed condensation of halogenocyclophosphazene with catechol.[214] Hellwinkel[215] has developed a simple access to phosphorus(VI) compounds (188) starting from phosphoranes (187). This method has been developed further by Denney,[216] showing that pentaoxyphosphoranes of

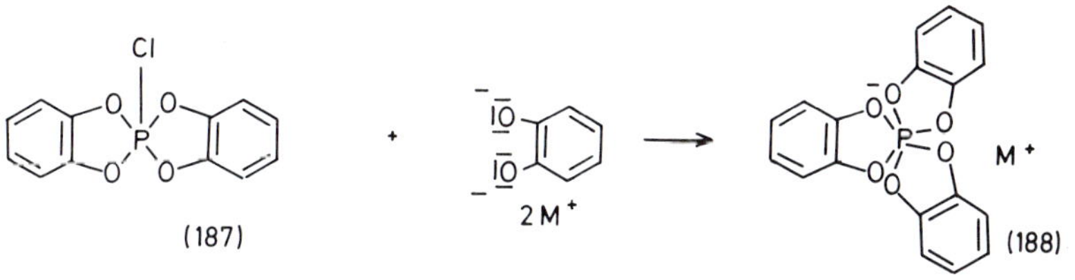

Scheme 77

type (189) react readily with 1,2- and 1,3-diols to give phosphorus(VI) derivatives (190). This strategy is now widely used for synthesis of phosphorus(VI) compounds.[217]

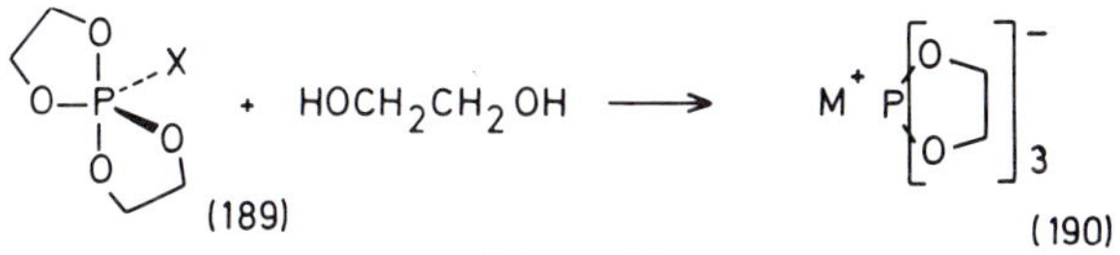

Scheme 78

21. References

For earlier reviews see 1–5.

1. F. Ramirez, (a) *Pure Appl. Chem.*, 1964, **9**, 337; (b) *Bull. Soc. Chim. France*, 1966, 2443; (c) *Accounts Chem. Res.*, 1968, **1**, 168; (d) *Trans. New York Acad. Sci.*, 1968, **30**, 410; (e) *Colloques Intern. du C.N.R.S., Bull. Soc. Chim. France*, 1970, **182**, 67; (f) *Bull. Soc. Chim. France*, 1970, 3491; (g) *Synthesis*, 1974, 90.
2. (a) K. D. Berlin and D. M. Hellwege, *Topics Phosphorus Chem.*, 1969, **6**, 83. (b) D. Hellwinkel, *Colloques Intern. du C.N.R.S., Bull. Soc. Chim. France*, 1970, **182**, 177. (c) A. N. Hughes and C. Srivanavit, *J. Heterocyclic Chem.*, 1970, **7**, 1. (d) G. Wittig, *Bull. Soc. Chim. France*, 1966, 1162.
3. D. Hellwinkel, *Organic Phosphorus Compounds*, Vol. 3 (ed. G. M. Kosolapoff and L. Maier), J. Wiley and Sons, 1972, p. 185.
4. J. Emsley and D. Hall, *The Chemistry of Phosphorus*, Harper and Row, London, 1976, p. 228.
5. Specialist Periodical Reports, *Organophosphorus Chemistry* (ed. S. Trippett), The Chemical Society, London, Vols. 1–8; ongoing.
6. (a) R. Schmutzler, *Adv. Fluorine Chem.*, 1965, **5**, 31. (b) *Idem, Angew. Chem. Internat. Ed.*, 1965, **4**, 496. (c) *Idem, Halogen Chemistry*, Vol. 2, Academic Press, New York, 1967, p. 31. (d) E. L. Muetterties, *Accounts Chem. Res.*, 1970, **3**, 266.
7. G. E. Kimball, *J. Chem. Phys.*, 1940, **8**, 188.
8. (a) D. P. Craig, A. Maccoll, R. S. Nyholm, L. E. Orgel, and L. E. Sutton, *J. Chem. Soc.*, 1954, 332. (b) R. J. Gillespie and R. S. Nyholm, *Quart. Rev.*, 1957, **11**, 339. (c) R. J. Gillespie, *Canad. J. Chem.*, 1960, **38**, 818. (d) *Idem, ibid.*, 1961, **39**, 318. (e) *Idem, J. Chem. Soc.*, 1963, 4672, 4679. (f) *Idem, Angew. Chem. Internat. Ed.*, 1967, **6**, 819.

9. (a) J. A. Howard, D. R. Russel, and S. Trippett, *J. Chem. Soc., Chem. Comm.*, 1973, 856. (b) R. R. Holmes, *J. Amer. Chem. Soc.*, 1974, **96**, 4143. (c) *Idem, ibid.*, 1975, **97**, 5379. (d) H. Wunderlich, D. Mootz, R. Schmutzler, and M. Wieber, *Z. Naturforsch.*, 1974, **29b**, 32. (e) H. Wunderlich, *Acta Cryst.*, 1974, **30B**, 939.
10. J. Heller, *Chimia*, 1969, **23**, 351.
11. I. Ugi, D. Marquarding, H. Klusacek, P. Gillespie, and F. Ramirez, *Accounts Chem. Res.*, 1971, **4**, 288.
12. P. Gillespie, P. Hoffmann, H. Klusacek, D. Marquarding, S. Pfohl, F. Ramirez, E. A. Tsolis, and I. Ugi, *Angew. Chem. Internat. Ed.*, 1971, **10**, 687.
13. K. A. R. Mitchell, *Chem. Rev.*, 1969, **69**, 157.
14. F. Ramirez and I. Ugi, *Adv. Phys. Org. Chem.*, 1971, **9**, 25.
15. I. Ugi and F. Ramirez, *Chem. Brit.*, 1972, **8**, 198.
16. H. Kwart and K. G. King, *d-Orbitals in the Chemistry of Silicon, Phosphorus and Sulfur*, Springer-Verlag, Berlin-Heidelberg-New York, 1977, and literature cited therein.
17. (a) S. Bone, S. Trippett, and P. J. Whittle, *J. Chem. Soc., Perkin I*, 1974, 2125. (b) K. E. DeBruin, A. G. Padilla, and M.-T. Campbell, *J. Amer. Chem. Soc.*, 1973, **95**, 4681.
18. E. L. Muetterties and R. A. Schunn, *Quart. Rev.*, 1966, **20**, 245.
19. P. Gillespie, F. Ramırez, I. Ugi, and D. Marquarding, *Angew. Chem. Internat. Ed.*, 1973, **12**, 91.
20. F. Ramirez, K. Tasaka, N. B. Desai, and C. P. Smith, *J. Amer. Chem. Soc.*, 1968, **90**, 751.
21. F. Ramirez, G. V. Loewengart, E. A. Tsolis, and K. Tasaka, *J. Amer. Chem. Soc.*, 1972, **94**, 3531.
22. F. Ramirez, K. Tasaka, and R. Hershberg, *Phosphorus*, 1972, **2**, 41.
23. R. S. Berry, *J. Chem. Phys.*, 1960, **32**, 933.
24. R. Luckenbach, *Dynamic Stereochemistry of Pentaco-ordinated Phosphorus and Related Elements*, Thieme, Stuttgart, 1973, and literature cited therein.
25. *Organo-Phosphorus Stereochemistry*, Part 2: P(V) Compounds (ed. W. E. McEwen and K. D. Berlin), Wiley, Chichester, 1975.
26. F. H. Westheimer, *Accounts Chem. Res.*, 1968, **1**, 70.
27. F. Ramirez, R. B. Mitra, and N. B. Desai, *J. Amer. Chem. Soc.*, 1960, **82**, 2651.
28. F. Ramirez, S. B. Bhatia, R. B. Mitra, Z. Hamlet, and N. B. Desai, *J. Amer. Chem. Soc.*, 1964, **86**, 4394.
29. F. Ramirez, N. B. Desai, and R. B. Mitra, *J. Amer. Chem. Soc.*, 1961, **83**, 492.
30. F. Ramirez, R. B. Mitra, and N. B. Desai, *J. Amer. Chem. Soc.*, 1960, **82**, 5763.
31. L. M. Stephenson and L. C. Falk, *J. Org. Chem.*, 1976, **41**, 2928.
32. (a) V. P. Evdakov, L. I. Mizrakh, and L. Yu. Sandalova, *Zhur. Obshch. Khim.*, 1967, **37**, 1818 (*Chem. Abs.*, 1968, **68**, 59507c). (b) B. A. Arbusov, N. A. Polezhaeva, and V. S. Vinogradova, *Izv. Akad. Nauk SSSR, Ser. Khim.*, 1968, 2525 (*Chem. Abs.*, 1969, **70**, 87682r).
33. F. Ramirez, O. P. Madan, and C. P. Smith, *J. Amer. Chem. Soc.*, 1965, **87**, 670.
34. D. Swank, C. N. Caughlan, F. Ramirez, O. P. Madan, and C. P. Smith, *J. Amer. Chem. Soc.*, 1967, **89**, 6503.
35. F. Ramirez, B. Hansen, and N. B. Desai, *J. Amer. Chem. Soc.*, 1962, **84**, 4588.

36. F. Ramirez, M. Nowakowski, and J. F. Marecek, *J. Amer. Chem. Soc.*, 1977, **99**, 4515.
37. (a) F. V. Bagrov, N. A. Razumova, and A. A. Petrov, *Zhur. Obshch. Khim.*, 1972, **42**, 792 (*Chem. Abs.*, 1972, **77**, 101764x). (b) N. A. Razumova, N. A. Kurshakova, and A. A. Petrov, *ibid.*, 1975, **45**, 702 (*Chem. Abs.*, 1975, **83**, 43440z).
38. R. Burgada, *Bull. Soc. Chim. France*, 1967, 347.
39. D. Bernard and R. Burgada, (a) *Compt. Rend.* (*C*), 1973, **277**, 433; (b) *Tetrahedron Lett.*, 1973, 3455.
40. F. Ramirez, A. J. Bigler, and C. P. Smith, *Tetrahedron*, 1968, **24**, 5041.
41. B. C. Chang, W. E. Conrad, D. B. Denney, D. Z. Denney, R. Edelmann, R. L. Powell, and D. W. White, *J. Amer. Chem. Soc.*, 1971, **93**, 4004.
42. D. Bernard and R. Burgada, *Phosphorus*, 1975, **5**, 285.
43. F. Ramirez and N. B. Desai, *J. Amer. Chem. Soc.*, 1960, **82**, 2652.
44. F. Ramirez, S. B. Bhatia, A. V. Patwardhan, E. H. Chen, and C. P. Smith, *J. Org. Chem.*, 1968, **33**, 20.
45. F. Ramirez, S. L. Glaser, A. J. Bigler, and J. F. Pilot, *J. Amer. Chem. Soc.*, 1969, **91**, 496.
46. L. I. Mizrakh, L. Yu. Sandalova, and V. P. Evdakov, *Dokl. Akad. Nauk SSSR*, 1966, **171**, 1116 (*Chem. Abs.*, 1967, **66**, 85732a).
47. F. Ramirez, S. L. Glaser, A. J. Bigler, and J. F. Pilot, *J. Amer. Chem. Soc.*, 1969, **91**, 5966.
48. F. Ramirez, S. B. Bhatia, A. J. Bigler, and C. P. Smith, *J. Org. Chem.*, 1968, **33**, 1192.
49. I. P. Gozman, *Zhur. Obshch. Khim.*, 1969, **39**, 1954 (*Chem. Abs.*, 1970, **72**, 31140m).
50. F. Ramirez, J. F. Marecek, S. L. Glaser, and P. Stern, *Phosphorus*, 1974, **4**, 65.
51. D. N. Harpp and P. Mathiaparanam, (a) *Tetrahedron Lett.*, 1970, 2089; (b) *J. Org. Chem.*, 1971, **36**, 2540; (c) *ibid.*, 1972, **37**, 1367.
52. F. Ramirez, S. L. Glaser, P. Stern, P. D. Gillespie, and I. Ugi, *Angew. Chem. Internat. Ed.*, 1973, **12**, 66.
53. (a) F. Ramirez and I. Ugi, *Bull. Soc. Chim. France*, 1974, 453. (b) F. Ramirez, S. L. Glaser, P. Stern, I. Ugi, and P. Lemmen, *Tetrahedron*, 1973, **29**, 3741. (c) F. Ramirez, J. F. Marecek, H. Tsuboi, and H. Okazaki, *J. Org. Chem.*, 1977, **42**, 771.
54. F. Ramirez, K. Tasaka, N. B. Desai, and C. P. Smith, *J. Org. Chem.*, 1968, **33**, 25.
55. (a) B. A. Arbusov, *Pure Appl. Chem.*, 1964, **9**, 307. (b) R. G. Harvey and E. R. DeSombre, *Topics Phosphorus Chem.*, 1964, **1**, 57.
56. G. Pfeiffer, E. Gaydou, and A. Guillemonat, *Compt. Rend.* (*C*), 1969, **268**, 529.
57. E. Corre and A. Foucaud, *J. Chem. Soc., Chem. Comm.*, 1971, 570.
58. F. Ramirez, N. Ramanathan, and N. B. Desai, *J. Amer. Chem. Soc.*, 1962, **84**, 1317.
59. F. Ramirez, N. B. Desai, and N. Ramanathan, *Tetrahedron Lett.*, 1963, 323.
60. F. Ramirez, S. L. Glaser, P. Stern, P. D. Gillespie, and I. Ugi, *Angew. Chem. Internat. Ed.*, 1973, **12**, 66.
61. F. Ramirez, A. V. Patwardhan, N. B. Desai, N. Ramanathan, and C. V. Greco, *J. Amer. Chem. Soc.*, 1963, **85**, 3056.
62. F. Ramirez, A. V. Patwardhan, N. Ramanathan, N. B. Desai, C. V. Greco, and S. R. Heller, *J. Amer. Chem. Soc.*, 1965, **87**, 543.

63. F. Ramirez, A. V. Patwardhan, and C. P. Smith, *J. Org. Chem.* (a) 1965, **30**, 2575; (b) 1966, **31**, 474;(c) 1966, **31**, 3159.
64. F. Ramirez, H. J. Kugler, and C. P. Smith, *Tetrahedron Lett.*, 1965, 261.
65. F. Ramirez, H. J. Kugler, and C. P. Smith, *Tetrahedron*, 1968, **24**, 1931.
66. F. Ramirez, S. B. Bhatia, and C. P. Smith, *J. Amer. Chem. Soc.*, 1967, **89**, (a) 3026; (b) 3030.
67. F. Ramirez, S. B. Bhatia, A. V. Patwardhan, and C. P. Smith, *J. Org. Chem.*, 1967, **32**, (a) 2194; (b) 3547.
68. F. Ramirez, H. J. Kugler, A. V. Patwardhan, and C. P. Smith, *J. Org. Chem.*, 1968, **33**, 1185.
69. F. Ramirez, H. J. Kugler, and C. P. Smith, *Tetrahedron*, 1968, **24**, 3153.
70. F. Ramirez, N. B. Desai, A. V. Patwardhan, S. B. Bhatia, S. L. Glaser, and P. Stern, to be published (cited in ref. 1g).
71. F. Ramirez and N. Ramanathan, *J. Org. Chem.*, 1961, **26**, 3041.
72. (a) A. N. Pudovik, I. V. Gur'yanova, S. P. Perevezentseva, and S. A. Terent'eva, *Zhur. Obshch. Khim.*, 1969, **39**, 337 (*Chem. Abs.*, 1969, **70**, 115071m). (b) A. N. Pudovik, I. V. Gur'yanova, L. A. Burnaeva, and E. Kh. Karimullina, *ibid.*, 1971, **41**, 1995 (*Chem. Abs.*, 1972, **76**, 45669k). (c) A. N. Pudovik, I. V. Konovalova, V. P. Kakurina, E. Kh. Ofitserova, and L. V. Rakova, *ibid.*, 1974, **44**, 267 (*Chem. Abs.*, 1974, **80**, 120843a).
73. M. Muroi, Y. Inouye, and M. Ohno, *Bull. Chem. Soc. Japan*, 1969, **42**, 2948.
74. F. Ramirez, V. A. V. Prasad, and H. J. Bauer, *Phosphorus*, 1972, **2**, 185.
75. M. J. Gallagher and I. D. Jenkins, *Topics Stereochem.*, 1968, **3**, 1.
76. F. Ramirez, S. B. Bhatia, C. D. Telefus, and C. P. Smith, *Tetrahedron*, 1969, **25**, 771.
77. F. Ramirez and C. D. Telefus, *J. Org. Chem.*, 1969, **34**, 376.
78. F. Ramirez, J. Bauer, and C. D. Telefus, *J. Amer. Chem. Soc.*, 1970, **92**, 6935.
79. R. Neidlein and R. Mosebach, *Arch. Pharm.*, 1976, **309**, 724.
80. R. Neidlein and W. Friedrich, *Arch. Pharm.*, 1977, **310**, 622.
81. F. Ramirez, C. D. Telefus, and V. A. V. Prasad, *Tetrahedron*, 1975, **31**, 2007.
82. E. Ware, *Chem. Rev.*, 1950, **46**, 403.
83. A. J. Kirby, *Tetrahedron*, 1966, **22**, 3001.
84. F. Ramirez and G. V. Loewengart, *J. Amer. Chem. Soc.*, 1969, **91**, 2293.
85. F. Ramirez and N. B. Desai, *J. Amer. Chem. Soc.*, 1963, **85**, 3252.
86. V. A. Kukhtin and K. M. Kirillova, *Dokl. Akad. Nauk SSSR, Ser. Khim.* 1961, **140**, 835 (*Chem. Abs.*, 1962, **56**, 4607g).
87. W. G. Bentrude, *J. Chem. Soc., Chem. Comm.*, 1967, 174.
88. (a) V. A. Kukhtin and K. M. Orekhova, *Zhur. Obshch. Khim.*, 1960, **30**, 1208 (*Chem. Abs.*, 1961, **55**, 358). (b) V. A. Kukhtin, T. N. Voskoboeva, and K. M. Kirillova, *ibid.*, 1962, **32**, 2333 (*Chem. Abs.*, 1963, **58**, 9127).
89. T. Mukaiyama and T. Kumamoto, *Bull. Chem. Soc. Japan*, 1966, **39**, 879.
90. T. Mukaiyama, H. Nambu, and T. Kumamoto, *J. Org. Chem.*, 1964, **29**, 2243.
91. J. I. G. Cadogan and R. K. Mackie, *Quart. Rev.*, 1974, **3**, 87.
92. T. Mukaiyama, H. Nambu, and M. Okamato, *J. Org. Chem.* 1962, **27**, 3651.
93. W. G. Bentrude and K. R. Darnall, *Tetrahedron Lett.*, 1967, 2511.
94. W. G. Bentrude and K. R. Darnall, *J. Chem. Soc., Chem. Comm.*, 1969, 862.
95. R. Neidlein and R. Mosebach, *Arch. Pharm.* 1974, **307**, 291.
96. R. Neidlein and R. Mosebach, *Arch. Pharm.*, 1976, **309**, 707.
97. W. G. Bentrude, *J. Amer. Chem. Soc.*, 1965, **87**, 4026.

98. F. Ramirez, A. V. Patwardhan, N. B. Desai, and S. R. Heller, *J. Amer. Chem. Soc.*, 1965, **87**, 549.
99. F. Ramirez, O. P. Madan, N. B. Desai, S. Meyerson, and E. M. Banas, *J. Amer. Chem. Soc.*, 1963, **85**, 2681.
100. A. N. Pudovik, I. V. Gur'yanova, and S. P. Perevezentseva, *Zhur. Obshch. Khim.*, 1969, **39**, 1532 (*Chem. Abs.*, 1969, **71**, 112336m).
101. F. Ramirez, S. B. Bhatia, and C. P. Smith, *Tetrahedron*, 1967, **23**, 2067.
102. F. Ramirez, N. B. Desai, and N. Ramanathan, *J. Amer. Chem. Soc.*, 1963, **85**, 1874.
103. F. Ramirez, N. Ramanathan, and N. B. Desai, *J. Amer. Chem. Soc.*, 1963, **85**, 3465.
104. F. Ramirez and N. Ramanathan, *J. Org. Chem.*, 1961, **26**, 3041.
105. (a) R. Burgada and H. Germa, *Compt. Rend.* (*C*), 1968, **267**, 270. (b) *Idem, Bull. Soc. Chim. France*, 1975, 2607. (c) C. Laurenco and R. Burgada, *Tetrahedron*, 1976, 2089.
106. C. Laurenco, D. Bernard, and R. Burgada, *Compt. Rend.* (*C*), 1974, **278**, 1301.
107. (a) N. P. Grechkin and G. S. Gubanova, *Izv. Akad. Nauk SSSR, Ser. Khim.*, 1970, 2803 (*Chem. Abs.*, 1971, **75**, 49221k). (b) C. Laurenco and R. Burgada, *Tetrahedron*, 1976, 2253.
108. R. Burgada, *Compt. Rend.* (*C*), 1976, **282**, 849.
109. P. Savignac, B. Richard, Y. Leroux, and R. Burgada, *J. Organometal. Chem.*, 1975, **93**, 331.
110. S. Trippett and P. J. Whittle, *J. Chem. Soc., Perkin I*, 1975, 1220.
111. (a) R. Burgada and D. Bernard, *Compt. Rend.* (*C*), 1971, **273**, 164. (b) D. Bernard, C. Laurenco, and R. Burgada, *J. Organometal. Chem.*, 1973, **47**, 113.
112. Y. Charbonnel and J. Barrans, *Compt. Rend.* (*C*), 1974, **278**, 355.
113. A. Munoz, M. Koenig, R. Wolf, and F. Mathis, *Compt. Rend.* (*C*), 1973, **277**, 121.
114. C. Laurenco and R. Burgada, *Compt. Rend.* (*C*), 1972, **275**, 237.
115. C. Malavaud and J. Barrans, *Tetrahedron Lett.*, 1975, 3077.
116. R. Burgada, *Bull. Soc. Chim. France*, 1975, 407.
117. D. Hellwinkel, *Chem. Ber.*, 1969, **102**, 528.
118. T. Mukaiyama, J. Kuwajima, and K. Ohno, *Bull. Chem. Soc. Japan*, 1965, **38**, 1954.
119. F. Ramirez, A. S. Gulati, and C. P. Smith, *J. Org. Chem.*, 1968, **33**, 13.
120. D. B. Denney and D. H. Jones, *J. Amer. Chem. Soc.*, 1969, **91**, 5821.
121. P. D. Bartlett, A. L. Baumstark, and M. E. Landis, *J. Amer. Chem. Soc.*, 1973, **95**, 6486.
122. P. D. Bartlett, M. E. Landis, and M. J. Shapiro, *J. Org. Chem.*, 1977, **42**, 1661.
123. D. B. Denney, R. L. Powell, A. Taft, and D. Twitchell, *Phosphorus*, 1971, **1**, 151.
124. V. Mark, *J. Amer. Chem. Soc.*, 1963, **85**, 1884.
125. G. W. Griffin, D. M. Gibson, and K. Ishikawa, *J. Chem. Soc., Chem. Comm.*, 1975, 595.
126. R. Boigegrain and B. Castro, (a) *Tetrahedron Lett.*, 1975, 3459; (b) *Tetrahedron*, 1976, **32**, 1283.
127. I. J. Borowitz, P. D. Readio, and P. Rusek, *J. Chem. Soc., Chem. Comm.*, 1968, 240.

128. A. C. Poshkus and J. E. Herweh, *J. Org. Chem.*, 1964, **29**, 2567.
129. I. J. Borowitz and M. Anschel, *Tetrahedron Lett.*, 1967, 1517.
130. F. Ramirez and C. P. Smith, *J. Chem. Soc., Chem. Comm.*, 1967, 662.
131. P. Petrellis and G. W. Griffin, *J. Chem. Soc., Chem. Comm.*, 1968, 1099.
132. E. M. Rokhlin, Yu. V. Zeifman, Y. A. Cheburkov, N. P. Gambaryan, and I. L. Knunyants, *Dokl. Akad. Nauk SSSR*, 1965, **161**, 1356 (*Chem. Abs.*, 1965, **63**, 4153g).
133. F. Ramirez, C. P. Smith, and J. F. Pilot, *J. Amer. Chem. Soc.*, 1968, **90**, 6726.
134. F. Ramirez, C. P. Smith, J. F. Pilot, and A. S. Gulati, *J. Org. Chem.*, 1968, **33**, 3787.
135. G. H. Birum and C. N. Matthews, (a) *J. Chem. Soc., Chem. Comm.*, 1967, 137; (b) *J. Org. Chem.*, 1967, **32**, 3554.
136. F. Ramirez, J. F. Pilot, C. P. Smith, S. B. Bhatia, and A. S. Gulati, *J. Org. Chem.*, 1969, **34**, 3385.
137. J. A. Gibson, G.-V. Röschenthaler, K. Sauerbrey, and R. Schmutzler, *Chem. Ber.*, 1977, **110**, 3214.
138. B. A. Arbusov, E. N. Dianova, and V. S. Vinogradova, *Izv. Akad. Nauk SSSR, Ser. Khim.*, 1969, 1109 (*Chem. Abs.*, 1969, **71**, 50070j).
139. B. A. Arbusov, V. S. Vinogradova, and O. D. Zolova, *Izv. Akad. Nauk SSSR, Ser. Khim.*, 1968, 2290 (*Chem. Abs.*, 1969, **70**, 29004k).
140. G. Buono and G. Pfeiffer, *Tetrahedron Lett.*, 1972, 149.
141. B. E. Ivanov, L. A. Valitova, L. A. Kudryavtseva, T. G. Bykova, K. A. Derstuganova, and E. I. Gol'dfarb, *Izv. Akad. Nauk SSSR, Ser. Khim.*, 1974, **23**, 636 (*Chem. Abs.*, 1974, **81**, 13599z).
142. B. A. Arbusov, N. A. Polezhaeva, and V. V. Smirnov, *Izv. Akad. Nauk SSSR, Ser. Khim.*, 1975, 686 (*Chem. Abs.*, 1975, **83**, 43445e).
143. H. J. Bestmann, T. Denzel, R. Kunstmann, and I. Lengyel, *Tetrahedron Lett.*, 1968, 2895.
144. J. Wulff and R. Huisgen, *Chem. Ber.*, 1969, **102**, 1841.
145. A. R. Hands and A. J. H. Mercer, *J. Chem. Soc.*, (*C*), 1968, 2448.
146. E. E. Schweizer, W. S. Creasy, J. C. Liehr, M. E. Jenkins, and D. L. Dalrymple, *J. Org. Chem.*, 1970, **35**, 601.
147. W. G. Bentrude and W. D. Johnson, *J. Amer. Chem. Soc.*, 1968, **90**, 5924.
148. L. Horner, A. Schaefer, and W. Ludwig, *Chem. Ber.* 1958, **91**, 75.
149. Q. E. Thompson, *J. Amer. Chem. Soc.*, 1961, **83**, 845.
150. (a) R. W. Murray and M. L. Kaplan, *J. Amer. Chem. Soc.*, 1969, **91**, 5358. (b) *Idem, ibid.*, 1968, **90**, 537. (c) P. D. Bartlett and G. D. Mendenhall, *J. Amer. Chem. Soc.*, 1970, **92**, 210. (d) P. D. Bartlett and A. P. Schaap, *ibid.*, 1970, **92**, 6055. (e) K. Koch, *Tetrahedron*, 1970, **26**, 3503.
151. M. E. Brennan, *J. Chem. Soc., Chem. Comm.*, 1970, 956.
152. L. M. Stephenson and D. E. McClure, *J. Amer. Chem. Soc.*, 1973, **95**, 3074.
153. A. P. Schaap, K. Kees, and A. L. Thayer, *J. Org. Chem.*, 1975, **40**, 1185.
154. A. P. Stewart and S. Trippett, *J. Chem. Soc., Chem. Comm.*, 1970, 1279.
155. K. Burger, J. Fehn, and E. Moll, *Chem. Ber.*, 1971, **104**, 1826.
156. J. Albanbauer, K. Burger, E. Burgis, D. Marquarding, L. Schabl, and I. Ugi, *Annalen*, 1976, 36.
157. W. Steglich, K. Burger, M. Dürr, and E. Burgis, *Chem. Ber.*, 1974, **107**, 1488.
158. K. Burger, J. Albanbauer, and F. Manz, *Chem. Ber.*, 1974, **107**, 1823.
159. K. Burger and R. Ottlinger, *Synthesis*, 1978, 44.
160. K. Burger and J. Fehn, *Tetrahedron Lett.*, 1972, 1263.

161. K. Burger and J. Fehn, *Chem. Ber.*, 1972, **105**, 3814.
162. K. Burger, K. Einhellig, W.-D. Roth, and L. Hatzelmann, (a) *Tetrahedron Lett.*, 1974, 2701; (b) *Chem. Ber.*, 1975, **108**, 2737.
163. K. Burger, W.-D. Roth, and K. Neumayr, *Chem. Ber.*, 1976, **109**, 1984.
164. K. Burger and K. Einhellig, *Chem. Ber.*, 1973, **106**, 3421.
165. K. Burger, H. Goth, and W.-D. Roth, *Z. Naturforsch.*, 1977, **32b**, 607.
166. K. Burger and W.-D. Roth, *Synthesis*, 1975, 731.
167. K. Burger, K. Einhellig, W.-D. Roth, and E. Daltrozzo, *Chem. Ber.*, 1977, **110**, 605.
168. K. Burger, H. Goth, and W. Höhenberger, *Chem. Ztg.*, 1977, **101**, 453.
169. K. Burger, J. Fehn, and E. Müller, *Chem. Ber.*, 1973, **106**, 1.
170. A. Gieren, K. Burger, and W. Thenn, *Z. Naturforsch.*, 1974, **29b**, 399.
171. K. Burger, K. Einhellig, G. Süß, and A. Gieren, *Angew. Chem., Internat. Ed.*, 1973, **12**, 156.
172. A. Padwa, A. Ku, A. Mazzu, and S. I. Wetmore, Jr., *J. Amer. Chem. Soc.*, 1976, **98**, 1048.
173. K. Burger, E. Burgis, and P. Holl, *Synthesis*, 1974, 816.
174. W.-D. Roth, Thesis 1976, Technische Universität München.
175. R. Huisgen and J. Wulff, (a) *Tetrahedron Lett.*, 1967, 917; (b) *Chem. Ber.*, 1969, **102**, 1833.
176. M. Masaki, K. Fukui, and M. Ohta, *J. Org. Chem.*, 1967, **32**, 3564.
177. (a) A. Umani-Ronchi, M. Acampora, G. Gaudiano, and A. Selva, *Chim. Ind.* (Milan), 1967, **49**, 388. (b) G. Gaudiano, R. Mondelli, P. P. Ponti, C. Ticozzi, and A. Umani-Ronchi, *J. Org. Chem.*, 1968, **33**, 4431.
178. H. J. Bestmann and R. Kunstmann, *Chem. Ber.*, 1969, **102**, 1816.
179. G. L'abbé, J. M. Borsus, P. Ykman, and G. Smets, *Chem. and Ind.*, 1971, 1491.
180. T. Sasaki, T. Yoshioka, and Y. Suzuki, *Yuki Gosei Kagaku Kyokai Shi*, 1970, **28**, 1054 (*Chem. Abs.*, 1971, **74**, 125528).
181. R. D. Gareev, E. E. Borisova, and I. M. Shermergorn, *Zhur. Obshch. Khim.*, 1975, **45**, 944 (*Chem. Abs.*, 1975, **83**, 28337c).
182. R. Appel and M. Halstenberg, *Chem. Ber.*, 1976, **109**, 814.
183. J. I. G. Cadogan, D. S. B. Grace, P. K. K. Lim, and B. S. Tait, (a) *J. Chem. Soc., Chem. Comm.*, 1972, 520; (b) *J. Chem. Soc., Perkin I*, 1975, 2376.
184. J. I. G. Cadogan, B. S. Tait, and N. J. Tweddle, *J. Chem. Soc., Chem. Comm.*, 1975, 847.
185. R. Huisgen and J. Wulff, (a) *Chem. Ber.*, 1969, **102**, 746; (b) *Angew. Chem. Internat. Ed.*, 1967, **6**, 457.
186. V. A. Ginsburg, M. N. Vasil'eva, S. S. Dubov, and A. Ya. Yakubovich, *Zhur. Obshch. Khim.*, 1960, **30**, 2854 (*Chem. Abs.*, 1961, **55**, 17477b).
187. B. A. Arbusov, N. A. Polezhaeva, and V. S. Vinogradova, *Izv. Akad. Nauk SSSR, Ser. Khim.*, 1968, **11**, 2525 (*Chem. Abs.*, 1969, **70**, 87682r).
188. A. Schmidpeter and J. Luber, *Chem. Ber.*, 1975, **108**, 820.
189. R. Hoffmann, J. M. Howell, and E. L. Muetterties, *J. Amer. Chem. Soc.*, 1972, **94**, 3047.
190. J. M. Howell, *J. Amer. Chem. Soc.*, 1977, **99**, 7447.
191. C. D. Hall, J. D. Bramblett, and F. F. S. Lin, *J. Amer. Chem. Soc.*, 1972, **94**, 9264.
192. T. J. Katz and E. W. Turnblom, *J. Amer. Chem. Soc.*, 1970, **92**, 6701.
193. D. B. Denney and L. S. Shih, *J. Amer. Chem. Soc.*, 1974, **96**, 317.

194. G. Wittig and A. Maercker, *Chem. Ber.* 1964, **97**, 747.
195. (a) D. Hellwinkel, *Chem. Ber.*, 1965, **98**, 576. (b) D. Hellwinkel and W. Lindner, *ibid.*, 1976, **109**, 1497.
196. D. B. Denney and L. Saferstein, *J. Amer. Chem. Soc.*, 1966, **88**, 1839.
197. (a) D. B. Denney, D. Z. Denney, B. C. Chang, and K. L. Marsi, *J. Amer. Chem. Soc.*, 1969, **91**, 5243. (b) D. I. Phillips, I. Szele, and F. H. Westheimer, *ibid.*, 1976, **98**, 184. (c) C. L. Lerman and F. H. Westheimer, *ibid.*, 1976, **98**, 179.
198. W. G. Voncken and H. M. Buck, *Rec. Trav. Chim. Pays-Bas*, 1974, **93**, 14, 210.
199. F. Ramirez, A. J. Bigler, and C. P. Smith, *J. Amer. Chem. Soc.*, 1968, **90**, 3507.
200. F. Ramirez, A. J. Bigler, and C. P. Smith, *Tetrahedron*, 1968, **24**, 5041.
201. F. Ramirez, S. Lee, P. Stern, I. Ugi, and P. D. Gillespie, *Phosphorus*, 1974, **4**, 21.
202. G. A. Razuvaev, N. A. Osanova, and I. K. Grigor'eva, *Izv. Akad. Nauk SSSR, Ser. Khim.*, 1969, 2234 (*Chem. Abs.*, 1970, **72**, 31952c).
203. G. Wittig and G. Geissler, *Annalen*, 1953, **580**, 44.
204. (a) Houben-Weyl, *Methoden der Organischen Chemie*, Thieme, Stuttgart, 1964, Vol. 12/1, p. 433; (b) *ibid.*, Vol. 12/2, p. 80; and literature cited therein.
205. I. L. Knunyants, V. V. Tyuleneva, E. Ya. Pervova, and R. N. Sterlin, *Izv. Akad. Nauk SSSR, Ser. Khim.*, 1964, 1797 (*Chem. Abs.*, 1965, **62**, 2791d).
206. A. Skowrońska, J. Mikolajczak, and J. Michalski, *J. Chem. Soc., Chem. Comm.*, 1975, (a) 791; (b) 986.
207. G. Bauer and G. Hägele, *Angew. Chem. Internat. Ed.*, 1977, **16**, 477.
208. T. Saegusa, Y. Kimura, N. Ishikawa, and S. Kobayashi, *Macromolecules*, 1976, **9**, 724.
209. T. Saegusa, S. Kobayashi, and J. Furukawa, *Macromolecules*, 1977, **10**, 73.
210. T. Saegusa, S. Kobayashi, and Y. Kimura, *Macromolecules*, 1977, **10**, 64.
211. T. Saegusa, S. Kobayashi, and Y. Kimura, *J. Chem. Soc., Chem. Comm.*, 1976, 443.
212. T. Saegusa, *Angew. Chem. Internat. Ed.*, 1977, **16**, 826.
213. T. Saegusa, S. Kobayashi, Y. Kimura, and T. Yokoyama, *J. Org. Chem.*, 1977, **42**, 2797.
214. (a) H. R. Allcock, *J. Amer. Chem. Soc.*, 1963, **85**, 4050. (b) *Idem., ibid.*, 1964, **86**, 2591. (c) H. R. Allcock and R. L. Kugel, *J. Chem. Soc., Chem. Comm.*, 1968, 1606. (d) H. R. Allcock and E. C. Bissel, *ibid.*, 1972, 676. (e) *Idem, J. Amer. Chem. Soc.*, 1973, **95**, 3154.
215. D. Hellwinkel and H. J. Wilfinger, *Chem. Ber.*, 1970, **103**, 1056.
216. P. C. Chang, D. B. Denney, R. L. Powell, and D. W. White, *J. Chem. Soc., Chem. Comm.*, 1971, 1070.
217. M. J. Gallagher, *Stereochemistry of Heterocyclic Compounds*, Part 2. (ed. W. L. F. Armarego), Wiley-Interscience, New York, 1977, p. 429; and literature cited therein.

12
Organophosphorus Reagents in the Synthesis of Peptides

R. RAMAGE

1. Amino protection, 511
2. Carbonyl activation and amide formation, 518
3. References, 533

Organophosphorus reagents have found useful application in protection of amino functions and activation of carboxylic acid groups in peptide synthesis which can be expressed in simple terms as follows:

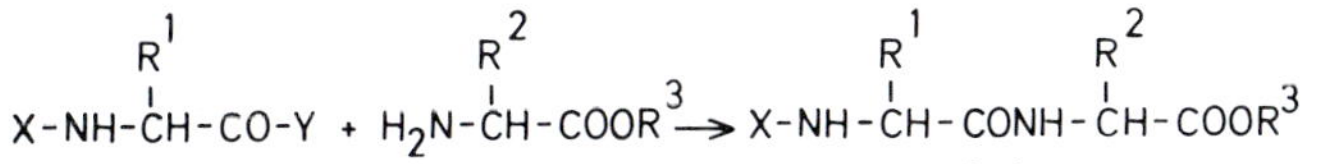

$$\underset{(1)}{X-NH-\underset{\underset{R^1}{|}}{CH}-CO-Y} + \underset{(2)}{H_2N-\underset{\underset{R^2}{|}}{CH}-COOR^3} \longrightarrow \underset{(3)}{X-NH-\underset{\underset{R^1}{|}}{CH}-CONH-\underset{\underset{R^2}{|}}{CH}-COOR^3}$$

Scheme 1

In addition to the groups X and R^3, which effectively direct the coupling to give (3) exclusively, the side-chain substituents of natural aminoacids, R^1 and R^2 (to R^n in a polypeptide), may contain functionality requiring protection to avoid interference with the desired amide formation. The α-amino protection, X, must be amenable to selective removal in the presence of side-chain protecting groups to allow subsequent amide formation solely at the amino terminus of (3). It is in this subtle interplay of protecting groups that organophosphorus derivatives have much to offer in the development of peptide methodology.[1]

1. Amino Protection

The most common amino protecting groups in peptide synthesis are urethanes of two general types (4) and (5) which differ in the conditions required for

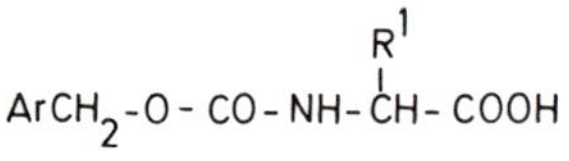

$$\text{ArCH}_2\text{-O-CO-NH-}\underset{\underset{R^1}{|}}{\text{CH}}\text{-COOH} \qquad R_3\text{-}\underset{\underset{R^4}{|}}{\overset{\overset{R^2}{|}}{C}}\text{-O-CO-NH-}\underset{\underset{R^1}{|}}{\text{CH}}\text{-COOH}$$

$$\text{(4)} \qquad\qquad\qquad\qquad \text{(5)}$$

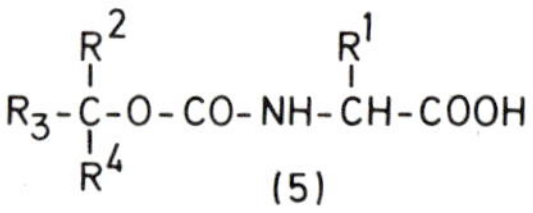

deprotection. Urethanes of type (4a; Ar = Ph; benzyloxycarbonyl or Z protection) can be cleaved by hydrogenolysis, or HBr in acetic acid, whereas (5a; $R^2 = R^3 = R^4 = CH_3$; t-butoxycarbonyl or Boc protection) may be deprotected by mild acid and is stable to hydrogenation. Substitution of R^4 by a biphenyl group produces increased acid lability of the urethane (5b; $R^2 = R^3 = CH_3$, $R^4 =$ biphenyl; biphenylisopropoxycarbonyl or Bpoc protection). Useful working dual combinations of (4), (5a), and (5b) are given in Table 1.

Table 1. Amino protecting groups

Protecting group	Cleavage conditions		Environment of amino function	
(4a)	H_2–Pd	α	side chain	
(5a)	mild acid	side chain	α	side chain
(5b)	v. mild acid			α

Acid cleavage of urethanes based on (5) gives rise to intermediate carbenium ions which may participate in side reactions with the side chains of sulphur-containing aminoacids (methionine and cysteine) or aminoacids having aromatic ring systems which undergo ready electrophilic substitutions, e.g. tryptophan and tyrosine. This necessitates the use of cation scavengers during acidolytic deprotection, but the presence of such scavengers can reduce the desired selectivity of cleavage.[2]

A most important consequence of the use of urethane protecting groups for the α-amino function is the suppression of racemization during activation prior to peptide bond formation (Scheme 1). When X is acyl, i.e. where (1) is a polypeptide, neighbouring group participation may produce an oxazolone (6) capable of facile racemization due to increased acidity of the α-methine bond whereas urethanes are found to couple without loss of stereochemical integrity.

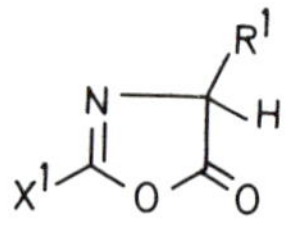

(6)

1 Peptide Synthesis

To summarize, amino protecting groups for aminoacids should be stable except under specific mild cleavage conditions, and should not facilitate racemization via oxazolone intermediates during the amide bond formation. Organophosphorus reagents have great potential for amino protection due to the inherent acid lability of the P–N bond.

1.1. Phosphoric Acid Derivatives

Zervas[3] investigated the use of substituted dibenzyl and diaryl phosphoramidates in peptide synthesis. The phosphorochloridate reagents required for the dibenzyl series could be prepared by the routes shown in Scheme 2.[3,4]

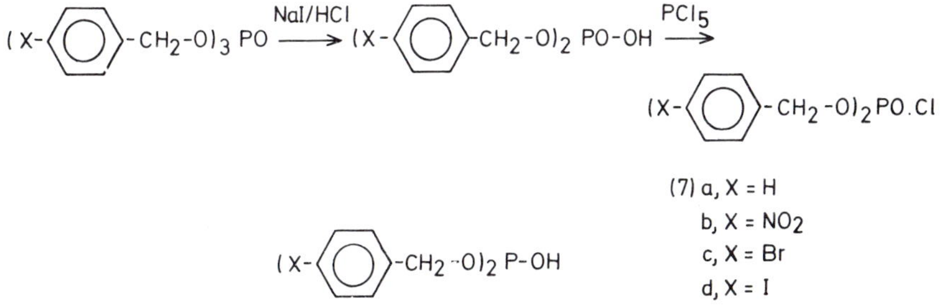

$$(X-\text{C}_6\text{H}_4-CH_2-O)_3 PO \xrightarrow{NaI/HCl} (X-\text{C}_6\text{H}_4-CH_2-O)_2 PO-OH \xrightarrow{PCl_5}$$

$$(X-\text{C}_6\text{H}_4-CH_2-O)_2 PO.Cl$$

$$(X-\text{C}_6\text{H}_4-CH_2-O)_2 P-OH$$

(7) a, X = H
b, X = NO$_2$
c, X = Br
d, X = I

Scheme 2

These reagents were found to react with esters of aminoacids to produce the corresponding dibenzyl phosphoryl aminoacid esters which could be hydrolysed by alkali to the desired acids (8) without cleavage of the phosphoramidate function. The aminoacid derivatives from alanine, glycine, leucine, phenylalanine, and tyrosine were prepared in this way.

Activation of the acids (8) using dicyclohexylcarbodiimide (DCCI) or the mixed diphenylphosphoric anhydride followed by reaction with an aminoacid ester, according to Scheme 1, gave racemization-free coupling. The dipeptide ester (9) could be transformed into the hydrazide using hydrazine hydrate and then, via an azide, into the corresponding anilide.

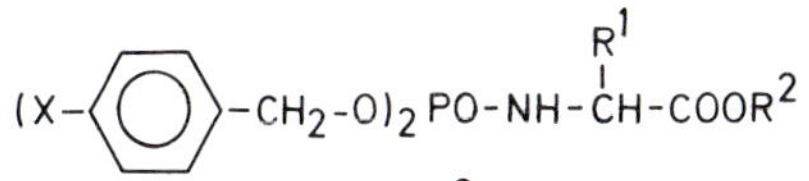

$$(X-\text{C}_6\text{H}_4-CH_2-O)_2 PO-NH-\overset{R^1}{\underset{|}{C}H}-COOR^2$$

(8) R^2 = H

$$(X-\text{C}_6\text{H}_4-CH_2-O)_2 PO-NH-\overset{CH_3}{\underset{|}{C}H}-CONH-\overset{CH_2-CH(CH_3)_2}{\underset{|}{C}H}-COOCH_3$$

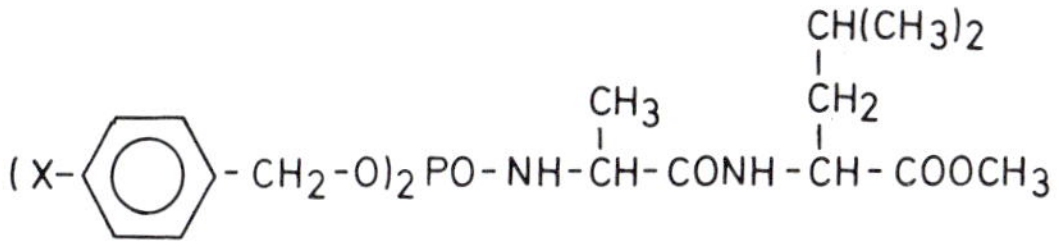

(9)

By this series of reactions Zervas showed that the phosphoramidates were relatively stable to alkaline hydrolysis, did not react adversely with hydrazine (therefore were compatible with the important azide coupling technique) and, furthermore, that the method of protection did not interfere with the normal activation of the carboxylic acid function in (8). Deprotection of the derivatives was effected by hydrogenolysis followed by cleavage of the P–N bond at pH 4 or, alternatively, direct treatment of the phosphoramidates with hydrogen bromide in chloroform. These conditions do not offer real advantage over the benzyloxycarbonyl protection (4a) mentioned earlier.

In the diphenyl phosphoramidate series,[5] hydrolysis of the ester (10a) using barium hydroxide solution at room temperature resulted in concomitant hydrolysis of one phenoxy group to give (11). Hydrogenolysis of the benzyl ester (10b) produced the cyclic anhydride (13) presumably via the desired acid (12) which could not be isolated. Interconversion of (11) and (13) may be accomplished as shown in Scheme 3.

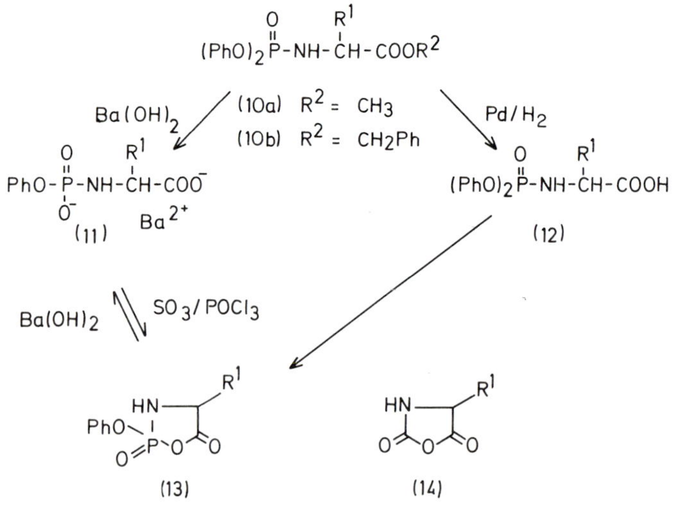

Scheme 3

It is noteworthy that the anhydrides (13) have not been thoroughly investigated[6] as synthons for polypeptide synthesis notwithstanding the close structural similarity with the N-carboxyanhydrides (14) which have had extensive application. An interesting approach[7] to the synthesis of unsymmetrical cystine peptides involves selective cleavage of one benzyloxy residue in (8) by sodium iodide to give (15) which could be coupled via the chloride to another aminoacid ester to produce (16) in which both of the amino functions are protected in the alkyl phosphorodiamidate structure.

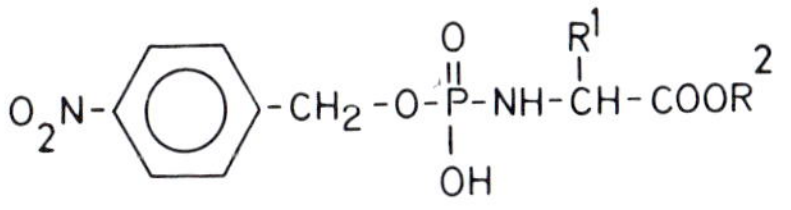

$O_2N-\langle O \rangle-CH_2-O-\overset{O}{\underset{OH}{P}}-NH-\overset{R^1}{\underset{}{CH}}-COOR^2$

(15)

$O_2N-\langle O \rangle-CH_2-O-P\overset{O}{\underset{}{}}\begin{smallmatrix}NH-CH(R^1)-COOR^2\\NH-CH(R^3)-COOR^4\end{smallmatrix}$

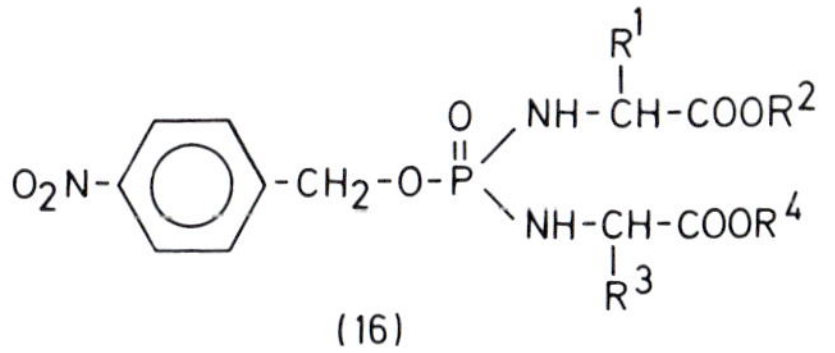

(16)

1.2. Phosphinic Acid Derivatives and Thio Analogues

Another approach to amino protection takes advantage of the remarkable acid lability of the P–N bond in phosphinamides (17).[8] Mechanistic investigations[9] would indicate that acid-catalysed solvolysis of (17) involves initial protonation on nitrogen followed by nucleophilic attack and subsequent fragmentation of the trigonal bipyramidal intermediate (Scheme 4). In this A-2 mechanism for the solvolysis no reactive intermediates are produced that may react with typical side-chain functionality found in peptides.

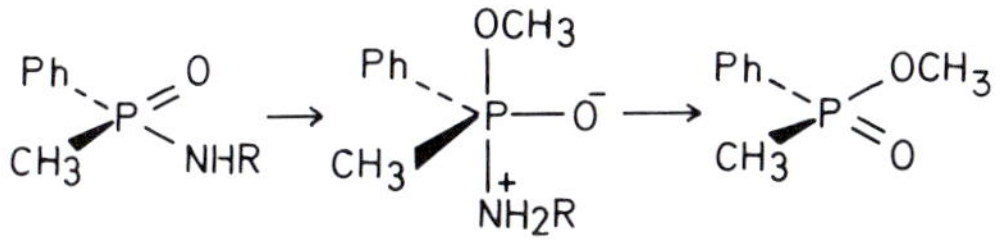

Scheme 4

Examination of the structures of $Ph_2PO.NMe_2$ and $Ph_2PO.NMe.CH_2-CH_2Ph$ by X-ray diffraction[10] shows that the nitrogen geometry is non-planar in contrast to carboxylic acid amide structures. The diminished delocalization of the lone pair of electrons on nitrogen with the P=O bond must be reflected in the relative rates of acid hydrolysis of phosphinamides and amides. Comparison of phosphoramidates (8) and (12) with the corresponding phosphinyl aminoacids (18) would suggest that, in the phosphinamide case, substituent variation should have a greater effect on the chemistry of the P–N bond. This should allow the development of a range of acid-labile phosphinamide derivatives comparable to that illustrated in Table 1 for the urethanes (5a) and (5b). Indeed it has been found[11] that, whereas the $Ph_2PO.NHR$ group is more acid-labile than $Bu^tO.CONHR$, the corresponding $Me_2PO.NHR$ series are so acid-labile that they defy isolation in most cases.

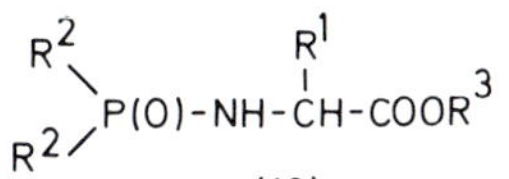

$\overset{R^2}{\underset{R^2}{\diagdown}}P(O)-NH-\overset{R^1}{\underset{}{CH}}-COOR^3$

(18)

Dpp-Gly-Gly-OBu^t

(19)

Dpp-Lys(Z)-Gly-OMe

(20)

The series of stable crystalline diphenylphosphinyl aminoacids shown in Table 2 were prepared[11] by the action of diphenylphosphinyl chloride on the

Table 2. Diphenylphosphinoyl (Dpp) derivatives of aminoacids

Derivative	M.p. (°C)	$[\alpha]^{25}_D$ (deg.)
Dpp.Gly.OH	132	—
Dpp.Ala.OH	152–153	−21·4
Dpp.Val.OH	103	−15·2
Dpp.Leu.OH	131–134	−20·0
Dpp.Ile.OH	113–114	−7·2
Dpp.Lys(Z).OH	153–154	+13·3
Dpp.Met.OH	141–142	−14·0
Dpp.Trp.OH	165–169	−60·5
Dpp.Pro.OH	170–173	−16.2
Dpp.Phe.OH	133	−40.1

corresponding methyl or benzyl ester followed by hydrolysis or hydrogenolysis respectively. As in the case of (9) the diphenylphosphinyl aminoacid methyl esters may be converted into hydrazides for subsequent azide coupling.

Deprotection of the diphenylphosphinamide (Dpp) derivatives may be accomplished by mild acid faster than the Boc group (5a). Furthermore, acid treatment of the dipeptides (19) and (20) showed selective cleavage of the Dpp group in the presence of either a t-butyl ester or side-chain benzyloxycarbonyl amino protection of great importance in solid-phase peptide methodology.

From the stereoelectronic considerations arising from the X-ray data already discussed, it seems unlikely that activated Dpp aminoacid derivatives would suffer racemization via the phosphorus analogue (21) of an oxazolone (6). Confirmation of this came from the synthesis of Dpp-Ile-Gly-OBu^t using DCCI to achieve the coupling of Dpp-Ile-OH with H-Gly-OBu^t. Aminoacid analysis could not detect *allo*-Ile which would have resulted from racemization via (21). As a further test, Dpp-Leu-Ala-OBzl and Z-Leu-Ala-OBzl prepared by three coupling procedures (DCCI, pivalic mixed anhydride, DCCI–HONSu) both gave diastereoisomerically pure H-Leu-Ala-OH after deprotection.

(21)

An excellent test of the compatibility of the Dpp group cleavage with tryptophan and methionine side chains is the synthesis of the partly protected C-terminal tetrapeptide of gastrin (22) by a stepwise synthesis from H-Asp-

(OBu^t)-Phe-NH$_2$ using Dpp-Met-OH and Dpp-Trp-OH in successive couplings. The intermediate and final deprotection of the Dpp groups required no carbenium ion scavengers. In this example the t-butoxycarbonyl amino protection is not satisfactory owing to impurities derived from the t-butyl carbenium ion intermediate produced during the deprotection stages.

$$\text{H-Trp-Met-Asp(OBu}^t\text{)-Phe-NH}_2$$
(22)

A similar approach[12] to α-amino protection of aminoacids using diphenylphosphinothioyl chloride to form the diphenylphosphinothioyl (Ppt) derivatives (23) has the advantage that derivatization may be effected in aqueous alkaline solution directly on the aminoacid without requiring the intermediacy of the corresponding esters as in the case of the phosphoramidates or phosphinamides. A list of Ppt derivatives of amino acids is given in Table 3.

$$\text{Ph}_2\text{P(S)-NH-CH(R}^1\text{)-COOH}$$
(23)

Table 3. Diphenylphosphinothioyl (Ppt) derivatives of aminoacids

Derivative	M.p. (°C)	$[\alpha]_D$ (deg.)
Ppt.Gly.OH	118–119	—
Ppt.Ala.OH, DCHA salt	177–178	−3·7
Ppt.Leu.OH, DCHA salt	137–138	−15·0
Ppt.Val.OH, DCHA salt	149–151	−10·0
Ppt.Phe.OH, DCHA salt	190–191	+8·7
Ppt.Pro.OH, DCHA salt	194–195	−40·0
Ppt.Met.OH, DCHA salt	145–146	−1·2
Ppt.Cys(Bzl)OH, DCHA salt	170–171	+22·5
Ppt.Asn.OH	163–164	−5·0
Ppt.Gln.OH, DCHA salt	172–174	+8·7
Ppt.Trp.OH, DCHA salt	187–191	+7·5

The Ppt derivatives are less acid-labile than the corresponding phosphinamides but were still found to be cleaved faster than the Boc urethane protecting group (5a). No racemization could be observed in the hindered coupling to give Ppt-L-Val-L-Val-OMe, in contrast with the situation found for the coupling employing benzoyl-L-valine where oxazolone formation could be expected to lead to racemization. Using Ppt protection the protected tripeptide Ppt-Trp-Trp-Trp-OMe was synthesized[13] using the solid-phase method without being detrimental to the sensitive tryptophanyl residues.

Another aspect of the chemistry of diphenylphosphinyl aminoacids is the base cleavage of the acidic N–H bond and subsequent methylation to give diphenylphosphinyl-*N*-methyl aminoacids (24) (Table 4).[14] Crystalline, acid-labile derivatives of *N*-methyl aminoacids are not readily available by other methods. The corresponding tosyl-*N*-methyl aminoacids (25) have limited utility owing to the difficulty encountered in cleaving the sulphonamide group.

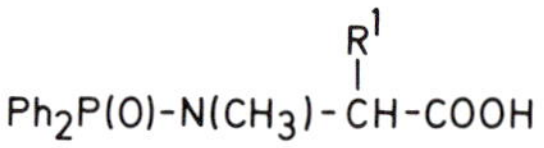

$$Ph_2P(O)-N(CH_3)-\overset{R^1}{\underset{|}{C}}H-COOH \qquad CH_3-\langle\bigcirc\rangle-SO_2-N(CH_3)-\overset{R^1}{\underset{|}{C}}H-COOH$$

$$(24) \qquad\qquad\qquad (25)$$

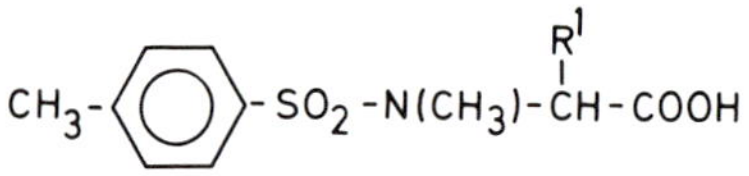

Alkylation of the N–H bond is dibenzylphosphoryl aminoacids (8) has not been investigated for the preparation of *N*-methyl aminoacid derivatives. However, the general synthetic procedure involving alkylation of the anion derived from phosphoramidates has been investigated extensively by Savignac.[15]

Table 4. Diphenylphosphinyl (Dpp) derivatives of *N*-methyl aminoacids

Derivative	M.p. (°C)	$[\alpha]_D$ (deg.)
Dpp(Me)Gly.OH	147–149	—
Dpp(Me)Ala.OH	148–149	−29·6
Dpp(Me)Leu.OH	161–163	−14·3
Dpp(Me)Ile.OH	150–152	−12·3
Dpp(Me)Met.OH	145–155	−24·0
Dpp(Me)Phe.OH	182–184	−87·1
Dpp(Me)Val.OH	150–151	−26·1

2. Carbonyl Activation and Amide Formation

Peptide synthesis is essentially concerned with the formation of the amide bond. However, this simplistic view does not take into account the number of amide linkages to be formed in the synthesis of a moderate-sized polypeptide and hence the efficiency demanded for each step. Other crucial considerations relate to the problems of racemization mentioned earlier in terms of oxazolone intermediates and the presence of side-chain functionality capable of interference with the crucial amide formation. In the continuing quest for reagents capable of meeting the stringent conditions of peptide synthesis, it is not surprising that organophosphorus reagents have received much attention,

12 Peptide Synthesis

since the thermodynamic aspects of carbonyl activation in mixed anhydrides of the type (26) and subsequent nucleophilic attack by an amine (Scheme 5) should be regiospecific and favour amide formation with concomitant generation of a new phosphorus–oxygen bond.

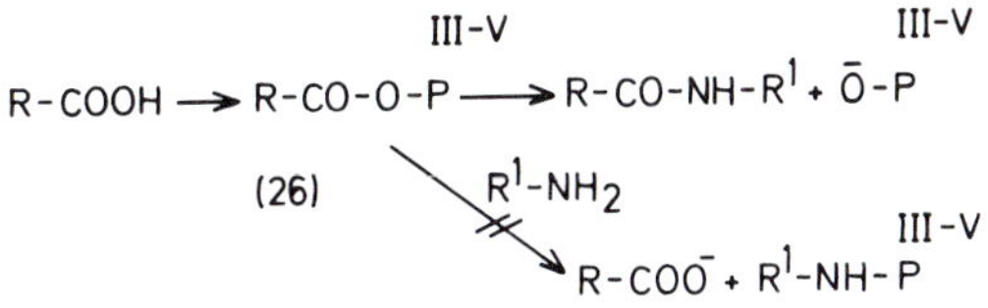

$$\text{R-COOH} \rightarrow \underset{(26)}{\text{R-CO-O-P}} \xrightarrow{\text{III-V}} \text{R-CO-NH-R}^1 + \bar{\text{O}}\text{-P}$$

$$\underset{R^1\text{-NH}_2}{\searrow} \text{R-COO}^- + \text{R}^1\text{-NH-P} \quad \text{III-V}$$

Scheme 5.

This situation should be compared with the problems associated with the use of mixed carboxylic anhydrides (27),[16] particularly in slow coupling reactions between hindered aminoacid residues, e.g. Val . . . Val, where the balance of regiospecificity in the nucleophilic attack by the amine is controlled by steric and electronic effects at the two possible target carbonyl groups (Scheme 6). A serious difficulty with the use of mixed carboxylic anhydrides is the facile thermal disproportionation to symmetrical structures which necessitates the reactions being operated at low temperatures ($-10\ °\text{C}$).

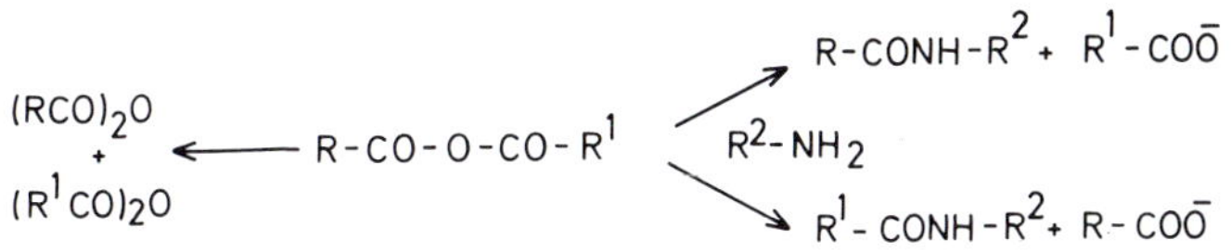

$$(\text{RCO})_2\text{O} + (\text{R}^1\text{CO})_2\text{O} \longleftarrow \text{R-CO-O-CO-R}^1 \xrightarrow{\text{R}^2\text{-NH}_2} \begin{array}{l} \text{R-CONH-R}^2 + \text{R}^1\text{-COO}^- \\ \text{R}^1\text{-CONH-R}^2 + \text{R-COO}^- \end{array}$$

Scheme 6.

2.1. Anhydrides with Phosphoric Acid Derivatives

Research in this area was stimulated by an early hypothesis by Lipmann[17] that protein biosynthesis involved mixed carboxylic-phosphoric anhydrides. This led to Chantrenne[18] and, later, Sheehan[19] studying the potentialities of the mixed anhydrides (27) and (28) respectively. It was found that (28) was capable of acylating aminoacid esters at pH 7·4 and 37°C, i.e. approaching physiological conditions. In an alternative approach[20] it was found that the anhydride (29) (formed from phenyl phosphorodichloridate and 2 equivalents of an acyl aminoacid) suffered from rapid disproportionation to the symmetrical anhydrides.

Reaction of acylamino acids with the enol phosphate (30) was found to produce the *O,O*-diethylphosphoric mixed anhydride (31) which could be used in peptide synthesis.[21] In this case disproportionation was not observed although the initial reaction occurred at 70°C.

$$\underset{(27)}{\text{Z-NH-CH(R}^1\text{)-CO-O-P(=O)(O}^-\text{)-OPh}}$$

$$\underset{(28)}{\text{Phth-N-CH(R}^1\text{)-CO-O-P(=O)(OBzl)}_2}$$

$$\underset{(29)}{(\text{AcylNH-CH(R}^1\text{)-CO-O})_2\text{P(O)-OPh}}$$

$$\underset{(30)}{(\text{EtO})_2\text{P(O)-O-CH(OEt)-CH(H)-COOEt}}$$

$$\underset{(31)}{\text{AcylNH-CH(R}^1\text{)-CO-O-PO(OEt)}_2}$$

$$\underset{(32)}{(\text{BzlO})_2\text{P(O)-NH-CH(R}^1\text{)-CO-O-PO(OPh)}_2}$$

Zervas,[3] in his study of phosphoramidate derivatives of aminoacids, used diphenyl phosphorochloridate to form the mixed anhydride (32) which gave racemization-free coupling in the preparation of the dipeptides H-Met-Met-OH and H-Phe-Gly-OH. In the latter case there was a by-product which could be attributed to the alternative fission of the mixed anhydride (32; $R^1 = CH_2Ph$).

Shiori and Yamada have introduced the reagents (33)[22] and (34)[23] into peptide methodology. Reaction of diphenyl phosphorochloridate with sodium azide in acetone gives diphenyl phosphorazidate (33). An Arbusov reaction between triethyl phosphite and cyanogen bromide gives diethyl phosphorocyanidate (34).

$$\underset{(33)}{(\text{PhO})_2\text{P(O)-N}_3} \qquad \underset{(34)}{(\text{EtO})_2\text{P(O)-CN}}$$

The coupling of acylaminoacids or acylpeptides with aminoacid esters or peptide amino components can be effected using (33) at 0°C in dimethylformamide containing 2 equivalents of triethylamine. No serious problems were encountered with functional side chains such as asparagine, glutamine, serine, threonine, tyrosine, histidine, methionine, and tryptophan. However, using (33) in the Young test,[24] i.e. the formation of Bz-Leu-Gly-OEt in which the oxazolone (6; $R^1 = CH_2CHMe_2$, $X^1 = Ph$) might be expected to be an intermediate, the optical purity of the product was 89%. The Izumiya test[25] involving condensation of Boc-Gly-Ala-OH with H-Leu-Resin was chosen to check the viability of the reagent in the coupling of fragments in solid-phase peptide synthesis. After deprotection to give H-Gly-Ala-Leu-OH it was found that the extent of racemization, $[D,L/(D,L + L,L)] \times 100\%$, was 2·5% at 20°C and 2% at 0°C. In the case of the reagent (34) even less racemization was observed in the Young test[24] (96% optically pure) and in the Izumiya test[25] (extent of racemization 1% at 20°C and 0·5% at 0°C) which is an excellent model for the true peptide coupling step. Encouraged by these results the Japanese workers synthesized [26] porcine motilin (35) using a combination of solution and solid-phase methods with (33) and (34) for the

H-Phe-Val-Pro-Ile-Phe-Thr-Tyr-Gly-Glu-Leu-Gln-
Arg-Met-Gln-Glu-Lys-Glu-Arg-Asn-Lys-Gly-Gln-OH

(35)

coupling reactions. Consideration of yields and optical purity indicates that (34) is slightly superior to (33) in peptide synthesis. Although the reagents (33) and (34) have been usefully applied, the mechanism of action, particularly of (33), is not certain (Scheme 7). It is not known whether the intermediate (36) reacts intermolecularly with the amino component or dissociates to give the acid azide (37) which would be expected to undergo bimolecular nucleophilic attack to give the amide product.

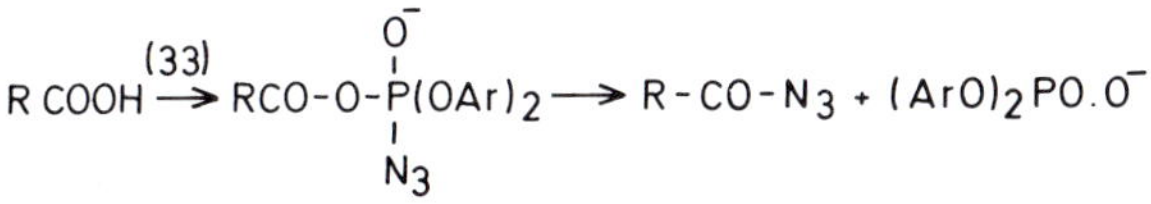

$$\text{R COOH} \xrightarrow{(33)} \text{RCO-O-}\underset{\underset{N_3}{|}}{\overset{\overset{O^-}{|}}{P}}(OAr)_2 \longrightarrow \text{R-CO-N}_3 + (ArO)_2 PO.O^-$$

(36)   (37)

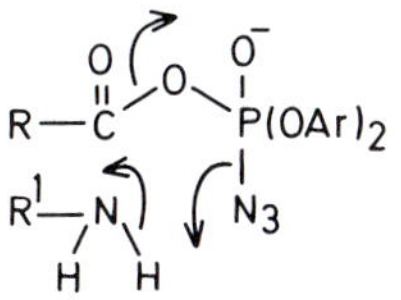

Scheme 7

Hexachlorotriphosphatriazene (38), the trimer of phosphonitrilic chloride, can be used to activate carboxylic acids for amide formation (Scheme 8).[27] The reagent reacts with 2 equivalents of the carboxylic acid in the presence of triethylamine or N-methylmorpholine to give an intermediate postulated as (39). Although the method is simple it does give some racemization and has the complication that the side-chain amides of glutamine and asparagine are dehydrated to the corresponding nitriles. It is possible to use (38) for the

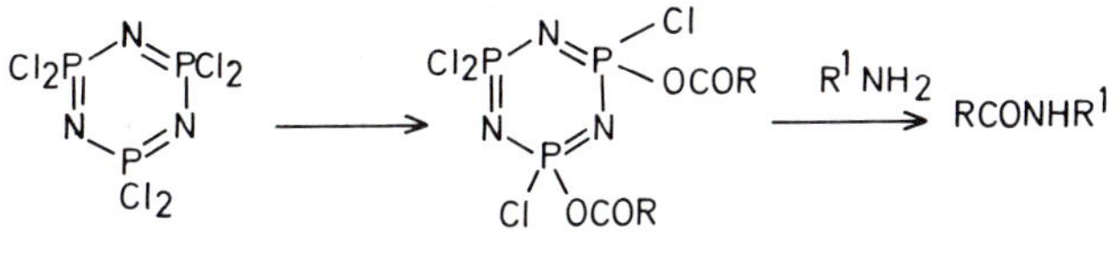

(38) → (39) $\xrightarrow{R^1 NH_2}$ RCONHR^1

Scheme 8

preparation of activated esters of protected aminoacids.[28] The phagocytosis-stimulating tetrapeptide tuftsin (40) has been synthesized[29] using (38) in stepwise activation of the protected aminoacids and also by the *o*-nitrophenyl active ester procedure.

$$\text{H-Thr-Lys-Pro-Arg-OH}$$
$$(40)$$

Perhaps the simplest case of formation of a mixed phosphoric-carboxylic anhydride would be the reaction of 3 moles of an *N*-acylaminoacid with 1 mole of phosphorus oxychloride; but this approach was unsuccessful. However, the monoanhydride (41) proved to be an effective acylating agent in anhydrous media provided that the reaction temperatures were kept at $-15°C$ to prevent disproportionation.[30] In this method the amino component should be present during activation of the carboxylic acid since it was found that the mixed anhydride derived from Z-Gly-OH–POCl$_3$ reacted with Z-Gly-OH to give Z-(Z-Gly)-Gly-OH (42) which is a known rearrangement product of the symmetrical anhydride of Z-Gly-OH. It is of interest that in the competition between the carboxylate and amine for phosphorus oxychloride the formation of the new P–O bond preferentially allows formation of (41). This method requires protection of thiol and probably hydroxyl side-chain functionality.

$$\text{R-CO-O-POCl}_2 \qquad \begin{array}{l}\text{Z-NH-CH}_2\text{-CO}\\ \quad\quad\quad\quad\quad | \\ \quad\quad\text{Z-N-CH}_2\text{COOH}\end{array}$$
$$(41) \qquad\qquad\qquad (42)$$

The amide groups of glutamine and asparagine give side reactions with the phosphorus oxychloride and correspondingly lower yields. In addition, the method is only suitable for stepwise elongation using urethane-protected aminoacids since coupling of Z-Phe-Val-OH with H-Phe-OBu^t produced considerable amounts of Z-Phe-D,L-Val-Phe-OBu^t.[31]

2.2. Anhydrides with Phosphinic Acid Derivatives

In considering the choice of mixed anhydride for activation of a carboxylic acid the following factors must be considered: the formation of the anhydride should be fast; the anhydride should, if possible, be thermally stable; there should be no side reactions due to lack of regiospecificity in the concluding nucleophilic step to give the product.

Comparison of the solvolysis data on phosphorochloridates (43) and phosphinyl chlorides (44) shows that the latter are more reactive towards oxygen nucleophiles.[32] The change from R = Me to R = OMe causes a 300-fold decrease in the rate of solvolysis in absolute alcohol. In addition, the mixed anhydrides (45) prepared from (44) should be more effectively cleaved

12 Peptide Synthesis

by nucleophilic attack than (46) derived from phosphorochloridates, examples of which were discussed earlier, e.g. (32), since hydrolysis of tetra-alkyl pyrophosphates is much slower than that of the corresponding anhydrides of dialkylphosphinic acids. Furthermore, any nucleophilic substitution of R^1O at phosphorus in (46) is obviated in the phosphinic-carboxylic anhydride (45).

$(R^1O)_2P(O)-Cl$ $(R^1)_2P(O)-Cl$ $R-CO-O-PO(R^1)_2$
(43) (44) (45)

$R-CO-O-PO(OR^1)_2$
(46)

Although carboxylic-phosphinic anhydrides have not received much study, the cyclic system (47) shows an interesting change in regiospecificity depending on the nature of the nucleophile, in that aminolysis produces (48) whereas alcoholysis favours the formation of a new P–O bond to give (49).[33] This situation is entirely compatible with peptide synthesis.

(48) (47) (49)

The regiospecificity of aminolysis is sustained even in the case of $Bu^tCOO-POPh_2$ in which the steric effect might be expected to counter the electronic effects. A comparative study of the anhydrides (50) and (51) in peptide synthesis under standard conditions showed the latter to be more reactive, have less tendency towards disproportionation, and give improved product purity, especially in hindered couplings.[34]

$R-CO-O-CO-Bu^t$ $R-CO-O-PO(Ph)_2$
(50) (51)

2.3. Anhydrides with Phosphorous Acid Derivatives

Reaction of acylaminoacids with diethyl chlorophosphite (52) results in the formation of the mixed carboxylic-phosphorous anhydride (53)[35,36] which is susceptible to aminolysis to give the dipeptide as shown in Scheme 9. An interesting variant[37] of this procedure involves the prior reaction of the amino component with (52) to give the amidate (54) which may be converted into the dipeptide on reaction with the acylaminoacid. In this process the drive towards P–O bond formation probably produces an intermediate of the type (55) which could decompose to the amide as indicated.

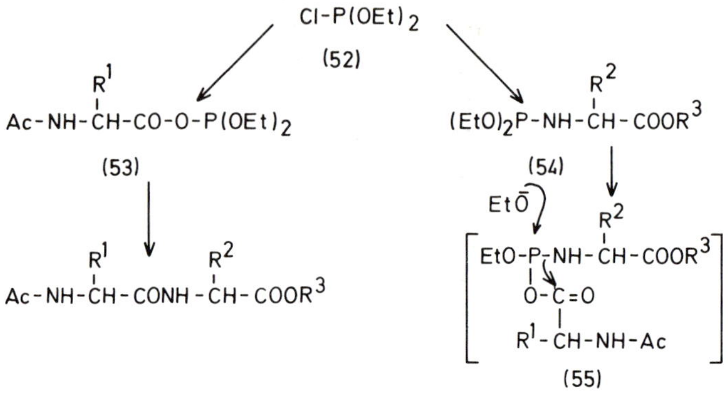

$$
\begin{array}{c}
\text{Cl-P(OEt)}_2 \\
(52)
\end{array}
$$

$$
\text{Ac-NH-} \underset{R^1}{\text{CH}} \text{-CO-O-P(OEt)}_2 \qquad (\text{EtO})_2\text{P-NH-}\underset{R^2}{\text{CH}}\text{-COOR}^3
$$

(53) \qquad\qquad (54)

$$
\text{Ac-NH-}\underset{R^1}{\text{CH}}\text{-CONH-}\underset{R^2}{\text{CH}}\text{-COOR}^3
$$

$$
\left[\begin{array}{c} \text{EtO-P-NH-}\underset{R^2}{\text{CH}}\text{-COOR}^3 \\ \text{O-C=O} \\ R^1\text{-CH-NH-Ac} \end{array} \right]
$$

(55)

Scheme 9

Later modifications[38–40] of this method led to the investigation of the phosphites (56)–(60) as superior reagents to (52). In an application[41] of the method to the synthesis of biologically active cyclic peptides, *o*-phenylene chlorophosphite was the preferred reagent because of accessibility and stability. Trimethyl phosphite or diethyl phosphite may be used to remove hydrogen chloride produced in the reaction instead of a tertiary amine.

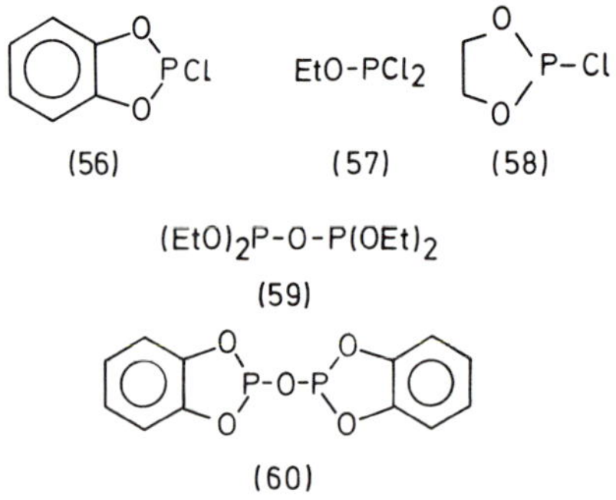

(56)   EtO-PCl$_2$ (57)   (58)

$(\text{EtO})_2\text{P-O-P(OEt)}_2$

(59)

(60)

For these reagents the side-chain functional groups of tyrosine, serine, and threonine should be protected whereas protonation of arginine is sufficient to avoid reaction at the guanidino function.

An interesting combination of P(V) and P(III) activation may be found in a report[42] on the use of the reagent (61) which reacts with a Z-aminoacid to give the carboxylic-phosphoric anhydride (62) [cf. (31),(32)] and then the dipeptide by nucleophilic attack as shown in Scheme 10. The initial attack of the carboxylate group on (61) obviously favours the trigonal bipyramidal intermediate whereas prior reaction of the amino component with (61) afforded the intermediate (54) resulting from direct nucleophilic displacement of the diethyl phosphate ion. As discussed previously (54) can be used in dipeptide formation by reaction with a Z-aminoacid.

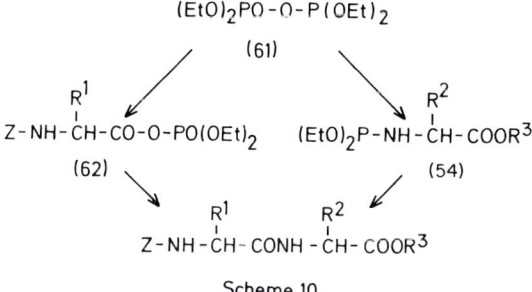

Scheme 10

The well established phosphorazo method of peptide synthesis shown in Scheme 11 has mechanistic similarities to the above methods involving intermediates of the type (54). Reaction of pure phosphorus trichloride with 2 moles of amino component together with 3 moles of a tertiary base gives the dimeric intermediates (63)[43] which must dissociate to the monomer (64) on reaction with an acylaminoacid to produce an intermediate of the type (65);

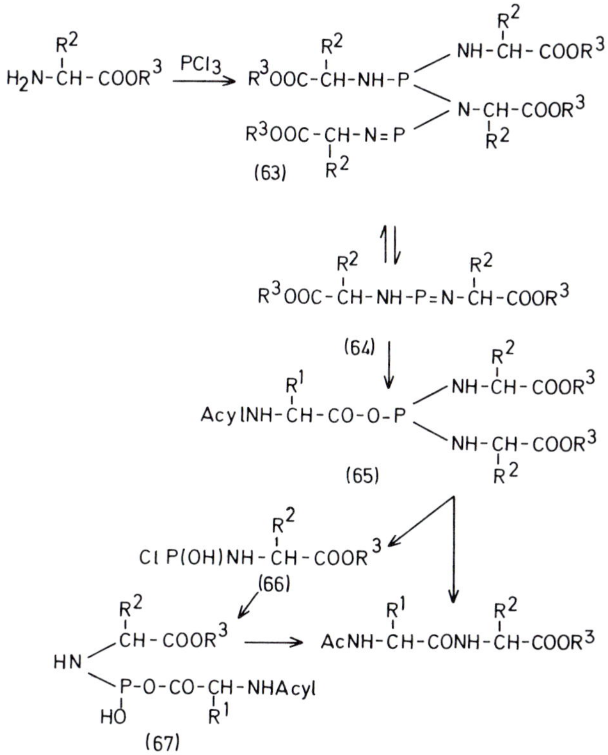

Scheme 11

cf. intermediate (55) in Scheme 9. A variant of the method[39] utilized ethyl phosphorodichloridate instead of phosphorus trichloride. Intramolecular amide formation within (65) gives the desired dipeptide together with (66) which can react with a further mole of carboxyl component to give (67) from which a second mole of dipeptide may be produced with the expulsion of the phosphorus by-product $(H-PO_2)_n$.

Although phosphorazo compounds react spontaneously with water, it was found[44] that Z-His(Bzl)OH,0·5H$_2$O was compatible with the method although the pathway would surely have deviated from that outlined in Scheme 11. The sterically demanding coupling to give Z-Val-Val-OEt could be achieved[45] in 85–90% yield by the phosphorazo method. Whereas secondary amines such as piperidine react with phosphorus trichloride to give the tris-amide of phosphorous acid, in the case of proline[46] steric effects allow only 2 moles of the proline to react to give an intermediate assumed to be (68). The method is very useful for the preparation of proline peptides, e.g. Z-Lys(Z)-Pro-OBzl (90%), but has found limited application.

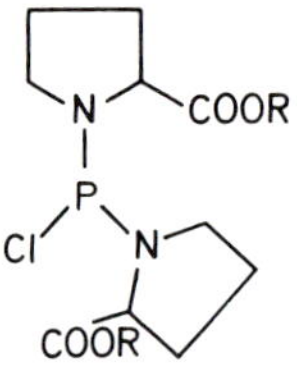

(68) 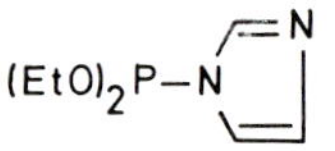 (69) 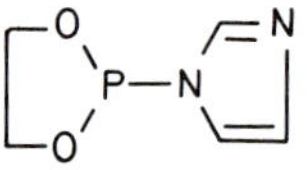 (70)

2.4. Combination of Phosphorus(III) Reagents with Amines

In an attempt to circumvent side-reactions possible with phosphonochloridates, e.g.

$$2 (RO)_2PCl \rightleftharpoons (RO)_3P + RO\,PCl_2$$

Brenner[47] investigated the application of RPCl$_2$ and R$_2$PCl to peptide synthesis. It was found that complete racemization occurred in the coupling of N-TFA-L-Pro-L-Val-OH to H-L-Pro-OMe in the Weygand test using RPCl$_2$ (R = Et, Ph, p-CNC$_6$H$_4$). However, it was found that reaction of Z-aminoacids with imidazole–PCl$_3$ in acetonitrile afforded an efficient preparation of the corresponding imidazolides (72). The degree of racemization, as measured by the Anderson test,[48] was found to be lower than when the imidazolides were prepared using carbonyldiimidazole. Addition of imidazole to coupling reactions using tetraethyl pyrophosphite (59) resulted in improved efficiency possibly due to the intermediacy of (69) which defied isolation,[49] although the corresponding ethylene glycol derivative (70) could be prepared and used for amide formation.

Condensation of N-acylaminoacids and aminoacid esters may be accomplished effectively by imidazole and triphenyl phosphite.[50] Omission of imidazole results in diminished yields and requires higher reaction temperatures. It is not certain whether the crucial reagent is (71), which may itself activate the carboxyl component in two ways via (53), or the acylimidazole (72). Application of the Izumiya test[25] (H-Gly-Leu-Ala-OH) to the method showed racemization to be solvent-dependent [acetonitrile 2·1%, dimethylformamide 3·7%, toluene 5·2%, ethyl acetate 5·7%, and pyridine 17·3%]. The method is applicable to most aminoacids, with suitable side-chain protection, but gives low yields generally with proline, asparagine, and glutamine.

$$\text{(71)} \quad \text{Im-N-P(OAr)}_2 \qquad \text{(72)} \quad \text{Im-N-CO-CH(R}^1\text{)-NHAcyl}$$

Phosphorous acid and its esters react with pyridine in the presence of mercuric chloride to give N-phosphonium salts of pyridine (73) which may be used for the preparation of simple peptides.[51]

$$H\text{-PO-(OPh)}_2 \xrightarrow[\text{Pyridine}]{HgCl_2} \underset{(73)}{\overset{Py^+}{\underset{Cl}{\overset{|}{O}-P}}}\!\!\!\begin{array}{c}OPh\\OPh\end{array}$$

$$\xrightarrow{RNH_2} \bar{O}-\overset{Py^+}{\underset{OPh}{P}}\!\!\!\begin{array}{c}NHR\\OPh\end{array}$$

$$\xrightarrow{RCO\bar{O}} \bar{O}-\overset{Py^+}{\underset{OPh}{P}}\!\!\!\begin{array}{c}O\text{-CO-R}\\OPh\end{array}$$

The postulated intermediate (73) may react with amines or carboxylic acids —a situation reminiscent of Scheme 9—to give species which may be substituted by carboxylate or amino components respectively in a coupling reaction. Since the procedure involves adding both the amino and carboxyl components to a pyridine solution of (73) at 45°C it is likely that the first reaction would involve formation of (74) which would be expected to suffer displacement of pyridine to give (75) with subsequent rearrangement to the dipeptide (Scheme 12). Alternatively the carboxylic-phosphoric anhydride may be involved either by decomposition of (74) or simply by formation from $(PhO)_2POCl$ derived *in situ* from (73).

An extension of this research[52] showed that the oxidant mercuric chloride may be omitted from the reaction mixture. Amides and also active esters, such as those derived from *p*-nitrophenol, can be prepared by this simplified procedure which is considered to involve the intermediate (77). Decomposition

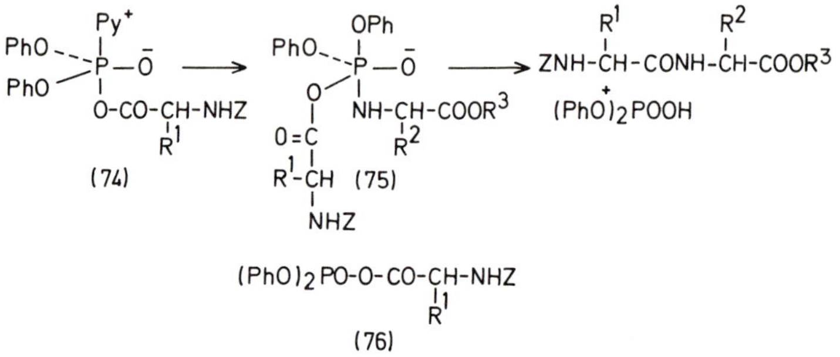

$$\text{PhO} \overset{Py^+}{\underset{PhO}{\succ}} \overset{|}{P} - \bar{O} \longrightarrow \text{PhO} \overset{OPh}{\underset{O}{\succ}} \overset{|}{P} - \bar{O} \longrightarrow ZNH-\underset{R^1}{CH}-CONH-\underset{R^2}{CH}-COOR^3$$

$$\begin{array}{c} | \\ O-CO-CH-NHZ \\ | \\ R^1 \\ (74) \end{array} \quad \begin{array}{c} O=\underset{|}{C} \quad NH-\underset{|}{CH}-COOR^3 \\ | \quad R^2 \\ R^1-CH \quad (75) \\ | \\ NHZ \end{array} \quad (PhO)_2POOH$$

$$(PhO)_2PO-O-CO-\underset{R^1}{CH}-NHZ$$
(76)

Scheme 12

of this trigonal bipyramidal intermediate could give the mixed anhydride (78) which would yield the amide or ester depending on the nature of the nucleophile (Scheme 13).

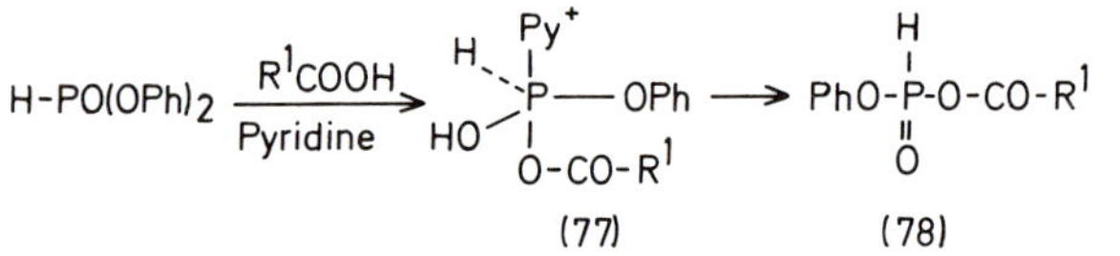

$$H-PO(OPh)_2 \xrightarrow[\text{Pyridine}]{R^1COOH} \begin{array}{c} H \overset{Py^+}{\underset{|}{\succ}} \\ \succ P-OPh \\ HO \overset{|}{\underset{O-CO-R^1}{}} \end{array} \longrightarrow \begin{array}{c} H \\ | \\ PhO-P-O-CO-R^1 \\ \overset{||}{O} \end{array}$$

(77)   (78)

Scheme 13.

2.5. Activation Methods Based on R_3PO Formation

Following the preparation[53] of acid chlorides using Ph_3P-CCl_4 (Chapter 9) this method of carboxyl activation was applied initially[54] with limited success to peptide synthesis; however, Appel[55] has established conditions which are useful for stepwise elongation of Z-aminoacids. There are no complications reported for serine, threonine, and tyrosine side-chains, and Z-Gln-Gly-OBzl could be prepared in satisfactory yield. Of great significance was the complete racemization found in the Young test[24] (Bz-Leu-Gly-OEt). The mechanism is thought to proceed via an acyloxyphosphonium salt (79) which may react to give the chloride or to give an amide by aminolysis. The dominant features in the sequence (Scheme 14) are the formation of a P–O bond in the intermediate and the generation of P^+-O^- in the nucleophilic displacement.

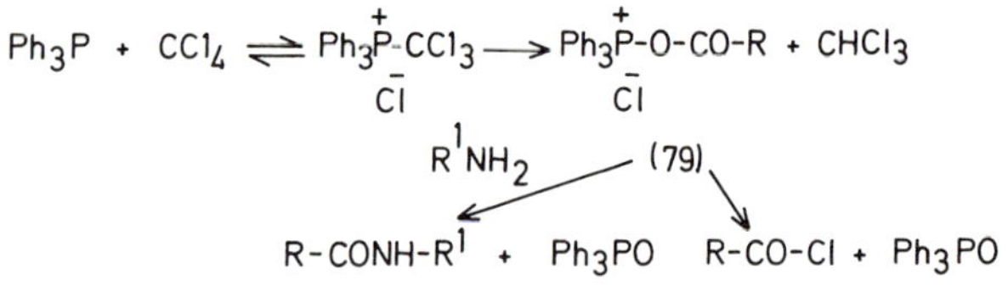

$$Ph_3P + CCl_4 \rightleftharpoons \underset{Cl}{Ph_3\overset{+}{P}-CCl_3} \longrightarrow \underset{Cl}{Ph_3\overset{+}{P}-O-CO-R} + CHCl_3$$

$$\underset{R-CONH-R^1 + Ph_3PO}{\xleftarrow{R^1NH_2}} \quad (79) \searrow \quad R-CO-Cl + Ph_3PO$$

Scheme 14

An analogous sequence of reactions has been investigated[56] incorporating tris(dimethylamino)phosphine and carbon tetrachloride. In the absence of a nucleophile the symmetrical anhydride is formed, possibly from the fugitive intermediate acyloxyphosphonium salt:

$$CCl_4 + (R_2N)_3P + RCOOH \rightarrow (RCO)_2O + (R_2N)_3PO$$

This approach has the advantage of facile removal of the phosphoric triamide by-product in contrast to triphenylphosphine oxide. A series of reagents $(Me_2N)_3P^+X\ PF_6^-$ have been investigated.[57] Where X = Br or CN, benzoic acid affords the anhydride in 90% yield, but when X = N_3 the acid azide is formed. Peptide coupling mediated by $(Me_2N)_3P^+Cl\ ClO_4^-$ led to complete racemization (Young test[24]) and so this reagent must be limited to stepwise elongation of Z-aminoacids.

The development of the oxidation–reduction method by Mukaiyama[58] has great similarities to the methods just discussed but has found much more application in peptide synthesis. The earliest procedure[59] (Scheme 15) involved reaction of copper(II) salts of N-acylaminoacids and N-(2-nitrophenylsulphenyl)aminoacid esters with triphenylphosphine. The mechanism may again involve acyloxyphosphonium intermediates since racemization was observed.

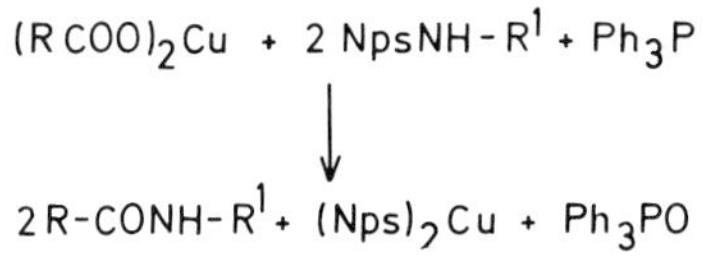

$$(RCOO)_2Cu + 2\ NpsNH-R^1 + Ph_3P$$
$$\downarrow$$
$$2\ R-CONH-R^1 + (Nps)_2Cu + Ph_3PO$$

Scheme 15

A modification of the method using Z-aminoacids, aminoacid ester, triphenylphosphine, a disulphide, and mercuric chloride as thiol scavenger[60] (Scheme 16) gave high yields with less racemization observed in the Young test (95% L-isomer). Further refinement by the Japanese group led to the elimination of thiol scavengers by the careful selection of the disulphide

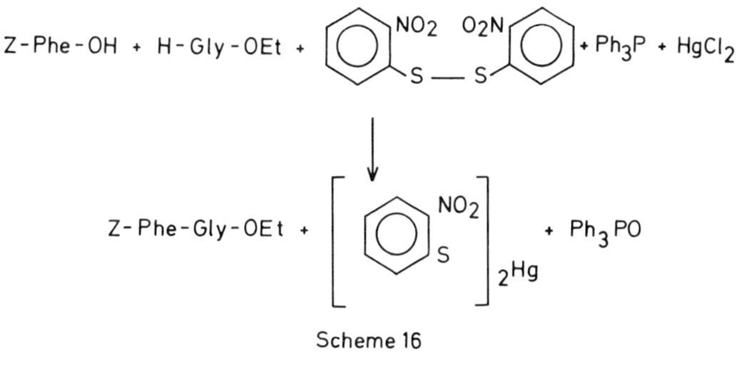

Z-Phe-OH + H-Gly-OEt + (2-O₂N-C₆H₄-S)₂ + Ph₃P + HgCl₂

↓

Z-Phe-Gly-OEt + [2-O₂N-C₆H₄-S]₂Hg + Ph₃PO

Scheme 16

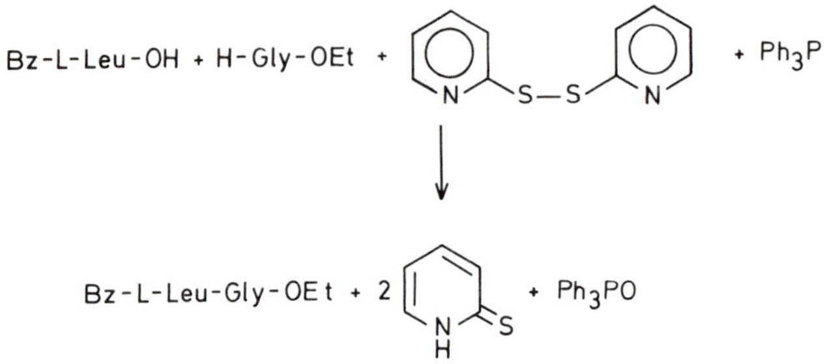

Bz-L-Leu-OH + H-Gly-OEt + [2,2'-dipyridyl disulfide] + Ph$_3$P

↓

Bz-L-Leu-Gly-OEt + 2 [pyridine-2-thione] + Ph$_3$PO

Scheme 17

additive.[61] In this case (Scheme 17) the thiol tautomerizes to the thione in solution, and is effectively removed without the use of mercuric ions.

According to the Young test,[24] racemization is low using this later technique which has great potential in peptide synthesis, being applicable to a variety of solvent systems. No difficulties were observed when methionine, cysteine, and tryptophan were the carboxyl components. The troublesome side-chain functionality of serine, threonine, and tyrosine may be used without protection. Furthermore, asparagine and glutamine did not participate appreciably in undesirable dehydration reactions to give the side-chain nitriles.

A combination of the principles embodied in Schemes 15 and 17 allows the elongation of two peptide bonds as shown in Scheme 18.

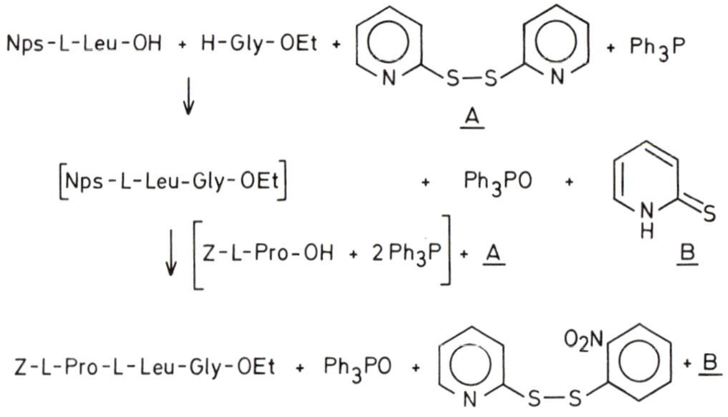

Nps-L-Leu-OH + H-Gly-OEt + [2,2'-dipyridyl disulfide] + Ph$_3$P

↓   **A**

[Nps-L-Leu-Gly-OEt]   +   Ph$_3$PO   +   [pyridine-2-thione]

↓ [Z-L-Pro-OH + 2Ph$_3$P] + **A**   **B**

Z-L-Pro-L-Leu-Gly-OEt + Ph$_3$PO + [2-nitrophenyl pyridyl disulfide] + **B**

Scheme 18

Acylaminoacid active esters may be prepared[62] by the oxidation–reduction method and subsequently be incorporated into a peptide-forming step as shown in Scheme 19.

Substitution of triphenylphosphine by the less nucleophilic triphenyl phosphite gives virtually optically pure Bz-Leu-Gly-OEt, and a study of substituent effects[63] suggests that tri-(p-bromophenyl) phosphite is the best

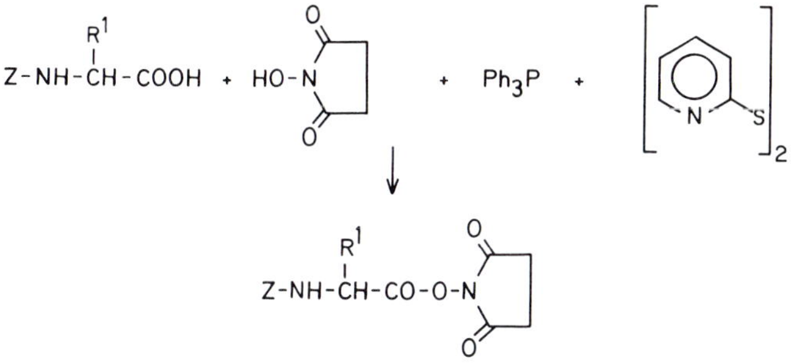

12 Peptide Synthesis

Scheme 19

reagent for the oxidation–reduction procedure. Further modifications of the reagents have been devised[64] to allow the non-peptide products to be removed from the reaction by virtue of acid solubility. Since the method has low racemization tendency, it is not likely that an acyloxyphosphonium salt is the intermediate. A feasible mechanism involves the intermediate (80).

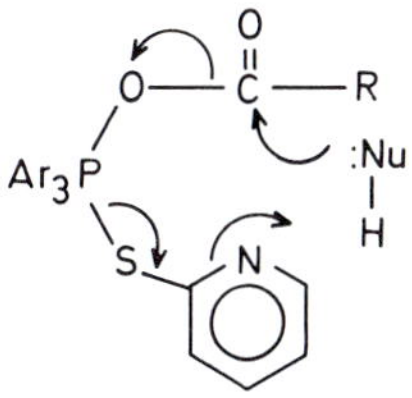

(80)

Mukaiyama has applied his method to the solid-phase synthesis of such important peptides as LH-RH[65] and ACTH(1–24),[66] so showing the great versatility and enormous potential of organophosphorus methodology to peptide synthesis.

In another attempt[67] to activate carboxylic acids in the form of acyloxyphosphonium salts, the reagent (81) was prepared from hexamethylphosphortriamide (HMPT) and p-toluenesulphonic anhydride followed by anion exchange with sodium tetrafluoroborate, and used as shown in Scheme 20. Reaction of (81) with Z-Gly-OH in dimethylformamide in the presence of N-methylmorpholine at 3°C gave a slow liberation of HMPT. Using [1-^{13}C]- and [2-^{13}C]-Z-Gly-OH and ^{31}P n.m.r. to detect new phosphorus-containing species by ^{13}C–^{31}P spin–spin coupling, it was not possible to identify the acyloxyphosphonium salt as an intermediate. When the Izumiya test[25] (Z-Gly-Ala-OH → Z-Gly-Ala-Leu-OBzl) was applied to the reagent (81) it was found that the degree of racemization varied according to the base strength (Et$_3$N 43%, N-methylmorpholine 16%, polymeric Hünig base 9%), which indicated the intermediacy of an oxazolone and was suggestive of a highly activated carboxyl function such as an acyloxyphosphonium salt.

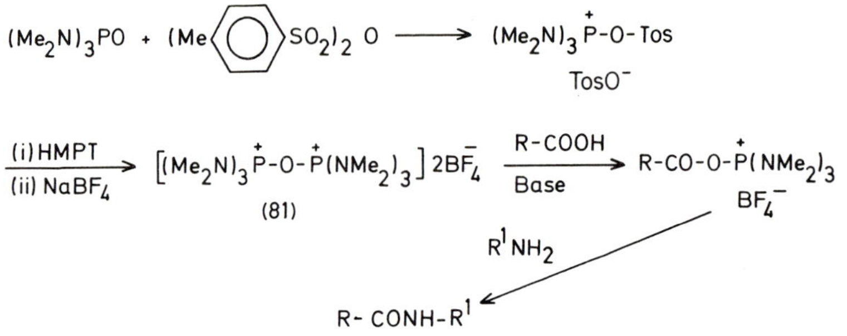

Scheme 20

In an attempt to take advantage of the fact that 1-hydroxybenzotriazole drastically reduces racemization in peptide coupling mediated by dicyclohexylcarbodiimide,[68] the reagent (81) was transformed into the derivative (82; X = BF_4^-).[69] In a similar approach essentially the same reagent (82; X = PF_6^-) was conveniently prepared[70] as shown in Scheme 21. Although high yields of peptides may be obtained using the hybrid reagent, unfortunately extensive racemization again occurred in the Izumiya test, strongly suggesting that nucleophilic attack of the carboxylate on (82) produces the highly reactive and fugitive acyloxyphosphonium system. The alternative mode of cleavage of (82) to give the desirable active ester (83) is precluded on stereoelectronic grounds; however, use of (82) in the presence of 1-hydroxybenzotriazole does

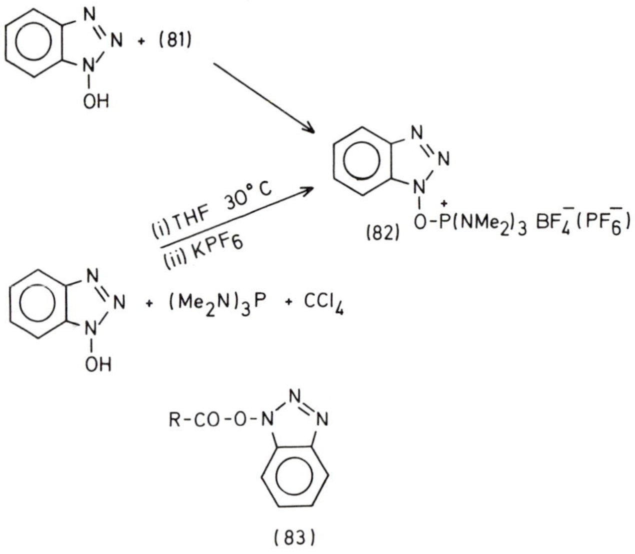

Scheme 21

decrease the racemization observed in the Izumiya test, presumably by rapid formation of the active ester (83). Castro has utilized the reagent (82) to prepare N_α-protected aminoacid phenyl esters in high yields.

3. References

1. *Methoden der Organischen Chemie* (Houben-Weyl), Vol. 15, Parts 1 and 2 (ed. E. Wünsch), Thieme, Stuttgart, 1974; *The Proteins*, 3rd Ed., Vol. 2 (ed. H. Neurath and R. L. Hill), Academic Press, London and New York, 1976.
2. D. F. Veber, S. F. Brady, and R. Hirschmann, Proc. 3rd American Peptide Symposium (Boston 1972), 315, Ann Arbor Science Publishers, Ann Arbor, Michigan, 1972.
3. L. Zervas and L. Dilaris, *J. Amer. Chem. Soc.*, 1955, **77**, 5354; L. Zervas and L. Dilaris, *Chem. Ber.*, 1956, **89**, 925; A. Cosmatos, I. Photaki, and L. Zervas, *ibid.*, 1961, **94**, 2644.
4. F. R. Atherton, H. T. Howard, and A. R. Todd, *J. Chem. Soc.*, 1948, 1106.
5. L. Zervas and P. G. Katsoyannis, *J. Amer. Chem. Soc.*, 1955, **77**, 5351.
6. H. Keller, H. Netter, and B. Niemann, *Z. Physiol. Chem.*, 1958, **313**, 244.
7. L. Zervas and I. Photaki, *J. Amer. Chem. Soc.*, 1962, **84**, 3887.
8. G. Tomashewski and G. Kuhn, *J. Prakt. Chem.*, 1968, **38**, 222; T. Koizumi and P. Haake, *J. Amer. Chem. Soc.*, 1973, **95**, 8073.
9. M. J. Harger, *J. Chem. Soc., Perkin I*, 1975, 514; M. J. Harger, A. J. Macpherson, and D. Pickering, *Tetrahedron Lett.*, 1975, 1797; M. J. Harger, *J. Chem. Soc., Chem. Comm.*, 1976, 520.
10. M. Hague and C. N. Caughlan, *J. Chem. Soc., Chem. Comm.*, 1966, 921; A. F. Cameron, private communication.
11. G. W. Kenner, G. A. Moore, and R. Ramage, *Tetrahedron Lett.*, 1976, 3623.
12. M. Ueki and S. Ikeda, *Chem. Lett.*, 1976, 827.
13. M. Ueki and S. Ikeda, *Chem. Lett.*, 1977, 869.
14. S. Coulton, G. A. Moore, and R. Ramage, *Tetrahedron Lett.*, 1976, 4005.
15. M. Dreux, P. Savignac, and J. Chenault, *Tetrahedron Lett.*, 1971, 4109; P. Savignac, M. Dreux and G. Plé, *J. Organometal. Chem.*, 1973, **60**, 103; B. Corbel, J.-P. Paugam, M. Dreux, and P. Savignac, *Tetrahedron Lett.*, 1976, 835.
16. P. Stelzel, in *Methoden der Organischen Chemie* (Houben-Weyl) (ed. E. Wünsch), Vol. 15/2, pp. 169–259, Thieme, Stuttgart, 1974; N. F. Albertson, *Org. Reactions*, 1962, **12**, 202.
17. F. Lipmann, *Adv. Enzymol.*, 1941, **1**, 99, 152.
18. H. Chantrenne, *Biochem. Biophys. Acta*, 1950, **4**, 484.
19. J. C. Sheehan and V. S. Frank, *J. Amer. Chem. Soc.*, 1950, **72**, 1312.
20. T. Wieland and H. Bernhard, *Annalen*, 1951, **572**, 190.
21. F. Cramer and K. G. Gärtner, *Chem. Ber.*, 1958, **91**, 1562.
22. T. Shiori, K. Ninomiya, and S. Yamada, *J. Amer. Chem. Soc.*, 1972, **94**, 6203; T. Shiori and S. Yamada, *Chem. Pharm. Bull.* (Japan), 1974, **22**, 849, 855, 859.
23. S. Yamada, K. Kasai, and T. Shiori, *Tetrahedron Lett.*, 1973, 1595.
24. M. W. Williams and G. T. Young, *J. Chem. Soc.*, 1963, 881.
25. N. Izumiya and M. Muraoka, *J. Amer. Chem. Soc.*, 1969, **91**, 2391.

26. S. Yamada, N. Ikota, T. Shiori, and S. Tachibana, *J. Amer. Chem. Soc.*, 1975, **97**, 7174.
27. L. Caglioti, M. Polone, and G. Rosini, *J. Org. Chem.*, 1968, **33**, 2979; J. Martinez and F. Winternitz, *Bull. Soc. Chim. France*, 1972, **12**, 4707.
28. J. Martinez and F. Winternitz, *Tetrahedron Lett.*, 1975, 2631.
29. J. Martinez and F. Winternitz, Proc. 14th European Peptide Symposium (Wépion 1976), 551, Éditions de l'Université de Bruxelles, 1976.
30. T. Wieland and B. Heinke, *Annalen*, 1956, **599**, 70.
31. E. Tascher, J. F. Biernat, B. Rzeszotarska, and C. Wasielewski, *Annalen*, 1961, **646**, 123.
32. I. Dostrovsky and M. Halmann, *J. Chem. Soc.*, 1953, 516.
33. P. C. Crofts, in *Organic Phosphorus Compounds* (ed. G. H. Kosolapoff and L. Maier), Vol. 6, Wiley-Interscience, New York, and references cited therein.
34. A. G. Jackson, G. W. Kenner, G. A. Moore, R. Ramage, and W. D. Thorpe, *Tetrahedron Lett.*, 1976, 3627.
35. G. W. Anderson and R. W. Young, *J. Amer. Chem. Soc.*, 1952, **74**, 5307.
36. G. W. Anderson, J. Blodlinger, and A. D. Welcher, *J. Amer. Chem. Soc.*, 1952, **74**, 5309.
37. G. W. Anderson, A. D. Welcher, and R. W. Young, *J. Amer. Chem. Soc.*, 1951, **73**, 501.
38. R. W. Young, K. H. Wood, R. J. Joyce, and G. W. Anderson, *J. Amer. Chem. Soc.*, 1956, **78**, 2126.
39. S. Goldschmidt and F. Obermeier, *Annalen*, 1954, **588**, 24.
40. P. C. Crofts, J. H. H. Markes, and H. N. Rydon, *J. Chem. Soc.*, 1959, 3610.
41. M. Rothe and W. Kreiss, Proc. 14th European Peptide Symposium (Wépion 1976), 71, Éditions de l'Université de Bruxelles, 1976.
42. M. Leplawy, J. Michalski, and J. Zabrocki, *Chem. and Ind.*, 1964, 835.
43. S. Goldschmidt and H. Lautenschlager, *Annalen*, 1953, **580**, 68; S. Goldschmidt and H. L. Krauss, *ibid.*, 1955, **595**, 193; W. Grassmann, E. Wünsch, and A. Riedel, *Chem. Ber.*, 1958, **91**, 455.
44. Ref. 16, p. 239.
45. E. Wünsch, *Angew. Chem.*, 1959, **71**, 743.
46. E. Wünsch, *Z. Physiol. Chem.*, 1963, **332**, 288.
47. M. Brenner, U. Giger, R. Nyfeler, P. Schenk, and A. Tschopp, Proc. 14th European Peptide Symposium (Wépion 1976), 65, Éditions de l'Université de Bruxelles, 1976.
48. G. W. Anderson and F. M. Callaghan, *J. Amer. Chem. Soc.*, 1958, **80**, 2902.
49. G. W. Anderson, A. C. McGregor, and R. W. Young, *J. Org. Chem.*, 1958, **23**, 1236.
50. Y. V. Mitin and O. V. Glinskaya, *J. Gen. Chem. (U.S.S.R.)*, 1971, **41**, 1152; Y. V. Mitin and E. E. Maximov, *ibid.*, 1973, **43**, 199; G. I. Shelykh, G. P. Vlasov, and Y. V. Mitin, *ibid.*, 1973, **43**, 369.
51. N. Yamazaki and F. Higashi, *Tetrahedron Lett.*, 1972, 415; N. Yamazaki and F. Higashi, *Bull. Chem. Soc. Japan*, 1973, **46**, 1235, 1239, 3821.
52. N. Yamazaki and F. Higashi, *Tetrahedron Letters*, 1972, 5047; *idem, Bull. Chem. Soc. Japan*, 1974, **30**, 1323; N. Yamazaki, F. Higashi, and S. Kazaryan, *Synthesis*, 1974, 436; N. Yamazaki, F. Higashi, and Y. Saito, *ibid.*, 1974, 495.
53. J. B. Lee, *J. Amer. Chem. Soc.*, 1966, **88**, 3440.
54. T. Wieland and A. Seeliger, *Chem. Ber.*, 1971, **104**, 3992; S. Yamada and Y. Takeuchi, *Tetrahedron Lett.*, 1971, 3595.

55. R. Appel, G. Baumer, and W. Strüver, *Chem. Ber.*, 1975, **108**, 2680; R. Appel, *Angew. Chem.*, 1975, **14**, 801.
56. B. Castro and J. R. Dormoy, *Bull. Soc. Chim. France*, 1971, 3034; *idem., Tetrahedron Lett.*, 1972, 4747.
57. B. Castro and J. R. Dormoy, *Tetrahedron Lett.*, 1973, 3243.
58. T. Mukaiyama, *Internat. J. Sulphur Chem.* (*B*), 1972, **7**, 173; *idem, Phosphorus and Sulphur*, 1976, **1**, 371; *idem, Angew. Chem.*, 1976, **15**, 94.
59. T. Mukaiyama, M. Ueki, H. Maruyama, and R. Matsueda, *J. Amer. Chem. Soc.*, 1968, **90**, 4490.
60. R. Matsueda, H. Maruyama, M. Ueki, and T. Mukaiyama, *Bull. Chem. Soc. Japan*, 1971, **44**, 1373.
61. T. Mukaiyama, R. Matsueda, and H. Maruyama, *Bull. Chem. Soc. Japan*, 1970, **43**, 1271.
62. T. Mukaiyama, K. Goto, R. Matsueda, and M. Ueki, *Tetrahedron Lett.*, 1970, 5293.
63. M. Ueki, S. Takashi, A. Hayashida, and T. Mukaiyama, *Chem. Lett.*, 1973, 733.
64. G. I. Shelykh, G. P. Vlasov, and Y. V. Mitin, *J. Gen. Chem.* (*U.S.S.R.*), 1973, **43**, 367.
65. R. Matsueda, H. Maruyama, E. Kitazawa, H. Takahagi, and T. Mukaiyama, *Bull. Chem. Soc. Japan*, 1973, **46**, 3240.
66. R. Matsueda, H. Maruyama, E. Kitazawa, H. Takahagi, and T. Mukaiyama, *J. Amer. Chem. Soc.*, 1975, **97**, 2573.
67. A. J. Bates, I. J. Galpin, A. Hallett, D. Hudson, G. W. Kenner, R. Ramage, and R. C. Shepherd, *Helv. Chim. Acta*, 1975, **58**, 688.
68. W. König and R. Geiger, *Chem. Ber.*, 1970, **103**, 788.
69. I. J. Galpin, P. F. Gordon, R. Ramage, and W. D. Thorpe, *Tetrahedron Lett.*, 1976, **32**, 2417.
70. B. Castro, J. R. Dormoy, G. Evin, and C. Selve, *Tetrahedron Lett.*, 1975, 1219; B. Castro, J. R. Dormoy, G. Evin, and C. Selve, Proc. 14th European Peptide Symposium (Wépion 1976), 79, Éditions de l'Université de Bruxelles, 1976.

Author Index

Aboul-Enein, H., 166(84c), 185(84c), *199*
Abramov, V. S., 419(176), *429*
Acampora. M., 494(177a), *509*
Achini, R. S., 138(349), 139(349), *153*
Achiwa, K., 131(332), 133(338), 135(338), *152*
Achremowicz, L., 460(147), *466*
Ackrell, J., 299(16), *346*
Adachi, M., 440(40b), *463*
Adam, W., 325(103, 104), 326(105), *348*
Adams, K. A. H., 272(7b), *292*
Adams, W., 357(28), *382*
Adler, V. E., 112(292), *151*
Agarwal, S. C., 309(48), 310(48, 50), *347*
Aguiar, A. M., 173(117), 194(117), *200*, 296(5), 322(5), *345*
Ahmed, M., 25(204), 85(204), 86(204), *148*
Ailman, D. E., 367(69), *384*
Airey, P. L., 422(198), *430*
Akashi, C., 446(72a), *464*
Akhtar, M., 24(368), *153*
Albanbauer, J., 492(156), 493(158), *508*
Albertson, N. F., 519(16), 526(44), *533*
Albright, T. A., 163(46), 194(46), *198*
Alekseev, Yu. E., 19(17), *142*, 184(208), *204*
Alekseeva, V. G., 19(17), *142*, 184(208), *204*
Alfredsson, G., 329(114a), 330(114a), 335(114a), *349*
Alkaysi, H. N., 443(54), *464*
Allcock, H. R., 502(214a, 214b, 214c, 214d), *510*
Allen, J. F., 454(105), *465*

Alley, W. D., 419(177), *429*
Almog, J., 164(56, 58), 173(56), 195(56), *198, 199*
Altaf-ur-Rahman, M., 299(16), *346*
Al'tmark, E. M., 164(62), *199*
Ambrus, G., 108(277), *150*
Ames, D. E., 272(10), *292*
Amin, S. G., 415(130), *428*
Ammon, H. L., 453(101), *465*
Amonoo-Neizer, E. H., 301(23), *346*, 363(56), *383*, 445(68), *464*
Anatol, J., 422(216, 217, 218), *430*
Ander, I., 275(22), *292*
Anderson, A. G., 443(55, 56), *464*
Anderson, A. G., Jr., 47(123), *145*, 442(53), *464*
Anderson, G. W., 523(35, 36, 37), 524(38), 526(48, 49), *534*
Anderson, J. E., 433(2), *462*
Anderson, N. H., 180(180), 195(180), *202*
Anderson, R. J., 24(130), 48(130), 49(130), 113(293), 114(293), 138(349), 139(349), *145, 151, 153*
Anderson, W. K., 356(25), *382*
Ando, K., 184(213), *204*
Andrewes, A. G., 37(94), 38(80), 40(94), *144*
Andrews, G. C., 364(61, 62, 63), *383, 384*
Andrews, S. D., 164(65), *199*
Aneja, R., 329(113), 330(115), *349*, 400(46, 52), *426*
Anh, N. T., 6(4b), *14*, 166(80), *199*
Anoshina, N. P., 314(63), *347*
Anschel, M., 310(53), *347*, 486(129), *508*

Antczak, S., 12(18a), *15*
Appel, H., 388(5), *425*
Appel, R., 212(16), 216(37), *221, 222,* 245(52), 246(54), *267,* 317(76), *348,* 388(6, 7, 8, 9, 10), 389(6, 9, 12, 13), 390(14, 15), 392(28, 30), 393(6, 28), 394(28), 395(7), 397(8, 9, 32, 33), 398(34), 399(14, 36), 400(28, 30, 32, 33, 34, 40, 50), 401(40), 403(6), 404(80), 405(83, 84), 406(86, 87, 88), 407(88, 91), 408(92, 93, 95), 409 (95, 97, 98, 99), 410(103, 105, 106, 107), 412(30, 109, 113), 413(114, 115, 116, 118, 119, 120, 121), 414 (106), 415(128, 129, 131), 416(137), 417(142), 421(188), 422(194, 195, 200), 423(236), 424(236, 242, 243, 247, 248, 249), *425, 426, 427, 428, 429, 430, 431,* 495(182), *509,* 528(55), *535*
Appelbaum, J., 195(237), *205*
Araki, M., 373(105), *385*
Arbusov, A. E., 156(4), 194(4), *197,* 418(152), *428*
Arbusov, B. A., 159(27), *198,* 300(20), *346,* 470(32b), 475(55a), 488(138, 139), 489(142), 496(187), *504, 505, 508, 509*
Aries, R., 180(185), *203*
Armour, M. A., 270(6), 278(36d), *292, 293*
Armsen, R., 226(7a), *266*
Arnason, B., (32), *267*
Arnold, W., 409(96), *427*
Arnold, Z., 180(181), *203*
Arora, P., 177(160), 194(160), *202*
Arutyunyan, E. A., 10(11a, 11b), *15*
Asato, A. E., 177(164), 194(164), *202*
Ash, D. K., 370(89), 378(132), *384, 385*
Assaf, Y., 332(126), *349*
Atherton, F. R., 287(63), *294,* 387(1), 422(203), *425, 430,* 513(4), *533*
Atkinson, J. G., 24(376), *153*
Atkinson, R. E., 418(165), 421(165), *429*
Avery, N. L., 312(60), *347*
Avonda, F. P., 455(108), *465*
Axelrod, E. H., 139(350), *153,* 208(2b), *221,* 400(42), *425*
Axen, U., 25(146), 53(146, 147), *146,* 455(115), *465*

Babad, H., 91(224), *148*
Babler, J. H., 170(111), 173(111), 193(111), *200*
Babouléne, M., 161(37), 189(226), *198, 204*
Bach, R. D., 8(6), *14*
Back, T. G., 374(109), *385*
Bacon, R. G. R., 277(34), 287(34), 288(34), *293*
Badar, Y., 33(64), *144*
Baeckstrom. P., 25(213), 88(213), *148*
Bagrov, F. V., 471(37a), *505*
Baker, D. A., 184(207), *204*
Bakker, E., 372(98), *384*
Bakuzis, M. L. F., 25(328), 130(328), *152*
Bakuzis, P., 25(328), 130(328), *152*
Bal, K., 181(198), 182(198), *203*
Balasubramanian, K. K., 275(21), *292*
Baldwin, J. E., 210(9), *221,* 364(59), 369(83), *383, 384,* 446(72b), *464*
Banas, E. M., 482(99), *507*
Banerjee, A. K., 139(356), 140(356), *153*
Bannet, D. M., 159(21), 171(21), 194 (21), *197*
Barakat, M. Z., 368(74), 371(74), *384*
Barclay, L. R. C., 272(7b), *292*
Barhardt, R. G., 220(50), *222*
Barley, G. C., 25(204), 85(204), 86 (204), 90(221), 98(221), *148*
Barlow, L., 24(199), 81(199), 88(199), 126(321), *147, 152*
Barneis, Z., 334(132), 335(132), *349*
Barnes, G. K., 6(5b), 7(5b), *14*
Barnett, W. E., 338(142), *349*
Barnette, W. E., 108(249), *149*
Barrans, J., 485(112, 115), *507*
Barstow, L. C., 415(123), *428*
Barsukov, L. I., 25(110), 27(36), 41(99), 46(110), 78(110), 79(110), 83(110), 91(225, 226), 92(226), *143, 145, 148*
Barta, I., 108(277), *150*
Bartlett, P. A., 138(346), *152*
Bartlett, P. D., 317(76), 336(138), 338 (138, 143), 340(138, 143, 145, 146), *348, 349,* 351(3), *382,* 486(121, 122), 491(150c, 150d), *507, 508*

Bartmann, W., 102(253), *149*
Barton, D. H. R., 25(360), 141(360), 142(361), *153*, 304(32), *346*, 359(36), 360(40), 364(60), *383*
Bartsch, W., 163(48), 184(210, 211), 193(48, 210), 194(210), *198*, *204*
Basedow, O. H., 311(58), 318(80), *347*, *348*, 360(43), 367(43), *383*
Batchelor, F. W., 195(235), *204*
Bates, A. J., 531(67), *535*
Bates, G. S., 28(41), *143*
Batlle de Pabon, H. N., 24(163), 63(163), 79(163), *146*
Bauer, D. P., 356(24), *382*
Bauer, E., 243(47), 244(47, 48), 257(87), *267*, *268*
Bauer, G., 501(207), *510*
Bauer, H. J., 476(74), 478(74), 479(74), *506*
Bauer, J., 476(78), 478(78), *506*
Baumann, M., 50(137), 132(335), *146*, *152*
Bäumer, G., 415(128, 129), *428*, 528(55), *535*
Baumgarten, M. E., 296(4), *345*
Baumstark, A. L., 317(76), *348*, 486(121), *507*
Bax, H. J., 34(69), 36(69), *144*
Bayet, P., 132(333), *152*, 218(42), *222*
Bayless, A. V., 456(117), *465*
Beasley, G. H., 179(176), 180(186), 196(186), *202*, *203*
Bebikh, G. F., 421(191), *430*
Beck, G., 102(253), *149*
Beck, P., 20(21), 24(21), *142*, 310(51), *347*
Becke, J., 24(369), *153*
Becker, K. B., 19(20), 139(20), *142*
Becker, T., 160(30), *198*
Beckmann, R., 110(283), *150*
Bee, L. K., 360(37), *383*
Beeby, J., 360(37), *383*
Beevor, P. S., 94(234), 116(234), 117(300, 301), 118(301), *149*, *151*
Bejer, B., 335(133), *349*
Belbéoch, A., 161(37), *198*
Bell, W. J., 24(143), 52(143), *146*
Bellin, S. A., 34(72), 36(72), 38(72), *144*

Belyaev, A. N., 85(203), *148*
Belyaev, N. N., 85(203), *148*
Bengelsdorf, I. S., 318(83), *348*
Bengelsdorf, J. S., 418(160), *429*
Bennett, F. W., 422(196), *430*
Bensel, N., 178(172), *202*
Bentley, P. H., 100(248), *149*
Bentley, R. K., 85(206), 86(206), *148*
Bentrude, W. G., 480(87, 93, 94), 481(97), 491(147), *506*, *508*
Berchtold, G. A., 322(97), 337(141), *348*, *349*
Bercz, P. J., 176(143), 194(143), *201*
Bergelson, L. D., 19(13), 25(110), 27(36), 33(60, 61), 41(98, 99), 46(110), 78(61, 110), 79(110), 83(110), 90(217, 218, 220), 91(220, 225, 226), 92(226, 227), 93(217, 229), 94(217, 232, 235), 95(236, 237), 96(237, 238, 239), 97(239), *142*, *143*, *144*, *145*, *148*, *149*
Bergman, R. G., 438(33), *463*
Bergmann, E. D., 164(56, 59), 166(59, 81), 173(56), 175(59), 181(196), 194(81), 195(56, 237), *198*, *199*, *203*, *205*
Berlin, K. D., 160(32), 161(32), *198*, 418(171), *429*, 467(2a), *503*
Berman, S. T., 208(3), *221*
Bernard, D., 471(39, 42), 484(106), 485(111a, 111b), *505*, *507*
Bernhard, H., 519(20), *533*
Berninger, C. J., 25(175), 68(175), *147*, 252(74), *268*
Bernot, K. G., 158(20), *197*
Beroza, M., 25(289, 299), 112(289, 292), 116(298, 299), *151*
Berry, R. S., 469(23), *504*
Bershas, J. P., 25(192), 76(189), 76(192), *147*
Bertele, E., 136(341), 137(341), *152*
Berustein, S., 180(191), *203*
Bestmann, H. J., 20(21), 23(28), 24(21), 25(117, 118, 131, 233, 287, 288, 290, 291, 296, 377), 29(52), 30(52), 46(117, 118), 47(117), 48(128, 131), 49(131), 79(131), 82(52), 94(233), 95(117, 118, 131), 110(287), 111(131, 187, 188), 112(118, 290, 291), 113

Bestmann, H. J., (*continued*)
(291), 115(118, 295, 296), 116(233, 297), *142, 143, 145, 146, 149, 150, 151, 153,* 189(224), *204,* 208(4), 209(6), 210(8), 211(11, 12, 13, 14), 212(15, 20), 213(22, 25, 26), 214(27), 217(39), 219(15), *221, 222,* 223(1, 4a), 225(5, 6), 226(7a, 7b, 7c), 227(7b, 7c), 228(10), 230(14), 231(17), 236(33), 265(103), *266, 267, 268,* 341(148), *350,* 399(35), 400(35), *425,* 441(47), 442(50), 446(73), 460(141), *464, 466,* 490(143), 494(178), 495(178), *508, 509*
Bezzubov, A. A., 92(227), 94(232, 235), *148, 149*
Bhacca, N. S., 308(45), *347,* 438(36), *463*
Bhalerao, U. T., 24(139), 50(139), 51(139), 139(351), *146, 153*
Bhatai, S. B., 305(37), 311(57), *346, 347,* 469(28), 472(44,) 473(48), 474 (48), 475(66, 67, 70), 476(66, 76), 478(76), 482(101), 483(67), 484(66), 486(67), 488(136), *504, 505, 506, 507, 508*
Bierl, B. A., 25(289), 112(289, 292), 116(298), *151*
Biernat, J. F., 522(31), *534*
Bigler, A. J., 471(40), 472(45), 473 (45, 47, 48), 474(48), 500(199, 200), *505, 510*
Billman, J. H., 10(8), *15*
Billmann, W., 25(287), 110(287), 111 (287), *150*
Bindra, J. S., 100(247), *149,* 180(184), *203*
Bindra, R., 100(247), *149*
Birch, A. J., 129(326), 130(326), *152*
Bird, C. W., 305(40), 312(59, 61), 313 (61), *346, 347*
Birkenmeyer, R. D., 400(55), *426,* 438 (37), *463*
Birum, G. H., 26(29), *143,* 487(135a, 135b), *508*
Bisagni, E., 175(127), *201*
Bissel, E. C., 502(214d), *510*
Bissing, D. E., 24(50, 51), 28(50), 29 (51), 30(50, 51), 40(51), 41(51), *143,* 167(96), *200,* 317(72, 73), *347*
Bisson, R., 177(160), 194(160), *202*
Bittner, S., 332(125, 126), 334(132), 335(132), *349*
Black, D. K., 434(6), *462*
Black, D. S. C., 190(229), *204*
Blackburn, G. M., 369(84), *384*
Bladé-Font, A., 214(31), *222*
Blagoveshchenskii, V. S., 423(221), *431*
Blanchard, J., 162(39), *198*
Blaschke, H., 25(166), 64(166), *146*
Blaser, B., 412(113), 413(114), *427*
Blatcher, P., 176(142), 177(158b), 192 (142, 158b), *201, 202*
Block, E., 341(147), *350,* 360(44), *383*
Blodlinger, J., 523(36), *534*
Blomquist, A. T., 28(49), *143*
Blount, J. F., 281(49), *293*
Blum, S., 309(47), *347*
Boden, R. M., 49(133), 78(133), *146*
Boeckman, R. K., Jr., 438(32), *463*
Boekelheide, V., 25(166), 64(166), *146,* 361(46), *383*
Boer, J. J. J., de, 186(215), *204*
Boer, T., de, 287(62), 288(62), *294*
Boer, T. J., de, 165(68), 175(68), *199*
Bohlmann, F., 18(7), 19(11), 34(69), 36(69), 44(106), 73(186), 74(186), 86(207), 99(244), 128(7), *142, 144, 145, 147, 148, 149,* 213(23), *222*
Bohm, H., 319(87), *348*
Bohme, E., 184(206), 193(206), *204*
Bohn, B., 24(375), 28(39), *143, 153*
Boigegrain, R., 392(27), 400(27), 401 (62), *425, 426,* 486(126a, 126b), *507*
Boland, W., 115(294), *151*
Bollinger, P., 107(273), *150*
Bondinell, W., 50(136), *146*
Bone, S. A., 12(18a), *15,* 468(17a), *504*
Bonnafous, J.-C., 34(73), 36(73), *144*
Bordwell, F. G., 317(75), *347,* 352(5), *382,* 453(99, 100), *465*
Borisova, E. E., 495(181), *509*
Borne, R., 166(84c), 185(84c), *199*
Borowitz, I. J., 310(53), *347,* 447(74, 75, 78, 79), 448(80, 84), 455(78, 109), 457(128), *464, 465, 466,* 486(127, 129), *507, 508*
Borsus, J. M., 495(179), *509*
Bose, A. K., 34(68), 38(81), *144,* 181

(193, 197), *203*, 329(111), 332(127), 333(127), *348, 349*, 415(130), *428*, 456(116), *465*
Boskin, M. J., 316(70), *347*, 352(6), *382*
Both, W., 210(8), *221*
Botteghi, C., 231(15), *266*
Bottin-Strzalko, T., 39(97), 80(197), *145, 147*, 163(46), 167(89), 168(104), 169(89), 194(46), *198, 200*
Boulos, L. S., 352(7), *382*
Boulton, A. J., 281(48), *293*, 299(15, 16), *346*
Boutagy, J., 23(24), 141(24), *143*, 180 (189), *203*
Boyer, J. H., 308(43), *347*
Bracelos, F., 180(185), *203*
Bradshaw, R. W., 97(241), 98(241), *149*
Brady, S. F., 52(142), *146*, 512(2), *533*
Brain, E. G., 139(355), 140(355), *153*
Bramblett, J. D., 498(191), *509*
Branca, Q., 373(102), *385*
Branchaud, B., 180(185), *203*
Brand, W. W., 301(22), *346*
Braun, B. H., 133(337), *152*
Braun, H., 71(180), 72(180), *147*
Braun, J., v, 407(89), *427*
Brawner Floyd, M., 101(250), *149*
Breazeale, R. D., 47(123), *145*
Brennan, M. E., 342(149), *350*, 491 (151), *508*
Brenner, M., 526(47), *534*
Bretschneider, H., 460(148), *466*
Brett, D., 400(41), 401(65), *425, 426*
Breuer, E., 159(21), 171(21), 190(229), 194(21), *197, 204*
Brewer, J. D., 24(169), 65(169), *147*
Bridges, A. J., 171(115), *200*
Brierley, J., 12(18a), *15*
Brightwell, N. E., 308(45), *347*
Brink, M., 361(47), *383*
Broadhurst, M. D., 179(176), 180(186), 196(186), *202, 203*
Broadhurst, M. J., 362(50), *383*
Brodsky, L., 73(185), *147*
Broekhof, N. L. J. M., 175(138), 194(138), *201*
Brooke, P. K., 270(4), *292*
Brooks, T. W., 25(100), 42(100), 128 (100), *145*
Brouwer, R. J., de, 138(345), *152*

Brown, C., 80(196), *147*, 422(201), *430*
Brown, G. W., 218(44), 219(44), *222*, 245(50), *267*
Brown, R. C., 299(16), *346*
Brück, D., vor der, 245(51), *267*
Brückner, K., 75(191), 176(191), *147*
Bruhin, J., 361(45), *383*
Bruin, J. W., 180(185), *203*
Brunelle, D. J., 175(139), 193(139), *201*, 373(107), 374(110), *385*
Brunn, E., 327(107), *348*
Bruntrup, G., 335(134), *349*
Brydon, D. I., 323(101), *348*
Brylikowska-Piotrowich, J., 422(215), *430*
Büchi, G., 73(184), *147*
Buck, H. M., 138(342, 345), *152*, 499 (198), 501(198), *510*
Buddrus, J., 24(45, 46, 47), 27(45, 46, 47), 29(45, 46, 47), 45(47), *143*
Buendia, J., 180(187), *203*
Buerger, W., 176(147), 195(147), *201*
Bugerenko, E. F., 422(211), *430*
Büjuktür, G., 139(356), 140(356), *153*
Bukowick, P. A., 177(155), 191(155), 194(155), *201*
Bulina, V. M., 46(116), 90(116), *145*
Bulka, E., 362(54), *383*
Bünger, H., 177(156), *201*
Bunyan, P. J., 305(39), 323(100), 324 (100), *346, 348*, 418(169), 420(169, 182, 183), 421(169), *429*
Buono, G., 488(140), *508*
Burden, R. S., 177(157), 194(157), *202*
Burgada, R., 402(69, 70), *426*, 450(89), *464*, 471(38, 39, 42), 484(105a, 105b, 105c, 106, 107b, 108, 109), 485 (111a, 111b, 114, 116), *505, 507*
Burger, K., 492(155, 156, 157), 493 (158, 159, 160, 161, 162, 163, 164, 165, 166, 167, 168, 169, 170, 171, 173), 494(162), *508, 509*
Burgis, E., 492(156, 157), 493(173), *508, 509*
Burgstahler, A. W., 24(143), 52(143), *146*
Burke, S. D., 180(182), 195(182), *203*
Burn, A. J., 323(100), 324(100), *348*, 400(38), 420(38, 182, 183, 185), 421 (38), *425, 429*

Burnaeva, L. A., 476(72b), *506*
Burrell, R. A., 297(11), *346*
Burton, D. J., 45(109), *145*, 314(64), 315(64), *347*, 403(71), 418(172), 423 (233, 234, 235), *426*, *429*, *431*, 442 (52), 449(88), *464*
Butenandt, A., 110(283, 284, 285), *150*
Butler, G. B., 25(100), 42(100), 128 (100), *145*, 160(32), 161(32), *198*
Buzas, A., 166(85), 185(85), 193(85), *199*
Bykova, T. G., 489(141), *508*

Caccia, G., 231(15), *266*
Cadogan, J. I. G., 10(9), 12(18b), 13 (18b, 19), *15*, 223(3), *266*, 269(1), 270(2, 3, 6, 7a), 272(2, 8), 273(1, 8, 16), 274(8, 16), 275(27), 276(31), 278(8, 36a, 36b, 36d, 38), 279(40), 280(8, 36a), 281(27, 47), 286(59, 60a, 60b, 60c, 60d), 287(60d, 62), 288(62, 65, 66a), 289(8, 40, 66a, 67, 68), 290 (59), *292*, *293*, *294*, 296(6), 297(7), 300(6), 305(39), 323(100, 101), 324 (100), *345*, *346*, *348*, 387(2), 400(38), 418(163, 164, 165, 169), 420(38, 163, 169, 182, 183, 185), 421(38, 163, 165, 169, 189), *424*, *425*, *429*, 480(91), 496(183a, 183b, 184), *506*, *509*
Caglioti, L., 521(27), *534*
Calagno, M. A., 265(102), *268*
Calder, I., 25(166), 64(166), *146*
Callaghan, F. M., 526(48), *534*
Cama, L. D., 187(220), 195(220), *204*
Cameron, A. F., 515(10), *533*
Cameron-Wood, M., 269(1), 272(8), 273(1, 8), 274(8), 278(8), 280(8), 289(8), *292*, 296(6), 300(6), *345*
Campbell, M.-T., 468(17b), *504*
Campbell, T. W., 24(371), *153*
Camps, F., 79(195), *147*, 164(60), 178 (60), *199*
Cane, D. E., 157(14), 164(14), 168(14), 170(14), 172(14), 178(14), 196(14), *197*
Carboni, R. A., 281(42), *293*
Carey, F. A., 184(210), 193(210), 194 (210), *204*, 355(15), 356(15), *382*
Carlon, F. E., 319(86), *348*
Carlsen, D., 10(12a), 11(12c), *15*
Carlson, D. A., 116(298), *151*

Carlson, S. C., 403(73), *426*
Carpentier, M., 167(92), *200*
Carpino, L. A., 453(102), *465*
Carrie, R., 167(86, 89, 94), 169(89, 107), 194(107), 195(86), *199*, *200*, 234(31), 235(31b), *267*
Carson, J. F., 303(31), *346*
Casper, E. W. R., 447(74), *464*
Caspi, E., 436(23), *463*
Cassidy, F., 139(355), 140(355), *153*
Castells, J., 79(195), *147*
Castro, B., 392(22, 26, 27), 400(22, 26, 27, 54), 401(60, 61, 62, 64), 402 (69, 70), 405(82), 417(139, 140), *425*, *426*, *427*, *428*, 450(89), 456(120, 121, 122, 123), *464*, *465*, 486(126a, 126b), *507*, 529(56, 57), 532(70), *535*
Catsoulacos, P., 184(207), *204*
Caughlan, C. N., 470(34), *504*, 515(10), *533*
Cava, M. P., 290(70b), *294*
Cavill, G. W. K., 37(91, 92), 40(91, 92), 133(92), *144*, 178(167), 194(167), *202*
Cernigliaro, G., 400(49), *426*
Challener, F., 323(98), *348*
Challenger, F., 369(80), *384*
Chang, B. C., 471(41), 486(41), 499 (197a), 500(41), *505*, *510*
Chang, P. C., 502(216), *510*
Chantrenne, H., 519(18), *533*
Chao, T. H., 11(13), *15*
Chapleur, Y., 400(54), 401(60, 64), *426*, 456(120, 123), *465*
Chaplin, E. C., 457(126), *466*
Charbonnel, Y., 485(112), *507*
Chasar, D. W., 303(29), *346*
Chattha, M. S., 173(117), 194(117), *200*
Chauser, E. G., 440(40a), *463*
Chaykovsky, M., 24(113), 46(113), *145*
Chebotarev, A. S., 178(166), 195(166), *202*
Cheburkov, Y. A., 487(132), *508*
Chen, E. H., 311(57), *347*, 472(44), *505*
Chen, W. Y., 290(71), *294*
Chenault, J., 518(15), *533*
Chernack, J., 10(8), *15*
Chin, C. G., 25(380), *153*
Chippendale, K. E., 272(13), 275(20), 278(37a), 279(37a), *292*, *293*

Chiu, N. W. K., 37(87), *144*, 168(99), 194(99), *200*
Choi, S. C., 364(59), *383*
Chopard, P. A., 24(370), *153*, 172(116), *200*, 212(19), *221*, *424*, 459(140), *466*
Chopard-dit-Jean, L. H., 123(317), *151*
Christensen, B. G., 187(220), 195(220), *204*
Christmann, K. F., 25(33, 155), 26 (32, 33), 27(32), 48(33, 126), 50(33, 126), 55(33, 151, 153), 56(33, 151, 153, 155, 157), 57(155, 157), 58(155, 157), 62(157), 78(33), 79(33), 83(33), *143*, *145*, *146*, 164(55), 170(114), *198*, *200*
Christol, H., 460(142, 143, 144), *466*
Chugh, O. P., 178(174), *202*
Chung, B. C., 416(133), 423(133), *428*, 437(28), 440(28), *463*
Ciganek, E., 220(49), *222*
Clark, R. D., 161(35), 187(217), 194 (35), 196(217), *198*, *204*
Clausen, K., 11(14c), *15*
Clement, B. A., 423(231), *431*
Clifford, D. B., 403(72), *426*
Clive, D. L. J., 190(230, 231), 193 (230, 231), *204*, 379(133), 381(145, 146), *385*, *386*
Coates, H., 422(201), *430*
Coe, D. G., 433(4, 5), *462*
Coffinet, D., 56(157), 57(157), 58 (157), 60(160), 61(160, 161), 62(157, 161), 63(162), *146*
Cohen, M. P., 229(12), 238(12), *266*
Colburn, V. M., 275(23), 278(23), 279(37b), *292*, *293*
Cole, R. A., 117(301), 118(301), *151*
Coll, J., 164(60), 178(60), *199*
Collier, C. W., 25(289), 112(289), *151*
Collignon, N., 162(39), *198*
Collington, E. W., 25(108), 44(108), *145*
Colton, F. B., 108(278), *150*
Colvin, E., 400(45), *426*
Colvin, E. W., 177(153), 193(153), *201*
Combret, J. C., 402(67, 68), *426*, 450 (89), *464*
Conia, J. M., 320(88), *348*
Conrad, W. E., 471(41), 486(41), 500 (41), *505*
Cookson, R. C., 245(50), *267*

Cooper, A., 270(7a), *292*
Cooper, R. D. G., 304(33), *346*, 363 (58), *383*
Corbel, B., 518(15), *533*
Corey, E. J., 3(1), *14*, 24(105, 113, 114, 246), 43(105), 46(113, 114), 56(156), 57(156), 58(156, 158), 59(158, 159), 60(158, 159), 61(158), 72(183), 73 (183), 84(202), 85(202), 89(214), 99 (245, 246), 100(246), 100(251, 252), 104(262), 106(268, 270), 109(281), 117 (114), 131(332), 133(338, 339), 134 (158, 340), 135(338, 339, 340), 136 (158), *145*, *146*, *147*, *148*, *149*, *150*, *152*, 157(11, 12, 13, 14), 163(47), 164 (11, 12, 13, 14, 63), 167(47), 168(11, 12, 13, 14, 47), 170(11, 12, 13, 14), 172(12, 13, 14), 175(63), 176(141), 178(14, 169), 179(177), 180(180, 181), 192(13), 193(47, 177), 194(141, 177), 195(169, 180), 196(11, 12, 13, 14), *197*, *198*, *199*, *201*, *202*, *203*, 336(136), *349*, 355(14, 15, 20), 356(15), 359(35), 360(44), 373(107, 108), 374(110), *382*, *383*, *385*, 423(229), *431*, 433(2), *462*
Corey, H. S., 92(228), *148*
Corfield, P. W. R., 453(99, 100), *465*
Corre, E., 450(96), *465*, 475(57), *505*
Cortez, C., 309(48), 310(48), *347*
Corvers, A., 138(342), *152*
Cosmatos, A., 513(3), 520(3), *533*
Cossentini, M., 166(80), *199*
Costisella, B., 176(148), 194(148), *201*
Costrillon, J. P. A., 302(25), 306(25), *346*
Cotton, F. A., 163(45), 194(45), *198*
Coulton, S., 518(14), *533*
Court, A. S., 184(210), 193(210), 194 (210), *204*
Coutrot, P., 163(49), 173(49), 174(134), 193(49), 193(134), 195(236), *198*, *201*, *205*, 450(89), *464*
Cox, E. F., 351(3), *382*
Cox, J. M., 297(11), *346*
Cox, O., 301(24), *346*
Cox, P. J., 139(355), 140(355), *153*
Crabbé, P., 25(267), 101(252), 105 (267), *149*, *150*
Craig, D. P., 468(8a), *503*
Cramer, F., 450(92), *465*, 519(21), *533*

Creasy, W. S., 455(115), *465*, 490(146), *508*
Creese, M. W., 443(54), *464*
Cresp, T. M., 80(196), *147*
Crim, F. F., 403(72), *426*
Cristau, H.-J., 460(142, 143, 144), *466*
Cristol, S. J., 453(104), *465*
Crofts, P. C., 420(184), *429*, 523(33), 524(40), *534*
Crombie, L., 66(172), 67(172), 128 (322, 323, 324), 130 (172, 330), 131 (330), 132(330), *147*, *152*, 177(157), 194(157), *202*
Crouch, R. K., 457(128), *466*
Cseh, G., 108(277), *150*
Currie, J. O., Jr., 47(123), *145*
Czuba, L. J., 107(274), *150*

Dahill, R. T., 38(81), *144*, 181(193), *203*
Dahl, A. R., 453(104), *465*
Dahl, T., 438(36), *463*
Dahm, D. J., 314(62), *347*, 357(31), *382*
Dahm, K. H., 37(90), *144*, 178(168), *202*
Dahms, G., 183(205), 194(205), *204*
D'alelio, G. F., 418(144, 147), 421 (144, 147), *428*
Dalla Croce, P., 233(28, 29, 30), 234(30), *267*
Dalrymple, D. L., 490(146), *508*
Daltrozzo, E., 493(167), *509*
Dandarova, M., 169(107), 173(107), 194(107), *200*
Daniel, H., 219(47), *222*
Danion, D., 167(86, 89, 94), 169(89, 107), 194(107), 195(86), *199*, *200*, 234(31), 235(31b), *267*
Danion-Bougot, R., 234(31), 235(31b), *267*
Dansette, P., 310(49), *347*
Darling, S. D., 459(139), *466*
Darnall, K. R., 480(93, 94), *506*
Das, B. P., 287(64), *294*
Dauben, W. G., 24(187), 25(190), 74 (187), 75(187, 188, 190), 76(190), *147*, 179(176), 180(186), 196(186), *202*, *203*, 251(67), *267*
Davidson, A. H., 166(82), 168(105),170 (105), 177(105, 158a), 192(105, 158a), 193(105), *199*, *200*, *202*

Davidson, R. S., 372(101), *385*
Davies, A., 319(85), *348*
Davies, A. G., 318(78), *348*
Davies, A. P., 329(113), 330(115), *349*, 400(46, 52), *426*
Davies, C. S., 278(36c), *293*
Davies, P. J., 25(360), 141(360), *153*
Davies, W. C., 379(136), *386*
Davis, C. E., 10(8), *15*
Davis, R. E., 351(3, 4), *382*
Davis, V. C., 190(229), *204*
Dawson, M., 24(115), 46(115), *145*
Dawson, M. I., 24(115), 46(115), *145*
Day, A. C., 97(241), 98(241), *149*, 164 (65), *199*
Dean, F. M., 275(24), *293*
De'ath, N. J., 370(87), *384*
De Bruin, K. E., 468(17b), *504*
Decorzant, R., 24(103), 25(103), 43 (103), 68(103), 69(103), *145*
Degterev, E. V., 422(206), *430*
Dehnel, A., 176(145), 194(145), *201*
Denney, D. B., 24(366), *153*, 316(70), 318(81), 319(81), 320(90, 91), 321 (91, 92), 323(99), 324(102), 325(102), *347*, *348*, 352(6), 370(87), *382*, *384*, 439(39), 448(85), *463*, *464*, 471(41), 486(41, 120, 123), 498(193), 499 (196, 197a), 500(41, 123), 502(216), *505*, *507*, *509*, *510*
Denney, D. Z., 320(91), 321(91), *348*, 471(41), 486(41), 499(197a), 500(41), *505*, *510*
Denyer, C. V., 379(133), *385*
Denzel, T., 490(143), *508*
Depew, C., 331(121, 122), *349*
Després, J. P., 25(267), 105(267), *150*
Dershowitz, S., 457(127), *466*
Derstuganova, K. A. 489(141), *508*
Desai, N. B., 315(66), *347*, 423(228), *431*, 469(20, 27, 28, 29, 30), 470(35), 471(20), 472(43), 474(43, 54), 475 (58, 59, 61, 62, 70), 476(20, 58, 59, 61, 62), 479(85), 482(59, 61, 62, 98, 99, 102, 103), 483(58), *504*, *505*, *506*, *507*
Deschamps, B., 39(96), 41(96), *145*, 159(22), 166(80), 167(93), 168(104), 171(22), 172(22), 173(22), 192(22), 194(104), *197*, *199*, *200*

Descoins, C., 25(173), 66(173), 67(173), *147*
Desmarchelier, J. M., 456(119), *465*
De Sombre, E. R., 159(27), *198*, 475(55b), *505*
Dever, J. L., 169(109), 175(109), 194(109), *200*
Devitt, F. H., 212(19), *221*
Devlin, C. J., 453(103), 458(129), *465, 466*
De Voe Goff, S., 253(77, 78), 254(78), 265(101), *268*
Devos, M. J., 132(333), *152*, 218(42), *222*
Dewey, R. H., 356(25), *382*
Dey, A. S., 422(213), *430*
Dianova, E. N., 300(20), *346*, 488(138), *508*
Dickstein, J. I., 455(110, 111), *465*
Diehl, J. W., 325(104), *348*
Dilaris, L., 513(3), 520(3), *533*
DiLeone, R. R., 439(39), *463*
Dimsdale, M. J. 105(266), *150*
D'Incan, E., 161(38), 165(74), 175(74), 193(38), *198, 199*
Dinizo, S. E., 174(126), 195(126), *201*
Dirks, J. E., 296(4), *345*
Dirlam, J. P., 298(12), *346*
Distler, W., 226(7c), 227(7c), 230(14), 266, 341(148), *350*
Dittmer, D. C., 363(55), *383*
Djerassi, C., 141(357, 359), *153*
Doherty, C. F., 130(330), 131(330), 132(330), *152*
Dolce, D. L., 400(51), *426*
Domagk, G. F., 75(191), 76(191), *147*
Doolittle, R. E., 116(298), *151*
Doomes, E., 453(99), *465*
Dormoy, J. R., 405(82), 417(139, 140), *427, 428*, 529(56, 57), 532(70), *535*
Dorn, C. R., 108(278), *150*
Dornand, J., 122(313), *152*
Dornauer, H., 211(14), *221*
Dostrovsky, I., 522(32), *534*
Douglas, B. E., 24(372), *153*
Dourtoglou, B., 417(140), *428*
Downie, I. M., 3(2), *14*, 392(17, 19, 21), 400(17, 19, 21, 41), 401(63, 65), 420(17, 184), *425, 426, 429*
Dradi, E., 102(255), *149*

Drefahl, G., 24(167, 354), 64(167), 135(354), *147, 153*, 178(166), 180(192), 195(166), *202, 203*
Dreux, M., 195(236), *205*, 303(28), *346*, 445(70b), *464*, 518(15), *533*
Duax, W. L., 436(23), *463*
DuBois, G. E., 138(347), *153*
Dubov, S. S., 496(186), *509*
Dubs, P., 354(12), *382*
Dudock, B. S., 263(97), *268*
Duffner, P., 48(127), *145*
Dunham, L., 178(171), *202*
Dunin, A. A., 156(4), 194(4), *197*
Dupré, M., 237(34, 36, 37, 38), 238(37), *267*
Dürr, M., 492(157), *508*
Durrant, G., 164(61), 167(61), 173(61), *199*, 214(28), *222*
Dusza, J. P., 25(101), 42(101), *145*, 180(191), *203*
Dutkey, S. D., 453(101), *465*
Dyatlovitskaya, E. V., 33(60), 96(238, 239), 97(239), *144, 149*

Earnshaw, C., 163(50), 166(82), 192(50), *198, 199*
Eberlein, W., 183(205), 194(205), *204*
Ebetino, F. F., 272(15), 275(19d), 276(19d), 283(55a), *292, 294*
Eckes, H., 177(154), *201*
Edelmann, R., 471(41), 486(41), 500(41), *505*
Edmundson, R. S., 422(207), *430*
Edwards, J. A., 161(162), 184(206), 193(206), *202, 204*
Efimova, V. D., 418(154, 167, 168), *428, 429*
Egg, H., 460(148), *466*
Eggelte, T. A., 180(188), 195(188), *203*
Eggers, H., 48(127), *145*
Egorova, L. P., 417(143), 418(143), *428*
Eguchi, M., 328(110), 330(110), *348*, 374(114), *385*
Eichenhofer, W., 422(200), *430*
Eichler, S., 34(66, 67), 36(67), *144*
Einhellig, K., 493(162, 164, 167, 171), 494(162), *509*
Einig, H., 404(80), 411(106), 413(121), 414(106, 122), 415(131), *427, 428*

Eiter, K., 19(12), 24(171), 29(12), 65(171), 68(12), 69(12), 85(171), 110(12), *142*, *147*
Elix, J. A., 24(169), 65(169), *146*
Elliot, M., 32(57), *143*
ElSanadi, N., 25(144), 52(144), 53(144), *146*
El Sawi, M. M., 435(13), *463*
Emeléus, H. J., 422(196), *430*
Emerson, T. R., 295(2), 296(2), 297(8), 300(2), 318(8), 322(2), *345*, *346*
Emmons, W. D., 23(26), 35(26), *143*, 156(3), 157(9), 162(9), 164(3), 165(3), 167(3), 173(3), 188(9), 189(222), 191(3), 194(3), 196(9, 222), *197*, *204*
Emsley, J., 14(23), *15*, 467(4), *503*
Endo, T., 373(104), *385*
Engel, N., 411(108), *427*
Engelhard, H., 28(37), *143*, 450(91), *465*
Engler, E. M., 357(30), *382*
Erbelding, G., 398(34), 400(34), *425*
Erdmann, H.-M., 75(191), 76(191), *147*
Erickson, B. W., 72(183), 73(183), *147*
Erman, W. F., 33(62, 63), 34(63), 134(62), 136(62), *144*
Ertel, A., 157(10), 193(10), *197*
Ertel, H., 159(23), 189(223), 192(23), 193(223), *198*, *204*
Eschenmoser, A., 354(10, 11, 12), 362(49), *382*, *383*
Ess, R. J., 360(41), *383*
Ettel, V., 422(205), *430*
Evans, B. E., 165(69), 186(69), *199*
Evans, D. A., 364(61, 62, 63), *383*, *384*
Evans, E. A., 47(121, 122), *145*
Evans, M. B., 369(82), *384*
Evdakov, V. P., 470(32a), 473(46), *504*, *505*
Everett, J. W., 360(37), *383*
Evin, G., 417(139, 140), *428*, 532(70), *535*
Evstigneeva, R. P., 46(116), 90(116), *145*
Eyles, C. T., 208(2a), *221*

Fackler, J. P., Jr., 368(79), *384*
Fajkos, J., 177(162), *202*
Faler, G. R., 360(37), *383*
Falk, L. C., 470(31), *504*

Fallis, A. G., 85(205), 98(205), 99(205), *148*
Farley, C. E., 373(103), *385*
Farnham, W. B., 356(23), *382*
Farrand, R., 356(26), *382*
Farrell, I. W., 99(242), *149*
Faruk, A. E., 24(368), *153*
Faulkner, D. J., 90(216), *148*
Fehn, J., 492(155), 493(160, 161, 169), *508*, *509*
Feinberg, A., 263(97), *268*
Feld, R., 319(85), *348*
Feldstein, G., 400(49), *426*
Felix, S., 334(132), 335(132), *349*
Fetchin, J. A., 368(79), *384*
Fétizon, M., 49(132), *146*
Feuer, B. I., 276(30), *293*
Fields, D. L., 461(149), *466*
Fields, E. K., 272(9), 273(9), *292*
Fields, E. S., 407(90), *427*
Filler, R., 455(108), *465*
Finch, N., 109(280), *150*
Findlay, J. A., 24(141), 51(141), 52(141), 133(141), *146*
Finet, J. P., 176(145), 194(145), *201*
Finkelhor, R. S., 218(42), *222*, 433(2), *462*
Firl, J., 71(180), 72(180), *147*
Fischler, H-M., 355(18), *382*
Fischli, A., 362(49), 373(102), *383*, *385*
Fisher, J., 304(33), *346*
Fisher, M. H., 24(376), *153*
Fitt, J. J., 109(280), *150*
Fitzpatrick, J. M., 373(108), *385*
Flatow, A., 361(48), *383*
Fleming, I., 177(158c), 192(158c), *202*
Fleming, R. H., 210(9), *221*
Fliszar, S., 27(34), 34(70), *143*, *144*
Flitsch, W., 266(106), *268*
Flynn, R. M., 418(172), *429*
Fodor, G., 23(27), 34(27), 38(27), *143*
Ford, J. A., Jr., 37(84), *144*, 168(102), 174(102), 195(102), *200*
Ford, M. E., 352(9), 353(9), *382*
Forgione, A., 102(255), *149*
Forsman, J. P., 433(3), *462*
Förster, H., 452(97), *465*
Foster, C. H., 322(97), 337(141), *348*, *349*
Foster, S. A., 248(56a), *267*

Foster, W. R., 418(163), 420(163), 421(163), *429*
Foucaud, A., 403(74), *426*, 450(96), *465*, 475(57), *505*
Frank, G. A., 24(85), 25(85), 37(85), 41(85), 78(85), 83(85), *144*
Frank, V. S., 519(19), *533*
Freeman, J. P., 218(43), *222*, 440(43), *463*
Freenor, F. J., 443(55, 56), *464*
Freerksen, R. W., 174(126), 195(126), *201*
Freyschlag, H., 19(16), *142*, 459(135), *466*
Fried, J., 102(256), *149*
Fried, J. H., 177(162), 184(206), 193(206), *202, 204*
Friederang, A. W., 400(58), 401(58), *426*
Friedrich, K., 380(142), *386*
Friedrich, W., 478(80), 481(80), *506*
Friemann, H., 413(117), *427*
Friemann, K., 411(105), *427*
Fritsch, W., 181(194), 183(194, 201), *203*
Frommeld, H. D., 300(21), *346*
Fruchey, O. S., 400(59), *426*
Fryberg, M., 142(362, 363), *153*
Fryer, R. I., 281(49), *293*
Fuchs, P. L., 76(189), *147*, 423(229), *431*
Fugiwara, K., 156(8), 173(8), 193(8), 194(8), *197*
Fuhrer, W., 354(11), *382*
Fujimoto, T. T., 364(62, 63), *383, 384*
Fujimoto, Y., 292(74), *294*
Fukami, J., 37(93), *144*, 178(167), 194(167), *202*
Fukomoto, K., 275(19d), 276(19d), 283(55a), 290(70a), *292, 294*
Fukuhara, G., 283(56), *294*
Fukui, K., 441(49), 452(49), *463*, 494(176), *509*
Fukuto, T. R., 456(119), *465*
Funamizu, M., 263(97), *268*
Furlenmeier, A., 345(158), *350*
Furst, A., 345(158), *350*
Furukawa, J., 502(209), *510*
Fusco, R., 233(28), *267*
Fyles, T. M., 119(303), 120(303, 304), *151*

Gabel, E., 424(247), *431*
Gadreau, C., 403(74), *426*
Gaede, B., 102(256), *149*
Gajda, T., 423(224), *431*
Gakis, N., 254(80), *268*
Gallagher, G., Jr., 168(98), 169(98), 173(98), 194(98), *200*
Gallagher, M. J., 169(110), *200*, 310(52), *347*, 476(75), 503(217), *506, 510*
Gallagher, P. T., 279(37b), *293*
Galle, J. E., 249(58), *267*
Galpin, I. J., 417(141), *428*, 531(67), 532(69), *535*
Galunski, B., 165(67), *199*
Gambaryan, N. P., 487(132), *508*
Ganba, A. L., 173(123), 194(123), *201*
Gandhi, C. S., 354(13), *382*
Gandolfi, C., 102(255), *149*
Garanti, L., 110(286), *150*
Garbers, C. F., 82(200), 83(200), 126(200), 127(200), *147*
Garcia, E. C., 274(18), *292*
Gardner, J. N., 319(86), *348*
Gareev, R. D., 495(181), *509*
Garegg, P. J., 329(114a), 330(114a), 335(114a), *349*
Garner, A. Y., 457(126), *466*
Garner, G. V., 277(33) *293*
Garner, R., 277(33), *293*
Garratt, P. J., 227(8a, 8b, 9), *266*, 360(37), *383*
Garst, M. E., 251(66), *267*
Gärtner, K. G., 519(21), *533*
Gash, V. W., 458(132), *466*
Gasic, G. P., 108(249), *149*
Gaudiano, G., 494(177a, 177b), *509*
Gaydou, E., 475(56), *505*
Gedge, R. N., 177(160), 194(160), *202*
Geierhaas, H., 423(237), *431*
Geiger, R., 532(68), *535*
Geissler, G., 3(3), *14*, 17(1), 25(1), *142*, 501(203), *510*
Gensler, W. J., 90(219), *148*
Georghiou, P. E., 28(41), *143*
Georgoulis, C., 400(47), *426*
Gerhard, W., 400(39), 401(39), 422(193), *425, 430*
Gerkin, R. M., 215(33), *222*
Gerlach, H., 373(106, 108), *385*
Germa, H., 484(105a, 105b), *507*

Gernayova, M., 169(107), 173(107), 194(107), *200*
Gesing, E. R., 266(106), *268*
Ghirardelli, R. G., 189(227), *204* 216, (34), *222*
Ghosez, L., 442(51), *464*
Gibson, D. M., 308(46), 309(46), *347*, 486(125), *507*
Gibson, J. A., 488(137), *508*
Gieren, A., 493(170, 171), *509*
Giesen, K., 408(92, 93), *427*
Giger, R., 6(4d), *14*
Giger, U., 526(47), *534*
Gilak, A., 407(91), *427*
Gilani, S. S. H., 392(20), 400(20), *425*
Gilbert, H. E., 24(365), *153*
Giles, R. G. F., 80(196), *147*
Gillespie, P., 468(11, 12, 19), 469(12, 19), *504*
Gillespie, P. D., 474(52), 475(60), 500 (201), *505, 510*
Gillespie, R. J., 468(8b, 8c, 8d, 8e, 8f), *503*
Gilman, N. W., 178(169), 195(169), *202*
Gilman, S., 380(143), *386*
Gilmore, W. F., 157(16), 193(16), *197*
Ginsburg, V. A., 496(186), *509*
Giolito, S. L., 418(146), *428*
Gladiali, S., 231(15), *266*
Glaser, S. L., 472(45), 473(45, 47, 50), 474(52, 53b), 475(60, 70), *505, 506*
Gleason, J., 354(11), *382*
Gleason, J. G., 368(75), 369(86), 370 (75, 88), 378(132), *384, 385*
Glinskaya, O. V. 527(50), *534*
Gloyna, D., 158(20), *197*
Gnoj, O., 319(86), *348*
Goerdeler, J., 378(129), *385*, 410(102), *427*
Goering, H. L., 400(48), *426*
Goetz, H., 167(96), *200*
Goff, S. D., 8(7b), *15*
Goh, S. H., 309(48), 310(48), *347*
Gol'dfarb, E. I., 489(141), *508*
Goldschmidt, S., 524(39), 525(43), 526 (39), *534*
Goldstein, B., 318(81), 319(81), 324 (102), 325(102), *348*
Golobov, Y. G., 423(227), *431*
Gombos, J., 373(108), *385*
Gomi, M., 297(159), *350*
Gompper, R., 176(146), 195(146), *201*
Goodbrand, H. B., 251(68), *267*, 459 (133), *466*
Goodyear, W. F., 318(81), 319(81), 324(102), 325(102), *348*
Gordon, P. F., 417(141), *428*, 532(69), *535*
Gosney, I., 13(19), *15*
Goth, H., 493(165, 168), *509*
Goto, G., 24(174), 67(174), 117(174), 118(174), *147*
Goto, K., 530(62), *535*
Goto, M., 263(97), *268*
Goto, S., 437(26b), *463*
Götschi, E., 354(11, 12), *382*
Gottfried, N., 139(353), *153*
Gough, S. T. D., 162(41), 176(41), 193(41), *198*, 212(18), *221*, 421(187), *429*
Gozman, I. P., 473(49), *505*
Grabley, F., 163(51), *198*
Grace, D. S. B., 270(6), 278(36d), 288 (66a), 289(66a), *292, 293, 294*, 496 (183a, 183b), *509*
Gracheva, E. P., 10(11a), *15*
Graf, G., 211(13), 212(20), *221*
Granem, B., 438(32), *463*
Granoth, I., 363(57), *383*
Grassberger, M. A., 48(129), *145*
Grassmann, W., 525(43), *534*
Grassner, H., 19(16), *142*, 459(135), *466*
Gravestock, M. B., 138(343), *152*
Gray, A. C. G., 299(15), *346*
Graydou, E., 455(112), *465*
Grayson, J. I., 166(82), 176(142), 177 (158b), 192(142, 158b), *199, 201, 202*
Grayson, M., 250(60), *267*, 373(103), *385*
Grechkin, N. P., 422(219), *430*, 484 (107a), *507*
Greco, C. V., 475(61, 62), 476(61, 62), 482(61, 62), *505*
Green, M., 164(63), 175(63), *199*
Greenbaum, M. A., 323(99), *348*
Greene, A. E., 25(267), 105(267), *150*
Greenwald, J. R., 403(71), *426*, 449(88), *464*

Greenwald, R., 24(113), 46(113), *145*
Greenwood, D., 369(80), *384*
Greenwood, R. A., 11(13), *15*
Grell, W., 164(66), *199*
Gretz, G. H., 116(298), *151*
Grewal, M. S., 173(123), 194(123), *201*
Grieco, P. A., 102(254), 105(265), 129 (327), 130(327), *149*, *150*, *152*, 161 (36), 180(182), 187(219), 194(36), 195(182), *198*, *203*, *204*, 218(42), *222*, 380(143, 144), *386*, 433(2), *462*
Griffin, C. E., 24(372, 373), *153*, 418 (162), 420(162, 181), *429*, 460(146), *466*
Griffin, G. W., 308(45, 46), 309(46), *347*, 486(125), 487(131), *507*, *508*
Griffiths, N. D., 272(10), *292*
Grigg, R., 362(50), *383*
Grigoreva, I. K., 482(202), 501(202), *510*
Griller, D., 318(78), *348*
Grinberg, S., 332(125), *349*
Grochowski, E., 331(123), 332(124), *349*
Gronowitz, S., 275(22), *292*
Gross, B., 400(54), 401(60, 64), *426*, 456(120, 123), *465*
Gross, H., 176(147, 148), 194(148), 195(147), *201*
Grossman, L. I., 448(80), *464*
Gruber, L., 392(29), 394(29), 400(29), 403(76), *425*, *426*
Grundmann, C., 298(13), 300(21), *346*
Grynkiewiez, G., 331(120), *349*
Grzejszczak, S., 164(64), 165(75, 76), 166(77), 175(64, 76), *199*
Gubanova, G. S., 484(107a), *507*
Guermont, J. P., 90(222, 223), *148*
Guillemonat, A., 455(112), *465*, 475 (56), *505*
Gulati, A. S., 307(34), 310(54), 311(54), *346*, *347*, 486(119), 487(134), 488(134, 136), *507*, *508*
Gunar, V. I., 10(11a, 11b), *15*
Gunstone, F. D., 93(231), 110(286), *149*, *150*
Gupta, B. G. B., 302(26), *346*, 445(69), *464*
Gupta, K. C., 25(201), 84(201), *147*, 173(123), 194(123), *201*

Guryanova, I. V., 314(63), *347*, 476 (72a, 72b), 482(100), *506*, *507*
Gustafson, D., 33(62), 134(62), 136 (62), *144*
Gustowski, W., 181(195), 195(195), *203*
Gut, M., 139(356), 140(356), *153*
Guthikonda, R. N., 187(220), 195(220), *204*
Gutmann, H., 24(367), *153*
Gutsch, P., 444(65), *464*
Guziec, F. S., Jr., 360(40), *383*
Guzman, A., 101(252), *149*

Haag, J., 378(129), *385*
Haag, W., 17(3), 18(3), 24(3), 79(3), 82(3), *142*, 317(71), *347*
Haake, P., 515(8), *533*
Häberlein, H., 211(11), *221*, 225(6), *266*
Habib, M. S., 297(10), *346*
Haede, W., 183(201), *203*
Haenel, M. W., 361(48), *383*
Hafferl, W., 178(171), *202*
Hafner, K., 291(73), *294*
Hägle, G., 501(207), *510*
Hague, M., 515(10), *533*
Haines, A. H., 355(17), *382*
Hall, C. D., 320(91), 321(91), *348*, 498(191), *509*
Hall, D., 14(23), *15*, 467(4), *503*
Hall, D. R., 117(300), *151*
Hallett, A., 531(67), *535*
Halmann, M., 522(32), *534*
Halstenberg, M., 216(37), *222*, 495 (182), *509*
Hamanaka, E., 24(105), 43(105), *145*
Hamelin, J., 234(31), 235(31b) *267*
Hamill, B. J., 177(153), 193(153), *201*
Hamilton, S. D., 277(34), 288(34), *293*
Hamlet, Z., 176(150), *201*, 469(28), *504*
Hamme, M. J., 423(235), *431*
Handa, V. K., 178(174), *202*
Hands, A. R., 25(107), 44(107), *145*, 490(145), *508*
Hanessian, S., 25(271), 106(271), 108 (271), 109(271), *150*, 456(118a), *465*
Hansen, B., 470(35), *504*
Hansen, H-J., 25(175), 68(175), *147*
Hansen, K. J., 272(10), *292*

Hantzsch, A., 379(136), *386*
Hanzel, R. S., 360(41), *383*
Harding, K., 180(180), 195(180), *202*
Harding, K. E., 173(124), *201*
Harger, M. J., 515(9), *533*
Hargis, J. H., 419(177), *429*
Harigaya, Y., 108(279), *150*
Harkes, A., 330(115), *349*
Harper, S. H., 128(322, 323), *152*
Harpp, D. N., 368(75), 369(86), 370 (75, 88, 89), 377(125), 378(132), *384*, *385*, 473(51a, 51b), *505*
Harris, C. J., 24(368), *153*
Harrison, C. R., 424(241), *431*
Harrison, I. T., 69(178), 70(178), *147*
Hart, D. J., 25(190), 75(190), 76(190), *147*, 251(67), *267*
Härtel, M., 87(212), *148*
Hartford, T. W., 436(25), *463*
Hartmann, V., 177(156), *201*
Hartmann, W., 355(18, 19), *382*
Hartung, H., 211(13), 213(25), *221*, *222*
Hartzler, H. D., 379(137), 380(138), *386*
Haruby, V. J., 415(123), *428*
Harvey, G. A., 231(16), 232(16), *266*
Harvey, G. R., 159(27), *198*
Harvey, J. A., 316(68), *347*
Harvey, R. G., 309(48), 310(48), *347*, 365(66), 367(70), *384*, 475(55b), *505*
Hase, T., 441(46), *463*
Hashimato, H., 401(66), *426*
Hashimoto, M., 375(116, 117, 119, 120, 121), *385*
Haslinger, E., 248(57), *267*, 373(108), *385*
Hasova, B., 169(107), 173(107), 194(107), *200*
Hassner, A., 249(58), *267*
Haszeldine, R. N., 422(196), *430*
Hata, T., 375(120), *385*, 450(92, 94), *465*
Hatfield, L. D., 304(33), *346*
Hattori, M., 176(144), 195(144), *201*
Hatzelmann, L., 493(162), 494(162), *509*
Hauser, C. F., 25(100), 42(100), 128(100), *150*
Hauser, E., 213(24), 219(48), *222*

Hauss, A., 212(16), *221*
Hawkins, J. M., 371(92), *384*
Hay, J., 289(68), *294*
Hayase, Y., 50(138), *146*, 173(119), 194(119), 195(234), *201*, *204*
Hayashi, M., 103(257), 104(264), 107 (272, 275), 108(276), *149*, *150*
Hayashi, R., 297(159), *350*
Hayashida, A., 530(63), *535*
Heaney, H., 401(63), *426*
Hearn, M. T. W., 25(204), 85(204, 205), 86(204), 98(205), 99(205, 243), *148*, *149*
Heathcock, C. H., 161(35), 187(217), 194(35), 196(217), *198*, *204*
Heck, R. F., 25(379), *153*
Hecker, E., 110(283, 284, 285), *150*
Heckman, R. A., 34(72), 36(72), 38(72), *144*
Heeres, G. J., 24(168), 64(168), *147*
Hege, B. P., 34(72), 36(72), 38(72), *144*
Heimgartner, H., 254(80), *268*
Heine, H-G., 355(18), *382*
Heine, H. W., 216(38), *222*
Heinke, B., 522(30), *534*
Heinzelmann, W., 423(236), 424(236), *431*
Heiss, J., 175(131), 195(131), *201*
Heitz, W., 25(154), 56(154), 79(154), *146*, 423(238), 424(244), *431*
Hejno, K., 177(165), 194(165), *202*
Heller, C. I., 312(60), *347*
Heller, J., 468(10), *504*
Heller, S. R., 475(62), 476(62), 482(62, 98), *505*, *507*
Hellwege, D. M. 467(2a), *503*
Hellwinkel, D., 467(2b, 3), 485(117), 499(195a, 195b), 502(215), *503*, *507*, *510*
Hemesley, P., 66(172), 67(172), 128(324), 130(172), *147*, *152*
Hendrikson, J. B., 218(45), *222*
Hengartner, U., 6(4c), *14*
Henglein, A., 422(197), *430*
Hengstberger, H., 239(43), 257(86), *267*, *268*
Henning, H. G., 158(20), 162(40), 193(40), *197*, *198*, 459(136), *466*
Henrick, C. A., 24(130), 25(173), 48(130), 49(130), 66(173), 67(173),

110(282), 113(293), 114(293), 118 (282), *145*, *147*, *150*, *151*, 184(206), 193(206), *204*
Henry, E., 13(19), *15*
Herald, C. L., 181(199), 183(203), *203*
Herb, S. F., 110(286), *150*
Herbert, R. B., 270(4), *292*
Herbst, P., 19(11), 44(106), 86(207), *142*, *145*, *148*
Hercouet, A., 210(10), *221*, 265(105), *268*
Herisson, C., 166(85), 185(85), 193(85), *199*
Herkes, F. E., 45(109), *145*, 314(64), 315(64), *347*
Herr, M. E., 438(31), *463*
Herron, D. K., 133(338), 135(338), *152*
Hershberg, R., 469(22), 471(22), 476(22), *504*
Hershkowitz, R. L., 416(133, 134), 423(133, 134), *428*, 437(28, 29), 438(29), 440(28), *463*
Herweg, G., 445(70a), *464*
Herweh, J. E., 305(38), 310(38), *346*, 365(67), *384*, 486(128), *508*
Hess, H. J. E., 107(274), *150*, 180(184), *203*
Hesson, D. P., 446(72b), *464*
Heublein, G., 54(148), *146*
Heuer, G. E., 176(143), 194(143), *201*
Heusler, K., 344(153), *350*
Hevesi, L., 132(333), *152*, 218(42), *222*
Hewson, K., 184(209), *204*
Heymons, A., 407(89), *427*
Hiatt, R., 318(84), 319(84), *348*
Hibbert, H., 379(136), *386*
Hickey, H. R., 195(239), *205*
Hickey, K. R., 174(137), 195(137), *201*
Higashi, F., 434(10), 437(150), *463*, *466*, 527(51, 52), *534*
Higgins, G. M. C., 369(82), *384*
Higgins, J., 440(44), *463*
Higham, C. A., 85(206), 86(206), *148*
Higuchi, K., 418(170), *429*
Hill, D. T., 290(70b), *294*
Himizu, J., 108(279), *150*
Hiraga, K., 24(174), 67(174), 117(174), 118(174), *147*

Hirai, K., 352(8), *382*
Hirata, Y., 437(26b), *463*
Hirayama, M., 164(57), 196(57), *199*
Hirrichs, E., von, 335(133), *349*
Hirschmann, R., 512(2), *533*
Ho, T. L., 440(41), *463*
Hocking, M. B., 172(116), *200*
Hocks, P., 345(158), *350*
Hodge, P., 424(240, 241), *431*
Hoerhold, H. H., 178(166), 195(166), *202*
Hoffman, W. A., 329(111), *348*
Hoffman, W. H., 332(127), 333(127), *349*
Hoffmann, A. K., 323(99), *348*
Hoffmann, F. W., 352(2), 360(41), *382*, *383*
Hoffmann, H., 35(75), *144*, 155(1), 156(1, 5, 6), 157(10), 158(18), 159(23), 162(40), 163(1, 5), 167(1, 5), 174(1, 6), 184(212), 189(223, 225), 192(1, 23, 225), 193(5, 6, 10, 18, 40, 223), *197*, *198*, *204*, 216(36), *222*, 300(18), *345*, *346*, 360(38), *383*, 415(132), 416(132), 423(132), *428*, 437(27), 440(27, 45), 452(97), 461(45), *463*, *465*
Hoffmann, P., 468(12), 469(12), *504*
Hoffmann, R., 497(189), *509*
Hoffmann, W., 50(137), 132(335), *146*, *152*
Hoffsommer, R. D., 180(180), 195(180), *202*
Höfle, G., 364(59), 369(83), *383*, *384*
Hofmann, A. W., 379(135), *386*
Höhenberger, W., 493(168), *509*
Höhle, A., 405(85), *427*
Holcombe, F. O., 166(84a), 177(155), 185(84a), 191(155), 194(84a, 155), *199*, *201*
Holl, P., 493(173), *509*
Holley, R. W., 263(97), *268*
Holliman, F. G., 270(4), *292*
Holmes, A. B., 227(9), *266*
Holmes, J. B., 3(2), *14*, 392(17), 400(17), 420(17), *425*
Holmes, R. R., 468(9b, 9c), *504*
Holtz, H. D., 320(89), 321(89), *348*
Honkomp, L. J., 278(36c), *293*
Hooz, J., 392(20), 400(20), *425*

Hopkins, T. L., 421(190), *430*
Hopp, M., 110(285), *150*
Horley, D., 24(376), *153*
Horner, L., 35(75), *144*, 155(1), 156 (1, 5), 157(10), 158(18), 159(23), 163(1, 5), 167(1, 5), 168(103), 174(1, 103), 189(223, 225), 192(1, 23, 225), 193(5, 10, 18, 103, 223), *197*, *198*, *200*, *204*, 216(36), *222*, 300(18), 303 (30), 310(51), 318(79), 320(79), 322 (79), 323(79), *345*, *346*, *347*, *348*, 378(131), *385*, 415(132), 416(132, 136), 423(132), *428*, 437(27), 440 (27, 45), 444(64), 445(67), 459 (137), 461(45), 491(148), *508*
Horspool, W. M., 308(42), 310(42), *346*
Hörster, H. G., 360(37), *383*
Horton, D., 355(16), *382*
Hosokami, T., 287(61), *294*
House, H. O., 19(15e), 24(85), 25(85), 33(58), 37(85), 41(85), 78(85), 83(85), 91(224), *142*, *143*, *144*, *148*
Howard, E., 297(9), *346*
Howard, H. T., 513(4), *533*
Howard, J. A., 468(9a), *504*
Howe, R. K., 70(179), 71(179), *147*
Howell, J. M., 497(189, 190), *509*
Hruby, V. J., 28(49), *143*
Huber, J. W., 157(16), 193(16), *197*
Huber, W., 24(246), 99(246), 100(246), *149*
Hudson, D., 531(67), *535*
Hudson, H. R., 434(8, 9), 440(8), *462*
Hudson, R. F., 14(24), *15*, 24(370), 27(34), 34(70), *143*, *144*, *153*, 172 (116), *200*, 422(201), *430*, 447(77), *464*
Huff, R. K., 168(100), 169(100), 173 (100), 194(100), *200*
Hug, R., 25(175), 68(175), *147*
Hughes, A. N., 467(2c), *503*
Hughes, C. T., 25(175), 68(175), *147*
Hugl, E., 258(89), 260(91), 261(91), 262(93), 263(89, 93, 95), *268*
Huisgen, R., 12(17), *15*, 215(32), *222*, 233(27), 236(27), 243(27), 245(27), *266*, 327(107), *348*, 490(144), 494 (175a, 175b), 496(185a, 185b), *508*, *509*
Huisman, H. O., 177(161), 179(179), 180(185, 188), 195(179, 188), *202*, *203*
Hull, R., 356(26), *382*
Humphrey, R. E., 371(92), 372(93), *384*
Humphris, K. J., 318(82), *348*
Hung, Y. Y., 280(41), *293*
Hunkeler, W., 354(11), *382*
Hunt, L. B., 400(56), *426*
Huppertz, M., 390(15), *425*
Hutchins, M. G., 436(24), *463*
Hutchins, R. F. N., 141(358), *153*
Hutchins, R. O., 437(26a), *463*
Hutton, J., 180(181), *203*
Huynh, T. B., 164(54), 177(54), 192 (54), 193(54), *198*

Ichimura, K., 249(59), *267*, 411(104), *427*, 440(42), *463*
Iddon, B., 272(13), 275(20, 23), 278(23, 37a), 279(37a, 37b), *292*, *293*
Iguchi, S., 103(257), *149*
Iguchi, Y., 107(275), *150*
Ikeda, S., 517(12, 13), *533*
Ikenaga, S., 373(104), *385*
Ikota, N., 521(26), *534*
Iljina, N. A., 419(176), *429*
Ilton, M. A., 52(142), *146*
Imagilov, R. K., 423(226), *431*
Inamoto, N., 357(27), *382*
Indo, M., 25(111), 46(111), 130(111), *145*
Inhoffen, E., 19(11), 44(106), *142*, *145*
Inhoffen, H. H., 75(191), 76(191), *147*
Inoue, I., 24(165), 63(165), *146*
Inoue, S., 24(140), 51(140), 56(140), *146*
Inoue, T., 164(57), 196(57), *199*
Inouye, Y., 476(73), *506*
Inwood, R. N., 164(65), *199*
Ipaktschi, J., 24(187), 25(190), 74(187), 75(187, 190), 76(190), *147*
Ireland, R. E., 6(4a, 4c, 4d), *14*
Irving, K. C., 216(38), *222*
Isaacs, N. S., 403(75, 77), 404(77), *426*, 443(61), *464*
Isbell, A. F., 419(178), *429*
Ishihara, T., 371(90), *384*
Ishikawa, K., 308(45, 46), 309(46), *347*, 486(125), *507*

Ishikawa, N., 502(208), *510*
Isler, O., 24(367), 120(305), 123(317, 319), *151, 152, 153*
Isoe, S., 50(138), *146*
Ito, M., 177(163), 196(163), *202*
Ito, Y., 322(95, 96), 323(96), *348*
Itoh, I., 356(23), *382*
Itoh, K., 278(38), *293*
Ivancsics, C., 262(94), 263(94, 96), 264(96, 98), *268*
Ivanov, B. E., 489(141), *508*
Ivanov, L. L., 46(116), 90(116), *145*
Izumiya, N., 520(25), 527(25), 531(25), *533*
Izydore, R. A., 189(227), *204*, 216(34), *222*

Jackson, A. G., 523(34), *534*
Jackson, A. H., 272(11), *292*
Jackson, R. W., 106(269), *150*
Jacobs, R. L., 351(4), *382*
Jacobson, H. I., 365(66), 367(70), *384*
Jacobson, M., 118(302), 133(337), *151, 152*
Jaenicke, L., 115(294), *151*
Jambotkar, D., 29(44), 77(44), *143*
James, B. G., 24(135, 199), 50(135), 51(135), 81(135, 199), 88(199), *146, 147*
James, J. D., 116(298), *151*
Janes, N. F., 32(57), *143*
Janssen, H., 424(249), *431*
Janssen, M. J., 24(168), 64(168), *147*
Janzso, G., 174(125), *201*
Jarell, H. C., 331(122), 332(128), *349*
Jarolim, V., 177(165), 178(170), 194(165, 170), *202*
Jarvis, B. B., 453, (98, 100, 101), *465*
Jary, J., 435(17), *463*
Jaunin, R., 409(96), *427*
Jayawant, M., 444(65), *464*
Jefford, C. W., 321(93), *348*
Jenkins, I. D., 169(110), *200*, 310(52), *347*, 476(75), *506*
Jenkins, J. K., 85(206), 86(206), *148*
Jenkins, M. E., 490(146), *508*
Jenny, W., 361(45), *383*
Jensen, E. V., 365(66), 367(70), *384*
Jerina, D. M., 310(49), *347*
Jie, M. L. K., 110(286), *150*

Johansen, J. E., 68(176), *147*
Johncock, P., 458(130), *466*
Johnson, A. W., 19(15d), 20(15d), 21(23), 23(15d), 25(193), 30(23), 77(193), 78(193), *142, 143, 147*, 158(19), 161(34), 165(70), 167(88), *197, 198, 199*, 362(50), *383*
Johnson, C. R., 352(8), *382*
Johnson, D. N., 272(11), *292*
Johnson, M. R., 180(184), *203*
Johnson, R. A., 438(31), *463*
Johnson, T., 25(213), 88(213), *148*
Johnson, W. D., 491(147), *508*
Johnson, W. L., 400(59), *426*
Johnson, W. S., 52(142), 138(343, 344, 346, 347), *146, 152, 153*
Johnston, G. A. R., 435(20), *463*
Johnston, J. A., 403(72), *426*
Jones, D. H., 321(92), *348*, 486(120), *507*
Jones, E. R. H., 25(204), 85(204, 205, 206), 86(204, 206), 90(221), 97(241), 98(205, 221, 241), 99(205, 243), *148, 149*
Jones, G., 37(95), *144*, 169(106), 173(106), 193(106), *200*, 288(75), 291(72), *294*
Jones, J. K. N., 331(119, 121, 122), 332(128), *349*
Jones, M. E., 82(198), *147*
Jones, R. A., 254(81), *268*
Jones, V. K., 24(85), 25(85), 37(85), 41(85), 78(85), 83(85), *144*
Jones, W. J., 379(136), *386*
Jørgensen, K. A., 11(14d), *15*
José, F. L., 304(33), *346*, 363(58), *383*
Joseph, J. P., 180(191), *203*
Josephson, A. S., 372(99), *385*
Jostes, F., 407(89), *427*
Joubert, J. P., 460(142, 143), *466*
Joyce, R. J., 524(38), *534*
Jung, M. E., 400(45), *426*
Jurezak, J., 331(120, 123), 332(124), *349*
Jurgeleit, W., 318(79), 320(79), 322(79), 323(79), *348*
Jurian, M., 6(4b), *14*
Just, G., 53(147), *146*

Kabachnik, M. J., 250(63), *267*
Kaehne, R., 159(26), *198*

Kagan, F., 438(37), *463*
Kaiser, J. K., 360(39), *383*
Kaiser, W., 291(73), *294*
Kakase, F., 378(126), *385*
Kakurina, V. P., 314(63), *347*, 476 (72c), *506*
Kalir, A., 363(57), *383*
Kalmus, A., 165(69), 186(69), *199*
Kamai, G., 417(143), 418(143, 149, 150, 151, 155, 156, 157, 159, 161), 419 (149, 155, 156, 159, 161, 173, 174, 175, 179), 420(161, 179), 423(226), *428, 429, 431*
Kamata, S., 6(4d), *14*
Kametani, T., 272(14, 15), 274(17), 275(19d, 26), 276(19d, 29), 282(17), 282(50, 51a, 51b), 283(53, 54, 55a), 284(57), 285(58), 290(69, 70a), 292 (74), *292, 293, 294*
Kandasamy, D., 437(26a), *463*
Kando, K., 175(140), 194(140), *201*
Kaneko, C., 297(159), *350*
Kaneko, H., 169(112b), 170(112b), 180(192), 181(112b), 182(112b), 193 112b), *200, 203*
Kantardjiew, I., 25(288), 111(288), *150*
Kaplan, L., 416(135), 423(135), *428*
Kaplan, M. L., 336(137), 337(140), 338(137), 339(137), *349*, 491(150a, 150b), *508*
Karimullina, E. Kh., 476(72b), *506*
Kartoon, I., 332(125), *349*
Kasai, H., 263(97), *268*
Kasai, K., 520(23), *533*
Katinoki, H., 437(150), *466*
Kato, K., 378(126, 127, 128), *385*
Katritzky, A. R., 295(1), 299(15), *345, 346*
Katsoyannis, P. G., 514(5), *533*
Katsube, J., 103(260), *149*
Katz, G., 263(97), *268*
Katz, T. J., 498(192), *509*
Katzenellenbogen, J. A., 131(332), *152* 178(169), 195(169), *202*
Katzman, S. M., 298(14), *346*
Kauer, J. C., 281(42), *293*
Kaur, J., 178(174), *202*
Kawabe, N., 454(107), *465*
Kawamura, M., 108(276), *150*
Kawamura, S., 371(90, 91), *384*

Kawasaki, T., 440(40b), *463*
Kaye, H., 405(81), *427*
Kazaryan, S. A., 434(10), *463*, 527(52), *534*
Kearns, D. R., 336(139), 338(139), 340(139), *349*
Keates, C., 288(75), *294*
Kees, K., 342(150), *350*, 491(153), *508*
Kefurt, K., 435(17), *463*
Keller, H., 514(6), *533*
Kellogg, R. M., 360(39), *383*
Kemp, G., 401(63), *426*
Kempe, U. M., 25(360), 141(360), *153*
Kenner, G. W., 160(31), 163(31), *198*, 515(11), 523(34), 531(67), *533, 534, 535*
Keogh, K. S., 129(326), 130(326), *152*
Keough, P. T., 250(60), *267*
Kerb, U., 345(158), *350*
Kesling, H. S., 423(233), *431*
Ketcham, R., 29(44), 77(44), *143*
Khalil, A. H., 177(160), 194(160), *202*
Khalil, H., 252(72), 265(104), *268*
Khan, N., 68(177), 70(177), 126(177), *147*
Kharasch, M. S., 318(83), *347*, 418 (160), *429*
Kharrasova, F. M., 418(150, 151, 153, 154, 161, 167, 168), 419(161, 173, 174, 179), 420(161, 179), *428, 429*
Khazanie, P. G., 272(7b), *292*
Kheifets, L. Ya., 175(133), *201*
Khim, C. U., 28(41), *143*
Khisamova, S. L., 418(156), 419(156), *429*
Khorana, U. G., 263(97), *268*
Kidwell, R. L., 459(139), *466*
Kiel, W. A., 190(230), 193(230), *204*, 381(145), *386*
Kienzle, F., 123(316), *152*
Kikuchi, S., 164(57), 196(57), *199*
Kim, C. S., 247(55), 254(81, 82), 255 (83), *267, 268*
Kimball, G. E., 468(7), *503*
Kimura, J., 330(114b, 116), 335(114b), *349*
Kimura, Y., 502(208, 210, 211, 213), *510*
King, K. G., 468(16), *504*
Kinnick, M. D., 173(120), 193(120), *201*

Kinoshita, M., 10(10), *15*
Kinstle, T. H., 37(88), 39(88), *144*, 169(107), 173(107), 194(107), *200*
Kirby, A. J., 14(25), *15*, 368(72), *384*, 479(83), *506*
Kirby, K. C., Jr., 447(74, 78, 79), 455(78), *464*
Kirillova, K. M., 480(86, 88b), *506*
Kirilov, M., 38(79), *144*, 165(67), 169(113), 193(233), 195(113), *199*, *200*, *204*
Kirkpatrick, A., 372(97), *384*
Kirkpatrick, D., 403(75, 77), 404(77), *426*, 443(61), *464*
Kirschenbaum, D. M., 372(98, 99), *384*, *385*
Kishida, Y., 352(8), *382*
Kisielowski, L., 226(7c), 227(7c), *266*, 341(148), *350*
Kita, Y., 440(40b), *463*
Kitahana, Y. K., 322(95, 96), 323(96), *348*
Kitao, T., 371(90, 91), *384*
Kitazawa, E., 531(65, 66), *535*
Klabunde, K. J., 45(109), *145*, 314(64), 315(64), *347*
Klado, I., 288(75), *294*
Klahre, G., 156(5), 163(5), 167(5), 189(223), 193(5, 223), *197*, *204*
Kleinstuck, R., 317(76), *348*, 408(95), 409(95, 97, 98, 99), 410(103), 412(109, 113), 413(118, 120), *427*, *428*
Klingler, T. C., 301(22), *346*
Klink, W., 157(10), 158(18), 159(23), 168(103), 174(103), 192(23), 193(10, 18, 103), *197*, *198*, *200*
Klitgard, N. A., 230(13), *266*
Kloesters, W., 214(27), *222*
Klusacek, H., 468(11, 12), 469(12), *504*
Knaggs, J. A., 329(113), 330(115), *349*, 400(46, 52), *426*
Knight, J. C., 181(199), *203*
Knoll, F., 388(6, 8, 10), 389(6), 393(6), 397(6), 403(6), 412(109), *425*, *427*
Knolle, J., 106(270), *150*
Knowles, W. S., 343(152), *350*
Knunyants, I. L., 487(132), 501(205), *508*, *510*
Kobayashi, S., 502(208, 209, 210, 211, 213), *510*
Kobayashi, Y., 446(72a), *464*
Koch, K., 491(150e), *508*
Koch, W., 110(285), *150*
Kochetkov, N. K., 435(14, 15, 16), 438(15), *463*
Kochi, J. K., 318(78), *348*
Kochmann, W., 422(193), *430*
Kocienski, P. J., 400(49), *426*
Kocor, M., 181(195), 195(195), *203*
Koekemoer, J. M., 103(258), *149*
Koelliker, U., 24(246), 99(246), 100(246), *149*
Koenig, M., 485(113), *507*
Koeppel, H., 158(20), *197*
Kohl, H., 183(201), *203*
Koizumi, T., 515(8), *533*
Kojima, K., 104(261), *149*
Kolewa, S., 212(20), *221*
Kolodyazhnyi, O. J., 423(227), *431*
Kolonits, P., 38(82), *144*, 184(214), *204*
Kondo, K., 209(5), *221*, 365(65), *384*
Konig, W., 532(68), *535*
Koning, H., de, 177(161), 179(179), 180(185, 188), 195(179, 188), *202*, *203*
Konishi, Y., 108(276), *150*
Kono, D. H., 442(53), *464*
Konovalova, I. V., 476(72c), *506*
Kopay, C. M., (79), *268*
Koppel, G. A., 173(120), 193(120), *201*
Koppes, W. M., 442(52), *464*
Kori, S., 103(257), 104(264), 107(275), 108(276), *149*, *150*
Korte, F., 368(71), 369(85), *384*
Koschatzky, K. H., 25(291), 112(291), 113(291), *151*
Kosolapoff, G. M., 14(20), *15*, 191(240), *205*, 418(158), *429*
Köster, R., 48(129), *145*
Kostetskii, P. V., 130(329), *152*
Koszmehl, G., 24(375), 28(39), *143*, *153*
Kothe, N., 405(85), *427*
Kotin, S. M., 363(55), *383*
Kovac, J., 169(107), 173(107), 194(107), *200*
Kovacs, J., 412(112), *427*
Kovalenko, N. P., 175(133), *201*
Kovalev, B. G., 164(62), *199*
Kovaleva, A. S., 46(116), 90(116), *145*
Kovtun, V. Yu., 90(220), 91(220), *148*

Kozar, L. G., 161(35), 187(217), 194 (35), 196(217), *198*, *204*
Kozikowski, A. P., 25(190), 75(188, 190), 76(190), *147*
Kozuka, S., 302(27), 303(27), *346*, 363 (56), *383*
Kranz, E., 25(377), *153*, 225(5), *266*
Kratzer, O., 29(52), 30(52), 82(52), *143*, 189(224), *204*, 225(6), 226(7b), 227(7b), *266*
Krauss, H. L., 525(43), *534*
Krawczyk, H., 176(151), 193(151), *201*
Kreiser, W., 181(200), 183(204), 195 (200), *203*
Kreiss, W., 524(41), *534*
Kresze, G., 71(180), 72(180), *147*, 176 (149), *201*
Kretchmer, R. A., 344(154), *350*
Kretschmar, H., 33(63), 34(63), *144*
Kreutzkamp, N., 212(17), *221*
Krief, A., 132(333, 334), *152*, 218(42), *222*
Krolikiewicz, K., 362(51), *383*
Krubiner, A. M., 139(352, 353), 140 (352), *153*
Kruger, G., 183(205), 194(205), *204*
Kruse, C. G., 175(138), 194(138), *201*
Krusic, P. J., 318(78), *348*
Ku, A., 493(172), *509*
Kübel, B., 411(108), *427*
Kucherov, V. F., 37(89), 38(78), *144*
Kudryavtseva, L. A., 489(141), *508*
Kugel, R. L., 502(214c), *510*
Kugler, H. J., 311(56), *347*, 475(64, 65, 68, 69), 482(64, 65, 68, 69), 483 (64, 65), *506*
Kühlein, K., 180(183), *203*
Kuhn, G., 515(8), *533*
Kukhtin, V. A., 480(86), 480(88a, 88b), *506*
Kulik, S., 10(9), *15*, 270(3), 286(60b, 60c, 60d), 287(60d), *292*, *294*
Kumamoto, T., 315(67), 316(67), *347*, 480(89, 90), *506*
Kunau, W. H., 93(230), *148*
Kunstmann, R., 115(295), *151*, 236 (33), *267*, 490(143), 494(178), 495 (178), *508*, *509*
Kunzek, H., 459(138), *466*
Kuo, C. H., 180(180), 195(180), *202*

Kurshakova, N. A., 471(37b), *505*
Kurtz, D. W., 25(213), 88(213), *148*
Kurz, W., 365(64), *384*
Kurze, R., 422(202), *430*
Kuwajima, I., 308(44), 333(130), *347*, *349*, 374(112, 113), *385*, 486(118), *507*
Kvantrishvili, V. B., 422(220), *431*
Kwart, H., 468(16), *504*
Kwiatkowski, G. T., 24(114), 46(114), 89(214), 117(114), *145*, *148*, *157* (11, 12, 13), 163(47), 164(11, 12, 13), 167(47), 168(11, 12, 13, 47), 170 (11, 12, 13), 172(12, 13), 192(13), 193(47), 196(11, 12, 13), *197*, *198*
Kyllingstad, V. L., 25(193), 77(193), 78(193), *147*
Kyotani, Y., 437(26b), *463*

Labar, R. A., 47(123), *145*
Labaw, C. S., 247(55), 254(82), 266 (107), *267*, *268*
L'abbé, G., 231(17, 18, 19, 20, 21), 232 (20, 23, 24, 25), *266*, 495(179), *509*
LaCount, R. B., 21(23), 30(23), *143*
Lagowski, J. M., 295(1), *345*
Laing, D. G., 37(92), 40(92), 133(92), *144*, 178(167), 194(167), *202*
Lake, A. W., 139(355), 140(355), *153*
Lal, A., 178(174), *202*
Lal, B., 329(111), 332(127), 333(127), *348*, *349*, 456(116), *465*
Lambert, R. W., 287(63), *294*
Lampin, J. P., 159(22), 171(22), 172 (22), 173(22), 192(22), *197*
Landauer, S. R., 433(4), 435(11), *462*, *463*
Landis, M. E., 317(76), *348*, 486(121, 122), *507*
Landor, P. D., 166(79), 176(79), 178 (79, 173), 194(79), *199*, *202*
Landor, S. R., 166(79), 176(79), 178 (79, 173), 194(79), *199*, *202*, 434(6), *462*
Lang, F., 403(73), *426*
Lang, H. J., 228(10), 230(14), *266*
Langemann, A., 345(158), *350*
Lapitskii, G. A., 178(166), 195(166), *202*
Laszlo, H., 132(334), *152*

Laurenco, C., 195(236), *205*, 484(105c, 106, 107b), 485(111b, 114), *507*
Lautenschlager, H., 525(43), *534*
Lavallee, P., 25(271), 106(271), 108(271), 109(271), *150*
Lavielle, G., 166(83), 167(92), 168(101), 176(145), 177(83), 194(83, 145), 195(83, 232), 196(101), *199, 200, 201, 204,* 402(67, 68, 69, 70), *426,* 450(89), *464*
Lavrinenko, E. S., 164(62), *199*
Lawesson, S-O., 11(14a, 14b, 14c, 14d), *15*
Lawton, E. L., 410(100), *427*
Leader, G. R., 169(109), 175(109), 194(109), *200*
Lebedeva, N. M., 166(78), 194(78), *199*
Leblanc, R., 450(96), *465*
Lecher, H. Z., 11(13), *15*
Lechmann, H. G., 183(205), 194(205), *204*
Le Corre, M., 210(10), *221*, 265(105), *268*
Lee, D. P., 424(239), *431*
Lee, J. B., 3(2), *14*, 392(17, 18, 19, 21), 400(17, 18, 19, 21, 41, 53), 401(65), 404(79), 420(17, 53), *425, 426*, 435(13), *463*, 528(53), *534*
Lee, J. H., 281(46), *293*
Lee, M. S., 102(256), *149*
Lee, S., 500(201), *510*
Lee, T. V., 105(266), *150*
Lefebvre, G., 39(96), 41(96), *145*, 167(90, 93), 168(104), 170(90), 194(104), *200*
Lefebvre, M. G., 169(112a), 170(112a), 182(112a), 193(112a), *200*
Le Guern, D., 450(96), *465*
Lehmann, H. A., 422(202), *430*
Lemmen, P., 474(53b), *505*
Le Moing, M. A., 450(96), *465*
Lengyel, I., 490(143), *508*
Le Perchec, P., 320(88), *348*
Leplawy, M., 524(42), *534*
Lerch, U., 102(253), *149*
Lerman, C. L., 499(197c), *510*
Leroux, Y., 303(28), *346*, 445(70b), *464*, 484(109), *507*
Lester, R., 117(300, 301), 118(301), *151*

Letourneux, Y., 139(356), 140(356), *153*
Levison, M. E., 372(99), *385*
Levron, J. C., 90(222, 223), *148*
Levy, D., 438(34, 35), 441(34), *463*
Levy, S., 21(22), 142(22), *143*
Lewis, R. G., 33(62), 134(62), 136(62), *144*
Leyshon, L., 283(52b), 285(52b), *293*
Leyshon, L. J., 248(56a, 56b), *267*, 270(5b), *292*
Leznoff, C. C., 119(303), 120(303, 304), *151*
Liaaen-Jensen, S., 37(94), 38(80), 40(94), 68(176), *144, 147*
Liang, Y. F., 358(34), *383*
Liehr, J. C., 490(146), *508*
Liehr, J. G., 250(64), 252(73), *267, 268*
Lienert, J., 442(50), 446(73), *463, 464*
Light, K. K., 8(7a), *15*, 250(61), 252(61), *267*, 455(115), *465*
Lim, P. K. K., 288(66a), 289(66a), *294*, 496(183a, 183b), *509*
Lim, W. Y., 453(104), *465*
Lin, F. F. S., 498(191), *509*
Lin, J. W-P., 336(137), 338(137), 339(137), *349*
Lincoln, F. H., 25(146), 53(146, 147), *146*
Lindlar, H., 123(319), *152*
Lindner, C., 378(129), *385*, 410(102), *427*
Lindner, W., 499(195b), *510*
Linkies, A., 180(183), *203*
Lipkin, D., 433(3), *462*
Lipmann, F., 519(17), *533*
Lisin, A. F., 300(20), *346*
Liu, J-C., 357(28), *382*
Liu, R. S. H., 177(164), 194(164), *202*
Liu, Y., 175(140), 194(140), *201*
Lockley, W. J. S., 33(64), *144*
Loeber, D. E., 68(177), 70(177), 126(177), *147*
Loewengart, G. V., 469(21), 471(21), 476(21), 479(84), 487(21), *504, 506*
Loibner, H., 329(112), 334(131), 335(112), *349*
Loiseau, G. P. M., 422(216, 218), *430*
Loman, A. A., 195(235), *204*
Loomis, G. L., 423(230), *431*

Lorenz, D., 24(167), 64(167), *147*
Lorenz, O., 344(155, 156), *350*
Losch, R., 378(129), *385*
Louisfert, J. A., 175(127), *201*
Lourens, G. J., 103(258), *149*
Lowrie, G. B., 216(38), *222*
Luber, J., 497(188), *509*
Luckenbach, R., 14(26), *15*, 445(70a), *464*, 469(24), *504*
Luddy, F. E., 110(286), *150*
Ludwig, W., 491(148), *508*
Lüttke, W., 438(30), *463*
Ly, M., 456(122), *465*
Lyapina, E. K., 178(166), 195(166), *202*
Lynch, B. M., 280(41), *293*
Lythgoe, B., 69(178), 70(178), *147*, 164(53), 177(53), 192(53), *198*

Maassen, J. A., 165(68), 175(68), *199*
McCarry, B. E., 138(343, 344), *152*
McClure, D. E., 342(151), 343(151), *350*, 491(152), *508*
Macco, A. A., 138(345), *152*
McColeman, C., 318(84), 319(84), *348*
Maccoll, A., 468(8a), *503*
McCombie, H., 433(1), *462*
McCormick, J. R. D., 92(228), *148*
McCracken, J. A., 102(256), *149*
McDonald, R. N., 24(371), *153*
McEwen, W. E., 214(31), 220(50), *222*, 455(114), *465*
McFarland, J. W., 298(12), *346*
McGarrity, J. F., 25(360), 141(360), *153*
McGreer, D. E., 37(87), *144*, 168(99), 194(99), *200*
McGregor, A. C., 526(49), *534*
McGuire, H. M., 179(178), 193(178), 194(178), *202*
Machinek, R., 438(30), *463*
Machleidt, H., 164(66), 177(156), 183(205), 194(205), *199, 201, 204*
McIntosh, J. M., 251(68, 69), 252(70, 71, 72), 265(104), *267, 268*, 459(133), *466*
Mackay, W. D., 24(141), 51(141), 52(141), 133(141), *146*
McKelvie, N., 423(228), *431*
Mackenzie, K., 28(48), *143*

Mackie, R. K., 272(8), 273(8), 274(8), 278(8, 36b), 280(8), 281(47), 286(60a), 288(65), 289(8), *292, 293, 294*, 296(6), 300(6), *345*, 387(2), *424, 425*, 480(91), *506*
McKinley, S. V., 79(194), *147*
McKinley, W. H., 291(72), *294*
McLamone, W. M., 344(153), *350*
Maclaren, J. A., 372(94, 95, 97), *384*
McNeilly, S. T., 308(42), 310(42), *346*
Macomber, R. S., 356(24), *382*
Macpherson, A. J., 515(9), *533*
MacSweeny, D. F., 400(56), *426*
McWilliam, H. M., 287(62), 288(62), *294*
Madan, O. P., 470(33, 34), 482(99), *404, 507*
Madan, P. D., 281(49), *293*
Maddoz, M. L., 177(162), *202*
Madersteig, H., 416(138), *428*
Maercker, A., 19(15a), *142*, 214(29), *222*, 499(194), *510*
Magid, R. M., 400(59), *426*
Magidman, P., 110(286), *150*
Mahan, J. E., 320(89), 321(89), *348*
Mahler, W., 318(78), *348*
Mahran, M. R., 352(7), *382*
Maier, L., 14(20), *15*, 422(199), *430*
Maisey, R. F., 37(95), *144*
Majetich, G., 180(182), 195(182), *203* 380(144), *386*
Maki, Y., 287(61), *294*
Makin, S. M., 178(166), 195(166), *202*
Malavaud, C., 485(115), *507*
Malkes, L. Ya, 175(133), *201*
Mallo, G. N., 177(161), *202*
Malmsten, C., 106(268), *150*
Malone, G. R., 24(374), *153*, 173(118), 195(118), *200*
Mamdapur, V. R., 129(326), 130(326), *152*
Mandanas, B. Y., 37(88), 39(88), *144*, 169(107), 173(107), 194(107), *200*
Manecke, G., 87(212), *148*
Manhas, M. S., 34(68), *144*, 329(111), 332(127), 333(127), *348, 349*, 415(130), *428*
Mann, J. S., 173(123), 194(123), *201*
Mannhardt, H-J., 18(7), 128(7), *142*
Manz, F., 493(158), *508*

Marchand, A. P., 443(57), *464*
Marchesini, A., 110(286), *150*
Marcus, R., 388(4), 403(4), *425*
Marecek, J. F., 470(36), 473(50), 474 (53c), *505*
Mark, V., 304(35), 306(41), 307(35, 41), *346*, 486(124), *507*
Markes, J. H. H. 524(40), *534*
Markezich, R. L., 138(344), *152*
Markgraf, J. H., 312(60), *347*
Märkl, G., 24(164), 28(40), 63(164), 79(164), 83(164), 84(164), *143*, *146*, 359(35), *382*
Marmor, R. S., 174(135), 193(135), *201*
Marquarding, D., 468(11, 12, 19), 469(12, 19), 492(156), *504*, *508*
Marschall, H., 178(172), *202*
Marshall, J. L., 275(25), 278(25), *293*
Marshall, R., 275(27), 281(27), *293*
Marsheck, W. J., 108(278), *150*
Marsi, K. L., 320(91), 321(91), *348*, 499(197a), *510*
Marszak, M. B., 176(152), *201*
Martin, K. R., 24(372), *153*
Martin, S. F., 176(146), 195(146), *201*, 209(7), *221*
Martinelli, L., 29(44), 77(44), *143*
Martinez, J., 521(27), 522(28, 29), *534*
Martinov, A. A., 418(168), *429*
Martynuk, A. P., 412(110), *427*
Maruyama, H., 374(115), 377(124), *385*, 529(59, 60), 530(61), 531(65, 66), *535*
Maryanoff, B. E., 437(26a), *463*
Maryanoff, C. A., 437(26a), *463*
Masahara, R., 177(163), 196(163), *202*
Masaki, M., 424(246), *431*, 441(49), 452(49), *463*, 494(176), *509*
Masaki, Y., 180(182), 195(182), *203*
Masamune, S., 28(41), *143*
Masilamani, D., 437(26a), *463*
Masse, G. M., 251(68), 252(71), *267*, *268*
Massey, R. F., 169(106), 173(106), 193 (106), *200*
Masuoka, Y., 24(174), 67(174), 117 (174), 118(174), *147*
Masuya, H., 24(174), 67(174), 117 (174), 118(174), *147*

Mathey, F., 159(22), 162(42), 171(22), 172(22), 173(22), 192(22), 193(42), *197*, *198*
Mathiaparanam, P., 473(51a, 51b), *505*
Mathieu, J., 19(15f), *142*
Mathis, F., 485(113), *507*
Mathys, G., 232(25), *266*
Matough, M. F. S., 392(21), 400(21, 41), *425*
Matsuda, H., 352(8), *382*
Matsueda, R., 374(115), 377(124), *385*, 529(59, 60), 530(61, 62), 531(65, 66), *535*
Matsui, M., 131(331), *152*
Matsui, T., 25(111), 46(111), 130(111), *145*
Matsumoto, A., 281(46), *293*
Matsumoto, M., 361(48), *383*
Matsuo, H., 37(76), *144*
Matta, K. L., 178(174), *202*
Matthews, C. N., 26(29), *143*, 487 (135a, 135b), *508*
Matthews, R., 187(218), *204*
Maurer, L. De, B., 345(157), *350*
Maximov, E. E., 527(50), *534*
Mayer, H., 120(305), *151*
Mayer, M. S., 116(298), *151*
Mazzu, A., 493(172), *509*
Meana, M. C., 25(267), 105(267), *150*
Mechoulan, R., 218(41), *222*
Medved, T. Y., 250(63), *267*
Meguerian, G., 351(3), *382*
Mehesfahvi, C., 108(277), *150*
Meinel, L., 416(138), *428*
Meinwald, J., 175(132), 196(132), *201*
Melby, L. R., 380(138), *386*
Melvin, L. S., Jr., 373(107), *385*
Menchen, S. M., 190(230, 231), 193 (230, 231), *204*, 381(145, 146), *386*
Mendenhall, G. D., 336(138), 338(138, 143), 340(138, 143), *349*, 491(150c), *508*
Mercer, A. J. H., 25(107), 44(107), *145*, 490(145), *508*
Merwe, J. P., van der, 82(200), 83 (200), 126(200), 127(200), *147*
Merz, A., 24(164), 63(164), 79(164), 83(164), 84(164), *146*
Messegner, A., 164(60), 178(60), *199*

Messmer, A., 412(112), *427*
Meth-Cohn, O., 278(35), 280(35), *293*
Meyer, E., 33(65), 35(65), *144*
Meyer, F., 159(28), *198*
Meyers, A. I., 24(374), 25(38, 108), 28 (38), 44(108), *143*, *145*, *153*, 173 (118), 195(118), *200*, 352(9), 353(9), *382*
Meyerson, S., 482(99), *507*
Micetich, R. G., 25(380), *153*
Michael, D., 440(45), 461(45), *463*
Michaelis, A., 159(26), 160(30), *198*
Michaelis, H., 167(96), *200*
Michalski, J., 157(15), 166(78), 181 (195), 194(15, 78), 195(195), *197*, *199*, *203*, 378(130), *385*, 501(206), *510*, 524(42), *534*
Michel, W., 388(6), 389(6), 393(6), 397(32), 400(32), 403(6), *425*
Michels, R., 25(154), 56(154), 79(154), *146*, 423(238), 424(244), *431*
Midura, W., 165(75), 166(77), *199*
Mikolajczak, J., 501(206), *510*
Mikolajczyk, M., 164(64), 165(75, 76), 166(77), 175(64, 76), *199*
Miles, M. G., 314(62), *347*, 357(31), 358(32), *382*, *383*
Miles, M. L., 25(100), 42(100), 128 (100), *145*
Milewski, C. A., 436(24), *463*
Milker, R., 389(12), 422(194, 195), *425*, *430*
Miller, B., *424*
Miller, J. A., 308(42), 310(42), *346*
Miller, J. B., 461(149), *466*
Miller, J. G., 365(64), *384*
Miller, R. D., 400(51), *426*
Miller, R. W., 25(299), 116(299), *151*
Miller, S. I., 455(110, 111), *465*
Milne, G. M., 138(349), 139(349, 350), *153*, 208(2b), *221*, 400(42), *425*
Mintz, M. J., 423(223), *431*
Mironov, V. F., 422(211), *430*
Mironova, N. V., 422(211), *430*
Mirth, D. B., 422(213), *430*
Missol, U., 319(87), *348*
Misumi, S., 361(48), *383*
Mitchell, K. A. R., 468(13), *504*
Mitchell, M. J., 290(70b), *294*
Mitchell, R. H., 362(52, 53), *383*

Mitin, Y. V., 527(50), 531(64), *534*, *535*
Mitra, A., 100(247), *149*
Mitra, R. B., 315(66), *347*, 469(27, 28, 29, 30), *504*
Mitsui, T., 37(93), *144*, 178(167), 194 (167), *202*
Mitsunobu, O., 327(108), 328(109, 110), 330(110, 114b, 116, 117, 118), 331 (108, 117, 118), 333(108), 334(118), 335(114b, 118), *348*, *349*, 374(114), 378(126, 127, 128), *385*, 450(93, 94, 95), 455(95), *465*
Miyake, H., 103(257), 104(264), 107 (272), *149*, *150*
Miyano, M., 108(278), *150*
Miyano, S., 401(66), *426*
Miyashita, M., 105(265), *150*, 180(182), 195(182), *203*
Mizrakh, L. I., 470(32a), 473(46), *504*, *505*
Mizushima, M., 292(74), *294*
Mizutani, M., 380(140, 141), *386*
Mlotkowska, B., 165(76), 175(76), *199*
Modro, T., 378(130), *385*
Moffat, J., 298(14), *346*
Moffatt, J. G., 400(57), *426*, 435(18, 19, 21, 22), 436(18, 21, 22), 439(22), 456 (22), *463*
Mogel, L. G., 439(38), *463*
Moinet, G., 101(251, 252), *149*
Moll, E., 492(155), *508*
Momchilova, S., 165(67), *199*
Monaco, D. J., 252(73), *268*
Mondelli, R., 494(177b), *509*
Mondon, A., 18(8), *142*
Mondron, P. J., 440(43), *463*
Montavon, M., 24(367), 123(317, 319), *152*, *153*
Montgomery, J. A., 184(209), *204*
Moore, C. G., 369(81), 370(81), *384*
Moore, G. A., 515(11), 518(14), 523 (34), *533*, *534*
Moore, T. R., 351(2), *382*
Mootz, D., 468(9d), *504*
Moppett, C. E., 168(100), 169(100), 173(100), 194(100), *200*
Moran, T. A., 164(53), 177(53), 192 (53), *198*

Morbach, W., 388(6, 7), 389(6), 393(6), 395(7), 397(32, 33), 400(32, 33), 403(6), *425*
Morel, G., 450(96), *465*
Mori, K., 37(93), *144*, 178(167), 194(167), *202*
Moriarty, R. M., 310(49), *347*
Morinaga, K., 446(72a), *464*
Moritani, I., 54(150), *146*
Moroe, T., 25(111), 46(111), 130(111), *145*
Morris, S. G., 110(286), *150*
Morrison, D. C., 376(122), *385*
Morrow, C. J., 173(117), 194(117), *200*
Morse, A. T., 24(376), *153*
Mortell, T. R., 170(111), 173(111), 193(111), *200*
Mosebach, R., 476(79), 479(79), 481(79, 95, 96), 496(79, 96), *506*
Mosher, R. A., 318(83), *348*
Moss, G. P., 24(368), *153*
Mott, L., 441(47), 442(50), 446(73), *463*, *464*
Mousseron-Canet, M., 34(73), 36(73), 122(313), *144*, *152*
Muchmore, D. C., 6(4c), *14*
Muetterties, E. L., 467(6d), 468(18), 469(18), 497(189), *503*, *504*, *509*
Mukai, T., 299(17), 300(17), *346*
Mukaiyama, T., 300(19), 308(44), 315(19, 65, 67), 316(19, 67), 327(106), 328(106), 330(116), 333(130), *345*, *346*, *347*, *348*, *349*, 373(104, 105), 374(111, 112, 113, 115), 375(116, 117, 119, 120, 121), 376(123), 377(124), *385*, 450(93, 94, 95), 455(95), *465*, 480(89, 90, 92), 486(118), *506*, *507*, 529(58, 59, 60), 530(61, 62, 63), 531(65, 66), *535*
Mukasa, S., 178(173), *202*
Muller, B., 179(176), 180(186), 196(186), *202*, *203*
Müller, E., 175(131), 195(131), *201*, 493(169), *509*
Müller, G., 48(126), 50(126), *145*
Mulzer, J., 335(134), *349*
Mundy, B. W., 10(8), *15*
Munoz, A., 485(113), *507*
Muraoka, M., 520(25), 527(25), 531(25), *533*
Murdock, L. L., 421(190), *430*
Muroi, M., 476(73), *506*
Murray, A. W., 281(45), *293*
Murray, R. W., 336(137), 337(140), 338(137), 339(137), 341(147), *349*, *350*, 491(150a, 150b), *508*
Murray, W. P., 8(7b), *15*, 247(55), 253(77), *267*, *268*
Musierowicz, S., 157(15), 166(78), 176(151), 193(151), 194(15, 78), *197*, *199*, *201*
Mychajlowskij, W., 176(150), *201*

Naae, D. G., 423(233, 234), *431*
Näf, F., 24(103, 240), 25(103), 43(103), 68(103), 69(103), 97(240), 118(240), 119(240), *145*, *149*
Nagabhushan, T. L., 355(17), *382*
Nagamatsu, N., 333(129), *349*
Nagase, H., 437(26b), *463*
Nagata, W., 173(119), 194(119), 195(234), *201*, *204*
Naisby, T., 13(19), *15*
Nakabayashi, T., 371(90, 91), *384*
Nakada, Y., 375(120), *385*
Nakagawa, I., 375(120), *385*
Nakamura, N., 104(263), *150*
Nakanishi, A., 302(27), 303(27), *346*, 352(8), 363(56), *382*, *383*
Nakanishi, K., 263(97), *268*
Nakanishi, N., 352(8), *382*
Nakasato, S., 418(170), 429
Nakeya, N., 424(246), *431*
Nambu, H., 300(19), 315(19, 67), 316(19, 67), *346*, *347*, 374(112), *385*, 480(90, 92), *506*
Nambudiry, M. E. N., 164(53), 177(53), 192(53), *198*
Napier, R. P., 418(145), 421(145, 187), *428*, *429*
Nara, K., 165(71), 175(71), *199*
Narang, S. C., 302(26), *346*, 445(69), *464*
Narasaka, K., 104(262), *150*
Narita, M., 357(29), 358(34), 380(139), *382*, *383*, *386*
Narula, S. S., 354(13), *382*
Nay, B., 13(19), *15*
Needham, L. L., 338(142), *349*
Neef, G., 181(200), 183(204), 195(200), *203*

Negishi, A., 209(5), *221*
Neidlein, R., 476(79), 478(80), 479 (79), 481(79, 80, 95, 96), 496(79, 96), *506*
Nelson, N. A., 106(269), *150*
Nerdel, F., 167(96), *200*
Nesbitt, B. F., 117(300, 301), 118(301), *151*
Nesmeyanov, N. A., 208(3), *221*
Nesterov, L. V., 418(168), *429*
Netter, H., 514(6), *533*
Neumayr, K., 493(163), *509*
Neureiter, N. P., 317(75), *347*, 352(5), *382*
Newman, F. C., 128(323), *152*
Newman, M. S., 309(47), *347*
Newton, R. F., 105(266), *150*
Nguyen, L., 184(207), *204*
Nickel, H., 303(30), *346*, 378(131), *385*
Nickl, J., 183(205), 194(205), *204*
Nicolaides, D. N., 227(8b), *266*
Nicolaou, K. C., 106(268), 108(249), *149*, *150*, 373(107), 374(109), *385*
Niemann, B., 514(6), *533*
Nifantev, E. E., 421(191), 422(210, 220), 423(221), *430*, *431*
Nikolenko, L. N., 422(206), *430*
Nikonorov, K. V., 418(152), *428*
Nikonorova, L. K., 422(219), *430*
Nilsson, N. H., 11(14a), *15*
Ning, R. Y., 281(49), 290(71), *293*, *294*
Ninomiya, K., 520(22), *533*
Nishida, S., 54(150), *146*
Nishiwaki, T., 283(56), *294*
Nishizawa, M., 180(182), 195(182), *203*, 380(143, 144), *386*
Nitta, M., 299(17), 300(17), *346*
Noel, Y., 400(45), *426*
Noguchi, K., 108(279), *150*
Nolan, T. J., 392(18), 400(18, 53), 420 (53), *426*
Normant, H., 162(39), 168(98), 169 (98), 173(98), 194(98), *198*, *200*
Normant, J. F., 162(43), 164(52), 174 (136), 193(52, 136), *198*, *201*, 213 (21), *222*
North, R. A., 289(68), *294*
Nose, M., 371(91), *384*
Noteboom, M., 360(39), *383*
Nöth, H., 416(138), *428*

Nowakowski, M., 470(36), *505*
Nunn, A. J., 278(39), 281(39), *293*
Nürrenbach, A., 19(16), 121(307b), *142*, *151*, 459(135), *466*
Nycz, D. M., 265(100), *268*
Nyfeler, P., 526(47), *534*
Nyholm, R. S., 468(8a, 8b), *503*
Nyu, K., 272(15), 274(17), 275(26), 276(29), 282(17, 50, 51b), 283(53, 54), *292*, *293*

Oae, S., 302(27), 303(27), *346*, 363(56), *383*
Obata, T., 450(93, 94, 95), 455(95), *465*
Obermeier, F., 524(39), 526(39), *534*
Obrycki, R., 460(146), *466*
Ochai, E., 296(3), *345*
Oda, M., 322(95, 96), 323(96), *348*
Odom, H. C., Jr., 179(178), 193(178), 194(178), *202*
Odyek, O., 166(79), 176(79), 178(79), 194(79), *199*
Oediger, H., 24(171), 65(171), 85(171), *147*, 415(132), 416(132, 136), 423 (132), *428*, 437(27), 440(27), 444(64), 445(67), *463*, *464*
Oehlschlager, A. C., 142(362, 363), *153*, 404(78), *426*, 443(60), *464*
Oertel, M., 422(202), *430*
Oertle, K., 373(108), *385*
Ofitserova, E. Kh., 476(72c), *506*
Ofner, A., 121(309), *151*
Ogasawara, K., 272(14), 274(17), 282 (17, 50, 51a), 284(57), 285(58), 290 (69, 70a), *292*, *293*, *294*
Ogata, Y., 380(140, 141), *386*
Ogliarso, M. A., 316(68), *347*
Ohashi, T., 411(107), *427*
Öhler, E., 248(99), *268*
Ohloff, G., 24(103, 240), 25(103, 104), 43(103, 104), 68(103), 69(103), 97 (240), 118(240), 119(240), 132(104), 133(104), *145*, *149*, 345(157), *350*
Ohno, K., 308(44), *347*, 486(118), *507*
Ohno, M., 441(48), 454(48, 107), *463*, *465*, 476(73), *506*
Ohta, M., 156(8), 173(8), 193(8), 194 (8), *197*, 441(49), 452(49), *463*, 494 (176), *509*

Ohtaki, T., 37(93), *144*, 178(167), 194 (167), *202*
Okada, I., 249(59), *267*, 411(104), *427*, 440(42), *463*
Okada, T., 450(90), *464*
Okamoto, M., 300(19), 315(19), 316(19), *346*, 480(92), *506*
Okawa, K., 357(27), *382*
Okawara, R., 450(90), *464*
Okazaki, H., 474(53c), *505*
Okazaki, M., 169(112b), 170(112b), 180(192), 181(112b), 182(112b), 193(112b), 195(238), *200, 203, 205*
Okazaki, R., 357(27), *382*
Okonogi, T., 24(165), 63(165), *146*
Okorodudu, A. O. M., 422(212), *430*
Okuniewicz, F. J., 102(254), *149*
Olah, G. A., 302(26), *346*, 445(69), *464*
Olivé, J-L., 122(313), *152*
Oliver, J. E., 443(58, 62), *464*
Oliveto, E. P., 139(352, 353), 140(352), *153*
Ollis, W. D., 369(84), *384*
Olszewski, W. F., 297(7), *346*
Openshaw, H. T., 34(71), 38(83), *144*, 387(1), *425*
Orekhova, K. M., 480(88a), *506*
Orgel, L. E., 468(8a), *503*
Oritani, T., 34(74), 36(74), *144*
Orwig, B. A., 377(125), *385*
Orzech, C. E., 455(110), *465*
Osanova, N. A., 482(202), 501(202), *510*
Ota, S., 24(140), 51(140), 56(140), *146*
Otsubo, T., 361(48), *383*
Ottenheym, H. C. J., 368(78), *384*
Ottlinger, R., 493(159), *508*
Overman, J. D., 372(100), *385*
Overman, L. E., 372(100), *385*

Pabst, W. E., 174(126), 195(126), *201*
Padilla, A. G., 468(17b), *504*
Padwa, A., 73(185), *147*, 493(172), *509*
Paetsch, J., 219(47), *222*
Page, C. B., 97(241), 98(241), *149*
Page, G., 304(32), *346*, 364(60), *383*
Pagnoni, U. M., 110(286), *150*
Pailer, M., 248(57), *267*
Palovcik, R., 169(107), 173(107), 194(107), *200*

Pande, K. C., 457(124), *466*
Panse, P., 34(66, 67), 36(67), *144*
Papanthanasopoulos, N., 10(11c), *15*
Paquette, L. A., 355(19), 356(23), *382*
Parks, C. R., 344(156), *350*
Partos, R. D., 448(81, 82), 450(82), *464*
Patai, S., 156(7), 166(7), 194(7), *197*
Patampongse, C., 275(24), *293*
Patchett, A. A., 355(17), *382*
Patel, A. N., 434(6), *462*
Patel, V. V. 357(30), *382*
Pattenden, G., 24(135, 199), 29(42), 33(59), 50(135), 51(135), 66(172), 67(172), 72(181, 182), 73(181, 182), 78(181), 81(135, 199), 82(182), 83(181, 182), 88(199), 90(181), 91(181), 124(42), 126(181, 182, 321), 127(181), 128(324), 129(325), 130(172), 130(330), 131(330), 132(330), *143, 146, 147, 152*, 177(159, 160), 192(159), 194(160), *202*
Patwardhan, A. V., 311(37, 56), *347*, 472(44), 475(61, 62, 63, 67, 68, 70), 476(61, 62, 63), 482(61, 62, 63, 68, 98), 483(67), 486(67), *505, 506, 507*
Paugam, J.-P., 518(15), *533*
Paulsen, H., 163(48), 184(210, 211), 193(48, 210), 194(210), *198, 204*, 459(134), *466*
Paulus, H., 25(287), 110(287), 111(287), *150*
Pawlak, M., 24(240), 25(104), 43(104), 97(240), 118(240), 119(240), 132(104), 133(104), *145, 149*
Pearce, A., 177(158c), 192(158c), *202*
Pearson, M. S., 367(68), *384*
Pedersen, E. B., 10(12a), 11(12c), *15*
Pedersen, B. S., 11(14a, 14b, 14c), *15*
Pelah, Z., 363(57), *383*
Pella, E., 102(255), *149*
Pellatt, M. G., 99(243), *149*
Pellegata, R., 102(255), *149*
Peppard, D. J., 179(176), 180(186), 196(186), *202, 203*
Perevezentseva, S. P., 476(72a), 482(100), *506, 507*
Perkow, W., 159(28), *198*
Perner, D., 422(197), *430*
Perriot, P., 164(52), 174(136), 193(52, 136), *198, 201*, 213(21), *222*

Pervova, E. Ya., 501(205), *510*
Pesnelle, P., 179(176), 180(186), 196 (186), *202*, *203*
Pesso, J., 190(229), *204*
Peter, M. G., 179(175), *202*
Peters, G. M., 177(161), *202*
Peters, V. M., 457(125), *466*
Peterson, D. J., 184(210), 193(210), 194(210), *204*
Petragnani, N., 24(124), 47(124), 95(124), *145*
Petrellis, P., 487(131), *508*
Petrov, A. A., 85(203), *148*, 471(37a, 37b), *505*
Petrov, G., 38(79), *144*, 193(233), *264*
Petrova, J., 165(67), 169(113), 195(113, 236), *199*, *200*, *205*
Pettit, G. R., 181(199), 183(203), *203*
Pfeiffer, G., 455(112), *465*, 475(56), 488(140), *505*, *508*
Pfeiffer, W. D., 362(54), *383*
Pfister, G., 6(4a), *14*
Pfohl, S., 468(12), 469(12), *504*
Philips, D. D., 369(85), *384*
Philips, J. C., 355(19), *382*
Phillips, D. I., 499(197b), *510*
Photaki, I., 513(3), 514(7), 520(3), *533*
Pichat, L., 90(222, 223), *148*
Pickering, D., 515(9), *533*
Pickut, S., 181(198), 182(198), *203*
Piechucki, C., 165(72, 73), 175(72, 73), 177(72), *199*
Pierson, G. O., 322(94), *348*
Pike, J. E., 53(147), *146*, 455(115), *465*
Pilette, J. F., 24(364), 141(364), 142(364), *153*
Pilgram, K., 368(71), 369(85), *384*
Pilot, J. F., 472(45), 473(45, 47), 487(133, 134), 488(134, 136), *505*, *508*
Pils, I., 211(11), *221*
Pinck, L. A., 24(365), *153*
Pinder, A. R., 179(178), 193(178), 194(178), *202*
Pinker, I., 412(112), *427*
Piraux, M., 24(364), 141(364), 142(364), *153*
Piskala, A., 26(31), 55(153), 56(153, 157), 57(157), 58(157), 62(157), *143*, *146*

Pittman, C. U., Jr., 357(29), 358(34), 380(139), *382*, *383*, *386*
Plantema, O. G., 179(179), 195(179), *202*
Platt, A. E., 458(131), *466*
Plattner, J. J., 24(139), 50(139), 51(139), *146*
Platz, H., 116(297), *151*
Ple, G., 518(15), *533*
Plenchette, A., 25(233), 94(233), 116(233), *149*
Plieninger, H., 33(65), 35(65), *144*, 245(51), *267*
Plotch, S., 372(98), *384*
Pocar, D., 233(30), 234(30), *267*
Podimuang, V., 275(24), *293*
Pogonowski, C. S., 105(265), *150*, 161(36), 180(182), 187(219), 194(36), 195(182), *198*, *203*, *204*
Polezhaeya, N. A., 470(32b), 489(142), 496(187), *504*, *508*, *509*
Polikarpov, Y. M., 250(63), *267*
Polone, M., 521(27), *534*
Pommer, H., 19(16, 18), 50(137), 87(208, 209, 210, 211), 109(211), 121(210, 307a, 307d, 308a, 308b, 308c, 310a, 310b, 311a), 122(209, 312), 123(308a, 310a, 310b, 314), 127(210, 311a), *142*, *146*, *148*, *151*, *152*, 227(8c), *266*, 459(135), *466*
Ponpipom, M. M., 456(118a), *465*
Ponsold, K., 24(354), 139(354), *153*, 180(192), *203*
Ponti, P. P., 494(177b), *509*
Pope, P., 404(78), *426*, 443(60), *464*
Popoff, I. C., 169(109), 175(109), 194(109), *200*
Popov, A. J., 422(211), *430*
Poppi, R. G., 117(300, 301), 118(301), *151*
Popplestone, C. R., 183(202), *203*
Portnoy, N. A., 173(117), 194(117), *200*
Poshkus, A. C., 305(38), 310(38), *346*, 365(67), *384*, 486(128), *508*
Posner, G. H., 175(139), 193(139), *201*, 423(230), *431*
Potter, J. L., 372(93), *384*
Potts, K. T., 275(25), 278(25), *293*
Powell, R. L., 471(41), 486(41, 123), 500(41, 123), 502(216), *505*, *507*, *510*

Prasad, V. A. V., 476(74), 478(74, 81), 479(74), *506*
Pratt, T. W., 303(29), *346*
Predvoditelev, D. A., 422(220), *431*
Prevost, C., 238(39, 41), *267*
Price, R. D., 166(84b), 185(84b), 194(84b), *199*
Prikota, T. I., 178(172), *202*
Prinzbach, H., 360(37), *383*
Proskauer, S., 457(127), *466*
Puchalski, C., 229(12), 238(12), *266*
Pudovik, A. N., 166(78), 194(78), *199*, 314(63), *347*, 422(214), *430*, 476 (72a, 72b, 72c), 482(100), *506, 507*
Pudovik, M. A., 422(214), *430*
Pulman, D. A., 32(57), *143*
Puskas, I., 272(9), 273(9), *292*
Pyatnova, Yu. B., 46(116), 90(116), *145*

Quin, L. D., 12(15), *15*, 166(84b), 185 (84b), 194(84b), *199*
Qureshi, A. R., 434(9), *462*

Raap, R., 25(380), *153*
Rabinovitz, M., 175(129), 195(129), *201*
Rabinowitz, R., 318(78), *348*, 360(41, 42), 366(42), *383*, 388(4), 403(4), 420(180), *425, 429*
Radics, L., 392(29), 394(29), 400(29), 403(76), *425, 426*
Radike, A., 10(8), *15*
Radley, P., 288(75), *294*
Radscheit, K., 183(201), *203*
Ragoonanan, D., 434(9), *462*
Rahman, A. U., 281(48), *293*
Rainey, D. K., 105(266), *150*
Rakhimova, G. I., 418(151, 153), *428*
Rakova, L. V., 476(72c), *506*
Rakshys, J. W., Jr., 79(194), *147*
Ram, B., 178(174), *202*, 415(130), *428*
Ramage, R., 400(56), 417(141), *426, 428*, 515(11), 518(14), 523(34), 531 (67), 532(69), *533, 534, 535*
Ramanathan, N., 311(55), *347*, 475(58, 59, 61, 62, 71), 476(58, 59, 61, 62), 482(59, 61, 62, 102, 103), 483(58, 104), *505, 506, 507*
Ramer, R. M., 34(68), *144*, 181(197), *203*
Ramey, C. E., 25(166), 64(166), *146*

Ramirez, F., 12(16), *15*, 21(22), 142 (22), *143*, 223(2), *266*, 296(5), 304(36), 305(37), 307(34), 310(54), 311(54, 55, 56, 57, 58), 315(66), 322(5), *345, 346, 347*, 423(228), *431*, 467(1a, 1b, 1c, 1d, 1e, 1f, 1g), 468(1e, 11, 12, 14, 15, 19), 469(12, 14, 15, 19, 20, 21, 22, 27, 28, 29, 30), 470(33, 34, 35, 36), 471(20, 21, 22, 40), 472(43, 44, 45), 473(45, 47, 48, 50), 474(43, 48, 52, 53a, 53b, 53c, 54), 475(58, 59, 60, 61, 62, 63, 64, 65, 66, 67, 68, 69, 70, 71), 476(20, 21, 22, 58, 59, 61, 62, 63, 66, 74, 76, 77, 78), 478(74, 76, 78, 81), 479(74, 77, 84, 85), 481(1g), 482 (59, 61, 62, 63, 64, 65, 68, 69, 98, 99, 101, 102, 103), 483(58, 64, 65, 67, 104), 484(66), 486(67, 119, 130), 487(21, 133, 134), 488(134, 136), 500(199, 200, 201), *503, 504, 505, 506, 507, 508, 510*
Ramirez, R. J., 325(103), *348*
Range, P., 25(296), 115(295, 296), *151*
Rangoonwala, R., 46(112), 47(112), *145*
Rapoport, H., 24(139), 50(136, 139), 51(139), 139(351), *146, 153*
Rasberger, M., 6(5a), 7(5a), *14*, 257 (85, 86), *268*
Rasmussen, J. B., 11(14d), *15*
Rasmusson, G. H., 33(58), *143*
Ratcliffe, R. W., 187(220), 195(220), *204*
Ratts, K. W., 27(35), 32(35), *143*, 391 (16), *425*
Ray, S. K., 301(23), *346*, 363(56), *383*, 445(68), *464*
Raymond, M. A., 25(100), 42(100), 128(100), *145*
Razumov, A. I., 156(4), *197*
Razumova, N. A., 471(37a, 37b), *505*
Razuvaev, G. A., 482(202), 501(202), *510*
Readio, P. D., 310(53), *347*, 455(109), *465*, 486(127), *507*
Rebafka, W., 361(47), *383*
Redjel, A., 167(91), 168(91, 104), 169 (91), 170(91), 173(91), 194(91, 104), *200*
Redmore, D., 175(130), 196(130), *201*, 422(208), *430*

Reed, J. S., 94(234), 116(234), *149*
Rees, C. W., 295(2), 296(2), 297(8, 10), 300(2), 318(8), 322(2), *345*, *346*
Rees, R., 218(45), *222*
Reetz, K. P., 412(111), *427*
Regen, S. L., 424(239), *431*
Regitz, M., 177(154), *201*
Reid, E., 272(7b), *292*
Reid, W., 121(307b), *151*, 388(5), *425*
Reif, W., 19(16), *142*, 459(135), *466*
Rein, B. M., 416(133, 134), 423(133, 134), *428*, 437(28, 29), 438(29), 440(28), *463*
Reingold, I. D., 361(46), *383*
Reisdorf, D., 168(101), 196(101), *200*
Reliquet, A., 162(43), *198*
Relles, H. M., 320(90), *348*, 424(245), *431*, 445(71), *464*
Repina, L. A., 423(227), *431*
Reucroft, J., 90(215), *148*
Reuschling, D., 180(183), *203*
Reutov, D. A., 208(3), *221*
Richard, B., 484(109), *507*
Richardson, G., 424(240), *431*
Richter, H., 422(202), *430*
Richter, W., 460(148), *466*
Rickborn, B., 215(33), *222*
Rieber, M., 31(53), *143*, 207(1), *221*
Ried, W., 405(85), *427*
Riedel, A., 525(43), *534*
Rieser, J., 380(142), *386*
Rimbault, C. G., 321(93), *348*
Rivalle, C., 175(127), *201*
Robbie, I. M., 278(35), 280(35), *293*
Roberts, B. P., 318(78), *348*
Roberts, D. L., 34(72), 36(72), 38(72), *144*
Roberts, S. M., 105(266), *150*
Roca, A., 164(60), 178(60), *199*
Röder, H., 459(137), *466*
Rodionov, E. S., 422(211), *430*
Roedig, A., 28(40), *143*
Rogers, W. J., 424(241), *431*
Rogoff, W. M., 116(298), *151*
Rois, A., 326(105), *348*
Rokhlin, E. M., 487(132), *508*
Röller, H., 37(90), *144*, 178(168), *202*
Roman, S. A., 178(169), 195(169), *202*
Ronen-Braunstein, I., 190(229), *204*
Röschenthaler, G-V., 488(137), *508*

Rösel, P., 25(288), 111(288), *150*
Rosenthal, A., 184(207), *204*
Rosini, G., 521(27), *534*
Rosowsky, A., 10(11c), *15*
Ross, B., 412(111), *427*
Ross, S. T., 24(366), *153*
Rostock, K., 211(14), *221*
Roth, M., 354(12), *382*
Roth, W-D., 493(162, 163, 165, 166, 167, 174), 494(162, 174), *509*
Rothe, M., 524(41), *534*
Rousseau, G., 320(88), *348*
Rowell, F. J., 278(39), 281(39), *293*
Rowley, A. G., 287(62), 288(62), 289(68), *294*
Roy, N. K., 418(171), *429*
Ruch, E., 176(149), *201*
Rüchardt, C., 34(66, 67), 36(67), *144*
Ruden, R. A., 84(202), 85(202), *148*
Rudinger, J., 372(96), *384*
Rudler Chauvin, M. C., 138(349), 139(349), *153*
Rüegg, R., 24(367), 123(317, 319), *152*, *153*
Rüegg, U. Th., 372(96), *384*
Rueggburg, W. H. C., 10(8), *15*
Rühlmann, K., 459(138), *466*
Runge, W., 176(149), *201*
Runquist, O. A., 322(94), *348*
Ruppert, I., 389(12), *425*
Ruschig, H., 181(194), 183(194, 201), *203*
Rusek, P. E., 447(74, 79), 448(84), 455(109), *464*, *465*, 486(127), *507*
Russel, D. R., 468(9a), *504*
Russell, J. W., Jr., 166(84b), 185(84b), 194(84b), *199*
Russell, S. W., 24(368), *153*
Rust, M. K., 24(143), 52(143), *146*
Ruston, S., 164(53), 177(53), 192(53), *198*
Ruyle, W. V., 355(17), *382*
Rydon, H. N., 433(4, 5), 434(7), 435(7, 11, 12), *462*, *463*, 524(40), *534*
Ryser, G., 24(367), *153*
Ryzlak, M. T., 139(356), 140(356), *153*
Rzeszotarska, B., 522(31), *534*

Saegusa, T., 502(208, 209, 210, 211, 212, 213), *510*

Safe, S., 368(77), 370(77), *384*
Saferstein, L., 499(196), *510*
Saijo, S., 108(279), *150*
Saito, M., 186(215), *204*
Saito, Y., 527(52), *534*
Sakai, I., 441(48), 454(48), *463*
Sakai, K., 104(261, 263), *149*, *150*
Sakan, T., 50(138), *146*
Sakata, Y., 361(48), *383*
Sakodynskaja, T. P., 421(191), *430*
Salakhutdinov, R. A., 418(153, 154, 167, 168), *428*, *429*
Saleh, G., 413(116, 119), *427*
Salmond, W. G., 423(232), *431*
Salomon, R. G., 25(144), 52(144), 53(144), *146*
Salvadori, G., 27(34), 34(70), *143*, *144*, 447(77), *464*
Sam, T. W., 339(144), *349*
Samek, Z., 435(17), *463*
Sammes, P. G., 90(215), *148*
Samuelsson, B., 106(268), *150*
Sandalova, L. Yu., 470(32a), 473(46), *504*, *505*
Sando, T., 330(117, 118), 331(117, 118), 334(118), 335(118), *349*
Sandris, C., 355(22), *382*
Santaniello, E., 436(23), *436*
Sap, M. M. E., 138(342), *152*
Sargent, M. V., 80(196), *147*
Sarnecki, W., 19(16), 87(210, 211), 109(211), 121(210, 307a, 307b, 308a, 308b, 308c, 310b, 311a), 122(312), 123(310b, 314), 127(210, 311a), *142*, *148*, *151*, *152*, 459(135), *466*
Sasaki, K., 37(86), *144*, 169(108), 177(108), 191(108), 194(108), *200*
Sasaki, T., 184(213), *204*, 495(180), *509*
Sasse, K., 159(24), 160(24), 161(24), *198*
Sato, K., 24(140), 51(140), 56(140), *146*, 164(57), 196(57), *199*
Sato, S., 176(144), 195(144), *201*
Satra, S. K., 308(45), *347*
Sauerbier, M., 175(131), 195(131), *201*
Sauerbrey, K., 488(137), *508*
Saukaitis, J. C., 453(98), *465*
Saunders, B. C., 433(1), *462*
Saunders, D. G., 248(56a, 56b), *267*, 270(5a, 5b), 283(52a, 52b), 285(52a, 52b), *292*

Sauter, H., 360(37), *383*
Savas, E. S., 360(43), 367(43), *383*
Savignac, P., 162(39, 42), 163(49), 167(92), 173(49), 174(134), 193(42, 49, 134), 195(236), *198*, *200*, *201*, *205*, 303(28), *346*, 445(70b), *464*, 484(109), *507*, 518(15), *533*
Saville, B., 369(82), *384*
Savins, E. G., 297(11), *346*
Savos, E. S., 318(80), *348*
Sawaya, H., 423(230), *431*
Scanlon, P. M., 457(126), *466*
Schaaf, T. K., 24(246), 99(246), 100(246), 107(274), 109(281), *149*, *150*, 180(184), *203*
Schaap, A. P., 336(138), 338(138), 340(138, 145, 146), 342(150), *349*, *350*, 360(37), *383*, 491(150d, 153), *508*
Schabl, L., 492(156), *508*
Schadow, H., 422(202), *430*
Schaefer, A., 491(148), *508*
Schaefer, J. P., 440(44), *463*
Schafer, W. M., 344(154), *350*
Schalbar, J., 180(187), *203*
Schaller, H., 28(40), *143*
Schaper, W., 265(103), *268*
Scharf, D. J., 455(113), *465*
Schauble, J. H., 368(76), *384*
Schaumann, E., 163(51), *198*
Scheibye, S., 11(14a, 14b, 14c), *15*
Schenk, P., 526(47), *534*
Scherowsky, G., 358(33), *383*
Schick, H., 24(354), 139(354), *153*, 180(192), *203*
Schiemenz, G. P., 24(369), 28(37), *143*, *153*, 450(91), *465*
Schill, G., 24(124), 47(124), 95(124), *145*
Schlosser, M., 19(14), 25(33, 155), 26(31, 32, 33), 27(32), 31(56), 48(33, 126), 50(14, 33, 126, 134), 55(33, 151, 152, 153), 56(14, 33, 151, 153, 155, 157), 57(155, 157), 58(14, 155, 157), 60(160), 61(160, 161), 62(157, 161), 63(162), 65(170), 78(33), 79(14, 33), 82(14), 83(33), *142*, *143*, *145*, *146*, *147*, 164(54, 55), 170(114), 177(54), 192(54), 193(54), *198*, *200*
Schluenz, R. W., 424(245), *431*, 445(71), *464*

Schmid, H., 25(175), 68(175), *147*, 179 (175), *202*, 254(80), *268*
Schmidpeter, A., 497(188), *509*
Schmidt, U., 248(99), *268*, 373(108), *385*
Schmit, J. P., 24(364), 141(364), 142 (364), *153*
Schmutzler, R., 467(6a, 6b, 6c), 468 (9d), 488(137), *503*, *504*, *508*
Schnabel, K. H., 460(141), *466*
Schneider, D. F., 82(200), 83(200), 126 (200), 127(200), *147*
Schneider, M., 400(51), *426*
Schneider, P., 354(11), *382*
Schneider, W. P., 53(145, 147), 82(145), *146*
Schnitt, G., 24(167), 64(167), *147*
Schöler, H-F., 388(10), 389(13), 416 (137), *425*, *428*
Schöllkopf, U., 17(2), 18(2), 25(2, 378), 65(2), 79(2), *142*, *143*, 156(2), 173 (122), 193(122), *197*, *201*
Schönberg, A., 368(73, 74), 371(74), *384*
Schörkhuber, W., 233(26), *266*
Schrader, L., 355(19), *382*
Schröder, R., 173(122), 193(122), *201*
Schroeder, J. P., 457(125), *466*
Schudel, P., 136(341), 137(341), *152*
Schuetz, R. D., 351(4), *382*
Schultz, G., 263(95), *268*
Schulz, H., 133(336), 134(336), *152*, 208(4), 211(12), *221*
Schulz, M., 319(87), *348*
Schumacher, R. J., 174(137), 195(137), *201*
Schunn, R. A., 163(45), 194(45), *198*, 468(18), 469(18), *504*
Schupp, O. E., 37(84), *144*, 168(102), 174(102), 195(102), *200*
Schwan, T. J., 278(36c), *293*
Schwartz, A., 156(7), 166(7), 194(7), *197*
Schwarz, M., 133(337), *152*, 443(58), *464*
Schwarz, V., 177(162), *202*
Schweitzer, R., 405(85), *427*
Schweizer, E. E., 8(6, 7a, 7b), *14*, *15*, 25(149, 175), 54(149), 68(175), *146*, *147*, 163(46), 194(46), *198*, 247(55), 250(61, 62, 64, 65), 252(61, 73, 74, 75), 253(76, 77, 78), 254(78, 81, 82), 255 (83), 265(100, 101, 102), *267*, *268*, 455(115), 465, 490(146), *508*
Scott, C. B., 317(74), *347*
Scott, G., 318(82), *348*
Searle, R. J. G., 172(116), *200*, 212(19), *221*, 272(8), 273(8), 274(8), 278(8, 36a), 280(8, 36a), 289(8), *292*, *293*, 296(6), 300(6), *345*
Sears, D. J., 288(65), *294*, 297(7), *346*
Seeliger, A., 415(125), *428*, 528(54), *534*
Seguin, F. P., 252(70), *268*
Seitz, G., 46(112), 47(112), *145*, 184 (212), *204*, 360(38), *383*
Sellstedt, T. H., 173(121), *201*
Selva, A., 494(177a), *509*
Selvarajan, R., 308(43), *347*
Selve, C., 392(22, 27), 400(22, 27), 401 (61, 64), 417(139, 140), *425*, *426*, *428*, 456(121, 122), *465*, 532(70), *535*
Seng, F., 217(39), *222*
Senyavina, L. B., 90(220), 91(220), *148*
Servi, S., 373(108), *385*
Seuleiman, A., 176(152), *201*
Seus, E. J., 37(84), *144*, 168(102), 174 (102), 195(102), *200*
Seyden-Penne, J., 39(96), 41(96), 80 (197), *145*, *147*, 161(38), 163(46), 166(80), 167(90, 91, 93), 168(91, 104), 169(91, 112a), 170(90, 91, 112a), 173(91), 182(112a,) 193(38, 112a), 194(46, 91, 104), *198*, *199*, *200*
Seyferth, D., 174(135), 193(135), *201*, 351(1), *382*
Shaffer, E. T., 25(175), 68(175), *147*, 455(115), *465*
Shafigullina, R. D., 418(167), *429*
Shahak, I., 164(56, 58), 173(56), 195 (56, 237), *198*, *199*, *205*, 360(40), *383*
Shannon, P. V. R., 272(11), *292*
Shapiro, M. J., 486(122), *507*
Sharif-Nassirian, F., 33(65), 35(65), *144*
Sharp, J. T., 418(164, 165), 421(165), *429*
Shavel, J., 229(12), 238(12), *266*
Shaw, R. A., 301(23), *346*, 363(56), *383*, 445(68), *464*
Shechter, H., 229(11), 242(46), *266*, *267*

Sheehan, J. C., 519(19), *533*
Sheikh, Y. M., 141(359), *153*
Sheludyakov, V. D., 422(211), *430*
Shelykh, G. I., 527(50), 531(64), *534, 535*
Shemyakin, M. M., 19(13), 25(110), 27(36), 33(60, 61), 41(98, 99), 46(110), 78(61, 110), 79(110), 83(110), 90(217, 218, 220), 91(220, 225, 226), 92(226, 227), 93(217, 229), 94(217, 232, 235), 95(236, 237), 96(237, 238, 239), 97(239), *142, 143, 144, 145, 148, 149*
Shen, T. Y., 355(17), *382*
Shepherd, R. C., 531(67), *535*
Sheppard, W. A., 379(134), 380(138), *385, 386*
Shermergorn, I. M., 495(181), *509*
Shibasaki, M., 104(262), 106(268, 270), *150*
Shih, L. S., 498(193), *509*
Shima, T., 24(174), 67(174), 117(174), 118(174), *147*
Shimokawa, J., 330(114b), 335(114b), *349*
Shimomura, H., 103(260), *149*
Shioiri, T., 142(361), *153*, 520(22, 23), 521(26), *533, 534*
Shirahama, H., 109(281), *150*
Shook, H. E., Jr., 166(84b), 185(84b), 194(84b), *199*
Shulman, J. I., 56(156), 57(156), 58(156), *146*, 164(63), 175(63), 176(141), 194(141), *199, 201*, 355(20), *382*
Sicher, J., 355(21), *382*
Sidky, M. M. 352(7), *382*
Siedle, A. R., 314(62), *347*, 358(32), *383*
Siegemund, G., 413(114), *427*
Siegmund, G., 246(54), *267*
Sieper, H., 281(43, 44), *293*
Sih, J. C., 102(256), *149*
Silhacek, D. L., 116(298), *151*
Silver, E. H., 29(43), 83(43), *143*
Sim, G. A., 139(355), 140(355), *153*
Simalty, M., 176(152), *201*, 237(36, 37, 38), 238(37, 41), *267*
Simalty-Siemiatycki, M., 237(35), 238(35, 39), *267*
Simamura, O., 281(46), *293*
Simic, D., 48(129), *145*

Simmons, T. C., 360(41), *383*
Simonnin, M-P., 163(46), 194(46), *198*
Simonovitch, C., 53(147), *146*
Sims, C. L., 364(62), *383*
Singh, G., 178(174), *202*, 444(65, 66), *464*
Singh, H., 354(13), *382*
Sinha, N. D., 180(185), *203*
Skowronska, A., 501(206), *510*
Skuballa, W., 99(244), *149*, 213(23), *222*
Skyarskii, L. S., 423(221), *431*
Slates, H. L., 180(180), 195(180), *202*
Smetana, R. D., 341(147), *350*
Smets, G., 231(18, 19, 20, 21), 232(20, 23, 24), *266*, 495(179), *509*
Smirnov, V. V., 489(142), *508*
Smissman, E. E., 443(54), *464*
Smith, B. C., 301(23), *346*, 363(56), *383*, 445(68), *464*
Smith, C., 369(84), *384*
Smith, C. P., 305(37), 307(34), 310(54), 311(54, 56, 57), *346, 347*, 469(20), 470(33, 34), 471(20, 40), 472(44), 473(48), 474(48, 54), 475(63, 64, 65, 66, 67, 68, 69), 476(20, 63, 66, 76), 478(76), 482(63, 64, 65, 68, 69, 101), 483(64, 65, 67), 484(66), 486(67, 119, 130), 487(133, 134), 488(134, 136), 500(199, 200), *504, 505, 506, 507, 508, 510*
Smith, D. M., 275(27), 281(27), 288(65), 293, 294, 297(7), *346*
Smith, D. O'N., 80(196), *147*
Smith, E. H., 359(36), *383*
Smith, J. A., 368(79), *384*
Smith, L. C., 448(85), 449(86), *464*
Smith, R. W., 287(64), *294*
Smolinsky, G., 276(30), *293*
Smoot, J., 372(100), *385*
Smucker, L. D., 253(76), *268*
Smythe, R. J., 318(84), 319(84), *348*
Snoble, K. A. J., 25(30), 52(30), 53(30), 82(30), *146*, 167(95), *200*
Snoussi, M., 163(49), 173(49), 193(49), *198*
Snowden, R. L., 177(158c), 192(158c) *202*
Snowdon, L. R., 195(235), *204*
Snyder, C. D., 50(136), *146*

Snyder, E. I., 392(23, 24, 25), 400 (23, 24, 25, 44), *425*, *426*
Snyder, E. J., 400(43), *426*
Snyder, J. P., 369(86), *384*
Sokurenko, A. M., 423(221), *431*
Soleiman, M., 460(143, 144), *466*
Solodovnik, V. D., 95(236, 237), 96 (237), *149*
Solomon, P. W., 320(89), 321(89), *348*
Solomonovici, A., 164(59), 166(59, 81), 175(59, 129), 181(196), 194(81), 195 (129), *199*, *201*, *203*
Soloway, A. H., 29(43), 83(43), *1 43*
Soloway, B., 422(213), *430*
Sondheimer, F., 218(41), *222*, 227(9), *266*, 344(153), *350*
Sonnet, P. E., 24(125), 25(119, 299), 46(119, 120), 47(125), 95(125), 112 (292), 113(119, 120), 114(120), 116 (119, 299), 117(119), 133(337), *145*, *151*, *152*, 443(58, 59, 62), *464*
Sonveaux, E., 442(51), *464*
Sorensen, A. K., 230(13), *266*
Sorm, F., 178(170), 194(170), *202*
Soulen, R. L., 403(72, 73), 423(231), *426*, *431*
Spande, T. F., 368(78), *384*
Spangler, C. W., 436(25), *463*
Speckamp, W. N., 186(215), *204*
Spencer, Th. A., 251(66), *267*
Spero, G. B., 53(147), *146*
Spessard, G. O., 28(41), *143*
Speziale, A. J., 27(35), 28(50), 30(50), 32(35), *143*, 167(96), *200*, 317(72, 73), *347*, 391(16), *425*, 448(81, 82), 449 86, 87), 450(82), *464*
Spitaler, R., 351(2), *382*
Springer-Fidder, A., 177(161), *202*
Sprung, I., 133(336), 134(336), *152*
Srinivasan, K., 275(21), *292*
Srinivasan, K. G., 275(21), *292*
Srivanavit, C., 467(2c), *503*
Staab, H. A., 361(47, 48), *383*
Stacey, G. J. 433(1), *462*
Stache, U., 181(194), 183(194, 201), *203*
Stadnichuk, M. D., 85(203), *148*
Stafforst, D., 173(122), 193(122), *201*
Stamm, D., 110(283), *150*
Stanishevskii, L. S., 178(172), *202*
Staten, R. T., 112(292), *151*

Staudinger, H., 213(24), 219(48), *222*
Stealy, M. A., 108(278), *150*
Steevensz, R. S., 251(68, 69), *267*, *268*
Steglich, W., 411(108), *427*, 492(157), *508*
Steinberg, G. M., 422(204), 423(204), *430*
Stelzel, P., 519(16), 526(44), *533*
Stephani, R., 109(280), *150*
Stephenson, L. M., 342(151), 343(151), *350*, 470(31), 491(152), *504*, *508*
Sterlin, R. N., 501(205), *510*
Stern, P., 473(50), 474(52, 53b), 475 (60, 70), 500(201), *505*, *506*, *510*
Sternbach, L. H., 290(71), *294*
Stevens, I. D. R., 245(50), *267*
Stevenson, R., 438(34, 35, 36), 441 (34), *463*
Stewart, A. P., 492(154), *508*
Stewart, N. J., 12(18b), 13(18b, 19), *15*
Stöckigt, J., 25(369), *153*
Storer, R., 129(325), *152*
Stork, G., 187(218), *204*, 365(64), *384*, 400(45), *426*
Stork, P. J., 373(107), *385*
Strandtmann, M. von, 229(12), 238 (12), *266*
Stransky, W., 25(117, 131, 287, 288, 290, 291, 296), 46(117), 47(117), 48 (131), 49(131), 79(131), 95(117, 131), 110(287), 111(131), 187, 188), 112 (290, 291), 113(291), 115(296), *145*, *146*, *150*, *151*
Strecker, W., 351(2), *382*
Streichfuss, D., 175(131), 195(131), *201*
Strickland, R. C., 25(38), 28(38), *143*
Stroh, J., 232(22), 241(44), 242(45), 243(47), 244(45, 47), *266*, *267*
Strüver, W., 415 (128, 129), 424(242), *428*, *431*, 528(55), *535*
Strzelecka, H., 237(34, 35, 36, 37, 38), 238(35, 37, 39, 40, 41, 42), *267*
Stuart, R. S., 24(376), *153*
Stuber, F. A., 177(161), *202*
Sturtz, G., 161(37), 166(83), 168(97, 98), 169(98), 173(98), 177(83), 189 (226), 194(83, 98), 195(83, 232), *198*, *199*, *200*, *204*
Süb, G., 493(171), *509*

Subramanian-Erhart, K. E. C., 177 (161), *202*
Subramanyam, V., 29(43), 83(43), *143*
Sudo, R., 249(59), *267*, 411(104), *427*, 440(42), *463*
Suffness, M. I., 138(349), 139(349), *153*
Sugahara, T., 106(270), *150*
Sugasawa, S., 37(76), *144*, 408(94), *427*
Sugie, A., 103(260), *149*
Sugimae, T., 10(10), *15*
Sullivan, W. W., 229(11), *266*
Sultanova, R. B., 418(151), *428*
Sundberg, R. J., 166(84a), 177(155), 185(84a), 191(155), 194(84a, 155), *199*, *201*, 275(19a, 28), 276(28), 287(64), *292*, *293*, *294*
Surmatis, J. D., 121(309), 123(318), 124(318), *151*, *152*, 177(163), 196(163), *202*
Suschitzky, H., 272(13), 275(20, 23), 277(32, 33), 278(23, 35, 37a), 279(37a, 37b), 280(35), *292*, *293*
Suter, C., 179(176), 180(186), 196(186), *202*, *203*
Sutherland, I. O., 369(84), *384*
Sutherland, J. K., 164(61), 167(61), 168(100), 169(100), 173(61, 100), 194(100), *199*, *200*, 214(28), *222*, 339(144), *349*
Sutton, L. E., 468(8a), *503*
Sutton, M. E., 277(32), *293*
Suzuki, M., 287(61), *294*
Suzuki, Y., 495(180), *509*
Svoboda, J. A., 141(358), *153*
Swank, D., 470(34), *504*
Sweetman, B. J., 372(94, 95), *384*
Swensen, W. E., 92(228), *148*
Synnes, E., 24(376), *153*
Szantay, C., 38(82), *144*, 184(214), *204*
Szarek, W. A., 331(119, 121, 122), 332(128), *349*
Szele, I., 499(197b), *510*
Szmant, H., 301(24), *346*
Szmant, H. H., 302(25), 306(25), *346*

Tachibana, S., 521(26), *534*
Taft, A., 486(123), 500(123), *507*
Tagaki, W., 24(165), 63(165), *146*
Taglieber, V., 361(48), *383*
Tait, B. S., 279(40), 288(66a), 289(40, 66a), *293*, *294*, 496(183a, 183b, 184), *509*
Takahagi, H., 531(65, 66), *535*
Takahashi, H., 156(8), 173(8), 193(8), 194(8), *197*
Takahashi, T., 283(56), *294*
Takahatake, Y., 365(65), *384*
Takaichi, O., 108(279), *150*
Takano, S., 275(26), *293*
Takashi, S., 530(63), *535*
Takebayashi, N., 370(120), *385*
Takei, H., 373(105), 374(111), *385*
Takeuchi, Y., 415(124, 126, 127), 418(124), *428*, 528(54), *534*
Tamura, Y., 440(40b), *463*
Tanaka, A., 272(12), *292*
Tanaka, K., 352(8), *382*
Tanio, M., 440(40b), *463*
Tarbell, D. S., 400(58), 401(58), *426*
Tarchini, C., 26(31), *143*
Tasaka, K., 450(94), *465*, 469(20, 21, 22), 471(20, 21, 22), 474(54), 476(20, 21, 22), 487(21), *504*, *505*
Tascher, E., 522(31), *534*
Taub, D., 180(180), 195(180), *202*, 344(153), *350*
Tavs, P., 281(44), *293*, 460(145), *466*
Taylor, A., 368(77), 370(77), *384*
Taylor, E. C., 165(69), 186(69), *199*, 209(7), *221*
Taylor, E. E., 274(18), *292*
Taylor, E. K., 263(97), *268*
Taylor, L. J., 449(87), *464*
Taylor, W. C., 336(136), *349*
Tchoubar, B., 80(197), *147*
Tebby, J. C., 162(44), *198*
Tefizon, M., 6(4b), *14*
Teichmann, H., 387(3), 400(39), 401(39), 410(101), 422(193), *425*, *427*, *430*
Telefus, C. D., 476(76, 77, 78), 478(76, 78, 81), 479(77), *506*
Templeton, J. F., 218(45), *222*
Tennant, G., 6(5b), 7(5b), *14*
Terahara, A., 263(97), *268*
Teraji, T., 54(150), *146*
Terashima, S., 109(281), *150*
Terent'eva, S. A., 476(72a), *506*
Tew, L. B., 457(125), *466*
Tewari, R. S., 25(201), 84(201), *147*

Texier, F., 234(31), *267*
Thai, A., 410(101), *427*
Thakore, A. N., 404(78), *426*, 443 (60), *464*
Thaller, V., 25(204), 85(204, 205, 206), 86(204, 206), 90(221), 97(241), 98 (205, 221, 241), 99(205, 242, 243), *148*, *149*
Thalmann, A., 373(106, 108), *385*
Thayer, A. L., 342(150), *350*, 491(153), *508*
Thenn, W., 493(170), *509*
Theysohn, W., 214(29), *222*
Thiel, L., 372(98), *384*
Thiem, J., 159(24), 160(24), 161(24), 184(211), *198*, *204*, 459(134), *466*
Thomas, D. G., 10(8), *15*
Thomas, R., 23(24), 141(24), *143*, 175 (131), 180(189), 195(131), *201*, *203*
Thommen, R., 177(163), 196(163), *202*
Thommen, W., 24(103), 25(103), 43 (103), 68(103), 69(103), *145*
Thompson, J. G., 25(149), 54(149), *146*, 252(74), *268*
Thompson, J. L., 25(146), 53(146), *146*, 455(115), *465*
Thompson, M. J., 141(358), *153*
Thompson, Q. E., 335(135), 336(135), 341(135), 343(152), *349*, *350*, 491 (149), *508*
Thomson, Q., 10(9), *15*, 286(60c), *294*
Thorpe, W. D., 417(141), *428*, 523(34), 532(69), *534*, *535*
Tichy, M., 355(21), *382*
Ticozzi, C., 494(177b), *509*
Tideswell, J., 164(53), 177(53), 192 (53), *198*
Tindall, C. G., 355(16), *382*
Tischenko, I. G., 178(172), *202*
Tittle, B., 458(131), *466*
Todd, A. R., 387(1), 422(203), *425*, *430*, 513(4), *533*
Todd, L., 405(81), *427*
Todd, M. J., 10(9), *15*, 270(3), 273 (16), 274(16), 275(27), 281(27, 47), 286(60a, 60b, 60c), 287(62), 288(62, 65), *292*, *293*, *294*
Toke, L., 38(82), *144*, 184(214), *204*
Tolney, P., 108(277), *150*
Tomashewski, G., 515(8), *533*

Tomoskozi, I., 23(27), 34(27), 38(27), *143*, 174(125), 189(226, 227, 228), *201*, *204*, 213(26), 216(35), *222*, 392 (29), 394(29), 400(29), 403(76), *425*, *426*
Tonge, B. L., 433(5), *462*
Toscano, V. G., 157(10), 159(23), 189 (223, 225), 192(23, 225), 193(10, 223), *197*, *198*, *204*, 216(36), *222*, 310(51), *347*
Toube, T. P., 33(64), 68(177), 70(177), 126(177), *144*, *147*
Trampe, G., 457(124), *466*
Trave, R., 110(286), *150*
Traynard, J-C., 455(112), *466*
Trego, B. R., 369(81), 370(81), *384*
Trenbeath, S. L., 400(48), *426*
Trippett, S., 12(18a), *15*, 18(9), 19 (9, 15a, 15b), 25(9), 31(54, 55), 38 (77), 40(15b), 82(198), *142*, *143*, *144*, *147*, 157(17), 162(41), 176(41), 193 (41), *197*, *198*, 208(2a), 212(18), 214 (30), 217(40), *221*, *222*, 316(69), *347*, 447(76), 448(83), 454(106), 455(76), *464*, *465*, 468(9a), 468(17a), 484(110), 492(154), *504*, *507*, *508*
Trost, B. M., 37(90), *144*, 178(168), *202*
Truce, W. E., 301(22), *346*
Truscheit, E., 19(12), 29(12), 68(12), 69(12), 110(12), *142*
Tsai, S. C., 325(103), *348*
Tschopp, A., 526(47), *534*
Tseng, C-Y., 173(124), *201*
Tsolis, A. K., 320(90), *348*, 455(114), *465*
Tsolis, E. A., 468(12), 469(12, 21), 471 (21), 476(21), 487(21), *504*
Tsuboi, H., 474(53c), *505*
Tsuchiya, K., 25(111), 46(111), 130 (111), *145*
Tsuda, E., 54(150), *146*
Tsukida, K., 177(163), 196(163), *202*
Tsurugi, J., 371(90, 91), *384*
Tunemoto, D., 175(140), 194(140), *201*, 209(5), *221*, 365(65), *384*
Tuong, H. B., 26(31), 63(162), *143*, *146*, 164(54), 177(54), 192(54), 193(54), *198*
Turesky, S. S. 422(213), *430*
Turnblom, E. W., 498(192), *509*

Turner, J. L., 85(205), 98(205), 99(205, 242, 243), *148*, *149*
Tuttle, M., 361(46), *383*
Tweddle, N. J., 12(18b), 13(18b, 19), *15*, 289(67), *294*, 496(184), *509*
Twitchell, D., 486(123), 500(123), *507*
Tyuleneva, V. V., 501(205), *510*
Tyvorskii, V. I., 178(172), *202*

Uebel, E. C., 25(299), 116(299), *151*
Ueda, K., 131(331), *152*
Ueki, M., 374(115), 376(123), 377(124), *385*, 517(12, 13), 529(59, 60), 530 (62, 63), *533*, *535*
Ugi, I., 335(133), *349*, 468(11, 12, 14, 15, 19), 469(12, 14, 15, 19), 474(52, 53a, 53b), 475(60), 492(156), 500 (201), *504*, *505*, *508*, *510*
Ui, M., 371(91), *384*
Ullerich, K., 159(28), *198*
Ullmann, D., 229(11), *266*
Ulrich, H., 177(161), *202*
Ulrich, P., 59(159), 60(159), *146*, 373 (108), *385*
Ulrich, T. A., 25(149), 54(149), *146*
Umani-Ronchi, A., 494(177a, 177b), *509*
Umemoto, T., 365(65), *384*
Unrau, A. M., 142(362, 363), *153*, 183(202), *203*
Untch, K. G., 365(64), *384*
Ushiyama, H., 401(66), *426*
Usov, A. I., 435(14, 15, 16), 438(15), *463*

Valadon, L. R. Guy, 33(64), *144*
Valetdinov, R. K., 423(226), *431*
Valitova, L. A., 489(141), *508*
Van der Gen, A., 175(138), 194(138), *201*
Vanderhaar, R. W., 423(234), *431*
Vanderwerf, C. A., 214(31), *222*, 455 (114), *465*
Van Duuren, B. L., 309(48), 310(48, 50), *347*
Van Mil, J. H., 138(342), *152*
Van Reijendam, J. W., 24(168), 64 (168), *147*
Van Tamelen, E. E., 138(348, 349), 139 (349, 350), 208(2b), *221*, 400(42), *425*

Vasil'eva, M. N., 496(186), *509*
Vasser, M., 24(115), 46(115), *145*
Vaughan, K., 281(45), *293*
Vaultier, M., 234(31), 235(31b), *267*
Vaver, V. A., 27(36), 41(98, 99), 90 (220), 91(220, 225, 226), 92(226, 227), 93(229), 94(232, 235), *143*, *145*, *148*, *149*
Veber, D. F., 512(2), *533*
Vecchia, L. D., 109(280), *150*
Vedejs, E., 25(30, 192), 26(30), 52(30), 53(30), 76(189), 76(192), 82(30), *143*, *146*, *147*, 167(95), *200*
Vela, F., 79(195), *147*
Veltmann, H., 388(6, 8, 9), 389(6, 9, 11), 393(6), 397(8, 9), 403(6), *425*
Venkateswarlu, A., 59(159), 60(159), 109(281), *146*, *150*
Vera, M., 101(252), *149*
Vere Hodge, R. A., 90(221), 97(241), 98(221, 241), *148*, *149*
Verenikina, S. G., 440(40a), *463*
Verheyden, J. P. H., 500(57), *426*, 435(18, 19, 21, 22), 436(18, 21, 22), 439(22), 456(22), *463*
Vial, C., 24(240), 97(240), 118(240), 119(240), *149*
Vidalenc, H., 422(218), *430*
Vidalenc, H. M., 422(216), *430*
Vig, A. K., 173(123), 178(174), 194 (123), *201*, *202*
Vig, O. P., 173(123), 178(174), 194(123), *201*, *202*
Ville, G., 400(47), *426*
Villieras, J., 162(43), 164(52), 174(136), 193(52, 136), *198*, *201*, 213(21), *222*, 402(67, 68, 69, 70), *426*, 450(89), *464*
Vilsmeier, E., 212(20), *221*
Vinogradova, V. S., 300(20), *346*, 470 (32b), 488(138, 139), 496(187), *504*, *508*, *509*
Virkhaus, R., 447(75, 78, 79), 448(84), 455(78), *464*
Vlasov, G. P., 527(50), 531(64), *534*, *535*
Vlattas, I., 109(280), *150*, 180(180), 195(180), *202*
Vollhardt, K. P. C., 19(19), *142*, 225 (4b), 227(8a, 9), *266*
Volz, P., 245(52), *267*, 413(115), *427*

Voncken, W. G., 499(198), 501(198), *510*
Vorbrüggen, H., 362(51), *383*
Voskoboeva, T. N., 480(88b), *506*
Vostrowsky, O., 25(118, 131, 233, 287, 288, 290, 291, 296), 46(118), 48(131), 49(131), 79(131), 94(233), 95(118, 131), 110(287), 111(131, 187, 188), 112(118, 290, 291), 113(291), 115 (118, 296), 116(233, 297), *145*, *146*, *149*, *150*, *151*

Wada, M., 108(279), *150*, 327(108), 330 (117, 118), 331(108, 117, 118), 333 (108), 334(118), 335(118), *348*, *349*, 378(127), *385*
Wadsworth, D. H., 37(84), *144*, 168 (102), 174(102), 195(102), *200*
Wadsworth, W. S., Jr., 23(25, 26), 35 (26), 131(25), 141(25), *143*, 156(3), 157(9), 162(9), 164(3), 165(3), 167(3, 87), 168(87), 173(3), 188(9), 189(222), 191(3), 194(3), 196(9, 222), *197*, *199*, *204*
Wagenecht, J. H., 314(62), *347*, 357 (31), *382*
Wager, J. S., 314(62), *347*, 358(32), *383*
Wagner, H., 225(6), 226(7a), *266*
Wailes, P. C., 19(10), 25(10), *142*
Wajer, T. A. J., 165(68), 175(68), *199*
Wakabayashi, N., 133(337), *152*, 407 (90), *427*
Wakabayashi, T., 186(215, 216), *204*
Wakatsuka, H., 108(276), *150*
Waldvogel, G., 345(158), *350*
Walker, B. J., 14(22), *15*, 159(29), *198*, 453(103), 458(129), *465*, *466*
Walker, C. C., 242(46), *267*
Walker, D. M., 31(54, 55), 38(77), *143*, *144*, 157(17), *197*, 316(69), *347*, 448(83), 454(106), *464*, *465*
Walker, K. A., 456(118b), *465*
Walling, C., 318(78, 80), *348*, 360(41, 42, 43), 366(42), 367(43, 68), *383*, *384*, 420(180), *429*
Wallis, C. J., 163(50), 192(50), *198*
Walters, M. E., 6(4e), *14*
Wang, C-L. J., 102(254), *149*, 180 (182), 195(182), *203*
Wang, C. S., 423(223), *431*

Ware, E., 479(82), *506*
Warnecke, H-U., 183(204), *203*
Warner, V. D., 422(213), *430*
Warning, K., 392(28, 30), 393(28), 394(28), 400(28, 30, 40), 401(40), 405(83, 84), 406(86, 87, 88), 407(88, 91), 412(30), *425*, *427*
Warning, U., 421(188), 422(192), *429*, *430*
Warren, S., 163(50), 166(82), 168(105), 170(105), 176(142), 177(105, 158a, 158b), 192(50, 105, 142, 158a, 158b), 193(105), *198*, *199*, *200*, *201*, *202*
Warren, S. G., 14(25), *15*
Warten, G. A., 422(201), *430*
Wasielewski, C., 522(31), *534*
Watanabe, K., 186(216), *204*
Waters, R. M., 133(337), *152*, 407(90), *427*
Watson, A. A., 369(82), *384*
Watt, D. S., 174(126), 179(177), 193 (177), 194(177), 195(126), *201*, *202*
Way, J. E., 29(42), 124(42), *143*
Weatherston, J., 119(303), 120(303, 304), *151*
Webb, C. F., 105(266), *150*
Webb, R. L., 168(98), 194(98), *200*, 266(107), 268
Weedon, B. C. L., 24(368), 29(42), 33(59, 64), 68(177), 70(177), 72(181, 182), 73(181, 182), 78(181), 82(182), 83(181, 182), 90(181), 91(181), 120 (306), 123(315), 124(42), 125(320), 126(177, 181, 182, 320), 127(181), *143*, *144*, *147*, *151*, *152*, *153*, 177(159), 192(159), *202*
Weeks, C. M., 436(23), *463*
Wehman, A. T., 265(100), *268*
Weidmann, E., 33(65), 35(65), *144*
Weigel, L. O., 24(143), 52(143), *146*
Weigmann, H-D., 31(56), *143*
Weiland, J., 358(33), *383*
Weiler-Feilchenfeld, H., 175(129), 195 (129), *201*
Weill-Raynal, J., 19(15f), *142*
Weimar, W. R., Jr., 443(57), *464*
Weinshenker, N. M., 24(246), 99(246), 100(246), *149*
Weis, C. D., 318(77), *348*
Weiss, D., 457(128), *466*

Weiss, J., 399(37), *425*
Weiss, R. G., 392(23, 24), 400(23, 24, 43), *425, 426*
Welch, C. A., 455(110), *465*
Welch, D. E., 351(1), *382*
Welch, S. C., 6(4e), *14*
Welcher, A. D., 523(36, 37), *534*
Wellman, G. R., 266(107), *268*
Wells, D., 364(62, 63), *383, 384*
Wendisch, D., 355(19), *382*
Wendler, N. L., 180(180), 195(180), *202*
Werner, E., 256(84), *268*
Westaway, K. C., 177(160), 194(160), *202*
Westheimer, F. H., 469(26), 499(197b, 197c), *504, 510*
Wetmore, S. I., Jr., 493(172), *509*
Weyerstahl, P., 178(172), *202*
Wheeler, O. H., 24(163), 63(163), 79(163), *146*
White, D. W., 471(41), 486(41), 500(41), 502(216), *505, 510*
Whitehouse, K. C., 11(13), *15*
Whiter, P. F., 434(6), *462*
Whitham, G. H., 171(115), *200*
Whitlock, H. W., Jr., 187(221), 195(221), *204*
Whittaker, N., 34(71), 38(83), *144*
Whittle, P. J., 468(17a), 484(110), *504, 507*
Wicha, J., 181(198), 182(198), *203*
Widdowson, D. A., 25(360), 141(360), 142(361), *153*, 304(32), *346*, 364(60), *383*
Wieber, M., 468(9d), *504*
Wiechert, R., 183(205), 194(205), *204*, 345(158), *350*
Wieczorkowski, J., 378(130), *385*
Wieland, T., 415(125), *428*, 519(20), 522(30), 528(54), *533, 534*
Wihler, H. D., 388(6, 10), 389(6), 390(14), 393(6), 399(14, 36), 400(50), 403(6), *425, 426*
Wijsman, A., 175(138), 194(138), *201*
Wild, H-J., 354(11), *382*
Wildner, H., 178(166), 195(166), *202*
Wiley, G. A., 416(133, 134), 423(133, 134), *428*, 437(28, 29), 438(29), 440(28), *463*

Wilfinger, H. J., 502(215), *510*
Willhalm, B., 24(103), 25(103), 43(103), 68(103), 69(103), *145*
Williams, J. C., 173(117), 194(117), *200*
Williams, J. D., 368(76), *384*
Williams, J. R., 453(102), *465*
Williams, M. W., 520(24), 528(24), 529(24), 530(24), *533*
Williams, P. J., 37(91, 92), 40(91, 92), 133(92), *144*, 178(167), 194(167), *202*
Willis, B. J., 359(36), *383*
Willms, L., 417(142), 424(242, 243, 248), *428, 431*
Willy, W. E., 138(344), *152*
Wilson, J. D., 314(62), *347*, 357(31), 358(32), *382, 383*
Wilson, K. E., 28(41), *143*
Wilson, V. K., 323(98), *348*
Winkler, H., 168(103), 174(103), 193(103), *200*
Winter, R. A. E., 3(1), *14*, 355(14, 15), 356(15), *382*
Winter, W., 175(131), 195(131), *201*
Winternitz, F., 521(27), 522(28, 29), *534*
Wintersberger, K., 423(237), *431*
Wippel, H. G., 35(75), *144*, 155(1), 156(1, 5), 163(1, 5), 167(1, 5), 174(1), 192(1), 193(5), *197*, 440(45), 461(45), *463*
Wit, J., de, 444(63), *464*
Witkop, B., 368(78), *384*
Witold, W. E., 175(128), 195(128), *201*
Witschard, G., 24(373), *153*
Wittenberg, D., 25(102), 42(102), *145*
Wittig, G., 3(3), *14*, 17(1, 2, 3, 4, 5, 6), 18(2, 3), 24(3), 25(1, 2, 102, 378), 31(53, 56), 42(102), 48(127), 65(2), 79(2, 3), 82(3), 87(208), 121(308a, 310a), 123(308a, 310a), *142, 143, 145, 148, 151, 153*, 156(2), *197*, 207(1), 221, 317(71), *347*, 467(2d), 499(194), 501(203), *503, 510*
Woggon, W-D., 179(175), *202*
Wolf, R., 485(113), *507*
Wolkoff, P., 393(31), 400(31), *425*
Wolloch, A., 244(49), *267*
Wong, C. K., 190(230), 193(230), *204*, 385(145), *386*
Wong, D. Y., 312(59, 61), 313(61), *347*

Wong, F. F., 303(31), *346*
Wood, K. H., 524(38), *534*
Woodward, R. B., 344(153), *350*
Wright, P. W., 164(53), 177(53), 192 (53), *198*
Wróblewski, A., 176(151), 193(151), *201*
Wüest, H., 73(184), *147*
Wulff, J., 12(17), *15*, 215(32), *222*, 233(27), 236(27), 243(27), 245(27), *266*, 490(144), 494(175a, 175b), 496 185a, 185b), *508, 509*
Wunderlich, H., 468(9d, 9e), *504*
Wünsch, E., 525(43), 526(45, 46), *534*
Wunsch, G., 423(237), *431*
Wyllie, S. G., 141(357), *153*
Wynberg, H., 444(63), *464*

Yagi, H., 274(17), 282(17, 50), *292, 293*
Yakota, Y., 315(65), *347*
Yakubovich, A. Ya., 496(186), *509*
Yakushijin, K., 272(12), *292*
Yamada, K., 437(26b), *463*
Yamada, M., 328(109), 330(116), *348, 349*
Yamada, S., 415(124, 126, 127), 418 (124), *428*, 520(22, 23), 521(26), 528 (54), *533, 534*
Yamaguchi, A., 195(238), *205*
Yamaguchi, K., 10(10), *15*
Yamaguchi, M., 437(150), *466*
Yamamori, M., 297(159), *350*
Yamamoto, A., 297(159), *350*
Yamamoto, H., 56(156), 57(156), 58 (156, 158), 59(158), 60(158), 61 (158), 103(260), 109(281), 133(338, 339), 134(158, 340), 135(338, 339, 340), 136(158), *146, 149, 150, 152*
Yamanaka, T., 272(14, 15), 274(17), 275(26), 276(29), 282(17, 50, 51a, 51b), 283(53, 54), 284(57), 285(58), 290(69, 70a), *292, 293, 294*
Yamanaka, H., 311(58), *347*
Yamashita, K., 34(74), 36(74), *144*
Yamashita, M., 380(140, 141), *386*
Yamato, E., 408(94), *427*
Yamauchi, K., 10(10), *15*
Yamazaki, N., 434(10), 437(150), *463, 466*, 527(51, 52), *534*

Yamazaki, T., 275(28), 276(28), *293*
Yano, Y., 24(165), 63(165), *146*
Yanovskaya, L. A., 37(89), 38(78), *144*
Yardley, J. P., 183(203), *203*
Yates, J. A., 25(204), 85(204), 86(204), *148*
Yeh, C-L., 24(115), 46(115), *145*
Ykman, P., 231(18, 19, 20, 21), 232 (20, 23, 24, 25), *266*, 495(179), *509*
Yokoyama, T., 502(213), *510*
Yoneda, F., 333(129), *349*
Yong, D. Y., 305(40), *346*
Yoshida, M., 281(46), *293*
Yoshikawa, Y., 102(256), *149*
Yoshina, S., 272(12), *292*
Yoshioka, T., 495(180), *509*
Young, G. T., 520(24), 528(24), 529 (24), 530(24), *533*
Young, I. M., 308(42), 310(42), *346*
Young, J. W., 175(132), 196(132), *201*
Young, R. W., 523(35, 37), 524(38), 526(49), *534*
Youssfyeh, R. D., 165(69), 186(69), *199*
Yudena, K. S., 250(63), *267*
Yur'ev, Yu. K., 130(329), *152*
Yurkevich, A. M., 439(38), 440(40a) *463*

Zabrocki, J., 524(42), *534*
Zak, H., 373(108), *385*
Zamojski, A., 331(119, 120), *349*
Zatorski, A., 164(64), 165(75, 76), 166 (77), 175(64, 76), *199*
Zavalisina, A. I., 422(210), *430*
Zavyalov, S. I., 10(11a, 11b), *15*
Zbaida, S., 190(229), *204*
Zbiral, E., 6(5a), 7(5a), *14*, 219(46), *222*, 232(22), 233(26), 239(43), 241 (44), 242(45), 243(47), 244(45, 47, 48, 49), 246(53), 256(84), 257(85, 86, 87), 258(88), 260(91), 261(91), 262 (93, 94), 263(93, 94, 95, 96), 264(96, 98), *266, 267, 268*, 329(112), 334(131), 335(112), *349*
Zbirovsky, M., 422(205), *430*
Zdero, C., 73(186), 74(186), *147*
Zefirov, N. S., 130(329), *152*
Zeifman, Yu. V., 487(132), *508*
Zelawski, Z. S., 180(180), 195(180), *202*

Zeller, P., 24(367), 123(317, 319), *152, 153*
Zervas, L., 513(3), 514(5), 514(7), 520(3), *533*
Zhdanov, Yu. A., 19(17), 139(20), *142*, 184(208), *204*
Zhumovova, I. N., 412(110), *427*
Ziegler, G. R., 455(110), *465*
Ziegler, J. G., 417(140), *428*
Ziehn, K. D., 400(40), 401(40), 405 (83, 84), 406(86, 87), 407(91), 408 (95), 409(95, 97, 98, 99), 412(109, 113), 413(118, 120), *425, 427, 428*
Zimmer, H., 174(137), 176(143), 194 (143), 195(137, 239), *201, 205*, 444 (65, 66), 456(117), *464, 465*
Zimmermann, H. E., 25(213), 88(213), *148*
Zimmermann, M., 65(170), *147*
Zimmermann, R., 23(28), *143*, 212 (15), 219(15), *221*, 223(1), *266*
Zimnicki, J., 175(128), 195(128), *201*
Zolova, O. D., 488(139), *508*
Zoretic, P. A., 180(185), *203*
Zountsas, G., 175(131), 195(131), *201*
Zurflueh, R., 178(171), *202*
Zwierzack, A., 422(215), 423(224), *430, 431*
Zykova, T. V., 418(153, 154, 167), *428, 429*

Subject Index

Abscisic acid, 34
Acenaphthoquinone, ring expansion with (MeO)$_3$P, 311
Acetyl bromide, reaction with Ph$_3$P, 460
Acetylene carboxylates, from alkyl halides, 209
  via thermal decomposition of ylides, 213
Acetylene equivalent, via desulphurization of thionocarbonate, 356
Acid catalysis, of hydroperoxide deoxygenation, 318
Acraldehyde, reaction with phosphorus ylides, 128
Acridinium betaines, from alkylidenephosphoranes, 230
Acridones, from anthranils, 281
ACTH (1–24), 531
Activated C=C, reactions with alkylidenephosphoranes, 217
Active hydrogen compounds, alkylation, 333
Acyl bromides, from acids, 441
Acyl chlorides from carboxylic acids, 404, 528
  from phosphinic acids, 404
  from phosphoric esters, 404
Acyl halides, from esters, 442
Acyl ylides, 211
β-Acylacrylic acids, from halogenoketones, 211
Acylated alkylidenephosphoranes, electrolytic or reductive cleavage, 213
Acylated enediol phosphates, from dioxaphospholens, 473

Acylation, of β-oxido ylides, 58
N-Acylcarbamates, alkylation by alcohols, 331
2-Acylindoles, 282
Acyloin condensation products, into monoketones, 470
Acyloins, in furan formation, 250
Acyloxyphosphonium intermediate, in carboxyl activation, 531, 532
  in peptide synthesis, 529
  racemization via, in peptide synthesis, 528
Acyloxyphosphonium salt intermediate, 414
N-Acylphenylglycine esters, dehydration and cyclization, 411
β-Acylvinylphosphonium salts in synthesis: heterocyclic-fused imidazoles, 261
  pyrazoles, 257
  pyrroles and imidazoles, 260
  1,3-thiazoles, 258
β-Acylvinyltriphenylphosphonium bromide, in nucleic acid and nucleoside modification, 263
  in 1,2,3-triazole synthesis, 256
α,β-Addition of P(III) reagents to nitroalkenes, 289
Adenine, 3-methyladenine, and adenosine, modification by β-acylvinyltriphenylphosphonium bromide, 263
Alcohols, activation to nucleophilic substitution by DEADC, 327
  alkylation via, 331
  as alkylating agents via DEADC, 327

Subject Index

chlorination, 391
chlorination via Ph_3P–CCl_4, 400
esterification via Ph_3P–CCl_4, 414
from alkanes, 319
from dialkyl peroxides, 320
from hydroperoxides, 318, 319
in alkylations, 332
in nucleophilic substitutions, 333
into alkyl chlorides, 3
into amines using DEADC–TPP, 330
into diesters, 405
into halides, 437
into hydrocarbons, 437
iodination, 401
phosphorylation via Ph_3P and di-2-pyridyl disulphide, 375
phosphorylation via vinylphosphonium salt, 450
Aldehydes, from acids, 459
reductive diacylation, 483
reductive monoacylation, 483
Vilsmeier synthesis using Ph_3PBr_2, 442.
Aldonitrones, reduction by tertiary phosphines, 300
Aldoximes, dehydration, 409
Alicyclic lactones, reaction with Ph_3P, 443
Alkanes, from thiols, 360
cyclic, from alkyl dihalides, 210
Alkanoic acids, from ketene thioketals, 359
Alkene epoxide, from dioxaphospholan, 306
Alkenes, 1,2-addition of singlet oxygen, 340
cleavage by ozonolysis, 343
cyclic, from alkyl dihalides, 210
from alcohols, 435, 436
from allyl halides, 209
from aryl trifluoromethyl ketones, 314
from α-bromoketones, 5
from carbonyl compounds, 314
from 1,2-dibromoalkanes, 457
from α,α'-dihalogenosulphones, 453
from vic-diols, 354
from dioxaphospholan, 306
from enol phosphates, 6
from enolizable ketones, 164
from episulphides, 351
from epoxides, 215, 316, 381
from β-hydroxyacids, 335
from imines, 165
from β-keto and ester stabilized ylides, 157
from ketones,
from 1,3-oxathiolan-5-ones and azosulphides (two-fold extrusion), 359
from oxosteroids, 181
from PO- and PS-activated reagents, 157
from PO-stabilized carbanions, 155
from primary alcohols via selenides, 380
from Schiff bases, 219
from secondary alcohols, 400
from thioncarbonates, 2, 354
from ylides via $(PhO)_3P$–O_3
hindered, 359
isomerization, 172
Z- to E-isomerization via 1,2-trithiocarbonates, 356
stereoselectively via sulphoxides, 364
stereoselectively via Wittig reaction, 17–142
stereospecifically from epoxides, 437
via reductive dimerization of carbonyl compounds, 310
via Wittig reaction, 3, 17
$trans$-Alkenes, from allylphosphonate alkylation, 209
Alk-1-enylphosphonium salts, in heterocyclic synthesis, 250
Alkoxy radicals, deoxygenation, 318
Alkoxycarbonylalkylidene-phosphoranes, 214
5-Alkoxyoxazole, from N-acylphenylglycine esters, 411
Alkoxyphosphonium salts, dechlorination, 403
N-Alkoxyphthalimides, from N-hydroxyimides, 331
Alkoxy-tris(dimethylamino)-phosphonium salts of secondary alcohols and propane-1,3-diols, 401
Alkyl alcohols, regio- and stereo-selective chlorination, 400

Alkyl aryl ethers, from phenols, 332
  via phosphonium salts, 401
Alkyl azides, from alcohols, 334
Alkyl bromides, from alcohols, 443
Alkyl chlorides, from alcohols, 3, 400
  from chloroformate esters, 459
  from thiocarbamic O-esters with Ph$_3$P–CCl$_4$, 408
  from thiols, 379
Alkyl halides, from alcohols, 334, 401, 435, 438, 448, 450, 455
  via triaryl phosphites, 433
  optically active, from phosphites and hydrogen halide, 434
Alkyl iodides, from alcohols, 433, 437
  from phosphoroiodates, 433
  via methyltriphenoxyphosphonium iodide, 435
Alkyl nitrates, from alcohols, 334
Alkyl nitrites, from alcohols, 334
Alkyl phosphorodiamidate, in amino function protection, 514
Alkyl thiocyanates, Arbusov reaction, 379
  from alcohols, 334
Alkylation, by alcohols, 327
  of cysteine, 10
  of ester stabilized ylides, 208
  of N-heterocycles by (RO)$_3$PO, 10, 272
  of β-oxido ylides, 58
  of phenothiazines, 10
  of ylides, 207
  using trialkyl phosphates, 10, 272
O-Alkylation of acylphosphoranes, 208
11-Alkyl-11H-benzimidazolo[1,2-b]-indazole, from 2-nitrobenzaldazine, 278
Alkylidenephosphoranes, acylation, 211, 213
  addition to nitrile oxides, 236
  from phosphonium halides, 207
  intramolecular C-alkylation, 209
  reactions with activated C═C, 217
  reactions with C≡C, 218
Alkylidenephosphoranes in synthesis:
  acridinium betaines, 230
  azirines, 236
  coumarins, 227
  dibenzo[b, f]oxepins, 225
  2,3-dihydro-1-benzofurans, 227
  heterocyclic compounds, 223
  1,2-oxazoles, 234
  pyrazoles, 234
  α-pyrones, 227
  pyrroles, 234
  3-pyrrolines, 234
  substituted pyridines, 231
  tetrahydropyrans, 225
  thiacyclohexanes, 225
  thiacyclononatetraenes, 227
  thiophens, 227, 268
  1,2,3-triazoles, 231
Alkylmalondinitriles, from dicyanosulphonium ylides, 380
β-Alkyl-β-phenyl-β-peroxypropio-lactones, reaction with Ph$_3$P, 325
N-Alkylphthalimides, in formation of amines, 330
α-Alkylstyrenes, from β-alkyl-β-phenyl-β-peroxypropiolactones, 325
  from peroxylactones, 325
N-Alkylsuccinimides, from succinimides and alcohols, 331
Alkyltrimethylsilyl peroxide, in preparation of silyl alkyl ethers, 321
Alkynes, from 1,2-diketones, 316
  from α-halogenoketones, 316
  from halogenoalkylphosphonates, 174
  from ketenes, 316
  from tetrabromosulphones, 453
  into tetrathiafulvenes, 379
  ozonolysis, 344
  via phosphinates, 176
Alkynyl dibromides, from Ph$_3$PBr$_2$, 438
Allenes, from imines, 219
  from ketenes, 164
  via phosphinates, 176
Allenic esters, from acyl chlorides, 213
(Z)-Alloxanthin, 125
Allylic alcohols, via β-oxido ylides, 58–63, 109
Allylic hydroperoxides, "ene" reaction in formation, 338
Allylic ylides, 18, 66–77
Amides, dehydration by Ph$_3$P–CCl$_4$, 408

Amides, (*continued*)
  formation in peptide synthesis, 511
  from acids, 434
  from cycloalkanone oximes, 441
  from imidoyl chlorides, 407
  into thioamides, 11
  phosphonium salts in promotion of synthesis, 437
  reactions with P(III) reagents, 450
  using organophosphorus anhydrides, 518
Amidines, from imidoyl halides, 406
Amination, of 6-methylisocytosine, methyluracil, 10
  of quinazolin-4-one, 10
  of tautomeric oxohydroxy groups by phenyl phosphorodiamidate, 10
  using $(Me_2N)_3PO$, 10
  using $PhOP(O)(NEt_2)_2$, 10
Amines, alkylation by $(RO)_3PO$, 10
  from alcohols, 330
  from *N*-alkylphthalimides, 330
  from *N*-oxides using triethyl phosphite, 297
  phosphorylation via Ph_3P and di-2-pyridyl disulphide, 375
  secondary, from primary amines via phosphinimines, 444
Amino group protection, 511
Aminoacids, activation by diphenylphosphoric anhydride, 513
  diphenylphosphinyl derivatives, 516
  diphenylthiophosphinyl derivatives, 517
  elongation using Ph_3P-CCl_4, 528
α-Aminoacids, from hydantoins, 479
  esters, from α-hydroxyacid esters, 331
Aminoalcohols, in N-heterocyclic synthesis, 410
Aminoalkylene-malonamic esters, from isocyanates and secondary enamines, 410
Amino(oxy)phosphoranes, from aromatic nitro compounds, 289
  from 1,2-diaryl-1-nitroethanes, 289
Aminophenols, reactions with P(III) esters, 12

2-Aminophenylcarbonyl compounds, in quinoline synthesis, 253
Aminophosphonium chlorides, preparation, 412
  from sulphur amides, 413
2-Aminopyridine, reaction with triphenyl-β-cyanovinylphosphonium bromide, 264
2-Aminopyrimidine, reaction with triphenyl-β-cyanovinylphosphonium bromide, 264
Aminopyrimidines, formation, 10
4-Aminoquinazoline, from 4-quinazolone, 10
Aminosugars, from sugars, 331
  phosphorylation, 412
Aminothiazoles, from β-acylvinylphosphonium salts, 259
2-Amino-1,3-thiazoles, 239
Amino(thiyl)phosphoranes, synthesis, 289
Ammonia and derivatives, reaction with R_3P-CCl_4, 411
Anabolic steroids, 180
Anderson test, 416
Anhydrides, from acids via tris(dimethylamino)phosphine, 529
  from carboxylic acids, 333
  from diaroyl peroxides, 323
Anthracene-9-aldehyde, in substituted ethylene synthesis, 310
Anthranils, into acylnitrenes, 281, 290
Anthranils (1,2-benzoxazoles) from *o*-acylnitrobenzenes, 281
9-Anthroxyl, in alcohol OH protection, 338
Apicophilicity of phosphoranes, 468
Aralkyl sulphoxides, deoxygenation, 302
Arbusov reaction, 4, 455, 459, 489
Aromatic endoxides, deoxygenation, 318
Aromatic epoxides, from aryl aldehydes, 309
Aromatization, via endoxide cleavage, 444
Aroyl compounds, reductive condensation, 314

Aryl alkyl ketones, from
  β-alkyl-β-phenyl-β-propiolactone,
  325
Aryl allyl sulphoxides, rearrangement
  to allyl sulphenates, 364
Aryl benzyl sulphoxides,
  deoxygenation, 303
Aryl bromides, from aryl ethers, 443
Aryl halides, from phenols, 440
  via triaryl phosphites, 433
Aryl iodides, from phenols, 434
Aryl isocyanates, deoxygenation, 315
Aryl trifluoromethyl ketones, into
  alkenes, 314
2-Aryl-[1]benzothieno[3,2-c]pyrazoles
  and [2,3-c] isomers, 278
2-Aryl-1H-[1]benzothieno[3,2-b]-
  pyrroles and [2,3-b] isomers, 275
2-Aryl-2H-benzotriazoles from
  2-nitroazoarenes, 280
O-(N-Arylcarbamoyl)hydroxamate,
  332
3-Aryl-2,3-dihydro-1,3,2-benzoxaza-
  phosph(v)oles, 288
Arylglyoxylic methyl esters, reaction
  with P(III) reagents, 308
N-Aryl-phosphoramidites,
  dehydrohalogenation, 421
Arylsulphonic acids, into thiols, 303
12-Aryl-1,2,3,4-tetrahydro-6H-
  indazolo[2,1-a]benzotriazole, 281
Ascaridole, deoxygenation, 322
Asymmetric induction, in olefin
  synthesis via phosphonates, 174
Azaberbinone, from nitropapaverine,
  290
Azanorcaradienes, possible
  intermediates in deoxygenation
  of $ArNO_2$, 286
3H-Azepines, from nitrobenzene
  derivatives, 287
Azepinobenzimidazoles, 288
Azepinocarbazoles, 272
Azepinoindoles, 288
Azepinyl phosphonates, 288
Azetidines, from 3-aminopropan-1-ols,
  440
Azides, from alcohols, 334
  in 1,3-dipolar cycloadditions of
  P=C, 232

in phosphorane synthesis, 13
  reaction with
  2-oxoalkylidenephosphoranes, 232
Azidocarbohydrates, from
  carbohydrates, 456
Azidoiodoalkanes, reaction with
  phosphines or phosphites, 249
α-Azido-α-H-ketones, in 1,3-oxazole
  synthesis, 243
  reaction with Ph_3P, 241
5-Azido-2-oxovaleric methylamide,
  in 1,2-dehydroproline
  methylamide synthesis, 248
Aziridines, cleavage by ylides, 216
  reaction with phosphonium salts,
  268
  from amines, 216
  from amino-alcohols, 411
  from 2-aminoalcohols, 249
  from 2-aminoethanols, 440
  from β-aminosulphonic acids, 410
  from epoxides, 216
  from β-halogeno-amines, 410
  from iminophosphoranes, 249
  from nitrones, 190
  from $1,3,2\lambda^5$-oxazaphospholidines,
  495
  from phosphoranes, 500
  into pyrrolines, 216
Aziridinylphosphonium salts, 268
Azirines, from alkylidenephosphoranes,
  236
  from nitro compounds, 283
  from oxazaphospholines, 494
Azosulphides, desulphurization, 359
p-Azotoluene, from
  N-p-tolylhydroxylamine, 411
Azoxy compounds, deoxygenation, 300

Beckmann rearrangement, catalysis by
  Ph_3PBr_2, 441
  of O-phosphorylated ketoximes, 406
Benzaldehyde, deuteration, 441
  into stilbene, 310
  into stilbene oxide, 308
Benzene di-epoxide, 322
Benzene monoxide, into trans-benzene
  trioxide, 337
trans-Benzene trioxide, from benzene
  monoxide, 337

Subject Index

Benzenesulphinic acid, deoxygenation, 303
Benzil, into diphenylacetylene, 315
  adduct, thermolysis, 480
Benzils, from benzoins, 440
Benzimidazoles, 268, 275, 277
  from phosphonates, 184
7H-Benzocarbazole, 272
Benzocyclobutene epoxide, attempted phthalaldehyde cyclization, 310
Benzofurazans, 299
Benzo[c]naphthyridines, from 2-nitro-2'-carboxyalkylbiaryls, 284
Benzonitrile, from benzamide or benzaldoxime, 441
  from phenylnitromethane, 445
Benzophenone, in substituted ethylene synthesis, 310
Benzotetrazapentalenes, 281
Benzothiazoles, 254, 268
Benzothiazolo[3,2-b]indazole, 278
10H-[1]Benzothieno[3,2-b]indole, 272
1H-[1]Benzothieno[3,2-b]pyrroles and [2,3-b] isomers, 275
Benzo[b]thiophen-3-carbonitrile, from 3-nitrobenzothiophen-2-anil, 279
1,2,4-Benzotriazines, from iminophosphoranes, 242
Benzotriazolo[2,1-a]naphtho[1,8-b,e]-triazine and [1,2-a] isomer, 281
Benzoxazoles, 254, 268
1,2-Benzoxazoles, 281
1,3-Benzoxazoles, from o-azidophenyl benzoate, 285
  from o-nitrophenyl benzoate, 285
Benzoxazolo[3,2-b]indazole, 278
Benzoyl cyanide, reaction with triethyl phosphite, 308
Benzoyl halides, from benzoic acid, 441
Benzoylation, selective, 329
3-Benzylidene-9-oxabicyclo[3.3.1]-nonane, 225
Benzyne, from oxazaphospholidines, 496
  reaction with ylides, 219
Bestmann syntheses, 223
Betaine, decomposition, 170
  from reaction of diethyl azodicarboxylate with Ph_3P, 327
  intermediate, in deoxygenation of epoxides, 316
  reversibility in olefin synthesis, 169, 170
Betaines, diastereoisomeric, 170
Bibenzyl, from trioctyl phosphite and dibenzyl sulphide, 361
Bicyclo[4.1.0]heptenes, via intramolecular Wittig reaction, 75
Bifluorenylidene, from 9,9-dibromo-fluorene, 455
  from fluorenone, 310
Bifurandiones, from substituted maleic anhydrides, 313
2,2'-Bi-2H-indazolyl from 2-nitrobenzaldazine, 278
Bi-indolyl, 276
9-Biphenylenephenanthrone, from fluorenone, 310
Biphilic attack, by P(III) reagents, 12
Biphthalyls, from phthalic anhydride, 311
Bis-diphenylacetylenes, from 1,2-dicarbonyl compounds, 316
Bis(hydroxyethyl) disulphide, reaction with Ph_3P, 373
Bis(iminophosphoranes), reaction with 1,3-dicarbonyl compounds, 245
Bisphosphonates, in cyclic diene preparation, 187
Bisphosphoranediylhydrazine, formation, 413
2,6-Bis(tosyloxymethyl)tetrahydropyran, 225
N,N'-Bistriphenylphosphoniohydrazine chlorides, from hydrazines, 412
Bis-Wittig reactions, in heterocyclic synthesis, 19
Bombykol, 110
N-Bromoacetamide, with Ph_3P in alcohol halogenation, 455
α-Bromoacetonitrile, reaction with Ph_3P, 450
ω-Bromoacyl bromides, from alicyclic lactones, 443
Bromoalkynes, dehalogenation in alcohol, 455

β-Bromocinnamaldehyde, from
 phenylacetylene, 442
α-Bromo-α-cyanoacetamide, reaction
 with Ph_3P, 450
2-Bromocyclohexanone, formation of
 ketophosphonium salt, 447
α-Bromodiphenylacetyl bromide,
 debromination by Ph_3P, 459
α-Bromoglycosides,
 dehydrobromination, 459
α-Bromoisobutyrophenone, formation
 of ketophosphonium salt, 447
α-Bromoketones, into alkenes via
 Perkow reaction, 5
α-Bromooximes, reactions with
 phosphines, 452
o,p-Bromophenols, reaction with
 Ph_3P, 461
Bufadienolides, via phosphonate
 carbanions, 181
But-2-ene epoxide, deoxygenation, 317
t-Butyl benzoate, from t-butyl
 perbenzoate, 324
t-Butyl perbenzoate, reduction to
 t-butyl benzoate, 324
Butyltriphenoxyphosphonium
 bromide, promotion of synthesis
 of amides, esters, ureas, and
 thioureas, 437
Butyne-1,4-diol diacetate, into
 1,4-diacetoxybutane-1,2-dione, 343

α-Camlolenic acid, 96
Carbamic esters, N-phosphonium
 salts, 408
Carbamoyl halides, deprotonation, 407
Carbanion formation, conditions for,
 190
Carbanion stabilization, by P(III)
 reagents, 310
 by phosphonate group, 156
Carbazoles, from 2-nitrobiaryls, 272
 via photolysis of phosphoranes, 496
Carbene dimer, from dioxaphospholen
 thermolysis, 480
Carbenes, from photocycloelimination,
 487
Carbenoid mesomer of nitrile ylides,
 493
Carbodiimides, synthesis, 424

from alkyl halides, 219
from N,N'-disubstituted ureas, 442
from N,N'-disubstituted ureas and
 thioureas, 409
from thioureas, 378
Carbodiphosphoranes, from
 (chloromethyl)triphenylphos-
 phonium chloride, 394
 via tertiary phosphine–CCl_4, 397
Carbohydrates, benzoylation, 329
 chloro derivatives, 400
 via phosphonate olefination, 184
Carbolines, 272
β-Carbolines, from
 2-nitro-2'-carboxyalkylbiaryls, 284
Carbon disulphide, reaction with
 phosphines, 379
Carbon tetrachloride, reaction with
 P(III) reagents, 387
 reaction with
 tris(dimethylamino)phosphine, 456
Carbon–carbon bond, via
 phosphoranes, 475
Carbonates, from peroxycarbonates,
 326
Carbonyl compounds, deoxygenation,
 304
 deoxygenation to indazolotriazoles,
 281
 reaction with Ph_3P–CCl_4, 403
 reaction with P(III) reagents, 306,
 475
 reductive condensations, 314
Carbonyl groups, reaction with
 phosphonate carbanions, 164
Carboxamides, dehydration by
 Ph_3P–CCl_4, 408
Carboximides, alkylation by alcohols,
 328
Carboxylic acid group, activation by
 phosphoric anhydrides, 511
Carboxylic acids, amide formation via
 Ph_3P–CCl_4, 415
 anhydrides from, 405
 from chloroformate esters, 214
 into acyl chlorides, 404
Carboxylic acid syntheses via Wittig
 reaction: branched chain, 91
 ω-hydroxy unsaturated, 96
 monoenic, 90–93

polyacetylenic, 97–99
polyenic, 93–96
Carboxylic esters, transesterification, 334
Cardenolides, via phosphonate carbanions, 183
Cardiotonics, 180
β-Carotene, via oxidative coupling of ylides, 227
via PO-stabilized carbanions, 177
Carotenoids and related polyenes, isomerization, 122–126
via Wittig reaction, 37, 120–127
Catalpa acid, 95
Cation effects, in olefination via carbanions, 165
Cecropia hormones, via phosphonate carbanions, 178
α-Cyclopropylidene acetaldehydes, from peroxyhemiacetals, 320
Cyclotri- and cyclotetra-λ^5-phosphazenes from chlorophosphines, 413
Cyclotri-λ^5-phosphazenes, from diethyl phosphoramidate or diphenylphosphinamide, 411
Cysteine, aqueous solution stabilization by Bu_3P, 372
L-Cysteine, into L-lanthionine, 370
S-methylation by $(MeO)_3PO$, 10
Cytidine and cytosine, modification via reaction with β-acylvinyltriphenylphosphonium bromide, 263

DEADC, 327
Dehalogenation, 459
retention of configuration, 453
of bromoalkynes, 455
of halogenophenols, 461
of trihalogenoacetic acids, 449
Dehydration, of amides by Ph_3P–CCl_4, 408
of N-substituted formamides, 409
of phosphoramidates, 411
of tolylhydroxylamine, 411
via dichlorophosphoranes, 416
Dehydrochlorination, via ylides, 397
Deoxygenation, catalysis by vanadium ions, 318

of nitro compounds, reagents for, 271
of nitro compounds, in heterocycle synthesis, 269
of N-oxides by Ph_3P/photolysis, 297
3-Deoxy-vitamin D_2, via PO-stabilized carbanions, 177
Desulphurization, 351
of sulphur chelates, 368
of thiirans, 351
via Ph_3P–Bu^tOK, 362
Detergents, 423
Deuteration, of benzaldehyde, 441
1,4-Diacetoxybutane-1,2-dione, from butyne-1,4-diol diacetate, 343
1,4-Diacylethylenes, from acyl ylides, 342
from phosphite rearrangement, 303
Dialkyl alkylphosphonates, synthesis and reaction, 4, 8
Dialkyl azodicarboxylate, see DEADC, 327
Dialkyl carbonate, from pyrocarbonate, 326
Dialkyl peroxides, deoxygenation, 320
Dialkyl sulphides, from thiolsulphinates, 304
Dialkyl sulphoxides, deoxygenation, 302, 303
Dialkylphenylphosphines, reaction with CCl_4, 389
Diallyl disulphides, desulphurization, 369
Diaroyl peroxides, deoxygenation, 323
Diaryl sulphides, from diaryl sulphoxides, 301
from thiolsulphinates, 304
Diaryl sulphoxides, deoxygenation, 303
Diarylacetylenes, from α,α-chlorodibenzyl sulphides, 362
3,4-Diaryl-4,5-dihydro-2-oxo-1,2,5-oxazaphosph(v)oles, from 1,2-diaryl-1-nitroethanes, 289
Diarylurea, from O-(N-arylcarbamoyl)hydroxamate rearrangement, 332
Diaza- or triaza-diphosphorines from diphosphines, 413
Diazo compounds, in pyrazoline synthesis, 254

Dibenzo[b,f]oxepins, from
  alkylidenephosphoranes, 225
Dibenzo[b,f]-1,3,4,6-tetraaza-
  pentalenes, 281
Dibenzoyl peroxide, with triethyl
  phosphite, 324
Dibenzyl sulphoxides, deoxygenation,
  303
Dibenzyl trisulphide, desulphurization,
  370
Dibenzylphosphoryl aminoacids,
  alkylation, 518
1,2-Dibromoethanes, elimination of
  bromine, 457
9,9-Dibromofluorene, formation of
  bifluorenylidene, 455
Dibromoketene, from trimethylsilyl
  tribromoacetate, 450
α,α-Dibromoketones from
  dioxaphospholens, 475
Dibromotriphenylphosphorane,
  reaction with 2-aminoalcohols, 249
Dibutylmalonic anhydride, from
  dibutylmalonyl peroxide, 325
Dibutylmalonyl peroxide, reaction
  with Ph_3P, 325
Di-t-butyl peroxide, reaction with
  Ph_3P, 318
1,2-Dicarbonyl compounds,
  alkylation, 333
  alkylation by alcohols, 328
  biphilic attack by P(III) reagents, 12
  deoxygenation, 315
1,3-Dicarbonyl compounds, reaction
  with bis(iminophosphoranes), 245
Dicarboxylic acid anhydrides,
  reductive dimerization, 311
1,1-Dichlorides, formation, 403
Dichloroalkanes, from epoxides, 403
1,1-Dichloroalkenes, from
  α-(trichloromethyl)-alcohols, 403
  from enolizable ketones, 403
  via alkoxyphosphonium salts, 402
cis-1,2-Dichlorocycloalkanes, from
  cycloalkene oxides, 403
Dichloromethylenephosphorane, as
  HCl acceptor, 392
  formation, 390
Dichloromethyltriphenylphosphonium
  chloride, formation, 394

Dichlorophosphorane resin,
  formation and regeneration, 424
Dichlorophosphoranes,
  as dehydrating agents, 409, 416
  covalent or ionic, 390
  formation, 403
  from (dichloromethyl)phosphonium
    chloride, 392
  from trialkyl and dialkylphenyl
    phosphines and CCl_4, 389
  hydrolysis, 396
  in chlorinations, 416
  in P–N linkage, 416
  in phosphorylations, 393
  intermediates in chloroformamidine
    formation, 406
  phosphorylating agent with
    nucleophiles, 392
α,α-Dichlorotoluene, from
  benzaldehyde, 440
Dichlorotriphenylphosphorane, from
  (trichloromethyl)phosphonium
  chloride, 388
N,N-Dichlorourethane, with $(PhO)_3P$
  in alcohol halogenation, 455
Dictamnin, 251
Dicyano disulphide, reaction with
  2-oxoalkylidenephosphoranes,
  239
2,2'-Dicyanoformylbiphenyl, into
  9,10-dicyanophenanthrene-
  9,10-oxide, 308
9,10-Dicyanophenanthrene-9,10-oxide,
  from 2,2'-dicyanoformylbiphenyl,
  308
1,2-Dicyanostilbene oxide, from
  benzoyl cyanide, 308
Dicyanosulphonium ylides, reaction
  with Ph_3P, 380
cis-Dideuterioethylene, via a
  1,1-dihydro-phosphiran, 498
Dienes, via PO-stabilized carbanions,
  177
1,5-Dienes, synthesis, 139
  from allyl bromide, 208
Diesters, from alcohols, 405
1,4a-Diethoxycarbonyl-4aH-
  phenothiazine, from 2,6-
  dicarboxyethylphenyl
  2-nitrophenyl sulphide, 287

Subject Index 587

1,2-Diethoxydioxetans, from
 1,2-diethoxyethylene, 334
1,2-Diethoxyethylene, into cis- and
 trans-diethoxydioxetans, 340
Diethyl 3H-azepin-7-yl phosphonates,
 via triethyl phosphite, 288
Diethyl 2-azidosuccinate, from
 diethyl malate, 334
Diethyl fumarate, from diethyl
 malate, 331
Diethyl malate, into diethyl
 2-azidosuccinate, 334
 into diethyl fumarate, 331
Diethyl malonate, alkylation, 333, 334
Diethyl methylphosphonite, in
 deoxygenation of nitro
 compounds, 287
Diethyl peroxide, order of reactivity
 with phosphites, 321
Diethyl phosphite, in HCl removal,
 524
O,O-Diethyl phosphoric anhydrides,
 from acylamino acids and enol
 phosphates, 519
Diethyl phosphorochloridate, in
 carboxylic-phosphorus anhydride
 formation, 523
Diethyl phosphorocyanidate, in
 peptide synthesis, 520
Diethyl phosphoro-selenoate and
 -telluroate, in deoxygenation of
 sulphoxides and epoxides, 190
Difarnesyl disulphide, into squalane,
 369
Difluorenylidene from
 9,9-dibromofluorene, 455
2,2′-Diformylbiphenyl, in formation
 of phenanthrene-9,10-oxide, 309
Digitoxigenin analogues, 180
α,α'-Dihalogenosulphones,
 1,3-eliminations, 453
α,α'-Dihalogenosulphoxide,
 1,3-eliminations, 453
Dihydroazulenes, from
 9-octalin-2-ones, 344
2,3-Dihydro-1-benzofurans, from
 alkylidenephosphoranes, 227
2,3-Dihydro-1-benzoxepins, from
 triphenylvinylphosphonium
 bromide, 252

2,2-Dihydro-1,4,2-dioxaphospholan,
 from P(III) compounds with
 hexafluoroacetone–
 pentafluorobenzaldehyde, 488
2,2-Dihydro-1,3,2-dioxaphospholen,
 photolysis, 480
2,2-Dihydro-1,3,2-dioxaphospholens,
 alcoholysis, 471
 C-acylation, 473
 fragmentation, 479
 from dioxaphospholens and ketene,
 476
 hydrogenation, 469
 O-acylation, 472
 reaction with acid chlorides and
 anhydrides, 472
 reaction with activated C=C, 475
 reaction with heterocumulenes, 476
 reaction with hydrogen halides, 472
 reaction with oxygen and ozone, 469
Dihydrofurans, 268
2,3-Dihydrofurans, from carboxylates
 and 1-ethoxycarbonyl-1-
 cyclopropyltriphenyl-
 phosphonium bromide, 251
2,5-Dihydrofurans, from acyloins, 250
Dihydroindoles, 276
3,4-Dihydro-5-methoxy[1,3]oxazino-
 [3,4-a]indol-1-one, 283
5, 11-Dihydro-4-methyldibenzo[b,e]-
 [1,4]thiazepine, from
 2,6-dimethylphenyl
 2-nitrophenyl sulphide, 287
2,2-Dihydro-1,2-oxaphosphetans, from
 2,2-dihydro-1,3,2-dioxaphos-
 pholans, 487
2,2-Dihydro-1,2-oxaphospholan,
 ylide behaviour, 490
3,3-Dihydro-1,3-oxaphospholan, from
 ketene, 491
2,2-Dihydro-1,2-oxaphospholen, into
 2,2-dihydro-1,2-thiaphospholen,
 489
2,2-Dihydro-1,2-oxaphospholens,
 thermolysis, 489
2,2-Dihydro-1,3,2-oxazaphospholans,
 photolysis, 496
 thermolysis, 495
2,2-Dihydro-1,4,2-oxazaphospholens,
 thermolysis and photolysis, 492

5,5-Dihydro-1,2,5-oxazaphospholens, thermolysis, 494
Dihydrophenazines, 287
1,1-Dihydro-phosphiran, in ethylene formation, 498
1,2-Dihydroproline methylamide, 248
Dihydropyrans, 268
Dihydropyrazolo[3,4-b]quinolin-4-yl phosphonates, from pyrazolin-5-ones, 283
2,3-Dihydropyridazines, from triphenylvinylphosphonium bromide, 254
5,6-Dihydro-2,3,9,10-tetramethoxy-benzimidazo[2,1-a]isoquinoline, from 1,2,3,4-tetrahydro-1-(2-nitrophenethyl)isoquinolines, 292
5,6-Dihydro-2,3,9,10-tetramethyl-benzo[a]carbazole, from 6-nitrolaudanosine, 290
2,2-Dihydro-1,2-thiaphospholen intermediate, 489
Dihydrothiophens, 268
2,5-Dihydrothiophens, from α-mercapto-ketones and -aldehydes, 251
2,3-Dihydroxy-1,4-diketones, from dioxaphospholans, 482
α,β-Dihydroxyketones, from dioxaphospholans, 482
Diisopropyl peroxydicarbonates, deoxygenation, 327
α,β-Diketones, from acetylenes, 345
β-Diketones, from dioxaphospholans, 483
 into β-halogeno-α,β-unsaturated ketones, 441
1,2-Diketones, by oxidation of phosphoranes, 258
1,3-Diketones, via desulphurization, 354
Dimerization, reductive, 310
2,4-Dimethoxycarbonylbut-2-ene, from methyl pyruvate, 314
1,2-Dimethoxycarbonylstilbene oxides, from arylglyoxylic acids, 308
Dimethyl azodicarboxylate, reaction with Ph_3P, 327

Dimethyl sulphoxide, reaction with Ph_3P, 302
o-Dimethylaminomethylphenols, into o-quinone methide, 238
2,3-Dimethylindole, from 2-nitro-β,β-dimethylstyrene, 276
8,10-Dimethylphenanthridine, from 2-nitro-2',4',6'-trimethylbiphenyl, 278
Dimroth rearrangement, in imidazopyridines, 262
Diol monophosphates, from dioxaphospholans, 482
1,2-Diols, from dioxaphospholans, 482
1,2-Dioxan, reaction with Bu_3P, 321
1,4-Dioxan, from phosphoranes, 500
Dioxaphospholan-type systems, into acetylenes, 315
Dioxaphospholans, decomposition, 306
 from o-quinones, 311
 intermediate in tetramethylethylene oxide synthesis, 317
 pinacol–pinacolone rearrangement, 311
Dioxaphosphorinan, in peroxylactone deoxygenation, 325
1,2-Dioxetans, deoxygenation, 317
 from alkenes by 1,2-cycloadditions, 340
Di-N-oxides, selective mono-deoxygenation, 297
4,8-Dioxoadamantane-2-carbonitrile, preparation, 409
Diphenyl phosphoramidates, in peptide synthesis, 514
Diphenyl phosphorazidate, in peptide synthesis, 520
Diphenyl phosphorochloridate, in anhydride formation, 520
Diphenylacetylene, from benzil, 315
 from diphenylketene, 316
1,2-Diphenyl-2-chloroethan-1-one, into diphenylacetylene, 316
2,3-Diphenylcoumaran derivatives, 228
Diphenylcyclopropenone, from α,α,α'-tribromodibenzyl ketone, 453
Diphenylketene, deoxygenation, 480
 from α-bromodiphenylacetyl bromide, 459

Subject Index 589

reductive condensation to diphenylacetylene, 316
Diphenylnitrone, deoxygenation, 300
Diphenylphosphine oxide anion, deoxygenating agent, 308
 in stilbene formation, 310
Diphenylphosphinyl chloride, in preparation of diphenylphosphinyl amino acids, 515
Diphenylphosphinyl derivatives, basic cleavage, 518
 deprotection, 512
 of amino acids, 515, 516
 of N-methylamino acids, 516
Diphenylphosphinyl-N-methyl amino acids, from diphenylphosphinyl amino acids, 518
Diphenylphosphoric anhydride, in amino acid activation, 513
Diphenylthiophosphinyl derivatives, of amino acids, 516
 protection by, in peptide synthesis, 517·
1,3-Dipolar addition, of ylides, 12
1,3-Dipolar species, dimerization, 493
1,3-Dipoles, in heterocyclic synthesis, 231
 reaction with 2-oxoalkylidene-triphenylphosphoranes, 225
Di-2-pyridyl disulphide, in nucleotide chemistry, 375
 in phosphorylation, 375
1,2-Disubstituted ethylene oxides, from aldehydes, 307
Disulphides, desulphurization, 365
 oxidation by $(PhO)_3P-O_3$, 341
 reduction to thiols, 371
 with $Ph_3P-HgCl_2$ in peptide synthesis, 529
1,2-Dithian, from tetrahydrothiophen sulphoxide, 11
1,2,4-Dithiazoles, into heterocumulene, 378
Dithiocarbonate, reductive condensation, 314
Dithioles, from acetylene dicarboxylic acids, 380
Double activation process, in hydroxy acid into lactone, 373

β-Elemenone, 380
α-Eleostearic acid, 95
Elimination of water, in carbohydrate benzoylation, 329
Emmons PO-activated olefin synthesis, 156
Enamines, from α-aminoalkylphosphonates, 176
β-Enaminocarbonyl compounds, reaction with β-acylvinyl-phosphonium salts, 260
Endocyclic peroxides, into exocyclic epoxide, 322
Endoperoxides, from 1,3-dienes, 337
 from furans, 340
Endoxides, cleavage by Ph_3PBr_2 deoxygenation, 318
Enol ethers, by alkylation via alkanols, 332
Enol phosphates, via Perkow reaction, 5, 159
 reduction to alkenes, 5
Enolizable ketones, alkenes from, 164
Episulphides, from sulphenyl chlorides, 446
 reaction with P(III) compounds, 351
Episulphoxides, from α,α'-dihalogenosulphoxides, 453
 sigmatropic rearrangement, 364
Epoxides, cis to trans isomerization via phosphoranes, 486
 deoxygenation, 190, 316
 from aroyl cyanides, 308
 from aryl aldehydes, 306
 from dialdehyde cyclization, 310
 from 1,2-diols, 486
 from dioxaphospholans, 483, 486
 from endoperoxide, 322
 from α-glycols, 401
 from phosphoranes, 486
 from trihalogenoacetic esters and carbonyl compounds, 449
 into aziridines, 216
 into dichloroalkanes, 403
 reaction with dichlorophosphoranes, 404
 reaction with P-stabilized carbanions, 214
 reaction with phosphonate carbanions, 216

reaction with ylides, 215
ring opening by phosphine dihalides, 443
stereospecific deoxygenation, 437
Ergosta-7,22,24(28)-trien-3β-ol, 141
Erucic acid, 90
Ester phosphonates, in olefin formation, 156
Ester-stabilized ylides, C-acylation, 212
Esters, from acids, 434
  from alcohols, 328
  from alkyl halides, 208
  hydrolysis using Ph_3P–HBr, 446
  into thioesters, 11
  phosphonium salts in promotion of synthesis, 437
  phosphorylation via Ph_3P and di-2-pyridyl disulphide, 375
Ether cleavage, 443, 446
Ethers, from alcohols, 332, 401
  from dialkyl peroxides, 320, 321
  from phenols, 332
C-Ethoxycarbonyl-C-imidoylketene imines, from (amino-alkylene) malonamic esters, 410
Ethyl acetoacetate, alkylation, 333
Ethyl cyanoacetate, alkylation, 333
Ethyl dimethylglycidate, deoxygenation, 317
Ethyl diphenylphosphinite, in deoxygenation of nitro compounds, 287
Ethyl 3-phenylthiazolo[3,4-b]indazole-1-carboxylate, 278
Ethyl phosphorodichloridate, in peptide synthesis, 526
Ethylene phosphonothioites, in thiiran formation, 492
1-Ethyl-4-phospha-3,5,8-trioxabicyclo[2,2,0]octane, ozonide, 491
4-Ethyl-2,6,7-trioxa-1-phosphabicyclo[2,2,2]octane, ozone adduct, 342
Exocyclic peroxide, from endocyclic peroxide, 322
Explosion hazard, with tris(hydroxymethyl)phosphine, 372

Farnesol, 134
  via phosphonate carbanions, 179
Fatty acids, see Carboxylic acids

Flame retardants, 423
Fluorenone, into bifluorenylidene, 310
Fluorination, via Ph_3PF_2, 446
Formamides, formation of isocyanides, 335
2-(2-Formylphenyl)naphthalene-3-carboxaldehyde, intramolecular condensation, 309
Fulvenes, from thio- and selenocarbonates, 357
Furans, fission on reaction with nitrenes, 291
  from α-hydroxyketones, 251
  from triphenylvinylphosphonium bromide, 250
Furazan, formation from N-oxide, 298
  ring opening, 298
Furazans, deoxygenation, 298, 299
Furo[3,2-c]carbazoles, from difuryl-o-nitrotoluenes, 290
4H-Furo[3,2-b]indole, 272
Furoxans, deoxygenation, 298

Gabriel synthesis, alternative, 377
Gastrin, C-terminal tetrapeptide of, 516
Geraniol, chloro derivative, 400
  via phosphonate carbanions, 179
Germacrene, reaction with $(PhO)_3P$–O_3, 339
γ-Globulin human, reduction, 372
Glutathione, aqueous solution, stabilization by Bu_3P, 372
Glycerides, chloro derivatives, 400
α-Glycols, condensation, 401
Glyoxylates, reaction with P(III) reagents, 314
Gossyplure, 112
Guanidines and biguanides, into phosphoramidates, 422
Guanine, 1- and 3-methyl, and guanosine, modification by β-acylvinyltriphenylphosphonium bromide, 263

Halogenation, of carbohydrates, nucleosides, and steroids, 435–436
α-Halogenoaldehydes, from phosphoranes, 472
α-Halogenocarbohydrates, from carbohydrates, 435

Subject Index

α-Halogenocarbonyl compounds, reaction with phosphines, 446
α-Halogeno-α-cyanoesters, reactions with P(III) reagents, 450
3-Halogenocyclohex-2-enone, from cyclohexane-1,3-dione, 441
α-Halogeno-α-hydroxyketones, from dioxaphospholens, 474
β-Halogeno-α-ketoesters, from trihalogenoacetic esters and carbonyl compounds, 449
α-Halogenoketones, from phosphoranes, 472
  in β-acylacrylic ester synthesis, 211
Halogenonucleosides, from nucleosides, 435, 439
Halogenopyridines, from hydroxypyridines, 440
Halogenoquinolines, from hydroxyquinolines, 440
Halogenosteroids, from hydroxysteroids, 436, 456
N-Halogenosuccinimides, reaction with Ph_3P, 455
Harman, 272
Heteroallenes, from phosphoramidate anions, 157, 188
Heterocumulenes, from thiazoles, 378
Heterocyclic sulphoxides, deoxygenation, 303
Heterocyclic synthesis, 223, 269
  via bis-Wittig reactions, 226
  via extrusion of methylenephosphorane, 268
N-Heterocyclics, from phosphonates, 184
Heterocyclization, of aromatic nitro compounds by $(RO)_3P$, 269
Hexachloroacetone, in chlorination with Ph_3P, 400
Hexachloroantimonates, of bis-phosphonium cation, 389
  of thiophosphonium salts, 401
Hexachloroethane, reactions with P(III) reagents, 387
Hexachlorotriphosphatriazene, in amide formation, 521
Hexacoordinate P(VI) species, 471, 476
Hexacoordinate phosphorus compounds, from phosphoranes, 502
trans, trans-Hexa-2,4-diene, from 1,1-diethoxy-1,2,5-trimethyl-1,1-dihydrophospholen-3, 498
N-Hexafluoroisopropylamides, from oxazaphospholines, 492
Hexafluorophosphate isolation of phosphonium ions, 401
5,6,7,8,9,10-Hexahydrocyclohepta[b]-indole, from cyclohexylidene-(2-nitrophenyl)methane, 276
Hexamethylphosphorous triamide, in deoxygenation of nitro compounds, 270, 271
Hexamethylphosphorus triamide, amination by, 11
  aminoheterocyclization by, 11
  carcinogenicity, 11
  in preparation of acyloxyphosphonium carboxylates, 531
  replacement by P_2O_5–Me_2NH, 11
Hexaphenylcarbodiphosphorane, isolation of, 397
  reaction with acyl chlorides, 214
Hofmann elimination, in phosphonium salts, 260
Horner PO-activated olefin synthesis, 156
Horner–Wittig, PO-activated olefin synthesis, 155
Hydantoins, from aryl isocyanates, 478
  from dioxaphospholens, 476
Hydrazines, alkylation, 444
  phosphorylation, 412
Hydrazones, in preparation of α-hydroxydiazo compounds, 319
Hydrazonyl halides, via Ph_3P–CBr_4, 393
Hydrobromic acid, in reactions with Ph_3P, 446
Hydrodisulphides, desulphurization, 371
Hydrogen chloride, removal via ylides, 397
Hydrogenation, of conjugated enynes, 110, 128
Hydroperoxides, deoxygenation, 318
  from alkenes, 338

Hydroxamic acids, rearrangement, 332
Hydroxy group, protection, 338
N-Hydroxy groups, alkylation, 331
α-Hydroxyacid esters, into α-amino acid esters, 331
α-Hydroxyaldehyde derivatives, 472
α-Hydroxyamides from 2-oxazolin-4-ones, 479
1-Hydroxybenzotriazole, in peptide coupling, 532
β-Hydroxycarboxylic acids, decarboxylative dehydration, 335
α-Hydroxydiazo compounds, from hydrazones, 319
α-Hydroxy-β-diketones, from dioxaphospholans, 482
phosphate esters, from dioxaphospholens, 473
N-Hydroxyimides, into N-alkoxyphthalimides, 331
α-Hydroxyketones, derivatives, 472
phosphate formation, 470
Hydroxylation of alkanes, 319
1-Hydroxymethyl-4,5-dimethoxy-7H-azirino[1,2-a]indole-7a-carboxylic acid, 283
4-Hydroxyoxazoles, from aroyl isocyanates, 478
β-Hydroxyphosphine oxides and phosphonates, isolation, 164
Hydroxyphosphorane–phosphate ester equilibrium, 470
17α-Hydroxypregnan-20-ones, 319
α-Hydroxythioamides, from 2-thiazolin-4-ones, 479

Imidazole, reaction with PCl_3 and Z-amino acids, 526
reaction with triphenyl phosphite, 527
Imidazoles, from β-acylvinylphosphonium salts, 260, 261
phosphorylation via vinylphosphonium salt, 450
Imidazolides, from imidazoles, 527
Imidazopyridines, Dimroth rearrangement in, 262
Imidazoquinolines, 277
Imide halides, via intermolecular elimination of water, 424
Imides, from N-hydroxyimides, 331
Imidoyl chlorides, from dialiphatic ketoximes, 407
hydrolysis sensitivity, 407
Imidoyl halides, from N-monosubstituted carboxamides, 406
Imines, from aldehydes, 157
via phosphoramidates, 188
via phosphoramidate anions, 157
Imino ethers, from imidoyl halides, 406
Iminophosphoranes, 219
acylation, 212
in primary into secondary amine conversion, 444
photochemical cyclization, 248
reactions, 9, 208
synthesis, 9, 412
Iminophosphoranes in synthesis:
aziridines, 216, 249
1,2,4-benzotriazines, 242
heterocyclics, 223, 241, 282
nigrifactin, 248
oxadiazoles, 244
1,3-oxazoles, 243
phthalazines, 246
pyrazines, 241
pyrazoles, 245
pyrazoles, pyridazines, and quinolines, 247
quinolines, 248
1,2,4,5-tetrazines, 242
tetrazoles, 241
1,3-thiazoles and 1,3-selenazoles, 246
4-Iminopyrans, from isocyanates, 238
from 2-oxoalkylidenephosphoranes, 238
Iminotriphenylphosphoranes, from aminophosphonium chlorides, 413
Indazoles, from benzylideneanilines, 271
Indazolotriazole, from 2-benzoyl-2'-nitroazobenzene cyclization, 281
Indene, ozonolysis, 343
Indole, from 2-nitrostyrene, 274
Indoles, from amino(oxy)phosphoranes, 289
from nitroso compounds, 272
via addition to conjugated $C=C$ bonds, 275

Subject Index

via insertion into olefinic and imino CH bonds, 273
Indolines, from 2-nitroalkyl benzenes, 276
Indolo[2,3-b]indole, 272, 273
Indolo[3,2-b]indole, 272
Insect metabolites, 178
Insecticides, bioresmethrin, 32
  pyrethric acid, 130
  pyrethrin esters, 127
Insertion into P–H bonds, 484
Insulin, reduction and aminoethylation, 372
Intermolecular dehydrations, via R_3P–CCl_4, 414
  using DEADC, 335
Intramolecular olefination, 183
Inversion, in halogenation of optically active alcohols, 400
Iodine, in sulphoxide deoxygenation, 302
Ion pair formation, in reaction between P(III) reagents and polyhalogenomethanes, 388
Ionization mechanism, in P(V) compound transformation, 469
Isobufadienolides, via phosphonate carbanions, 181
Isocyanates, deoxygenation, 315
  from carbamoyl chlorides, 407
  from N-substituted formamides, 442
  from thiocarbamic O-esters, 408
  in formation of cumulenes, 219
  reaction with 2-oxoalkylidenephosphoranes, 238
Isocyanides, synthesis, 424
  from formamides, 335, 409
  from isocyanates, 315
Isomerization (E-Z) of alkenes, 443
cis, trans-Isomerization of epoxides, 486
Isothiazoles, into heterocumulenes, 378
Isothiocyanates, in cumulene formation, 219
1-Isothiocyanato-2-thiocyanato-alkene, cyclization, 239
Isoxazoles, from β-ketoalkylidene-phosphoranes, 495
Izumiya test, 520, 527

Jasmolone, 127

Jasmone, 130
cis-Jasmone, from β,ε-diketophosphonate cyclization, 186
Juvenile hormone, via Wittig reaction, 37, 51, 133, 134

K region aromatic oxides, synthesis, 310
Karl Fischer reagent, alternative to, 397
Keratin (wool), reduction of S–S links, 372
Ketals, hydrolysis, 446
Ketene, reaction with phosphites, 491
Ketene dimer, from 3,3-dihydro-1,3-oxaphospholan, 491
Ketene thioacetals, from phosphonates, 175
Ketene thioketals, formation, 359
Ketenes, from aryl-(1-diazoalkyl)ketones, 238
  reaction with phosphonate anions, 164
Ketenimines, from primary acid amides, 410
  from amines, 219
  from aryl ketoximes and secondary amides, 441
  from oxazaphospholines, 494
  from thioamides, 378
  in lactam formation, 415
γ-Ketoacids, from α-halogenoacetates, 211
β-Ketoalkylidenephosphoranes, O-acylation, 212
β-Ketocarbanions, ^{31}P n.m.r. chemical shifts, 163
α-Ketoesters, reaction with P(III) reagents, 314
β-Ketoesters, from acid chlorides, 212
β-Keto-α-hydroxyamides, from carbamyldioxaphospholens, 478
Ketone transposition, 470
Ketones, from acyl chlorides, 213
  from acyl ylides, 342
  from alkenes, 343
  from alkynes, 345
  from 2,2-dihydro-1,3,2-dioxa-phospholens, 469

from ketals, 446
from β-ketoesters, 446
from lactones, 325
from nitriles, 220
from 2-pyridyl thioesters, 373
from thioesters, 213
from ylides via $(PhO)_3P-O_3$, 342
into alkenes, 5
into thioketones, 11
reaction with carbon tetrahalides, 423
synthesis of α,β-unsaturated, 63
Ketones (enolizable), reaction with Ph_3P-CCl_4 reagent, 394
β-Ketophosphonates, from α-halogenoketones, 159
β-Ketophosphonium salts, from alkylidenephosphoranes, 211
transylidation, 211
Ketosteroids, esterification complication, 329
Ketoximes, in imidoyl halide preparation, 406

β-Lactam, from N-sulphinylsulphonamide, 481
antibiotics, via phosphonate cyclization, 187
Lactams, from ketenimines and substituted acetic acids, 415
Lactone, by ring expansion of acenaphthoquinone, 311
Lactones, from hydroxy acids, 373
from phosphonate cyclization, 186
L-Lanthionine, from L-cysteine, 370
Lead tetraacetate, in oxidation of thio- and seleno-cyanates, 246
LH-RH, synthesis, 531
Ligand exchange, between phosphoranes, 500
Limonene diepoxide, into 1,2-epoxy-p-menth-8-ene, 381
Lincomycins, chloro derivatives, 400
Linoleic acid, 93
Lossen rearrangement, 332
Lubricants, 423

Macrocyclic lactones, 373
Macrolides, via Ph_3P, 373
Maleic anhydride, reaction with Ph_3P, 312

Malononitrile, alkylation, 333
α-Mercapto-amides, from 2-thiazolin-4-ones, 479
5-Mercaptocorrin zinc complex, desulphurization, 361
α-Mercaptoketones, into 2,5-dihydro-thiophens, 251
2-Mercapto-1,3-thiazoles, 239
Mercuric chloride, in synthesis of pyridine N-phosphonium salts, 527
Mesoxalates, reaction with P(III) reagents, 314
Methoxycarbonylmethylenetriphenyl-phosphorane, C-alkylation, 213
p-Methoxyphenylthionophosphine disulphide dimer, thiation by, 11
Methyl pyruvate, reaction with Bu_3P, 314
N-Methylamino acids, 518
S-Methylation, of L-cysteine, 10
2-Methyl-11H-[1]benzopyrano[3,2-b]-indol-11-one, 275
(Z)-Methylbixin, 124
20-Methylcholesterol, 139
N-Methylcorrole, from N-methylthiaphlorin, 362
Methylenephosphorane, dehydrochlorination, 397
Methylenephosphoranes, reactions with nitrile oxides, 494
Methylenetriphenylphosphorane extrusion, in heterocyclic synthesis, 268
3-[N-Methyl(indazol-2-yl)amino]-cyclohex-2-enone, 278
2-Methyl-3-phenylindole, from 2-phenyl-1-(2-nitrophenyl)-propene, 275
1-Methylthio-5-acetyl-5,10-dihydrophenazine, from N-acetyl-2-nitro-2'-methyl-thiodiphenylamine, 287
Methyltriphenoxyphosphonium iodide, in alkyl iodide synthesis, 435
Michaelis–Arbusov reaction, 4, 8, 158, 159, 160, 161, 190, 417, 435
phosphorane intermediates, 501
Michaelis–Becker reaction, 8, 160

Subject Index

Mo ions, deoxygenation catalysts, 318
Monophosphates, phosphorylation via Ph_3P and di-2-pyridyl disulphide, 375
Mukaiyama, oxidation-reduction condensations, 374
  route to peptides, 529
(S)-Muscone, 372

1-Naphthaldehyde, in substituted ethylene synthesis, 310
Naphthimidazole, 277
2-Naphthylaminotriphenylphosphonium selenocyanate and thiocyanide, oxidation with lead tetraacetate, 246
Nigrifactin, synthesis using iminophosphoranes, 248
Nitrenes, intermediates in nitro compound deoxygenation, 270
  in fission of anthranils, 290
  in fission of furans, 291
Nitrile imines, in 1,3-dipolar cycloadditions, 234
  intermediates in 1,2,4,5-tetrazine formation, 497
Nitrile oxides, deoxygenation, 300
  in 1,3-cycloadditions to P=C, 234
  reaction with methylenephosphoranes, 494
Nitrile ylides, carbene mesomer, 493
  from oxazaphospholines and thiazaphospholines, 493
  partly chlorinated, from phosphorane thermolysis, 493
Nitriles, from acyl halides, 213
  from alcohols, 401, 443
  from aldoximes, 408
  from alkyl thiocyanates, 379
  from 1-bromo-nitroalkanes, 454
  from carboxamides, 408
  from nitrile oxides, 300
  reactions with P(III) reagents, 450
  via fission of furazans, 298, 299
  via intermolecular elimination of water, 424
Nitro compound deoxygenation, 270
  relative reactivity of reagents, 271
Nitroalkenes, indoles from, 273, 289

2-Nitroazoarenes, into benzotriazoles, 280
o-Nitrobenzylideneaniline, deoxygenation by P(III) reagents, 271
Nitrobiaryls, deoxygenation, 270
Nitro-2,3-dihydrobenzimidazoles, 277
2-Nitro-β,β-dimethylstyrene, into 2,3-dimethylindole, 276
6-Nitrolaudanosine, deoxygenation, 290
Nitrones, deoxygenation, 300
  reaction with phosphonate carbanions, 190
Nitropapaverine, deoxygenation, 290
o-Nitrophenyl active ester procedure in peptide synthesis, 522
Nitrophenyl aryl ethers, deoxygenation, 288
Nitrophenyl aryl sulphides, deoxygenation, 286
p-Nitrophenyl phosphates, in peptide synthesis, 527
1-(2-Nitrophenyl)naphthalene, reduction to 7H-benzo[c]carbazole, 272
2-Nitrophenylsulphenyl esters, with Ph_3P in peptide synthesis, 529
Nitroso compounds, deoxygenation, 270
2-Nitrostilbene, into 2-phenylindole, 273
Nootkatane, via methylthioalkylphosphonates, 179
exo-Norbornyl bromide, from $endo$-norborneol, 438
Nuciferal, 364
Nucleophilic substitution, of alcohols, 327
  via alcohols, 333
Nucleosides,
  benzoylation, 330
  halogenation, 400
  phosphonylation, 330
  preparation, 331
  C-substituted, via desulphurization, 362
  N-triazolylriboside analogues, 234
  via Wittig reaction, 264

Nucleotides, from nucleosides, 330
  use of Ph_3P with di-2-pyridyl disulphide, 375
(+)-Occidentalol, via methylthioalkylphosphonates, 179
Olefins via phosphonates, asymmetric induction, 174
Oleic acid, 90
Organophosphorus reagents, order of reactivity in carbonyl compound condensation, 314
Oxadiazoles, from iminophosphoranes, 244
Oxalic esters, reaction with phosphonate anions, 164
Oxaphosphorans, hydrolysis, 470
1,3-Oxathiolan-5-ones, in alkene synthesis, 359
Δ^2-1,2,5λ^5-Oxazaphospholines, from nitrile oxides and phosphine alkylenes, 494
  intermediates in isoxazole formation, 495
  N-oxides, into phosphonates, 495
1,2-Oxazoles, from alkylidenephosphoranes, 234
1,3-Oxazoles, from iminophosphoranes, 243
  derivatives of steroids, 244
Oxazolines, from dioxaphospholans, 486
2-Oxazoline-4-thiones, from isothiocyanates, 478
4-Oxazoline-2-thiones, from sulphonyl isothiocyanates, 481
2-Oxazolin-4-ones, from aroyl isocyanates, 478
  from dioxaphospholens, 478
Oxazolone, formation in peptide synthesis, 517, 520
  phosphorus analogue, 516
Oxazolo[5,4-b]quinolines, 282
Oxepino-carbazoles, 272
Oxetan, 480
Oxidation, of ylides, 227
  selective, of alcohols, 438
Oxidation–reduction, condensations, 374
  route to peptides, 529

N-Oxides, deoxygenation using phosphines, phosphites, and phosphorus halides, 295
  deoxygenation using u.v. and Ph_3P, 297
  from di-N-oxides, 297
  inhibition of deoxygenation by HCl, 295
  photolytic deoxygenation with Ph_3P, 297
  selective deoxygenation, 298
Oximes, from oxazaphospholines, 494
Oxirans, see Epoxides
Oxo groups (tautomeric), thiation, 11
3-Oxoalk-1-enylphosphonium salts, in heterocyclic synthesis, 256
2-Oxoalkylidenephosphoranes, reaction with aryl azides, 232
  reaction with aryl 1-diazoalkyl ketones, 238
  reaction with isocyanates, 238
2-Oxoalkylidenephosphoranes in synthesis: fused pyrans, 238
  4-iminopyrans, 238
  4-(2-oxoalkylidene)pyrans, 236
  4-phenyliminopyrans, 236
  γ-pyrones, 236
  1,3-thiazoles, 239
4-(2-Oxoalkylidene)pyrans, from 2-oxoalkylidenephosphoranes, 236
2-Oxoalkylidenetriphenylphosphorane, reaction with 1,3-dipolarophile, 224
13-Oxobenzotriazolo[2,1-b]benzo-[1,2-e]triazine, 281
β-Oxoesters, from alkyl halides, 209
Oxonium salt intermediate in ligand exchange between 2,2-dihydro-1,3,2-dioxaphospholen and pentaphenoxyphosphorane, 500
2-Oxo-Δ^4-1,2,3-oxathiazolines, from N-sulphinylsulphonamides, 481
Oxygen heterocycles, deoxygenation, 317
Ozone, adducts with P(III) reagents, 335
  reaction with P(III) compounds, 343
Ozone–triphenyl phosphite adduct, 335
Ozonides, reduction with Ph_3P, 344
Ozonolysis, in steroid synthesis, 345

Subject Index

Papain, activation, 372
P–C bond formation, 484
Penicillin sulphoxides, deoxygenation, 304, 363
Pentaalkoxyphosphoranes, from phosphites and dialkyl peroxides, 321
  in alkylations, 499
Pentacoordinate phosphoranes, 11
  intermediates, 412
  in Wittig reaction, 9
Pentaphenoxyphosphorane, in heterocyclic synthesis, 500
  reaction with HCl, 501
Peptides, 418, 424, 511
  via disulphides, 511
  via Mukaiyama oxid.-red. route, 529
  via o-phenylene phosphorochloridates, 524
  via Ph_3P and aryl disulphides, 375
  via $Ph_3P–CCl_4$, 415
  via phosphinamidates, 515–517
  via N-phosphonium pyridines, 527
  via phosphoramidates, 513–514
  via phosphorazo method, 525, 526
  via sulphenamides and Ph_3P, 377
  via triphenyl phosphite, 530
  via tris(dimethylamino)phosphine–C_2Cl_6, 416
Perchlorate isolation of phosphonium ions, 401
N-Perfluoroisopropylidene-carboxamides, in Δ^4-1,4,2λ^5-oxazaphospholine synthesis, 492
Perfluorotetraphenylethane, from perfluorodiphenylmethane, 455
Perkow reaction, 5, 159, 160, 446, 454
Permutational isomerizations of phosphoranes, 469
Peroxy compounds, deoxygenation, 318
Peroxydicarbonates, deoxygenation, 326
Peroxyesters, deoxygenation, 324
Peroxyhemiacetals, into α-cyclopropylideneacetaldehydes, 320
Peroxylactones, deoxygenation, 324
Phase-transfer catalysis, in alkylation of carbanions, 161
  in PO-activated olefin synthesis, 165

Phenanthrene, ozonolysis, 343
Phenanthrene-9,10-oxide, from 2,2′-diformylbiphenyl, 309
Phenanthridines, 278
Phenazines, from N-methyl-2-nitrodiphenylamines, 277
Phenol, reaction with Ph_5P, 501
Phenols, into aryl iodides, 434
Phenothiazines, alkylation by $(RO)_3PO$, 10
  from 2-nitroaryl 4-substituted-aryl sulphides, 286
2-Phenoxy-1,3,2-benzodioxaphosph-(III)ole, in sulphoxide deoxygenation, 303
2-Phenoxyethyltriphenylphosphonium bromide, in ylide formation, 8
Phenoxythiin-10-oxide, deoxygenation, 301
Phenyl phosphorodiamidate, in amination of OH groups, 10
Phenyl phosphorodichloridate, in anhydride formation, 519
Phenyl tetraethylphosphorodiamidate, amination by, 10
  synthesis of diethylaminoquinolines, 10
Phenylacetylene, into β-bromocinnamaldehyde, 442
  oxidation to phenylglyoxal, 343
3-Phenyl-2H-azirine, reaction with triphenylvinylphosphonium bromide, 254
2-Phenylbenzimidazoles, from N-benzylidene-2-nitroanilines, 275
N-Phenyl-chlorothioformimidic esters, preparation, 408
o-Phenylene phosphorochloridate, in cyclic peptide synthesis, 524
Phenylglycidic esters, deoxygenation to cinnamic esters, 317
Phenylglyoxal, from phenylacetylene, 343
4-Phenyliminopyrans, from 2-oxoalkylidenephosphoranes, 236
2-Phenylindazole, from o-nitrobenzylideneaniline, 271
2-Phenylindole, from 2-nitrostilbene, 273

2-Phenyl-3-methyl-1,3,2-oxazaphospholan, in aryl isocyanate deoxygenation, 315
Phenylnitromethane, in benzonitrile formation, 445
2-Phenylthiazolo[5,4-*b*]indole, 275
Pheromones, via Wittig reaction, 67, 109–120
Phospha-allenylidenes, in heterocyclic synthesis, 223
Phosphacumulene ylides, from carbodiphosphoranes, 399
Phosphacumulenes, in heterocyclic synthesis, 223
1-Phospha-2,8,9-trioxaadamantane, ozone adduct, 342, 491
Phosphates, phosphorylation via vinylphosphonium salt, 450
Phosphinamides, dehydration, 411
  from phosphinic acids, 415
  in amino group protection in peptide formation, 515–517,
  solvolysis, 515
Phosphinates, 159
  from phosphinites, 418
  in olefination, 157
Phosphine dihalides, in epoxide ring opening, 443
Phosphine oxide carbanions, acylation, 212
Phosphine oxides, 159
  tertiary, via Grignard reagents, 160
Phosphines, reaction with ozone, 335
  and phosphites, reaction with sulphur, 351
Phosphinic anhydrides, in carboxyl activation, 522
Phosphinic-carboxylic anhydride, formation, 522
  in nucleophilic substitutions, 522
Phosphinochloridates, solvolysis, 522
Phosphiran, 1,1-dihydro-, intermediate, 498
Phosphites, oxidation to phosphates, 450
  reaction with ozone, 335
  reactions with hydrogen halides, 434
Phosphonamidates, in olefination, 157
  carbanions, with carbonyls, 88

Phosphonate- and phosphonyl-stabilized carbanions, 156
Phosphonate- or phosphine oxide-stabilized carbanions, reactions with carbonyl compounds, 163
Phosphonate and phosphine oxide carbanions, nucleophilic properties, 161
Phosphonate carbanions, see also Wadsworth–Emmons, Horner
  allylic, 73
  in insect metabolite synthesis, 178
  nucleophilicity, 35
  reactivity with carbonyl compounds, 23, 35
Phosphonate group, removal by LiAlH$_4$ reduction, 209
Phosphonates, 159
  from phosphites, 418
  in N-heterocycle synthesis, 184
  alkylation, 161
Phosphonium salts, 20, 91
  acidity, 23
  promoters in formation of amides, esters, ureas, and thioureas, 437
  base-catalysed decomposition, 91, 93
  Hofmann elimination of, 260
  into phosphorus ylides, 23, 47–49
  reductive cleavage, 138
Phosphonothionates, in alkene synthesis, 157
Phosphoramidates, 162
  alkylation of anion from, 518
  dehydration, 411
  from dialkyl hydrogen phosphites and amines, 422
  in peptide synthesis, 513–515
  of amino acids, 520
  reaction with carbonyl compounds, 188
Phosphoramidic esters, from phosphinic acids, 415
Phosphoranes, 12
  apicophilicity, 468
  bonding, 468
  stability, 468
  permutational isomerization, 469
  ionization mechanism in, 469

ligand exchange between, 500
C–C bond formation via, 475
intermediates in Michaelis–Arbusov reaction, 501
intermediates in reaction of nitro compounds with $(RO)_3P$, 286
intermediate in stilbene oxide formation, 308
in dicarbonate deoxygenation, 326
in synthesis, 467
in N-heterocyclic synthesis, 500
in oxa-heterocyclic synthesis, 500
fragmentation, 479
from S_N2 attack of phosphine on a peroxide, 320
from allylic ylides, 73
from aminophenols, 12
from azides, 13
from chloramines and diols, 12
from deoxygenation of nitro compounds, 288
from dienes, 12
from disulphides, 369
from 1-nitroalkenes, 289
via 1,3-dipolar addition, 12
spiro, from α-glycols, 401
Phosphorazo method, in peptide synthesis, 525, 526
Phosphoric-carboxylic anhydrides, thermal disproportionation, 519
from diethyl phosphorochloridate, 523
in peptide formation, 527, 531
in protein biosynthesis, 519
reactions with amino acids, 519
Phosphorobromidates, from aryl or aryl alkyl phosphites, 419
Phosphorochloridates, in anhydride formation, 522
from tetraethyl diphosphite, 419
solvolysis, 522
Phosphorohalidates, from halogens and trialkyl phosphites, 433
Phosphorotelluroates, in 1,2-epoxides into alkenes, 381
Phosphorus trichloride, in deoxygenation of N-oxides, 295
in peptide synthesis, 525
Phosphorus ylide classification, 22, 63
moderate, 63–89

reactive, 42–63
stable, 27–41
Phosphorus ylide reactivity, 20–23, 63
effect of P-substituents, 21, 29, 108
Phosphorus ylides, p–d π-overlap, 20
hydrolytic stability, 21
effect of carbanion substituents, 20
from phosphonium salts, 23, 47–49
intramolecular acylation, 91
abnormal Wittig reaction, 73
reaction with carbonyl compounds, see Wittig reaction
reaction with lactols, 82, 96, 99–106, 118, 126, 130
Phosphorus ylides (allylic) conformation, 70
γ-condensation with carbonyls, 73
in synthesis of 1,5-dienes, 139
stereochemistry in Wittig reaction, 18, 66–67, 126, 141
stereomutation, 68
with conjugated carbonyls, 74
Phosphorus ylides (benzylic), stereochemistry in Wittig reaction, 77–84
Phosphorus ylides (β-oxido), 55–63
alkylation, 58
stereospecific protonation; see also Wittig–Schlosser reaction, 55
with acyl halides, 58
with carbonyl compounds, 58–63, 109, 133–136
with electrophiles, 56
with nitriles, 63
Phosphorus ylides (polymer-supported), 56, 79, 119
Phosphorus ylides (propargylic), 84–86
Phosphorus ylides (γ,δ-unsaturated), isomerization, 93
Phosphorus(III) compounds, reaction with ozone, 343
Phosphorus(III) reagents, reaction with carbonyl compounds, 475
Phosphorus(VI) compounds, from phosphoranes, 502
Phosphorylation, using di-2-pyridyl disulphide, 375
using vinylphosphonium salts, 450
Photocycloelimination, [5 → 2 + 2 + 1] type, 487

Photolysis, assistance to Ph_3P
  deoxygenation of N-oxides, 297
Photolytic deoxygenation of N-oxides
  using Ph_3P, 297
Phthalazines, 246
  from iminophosphoranes, 246
Phthalic anhydride, formation of
  trans-3,3'-biphthalyl, 311
Phthalimide, condensation with
  alcohols, 330
O-Phthalimido-sugars, 332
Pinacol–pinacolone rearrangement of
  dioxaphospholans, 311
Piperazines, from
  2-N-alkylaminoethanols, 440
Piperidine, from phosphoranes, 500
P–N linking reactions, via P(III)
  reagents, 387
PO-stabilized carbanions, alkylation
  and acylation, 158
  reactions with carbonyl compounds,
  157
Polycyclic ozonides, stability, 491
Polyene synthesis via PO-stabilized
  carbanions, 177
Polyhalogenoalkanes, reaction with
  P(III) reagents, 387
Polyhalogenomethanes, reactions with
  phosphites, 417
Polyhydric alcohols, phosphonylation,
  330
Porcine motilin, 521
5α-Pregnane-3β, 20α-diol, 139
Primary acid amides, dehydration by
  Ph_3P-CCl_4, 410
Primary alcohols, selective oxidation,
  438
(±)-Progesterone, 138
Proline peptides, via phosphorazo
  method, 526
Propane-1,3-diols, alkoxy-tris-
  (dimethylamino)phosphonium
  salts, 401
Propylene oxide, deoxygenation, 317
Prop-2-ynyl-triphenylphosphonium
  bromide, in quinoline derivative
  synthesis, 253
Prostaglandins, via phosphonate
  carbanions, 179
  via Wittig reaction, 99–109

Z, E-isomerization via sulphenates,
  365
Protection of amino groups, 511
  by diphenylphosphinothioyl
  derivatives, 517
  via phosphoramidates, 517
Protein biosynthesis, via
  carboxylic-phosphoric anhydrides,
  519
Pseudorotation in pentacoordinate
  phosphoranes, 11
Pseudorutaecarpine, 276
Pteridines, from phosphonate
  carbanions, 186
Purine-thiol nucleosides,
  desulphurization, 362
Pyran derivatives, via
  2-oxoalkylidenephosphoranes,
  238
Pyrazines, from iminophosphoranes,
  241
Pyrazoles, 254
  from β-acylvinylphosphonium salts,
  257
  from iminophosphoranes, 245, 247
  from thiadiazines, 362
  from triphenylvinylphosphonium
  bromide, 254
  via alkylidenephosphoranes, 234
Pyrazolines, 254
Pyrazolo[1,2-a]benzotriazoles, 280
Pyrazolopyrazoles, 281
Pyrethric acid, 130
Pyrethrolone, 127
Pyridazines, from
  iminophosphoranes, 247
2H-Pyridazino[4,5-b]indoles, 283
Pyridine, N-phosphonium salts in
  peptide synthesis, 527
Pyridines, from
  alkylidenephosphoranes, 231
Pyrido[1,2-b]indazole, from
  2-nitrophenyl 2-pyridyl sulphide,
  279
  from α-2-nitrophenylpyridine,
  278
2-Pyridyl thioesters, from
  di-2-pyridyl disulphide, 373
Pyrimidinethiol nucleosides,
  desulphurization, 362

Pyrimido[1,2-*b*]indazole, from 2-nitrophenyl 2-pyrimidyl sulphide, 278
Pyrocarbonate, from peroxydicarbonate, 326
α-Pyrones, from alkylidenephosphoranes, 227
  from 1,3-diketones, 230
γ-Pyrones, from 2-oxoalkylidenephosphoranes, 236
Pyrophosphates, from phosphates, 333
Pyrrole-2-aldehyde, into 3*H*-pyrrolizine, 8
Pyrroles, 260
  from β-acylvinylphosphonium salts, 260
  from alkylidenephosphoranes, 234
  from triphenylvinylphosphonium bromide, 254
Pyrrolidines, asymmetric synthesis via phosphonates, 185
  from α-amino-δ-hydroxyalkanes, 411
2-*N*-Pyrrolidino-1,3-dimethyl-1,3,2-diazaphospholan, deoxygenating agent, 311
Pyrrolines, from aziridines, 216
3-Pyrrolines, from alkylidenephosphoranes, 234
3*H*-Pyrrolizine, from pyrrole-2-aldehyde, 8
Pyrrolizines, from triphenylvinylphosphonium bromide, 250
3*H*-Pyrrolizines, from 2-acylpyrroles, 250
Pyrrolo[1,2-*g*]-5-azacyclo[3,2,2]azine, 266
Pyrrolo[1,2-*a*]indole, 250
Pyrrolo[3,2-*a*]pyrimidine, 274
Pyruvates, reaction with P(III) reagents, 314
Pyrylium salts, from 4-(2-oxoalkylidene)pyrans, 238

Quasiphosphonium salt intermediates, 4
  in Arbusov reaction, 4
  in deoxygenations, 5
  in desulphurizations, 2
  in Perkow reaction, 5
  in reactions with CCl_4, 4
Quaternary phosphonium salts, formation, 4
Quinazoline derivatives, 282
Quinazolin-4-one, 268
Quinazolin-4(3*H*)-ones, into enol ethers, 332
Quinine and related alkaloids, via ylides, 209
Quinoline, diethylamino-, using $PhOP(O)(NEt_2)_2$, 11
Quinolines, 282
  from 2-aminobenzophenone and 2-aminobenzoic esters, 248
  from benzoylacetonitriles, 283
  from benzylidenecyanoacetates, 283
  from iminophosphoranes, 247, 248
  from phenyl α-phenylcinnamates, 283
  from triphenylvinylphosphonium bromide, 253
  halogeno-, formation, 440
Quinolines, thieno[3,2-*b*]-, 288
o-Quinone methide, from *o*-dimethylaminomethylphenols, 238
o-Quinones, into dioxaphospholans, 311

Racemization, in peptide synthesis, 415, 416, 517, 518, 527, 529, 530
Racemization-free synthesis of peptides, 520
Radical chain reactions, 420
Radicals, from spirophosphoranes, 485
Rearrangement, during Wittig reaction, 139
Reduction, of phosphonium salts, 138
Reductive diacylation of aldehydes, 483
Regioselectivity, in alkene formation, 18
  of $(PhO)_3P-O_3$ reagent, 339
Regiospecificity, in peptide synthesis via phosphorus reagents, 519
  in PO-activated olefination, 166
  loss on deoxygenation by PCl_3, 297
Rethrolones, via Wittig reaction, 127–133

cis-Retinoates and 7-cis via
  PO-stabilized carbanions, 177
[3 + 2]-Retrograde reaction, 480
[4 + 1]-Retrograde reaction, 480
Ribose, halogenation of OH, 400
Ring expansion of
  acenaphthoquinone, 311
Royal jelly, 173

α-Sanshool, 95
α-Santalol, 134
β-Santalol, 33, 134
Schiff bases, reaction with ylides, 219
Schweizer synthesis, 8, 250
Secondary amides, into ketenimines, 442
Selenazoles, from iminophosphoranes, 246
Selenides, from alcohols, 380
Selenocarbonates, deselenation, 357
Sesquiterpenes, via Wittig reaction, 133–136
Silyl alkyl ethers, from alkyl trimethylsilyl peroxide, 321
β-Sinensal, 136
Singlet oxygen, from trioxaphosphetan decomposition, 491
  from triphenyl phosphite–ozone, 336, 491
(±)-Sirenin, via Wittig reaction, 50
Sodium iodide, in sulphoxide deoxygenation, 302
Solid-phase chlorination reagents, 445
Solvent effects in stereochemical control, in PO-activated olefination, 169
Spiroindolene, 276
Spiroindolinones, from
  α-(o-nitrobenzoyl)butyrolactone, 276
Spiroketones, from dioxaphospholans, 486
Spirophosphoranes, as reducing agents, 485
  metallation, 484
Sporidesmin E, desulphurization, 370
Squalane, from difarnesyl disulphide, 369
Squalene, from phosphorus ylides, 19, 139

Staudinger reaction, of azides to give iminophosphoranes, 9
Stereochemistry, of PO-activated olefin synthesis, 158
  control in PO-activated olefin synthesis, 167
  of Wittig reaction, 26–39
Steric effects with carbanion reactivity, 164
Steroidal 1,3-oxazoles, 244
Steroids, via phosphonate carbanions, 180
  via Wittig reaction, 136–142
  esterification, 329
  inversion at chiral centres, 329
  reaction with phenol, 332
  regioselective benzoylation, 329
Stilbene, from benzaldehyde, 310
Stilbenes, via benzylphosphonate and phosphine oxide carbanions, 174
  from α-chlorodibenzyl sulphides, 362
Stilbene oxides, deoxygenation, 317
  from aromatic aldehydes, 306
Styrene oxide, deoxygenation, 317
α-Substituted trichloroethanols, via alkoxyphosphonium salts, 402
N-Substituted formamides, dehydration by Ph_3P–CCl_4, 409
N-Substituted phthalimides, attempted deoxygenation, 312
Substitution mechanism in P(v) compound transformation, 469
Sugars, amino-, phosphorylation, 412
  o-phthalimido-, 332
  unsaturated, via thionocarbonates, 355
Sulphates, from alcohols and dimethyl sulphate, 334
Sulphenates, desulphurization, 364
Sulphenic esters, desulphurization, 304
Sulphenyl chlorides, episulphides from, 446
  reaction with phosphites and phosphines, 376
Sulphenylthiosulphinates, reaction with phosphites, 378
Sulphides, from phosphonium salts, 401
  from sulphoxides, 301
  from thiosulphinates, 304

desulphurization, 360
oxidation via triphenyl
  phosphite–ozone, 341
Sulphones, from phosphonates, 175
  from sulphides, 336
  no deoxygenation by
    organophosphorus reagents, 301,
    303
Sulphoxides,
  acid-catalysed deoxygenation, 301
  deoxygenation, 190, 302, 363, 380,
    445
  deoxygenation using Ph_3P, iodine,
    and NaI, 302
  from phosphonates, 175
  from sulphides, 336
  reduction to sulphides, 301
Sulphur, reaction with phosphites and
  phosphines, 351
Sulphur chelates, desulphurization, 368
Sulphurization, using
  p-methoxyphenylthionophosphine
  sulphide dimer, 11

Tellurium, catalysis of epoxide
  deoxygenation, 381
Terpenes, via phosphonate
  carbanions, 178
Tervalent phosphorus reagents, order
  of reactivity in deoxygenation, 271
Tetraalkyl pyrophosphates,
  hydrolysis, 523
$\alpha,\alpha,\alpha',\alpha'$-Tetrabromosulphones, in
  cyclic sulphone formation, 453
Tetraethyl pyrophosphite, in
  couplings with imidazole, 526
Tetrahalogenomethanes, reaction with
  phosphines, 388
1,2,3,4-Tetrahydrocarbazole, 276
Tetrahydrofuran, from
  butane-1,4-diol, 500
Tetrahydroisoquinolines, from
  dihydrobenzimidazo[2,1-a]iso-
  quinolines, 292
Tetrahydropyran, from
  pentene-1,5-diol, 500
Tetrahydropyrans, from
  alkylidenephosphoranes, 225
Tetrahydropyran-4-one, from
  phosphoranes, 225

Tetrahydropyranyl ether, OH
  protection in alkyl bromide and
  nitrile synthesis, 443
$trans$-Tetrahydrothiophen-2,5-
  dicarboxylate, 370
2,3,9,10-Tetramethoxybenz[e]acridine,
  from 1,2,3,4-tetrahydro-1-
  (2-nitrophenethyl)isoquinolines,
  292
3,3,4,4-Tetramethyl-1,2-dioxetan, into
  tetramethylethylene oxide, 317
Tetramethylethylene, into its
  hydroperoxide, 338
Tetramethylethylene oxide, from
  3,3,4,4-tetramethyl-1,2-dioxetan,
  317
N,N,N',N'-Tetramethylphosphoro-
  diamidic protecting group, 6
Tetraphenylethylene, from
  benzophenone, 310
Tetraphenylphenoxyphosphorane,
  from Ph_5P and phenol, 501
Tetrasulphides, desulphurization to
  disulphides, 370
Tetrathiafulvenes, from alkynes, 379
Tetrathiin, desulphurization, 368
Tetrathioethylenes, from
  trithiocarbonates, 357
1,2,4,5-Tetrazines, from
  iminophosphoranes, 242
  from oxadiazaphospholines, 497
Tetrazoles, from
  iminophosphoranes, 241
Thiacyclohexanes, from
  alkylidenephosphoranes, 225
Thiacyclohexan-4-one, from
  phosphoranes, 225
Thiacyclononatetraenes, from
  alkylidenephosphoranes, 227
Thiadiazines, isomerization, 362
1,2,4-Thiadiazoles, into
  heterocumulenes, 378
Thiaphlorin, desulphurization, 362
Thiation, of oxo groups using
  p-methoxyphenylthionophosphine
  sulphide dimer, 11
Δ^4-1,4,2λ^5-Thiazaphospholines, in
  nitrile ylide production, 493
Thiazoles, from iminophosphoranes,
  246

1,3-Thiazoles, from
  β-acylvinylphosphonium salts, 258
  from iminophosphoranes, 246
  via 2-oxoalkylidenephosphoranes, 239
2-Thiazolin-4-ones, from dioxaphospholens, 478
Thiazolo[5,4-*b*]indoles, 275
Thieno[3,2-*c*]pyrazoles, 278
Thieno[2,3-*b*]pyridines, via $(Me_2N)_3PO$ cyclization, 11
Thieno[3,2-*b*]pyrroles and [2,3-*b*] isomers, 275
Thieno[3,2-*b*]quinolines, via deoxygenation of *o*-nitrophenyl-dithienylmethanes, 288
Thietan, desulphurization, 363
Thietan-2,4-diones, from acyl disulphides, 368
Thiiran, from bis(hydroxyethyl) disulphide, 373
Thiirans, formation 492
  desulphurization with P(III) compounds, 351
Thioamides, from amides, 11
  reaction with
    β-acylvinylphosphonium salts, 258
Thiobenzoates, from sulphenyl chlorides, 376
Thiobenzophenone, into tetraphenylethane and ethene, 380
  polymerization in presence of Bu_3P, 380
trans-3,3'-(2-Thio)biphthalyl, from 2-thiophthalic anhydride, 312
Thiocarbamic acid, *S*-esters, chlorination, 408
Thiocyanates, from alcohols, 334, 440
Thiocyanogen, formation of $Ph_3P(SCN)_2$ and RSCN from ROH, 440
Thiodehydrogliotoxin, desulphurization, 370
Thioethers, via phosphonium salts, 401
Thiofluorenone, desulphurization, 380
Thioketones, from ketones, 11
Thiols,
  chlorination via Ph_3P–CCl_4, 400
  desulphurization, 360
  from aryl sulphonic acids, 303
  from sulphenic acids, 304
  from sulphinic acids, 303
  from sulphonyl halides, 303
  into alkyl chlorides, 379
  phosphorylation via Ph_3P and di-2-pyridyl disulphide, 375
Thionocarbonates, desulphurization to alkenes, 2, 355
Thionoesters, from esters, 11
Thiophenetole, from benzenesulphenyl chloride, 376
Thiophenol, from benzenesulphinic acid, 303
Thiophens, from alkylidenephosphoranes, 227, 268
Thiophosphonates, in olefination, 157
Thiophosphonium salts, isolation as hexachloroantimonates, 401
  stability, 401
Thiophosphoranyl radicals, in thiol desulphurization, 360
2-Thiophthalic anhydride, into *trans*-3,3'-(2,2'-dithio)biphthalyl, 312
Thiosulphinates, from disulphides, 341
  deoxygenation, 303
Thiosulphonates, from disulphides, 341
  reaction with phosphites, 378
Thiourea, reaction with
  β-acylvinylphosphonium salts, 258
Thioureas, phosphonium salts in promotion of synthesis, 437
Thiyl radicals, reactivity towards P(III) reagents, 367
Thromboxane B2, 107
N-p-Tolylhydroxylamine, dehydration, 411
Tosylates, from alcohols and methyl tosylate, 334
Transesterification, of esters, 334
Transylidation, 208
  prevention, 212
Trialkyl phosphites, from white phosphorus with alkoxides, 422
  in deoxygenation of aromatic nitro compounds, 269
  in phosphonate formation, 158
  rearrangement to dialkyl alkylphosphonates, 303, 312
Trialkylphosphines, reaction with

Subject Index

α-halogenocarbonyl compounds, 446
reaction with CCl_4, 389
Trialkylphosphine oxides, syntheses and reactions, 8
Triaryl phosphites, 1:1 adducts with halogens, 433
quaternization via Cu catalysis, 5
in iodination, 5
Triarylphosphines, reaction with α-halogenocarbonyl compounds, 446
Triazoles, 278
1,2,3-Triazoles, from β-acylvinyltriphenylphosphonium bromides, 256
from alkylidenephosphoranes, 231
1,2,4-Triazoline-3-thiones, from isothiocyanates, 497
1,2,4-Triazolin-3-ones, from isocyanates, 496
Triazolo[3,4-a]quinoxalines, from o-nitrophenyltriazole, 285
1,2,3-Triazolo[5,1-c]-1,2,4-triazines, from ylides, 6
α,α,α′-Tribromodibenzyl ketone, 1,3-eliminations, 453
Tri-(p-bromophenyl) phosphite, in peptide synthesis, 530
Tributylphosphine, reaction with methyl pyruvate, 314
Trichloromethanide ion, from tris(dimethylamino)phosphine, 402
α-Trichloromethyl-alcohols, into 1,1-dichloroolefins, 403
Trichloromethylphosphonium chloride, formation, 388
from dichlorophosphorane, 389
intermediate, 388
reaction with amines and alcohols, 392
Trichlorosilane, in reductions, 424
Triethyl phosphite, deoxygenation of N-oxides, 297
in azepine synthesis, 287
reaction with arylglyoxylic esters, 308
reaction with azoxybenzene, 300
reaction with cyanogen bromide, 520
reaction with fluorenone, 310
reaction with phthalic anhydrides, 312
reactions with halogens, 434
Triethylphosphine, reaction with azoxybenzene, 300
Trifluoromethyl ketones, reaction with phosphonate carbanions, 164
Trihalogenoacetic esters, dehalogenation in ethanol, 449
Triisopropyl phosphite, reaction with fluorenone, 310
Trimethyl phosphite, in HCl removal, 524
reaction with acenaphthaquinone, 311
6,8,10-Trimethyl-10H-azepino[1,2-a]-indole, from 2-nitrophenyl 2,4,6-trimethylphenyl ether, 288
Trioxaphosphetans, 491
Triphenoxyphosphonium iodide, stereospecific conversion of epoxides into alkenes, 437
Triphenyl phosphate, from triphenyl phosphite, 336
Triphenyl phosphite, in condensation of N-acylaminoacids and amino acid esters, 527
in peptide synthesis, 530
oxidation to triphenyl phosphate, 336
reaction with N,N-dichlorourethane in alcohol halogenation, 455
reaction with imidazole, 527
Triphenyl phosphite dibromide, reaction with acetylenic and allenic alcohols, 433
Triphenylphosphine, adducts with halogens, 437
in aryl alkyl ether cleavage, 446
in debromination of α-bromodiphenylacetyl bromide, 459
in ester hydrolysis, 446
in N-oxide deoxygenation, 297
in ozonide reduction, 344
in peptide synthesis, 529
Triphenylphosphine reactions:
with acetyl bromide, 460
with α-bromoacetonitrile, 450
with 1-bromo-1-nitroalkanes, 454

with N-bromoacetamide in alcohol halogenation, 455
with bromophenols, 461
with CCl_4 and water, 394
with chloroformate esters, 459
with diaryl sulphoxide, 302
with dimethyl sulphoxide, 302
with N-halogenosuccinimides, 455
with trimethylsilyl tribromoacetate, 450
Triphenylphosphine–CCl_4 system, in chlorinations, 391
in dehydrations, 391
in multicomponent reactions, 391
in P–N linking reactions, 391
in synthesis of carbodiphosphoranes, 397
reactions with aldehydes, ketones, and epoxides, 402
Triphenylphosphine–dialkyl azodicarboxylate reagent (DEADC-TPP), 327
Triphenylphosphine dichloride, in ylide formation, 445
reaction with cycloalkanone oxime, 454
Triphenylphosphine difluoride, in fluorination, 446
Triphenylphosphine diiodide, in nucleoside iodination, 456
Triphenylphosphoranylidenesuccinic anhydride, from maleic anhydride, 312
Triphenylvinylphosphonium bromide in syntheses: chromens and 2,3-dihydro-1-benzoxepins, 252
furans, 250
pyrazoles, 254
pyrroles and 2,3-dihydropyridazines, 254
pyrrolizines, 250
quinolines, 253
Tris-(4-chlorophenyl) phosphite, in sulphoxide reduction, 302
Tris(dialkylamino)phosphines, reaction with aromatic aldehydes, 306
Tris(dimethylamino)phosphine, anhydrides via reaction with acids and CCl_4, 529

desulphurizing agent, 369
in dechlorinations, 397, 399
in ozonolysis, 345
in synthesis of epoxides from aryl aldehydes, 306
in tetraphenylethylene synthesis, 310
Tris(dimethylamino)phosphine reactions: with N-chloride–isopropylamine, 456
with 2-chlorobenzaldehyde, 307
with fluorenone, 310
with 1,3-oxathiolan-5-ones, 359
with polyhalogenoalkanes, 387
with 2-thiophthalic anhydride, 312
Tris(hydroxymethyl)phosphine, explosion hazard, 372
in reduction of human γ-globulin, 372
Trisporic acid, precursor via PO-stabilized carbanions, 177
N,N,N'-Trisubstituted hydrazines, from N,N-disubstituted hydrazines, 444
1,3-Trithiocarbonates, ylide formation with phosphites, 359
Trithiole, desulphurization, 368
1,2,4-Trithione, from dioxaphospholen and CS_2, 479
Tuberculostearic acid, 92
Tuftsin (tetrapeptide), 522
Turnstile rotation, 12
Twistene, via thionocarbonate, 355

Ultraviolet irradiation in furazan deoxygenation, 299
Unsaturated acids and derivatives, from phosphonates, 173
α,β-Unsaturated esters, from acyl chlorides, 213
from halogenoacetates, 211
Uracil, alkylation, 332
Ureas, phosphonium salts in promotion of synthesis, 437

Vaccenic acid, 90
Vanadium ions, deoxygenation catalysts, 318

Subject Index

Vilsmeier aldehyde synthesis, using Ph$_3$PBr$_2$, 442
Vinyl azides, 1,3-dipolar cycloaddition, 232
Vinyl ethers, from alcohols, 460
Vinyl halides, from alkylphosphonate carbanions, 174
Vinyl phosphates, by Perkow reaction, 5
Vinyl phosphonates, from ester phosphonates, 156
Vinyl sulphides, from phosphonates, 175
N-Vinylimidoyl chlorides, from chlorinated nitrile ylides, 493
Vinylketene imines, from allylamides, 442
Vinylogous amides, synthesis, 353
Vinylphosphine oxides, syntheses and reactions, 8
Vinylphosphonium salts, formation, 8
  in heterocyclic synthesis, 250
  in phosphorylation, 450
Vitamin A and derivatives, via Wittig reaction, 16, 86, 122, 127

Wadsworth–Emmons PO-activated olefin synthesis, 156
Wadsworth–Emmons reaction, comparison with Wittig reaction, 23, 35, 88
  stereochemistry, 37, 73, 88, 131
  phosphonamide modification, 89
Wadsworth–Emmons reaction in natural product synthesis:
  β,γ-carotene, 37
  juvenile hormone, *Hyalophora cecropia*, 37
  prostaglandins, 99
  steroids, 141
Water, determination in organic solvents, 396, 397
  in dialkyl peroxide deoxygenation, 320
Water-soluble phosphines, 372
White phosphorus, reactions, 422
Wittig reaction, 3, 17–153, 155, 157, 164, 358, 359
  abnormal, 81
  acid catalysis, 34

betaine decomposition, 21, 27, 31, 48
betaine equilibration, 27, 30, 48, 50, 55, 77, 80, 114
betaine formation, 21, 26
betaine stabilization, 40, 82, 122
betaine stereochemistry, 26, 27, 53
by phase-transfer catalysis, 84
industrial practice, 19
intramolecular, 8, 19, 74, 250
mechanism, 19, 26, 31
phosphonate modification, see Wadsworth–Emmons
rearrangement during, 139
salt-free, 48–50, 67, 78, 79, 115, 128, 131
Schlosser modification, see Wittig–Schlosser
scope, 19
Wittig reaction (stereochemistry), 30, 31, 53, 76, 82, 84
  effect of additives, 19, 40, 45, 80
  effect of α-substituents, 31–33, 54
  effect of carbonyl structure, 33, 50–53, 82
  effect of P substituents, 29, 77, 81
  effect of reaction conditions, 18, 45–49, 66, 69, 78–80
  effect of temperature, 18, 34, 40, 69, 111
  solvent effect, 33, 40, 80, 113, 116
  with moderate ylides, 63–89
  with reactive ylides, 42–63
  with stable ylides, 27–41, 99
Wittig reaction (syntheses), bicyclo[4,1,0]heptenes, 75
  bridgehead alkenes, 75
  conjugated dienes, 67, 111
  conjugated polyene carbonyls, 76
  cumulenes, 85
  cyclohexa-1,3-dienes, 74
  cyclopropanes, 132
  dienoic esters, 132
  enynes, 84–86, 110
  non-benzenoid aromatic ring systems, 19
  pentadienoic acids, 82
  stilbenes, 77–82
Wittig reaction (syntheses of natural products), abscisic acid, 34

Wittig reaction, (*continued*)
  bioresmethrin derivatives, 32
  carbohydrates, 19
  carotenoids and polyenes, 120–127
  isoprenoid chains, 51
  juvenile hormone, *Hyalophora cecropia*, 51, 133, 134
  pheromones, 67, 94, 109–120
  phytoenes, 68, 126
  polyacetylenic compounds, 85, 97–99
  prostaglandins, 99–109
  pyrethric acid, 130
  rethrolones, 127–133
  α-santalol, 134
  β-santalol, 33, 134
    reagents, 387
  sesquiterpenes, 133–136
  (±)-sirenin, 50
  trans-squalene, 19
  steroids, 136–142
  unsaturated fatty acids, 90–97
  vitamin A and derivatives, 19, 86, 122, 127
Wittig (bis-) reactions, in heterocyclic synthesis, 19
Wittig–Schlosser reaction, 55–63, 67, 79, 113, 136
Wolf–Kishner reduction, alternative to, 6

Ylide behaviour of 2,2-dihydro-1,2-oxaphospholan, 490
Ylides,
  alkaline hydrolysis, 211
  alkylation, 207
  bond lengths, 7
  1,3-dipolar addition, 12
  direct conversion into alkenes using $(PhO)_3P$–O_3, 341
  fragmentation, 234
  from active methylene compounds, 445
  from phosphonium salts, 7
  from 1,3-trithiocarbonates, 359
  from vinyl phosphonium salts, 8
  hydrolytically stable, 208
  in β-acylacrylic acid synthesis, 211
  in alkaloid synthesis, 209
  in alkene formation, 157
  in dehydrochlorination, 397
  in heterocyclic synthesis, 223
  in Wittig reaction, 3
  oxidative coupling, 227
  reactions with aziridines, 216
  reaction with benzyne, 219
  reaction with nuclear-chlorinated heterocycles, 209
  structure and reactivity, 6
Young test, 528